Modern Language Models and Computation

Alexander Meduna • Ondřej Soukup

Modern Language Models and Computation

Theory with Applications

 Springer

Alexander Meduna
Department of Computer Science
Brno University of Technology
Brno, Czech Republic

Ondřej Soukup
Department of Information Technology
Brno University of Technology
Brno, Czech Republic

ISBN 978-3-319-87473-9   ISBN 978-3-319-63100-4   (eBook)
DOI 10.1007/978-3-319-63100-4

Printed on acid-free paper

This Springer imprint is published by Springer Nature
The registered company is Springer International Publishing AG
The registered company address is: Gewerbestrasse 11, 6330 Cham, Switzerland

To Zora and to the Memory of Ivana

Words move
T. S. Eliot: *The Four Quartets*

Preface

To live, work, and prosper on the Earth, people need to communicate, and they do so by means of a broad variety of languages developed from the dawn of civilization up to today. First and foremost, human beings use natural languages, such as English. In essence, these natural languages represent systems of communication by written and spoken words, used by the people of a particular country or its parts. Furthermore, researchers often express their ideas, concepts, tests, and results in various artificially made languages introduced for specific purposes in their scientific disciplines. For instance, computer scientists have developed hundreds of programming languages in which they write their algorithms so they can be executed on computers. In addition, today's world is overflown with modern communication machines, such as mobile phones, which gave rise to developing brand new languages for man-machine and machine-machine communication. It thus comes as no surprise that the scientific development and study of languages and their processors fulfill a more important role than ever before.

Of course, we expect that the study of languages produces concepts and results that are solid and reliable. Therefore, we base this study upon mathematics as a systematized body of unshakable knowledge obtained by exact and infallible reasoning. Indeed, mathematics has developed a highly sophisticated theory that specifies languages quite rigorously and formally, hence the name of this theory—formal language theory or, briefly, language theory. From a mathematical viewpoint, this theory defines languages as sets of sequences consisting of symbols. This definition encompasses almost all languages as they are commonly understood. Indeed, natural languages are included in this definition. Of course, all artificial languages introduced by various scientific disciplines represent formal languages as well.

The strict formalization of languages necessitates an introduction of mathematical models that define them. Traditionally, these *language models* are based upon finitely many rules by which they sequentially rewrite sequences of symbols, called words, and that is why they are referred to as *rewriting systems*. They are classified into two basic categories—generating and accepting rewriting systems. Generating systems, better known as *grammars*, define strings of their language so their rewriting process generates them from a special start symbol. On the other hand, accepting systems, better known as *automata*, define strings of their language by rewriting process that starts from these strings and ends in a special set of strings, usually called final configurations.

However, apart from these traditional language models, language theory has also developed several modern language models based upon rewriting systems that work with words in a nontraditional way, and many of them have their great advantages over their traditional out-of-date counterparts. To give an insight into these advantages, we first need to understand the fundamental problems and difficulties the classical language models cope with. To start with, the traditional language-defining rewriting systems are defined quite generally. Unfortunately, from a practical viewpoint, this generality actually means that the rewriting systems work in a completely unpredictable way. As such, they are hardly implementable and, therefore, applicable in practice. Being fully aware of this pragmatic difficulty, formal language theory has introduced fully deterministic versions of these rewriting systems; sadly, their application-oriented perspectives are also doubtful. First of all, in an ever-changing environment in which real language processors work, it is naive, if not absurd, that these deterministic versions might adequately reflect and simulate real communication technologies applied in such real-world areas as various engineering techniques for language analysis. Furthermore, in many cases, this determinism decreases the power of their general counterparts, which represents another highly undesirable feature of this strict determinism.

Considering these difficulties and drawbacks, formal language theory has recently introduced new versions of rewriting systems, which avoid the disadvantages mentioned above. From a practical viewpoint, an important advantage of these newly introduced rewriting systems consists in controlling their language-defining process and, therefore, operating in a more deterministic way than classical systems, which perform their rewriting process in a quite traditional way. Perhaps even more significantly, the modern versions are stronger than their traditional counterparts. Considering these advantages, it comes as no surprise that formal language theory has paid an incredibly high attention to these modern versions of grammars and automata. Indeed, over the past quarter century, literally hundreds of studies were written about them, and their investigation represents a vivid trend within formal language theory. This investigation has introduced a number of alternative concepts of grammars and automata, and it has achieved many remarkable results. Nevertheless, all these concepts and results are only scattered in various conference and journal papers.

Modern versions of grammars and automata represent the principal *subject* of this book, whose main *focus* is on their concepts, properties, and applications in computer science. The book selects crucially important models and summarizes key results about them in a compact and uniform way. It always relates each of the selected models to a particular way of modern computation, such as computation in parallel or largely cooperating computation. The text explains how the model in question properly reflects and formalizes the corresponding way of computation, so it allows us to obtain a systematized body of mathematically precise knowledge concerning the computation under investigation. Apart from this obvious theoretical significance, from a more practical viewpoint, the book demonstrates and illustrates how the developers of new computational technologies can make use of this knowledge to build up and implement their modern methods and techniques in the most efficient way.

The text always starts the discussion concerning the language models under consideration by conceptualizing them and linking them to a corresponding form of computation. Then, it gives their mathematical definition, which is also explained intuitively and illustrated by many examples. After that, the text presents most computation-related topics about the models so it proceeds from their (i) theoretical properties through (ii) transformations up to (iii) applications as described next in a greater detail.

(i) The power of the models represents perhaps the most essential property concerning them. Therefore, the book always determines the language family that the models define. The text also includes many algorithms that modify the models so they satisfy some prescribed properties, which frequently simplify proofs demonstrating results about the models. Apart from this theoretical advantage, the satisfaction of these properties is often strictly required by language processors based on the models.

(ii) Various transformations of grammars and automata also represent an important investigation area of this book. Specifically, the transformations that reduce the specification of these language models are important to this investigation because the resulting reduced versions of the models define languages in a very succinct and elegant way. As obvious, this reduction simplifies the development of computational technologies, which then work economically and effectively. Of course, the same languages can be defined by different models, and as obvious, every computation-related investigation or application selects the most appropriate models for them under given circumstances. Therefore, whenever discussing different types of equally powerful language models, the book gives transformations that convert them to each other. More specifically, given a language model of one type, the text carefully explains how to transform it to another model so both the original system and the model produced by this transformation define the same language.

(iii) Finally, the book discusses the use of the models in practice. It describes applications and their perspectives from a general viewpoint. However, the text also covers several real-world applications with a focus on linguistics and biology.

As far as the *writing style* is concerned, we introduce all formalisms with enough rigor to make all results quite clear and valid because we consider this book primarily as a theoretically oriented treatment. Before every complicated mathematical passage, we explain its basic idea intuitively so that even the most complex parts of the book are relatively easy to grasp. We prove most of the results concerning the topics mentioned above effectively—that is, within proofs demonstrating them, we give algorithms that describe how to achieve these results.

For instance, we often present conversions between equally powerful systems as algorithms, whose correctness is then rigorously verified. In this way, apart from their theoretical value, we actually explain how to implement and use them in practice. Several worked-out examples and case studies illustrate this use.

Concerning the *use of the book*, from a general standpoint, this book is helpful to everybody who takes advantage of modern computational technologies based upon grammars or automata. Perhaps most significantly, all scientists who actually make these technologies, ranging from pure mathematicians through computational linguists up to computer engineers, might find this book useful for their work. Furthermore, the entire book can be used as a text for a two-term course in grammars and automata at a graduate level. The text allows the flexibility needed to select some of the discussed topics and, thereby, use it for a one-term course on this subject. Finally, serious undergraduate students may find this book helpful as an accompanying text for a course that deals with formal languages and their models.

Organization and Coverage

The text is divided into six parts, each of which consists of several chapters; altogether, the book contains 19 chapters. Each part starts with an abstract that summarizes its chapters.

Part I, consisting of Chaps. 1 and 2, gives an introduction to this monograph in order that the entire text of the book is completely self-contained. In addition, it places all the coverage of the book into scientific context and reviews important mathematical concepts with a focus on classical language theory.

Part II, which consists of Chaps. 3 through 6, presents an overview of modern grammatical models for languages and corresponding computational modes. Chapter 3 gives the fundamentals of grammars for regulated computation. In essence, these grammars regulate their language generation by additional mechanisms, based upon simple mathematical concepts, such as finite sets of symbols. Chapter 4 studies grammars for computation performed in parallel. These grammars generate their languages in parallel and, thereby, accelerate this generation significantly just like computation in parallel is usually much faster than that made in a sequential way. First, this chapter studies partially parallel generation of languages, after which, it investigates the totally parallel generation of languages. Chapter 5 explores grammars that work on their words in a discontinuous way, thus formalizing a discontinuous way of computation in a very straightforward way. Chapter 6 approaches grammatical models for languages and computation in terms of algebra. In particular, it examines grammatical generation of languages defined over free groups.

Part III consists of Chaps. 7 through 10. To some extent, in terms of automata, this part parallels what Part II covers in terms of grammars. Indeed, Chap. 7 gives the fundamentals of regulated computation formalized by automata. Similarly to grammars discussed in Chap. 5, Chap. 8 formalizes a discontinuous way of computation. However, Chap. 8 bases this formalization upon automata, which jump across the words they work on discontinuously. Chapter 9 discusses language models for computation based upon new data structures. More specifically, it studies deep pushdown automata, underlined by stacks that can be modified deeper than on their top. Finally, Chap. 10 studies automata that work over free groups, and in this way, it parallels Chap. 6, which studies this topic in grammatical terms.

Part IV, which consists of Chaps. 11 and 12, covers important language-defining devices that combine other rewriting systems, thus formalizing a cooperating way of computation. Chapter 11 untraditionally combines grammars and automata in terms of the way they operate. Specifically, it studies how to generate languages by automata although, traditionally, languages are always generated by grammars. Chapter 12 studies the generation of languages by several grammars that work in a simultaneously cooperative way.

Part V, consisting of Chaps. 13 through 15, discusses computer science applications of rewriting systems studied earlier in the book. First, Chap. 13 covers these computational applications and their perspectives from a rather general viewpoint. Then, more specifically, Chaps. 14 and 15 describe applications in computational linguistics and computational biology, respectively. Both chapters contain several case studies of real-world applications described in detail.

Part VI consists of a single chapter—Chap. 16, which closes the entire book by adding several remarks concerning its coverage. It briefly summarizes all the material covered in the text. Furthermore, it sketches many brand new investigation trends and longtime open problems. Finally, it makes several bibliographical and historical remarks. Further backup materials are available at http://www.fit.vutbr. cz/~meduna/books/mlmc.

Brno, Czech Republic   Alexander Meduna
Brno, Czech Republic   Ondřej Soukup

Acknowledgments

Parts of the manuscript for this book were used as lecture notes at various universities throughout the world. Most of them were based on papers published by us as well as other authors.

This work was supported by several grants—namely, BUT FIT grant FIT-S-11-2, European Regional Development Fund in the IT4Innovations Centre of Excellence (MŠMT CZ1.1.00/02.0070), research plan CEZ MŠMT MSM0021630528, and Visual Computing Competence Center (TE01010415).

Our thanks go to many colleagues from our home university for fruitful discussions concerning the subject of this book. We are grateful to Susan Lagerstrom-Fife and Dennis Pacheco at Springer for their invaluable assistance during the preparation of this book.

Brno, Czech Republic   Alexander Meduna
Brno, Czech Republic   Ondřej Soukup

Acknowledgments

Parts of the manuscript for this book were used as lecture notes at various universities throughout the world. Most of them were based on papers published by us as well as other authors.

This work was supported by several grants, namely BUT FEL grant FEKS-11-2, Internal Regional Development Fund in the IT4Innovations Centre of Excellence MSMT CZ.1.00/0.0/0.0 research plan CEZ MSMT MSM0021630528, and Visual Computing Competence Center (TE01010415).

Our thanks go to many colleagues from our home university for fruitful discussions concerning the subject of this book. We are grateful to Susan J. Agensmann-Filk and Dennis Biebera in Springer for their invaluable assistance during the preparation of this book.

Brno, Czech Republic   Alexander Medana
Brno, Czech Republic   Ondrej Staňa

Contents

Part I
Introduction

This part, consisting of Chaps. 1 and 2, reviews important mathematical concepts with a special focus on classical language theory. In this way, it places the entire coverage of the book into its scientific context. In addition, it guarantees that the whole text is completely selfcontained—that is, no other book is needed to follow its complete coverage.

Chapter 1 reviews rudimentary mathematical notions in order to speak clearly and accurately throughout the remaining chapters of this book. Then, Chap. 2 covers important concepts used in formal language theory. Apart from the classical rudiments, it includes several lesser-known areas of this theory, such as parallel grammars, because they are also needed to fully grasp some upcoming topics covered in this book.

Part I
Introduction

Chapter 1
Mathematical Background

This three-section chapter reviews rudimentary mathematical concepts, including key notions concerning sets (Sect. 1.1), relations (Sect. 1.2), and graphs (Sect. 1.3). For readers having background in these areas, this chapter can be skipped and treated as a reference for terminology used later in this book.

1.1 Sets and Sequences

This section outlines rudimentary concepts concerning sets and sequences.

1.1.1 Sets

In what follows, we assume that there exist primitive objects, referred to as *elements*, taken from some pre-specified *universe*, usually denoted by $\mathbb{U}$. We also assume that there are objects, referred to as *sets*, which represent collections of objects, each of which is an element or another set. If A contains an object a, then we symbolically write $a \in A$ and refer to a as a *member of A*. On the other hand, to express that a is not a member of A, we write $a \notin A$.

If A has a finite number of members, then A is a *finite set*; otherwise, it is an *infinite set*. The finite set that has no member is the *empty set*, denoted by $\emptyset$. The *cardinality of a finite set* A, $\mathrm{card}(A)$, is the number of members that belong to A; note that $\mathrm{card}(\emptyset) = 0$. A finite set A is customarily *specified by listing its members*; that is, $A = \{a_1, a_2, \ldots, a_n\}$, where a_1 through a_n are all members of A; as a special case, we have $\{\} = \emptyset$. An infinite set B is usually *defined by a property* π so that B contains all elements satisfying π; in symbols, this specification has the following general format: $B = \{a \mid \pi(a)\}$. Sometimes, an infinite set is *defined recursively* by

© Springer International Publishing AG 2017
A. Meduna, O. Soukup, *Modern Language Models and Computation*,
DOI 10.1007/978-3-319-63100-4_1

explicitly naming the first few values (typically, just one first value) in the set and then defining later values in the set in terms of earlier values.

In this book, we denote the set of all natural numbers by $\mathbb{N}$. In other words, $\mathbb{N}$ denotes the set of all positive integers, so

$$\mathbb{N} = \{1, 2, \ldots\}$$

Furthermore, $_0\mathbb{N}$ denotes the set of all non-negative integers, and $\mathbb{Z}$ denotes the entire set of all integers throughout.

Example 1.1.1. Take $\mathbb{U} = \mathbb{N}$. Let X be the set of all even positive integers defined as

$$X = \{i | i \in \mathbb{N}, \ i \text{ is even}\} \text{ or, alternatively, } X = \{j | j = 2i, \ i, j \in \mathbb{N}\}$$

Let Y be the set of all even positive integers between 1 and 9. Define

$$Y = \{i | i \in X, \ 1 \leq i \leq 9\} \text{ or, simply, } Y = \{2, 4, 6, 8\}$$

Observe that $\text{card}(Y) = 4$. Consider the next recursive definition of the set W

(i) 2 is in W;
(ii) if i is in W, then so is $2i$, for all $i \geq 2$.

By (i), W contains 2. Then, by (ii), it contains 4, too. By (ii) again, it includes 8 as well. Continuing in this way, we see that W contains $2, 4, 8, 16, \ldots$. In words, W consists of all positive integers that represent a power of two; mathematically,

$$W = \{j | j = 2^i, \ i, j \in \mathbb{N}\} \text{ or, briefly, } \{2^i | i \in \mathbb{N}\} \qquad \square$$

Let A and B be two sets. A is a *subset of* B, symbolically written as $A \subseteq B$, if each member of A also belongs to B. A is a *proper subset of* B, written as $A \subset B$, if $A \subseteq B$ and B contains a member that is not in A. By $A \nsubseteq B$, we express that A is not a subset of B. If $A \subseteq B$ and $B \subseteq A$, then A *equals* B, denoted by $A = B$; simply put, $A = B$ means that both sets are identical. By $A \neq B$, we express that A is not equal to B. To express that $A \nsubseteq B$ and $B \nsubseteq A$ we call A and B to be *incomparable*. The *power set of* A, denoted by 2^A, is the set of all subsets of A; formally, $2^A = \{B | B \subseteq A\}$.

For two sets, A and B, their *union*, *intersection*, and *difference* are denoted by $A \cup B$, $A \cap B$, and $A - B$, respectively, and defined as $A \cup B = \{a | a \in A \text{ or } a \in B\}$, $A \cap B = \{a | a \in A \text{ and } a \in B\}$, and $A - B = \{a | a \in A \text{ and } a \notin B\}$. If $A \cap B = \emptyset$, A and B are *disjoint*. More generally, n sets $C_1, C_2, \ldots, C_n$, where $n \geq 2$, are *pairwise disjoint* if $C_i \cap C_j = \emptyset$ for all $1 \leq i, j \leq n$ such that $i \neq j$. If A is a set over a universe $\mathbb{U}$, the *complement of* A is denoted by $\bar{\bar{A}}$ and defined as $\bar{\bar{A}} = \mathbb{U} - A$.

Sets whose members are other sets are usually called *classes* of sets rather than sets of sets.

Example 1.1.2. Consider the sets from Example 1.1.1. Observe that

$$2^Y = \{\emptyset,$$
$$\{2\}, \{4\}, \{6\}, \{8\},$$
$$\{2,4\}, \{2,6\}, \{2,8\}, \{4,6\}, \{4,8\}, \{6,8\},$$
$$\{2,4,6\}, \{2,4,8\}, \{2,6,8\}, \{4,6,8\},$$
$$\{2,4,6,8\}\}$$

Furthermore, $X \subset \mathbb{U}$ and $Y \subset X$. Set $W = \mathbb{U} - X$. In words, W is the set of all odd positive integers. As obvious, $X \cap W = \emptyset$, so X and W are disjoint. Notice that $Y \cup W = \{i | i \in \mathbb{U}, i \le 8 \text{ or } i \text{ is even}\}$ and

$$\bar{\bar{Y}} = \{1,3,5,7\} \cup \{i | i \in \mathbb{U}, \ i \ge 9\}$$

X, Y and W are not pairwise disjoint because $X \cap Y \ne \emptyset$. On the other hand, $\{2\}$, $\{8\}$, and $\{4,6\}$ are pairwise disjoint. Observe that $\{j | j \in \mathbb{U}, j \ne j\}$ and $\emptyset$ are identical, symbolically written as

$$\{j | j \in \mathbb{U}, j \ne j\} = \emptyset$$

Sets $\mathbb{U}$ and $(X \cup W)$ are identical, too.

To illustrate a class of sets, consider Δ defined as

$$\Delta = \{U | U \subseteq \mathbb{U}, 1 \in U\}$$

In words, Δ consists of all subsets of $\mathbb{U}$ that contain 1; for instance, $\{1\}$ and W are in Δ, but $\{2\}$ and Y are not. Notice that $\mathbb{U} \in \Delta$, but $\mathbb{U} \not\subseteq \Delta$; indeed, Δ contains sets of positive integers while $\mathbb{U}$ contains positive integers, not sets. □

1.1.2 Sequences

A *sequence* is a list of elements from some universe. A sequence is *finite* if it consists of finitely many elements; otherwise, it is *infinite*. The *length of a finite sequence x*, denoted by $|x|$, is the number of elements in x. The *empty sequence*, denoted by ε, is the sequence consisting of no element; that is, $|\varepsilon| = 0$.

Let s be a sequence. If s is finite, it is defined by listing its elements. If s is infinite, it is specified by using ellipses provided that this specification is clear. Alternatively, s is defined recursively by explicitly naming the first few values in s and then deriving later values in s by a property applied to earlier values in s.

1.2   Relations

The present section reviews several key concepts concerning relations (Sect. 1.2.1) and functions (Sect. 1.2.2).

1.2.1   Relations

For two elements, a and b, (a, b) denotes the *ordered pair* consisting of a and b in this order. Let A and B be two sets. The *Cartesian product of A and B*, $A \times B$, is defined as $A \times B = \{(a, b) \mid a \in A \text{ and } b \in B\}$. A *binary relation* or, briefly, a *relation*, ρ, from A to B is any subset of $A \times B$; that is, $\rho \subseteq A \times B$. If ρ represents a finite set, then it is a *finite relation*; otherwise, it is an *infinite relation*. The *domain of ρ*, denoted by domain(ρ), and the *range of ρ*, denoted by range(ρ), are defined as domain(ρ) $= \{a \mid (a, b) \in \rho \text{ for some } b \in B\}$ and range(ρ) $= \{b \mid (a, b) \in \rho \text{ for some } a \in A\}$. If $A = B$, then ρ is a *relation on A*. A relation σ is a *subrelation of ρ* if $\sigma \subseteq \rho$. The *inverse of ρ*, denoted by ρ^{-1}, is defined as $\rho^{-1} = \{(b, a) \mid (a, b) \in \rho\}$. Let $\chi \subseteq B \times C$ be a relation, where C is a set; the *composition of ρ with χ* is denoted by $\rho \circ \chi$ and defined as $\rho \circ \chi = \{(a, c) \mid (a, b) \in \rho, \ (b, c) \in \chi\}$.

As relations are defined as sets, the set operations apply to them, too. For instance, if ρ is a relation from A to B, its complement $\bar{\bar{\rho}}$ is defined as $(A \times B) - \rho$.

Example 1.2.1. Set

$$articles = \{a, an, the\} \text{ and } two\text{-}words = \{author, reader\}$$

The Cartesian product of *articles* and *two-words* is defined as

$$articles \times two\text{-}words = \{(a, author), (a, reader), (an, author),$$
$$(an, reader), (the, author), (the, reader)\}$$

Define the relation *proper-article* as

$$proper\text{-}article = \{(a, reader), (an, author), (the, author), (the, reader)\}$$

Observe that *proper-article* properly relates English articles to the members of *two-words*. Notice that

$$proper\text{-}article^{-1} = \{(reader, a), (author, an), (author, the), (reader, the)\} \quad \square$$

Let $\rho \subseteq A \times B$ be a relation. To express that $(a, b) \in \rho$, we sometimes write $a\rho b$. That is, we use $(a, b) \in \rho$ and $a\rho b$ interchangeably in what follows.

Let A be a set, and ρ be a relation on A. Then,

1. if for all $a \in A$, $a\rho a$, then ρ is *reflexive*;
2. if for all $a, b \in A$, $a\rho b$ implies $b\rho a$, then ρ is *symmetric*;
3. if for all $a, b \in A$, ($a\rho b$ and $b\rho a$) implies $a = b$, then ρ is *antisymmetric*;
4. if for all $a, b, c \in A$, ($a\rho b$ and $b\rho c$) implies $a\rho c$, then ρ is *transitive*.

Let A be a set, ρ be a relation on A, and $a, b \in A$. For $k \geq 1$, the *k-fold product of* ρ, ρ^k, is recursively defined as

1. $a\rho^1 b$ iff $a\rho b$, and
2. $a\rho^k b$ iff there exists $c \in A$ such that $a\rho c$ and $c\rho^{k-1} b$, for $k \geq 2$.

Furthermore, $a\rho^0 b$ if and only if $a = b$. The *transitive closure of* ρ, ρ^+, is defined as $a\rho^+ b$ if and only if $a\rho^k b$, for some $k \geq 1$; consequently, ρ^+ is the smallest transitive relation that contains ρ. The *reflexive and transitive closure of* ρ, ρ^*, is defined as $a\rho^* b$ if and only if $a\rho^k b$, for some $k \geq 0$.

Example 1.2.2. Let A be the set of all people who have ever lived. Define the relation *parent* so

$$(a, b) \in parent \text{ if and only if } a \text{ is a parent of } b, \text{ for all } a, b \in A$$

Observe that $parent^2$ represents the grandparenthood because $(a, b) \in parent^2$ if and only if a is a grandparent of b. Furthermore, $(a, b) \in parent^3$ if and only if a is a great-grandparent of b. Consequently, $parent^+$ corresponds to being an ancestor in the sense that $(a, b) \in parent^+$ iff a is an ancestor of b. Of course, $(a, a) \notin parent^+$ for any $a \in A$ because a cannot be an ancestor of a. On the other hand, notice that $(a, a) \in parent^*$ for all $a \in A$, so $(a, b) \in parent^*$ iff a is an ancestor of b or $a = b$. □

Let A be a finite set, $A = \{a_1, \ldots, a_n\}$, for some $n \geq 1$. Let ρ be a relation on A. A useful way to represent ρ is by its *adjacency matrix* $_\rho M$. That is, $_\rho M$ is an $n \times n$ matrix whose entries are 0s and 1s. Its rows and columns are both denoted by a_1 through a_n. For all $1 \leq i, j \leq n$, the entry $_\rho M_{ij}$ is 1 if and only if $(a_i, a_j) \in \rho$, so $_\rho M_{ij} = 0$ if and only if $(a_i, a_j) \notin \rho$.

From $_\rho M$, we can easily construct $_{\rho^+} M$, which represents the transitive closure of ρ, by using *Floyd-Warshall algorithm* (see Section 26.2 in [CLR90]). In essence, this algorithm is based upon the idea that if $_{\rho^+} M_{ik} = 1$ and $_{\rho^+} M_{kj} = 1$, then $_{\rho^+} M_{ij} = 1$. Starting from $_\rho M$, it repeatedly performs this implication until no new member can be added to the adjacency matrix.

Let Σ be a set, and let ρ be a relation on Σ. If ρ is reflexive, symmetric, and transitive, then ρ is an *equivalence relation*. Let ρ be an equivalence relation on Σ. Then, ρ partitions Σ into disjoint subsets, called *equivalence classes*, so that for each $a \in \Sigma$, the equivalence class of a is denoted by $[a]$ and defined as $[a] = \{b \mid a\rho b\}$. As an exercise, explain why for all a and b in Σ, either $[a] = [b]$ or $[a] \cap [b] = \emptyset$.

Example 1.2.3. Let $\equiv_n$ denote the *relation of congruence modulo n* on $_0N$, defined as

$$\equiv_n = \{(x, y) | x, y \in {}_0N, \ x - y = kn, \ \text{for some integer } k\}$$

Specifically, take $n = 3$. In other words,

$$\equiv_3 = \{(x, y) | x, y \in {}_0N, \ x - y \text{ is a multiple of 3}\}$$

Notice that $\equiv_3$ is reflexive because $i - i = 0$, which is a multiple of 3. Furthermore, $\equiv_3$ is symmetric because $i - j$ is a multiple of 3 iff $j - i$ is a multiple of 3. Finally, it is transitive because whenever $i - j$ is a multiple of 3 and $j - k$ is a multiple of 3, then $i - k = (i - j) + (j - k)$ is the sum of two multiples, so it is a multiple of 3, too. Thus, $\equiv_3$ represents an equivalence relation.

   Observe that

$$\{0, 3, 6, \ldots\}$$

forms an equivalence class because $3n \equiv_3 3m$ for all integers n and m. More generally, $\equiv_3$ partitions $_0N$ into these three equivalence classes

$$[0] = \{0, 3, 6, \ldots\}$$
$$[1] = \{1, 4, 7, \ldots\}$$
$$[2] = \{2, 5, 8, \ldots\}$$

Observe that [0], [1], [2] are pairwise disjoint and that $_0N = [0] \cup [1] \cup [2]$.   □

   Let Σ be a set, and let ρ be a relation on Σ. If ρ is reflexive, antisymmetric, and transitive, then ρ is a *partial order*. If ρ is a partial order satisfying either $a\rho b$ or $b\rho a$, for all $a, b \in \Sigma$ such that $a \neq b$, then ρ is a *linear order*. As an exercise, illustrate these relations by specific examples.

1.2.2   Functions

A *function* φ from A to B is a relation φ from A to B such that if $a\varphi b$ and $a\varphi c$, then $b = c$; in other words, for every $a \in A$, there is no more than one $b \in B$ such that $a\varphi b$. Let φ be a function from A to B. If domain$(\varphi) = A$, φ is *total*. If we want to emphasize that φ may not satisfy domain$(\varphi) = A$, we say that φ is *partial*.

Example 1.2.4.   Reconsider the relation *parent* from Example 1.2.2, defined as

$$(a, b) \in parent \text{ if and only if } a \text{ is a parent of } b, \text{ for all } a, b \in A$$

where A is the set of all people who have ever lived. Of course, *parent* is not a function because a parent may have two or more children. Neither is *parent*$^{-1}$ a function because every child has two parents. Consider *one-child-parent* as the subrelation of *parent* defined as

$(a, b) \in$ *one-child-parent*   if and only if a is a parent of a single child b,

for all $a, b \in A$

Clearly, *one-child-parent* is a partial function, but *one-child-parent*$^{-1}$ is not a function.

Finally, take B as the set of all mothers who have a single child, so $B \subseteq A$. Consider *parent* defined over B. Observe that *parent* redefined over B coincides with the following definition of relation *one-daughter-mother* over B

$(a, b) \in$ *one-daughter-mother*   if and only if a is a mother

of a single daughter b, for all $a, b \in B$

As obvious, *one-daughter-mother* and its inverse are both total functions. $\square$

Instead of $a\varphi b$, where $a \in A$ and $b \in B$, we often write $\varphi(a) = b$ and say that b is the *value* of φ for *argument* a. Let φ be a function from A to B. If for every $b \in B$, card($\{a \mid a \in A$ and $\varphi(a) = b\}) \le 1$, φ is an *injection*. If for every $b \in B$, card($\{a \mid a \in A$ and $\varphi(a) = b\}) \ge 1$, φ is a *surjection*. If φ is a total function that is both a surjection and an injection, φ represents a *bijection*.

Return to the set theory (see Sect. 1.1.1). Based upon bijections, we define the notion of a countable set. That is, if there is a bijection from an infinite set Ψ to an infinite set Ξ, then Ψ and Ξ have the *same cardinality*. An infinite set, Ω, is *countable* or, synonymously, *enumerable*, if Ω and $\mathbb{N}$ have the same cardinality; otherwise, it is *uncountable* (as stated in Sect. 1.1.1, $\mathbb{N}$ denotes the set of natural numbers).

1.3 Graphs

This section reviews the principal ideas and notions underlying directed graphs (Sect. 1.3.1) while paying a special attention to trees (Sect. 1.3.2).

1.3.1 Directed Graphs

Loosely speaking, a directed graph is a representation of a set with some pairs of its elements, called nodes, connected by directed links, called edges. Customarily, a graph is depicted as a set of dots for the vertices, joined by lines for the edges. A directed graph has its edges directed from one node to another.

More precisely, a *directed graph* or, briefly, a *graph* is a pair $G = (A, \rho)$, where A is a set and ρ is a relation on A. Members of A are called *nodes*, and ordered pairs in ρ are called *edges*. If $(a, b) \in \rho$, then edge (a, b) *leaves* a and *enters* b. Let $a \in A$; then, the *in-degree* of a and the *out-degree* of a are card($\{b \mid (b, a) \in \rho\}$) and card($\{c \mid (a, c) \in \rho\}$) and denoted by in-d($a$) and out-d($a$), respectively.

An n-tuple of nodes, $(a_0, a_1, \ldots, a_n)$, where $n \geq 0$, is a *sequence of length n* from a_0 to a_n if $(a_{i-1}, a_i) \in \rho$ for all $1 \leq i \leq n$; then (a_0) is a sequence of length 0. A sequence $(a_0, a_1, \ldots, a_n)$ of length n, for some $n \geq 0$, is a *path of length n* if $a_i \neq a_j$, for $0 \leq i \leq n, 0 \leq j \leq n, i \neq j$; if, in addition, $a_0 = a_n$, then $(a_0, a_1, \ldots, a_n)$ is a *cycle of length n*. A graph G is *acyclic* if and only if it contains no cycle.

In this book, we frequently *label* the edges of G with some attached information. Pictorially, we represent $G = (A, \rho)$ so we draw each edge $(a, b) \in \rho$ as an arrow from a to b possibly with its label as illustrated in the next example.

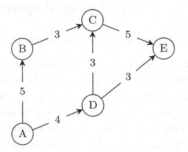

1.3.2 Trees

Let $G = (A, \rho)$ be a graph. If $(a_0, a_1, \ldots, a_n)$ is a path in G, then a_0 is an *ancestor of* a_n and a_n is a *descendant of* a_0; if in addition, $n = 1$, then a_0 is a *direct ancestor of* a_n and a_n a *direct descendant of* a_0. A *tree* is an acyclic graph $T = (A, \rho)$ such that A contains a specified node, called the *root of* T and denoted by troot(T), and every $a \in A - \{\text{troot}(T)\}$ is a descendant of troot(T) and its in-degree is one; in-degree of troot(T) is zero. If $a \in A$ is a node whose out-degree is 0, a is a *leaf*; otherwise, it is an *interior node*. In this book, a tree T is always considered as an *ordered tree* in which each interior node $a \in A$ has all its direct descendants, b_1 through b_n, where $n \geq 1$, ordered from the left to the right so that b_1 is the leftmost direct descendant of a and b_n is the rightmost direct descendant of a. At this point, a is the *parent* of its *children* b_1 through b_n, and all these nodes together with the edges connecting them, (a, b_1) through (a, b_n), are called a *parent-children portion of* T. Nodes b_1 through b_n are called *siblings*, while b_i is a *left sibling* of b_j and b_j is a *right sibling* of b_i, for $1 \leq i < j \leq n$; b_i is the *direct left sibling* of b_{i+1} and b_{i+1} is the *direct right sibling* of b_i. The *frontier of* T, denoted by frontier(T), is the sequence of T's leaves ordered from the left to the right. The *depth of* T, depth(T), is the length of the longest path in T. A tree $S = (B, v)$ is a *subtree of* T if $\emptyset \subset B \subseteq A$, $v \subseteq \rho \cap (B \times B)$, and in T, no node in $A - B$ is a descendant of a node in B; S is an *elementary subtree of* T if depth(S) = 1.

Like any graph, a tree T can be described as a two-dimensional structure. To simplify this description, however, we draw a tree T with its root on the top and with all edges directed down. Each parent has its children drawn from the left to the right according to its ordering. Drawing T in this way, we may omit all arrowheads.

Apart from this two-dimensional representation, however, it is frequently convenient to specify T by a one-dimensional representation, denoted by odr(T), in which each subtree of T is represented by the expression appearing inside a balanced pair of $\langle$ and $\rangle$ with the node which is the root of that subtree appearing immediately to the left of $\langle$. More precisely, odr(T) is defined by the following recursive rules:

(i) If T consists of a single node a, then odr(T) = a.
(ii) Let (a, b_1) through (a, b_n), where $n \geq 1$, be the parent-children portion of T, troot(T) = a, and T_k be the subtree rooted at b_k, $1 \leq k \leq n$, then

$$\text{odr}(T) = a \langle \text{odr}(T_1) \, \text{odr}(T_2) \ldots \text{odr}(T_n) \rangle$$

The next example illustrates both the one-dimensional odr-representation and the two-dimensional pictorial representation of a tree. For brevity, we prefer the former throughout the rest of this book.

Example 1.3.1. Consider the tree $T = (P, \rho)$, where $P = \{a, b, c, d, e\}$ and $\rho = \{(a, b), (a, c), (c, d), (c, e)\}$. Nodes a and c are interior nodes while b, d, and e are leaves. The root of T is a. We define b and c as the first child of a and the second child of a, respectively. A parent-children portion of T is, for instance, (a, b) and (a, c). Notice that frontier(T) = bde, and depth(T) = 2. Following (i) and (ii) above, we obtain the one-dimensional representation of T as

$$\text{odr}(T) = a \langle bc \langle de \rangle \rangle$$

Its subtrees are $a \langle bc \langle de \rangle \rangle$, $c \langle de \rangle$, b, d, and e. In Fig. 1.1, we pictorially describe $a \langle bc \langle de \rangle \rangle$ and $c \langle de \rangle$.

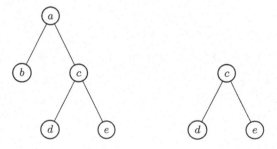

Fig. 1.1 Tree and subtree

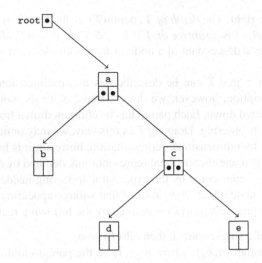

Fig. 1.2 Diagram of binary tree implementation

A *binary tree* is a tree in which every interior node has at most two children; for instance, T represents a binary tree. Next, we describe the implementation of a binary tree, in which every node is represented by a three-part structure. One part represents its name—that is, the symbol corresponding to the node. The others are blank if the node is a leaf; otherwise, they reference the two children of the node. As a whole, the tree implementation starts by a reference to its root node. Figure 1.2 illustrates this principle.   □

Chapter 2
Formal Language Theory: Basics

This chapter covers the basics of formal language theory. It covers all the notions that are necessary to follow the rest of this book. Apart from the classical rudiments, however, the chapter covers several lesser-known areas of this theory, such as parallel grammars, because these areas are also needed to fully grasp some upcoming topics discussed in this book. The chapter consists of four sections.

Section 2.1 introduces the very basics concerning strings, languages, and operations over them. Section 2.2 defines rewriting systems as fundamental language-defining devices. Section 2.3 overviews a variety of formal grammars, and Sect. 2.4 covers automata needed to follow the rest of this book.

Readers familiar with the classical concepts used in formal languages theory should primarily concentrate their attention on non-classical concepts, such as language-defining devices working in parallel.

2.1 Languages

An *alphabet* Σ is a finite, nonempty set of elements called *symbols*. If $\text{card}(\Sigma) = 1$, then Σ is a *unary alphabet*. A *string* or, synonymously, a *word* over Σ is any finite sequence of symbols from Σ. We omit all separating commas in strings; that is, for a string $a_1, a_2, \ldots, a_n$, for some $n \geq 1$, we write $a_1 a_2 \cdots a_n$ instead. The *empty string*, denoted by ε, is the string that is formed by no symbols, i.e. the empty sequence. By Σ^*, we denote the set of all strings over Σ (including ε). Set $\Sigma^+ = \Sigma^* - \{\varepsilon\}$.

Let x be a string over Σ, i.e. $x \in \Sigma^*$, and express x as $x = a_1 a_2 \cdots a_n$, where $a_i \in \Sigma$, for all $i = 1 \ldots, n$, for some $n \geq 0$ (the case when $n = 0$ means that $x = \varepsilon$). The *length* of x, denoted by $|x|$, is defined as $|x| = n$. The *reversal* of x, denoted by $\text{reversal}(x)$, is defined as $\text{reversal}(x) = a_n a_{n-1} \cdots a_1$. The *alphabet* of x, denoted by $\text{alph}(x)$, is defined as $\text{alph}(x) = \{a_1, a_2, \ldots, a_n\}$; informally, it is the set of symbols appearing in x. For $U \subseteq \Sigma$, $\#_U(x)$ denotes the number of occurrences

© Springer International Publishing AG 2017
A. Meduna, O. Soukup, *Modern Language Models and Computation*,
DOI 10.1007/978-3-319-63100-4_2

of symbols from U in x. If $U = \{a\}$, then instead of $\#_{\{a\}}(x)$, we write just $\#_a(x)$.
The *leftmost symbol* of x, denoted by $\mathrm{lms}(x)$, is defined as $\mathrm{lms}(x) = a_1$ if $n \geq 1$
and $\mathrm{lms}(x) = \varepsilon$ otherwise. The *rightmost symbol* of x, denoted by $\mathrm{rms}(x)$, is defined
analogously. If $n \geq 1$, then for every $i = 1, \ldots, n$, let $\mathrm{sym}(x, i)$ denote the ith symbol
in x. Notice that $|\varepsilon| = 0$, $\mathrm{reversal}(\varepsilon) = \varepsilon$, and $\mathrm{alph}(\varepsilon) = \emptyset$,

Let x and y be two strings over Σ. Then, xy is the *concatenation* of x and y. Note
that $x\varepsilon = \varepsilon x = x$. If x can be written in the form $x = uv$, for some $u, v \in \Sigma^*$, then
u is a *prefix* of x and v is a *suffix* of x. If $0 < |u| < |x|$, then u is a *proper prefix* of x;
similarly, if $0 < |v| < |x|$, then v is a *proper suffix* of x. Define $\mathrm{prefix}(x) = \{u \mid u$
is a prefix of $x\}$ and $\mathrm{suffix}(x) = \{v \mid v$ is a suffix of $x\}$. For every $i \geq 0$, $\mathrm{prefix}(x, i)$
is the prefix of x of length i if $|x| \geq i$, and $\mathrm{prefix}(x, i) = x$ if $|x| < i$. If $x = uvw$,
for some $u, v, w \in \Sigma^*$, then v is a *substring* of x. The set of all substrings of x is
denoted by $\mathrm{sub}(x)$. Moreover,

$$\mathrm{sub}(y, k) = \{x \mid x \in \mathrm{sub}(y), |x| \leq k\}$$

Let n be a nonnegative integer. Then, the *nth power* of x, denoted by x^n, is a string
over Σ recursively defined as

$$(1)\ x^0 = \varepsilon$$
$$(2)\ x^n = xx^{n-1} \text{ for } n \geq 1$$

Let $x = a_1 a_2 \cdots a_n$ be a string over Σ, for some $n \geq 0$. The set of all *permutations*
of x, denoted by $\mathrm{perm}(x)$, is defined as

$$\mathrm{perm}(x) = \{b_1 b_2 \cdots b_n \mid b_i \in \mathrm{alph}(x), \text{ for all } i = 1, \ldots, n, \text{ and }$$
$$(b_1, b_2, \ldots, b_n) \text{ is a permutation of } (a_1, a_2, \ldots, a_n)\}$$

Note that $\mathrm{perm}(\varepsilon) = \varepsilon$.

A *language* L over Σ is any set of strings over Σ, i.e. $L \subseteq \Sigma^*$. The set Σ^* is
called the *universal language* because it consists of all strings over Σ. If L is a finite
set, then it is a *finite language*; otherwise, it is an *infinite language*. The set of all
finite languages over Σ is denoted by $\mathrm{fin}(\Sigma)$. For $L \in \mathrm{fin}(\Sigma)$, $\mathrm{max\text{-}len}(L)$ denotes
the length of the longest string in L. We set $\mathrm{max\text{-}len}(\emptyset) = 0$. If $\mathrm{card}(\Sigma) = 1$, then
L is a *unary language*. The *empty language* is denoted by $\emptyset$.

The *alphabet* of L, denoted by $\mathrm{alph}(L)$, is defined as

$$\mathrm{alph}(L) = \bigcup_{x \in L} \mathrm{alph}(x)$$

The *permutation* of L, denoted by $\mathrm{perm}(L)$, is defined as

$$\mathrm{perm}(L) = \{\mathrm{perm}(x) \mid x \in L\}$$

The *reversal* of L, denoted by $\text{reversal}(L)$, is defined as

$$\text{reversal}(L) = \{\,\text{reversal}(x) \mid x \in L\,\}$$

For every $L \subseteq \Sigma^*$, where $\{\varepsilon\} \subseteq L$, and every $x \in \Sigma^*$, $\text{max-prefix}(x, L)$ denotes the longest prefix of x that is in L; analogously, $\text{max-suffix}(x, L)$ denotes the longest suffix of x that is in L.

Let L_1 and L_2 be two languages over Σ. Throughout the book, we consider L_1 and L_2 to be *equal*, symbolically written as $L_1 = L_2$, if $L_1 \cup \{\varepsilon\}$ and $L_2 \cup \{\varepsilon\}$ are identical. Similarly, $L_1 \subseteq L_2$ if $L_1 \cup \{\varepsilon\}$ is a subset of $L_2 \cup \{\varepsilon\}$.

As all languages are sets, all common operations over sets can be applied to them. Therefore,

$$L_1 \cup L_2 = \{x \mid x \in L_1 \text{ or } x \in L_2\}$$
$$L_1 \cap L_2 = \{x \mid x \in L_1 \text{ and } x \in L_2\}$$
$$L_1 - L_2 = \{x \mid x \in L_1 \text{ and } x \notin L_2\}$$

The *complement* of L, denoted by $\bar{L}$, is defined as

$$\bar{L} = \{x \mid x \in \Sigma^*, x \notin L\}$$

There are also some special operations which apply only to languages. The *concatenation* of L_1 and L_2, denoted by $L_1 L_2$, is the set

$$L_1 L_2 = \{x_1 x_2 \mid x_1 \in L_1 \text{ and } x_2 \in L_2\}$$

Note that $L\{\varepsilon\} = \{\varepsilon\}L = L$. For $n \geq 0$, the *nth power* of L, denoted by L^n, is recursively defined as

$$(1)\ L^0 = \{\varepsilon\}$$
$$(2)\ L^n = L^{n-1}L$$

The *closure* (*Kleene star*) of a language L, denoted by L^*, is the set

$$L^* = \bigcup_{i \geq 0} L^i$$

The *positive closure* of a language L, denoted by L^+, is the set

$$L^+ = \bigcup_{i \geq 1} L^i$$

The *right quotient* of L_1 with respect to L_2, denoted by L_1/L_2, is defined as

$$L_1/L_2 = \{y \mid yx \in L_1, \text{ for some } x \in L_2\}$$

Similarly, the *left quotient* of L_1 with respect to L_2, denoted by $L_2 \backslash L_1$, is defined as

$$L_2 \backslash L_1 = \{y \mid xy \in L_1, \text{ for some } x \in L_2\}$$

We also use special types of the right and left quotients. The *exhaustive right quotient* of L_1 with respect to L_2, denoted by $L_1 \mathbin{/\!/} L_2$, is defined as

$$L_1 \mathbin{/\!/} L_2 = \{y \mid yx \in L_1, \text{ for some } x \in L_2, \text{ and no } x' \in L_2$$
$$\text{such that } |x'| > |x| \text{ is a proper suffix of } yx\}$$

Similarly, the *exhaustive left quotient* of L_1 with respect to L_2, denoted by $L_2 \mathbin{\backslash\!\backslash} L_1$, is defined as

$$L_2 \mathbin{\backslash\!\backslash} L_1 = \{x \mid yx \in L_1, \text{ for some } y \in L_2, \text{ and no } y' \in L_2$$
$$\text{such that } |y'| > |y| \text{ is a proper prefix of } yx\}$$

Let $L_2 = \{\$\}^*$, where $\$$ is a symbol. Then, $L_1 \mathbin{/\!/} L_2$ is the *symbol-exhaustive right quotient* of L_1 with respect to $\$$, and $L_2 \mathbin{\backslash\!\backslash} L_1$ is the *symbol-exhaustive left quotient* of L_1 with respect to $\$$.

Let Σ be an alphabet. For $x, y \in \Sigma^*$, the *shuffle* of x and y, denoted by shuffle(x, y), is defined as

$$\text{shuffle}(x, y) = \{x_1 y_1 x_2 y_2 \cdots x_n y_n \mid x = x_1 x_2 \ldots x_n, y = y_1 y_2 \cdots y_n,$$
$$x_i, y_i \in \Sigma^*, 1 \le i \le n, n \ge 1\}$$

We extend the shuffle operation on languages in the following way. For $K_1, K_2 \subseteq \Sigma^*$,

$$\text{shuffle}(K_1, K_2) = \{z \mid z \in \text{shuffle}(x, y), x \in K_1, y \in K_2\}$$

Let Σ and Γ be two alphabets. Let K and L be languages over alphabets Σ and Γ, respectively. A *translation* from K to L is a relation ρ from Σ^* to Γ^* with domain$(\rho) = K$ and range$(\rho) = L$. A total function σ from Σ^* to 2^{Γ^*} such that $\sigma(uv) = \sigma(u)\sigma(v)$, for every $u, v \in \Sigma^*$, is a *substitution*. A substitution is *ε-free* if it is defined from Σ^* to 2^{Γ^+}. If $\sigma(a)$ for every $a \in \Sigma$ is finite, then σ is said to be *finite*. By this definition, $\sigma(\varepsilon) = \{\varepsilon\}$ and $\sigma(a_1 a_2 \cdots a_n) = \sigma(a_1)\sigma(a_2) \cdots \sigma(a_n)$, where $n \ge 1$ and $a_i \in \Sigma$, for all $i = 1, 2, \ldots, n$, so σ is completely specified by defining $\sigma(a)$ for each $a \in \Sigma$. For $L \subseteq \Sigma^*$, we extend the definition of σ to

$$\sigma(L) = \bigcup_{w \in L} \sigma(w)$$

A total function φ from Σ^* to Γ^* such that $\varphi(uv) = \varphi(u)\varphi(v)$, for every $u, v \in \Sigma^*$, is a *homomorphism* or, synonymously, a *morphism*. As any homomorphism is a special case of finite substitution, we specify φ by analogy with the specification of σ. For $L \subseteq \Sigma^*$, we extend the definition of φ to

$$\varphi(L) = \{\varphi(w) \mid w \in L\}$$

By analogy with substitution, φ is ε-free if $\varphi(a) \neq \varepsilon$, for every $a \in \Sigma$. By φ^{-1}, we denote the *inverse homomorphism*, defined as

$$\varphi^{-1}(u) = \{w \mid \varphi(u) = w\}$$

A homomorphism ω from Σ^* represents an *almost identity* if there exists a symbol $\# \in \Sigma$ such that $\omega(a) = a$, for every $a \in \Sigma - \{\#\}$, and $\omega(\#) \in \{\#, \varepsilon\}$. A homomorphism τ from Σ^* to Γ^* is a *coding* if $\tau(a) \in \Gamma$, for every $a \in \Sigma$.

Let L be a language over Σ, and let k be a positive integer. A homomorphism λ over Σ^* is a *k-linear erasing* with respect to L if and only if for each $y \in L$, $|y| \leq k|\lambda(y)|$. Furthermore, if $L \subseteq (\Sigma\{\varepsilon, c, c^2, \ldots, c^k\})^*$, for some $c \notin \Sigma$ and $k \geq 1$, and λ is defined by $\lambda(c) = \varepsilon$ and $\lambda(a) = a$, for all $a \in \Sigma$, then we say that λ is *k-restricted* with respect to L. Clearly, each k-restricted homomorphism is a k-linear erasing.

2.1.1 Language Families

By analogy with set theory, sets whose members are languages are called *families of languages*. A family of languages $\mathscr{L}$ is ε-free if for every $L \in \mathscr{L}$, $\varepsilon \notin L$. The family of finite languages is denoted by **FIN**.

Just like for languages, we consider two language families, $\mathscr{L}_1$ and $\mathscr{L}_2$, *equal* if and only if

$$\bigcup_{L \in \mathscr{L}_1} L \cup \{\varepsilon\} = \bigcup_{K \in \mathscr{L}_2} K \cup \{\varepsilon\}$$

If $\mathscr{L}_1$ and $\mathscr{L}_2$ are equal, we write $\mathscr{L}_1 = \mathscr{L}_2$. We also say that these two families *coincide*. $\mathscr{L}_1$ is a *subset* of $\mathscr{L}_2$, written as $\mathscr{L}_1 \subseteq \mathscr{L}_2$, if and only if

$$\bigcup_{L \in \mathscr{L}_1} L \cup \{\varepsilon\} \subseteq \bigcup_{K \in \mathscr{L}_2} K \cup \{\varepsilon\}$$

The closure of a language family under an operation is defined by analogy with the definition of the closure of a set. Next, we define three closure properties, discussed later in this book.

Definition 2.1.1. A language family $\mathscr{L}$ is *closed under linear erasing* if and only if for all $L \in \mathscr{L}$, $\lambda(L)$ is also in $\mathscr{L}$, where λ is a k-linear erasing with respect to L, for some $k \geq 1$. $\square$

Definition 2.1.2. A language family $\mathscr{L}$ is *closed under restricted homomorphism* if and only if for all $L \in \mathscr{L}$, $\lambda(L)$ is also in $\mathscr{L}$, where λ is a k-restricted homomorphism with respect to L, for some $k \geq 1$. $\square$

Definition 2.1.3. Let $\mathscr{L}$ be a language family. We say that $\mathscr{L}$ is *closed under endmarking* if and only if for every $L \in \mathscr{L}$, where $L \subseteq \Sigma^*$, for some alphabet Σ, $\# \notin \Sigma$ implies that $L\{\#\} \in \mathscr{L}$. $\square$

2.2 Rewriting Systems as Basic Language Models

Just like finite sets, finite languages can be specified by listing all the strings they contain. Of course, infinite languages, including almost all programming and natural languages, cannot be specified by an exhaustive enumeration of the strings they contain. Therefore, we customarily specify languages by suitable mathematical models so the models are of finite size even if the languages being specified are not. In the present section, based upon the mathematical notion of a relation, we define rewriting systems for this purpose.

Definition 2.2.1. A *rewriting system* is a pair, $M = (\Sigma, R)$, where Σ is an alphabet, and R is a finite relation on Σ^*. Σ is called the *total alphabet of M* or, simply, the *alphabet of M*. A member of R is called a *rule of M*, and accordingly, R is referred to as the *set of rules* in M.

The *rewriting relation* over Σ^* is denoted by $\vdash_M$ and defined so that for every $u, v \in \Sigma^*$, $u \vdash_M v$ in M iff there exist $(x, y) \in R$ and $w, z \in \Sigma^*$ such that $u = wxz$ and $v = wyz$. As usual, $\vdash_M^+$ and $\vdash_M^*$ denote the transitive and transitive and reflexive closure of $\vdash_M$, respectively. $\square$

Let $M = (\Sigma, R)$ be a rewriting system. Each rule $(x, y) \in R$ is written as $x \rightarrow y$ throughout this book. For $x \rightarrow y \in R$, x and y represent the *left-hand side of $x \rightarrow y$* and the *right-hand side of $x \rightarrow y$*, respectively. We drop M in $\vdash_M$ and, thereby, simplify $\vdash_M$ to $\vdash$ whenever M is automatically understood. By $u \vdash v \, [x \rightarrow y]$, where $u, v \in \Sigma^*$ and $x \rightarrow y \in R$, we express that M directly rewrites u as v according to $x \rightarrow y$. Of course, whenever the information regarding the applied rule is immaterial, we omit its specification; in other words, we simplify $u \vdash v \, [x \rightarrow y]$ to $u \vdash v$. By underlining, we specify the substring rewritten during a rewriting step if necessary. More formally, if $u = wxz$, $v = wyz$, $x \rightarrow y \in R$, where $u, v, x, y \in \Sigma^*$, then $w\underline{x}z \vdash wyz \, [x \rightarrow y]$ means that the x occurring behind w is rewritten during this step by using $x \rightarrow y$ (we usually specify the rewritten occurrence of x in this way when other occurrences of x appear in w and z).

To give a straightforward insight into the application of the defined notion, we now give the following examples. As the principle subject of this section, we discuss languages and their representations. It is thus only natural to illustrate our discussion by linguistically oriented examples.

Example 2.2.2. Let Δ denote the alphabet of English small letters (this alphabet is used in all examples of this section). In the present example, we introduce a rewriting system M that translates every digital string to the string in which every digit is converted to its corresponding English name followed by #; for instance, 010 is translated to *zero#one#zero#*.

First, we define the finite function h from $\{0, 1, \ldots, 9\}$ to Δ^* as

$$h(0) = zero,$$

$$h(1) = one,$$

$$h(2) = two,$$

$$h(3) = three,$$

$$h(4) = four,$$

$$h(5) = five,$$

$$h(6) = six,$$

$$h(7) = seven,$$

$$h(8) = eight,$$

$$h(9) = nine$$

In words, h translates every member of $\{0, 1, \ldots, 9\}$ to its corresponding English name; for instance, $h(9) = nine$. Based upon h, we define $M = (\Sigma, R)$ with $\Sigma = \{0, 1, \ldots, 9\} \cup \Delta \cup \{\#\}$ and $R = \{i \to h(i)\# \mid i \in \{0, 1, \ldots, 9\}\}$. Finally, we define the function $T(M)$ from $\{0, 1, \ldots, 9\}^*$ to $(\Delta \cup \{\#\})^*$ as

$$T(M) = \{(s, t) \mid s \vdash^* t, \ s \in \{0, 1, \ldots, 9\}^*, \ t \in (\Delta \cup \{\#\})^*\}$$

For instance, $T(M)$ contains $(911, nine\#one\#one\#)$. Indeed, M translates 911 to *nine#one#one#* as follows

$$9\underline{1}1 \qquad \vdash 9one\#1 \qquad [1 \to one\#]$$
$$9one\#\underline{1} \qquad \vdash 9one\#one\# \qquad [1 \to one\#]$$
$$\underline{9}one\#one\# \vdash nine\#one\#one\# \quad [9 \to nine\#]$$

Thus, $911 \vdash^* nine\#one\#one\#$ $[1 \to one\#, 1 \to one\#, 9 \to nine\#]$. Therefore, $(911, nine\#one\#one\#) \in T(M)$. Thus, M performs the desired translation. $\square$

Example 2.2.3. This example strongly resembles a very simple morphological study—in linguistics, *morphology* studies the structure of words. Indeed, it discusses re-structuring strings consisting of English letters, including strings that does not represent any English words, such as *xxuy*. More precisely, we introduce a rewriting system M that

(1) starts from non-empty strings consisting of small English letters delimited by angle brackets,

(2) orders the letters lexicographically, and
(3) eliminates the angle brackets.

For instance, M changes $\langle xxuy \rangle$ to $uxxy$.

Let Δ have the same meaning as in Example 2.2.2—that is, Δ denotes the alphabet of English lowercases. Let $_{\text{lex}}<$ denote the standardly defined lexical order over Δ—that is,

$$a \mathbin{_{\text{lex}}<} b \mathbin{_{\text{lex}}<} c \mathbin{_{\text{lex}}<} \cdots \mathbin{_{\text{lex}}<} y \mathbin{_{\text{lex}}<} z$$

We define $M = (\Sigma, R)$ with $\Sigma = \Delta \cup \{\langle, \rangle, 1, 2, 3\}$ and R containing the following rules

(i) $\langle \to 12, 12 \to 3$
(ii) $2\alpha \to \alpha2$ and $\alpha2 \to 2\alpha$ for all $\alpha \in \Delta$
(iii) $\beta2\alpha \to \alpha2\beta$ for all $\alpha, \beta \in \Delta$ such that $\alpha \mathbin{_{\text{lex}}<} \beta$
(iv) $3\alpha\beta \to \alpha3\beta$ for all $\alpha, \beta \in \Delta$ such that $\alpha \mathbin{_{\text{lex}}<} \beta$ or $\alpha = \beta$
(v) $3\alpha\rangle \to \alpha$ for all $\alpha \in \Delta$

Define the function $T(M)$ from $(\{\langle, \rangle\} \cup \Delta)^+$ to Δ^+ as

$$T(M) = \{(\langle s\rangle, t) \mid \langle s\rangle \vdash^* t, \text{ where } s, t \in \Delta^+\}$$

Observe that $(\langle s\rangle, t) \in TM)$ if and only if t is a permutation of s such that t has its letters lexicographically ordered according to $_{\text{lex}}<$. For instance, $T(M)$ contains $(\langle order\rangle, deorr)$. Indeed, M translates $\langle order\rangle$ to $deorr$ as follows

$$
\begin{array}{lll}
\langle order\rangle & \vdash 12order\rangle & [\langle \to 12] \\
1\underline{2o}rder\rangle & \vdash 1o2rder\rangle & [2o \to o2] \\
1o\underline{2r}der\rangle & \vdash 1or2der\rangle & [2r \to r2] \\
1or\underline{2d}er\rangle & \vdash 1od2rer\rangle & [r2d \to d2r] \\
1od\underline{2r}er\rangle & \vdash 1odr2er\rangle & [2r \to r2] \\
1odr\underline{2e}r\rangle & \vdash 1ode2rr\rangle & [r2e \to e2r] \\
1od\underline{e2}rr\rangle & \vdash 1od2err\rangle & [e2 \to 2e] \\
1o\underline{d2}err\rangle & \vdash 1o2derr\rangle & [d2 \to 2d] \\
1\underline{o2}derr\rangle & \vdash 1d2oerr\rangle & [o2d \to d2o] \\
1d\underline{2o}err\rangle & \vdash 1do2err\rangle & [2o \to o2] \\
1\underline{do}2err\rangle & \vdash 1de2orr\rangle & [o2e \to e2o] \\
1\underline{de}2orr\rangle & \vdash 1d2eorr\rangle & [e2 \to 2e] \\
1\underline{d2}eorr\rangle & \vdash 12deorr\rangle & [d2 \to 2d] \\
1\underline{2}deorr\rangle & \vdash 3deorr\rangle & [12 \to 3] \\
\underline{3d}eorr\rangle & \vdash d3eorr\rangle & [3de \to d3e] \\
d\underline{3e}orr\rangle & \vdash de3orr\rangle & [3eo \to e3o] \\
de\underline{3o}rr\rangle & \vdash deo3rr\rangle & [3or \to o3r] \\
deo\underline{3rr}\rangle & \vdash deor3r\rangle & [3rr \to r3r] \\
deor\underline{3r}\rangle & \vdash deorr & [3r\rangle \to r]
\end{array}
$$

Observe that M can translate $\langle order \rangle$ to *deorr* by a number of different sequences of rewriting steps. In fact, it can translate infinitely many members of $T(M)$ in various ways. In general, this phenomenon is referred to as non-determinism; accordingly, rewriting systems working in this way are said to be non-deterministic. In mathematics, we usually design the basic versions of rewriting systems so they work non-deterministically. In terms of their implementation, however, we obviously prefer using their deterministic versions. Therefore, we usually place a restriction on the way the rules are applied so the rewriting systems restricted in this way necessarily work deterministically; simultaneously, we obviously want that the deterministic restricted versions perform the same job as their original unrestricted counterparts. □

In the rest of this section, we focus on its key subject, which consists in using rewriting systems as language-defining models.

Whenever we use a rewriting system, $M = (\Sigma, R)$, as a language-defining model, then for brevity, we denote the language that M defines by $L(M)$. In principal, M defines $L(M)$ so it either *generates* $L(M)$ or *accepts* $L(M)$. Next, we explain these two fundamental language-defining methods in a greater detail. Let $S \in \Sigma^*$ and $F \in \Sigma^*$ be a *start language* and a *final language*, respectively.

(1) The *language generated by M* is defined as the set of all strings $y \in F$ such that $x \vdash^* y$ in M for some $x \in S$. M used in this way is generally referred to as a *language-generating model* or, more briefly, a *grammar*.
(2) The *language accepted by M* is defined as the set of all strings $x \in S$ such that $x \vdash^* y$ in M for some $y \in F$. M used in this way is referred to as a *language-accepting model* or, more briefly, an *automaton*.

Example 2.2.4. Let Δ have the same meaning as in Examples 2.2.2 and 2.2.3—that is, Δ denotes the alphabet of English lowercases. Let L be the language consisting of all even-length palindromes over Δ—a *palindrome* is a string that is the same whether written forwards or backward. For instance, *aa* and *noon* belong to L, but *ba* and *oops* do not. The present example introduces a grammar and an automaton that define L.

Let $G = (\Sigma, P)$ be the rewriting system with $\Sigma = \Delta \cup \{\#\}$ and

$$P = \{\# \rightarrow a\#a \mid a \in \Delta\} \cup \{\# \rightarrow \varepsilon\}$$

Set $S = \{\#\}$ and $F = \Delta^*$. Define the language generated by G as

$$L(G) = \{t \mid s \vdash^* t, \ s \in S, \ t \in F\}$$

In other words,

$$L(G) = \{t \mid \# \vdash^* t \text{ with } t \in \Delta^*\}$$

Observe that G acts as a grammar that generates L. For instance,

$$\# \vdash n\#n \vdash no\#on \vdash noon$$

in G, so *noon* $\in L(G)$.

To give an automaton that accepts L, introduce the rewriting system $A = (\Sigma, R)$ with $\Sigma = \Delta \cup \{\#\}$ and

$$R = \{a\#a \to \# \mid a \in \Delta\}$$

Set $S = \Delta^*\{\#\}\Delta^*$ and $F = \{\#\}$. Define the language accepted by A as

$$L(A) = \{st \mid s\#t \vdash^* u, \text{ where } s, t \in \Delta^*, \ u \in F\}$$

That is,

$$L(A) = \{st \mid s\#t \vdash^* \#, \text{ where } s, t \in \Delta^*\}$$

For instance, A accepts *noon*

$$no\#on \vdash n\#n \vdash \#$$

(as stated in the comments following Definition 2.2.1, the underlined substrings denote the substrings that are rewritten). On the other hand, consider this sequence of rewriting steps

$$no\#onn \vdash n\#nn \vdash \#n$$

It starts from *no#onn*, and after performing two steps, it ends up with *#n*, which cannot be further rewritten. Since $\#n \notin F$, which equals $\{\#\}$, M does not accept *no#onn*. As an exercise, based upon these observations, demonstrate that $L = L(A)$, so A acts as an automaton that accepts L.   □

Before closing this section, we make use of Example 2.2.4 to explain and illustrate the concept of equivalence and that of determinism in terms of rewriting systems that define languages.

2.2.1 Equivalence

If some rewriting systems define the same language, they are said to be *equivalent*. For instance, take G and A in Example 2.2.4. Both define the same language, so they are equivalent.

2.2.2 Determinism

Recall that Example 2.2.3 has already touched the topic of determinism in terms of rewriting systems. Notice that the language-defining rewriting system A from Example 2.2.4 works deterministically in the sense that A rewrites any string from

K by no more than one rule, where $K = S$, so $K = \Delta^*\{\#\}\Delta^*$. To express this concept of determinism more generally, let $M = (\Sigma, R)$ be a rewriting system and $K \subseteq \Sigma^*$. M is *deterministic over* K if for every $w \in K$, there is no more than one $r \in R$ such that $w \vdash v$ [r] with $v \in K$. Mathematically, if M is deterministic over K, then its rewriting relation $\vdash$ represents a function over K—that is, for all $u, v, w \in K$, if $u \vdash v$ and $u \vdash w$, then $v = w$. When K is understood, we usually just say that M is *deterministic*; frequently, we take $K = \Sigma^*$.

As already noted in Example 2.2.3, the basic versions of language-defining rewriting systems are always introduced quite generally and, therefore, non-deterministically. That is also why we first define the basic versions of these models in a non-deterministic way throughout this book. In practice, however, we obviously prefer their deterministic versions because they are easy to implement. Therefore, we always study whether any non-deterministic version can be converted to an equivalent deterministic version, and if so, we want to perform this conversion algorithmically. More specifically, we reconsider this crucially important topic of determinism in terms of finite automata in Sect. 2.4.

2.3 Grammars

In this section, based upon rewriting systems, we define grammars as formal devices that generate languages. Grammars play an important role throughout this book.

2.3.1 Grammars in General

Since grammars represent special cases of rewriting systems (see Sect. 2.2), we often use the mathematical terminology concerning these systems throughout the present section. Specifically, we apply the relations $\vdash$, $\vdash^n$, $\vdash^+$, and $\vdash^*$ to these grammars.

The notion of a *grammar* represents a rewriting system $G = (\Sigma, R)$, where

- Σ is divided into two disjoint subalphabets, denoted by N and T;
- R is a finite set of rules of the form $A \to x$, where $A \in N$ and $x \in \Sigma^*$.

N and T are referred to as the *alphabet of nonterminal symbols* and the *alphabet of terminal symbols*, respectively. N contains a special *start symbol*, denoted by S.

If $S \vdash^* w$, where $w \in \Sigma^*$, G *derives* w, and w is a *sentential form*. $F(G)$ denotes the set of all sentential forms derived by G. The *language generated by* G, symbolically denoted by $L(G)$, is defined as $L(G) = F(G) \cap T^*$. Members of $L(G)$ are called *sentences*. If $S \vdash^* w$ and w is a sentence, $S \vdash^* w$ is a *successful derivation* in G.

More customarily, however, the notion of a grammar or, more precisely that of a phrase-structure grammar is defined as follows.

Definition 2.3.1. A *phrase-structure grammar* is a quadruple

$$G = (V, T, P, S)$$

where

- V is a *total alphabet*;
- T is an alphabet of *terminals* such that $T \subseteq V$;
- P is a finite relation from $V^* - T^*$ to V^*;
- $S \in V - T$ is the *start symbol*.

The set $N = V - T$ is the set of *nonterminals* such that $N \cap T = \emptyset$.

Pairs $(u, v) \in P$ are called *rewriting rules* (abbreviated *rules*) or *productions*, and are written as $u \to v$. A rewriting rule $u \to v \in P$ satisfying $v = \varepsilon$ is called an *erasing rule*. If there is no such rule in P, then we say that G is a *propagating* (or *ε-free*) grammar.

The G-based *direct derivation relation* over V^* is denoted by $\Rightarrow_G$ and defined as

$$x \Rightarrow_G y$$

if and only if $x = x_1 u x_2, y = y_1 v y_2$, and $u \to v \in P$, where $x_1, x_2, y_1, y_2 \in V^*$. Since $\Rightarrow_G$ is a relation, $\Rightarrow_G^k$ is the kth power of $\Rightarrow_G$, for $k \geq 0$, $\Rightarrow_G^+$ is the transitive closure of $\Rightarrow_G$, and $\Rightarrow_G^*$ is the reflexive-transitive closure of $\Rightarrow_G$. Let $D: S \Rightarrow_G^* x$ be a derivation, for some $x \in V^*$. Then, x is a *sentential form*. If $x \in T^*$, then x is a *sentence*. If x is a sentence, then D is a *successful* (or *terminal*) *derivation*.

The *language* of G, denoted by $L(G)$, is the set of all sentences defined as

$$L(G) = \{w \in T^* \mid S \Rightarrow_G^* w\} \qquad \qquad \Box$$

Next, for every phrase-structure grammar G, we define two sets, $F(G)$ and $\Delta(G)$. $F(G)$ contains all sentential forms of G. $\Delta(G)$ contains all sentential forms from which there is a derivation of a string in $L(G)$.

Definition 2.3.2. Let $G = (V, T, P, S)$ be a phrase-structure grammar. Set

$$F(G) = \{x \in V^* \mid S \Rightarrow_G^+ x\}$$

and

$$\Delta(G) = \{x \in F(G)^* \mid x \Rightarrow_G^* y, \ y \in T^*\} \qquad \qquad \Box$$

For brevity, we often denote a rule $u \to v$ with a unique label r as $r: u \to v$, and instead of $u \to v \in P$, we simply write $r \in P$. The notion of rule labels is formalized in the following definition.

Definition 2.3.3. Let $G = (V, T, P, S)$ be a phrase-structure grammar. Let Ψ be a set of symbols called *rule labels* such that $\text{card}(\Psi) = \text{card}(P)$, and ψ be a bijection from P to Ψ. For simplicity and brevity, to express that ψ maps a rule, $u \to v \in P$, to r, where $r \in \Psi$, we write $r{:}u \to v \in P$; in other words, $r{:}u \to v$ means that $\psi(u \to v) = r$. For $r{:}u \to v \in P$, u and v represent the *left-hand side* of r, denoted by $\text{lhs}(r)$, and the *right-hand side* of r, denoted by $\text{rhs}(r)$, respectively. Let P^* and Ψ^* denote the set of all sequences of rules from P and the set of all sequences of rule labels from Ψ, respectively. Set $P^+ = P^* - \{\varepsilon\}$ and $\Psi^+ = \Psi^* - \{\varepsilon\}$. As with strings, we omit all separating commas in these sequences.

We extend ψ from P to P^* in the following way

(1) $\psi(\varepsilon) = \varepsilon$

(2) $\psi(r_1 r_2 \cdots r_n) = \psi(r_1)\psi(r_2) \cdots \psi(r_n)$

for any sequence of rules $r_1 r_2 \cdots r_n$, where $r_i \in P$, for all $i = 1, 2, \ldots, n$, for some $n \geq 1$.

Let $w_0, w_1, \ldots, w_n$ be a sequence of strings, where $w_i \in V^*$, for all $i = 0, 1, \ldots, n$, for some $n \geq 1$. If $w_{j-1} \Rightarrow_G w_j$ according to r_j, where $r_j \in P$, for all $j = 1, 2, \ldots, n$, then we write

$$w_0 \Rightarrow_G^n w_n \; [\psi(r_1 r_2 \cdots r_n)]$$

For any string w, we write

$$w \Rightarrow_G^0 w \; [\varepsilon]$$

For any two strings w and y, if $w \Rightarrow_G^n y \; [\rho]$ for $n \geq 0$ and $\rho \in \Psi^*$, then we write

$$w \Rightarrow_G^* y \; [\rho]$$

If $n \geq 1$, which means that $|\rho| \geq 1$, then we write

$$w \Rightarrow_G^+ y \; [\rho]$$

If $w = S$, then ρ is called the *sequence of rules (rule labels)* used in the derivation of y or, more briefly, the *parse*[1] of y. □

For any phrase-structure grammar G, we automatically assume that V, N, T, S, P, and Ψ denote its total alphabet, the alphabet of nonterminals, the alphabet of terminals, the start symbol, the set of rules, and the set of rule labels, respectively. Sometimes, we write $G = (V, T, \Psi, P, S)$ instead of $G = (V, T, P, S)$ with Ψ having the above-defined meaning.

---

[1]Let us note that the notion of a parse represents a synonym of several other notions, including a *derivation word*, a *Szilard word*, and a *control word*.

In the literature, a phrase-structure grammar is also often defined with rules of the form $x \rightarrow y$, where $x \in V^+$ and $y \in V^*$ (see, for instance, [Woo87]). Both definitions are interchangeable in the sense that the grammars defined in these two ways generate the same family of languages—the family of recursively enumerable languages.

Definition 2.3.4. A *recursively enumerable language* is a language generated by a phrase-structure grammar. The family of recursively enumerable languages is denoted by **RE**.   □

Throughout this book, in the proofs, we frequently make use of Turing-Church thesis (see [Chu36b, Chu36a, Tur36]), which we next state in terms of formal language theory. Before this, however, we need to explain how we understand the *intuitive notion of an effective procedure* or, briefly, a *procedure*, contained in this thesis. We surely agree that each procedure describes how to perform a task in an unambiguous and detailed way. We also agree that it consists of finitely many instructions, each of which can be executed mechanically in a fixed amount of time. When performed, a procedure reads input data, executes its instructions, and produces output data; of course, both the input data and the output data may be nil. We are now ready to state Turing-Church thesis in terms of **RE**—that is, the family of recursively enumerable languages, defined by phrase-structure grammars (see Definition 2.3.1).

Turing-Church Thesis. *Let L be a language. Then, $L \in$ **RE** if and only if there is a procedure that defines L by listing all its strings.*

All the grammars and automata discussed in this book obviously constitutes procedures in the above sense. Consequently, whenever grammars or automata of a new type are considered in this book, Turing-Church thesis automatically implies that the language family they define is necessarily contained in **RE**, and we frequently make use of this implication in the sequel.

Observe that Turing-Church thesis is indeed a thesis, not a theorem because it cannot be proved. Indeed, any proof of this kind would necessitate a formalization of our intuitive notion of a language-defining procedure so it can be rigorously compared with the notion of a phrase-structure grammar. At this point, however, there would be a problem whether this newly formalized notion is equivalent to the intuitive notion of a procedure, which would give rise to another thesis similar to Turing-Church thesis. Therefore, any attempt to prove this thesis inescapably ends up with an infinite regression. However, the evidence supporting Turing-Church thesis is hardly disputable because throughout its history, computer science has formalized the notion of a procedure in the intuitive sense by other language-defining models, such as Post systems (see [Pos43]) and Markov algorithms (see [Mar60]), and all of them have turned out to be equivalent with phrase-structure grammars. Even more importantly, nobody has ever come with a procedure that defines a language and demonstrated that the language cannot be generated by any phrase-structure grammar.

Originally, Turing-Church thesis have been stated in terms of Turing machines in [Tur36]. Indeed, Church and Turing hypothesized that any computational process which could be reasonably called as a procedure could be simulated by a Turing machine (see [Rog87] for an in-depth discussion concerning to Turing-Church thesis). In the present monograph, however, we do not need the notion of a Turing machine while we frequently make use of the notion of a phrase-structure grammar. Therefore, for the purposes of this book, we have reformulated Turing-Church thesis in the above way. As phrase-structure grammars and Turing machines are equivalent (see [Med00a]), this reformulation is obviously perfectly correct and legal from a mathematical viewpoint.

Any language models that characterize **RE** are said to be *computationally complete* because they are as strong as all possible procedures in terms of language-defining power according to Turing-Church thesis. Apart from them, however, this book also discusses many *computationally incomplete* language models, which define proper subfamilies of **RE**. For instance, the following special versions of phrase-structure grammars are all computationally incomplete.

Definition 2.3.5. A *context-sensitive grammar* is a phrase-structure grammar

$$G = (V, T, P, S)$$

such that every $u \to v$ in P is of the form

$$u = x_1 A x_2, \quad v = x_1 y x_2$$

where $x_1, x_2 \in V^*$, $A \in N$, and $y \in V^+$. A *context-sensitive language* is a language generated by a context-sensitive grammar. The family of context-sensitive languages is denoted by **CS**. □

The family of context-sensitive languages is also characterized by monotone phrase-structure grammars.

Definition 2.3.6. A *monotone phrase-structure grammar* is a phrase-structure grammar

$$G = (V, T, P, S)$$

such that $u \to v \in P$ satisfies $|u| \le |v|$. A *monotone recursively enumerable language* is a language generated by a monotone phrase-structure grammar. The family of monotone recursively enumerable languages is denoted by **MON**. □

Definition 2.3.7. A *context-free grammar* is a phrase-structure grammar

$$G = (V, T, P, S)$$

such that every rule in P is of the form

$$A \to x$$

where $A \in N$ and $x \in V^*$. A *context-free language* is a language generated by a context-free grammar. The family of context-free languages is denoted by **CF**. □

Definition 2.3.8. A *linear grammar* is a phrase-structure grammar

$$G = (V, T, P, S)$$

such that every rule in P is of the form

$$A \to xBy \text{ or } A \to x$$

where $A, B \in N$ and $x, y \in T^*$. A *linear language* is a language generated by a linear grammar. The family of linear languages is denoted by **LIN**.   □

Definition 2.3.9. A *regular grammar* is a phrase-structure grammar

$$G = (V, T, P, S)$$

such that every rule in P is of the form

$$A \to aB \text{ or } A \to a$$

where $A, B \in N$ and $a \in T$. A *regular language* is a language generated by a regular grammar. The family of regular languages is denoted by **REG**.   □

Alternatively, the family of regular languages is characterized by right-linear grammars, defined next.

Definition 2.3.10. A *right-linear grammar* is a phrase-structure grammar

$$G = (V, T, P, S)$$

such that every rule in P is of the form

$$A \to xB \text{ or } A \to x$$

where $A, B \in N$ and $x \in T^*$. A *right-linear language* is a language generated by a right-linear grammar. The family of right-linear languages is denoted by **RLIN**.   □

Definition 2.3.11. Define the following abbreviations. Let *PSG, MONG, CSG, CFG, LG, RLG,* and *RG,* denote phrase-structure, monotone phrase-structure, context-sensitive, context-free, linear, right-linear, and regular grammar, respectively. Let X denotes a specific type of grammar. Then, $X^{-\varepsilon}$ denotes its propagating variant. Additionally, for a specific type of grammar Y, Γ_Y denotes the set of all grammars of type Y.   □

Definition 2.3.12. Let $G = (V, T, P, S)$ be a phrase-structure grammar. To explicitly specify that G use a derivation relation $_d\Rightarrow$ to generate $L(G)$, we write

$$L(G, {}_d\Rightarrow) = \{x \in T^* \mid S \; {}_d\Rightarrow^* x\}$$

and say that $L(G, {}_d\Rightarrow)$ is the *language that G generates by using* ${}_d\Rightarrow$. For any $X \subseteq \Gamma_{PSG}$, set

$$\mathscr{L}(X, {}_d\Rightarrow) = \{L(G, {}_d\Rightarrow) \mid G \in X\} \qquad \square$$

Definition 2.3.13. Let $G = (V, T, P, S)$ be any grammar. Let $w \in T^*$ and

$$\alpha: S = w_0 \Rightarrow w_1 \Rightarrow w_2 \Rightarrow \cdots \Rightarrow w_n = w$$

be a derivation in G, for some $n \geq 0$. Then,

$$\text{Ind}(G, w, \alpha) = \max(\{\#_N(w_i) \mid 0 \leq i \leq n\})$$
$$\text{Ind}(G, w) = \min(\{\text{Ind}(G, w, \alpha) \mid \alpha \text{ is a derivation of } w \text{ in } G\})$$
$$\text{Ind}(G) = \sup(\{\text{Ind}(G, w) \mid w \in L(G)\})$$

is *index of derivation D of the string w in G*, *index of the string w in G*, and *index of G*, respectively. If there is a constant $k \geq 1$ such that $\text{Ind}(G) = k$, G is said to be of index k. $\qquad \square$

Set $\mathbf{CF}_k = \{L \mid L = L(G), G \text{ is CFG}, \text{Ind}(G) = k, k \geq 1\}$ and $\mathbf{CF}_{fin} = \{L \mid L \in \mathbf{CF}_i, \text{ for some } i \geq 1\}$. For further details concerning finite index of grammars, see Chapter 3 in [DP89].

To illustrate the above-introduced notation, let $G = (V, T, P, S)$ be a RLG; then, $L(G, \Rightarrow) = \{x \in T^* \mid S \Rightarrow^* x\}$, and $\mathscr{L}(\Gamma_{RLG}, \Rightarrow) = \{L(G, \Rightarrow) \mid G \in \Gamma_{RLG}\}$. To give another example, $\mathscr{L}(\Gamma_{CFG}, \Rightarrow)$ denotes the family of all context-free languages.

Notice that $\mathbf{REG} = \mathbf{RLIN} = \mathscr{L}(\Gamma_{RLG}, \Rightarrow) = \mathscr{L}(\Gamma_{RG}, \Rightarrow)$, $\mathbf{LIN} = \mathscr{L}(\Gamma_{LG}, \Rightarrow)$, $\mathbf{CF} = \mathscr{L}(\Gamma_{CFG}, \Rightarrow)$, $\mathbf{CS} = \mathbf{MON} = \mathscr{L}(\Gamma_{MONG}, \Rightarrow) = \mathscr{L}(\Gamma_{CSG}, \Rightarrow)$, and $\mathbf{RE} = \mathscr{L}(\Gamma_{PSG}, \Rightarrow)$.

Concerning the families of finite, regular, right-linear, linear, context-free of finite index, context-free, context-sensitive, monotone recursively enumerable, and recursively enumerable languages, the next important theorem holds true.

Theorem 2.3.14 (Chomsky Hierarchy, see [Cho56, Cho59]).

$$\mathbf{FIN} \subset \mathbf{REG} = \mathbf{RLIN} \subset \mathbf{LIN} \subset \mathbf{CF}_{fin} \subset \mathbf{CF} \subset \mathbf{CS} = \mathbf{MON} \subset \mathbf{RE}$$

Next, we recall canonical derivations in context-free grammars.

Definition 2.3.15. Let $G = (V, T, \Psi, P, S)$ be a context-free grammar. The relation of a *direct leftmost derivation*, denoted by ${}_{lm}\Rightarrow_G$, is defined as follows: if $u \in T^*$, $v \in V^*$, and $r: A \to x \in P$, then

$$uAv \,{}_{lm}\Rightarrow_G uxv \ [r]$$

Let $_{lm}\Rightarrow_G^n$, $_{lm}\Rightarrow_G^*$, and $_{lm}\Rightarrow_G^+$ denote the nth power of $_{lm}\Rightarrow_G$, for some $n \geq 0$, the reflexive-transitive closure of $_{lm}\Rightarrow_G$, and the transitive closure of $_{lm}\Rightarrow_G$, respectively. The *language that G generates by using leftmost derivations* is denoted by $L(G, _{lm}\Rightarrow)$ and defined as

$$L(G, _{lm}\Rightarrow) = \{w \in T^* \mid S\ _{lm}\Rightarrow_G^* w\}$$

If $S\ _{lm}\Rightarrow_G^* w\ [\rho]$, where $w \in T^*$, then ρ is the *left parse* of w.   □

By analogy with leftmost derivations and left parses, we define rightmost derivations and right parses.

Definition 2.3.16. Let $G = (V, T, \Psi, P, S)$ be a context-free grammar. The relation of a *direct rightmost derivation*, denoted by $_{rm}\Rightarrow_G$, is defined as follows: if $u \in V^*$, $v \in T^*$, and $r: A \to x \in P$, then

$$uAv\ _{rm}\Rightarrow_G uxv\ [r]$$

Let $_{rm}\Rightarrow_G^n$, $_{rm}\Rightarrow_G^*$, and $_{rm}\Rightarrow_G^+$ denote the nth power of $_{rm}\Rightarrow_G$, for some $n \geq 0$, the reflexive-transitive closure of $_{rm}\Rightarrow_G$, and the transitive closure of $_{rm}\Rightarrow_G$, respectively. The *language that G generates by using rightmost derivations* is denoted by $L(G, _{rm}\Rightarrow_G)$ and defined as

$$L(G, _{rm}\Rightarrow) = \{w \in T^* \mid S\ _{rm}\Rightarrow_G^* w\}$$

If $S\ _{rm}\Rightarrow_G^* w\ [\rho]$, where $w \in T^*$, then ρ is the *right parse* of w.   □

Without any loss of generality, in context-free grammars, we may consider only canonical derivations, which is formally stated in the following theorem.

Theorem 2.3.17 (See [Med00a]). *Let G be a context-free grammar. Then,*

$$L(G, _{lm}\Rightarrow) = L(G, _{rm}\Rightarrow) = L(G)$$

The following theorem gives a characterization of the family of recursively enumerable languages by context-free languages.

Theorem 2.3.18 (See [GGH67]). *For every recursively enumerable language K, there exist two context-free languages, L_1 and L_2, and a homomorphism h such that*

$$K = h(L_1 \cap L_2)$$

The next theorem says that if a phrase-structure grammar generates each of its sentences by a derivation satisfying a length-limited condition, then the generated language is, in fact, context sensitive.

Theorem 2.3.19 (Workspace Theorem, see [Sal73]). *Let* $G = (V, T, P, S)$ *be a phrase-structure grammar. If there is a positive integer* k *such that for every nonempty* $y \in L(G)$, *there exists a derivation*

$$D: S \Rightarrow_G x_1 \Rightarrow_G x_2 \Rightarrow_G \cdots \Rightarrow_G x_n = y$$

where $x_i \in V^*$ *and* $|x_i| \leq k|y|$, *for all* $i = 1, 2, \ldots, n$, *for some* $n \geq 1$, *then* $L(G) \in$ **CS.**

2.3.2 How to Prove Context-Freeness

Formal language theory has always intensively struggled to establish conditions under which phrase-structure grammars generate a proper subfamily of the family of recursively enumerable languages because results like this often significantly simplify proofs that some languages are members of the subfamily. To illustrate, consider the well-known workspace theorem for phrase-structure grammars, which fulfills a crucially important role in the grammatically oriented theory of formal languages as a whole (see Theorem III.10.1 in [Sal73]). This theorem represents a powerful tool to demonstrate that if a phrase-structure grammar H generates each of its sentences by a derivation satisfying a prescribed condition (specifically, this condition requires that there is a positive integer k such that H generates every sentence y in the generated language $L(H)$ by a derivation in which every sentential form x satisfies $|x| \leq k|y|$), then $L(H)$ is a member of the context-sensitive language family. Regarding the membership in the context-free language family, however, formal language theory lacks a result like this. To fill this gap, the present section establishes a tree-based condition so every phrase-structure grammar satisfying this condition generates a member of the context-free language family.

To give an insight into this result, we first sketch some terminology. Recall that a phrase-structure grammar G is in Kuroda normal form (see Definition 3.1.1) if any rule satisfies one of these forms

$$AB \rightarrow CD, A \rightarrow BC, A \rightarrow B, A \rightarrow a, \text{ or } A \rightarrow \varepsilon$$

where A, B, C, D are nonterminals, a is a terminal, and ε is the empty string. We define the notion of a derivation tree t graphically representing a derivation in G by analogy with this notion in terms of an ordinary context-free grammar (see Definition 6.8 on page 92 in [Med14]). In addition, however, we introduce context-dependent pairs of nodes in t as follows. In t, two paths are neighboring if no other path occurs between them. Let p and q be two neighboring paths in t. Let p contain a node k with a single child l, where k and l are labelled with A and C, respectively, and let q contain a node m with a single child n, where m and n are labelled with B and D, respectively. Let this four-node portion of t; consisting of k, l, m, and n; graphically represents an application of $AB \rightarrow CD$. Then, k and m are a context-dependent pair of nodes (see Fig. 2.1).

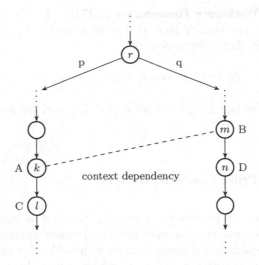

Fig. 2.1 Illustration of context dependency in t

As its main result, the present section proves that the language of G, $L(G)$, is context-free if there is a constant k such that every $w \in L(G)$ is the frontier of a derivation tree d in which any pair of neighboring paths contains k or fewer context-dependent pairs of nodes. Apart from its theoretical value, this result may be of some interest in practice, too. Specifically, some language processors, such as compiler parsers, frequently require that the languages processed by them are context-free. As obvious, the result stated above may fulfill a useful role during the verification of this requirement.

The section is organized as follows. First, we give all the necessary terminology. Then, we establish the main result of this section. Finally, we close this section by showing an application perspective of the main result.

Definitions and Examples

Definition 2.3.20. Let $t = (V, E)$ be a tree. Define a partial order relation $<$ over V as follows. For a path $\alpha = (m_0, m_1, \ldots, m_k)$, where $m_0 = \text{troot}(t)$, $m_i < m_k$, $0 \le i \le k - 1$. An ordered tree is called *labelled*, if there exists a set of labels $\mathcal{L}$ and a total mapping $l : V \to \mathcal{L}$. In what follows we substitute a node of a tree by its label if there is no risk of confusion.

Let t be an ordered tree with a node o. Let $\alpha = (o, m_1, m_2, \ldots, m_r)$ and $\beta = (o, n_1, n_2, \ldots, n_s)$ be two paths in t, for some $r, s \ge 1$, such that o is the parent of m_1 and n_1, where

(1) m_1 is the direct left sibling of n_1;
(2) m_i is the rightmost child of m_{i-1}, and n_j is the leftmost child of n_{j-1}, $2 \le i \le r$, $2 \le j \le s$.

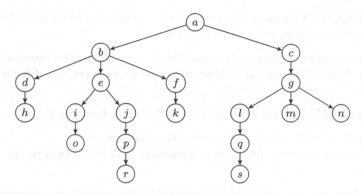

Fig. 2.2 Labelled ordered tree t

Then, α and β are two *neighboring paths* in t, α is a *left neighboring path* to β, and β is a *right neighboring path* to α. □

Let us demonstrate the tree-related notions by the following example.

Example 2.3.21. The following graph (Fig. 2.2.) represents labelled ordered tree t. The root node troot(t) is labelled a. It has no parent and two children b and c. Then, b is a sibling of c and c is a sibling of b. The leftmost child of b is d, while the rightmost is f. The node d is a left sibling of f, however, it is not the direct left sibling, which is e. The node f is the parent of k, but k has no child, so it is a leaf node. *horksmn* = frontier(t). Consider the node e. The nodes a and b are predecessors of e, while i, j, o, p, and r are e's descendants. The nodes c or d are not in predecessor relation with e, since they are neither predecessors of e, nor descendants of e. The sequence of nodes *bejpr* is a path in t. The path *bfk* is neighboring to *bejpr*; unlike *abfk*, *eio*, or *bdh*. □

Definition 2.3.22. Let $G = (V, T, P, S)$ be a phrase-structure grammar. G is in the *binary form* if any $p \in P$ has one of these forms,

$$AB \to CD, A \to BC, A \to X$$

where $A, B, C, D \in N$, $X \in V \cup \{\varepsilon\}$. In what follows, unless explicitly stated otherwise, we automatically assume that every PSG is in the binary form. □

Theorem 2.3.23. *A language L is context-sensitive iff $L = L(G)$, where G is a monotone phrase-structure grammar in the binary form.*

Proof. A language L is context-sensitive iff L is generated by monotone phrase-structure grammar in Kuroda normal form (see Definition 3.1.1). Every monotone PSG in Kuroda normal form is a special case of a monotone PSG in the binary form. □

Theorem 2.3.24. *A language L is recursively enumerable iff* $L = L(G)$, *where G is a phrase-structure grammar in the binary form.*

Proof. A language L is recursively enumerable iff L is generated by phrase-structure grammar. Every PSG can be converted to the binary form (see Chapter 4 in [RS97a]).   □

Definition 2.3.25. Let $G = (V, T, P, S)$ be a PSG in the binary form.

(1) For $p: A \rightarrow x \in P$, $A\langle x \rangle$ is the *rule tree* that represents p.
(2) The *derivation trees* representing derivations in G are defined recursively as follows:

> (a) One-node tree with a node labelled X is the derivation tree corresponding to $X \Rightarrow^0 X$ in G, where $X \in V$. If $X = \varepsilon$, we refer to the node labeled X as ε-*node* (ε-*leaf*); otherwise, we call it *non-ε-node* (*non-ε-leaf*).
> (b) Let d be the derivation tree with frontier$(d) = uAv$ representing $X \Rightarrow^*$ uAv $[\rho]$ and let $p: A \rightarrow x \in P$. The derivation tree that represents
>
> $$X \Rightarrow^* uAv \, [\rho] \Rightarrow uxv \, [p]$$
>
> is obtained by replacing the ith non-ε-leaf in d labelled A, with rule tree corresponding to p, $A\langle x \rangle$, where $i = |uA|$.
> (c) Let d be the derivation tree representing $X \Rightarrow^* uABv \, [\rho]$ with frontier$(d) = uABv$, and let $p: AB \rightarrow CD \in P$. The derivation tree that represents
>
> $$X \Rightarrow^* uABv \, [\rho] \Rightarrow uCDv \, [p]$$
>
> is obtained by replacing the ith and $(i+1)$th non-ε-leaf in d labelled A and B with $A\langle C \rangle$ and $B\langle D \rangle$, respectively, where $i = |uA|$.

(3) A *derivation tree* in G is any tree t for which there is a derivation represented by t (see (2) in this definition).

Note, after replacement in (c), the nodes A and B are the parents of the new leaves C and D, respectively, and we say that A and B are *context-dependent*, alternatively speaking, we say that there is a context dependency between A and B. In a derivation tree, two nodes are *context-independent* if they are not context-dependent.

Then, for any $p: A \rightarrow x \in P$, $_G\triangle(p)$ denotes rule tree corresponding to p. For any $A \Rightarrow^* x \, [\rho]$ in G, where $A \in N$, $x \in V^*$, and $\rho \in P^*$, $_G\triangle(A \Rightarrow^* x \, [\rho])$ denotes the derivation tree corresponding to $A \Rightarrow^* x \, [\rho]$. Just like we often write $A \Rightarrow^* x$ instead of $A \Rightarrow^* x \, [\rho]$, we sometimes simplify $_G\triangle(A \Rightarrow^* x \, [\rho])$ to $_G\triangle(A \Rightarrow^* x)$ in what follows if there is no danger of confusion. Let $_G\blacktriangle$ denotes the set of all derivation trees in G. Finally, by $_G\triangle_x \in {_G}\blacktriangle$, we mean a derivation tree whose frontier is x, where $x \in F(G)$.

If a node is labelled with a terminal, it is called a *terminal node*. If a node is labelled with a nonterminal, it is called a *nonterminal node*.

Let $\alpha = (o, m_1, m_2, \ldots, m_r)$ and $\beta = (o, n_1, n_2, \ldots, n_s)$ be two neighboring paths, where $r, s \geq 0$, α is the left neighboring path to β, and m_r and n_s are terminal nodes. Then, there is a t-tuple $\gamma = (g_1, g_2, \ldots, g_t)$ of nodes from α and t-tuple $\delta = (h_1, h_2, \ldots, h_t)$ of nodes from β, where $g_p < g_q$, for $1 \leq p < q \leq t$, $t < \min(r, s)$, and g_i and h_i are context-dependent, for $1 \leq i \leq t$. Let $\rho = p_1 p_2 \ldots p_t$ be a string of non-context-free rules corresponding to context dependencies between γ and δ. We call ρ the *right context of* α and the *left context of* β or the *context of* α *and* β. Consider a node m_i, where $1 \leq i \leq r$, and two $(t - k + 1)$-tuples of nodes $\sigma = (g_k, g_{k+1}, \ldots, g_t)$ and $\varphi = (h_k, h_{k+1}, \ldots, h_t)$, where k is a minimal integer such that $m_i < g_k$. Then, a string of non-context-free rules $\tau = p_k p_{k+1} \ldots p_t$ corresponding to context dependencies between σ and φ is called the *right descendant context of* m_i, for some $1 \leq k \leq t$. Analogously, we define the notion of the *left descendant context* of a node n_j in β, for some $1 \leq j \leq s$. □

Example 2.3.26. Let $G = (V, T, P, S)$ be a phrase-structure grammar, where $V = \{S, S_a, S_b, X, X_a, X_b, Z_a, Z_b, A, 1, 2, 3, A_x, \overline{a}, a, B, B_x, \overline{b}, b\}$, $T = \{a, b\}$, and P contains the following rules:

(1) $S \to S_a B_x$	(9) $X \to BX_a$	(17) $Z_b \to B$	(25) $A_x \to \overline{a}$
(2) $S \to S_b A_x$	(10) $X_a \to XA$	(18) $AB \to A_x B$	(26) $AA_x \to \overline{aa}$
(3) $S_a \to Z_a X$	(11) $X_b \to XB$	(19) $BA \to B_x A$	(27) $1A_x \to \overline{aa}$
(4) $S_b \to Z_b X$	(12) $Z_a A \to AZ_a$	(20) $BA_x \to B_x A_x$	(28) $2A_x \to \overline{aa}$
(5) $X \to XX$	(13) $Z_a B \to BZ_a$	(21) $AA \to \overline{a}1$	(29) $BB \to \overline{b}B_x$
(6) $X \to AB$	(14) $Z_b A \to AZ_b$	(22) $1A \to \overline{a}2$	(30) $B_x B \to \overline{b}B_x$
(7) $X \to BA$	(15) $Z_b B \to BZ_b$	(23) $2A \to \overline{a}3$	(31) $B_x B_x \to \overline{bb}$
(8) $X \to AX_b$	(16) $Z_a \to A$	(24) $3A_x \to \overline{aa}$	(32) $\overline{a} \to a$
			(33) $\overline{b} \to b$

At this point, let us make only an informal observation that $L(G)$ is the language of all nonempty strings above T consisted of an equal number of as and bs, where every sequence of as is of a length between 1 and 5 and every sequence of bs is longer or equal 3. A rigorous proof comes later.

The string $aabbba$ can be obtained by the following derivation:

$S \Rightarrow S_b A_x$	[(2)]		$\Rightarrow Z_b X A_x$	[(4)]
$\Rightarrow Z_b A X_b A_x$	[(8)]		$\Rightarrow Z_b A X B A_x$	[(11)]
$\Rightarrow Z_b A A B B A_x$	[(6)]		$\Rightarrow A Z_b A B B A_x$	[(14)]
$\Rightarrow A A Z_b B B A_x$	[(14)]		$\Rightarrow A A B B B A_x$	[(17)]
$\Rightarrow A A_x B B B A_x$	[(18)]		$\Rightarrow A A_x B B B_x A_x$	[(20)]
$\Rightarrow \overline{aa} B B B_x A_x$	[(26)]		$\Rightarrow \overline{aa} b B_x B_x A_x$	[(29)]
$\Rightarrow \overline{aabbb} A_x$	[(31)]		$\Rightarrow \overline{aabbb} a$	[(25)]
$\Rightarrow \overline{aabbb} a$	[(32)]		$\Rightarrow a\overline{abbb} a$	[(32)]
$\Rightarrow aa\overline{bbb} a$	[(33)]		$\Rightarrow aab\overline{bb} a$	[(33)]
$\Rightarrow aabbb\overline{a}$	[(33)]		$\Rightarrow aabbba$	[(32)]

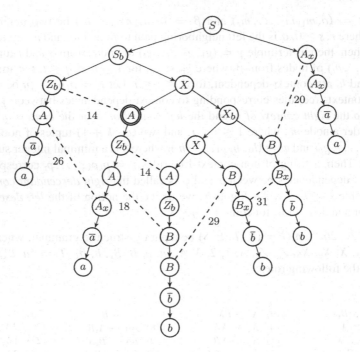

Fig. 2.3 $_G\triangle_{aabbba}$

A graph representing $_G\triangle(S \Rightarrow^* aabbba)$ is illustrated in Fig. 2.3.

Let us note that dashed lines, numbers, and double circle contour only denote the context dependencies, applied non-context-free rules, and a specific node, respectively, and are not the part of the derivation tree.

Pairs of context-dependent nodes are linked with dashed lines, all the other nodes are context-independent. Since $aabbba = \text{frontier}(_G\triangle_{aabbba})$, all the leafs are terminal nodes. Every other node is nonterminal node. For a pair of neighboring paths $\alpha = S_b Z_b A\bar{a}a$ and $\beta = S_b X A Z_b A A_x \bar{a}a$, a string $\rho = 14\ 26$ is their context, it is the left context of β and the right context of α. Consider the double circled node A. Then, $\tau = 26$ is the left descendant context of A and $\varphi = 14\ 18$ is the right descendant context of A.   □

Results

Theorem 2.3.27. *A language L is context-free iff there is a constant $k \geq 0$ and a phrase-structure grammar G such that $L = L(G)$ and for every $x \in L(G)$, there is a tree $_G\triangle_x \in {}_G\blacktriangle$ that satisfies:*

(1) any two neighboring paths contain no more than k pairs of context-dependent nodes;

(2) out of neighboring paths, every pair of nodes is context-independent.

Proof. Construction. Consider any $k \geq 0$. Let $G = (V, T, P, S)$ be a PSG such that $L(G) = L$. Set $N = V - T$. Let $P_{cs} \subseteq P$ denote the set of all non-context-free rules of G. Set

$$N' = \{A_{l|r} \mid A \in N,\ l, r \in (P_{cs} \cup \{\varepsilon\})^k\}$$

Construct a grammar $G' = (V', T, P', S_{\varepsilon|\varepsilon})$, where $V' = N' \cup T$. Set $P' = \emptyset$. Construct P' by performing (I) through (IV) given next.

(I)   For all $A \to B \in P, A, B \in N$, and $l, r \in (P_{cs} \cup \{\varepsilon\})^k$, add $A_{l|r} \to B_{l|r}$ to P';
(II) for all $A \to a \in P, A \in N, a \in (T \cup \{\varepsilon\})$, add $A_{\varepsilon|\varepsilon} \to a$ to P';
(III) for all $A \to BC \in P$, where $A, B, C \in N$, and $r, l, x \in (P_{cs} \cup \{\varepsilon\})^k$, add $A_{l|r} \to B_{l|x}C_{x|r}$ to P';
(IV) for all $p: AB \to CD \in P, A, B, C, D \in N, x, z \in (P_{cs} \cup \{\varepsilon\})^k$, and $y \in (P_{cs} \cup \{\varepsilon\})^{k-1}$, add $A_{x|py} \to C_{x|y}$ and $B_{py|z} \to D_{y|z}$ to P'.

Basic Idea. Notice nonterminal symbols. Since every pair of neighboring paths of G contains a limited number of context-dependent nodes, all of its context-dependencies are encoded in nonterminals. G' nondeterministically decides about all context-dependencies while introducing a new pair of neighboring paths by rules (III). A new pair of neighboring paths is introduced with every application of

$$A_{l|r} \to B_{l|x}C_{x|r}$$

where x encodes a new descendant context. Context dependencies are realized later by context-free rules (IV).

Since P' contains no non-context-free rule, G' is context-free. Next, we proof $L(G) = L(G')$ by establishing Claims 2.3.28 through 2.3.30. Define the new homomorphism $\gamma : V' \to V$, $\gamma(A_{l|r}) = A$, for $A_{l,r} \in N'$, and $\gamma(a) = a$ otherwise.

Claim 2.3.28. If $S \Rightarrow^m w$ in G, where $m \geq 0$ and $w \in V^*$, then $S_{\varepsilon|\varepsilon} \Rightarrow^* w'$ in G', where $w' \in V'^*$ and $\gamma(w') = w$.

Proof. We prove this by induction on $m \geq 0$.

Basis. Let $m = 0$. That is $S \Rightarrow^0 S$ in G. Clearly, $S_{\varepsilon|\varepsilon} \Rightarrow^0 S_{\varepsilon|\varepsilon}$ in G', where $\gamma(S_{\varepsilon|\varepsilon}) = S$, so the basis holds.

Induction Hypothesis. Suppose that there exists $n \geq 0$ such that Claim 2.3.28 holds for all $0 \leq m \leq n$.

Induction Step. Let $S \Rightarrow^{n+1} w$ in G. Then, $S \Rightarrow^n v \Rightarrow w$, where $v \in V^*$, and there exists $p \in P$ such that $v \Rightarrow w\,[p]$. By the induction hypothesis, $S_{\varepsilon|\varepsilon} \Rightarrow^* v'$, where $\gamma(v') = v$, in G'. Next, we consider the following four forms of p.

(I)   Let $p: A \to B \in P$, for some $A, B \in N$. Without any loss of generality, suppose l and r are a left descendant context and a right descendant context of A. By the construction of G', there exists a rule $p': A_{l|r} \to B_{l|r} \in P'$. Then, there exists a derivation $v' \Rightarrow w'\,[p']$ in G', where $\gamma(w') = w$.

(II) Let $p: A \to a \in P$, for some $A \in N$ and $a \in T \cup \{\varepsilon\}$. Since a is a terminal node, it has empty descendant contexts. By the construction of G', there exists a rule $p': A_{\varepsilon|\varepsilon} \to a \in P'$. Then, there exists a derivation $v' \Rightarrow w'\,[p']$ in G', where $\gamma(w') = w$.

(III) Let $p: A \to BC \in P$, for some $A, B, C \in N$. Without any loss of generality, suppose l and r are a left descendant context and a right descendant context of A, and $x \in (P_{cs} \cup \{\varepsilon\})^k$ is a context of neighboring paths beginning at this node. By the construction of G', there exists a rule $p': A_{l|r} \to B_{l|x}C_{x|r} \in P'$. Then, there exists a derivation $v' \Rightarrow w'\,[p']$ in G', where $\gamma(w') = w$.

(IV) Let $p: AB \to CD \in P$, for some $A, B, C, D \in N$. By the assumption stated in Theorem 2.3.27, A and B occur in two neighboring paths denoted by α and β, respectively. Without any loss of generality, suppose that a context of α and β is a string $c \in (P_{cs} \cup \varepsilon)^k$, where $c = pc_f$, and l is a left descendant context, r is a right descendant context of A, B, respectively. By the construction of G', there exist two rules

$$p'_l: A_{l|pc_f} \to C_{l|c_f}, \quad p'_r: B_{pc_f|r} \to D_{c_f|r} \in P'$$

Then, there exists a derivation $v' \Rightarrow^2 w'\,[p'_l p'_r]$ in G', where $\gamma(w') = w$.

Notice (IV). The preservation of the context is achieved by nonterminal symbols. Since the stored context is reduced symbol by symbol from left to right direction in both α and β, G' simulates the applications of non-context-free rules of G.

We covered all possible forms of p, so the claim holds.   □

Claim 2.3.29. Every $x \in F(G')$ can be derived in G' as follows.

$$S_{\varepsilon|\varepsilon} = x_0 \Rightarrow^{d_1} x_1 \Rightarrow^{d_2} x_2 \Rightarrow^{d_3} \cdots \Rightarrow^{d_{h-1}} x_{h-1} \Rightarrow^{d_h} x_h = x$$

for some $h \geq 0$, where $d_i \in \{1, 2\}$, $1 \leq i \leq h$, so that

(1) if $d_i = 1$, then $x_{i-1} = uA_{l|r}v$, $x_i = uzv$, $x_{i-1} \Rightarrow x_i\,[A_{l|r} \to z]$, where $u, v \in V'^*$, $z \in \{B_{l|r}, C_{l|x}D_{x|r}, a\}$, for some $A_{l|r}, B_{l|r}, C_{l|x}, D_{x|r} \in N'$, $a \in (T \cup \{\varepsilon\})$;

(2) if $d_i = 2$, then $x_{i-1} = uA_{x|py}B_{py|z}v$, $x_i = uC_{x|y}D_{y|z}v$, and

$$uA_{x|py}B_{py|z}v \Rightarrow uC_{x|y}B_{py|z}v\,[A_{x|py} \to C_{x|y}] \Rightarrow uC_{x|y}D_{y|z}v\,[B_{py|z} \to D_{y|z}]$$

for some $u, v \in V'^*$ and $A_{x|py}, B_{py|z}, C_{x|y}, D_{y|z} \in N'$.

Proof. Since G' is context-free, without any loss of generality in every derivation of G' we can always reorder applied rules to satisfy Claim 2.3.29.   □

Claim 2.3.30. Let $S_{\varepsilon|\varepsilon} \Rightarrow^{d_1} x_1 \Rightarrow^{d_2} \cdots \Rightarrow^{d_{m-1}} x_{m-1} \Rightarrow^{d_m} x_m$ in G' be a derivation that satisfies Claim 2.3.29, for some $m \geq 0$. Then, $S \Rightarrow^* w$ in G, where $\gamma(x_m) = w$.

Proof. We prove this by induction on $m \geq 0$.

Basis. Let $m = 0$. That is $S_{\varepsilon|\varepsilon} \Rightarrow^0 S_{\varepsilon|\varepsilon}$ in G'. Clearly, $S \Rightarrow^0 S$ in G. Since $\gamma(S_{\varepsilon|\varepsilon}) = S$, the basis holds.

Induction Hypothesis. Suppose that there exists $n \geq 0$ such that Claim 2.3.30 holds for all $0 \leq m \leq n$.

Induction Step. Let $S_{\varepsilon|\varepsilon} \Rightarrow^{d_1} x_1 \Rightarrow^{d_2} \cdots \Rightarrow^{d_{n-1}} x_{n-1} \Rightarrow^{d_n} x_n \Rightarrow^{d_{n+1}} x_{n+1}$ in G' be a derivation that satisfies Claim 2.3.29. By the induction hypothesis, $S \Rightarrow^* v$, $v \in V^*$, where $\gamma(x_n) = v$, in G. Divide the proof into two parts according to d_{n+1}.

(A) Let $d_{n+1} = 1$. By the construction of G', there exists a rule $p' \in P'$ such that $x_n \Rightarrow^{d_{n+1}} x_{n+1}$ $[p']$. Next, we consider the following three forms of p'.

   (A.I) Let $p': A_{l|r} \to B_{l|r} \in P'$, for some $A, B \in N$ and $l, r \in (P_{cs} \cup \{\varepsilon\})^k$. By the construction of G', rule p' was introduced by some rule $p: A \to B \in P$. Then, there exists a derivation $v \Rightarrow w$ $[p]$, where $\gamma(x_{n+1}) = w$.

   (A.II) Let $p': A_{\varepsilon|\varepsilon} \to a \in P'$, for some $A \in N$ and $a \in T \cup \{\varepsilon\}$. By the construction of G', rule p' was introduced by some rule $p: A \to a \in P$. Then, there exists a derivation $v \Rightarrow w$ $[p]$, where $\gamma(x_{n+1}) = w$.

   (A.III) Let $p': A_{l|r} \to B_{l|x}C_{x|r} \in P'$, for some $A, B, C \in N$ and $l, r, x \in (P_{cs} \cup \{\varepsilon\})^k$. By the construction of G', rule p' was introduced by some rule $p: A \to BC \in P$. Then, there exists a derivation $v \Rightarrow w$ $[p]$, where $\gamma(x_{n+1}) = w$.

(B) Let $d_{n+1} = 2$. Then, $x_n \Rightarrow^{d_{n+1}} x_{n+1}$ is equivalent to

$$u_1 A_{x|py} B_{py|z} u_2 \Rightarrow u_1 C_{x|y} B_{py|z} u_2 \; [p'_1] \Rightarrow u_1 C_{x|y} D_{y|z} u_2 \; [p'_2]$$

where $x_n = u_1 A_{x|py} B_{py|z} u_2$, $x_{n+1} = u_1 C_{x|y} D_{y|z} u_2$, and

$$p'_1: A_{x|py} \to C_{x|y}, \; p'_2: B_{py|z} \to D_{y|z} \in P'$$

for some $u_1, u_2 \in V'^*$ and $A_{x|py}, B_{py|z}, C_{x|y}, D_{y|z} \in N'$. By the construction of G', rules p'_1 and p'_2 were introduced by some rule $p: AB \to CD \in P$, Then, there exists a derivation $v \Rightarrow w$ $[p]$, where $\gamma(x_{n+1}) = w$.

We covered all possibilities, so the claim holds.   □

By Claims 2.3.28 and 2.3.30, $S \Rightarrow^* w$ in G iff $S_{\varepsilon|\varepsilon} \Rightarrow^* w'$ in G', where $\gamma(w') = w$. If $S \Rightarrow^* w$ in G and $w \in T^*$, then $w \in L(G)$. Since $\gamma(w') = w' = w$, for $w \in T^*$, $w' \in L(G')$. Therefore, $L(G) = L(G')$ and Theorem 2.3.27 holds.   □

Consider Theorem 2.3.27. Observe that the second condition is superfluous whenever G is monotone. Since a grammar is in the binary form and no symbol can be erased, all context dependencies are within pairs of neighboring paths.

Theorem 2.3.31. *A language L is context-free iff there is a constant $k \geq 0$ and a monotone phrase-structure grammar G such that $L = L(G)$ and for every $x \in L(G)$, there is a tree $_G\Delta_x \in {_G}\blacktriangle$, where any two neighboring paths contain no more than k pairs of context-dependent nodes.*

Proof. Prove this by analogy with the proof of Theorem 2.3.27.   □

Use

We close this section explaining how to apply the results achieved in the previous section in order to demonstrate the contextfreeness of a language, L. As a rule, this demonstration follows the next three-step proof scheme.

(1) Construct a phrase-structure grammar G in the binary form.
(2) Prove $L(G) = L$.
(3) Prove that G satisfies conditions from Theorems 2.3.27 or 2.3.31, depending on whether G is monotone.

Reconsider the grammar G from Example 2.3.26. Following the proof scheme sketched above, we next prove that $L(G) \in \mathbf{CF}$.

Consider G constructed in Example 2.3.26. Next, we show that for G,

$$L(G) = \{w \in (A \cup \{\varepsilon\})(BA)^*(B \cup \{\varepsilon\}) \mid \#_a(w) = \#_b(w),$$

$$A = \{a^i \mid 1 \le i \le 5\}, B = \{b^i \mid i \ge 3\}, \text{ and } |w| > 0\}$$

Without any loss of generality, every terminal derivation of G can be divided into the following 5 phases, where each rule may be used only in a specific phase:

(a) (1)–(4) (b) (5)–(11) (c) (12)–(17) (d) (18)–(31) (e) (32)–(33)

Next, we describe these phases in a greater detail.

(a) First, we generate one of the following two strings by rules (1) through (4).

$$Z_a X B_x, Z_b X A_x$$

Possibly applicable rule (25) may be postponed for phase (d) without affecting the derivation, since rules in the previous phases cannot rewrite A_x.

(b) The rules (5) through (11) are the only with X, X_a, or X_b on their left-hand sides, therefore we can group all their applications in a sequence to get a sentential form from

$$\{Z_a, Z_b\}\{A, B\}^*\{A_x, B_x\}$$

(c) The rules (12) through (17) possibly shift Z_a or Z_b to the right and rewrite it to A or B, respectively. Since these rules are the only with Z_a, Z_b on their left-hand sides, they can be always prioritized before the rest of rules without any loss of generality.

$$\{A, B\}^*\{A_x, B_x\}$$

(d) All the remaining rules may be applied in this phase. However, we can exclude rules (32) and (33), so we get a sentential form from

$$\{\overline{a}, \overline{b}\}^*$$

(e) Since rules (32) and (33) are context-free and produce terminal symbols, they can be always postponed until the end of any successful derivation.

$$\{a, b\}^* = T^*$$

Let us add a few remarks concerning (a) through (e).

Phase (a) is very straightforward. Only notice that it is decided whether the generated string finally ends with a or b and the paired symbol is stored in Z_a or Z_b for phase (c).

In phase (b) an arbitrary string of As and Bs is generated from the initial symbol X. However, for every A, one B is generated and vice versa, so their numbers are always kept equal.

In phase (a) the grammar decides about the last symbol and stores the paired one, which, however, need not to be the first one. Therefore phase (c) determines its final position, while possibly shifting it to the right and finally rewriting to A or B.

Phase (d) is the most tricky. It starts with a sentential form wc, where $w \in \{A, B\}^*$, $c \in \{A_x, B_x\}$. Informally speaking, it consists of the sequences of As which should be at most 5 symbols long, and Bs which should be at least 3 symbols long. Rules (18) through (31) are designed to ensure these restrictions. To give an example, suppose wc is as follows.

$$wc = AAAABBBBBABBBAA_x$$

First, by rules (18) through (20) the last symbol in every sequence is marked with index x. Otherwise, rules (24) through (28) and rule (31) never become applicable and all the unmarked sequences become permanent resulting into an unsuccessful derivation. The last sequence is already marked.

$$
\begin{aligned}
& AAAABBBBBABBBAA_x \\
&\Rightarrow AAAA_xBBBBBABBBAA_x \quad [(18)] \\
&\Rightarrow AAAA_xBBBBA_xBBBAA_x \quad [(18)] \\
&\Rightarrow AAAA_xBBBB_xA_xBBBAA_x \quad [(20)] \\
&\Rightarrow AAAA_xBBBB_xA_xBBB_xAA_x \quad [(19)]
\end{aligned}
$$

Notice, one symbol sequence of As is legal. Then, every sequence of As is processed in left-to-right direction by rules (21) through (24), but can be successfully rewritten earlier by rules (25) through (28), in the case it consists of less than 5 symbols. Thus, a longer sequence leads to an unsuccessful derivation.

$$
\begin{aligned}
& AAAA_xBBBB_xA_xBBB_xAA_x \\
&\Rightarrow \overline{a}1AA_xBBBB_xA_xBBB_xAA_x \quad [(21)] \\
&\Rightarrow \overline{aa}2A_xBBBB_xA_xBBB_xAA_x \quad [(22)] \\
&\Rightarrow \overline{aaaa}BBBB_xA_xBBB_xAA_x \quad [(27)] \\
&\Rightarrow \overline{aaaa}BBBB_x\overline{a}BBB_xAA_x \quad [(25)] \\
&\Rightarrow \overline{aaaa}BBBB_x\overline{a}BBB_x\overline{aa} \quad [(26)]
\end{aligned}
$$

If the processing does not start from the leftmost symbol in the current sequence, it remains permanent. Every sequence of Bs is processed by applying rule (29), zero or multiple times rule (30), and finally rule (31). It ensures the lengths of sequences of Bs are at least 3 symbols.

$$\overline{aaaa}BBBB_x\overline{a}BBB_x\overline{aa}$$
$$\Rightarrow \overline{aaaab}B_xBB_x\overline{a}BBB_x\overline{aa} \qquad [(29)]$$
$$\Rightarrow \overline{aaaabb}B_xB_x\overline{a}BBB_x\overline{aa} \qquad [(30)]$$
$$\Rightarrow \overline{aaaabbbba}BBB_x\overline{aa} \qquad [(31)]$$
$$\Rightarrow \overline{aaaabbbbab}B_xB_x\overline{aa} \qquad [(29)]$$
$$\Rightarrow \overline{aaaabbbbabbbaa} \qquad [(31)]$$

Notice, it depends on the order of applied rules only within one sequence. Multiple sequences may be processed at random without affecting the derivation.

In phase (e), a resulting terminal string is generated by rules (32) and (33).

$$\overline{aaaabbbbabbbaa} \Rightarrow^* aaaabbbbabbbaa$$

Therefore, if the derivation is terminating, we achieve a string with an equal number of as and bs, where every sequence of as is at most 5 symbols long and every sequence of bs is at least 3 symbols long.

Grammar G is obviously a monotone phrase-structure grammar in the binary form. Let us now show that for any $x \in L(G)$, there is $_G\triangle_x \in {_G}\blacktriangle$, where any two neighboring paths contain no more than 2 pairs of context-dependent nodes.

Every pair of context-dependent nodes in $_G\triangle_x$ corresponds to one non-context-free rule in $S \Rightarrow^* x$. Consider the six phases sketched above. Observe that phases (a), (b), and (e) contain only context-free rules, so we have only to investigate (c) and (d). On the other hand, (c) and (d) contain no rule of the form $A \to BC$, thus the number of neighboring paths remains unchanged.

In (c) by rules (12) through (17) the derivation may proceed in left-to-right direction through the whole sentence form (except the rightmost symbol) introducing a context dependency between every pair of neighboring paths.

In (d), first, the context dependency is introduced between all neighboring paths representing the borders between the sequences of As and Bs by rules (18) through (20). Second, every sequence of As or Bs is processed in the left-to-right direction by non-context-free rules (21) through (31) introducing a context dependency between all neighboring paths representing symbols inside the sequences of As and Bs.

No other non-context-free rule is applied, therefore no other context-dependent pair of nodes can occur. Then, every pair of neighboring paths may contain at most one context-dependent pair of nodes introduced in phase (c) and one introduced in phase (d).

Since G is a monotone PSG in the binary form, where for every $x \in L(G)$, there is $_G\Delta_x \in {}_G\blacktriangle$, where any two neighboring paths contain no more than 2 pairs of context-dependent nodes, by Theorem 2.3.31, $L(G) \in \mathbf{CF}$.

2.3.3 How to Disprove Context-Freeness

When examining complicated formal languages, we often need to demonstrate that they are non-context-free and, therefore, beyond the power of context-free grammars. The present section explains how to make a demonstration like this.

The Pumping Lemma for Context-Free Languages

The pumping lemma established in this section is frequently used to disprove that a language K is context-free. The lemma says that for every $L \in \mathbf{CF}$, there is a constant $k \geq 1$ such that every $z \in L$ with $|z| \geq k$ can be expressed as $z = uvwxy$ with $vx \neq \varepsilon$ so that L also contains $uv^m wx^m y$, for every $m \geq 0$. Consequently, to demonstrate the non-context-freeness of a language, K, by contradiction, assume that $K \in \mathbf{CF}$ and k is its pumping-lemma constant. Select a string $z \in K$ with $|z| \geq k$, consider all possible decompositions of z into $uvwxy$, and for each of these decompositions, prove that $uv^m wx^m y$ is out of K, for some $m \geq 0$, which contradicts the pumping lemma. Thus, $K \notin \mathbf{CF}$. Without any loss of generality, we prove the pumping lemma based on CFGs satisfying Chomsky normal form (see Definition 3.1.19). We also make use of some notions introduced earlier in this chapter, such as the derivation tree $\Delta(A \Rightarrow^* x)$ corresponding to a derivation $A \Rightarrow^* x$ in a CFG G, where $A \in N$ and $x \in T^*$. In addition, we use some related graph-theory notions introduced in Sect. 1.3, such as $\mathrm{depth}(\Delta(A \Rightarrow^* x))$, which denotes the depth of $\Delta(A \Rightarrow^* x)$.

Lemma 2.3.32. *Let $G = (V, T, P, S)$ be a CFG in Chomsky normal form. For every derivation $A \Rightarrow^* x$ in G, where $A \in N$ and $x \in T^*$, its corresponding derivation tree $\Delta(A \Rightarrow^* x)$ satisfies $|x| \leq 2^{\mathrm{depth}(\Delta(A \Rightarrow^* x))-1}$.*

Proof. (by induction on $\mathrm{depth}(\Delta(A \Rightarrow^* x)) \geq 1$).

Basis. Let $\mathrm{depth}(\Delta(A \Rightarrow^* x)) = 1$, where $A \in N$ and $x \in T^*$. Because G is in Chomsky normal form, $A \Rightarrow^* x$ $[A \to x]$ in G, where $x \in T$, so $|x| = 1$. For $\mathrm{depth}(\Delta(A \Rightarrow^* x)) = 1$, $2^{\mathrm{depth}(\Delta(A \Rightarrow^* x))-1} = 2^0$. As $2^0 = 1$, $|x| \leq 2^{\mathrm{depth}(\Delta(A \Rightarrow^* x))-1}$ in this case, so the basis holds true.

Induction Hypothesis. Suppose that this lemma holds for all derivation trees of depth n or less, for some $n \geq 0$.

Induction Step. Let $A \Rightarrow^* x$ in G with $\mathrm{depth}(\Delta(A \Rightarrow^* x)) = n + 1$, where $A \in N$ and $x \in T^*$. Let $A \Rightarrow^* x$ $[r\rho]$ in G, where $r \in P$ and $\rho \in P^*$. As G is in Chomsky normal form, $r : A \to BC \in P$, where $B, C \in N$. Let $B \Rightarrow^* u \; [\varphi]$, $C \Rightarrow^* v \; [\theta]$, $\varphi, \theta \in P^*$, $x = uv$, $\rho = \varphi\theta$ so that $A \Rightarrow^* x$ can be expressed in greater detail as $A \Rightarrow BC \; [r] \Rightarrow^* uC \; [\varphi] \Rightarrow^* uv \; [\theta]$. Observe that $\mathrm{depth}(\Delta(B \Rightarrow^* u \; [\varphi])) \leq \mathrm{depth}(\Delta(A \Rightarrow^* x)) - 1 = n$, so $|u| \leq 2^{\mathrm{depth}(\Delta(B \Rightarrow^* u)) - 1}$ by the induction hypothesis. Analogously, as $\mathrm{depth}(\Delta(C \Rightarrow^* v \; [\theta])) \leq \mathrm{depth}(\Delta(A \Rightarrow^* x)) - 1 = n$, $|v| \leq 2^{\mathrm{depth}(\Delta(C \Rightarrow^* v)) - 1}$. Thus, $|x| = |u| + |v| \leq 2^{\mathrm{depth}(\Delta(B \Rightarrow^* u)) - 1} + 2^{\mathrm{depth}(\Delta(C^* v)) - 1} \leq 2^{n-1} + 2^{n-1} = 2^n = 2^{\mathrm{depth}(\Delta(A \Rightarrow^* x)) - 1}$. □

Corollary 2.3.33. *Let $G = (V, T, P, S)$ be a CFG in Chomsky normal form. For every derivation $A \Rightarrow^* x$ in G, where $A \in N$ and $x \in T^*$ with $|x| \geq 2^m$ for some $m \geq 0$, its corresponding derivation tree $\Delta(A \Rightarrow^* x)$ satisfies $\mathrm{depth}(\Delta(A \Rightarrow^* x)) \geq m + 1$.*

Proof. This corollary follows from Lemma 2.3.32 and the contrapositive law. □

Lemma 2.3.34. Pumping Lemma for **CF**. *Let L be an infinite context-free language. Then, there exists $k \geq 1$ such that every string $z \in L$ satisfying $|z| \geq k$ can be expressed as $z = uvwxy$, where $0 < |vx| < |vwx| \leq k$, and $uv^m wx^m y \in L$, for all $m \geq 0$.*

Proof. Let $L \in \mathbf{CF}$, and $L = L(G)$, where $G = (V, T, P, S)$ be a CFG in Chomsky normal form. Let G have n nonterminals, for $n \geq 1$; in symbols, $\mathrm{card}(N) = n$. Set $k = 2^n$. Let $z \in L(G)$ satisfying $|z| \geq k$. As $z \in L(G)$, $S \Rightarrow^* z$, and by Corollary 2.3.33, $\mathrm{depth}(\Delta(S \Rightarrow^* z)) \geq \mathrm{card}(N) + 1$, so $\Delta(S \Rightarrow^* z)$ contains some subtrees in which there is a path with two or more nodes labeled by the same nonterminal. Express $S \Rightarrow^* z$ as $S \Rightarrow^* uAy \Rightarrow^+ uvAxy \Rightarrow^+ uvwxy$ with $uvwxy = z$ so that the derivation tree corresponding to $A \Rightarrow^+ vAx \Rightarrow^+ vwx$ contains no proper subtree with a path containing two or more different nodes labeled with the same nonterminal. To prove that $0 < |vx| < |vwx| \leq k$, recall that every rule in P has on its right-hand side either a terminal or two nonterminals because G is in Chomsky normal form. Thus, $A \Rightarrow^+ vAx$ implies $0 < |vx|$, and $vAx \Rightarrow^+ vwx$ implies $|vx| < |vwx|$. As the derivation tree corresponding to $A \Rightarrow^+ vAx \Rightarrow^+ vwx$ contains no subtree with a path containing two different nodes labeled with the same nonterminal, $\mathrm{depth}(\Delta(A \Rightarrow^* vwx)) \leq \mathrm{card}(N) + 1$, so by Lemma 2.3.32, $|vx| < |vwx| \leq 2^n = k$. Finally, we demonstrate that for all $m \geq 0$, $uv^m wx^m y \in L$. As $S \Rightarrow^* uAy \Rightarrow^+ uvAxy \Rightarrow^+ uvwxy$, $S \Rightarrow^* uAy \Rightarrow^+ uwy$, so $uv^0 wx^0 y = uwy \in L$. Similarly, since $S \Rightarrow^* uAy \Rightarrow^+ uvAxy \Rightarrow^+ uvwxy$, $S \Rightarrow^* uAy \Rightarrow^+ uvAxy \Rightarrow^+ uvvAxxy \Rightarrow^+ \cdots \Rightarrow^+ uv^m Ax^m y \Rightarrow^+ uv^m wx^m y$, so $uv^m wx^m y \in L$, for all $m \geq 1$. Thus, Lemma 2.3.34 holds true. □

Applications of the Pumping Lemma

We usually use the pumping lemma in a proof by contradiction to demonstrate that a given language L is not context-free. Typically, we make a proof of this kind in the following way.

(1) Assume that L is context-free.
(2) Select a string $z \in L$ whose length depends on the pumping-lemma constant k so that $|z| \geq k$ is necessarily true.
(3) For all possible decompositions of z into $uvwxy$ satisfying the pumping-lemma conditions, find $m \geq 0$ such that $uv^m wx^m y \notin L$, which contradicts Lemma 2.3.34.
(4) The contradiction obtained in (3) means that the assumption in (1) is incorrect; therefore, L is not context-free.

Example 2.3.35. Consider $L = \{a^n b^n c^n \mid n \geq 1\}$. Next, under the guidance of the recommended proof structure preceding this example, we demonstrate that $L \notin \mathbf{CF}$.

(1) Assume that $L \in \mathbf{CF}$.
(2) In L, select $z = a^k b^k c^k$ with $|z| = 3k \geq k$, where k is the pumping-lemma constant.
(3) By Lemma 2.3.34, z can be written as $z = uvwxy$ so that this decomposition satisfies the pumping-lemma conditions. As $0 < |vx| < |vwx| \leq k$, either $vwx \in \{a\}^*\{b\}^*$ or $vwx \in \{b\}^*\{c\}^*$. If $vwx \in \{a\}^*\{b\}^*$, $uv^0 wx^0 y$ has k cs but fewer than k as or bs, so $uv^0 wx^0 y \notin L$, but by the pumping-lemma, $uv^0 wx^0 y \in L$. If $vwx \in \{b\}^*\{c\}^*$, $uv^0 wx^0 y$ has k as but fewer than k bs or cs, so $uv^0 wx^0 y \notin L$, but by the pumping lemma, $uv^0 wx^0 y \in L$. In either case, we obtain the contradiction that $uv^0 wx^0 y \notin L$ and, simultaneously, $uv^0 wx^0 y \in L$.
(4) By the contradiction obtained in (3), $L \notin \mathbf{CF}$. □

Omitting some obvious details, we usually proceed in a briefer way than above when proving the non-context-freeness of a language by using Lemma 2.3.34.

Example 2.3.36. Let $L = \{a^n b^m a^n b^m \mid n, m \geq 1\}$. Assume that L is context-free. Set $z = a^k b^k a^k b^k$ with $|a^k b^k a^k b^k| = 4k \geq k$. By Lemma 2.3.34, express $z = uvwxy$. Observe that $0 < |vx| < |vwx| \leq k$ implies $uwy \notin L$ in all possible occurrences of vwx in $a^k b^k a^k b^k$; however, by Lemma 2.3.34, $uwy \in L$—a contradiction. Thus, $L \notin \mathbf{CF}$. □

Even some seemingly trivial unary languages are not context-free as shown next.

Example 2.3.37. Consider $L = \{a^{n^2} \mid \text{for some } n \geq 0\}$. To demonstrate $L \notin \mathbf{CF}$, assume that $L \in \mathbf{CF}$ and select $z = a^{k^2} \in L$ where k is the pumping-lemma constant. As a result, $|z| = k^2 \geq k$, so $z = uvwxy$, which satisfies the pumping-lemma conditions. As $k^2 < |uv^2 wx^2 y| \leq k^2 + k < k^2 + 2k + 1 = (k+1)^2$, we have $uv^2 wx^2 y \notin L$, but by Lemma 2.3.34, $uv^2 wx^2 y \in L$—a contradiction. Thus, $L \notin \mathbf{CF}$. □

2.3.4 Parallel Grammars

Parallel grammars represent modified versions of context-free grammars that rewrite their strings in parallel. Most often, this book represents a grammatical parallelism like this by *Extended tabled zero-sided Lindenmayer grammars* or, more briefly, ET0L grammars and their special cases, such as E0L grammars (see [PHHM96b, PL90a, RS80, RS86]). Originally, these grammars were introduced in connection with a theory proposed for the development of filamentous organisms. Developmental stages of cellular arrays are described by strings with each symbol being a cell. Rules correspond to developmental instructions with which organisms can be produced. They are applied simultaneously to all cell-representing symbols because in a growing organism development proceeds simultaneously everywhere. That is, all symbols, including terminals, are always rewritten to adequately reflect the development of real organisms that contain no dead cells which would remain permanently fixed in their place in the organism; disappearing cells are represented by ε. Instead of a single set of rules, ET0L grammars have a finite set of sets containing rules. Each of them contains rules that describe developmental instructions corresponding to a specific biological circumstances, such as environmental conditions concerning temperature or coexistence of other organisms. Naturally, during a single derivation step, rules from only one of these sets can be applied to the rewritten string.

Considering these biologically motivated features of ET0L grammars, we see the following three main conceptual differences between them and the previously discussed sequential grammars, such as context-free grammars (see Sect. 2.3).

(I) Instead of a single set of rules, they have finitely many sets of rules.
(II) The left-hand side of a rule may be formed by any grammatical symbol, including a terminal.
(III) All symbols of a string are simultaneously rewritten during a single derivation step.

Definition 2.3.38. An *ET0L grammar* is a $(t + 3)$-tuple

$$G = (V, T, P_1, \ldots, P_t, w)$$

where $t \geq 1$, and V, T, and w are the *total alphabet*, the *terminal alphabet* ($T \subseteq V$), and the *start string* ($w \in V^+$), respectively. Each P_i is a finite set of *rules* of the form

$$a \to x$$

where $a \in V$ and $x \in V^*$. If $a \to x \in P_i$ implies that $x \neq \varepsilon$ for all $i = 1, \ldots, t$, then G is said to be *propagating* (an *EPT0L grammar* for short).

Let $u, v \in V^*$, $u = a_1 a_2 \cdots a_q$, $v = v_1 v_2 \cdots v_q$, $q = |u|$, $a_j \in V$, $v_j \in V^*$, and $p_1, p_2, \ldots, p_q$ is a sequence of rules of the form $p_j = a_j \to v_j \in P_i$ for all $j = 1, \ldots, q$, for some $i \in \{1, \ldots, t\}$. Then, u *directly derives* v according to the rules p_1 through p_q, denoted by

$$u \Rightarrow_G v \; [p_1, p_2, \ldots, p_q]$$

If p_1 through p_q are immaterial, we write just $u \Rightarrow_G v$. In the standard manner, we define the relations $\Rightarrow_G^n$ ($n \geq 0$), $\Rightarrow_G^*$, and $\Rightarrow_G^+$.

The *language* of G, denoted by $L(G)$, is defined as

$$L(G) = \{y \in T^* \mid w \Rightarrow_G^* y\} \qquad \square$$

The families of languages generated by ET0L and EPT0L grammars are denoted by **ET0L** and **EPT0L**, respectively.

Definition 2.3.39. Let $G = (V, T, P_1, \ldots, P_t, w)$ be an ET0L grammar, for some $t \geq 1$. If $t = 1$, then G is called an *E0L grammar*. $\qquad \square$

The families of languages generated by E0L and propagating E0L grammars (*EP0L grammars* for short) are denoted by **E0L** and **EP0L**, respectively.

Definition 2.3.40. An *0L grammar* is defined by analogy with an E0L grammar except that $V = T$. $\qquad \square$

For simplicity, we specify an 0L grammar as a triple $G = (T, P, S)$ rather than a quadruple $G = (T, T, P, S)$. By **0L**, we denote the family of languages generated by 0L grammars.

Theorem 2.3.41 (See [RS80]).

$$\mathbf{CF} \subset \mathbf{E0L} = \mathbf{EP0L} \subset \mathbf{ET0L} = \mathbf{EPT0L} \subset \mathbf{CS}$$

ET0L grammars work in a totally parallel way because they simultaneously rewrite all symbols of the current sentential form during every single derivation step. Apart from them, however, there also exist semi-parallel grammars, which simultaneously change only selected symbols in the rewritten strings while keeping their rest unchanged. We close this section by defining one of them, namely, queue grammars. However, let us point out that we study several other semi-parallel grammatical mechanisms later in this book (see Sect. 4.1 and Chap. 12).

Queue Grammars

Queue grammars (see [KR83]) always rewrite the first and the last symbol in the strings in parallel; all the other symbols in between them remain unchanged. That is, as their name indicates, queue grammars rewrite strings in a way that resemble the standard way of working with an abstract data type referred to as a queue. Indeed, these grammars work on strings based upon the well-known first-in-first-out principle—that is, the first symbol added to the string will be the first one to be removed. More specifically, during every derivation step, these grammars attach a string as a suffix to the current sentential form while eliminating the leftmost symbol of this form; as a result, all symbols that were attached prior to this step have to be removed before the newly attached suffix is removed. Next, we define these grammars rigorously.

Definition 2.3.42. A *queue grammar* is a sixtuple

$$Q = (V, T, W, F, R, g)$$

where V and W are two alphabets satisfying $V \cap W = \emptyset$, $T \subset V$, $F \subset W$, $g \in (V - T)(W - F)$, and

$$R \subseteq V \times (W - F) \times V^* \times W$$

is a finite relation such that for each $a \in V$, there exists an element $(a, b, x, c) \in R$. If $u = arb$, $v = rxc$, and $(a, b, x, c) \in R$, $r, x \in V^*$, where $a \in V$ and $b, c \in W$, then Q makes a *derivation step* from u to v according to (a, b, x, c), symbolically written as

$$u \Rightarrow_Q v \; [(a, b, x, c)]$$

or, simply, $u \Rightarrow_Q v$. We define $\Rightarrow_Q^n$ ($n \geq 0$), $\Rightarrow_Q^+$, and $\Rightarrow_Q^*$ in the standard way. The *language* of Q, denoted by $L(Q)$, is defined as

$$L(Q) = \{x \in T^* \mid g \Rightarrow_Q^* xf, f \in F\} \qquad \square$$

As an example, consider a queue grammar $G = (V, T, W, F, s, P)$, where $V = \{S, A, a, b\}$, $T = \{a, b\}$, $W = \{Q, f\}$, $F = \{f\}$, $s = SQ$ and $P = \{p_1, p_2\}$, $p_1 = (S, Q, Aaa, Q)$ and $p_2 = (A, Q, bb, f)$. Then, there exists a derivation

$$s = SQ \Rightarrow AaaQ[p_1] \Rightarrow aabbf[p_2]$$

in this queue grammar, which generates *aabb*.

Theorem 2.3.43 (See [KR83]). *For every recursively enumerable language K, there is a queue grammar Q such that $L(Q) = K$.*

Next, we slightly modify the definition of a queue grammar.

Definition 2.3.44 (See [Med04, KM00, Med03c]). A *left-extended queue grammar* is a sixtuple

$$Q = (V, T, W, F, R, g)$$

where V, T, W, F, R, g have the same meaning as in a queue grammar; in addition, assume that $\# \notin V \cup W$. If $u, v \in V^*\{\#\}V^*W$ so $u = w\#arb$, $v = wa\#rzc$, $a \in V$, $r, z, w \in V^*$, $b, c \in W$, and $(a, b, z, c) \in R$, then

$$u \Rightarrow_Q v \; [(a, b, z, c)]$$

or, simply, $u \Rightarrow_Q v$. In the standard manner, extend $\Rightarrow_Q$ to $\Rightarrow_Q^n$, where $n \geq 0$. Based on $\Rightarrow_Q^n$, define $\Rightarrow_Q^+$ and $\Rightarrow_Q^*$. The *language* of Q, denoted by $L(Q)$, is defined as

$$L(Q) = \left\{v \in T^* \mid \#g \Rightarrow_Q^* w\#vf \text{ for some } w \in V^* \text{ and } f \in F\right\} \qquad \square$$

Less formally, during every step of a derivation, a left-extended queue grammar shifts the rewritten symbol over #; in this way, it records the derivation history, which represents a property fulfilling a crucial role in several proofs later in this book.

For example, consider a left-extended queue grammar, which has the same components as the previously mentioned queue grammar G. Then, there exists a derivation

$$\#s = \#SQ \Rightarrow S\#AaaQ[p_1] \Rightarrow SA\#aabbf[p_2]$$

in this left-extended queue grammar, which generates *aabb*. Moreover, this type of queue grammar saves symbols from the first components of rules which were used in the derivation.

Theorem 2.3.45. *For every queue grammar Q, there is an equivalent left-extended queue grammar Q' such that $L(Q') = L(Q)$.*

The proof is trivial and left to the reader.

Theorem 2.3.46. *For every recursively enumerable language K, there is a left-extended queue grammar Q such that $L(Q) = K$.*

Proof. Follows from Theorems 2.3.43 and 2.3.45.

2.4 Automata

In this section, based upon rewriting systems, we define automata as fundamental formal devices that accept strings of a given language (see [Med00a]). Specifically, we give the definitions of finite and pushdown automata, which accept the families of regular and context-free languages, respectively. As they are special cases of rewriting systems, introduced in Sect. 2.2, we straightforwardly apply the mathematical terminology concerning rewriting systems to finite automata. Perhaps most significantly, we apply relations $\vdash$, $\vdash^n$, $\vdash^+$, and $\vdash^*$ to them.

A *finite automaton* can be viewed as a rewriting system $M = (\Gamma, R)$, where

- the total alphabet Δ contains subsets Q, F, and Σ such that $\Delta = Q \cup \Sigma$, $F \subseteq Q$, and $Q \cap \Sigma = \emptyset$;
- R is a finite *set of rules* of the form $qa \rightarrow p$, where $q, p \in Q$ and $a \in \Sigma \cup \{\varepsilon\}$.

More commonly, however, the notion of a finite automaton is specified as follows.

Definition 2.4.1. A *general finite automaton* is a quintuple

$$M = (Q, \Sigma, R, s, F)$$

where

- Q is a finite set of *states*;
- Σ is an *input alphabet*;
- $R \subseteq Q \times \Sigma^* \times Q$ is a finite relation, called the set of *rules* (or *transitions*);
- $s \in Q$ is the *start state*;
- $F \subseteq Q$ is the set of *final states*.

Instead of $(p, y, q) \in R$, we write $py \to q \in R$. If $py \to q \in R$ implies that $y \neq \varepsilon$, then M is said to be *ε-free*.

A *configuration* of M is any string from $Q\Sigma^*$. The relation of a *move*, symbolically denoted by $\vdash_M$, is defined over $Q\Sigma^*$ as follows:

$$pyx \vdash_M qx$$

if and only if $pyx, qx \in Q\Sigma^*$ and $py \to q \in R$.

Let $\vdash_M^n$, $\vdash_M^*$, and $\vdash_M^+$ denote the nth power of $\vdash_M$, for some $n \geq 0$, the reflexive-transitive closure of $\vdash_M$, and the transitive closure of $\vdash_M$, respectively. The *language* of M is denoted by $L(M)$ and defined as

$$L(M) = \{w \in \Sigma^* \mid sw \vdash_M^* f, f \in F\} \qquad \Box$$

Next, we define three special variants of general finite automata.

Definition 2.4.2. Let $M = (Q, \Sigma, R, s, F)$ be a general finite automaton. M is a *finite automaton* if and only if $py \to q \in R$ implies that $|y| \leq 1$. M is said to be *deterministic* if and only if $py \to q \in R$ implies that $|y| = 1$ and $py \to q_1, py \to q_2 \in R$ implies that $q_1 = q_2$, for all $p, q, q_1, q_2 \in Q$ and $y \in \Sigma^*$. M is said to be *complete* if and only if M is deterministic and for all $p \in Q$ and all $a \in \Sigma$, $pa \to q \in R$ for some $q \in Q$. $\qquad \Box$

To make several definitions and proofs concerning finite automata more concise, we sometimes denote a rule $pa \to q$ with a unique label r as $r: pa \to q$. This notion of rule labels is formalized in the following definition.

Definition 2.4.3. Let $M = (Q, \Sigma, R, s, F)$ be a finite automaton. Let Ψ be an alphabet of *rule labels* such that $\text{card}(\Psi) = \text{card}(R)$, and ψ be a bijection from R to Ψ. For simplicity, to express that ψ maps a rule, $pa \to q \in R$, to r, where $r \in \Psi$, we write $r: pa \to q \in R$; in other words, $r: pa \to q$ means $\psi(pa \to q) = r$.

For every $y \in \Sigma^*$ and $r: pa \to q \in R$, M makes a *move* from configuration pay to configuration qy according to r, written as

$$pay \vdash_M qy \; [r]$$

Let χ be any configuration of M. M makes *zero moves* from χ to χ according to ε, symbolically written as

$$\chi \vdash_M^0 \chi \ [\varepsilon]$$

Let there exist a sequence of configurations $\chi_0, \chi_1, \ldots, \chi_n$ for some $n \geq 1$ such that $\chi_{i-1} \vdash_M \chi_i \ [r_i]$, where $r_i \in \Psi$, for $i = 1, \ldots, n$, then M makes n *moves* from χ_0 to χ_n according to $r_1 \cdots r_n$, symbolically written as

$$\chi_0 \vdash_M^n \chi_n \ [r_1 \cdots r_n]$$

Define $\vdash_M^*$ and $\vdash_M^+$ in the standard manner. $\qquad\qquad\square$

Sometimes, we specify a finite automaton as $M = (Q, \Sigma, \Psi, R, s, F)$, where Q, Σ, Ψ, R, s, and F are the set of states, the input alphabet, the alphabet of rule labels, the set of rules, the start state, and the set of final states, respectively.

Theorem 2.4.4 (See [Woo87]). *For every general finite automaton M, there is a complete finite automaton M' such that $L(M') = L(M)$.*

Finite automata accept precisely the family of regular languages:

Theorem 2.4.5 (See [Woo87]). *A language K is regular if and only if there is a complete finite automaton M such that $K = L(M)$.*

Pushdown automata represent finite automata extended by a potentially unbounded pushdown store. We first define their general version, customarily referred to as extended pushdown automata.

Definition 2.4.6. An *extended pushdown automaton* is a septuple

$$M = (Q, \Sigma, \Gamma, R, s, S, F)$$

where

- Q, Σ, s, and F are defined as in a finite automaton;
- Γ is a *pushdown alphabet*;
- $R \subseteq \Gamma^* \times Q \times (\Sigma \cup \{\varepsilon\}) \times \Gamma^* \times Q$ is a finite relation, called the set of *rules* (or *transitions*);
- S is the *initial pushdown symbol*.

Q and $(\Sigma \cup \Gamma)$ are always assumed to be disjoint. By analogy with finite automata, instead of $(\gamma, p, a, w, q) \in R$, we write $\gamma p a \to wq$.

A *configuration* of M is any string from $\Gamma^* Q \Sigma^*$. The relation of a *move*, symbolically denoted by $\vdash_M$, is defined over $\Gamma^* Q \Sigma^*$ as follows:

$$x \gamma p a y \vdash_M x w q y$$

if and only if $x \gamma p a y, x w q y \in \Gamma^* Q \Sigma^*$ and $\gamma p a \to wq \in R$.

Let $\vdash_M^k$, $\vdash_M^*$, and $\vdash_M^+$ denote the kth power of $\vdash_M$, for some $k \geq 0$, the reflexive-transitive closure of $\vdash_M$, and the transitive closure of $\vdash_M$, respectively. □

For an extended pushdown automaton, there exist three ways of language acceptance: (1) by entering a final state, (2) by emptying its pushdown, and (3) by entering a final state and emptying its pushdown. All of them are defined next.

Definition 2.4.7. Let $M = (Q, \Sigma, \Gamma, R, s, S, F)$ be an extended pushdown automaton. The *language accepted by M by final state* is denoted by $L_f(M)$ and defined as

$$L_f(M) = \{w \in \Sigma^* \mid Ssw \vdash_M^* \gamma f, f \in F, \gamma \in \Gamma^*\}$$

The *language accepted by M by empty pushdown* is denoted by $L_e(M)$ and defined as

$$L_e(M) = \{w \in \Sigma^* \mid Ssw \vdash_M^* q, q \in Q\}$$

The *language accepted by M by empty pushdown and final state*, denoted by $L_{ef}(M)$, is defined as

$$L_{ef}(M) = \{w \in \Sigma^* \mid Ssw \vdash_M^* f, f \in F\}$$ □

Let **EPDA**$_f$, **EPDA**$_e$, and **EPDA**$_{ef}$ denote the language families accepted by extended pushdown automata accepting by final state, by empty pushdown, and by final state and empty pushdown, respectively.

All of the three ways of acceptance are equivalent:

Theorem 2.4.8 (See [Med00a]). $\textbf{EPDA}_f = \textbf{EPDA}_e = \textbf{EPDA}_{ef}$

If an extended pushdown automaton rewrites a single symbol on its pushdown top during every move, we obtain a pushdown automaton, defined next.

Definition 2.4.9. Let $M = (Q, \Sigma, \Gamma, R, s, S, F)$ be an extended pushdown automaton. Then, M is a *pushdown automaton* if and only if $\gamma pa \rightarrow wq \in R$ implies that $|\gamma| = 1$. □

To make definitions and proofs concerning pushdown automata more readable, we sometimes denote a rule $Apa \rightarrow wq$ with a unique label r as $r \colon Apa \rightarrow wq$, which is formalized in the following definition.

Definition 2.4.10. Let $M = (Q, \Sigma, \Gamma, R, s, S, F)$ be a pushdown automaton. Let Ψ be an alphabet of *rule labels* such that $\text{card}(\Psi) = \text{card}(R)$, and ψ be a bijection from R to Ψ. For simplicity, to express that ψ maps a rule, $Apa \rightarrow wq \in R$, to r, where $r \in \Psi$, we write $r \colon Apa \rightarrow wq \in R$; in other words, $r \colon Apa \rightarrow wq$ means $\psi(Apa \rightarrow wq) = r$.

For every $x \in \Gamma^*$, $y \in \Sigma^*$, and $r: Apa \rightarrow wq \in R$, M makes a *move* from configuration $xApay$ to configuration $xwqy$ according to r, written as

$$xApay \vdash_M xwqy \ [r]$$

Let χ be any configuration of M. M makes *zero moves* from χ to χ according to ε, symbolically written as

$$\chi \vdash_M^0 \chi \ [\varepsilon]$$

Let there exist a sequence of configurations $\chi_0, \chi_1, \ldots, \chi_n$ for some $n \geq 1$ such that $\chi_{i-1} \vdash_M \chi_i \ [r_i]$, where $r_i \in \Psi$, for $i = 1, \ldots, n$, then M makes n *moves* from χ_0 to χ_n according to $r_1 \cdots r_n$, symbolically written as

$$\chi_0 \vdash_M^n \chi_n \ [r_1 \cdots r_n]$$

Define $\vdash_M^*$ and $\vdash_M^+$ in the standard manner.   □

Let $\mathbf{PDA}_f$, $\mathbf{PDA}_e$, and $\mathbf{PDA}_{ef}$ denote the language families accepted by pushdown automata accepting by final state, by empty pushdown, and by final state and empty pushdown, respectively.

Theorem 2.4.11 (See [Med00a]). $\mathbf{PDA}_f = \mathbf{PDA}_e = \mathbf{PDA}_{ef}$

As the next theorem states, pushdown automata characterize the family of context-free languages.

Theorem 2.4.12 (See [Med00a]).

$$\mathbf{CF} = \mathbf{EPDA}_f = \mathbf{EPDA}_e = \mathbf{EPDA}_{ef} = \mathbf{PDA}_f = \mathbf{PDA}_e = \mathbf{PDA}_{ef}$$

Finally, we define deterministic versions of PDAs.

Definition 2.4.13. Let $M = (Q, \Sigma, \Gamma, R, s, S, F)$ be a pushdown automaton. M is a *deterministic pushdown automaton* if

(1) for any $q \in Q$, $a \in \Sigma \cup \{\varepsilon\}$, $x \in \Gamma$, the set $\{(q, a, x) \mid (q, a, x) \in R\}$, has at most one element and
(2) if $\{(q, \varepsilon, x) \mid (q, \varepsilon, x) \in R\} \neq \emptyset$, then $\{(q, a, x) \mid (q, a, x) \in R, a \in \Sigma\} = \emptyset$ for every q and a.   □

Let $\mathbf{DPDA}$ denote the family of languages accepted by deterministic pushdown automata. Unlike finite automata, determinism in the case of pushdown automata decrease the acceptance power.

Theorem 2.4.14 (See [Med00a]).

$$\mathbf{REG} \subset \mathbf{DPDA} \subset \mathbf{CF}$$

Part II
Modern Grammars

This part, consisting of Chaps. 3 through 6, presents an overview of major modern types of grammars together with the corresponding computational modes formalized by them. Chapter 3 covers the most important grammars for regulated computation. In essence, these grammars regulate their language generation by additional mechanisms, based upon simple mathematical concepts, such as finite sets of symbols. Chapter 4 discusses grammatical models for computation in parallel. Accordingly, these grammars generate their languages in parallel and, thereby, accelerate the generation process enormously just like computation in parallel is usually much faster than that made in an ordinary sequential way. First, the chapter studies partially parallel generation of languages, after which it investigates the totally parallel generation of languages. Chapter 5 explores grammars that work on their words in a discontinuous way, thus reflecting and formalizing a discontinuous way of computation. Chapter 6 approaches grammatical models for languages and computation from an algebraic standpoint. In particular, it examines grammatical generation of languages defined over free groups.

Chapter 3
Regulated Grammars and Computation

In practice, computation is almost always regulated by some additional conditions and restrictions placed upon the way it is performed under given circumstances. To investigate computation regulated in this way as precisely as possible, language theory has formalized it by a variety of regulated grammars. In essence, all these grammars are based upon some restrictions placed upon their derivations and, thereby, properly express computational regulation. This chapter covers major types of these grammars.

More precisely, the present chapter, consisting of four sections, classifies regulated grammars into two categories—context-based regulated grammars (Sect. 3.1) and rule-based regulated grammars (Sects. 3.2 through 3.5).

Section 3.1 gives an extensive and thorough coverage of regulated grammars that generate languages under various context-related restrictions. First, it views classical grammars as context-based regulated grammars. Then, it studies context-conditional grammars and their variants, including random context grammars, generalized forbidding grammars, semi-conditional grammars, and simple semi-conditional grammars. They all have their rules enriched by permitting and forbidding strings, referred to as permitting and forbidding conditions, respectively. These grammars regulate the language generation process so they require the presence of permitting conditions and, simultaneously, the absence of forbidding conditions in the rewritten sentential forms.

Sections 3.2 through 3.5 study grammatical regulation underlain by restrictions placed on the use of rules. Three types of grammars regulated in this way are covered namely, state grammars (Sect. 3.2), grammars with control languages (Sect. 3.3), matrix grammars (Sect. 3.4), and programmed grammars (Sect. 3.5). State grammars regulate the use of rules by states in a way that strongly resembles the finite-state control of finite automata. Grammars with control languages regulate the use of rules by regular languages. Matrix grammars represent special cases of

© Springer International Publishing AG 2017
A. Meduna, O. Soukup, *Modern Language Models and Computation*,
DOI 10.1007/978-3-319-63100-4_3

regular-controlled grammars whose control languages have the form of the iteration
of finite languages. Finally, the regulation of programmed grammars is based upon
binary relations over the sets of rules.

3.1 Context-Based Grammatical Regulation

The present section discusses context-regulated grammars, which regulate their
derivations by placing context-related restrictions upon their rewritten sentential
forms. It consists of six subsections. Section 3.1.1 demonstrates that classical
grammars can be viewed as grammars regulated in this way. It concentrates its
attention on their normal forms and uniform rewriting in them. Section 3.1.2
introduces general versions of context-conditional grammars and establishes fundamental results about them. Then, the rest of Sect. 3.1 discusses special cases of
these general versions; namely, Sects. 3.1.3, 3.1.4, 3.1.5, and 3.1.6 cover random
context grammars, forbidding grammars, semi-conditional grammars, and simple
semi-conditional grammars, respectively.

3.1.1 Classical Grammars Viewed as Context-Regulated
Grammars

Classical grammars, such as context-sensitive and phrase-structure grammars (see
Sect. 2.3), can be seen, in a quite natural way, as context-regulated grammars.
Indeed, on the left-hand sides of their rules, they have strings—that is, sequences
of symbols, not single symbols. In effect, they thus regulate their derivations
by prescribing sequences of neighboring symbols that can be rewritten during a
derivation step; this kind of regulation is generally referred to as tight-context
regulation to distinct it from scattered-context regulation, in which the symbol-
neighborhood requirement is dropped (see Sect. 4.1).

In general, tight-context regulated grammars, represented by context-sensitive
and phrase-structure grammars in this section, may have rules of various forms,
and they may generate a very broad variety of completely different sentential forms
during the generation of their languages. As obvious, this inconsistency concerning
the form of rules as well as rewritten strings represents an undesirable phenomenon
in theory as well as in practice. From a theoretical viewpoint, the demonstration
of properties concerning languages generated in this inconsistent way usually lead
to unbearably tedious proofs. From a practical viewpoint, this kind of language
generation is obviously difficult to apply and implement. Therefore, we pay a special
attention to arranging these grammars so they generate their languages in a more
uniform way.

The present section consists of two subsections. First, it studies grammars with modified grammatical rules so they all satisfy some simple prescribed forms, generally referred to as normal forms. Second, it explains how to perform tight-context rewriting over strings that have a uniform form.

Normal Forms

In this section, we convert context-sensitive and phrase-structure grammars into several normal forms, including the Kuroda, Penttonen, and Geffert normal forms. We also reduce the number of context-free rules in these grammars. In addition, we describe the Greibach and Chomsky normal forms for context-free grammars.

Recall that for a grammar $G = (V, T, P, S)$, N denotes the set of all nonterminal symbols, where $N = V - T$.

Definition 3.1.1. Let $G = (V, T, P, S)$ be a phrase-structure grammar. G is in the *Kuroda normal form* (see [Kur64]) if every rule in P is of one of the following four forms

$$\text{(i) } AB \to CD \quad \text{(ii) } A \to BC \quad \text{(iii) } A \to a \quad \text{(iv) } A \to \varepsilon$$

where $A, B, C, D \in N$, and $a \in T$. □

Theorem 3.1.2 (See [Kur64]). *For every phrase-structure grammar G, there is a phrase-structure grammar G' in the Kuroda normal form such that $L(G') = L(G)$.*

Definition 3.1.3. Let $G = (V, T, P, S)$ be a phrase-structure grammar. G is in the *Penttonen normal form* (see [Pen74]) if every rule in P is of one of the following four forms

$$\text{(i) } AB \to AC \quad \text{(ii) } A \to BC \quad \text{(iii) } A \to a \quad \text{(iv) } A \to \varepsilon$$

where $A, B, C \in N$, and $a \in T$. □

In other words, G is in the Penttonen normal form if G is in the Kuroda normal form an every $AB \to CD \in P$ satisfies that $A = C$.

Theorem 3.1.4 (See [Pen74]). *For every phrase-structure grammar G, there is a phrase-structure grammar G' in the Penttonen normal form such that $L(G') = L(G)$.*

Theorem 3.1.5 (See [Pen74]). *For every context-sensitive grammar G, there is a context-sensitive grammar G' in the Penttonen normal form such that $L(G') = L(G)$.*

Observe that if G is a context-sensitive grammar in the Pentonnen normal form, then none of its rules is of the form (iv), which is not context-sensitive.

Theorems 3.1.4 and 3.1.5 can be further modified so that for every context-sensitive rule of the form $AB \to AC \in P$, where $A, B, C \in N$, there exist no $B \to x$ or $BD \to BE$ in P for any $x \in V^*$, $D, E \in N$:

Theorem 3.1.6. *Every context-sensitive language can be generated by a context-sensitive grammar* $G = (N_{CF} \cup N_{CS} \cup T, T, P, S)$, *where* N_{CF}, N_{CS}, *and* T *are pairwise disjoint alphabets, and every rule in* P *is either of the form* $AB \to AC$, *where* $B \in N_{CS}$, $A, C \in N_{CF}$, *or of the form* $A \to x$, *where* $A \in N_{CF}$ *and* $x \in N_{CS} \cup T \cup N_{CF}^2$.

Proof. Let $G' = (V, T, P', S)$ be a context-sensitive grammar in the Penttonen normal form (see Theorem 3.1.5). Then, let

$$G = (N_{CF} \cup N_{CS} \cup T, T, P, S)$$

be the context-sensitive grammar defined as follows:

$$N_{CF} = N$$
$$N_{CS} = \{\tilde{B} \mid AB \to AC \in P', A, B, C \in N\}$$
$$P \ \ = \{A \to x \mid A \to x \in P', A \in N, x \in T \cup N^2\} \cup$$
$$\{B \to \tilde{B}, A\tilde{B} \to AC \mid AB \to AC \in P', A, B, C \in N\}$$

Obviously, $L(G') = L(G)$ and G is of the required form, so the theorem holds. $\square$

Theorem 3.1.7. *Every recursively enumerable language can be generated by a phrase-structure grammar* $G = (N_{CF} \cup N_{CS} \cup T, T, P, S)$, *where* N_{CF}, N_{CS}, *and* T *are pairwise disjoint alphabets, and every rule in* P *is either of the form* $AB \to AC$, *where* $B \in N_{CS}$, $A, C \in N_{CF}$, *or of the form* $A \to x$, *where* $A \in N_{CF}$ *and* $x \in N_{CS} \cup T \cup N_{CF}^2 \cup \{\varepsilon\}$.

Proof. The reader can prove this theorem by analogy with the proof of Theorem 3.1.6. $\square$

The next two normal forms limit the number of nonterminals and context-sensitive rules in phrase-structure grammars.

Definition 3.1.8. Let G be a phrase-structure grammar. G is in the *first Geffert normal form* (see [Gef91]) if it is of the form

$$G = (\{S, A, B, C\} \cup T, T, P \cup \{ABC \to \varepsilon\}, S)$$

where P contains context-free rules of the following three forms

$$\text{(i)} \ S \to uSa \qquad \text{(ii)} \ S \to uSv \qquad \text{(iii)} \ S \to uv$$

where $u \in \{A, AB\}^*$, $v \in \{BC, C\}^*$, and $a \in T$. $\square$

Theorem 3.1.9 (See [Gef91]). *For every recursively enumerable language* K, *there exists a phrase-structure grammar* G *in the first Geffert normal form such that* $L(G) = K$. *In addition, every successful derivation in* G *is of the form* $S \Rightarrow_G^* w_1 w_2 w$ *by rules from* P, *where* $w_1 \in \{A, AB\}^*$, $w_2 \in \{BC, C\}^*$, $w \in T^*$, *and* $w_1 w_2 w \Rightarrow_G^* w$ *is derived by* $ABC \to \varepsilon$.

Definition 3.1.10. Let G be a phrase-structure grammar. G is in the *second Geffert normal form* (see [Gef91]) if it is of the form

$$G = \big(\{S, A, B, C, D\} \cup T, T, P \cup \{AB \to \varepsilon, CD \to \varepsilon\}, S\big)$$

where P contains context-free rules of the following three forms

$$\text{(i) } S \to uSa \qquad \text{(ii) } S \to uSv \qquad \text{(iii) } S \to uv$$

where $u \in \{A, C\}^*$, $v \in \{B, D\}^*$, and $a \in T$. ☐

Theorem 3.1.11 (See [Gef91]). *For every recursively enumerable language K, there exists a phrase-structure grammar G in the Geffert normal form such that $L(G) = K$. In addition, every successful derivation in G is of the form $S \Rightarrow_G^* w_1 w_2 w$ by rules from P, where $w_1 \in \{A, C\}^*$, $w_2 \in \{B, D\}^*$, $w \in T^*$, and $w_1 w_2 w \Rightarrow_G^* w$ is derived by $AB \to \varepsilon$ and $CD \to \varepsilon$.*

Next, we establish two new normal forms for phrase-structure grammars with a limited number of context-free rules in a prescribed form and, simultaneously, with non-context-free rules in a prescribed form. Specifically, we establish the following two normal forms of this kind.

(I) First, we explain how to turn any phrase-structure grammar to an equivalent phrase-structure grammar that has $2 + n$ context-free rules, where n is the number of terminals, and every context-free rule is of the form $A \to x$, where x is a terminal, a two-nonterminal string, or the empty string. In addition, every non-context-free rule is of the form $AB \to CD$, where A, B, C, D are nonterminals.

(II) In the second normal form, phrase-structure grammars have always only two context-free rules—that is, the number of context-free rules is reduced independently of the number of terminals as opposed to the first normal form. Specifically, we describe how to turn any phrase-structure grammar to an equivalent phrase-structure grammar that has two context-free rules of the forms $A \to \varepsilon$ and $A \to BC$, where A, B, C are nonterminals and ε denotes the empty string, and in addition, every non-context-free rule is of the form $AB \to CD$, where A, B, D are nonterminals and C is nonterminal or a terminal.

Theorem 3.1.12. *Let G be a phrase-structure grammar. Then, there is an equivalent phrase-structure grammar*

$$H = \big(V, T, P_1 \cup P_2 \cup P_3, S\big)$$

with

$$P_1 = \{AB \to CD \mid A, B, C, D \in N\}$$
$$P_2 = \{S \to S\#, \# \to \varepsilon\}$$
$$P_3 = \{A \to a \mid A \in N, a \in T\}$$

where $\# \in N$.

Proof. Let $G = (V, T, P, S)$ be a phrase-structure grammar. By Theorem 3.1.2, we assume that G is in the Kuroda normal form. Set $\bar{T} = \{\bar{a} \mid a \in T\}$. Without any loss of generality, we assume that $N, T, \bar{T}$, and $\{\#\}$ are pairwise disjoint. Construct the phrase-structure grammar

$$H = \left(V', T, P'_1 \cup P'_2 \cup P'_3, S\right)$$

as follows. Initially, set $V' = V \cup \bar{T} \cup \{\#\}$, $P'_1 = \emptyset$, $P'_2 = \{S \to S\#, \# \to \varepsilon\}$, and $P'_3 = \{\bar{a} \to a \mid a \in T\}$. Perform (1) through (5), given next.

(1) For each $AB \to CD \in P$, where $A, B, C, D \in N$, add $AB \to CD$ to P'_1.
(2) For each $A \to BC \in P$, where $A, B, C \in N$, add $A\# \to BC$ to P'_1.
(3) For each $A \to a \in P$, where $A \in N$ and $a \in T$, add $A\# \to \bar{a}\#$ to P'_1.
(4) For each $A \to \varepsilon \in P$, where $A \in N$, add $A\# \to \#\#$ to P'_1.
(5) For each $A \in N$, add $A\# \to \#A$ and $\#A \to A\#$ to P'_1.

Before proving that $L(H) = L(G)$, let us give an insight into the construction. We simulate G by H using the following sequences of derivation steps.

First, by repeatedly using $S \to S\#$, we generate a proper number of #s. Observe that if the number of #s is too low, the derivation can be blocked since rules from (2) consume # during their application. Furthermore, notice that only rules from (4) and the initial rule $S \to S\#$ increase the number of #s in sentential forms of H.

Next, we simulate each application of a rule in G by several derivation steps in H. By using rules from (5), we can move # in the current sentential form as needed. If we have # or B in a proper position next to A, we can apply a rule from (1) through (4). We can also apply $\# \to \varepsilon$ to remove any occurrence of # from a sentential form of H.

To conclude the simulation, we rewrite the current sentential form by rules of the form $\bar{a} \to a$ to generate a string of terminals. Observe that a premature application of a rule of this kind may block the derivation in H. Indeed, #s then cannot move freely through such a sentential form.

To establish $L(H) = L(G)$, we prove four claims. Claim 3.1.13 demonstrates that every $w \in L(H)$ can be generated in two stages; first, only nonterminals are generated, and then, all nonterminals are rewritten to terminals. Claim 3.1.14 shows that we can arbitrarily generate and move #s within sentential forms of H during the first stage. Claim 3.1.15 shows how derivations of G are simulated by H. Finally, Claim 3.1.16 shows how derivations of every $w \in L(H)$ in H are simulated by G.

Set $N' = V' - T$. Define the homomorphism τ from V'^* to V^* as $\tau(X) = X$ for all $X \in V$, $\tau(\bar{a}) = a$ for all $a \in T$, and $\tau(\#) = \varepsilon$.

Claim 3.1.13. Let $w \in L(H)$. Then, there exists a derivation $S \Rightarrow^*_H x \Rightarrow^*_H w$, where $x \in N'^+$, and during $x \Rightarrow^*_H w$, only rules of the form $\bar{a} \to a$, where $a \in T$, are applied.

Proof. Let $w \in L(H)$. Since there are no rules in $P'_1 \cup P'_2 \cup P'_3$ with symbols from T on their left-hand sides, we can always rearrange all the applications of the rules occurring in $S \Rightarrow^*_H w$ so the claim holds.   □

Claim 3.1.14. If $S \Rightarrow_H^* uv$, where $u, v \in V'^*$, then $S \Rightarrow_H^* u\#v$.

Proof. By an additional application of $S \rightarrow S\#$, we get $S \Rightarrow_H^* S\#^{n+1}$ instead of $S \Rightarrow_H^* S\#^n$ for some $n \geq 0$, so we derive one more # in the sentential form. From Claim 3.1.13, by applying rules from (5), # can freely migrate through the sentential form as needed, until a rule of the form $\bar{a} \rightarrow a$ is used. $\qquad \Box$

Claim 3.1.15. If $S \Rightarrow_G^k x$, where $x \in V^*$, for some $k \geq 0$, then $S \Rightarrow_H^* x'$, where $\tau(x') = x$.

Proof. This claim is established by induction on $k \geq 0$.

Basis. For $S \Rightarrow_G^0 S$, there is $S \Rightarrow_H^0 S$.

Induction Hypothesis. For some $k \geq 0$, $S \Rightarrow_G^k x$ implies that $S \Rightarrow_H^* x'$ such that $x = \tau(x')$.

Induction Step. Let $u, v \in N'^*$, $A, B, C, D \in N$, and $m \geq 0$. Assume that $S \Rightarrow_G^k y \Rightarrow_G x$. By the induction hypothesis, $S \Rightarrow_H^* y'$ with $y = \tau(y')$. Let us show the simulation of $y \Rightarrow_G x$ by an application of several derivation steps in H to get $y' \Rightarrow_H^+ x'$ with $\tau(x') = x$. This simulation is divided into the following four cases— (i) through (iv).

(i) *Simulation of $AB \rightarrow CD$:* $y' = uA\#^m Bv \Rightarrow_H^m u\#^m ABv \Rightarrow_H u\#^m CDv = x'$ using m derivation steps according to rules $A\# \rightarrow \#A$ from (5), and concluding the derivation by rule $AB \rightarrow CD$ from (1).

By the induction hypothesis and Claim 3.1.14, $y = \tau(u)A\tau(v)$ allows $y' = uA\#v$.

(ii) *Simulation of $A \rightarrow BC$:* $y' = uA\#v \Rightarrow_H uBCv = x'$ using rule $A\# \rightarrow BC$ from (2).

(iii) *Simulation of $A \rightarrow a$:* $y' = uA\#v \Rightarrow_H u\bar{a}\#v = x'$ using rule $A\# \rightarrow \bar{a}\#$ from (3).

(iv) *Simulation of $A \rightarrow \varepsilon$:* $y' = uA\#v \Rightarrow_H u\#\#v = x'$ using rule $A\# \rightarrow \#\#$ from (4). $\qquad \Box$

Claim 3.1.16. If $S \Rightarrow_H^k x'$, where $x' \in N'^*$, for some $k \geq 0$, then $S \Rightarrow_G^* x$ with $x = \tau(x')$.

Proof. This claim is established by induction on $k \geq 0$.

Basis. For $S \Rightarrow_H^0 S$, there is $S \Rightarrow_G^0 S$.

Induction Hypothesis. For some $k \geq 0$, $S \Rightarrow_H^k x'$ implies that $S \Rightarrow_G^* x$ such that $x = \tau(x')$.

Induction Step. Let $u, v, w \in N'^*$ and $A, B, C, D \in N$. Assume that $S \Rightarrow_H^k y' \Rightarrow_H x'$. By the induction hypothesis, $S \Rightarrow_G^* y$ such that $y = \tau(y')$. Let us examine the following seven possibilities of $y' \Rightarrow_H x'$.

(i) $y' = uSv \Rightarrow_H uS\#v = x'$: Then,

$$\tau(y') = y = \tau(uSv) \Rightarrow_G^0 \tau(uS\#v) = \tau(uSv) = x = \tau(x')$$

(ii) $y' = uABv \Rightarrow_H uCDv = x'$: According to (1),

$$y = \tau(u)AB\tau(v) \Rightarrow_G \tau(u)CD\tau(v) = x$$

(iii) $y' = uA\#v \Rightarrow_H uBCv = x'$: According to the source rule in (2),

$$y = \tau(u)A\tau(\#v) \Rightarrow_G \tau(u)BC\tau(\#v) = \tau(u)BC\tau(v) = x$$

(iv) $y' = uA\#v \Rightarrow_H u\#\#v = x'$: By the corresponding rule $A \to \varepsilon$,

$$y = \tau(u)A\tau(v) \Rightarrow_G \tau(u\#\#v) = \tau(uv) = x$$

(v) $y' = uA\#v \Rightarrow_H u\#Av = x'$ or $y' = u\#Av \Rightarrow_H uA\#v = x'$: In G,

$$y = \tau(uA\#v) = \tau(u)A\tau(\#v) \Rightarrow_G^0 \tau(u\#)A\tau(v) = x$$

or

$$y = \tau(u\#Av) = \tau(u\#)A\tau(v) \Rightarrow_G^0 \tau(u)A\tau(\#v) = x$$

(vi) $y' = u\#v \Rightarrow_H uv = x'$: In G,

$$y = \tau(u\#v) \Rightarrow_G^0 \tau(uv) = x$$

(vii) $y' = u\bar{a}v \Rightarrow_H uav = x'$: In G,

$$y = \tau(u\bar{a}v) = \tau(u)a\tau(v) \Rightarrow_G^0 \tau(u)a\tau(v) = x \qquad\qquad \square$$

Next, we establish $L(H) = L(G)$. Consider Claim 3.1.15 with $x \in T^*$. Then, $S \Rightarrow_G^* x$ implies that $S \Rightarrow_H^* x$, so $L(G) \subseteq L(H)$. Let $w \in L(H)$. By Claim 3.1.13, $S \Rightarrow_H^* x \Rightarrow_H^* w$, where $x \in N'^+$, and during $x \Rightarrow_H^* w$, only rules of the form $\bar{a} \to a$, where $a \in T$, are applied. By Claim 3.1.16, $S \Rightarrow_G^* \tau(x) = w$, so $L(H) \subseteq L(G)$. Hence, $L(H) = L(G)$.

Since H is of the required form, the theorem holds. $\qquad\qquad \square$

From the construction in the proof of Theorem 3.1.12, we obtain the following corollary concerning the number of nonterminals and rules in the resulting grammar.

Corollary 3.1.17. *Let $G = (V, T, P, S)$ be a phrase-structure grammar in the Kuroda normal form. Then, there is an equivalent phrase-structure grammar in the normal form from Theorem 3.1.12*

$$H = (V', T, P', S)$$

where

$$\text{card}\,(V') = \text{card}\,(V) + \text{card}\,(T) + 1$$

and

$$\text{card}\,(P') = \text{card}\,(P) + \text{card}\,(T) + 2\big(\text{card}(N) + 1\big) \qquad \square$$

If we drop the requirement on each symbol in the non-context-free rules to be a nonterminal, we can reduce the number of context-free rules even more.

Theorem 3.1.18. *Let G be a phrase-structure grammar. Then, there is an equivalent phrase-structure grammar*

$$H = \big(V, T, P_1 \cup P_2, S\big)$$

with

$$P_1 = \{AB \to XC \mid A, B, C \in N, X \in N \cup T\}$$
$$P_2 = \{S \to S\#, \# \to \varepsilon\}$$

where $\# \in N$.

Proof. Reconsider the proof of Theorem 3.1.12. Observe that we can obtain the new normal form by omitting the construction of P'_3 and modifying step (3) in the following way

(3) For each $A \to a \in P$, where $A \in N$ and $a \in T$, add $A\# \to a\#$ to P'_1.

The rest of the proof is analogical to the proof of Theorem 3.1.12 and it is left to the reader. $\square$

Next, we define two normal forms for context-free grammars—the Chomsky and Greibach normal forms (see [Cho59] and [Gre65]).

Definition 3.1.19. Let G be a context-free grammar. G is in the *Chomsky normal form* if every $A \to x \in P$ satisfies that $x \in NN \cup T$. $\square$

Theorem 3.1.20 (See [Cho59]). *For every context-free grammar G, there is a context-free grammar G' in the Chomsky normal form such that $L(G') = L(G)$.*

Definition 3.1.21. Let G be a context-free grammar. G is in the *Greibach normal form* if every $A \to x \in P$ satisfies that $x \in TN^*$. $\square$

Theorem 3.1.22 (See [Gre65]). *For every context-free grammar G, there is a context-free grammar G' in the Greibach normal form such that $L(G') = L(G)$.*

Finally, we define the following two normal forms for queue grammars and left-extended queue grammars.

Definition 3.1.23. Let $Q = (V, T, W, F, R, g)$ be a queue grammar (see Definition 2.3.42). Q satisfies the normal form, if $\{!,f\} \subseteq W$, $F = \{f\}$, and each $(a, b, x, c) \in R$ satisfies $a \in V - T$ and either

$$b \in W, x \in (V - T)^*, c \in W \cup \{!,f\} \quad \text{or} \quad b =!, x \in T, c \in \{!,f\} \qquad \square$$

Theorem 3.1.24. *For every queue grammar Q', there is a left-extended queue grammar Q in the normal form of Definition 3.1.23 such that $L(Q') = L(Q)$. Then, Q generates every $y \in L(Q) - \{\varepsilon\}$ in this way*

$$a_0 q_0 \Rightarrow_Q x_0 q_1 \qquad\qquad [(a_0, q_0, z_0, q_1)]$$

$$\vdots$$

$$\Rightarrow_Q x_{k-1} q_k \qquad\qquad [(a_{k-1}, q_{k-1}, z_{k-1}, q_k)]$$
$$\Rightarrow_Q x_k! \qquad\qquad [(a_k, q_k, z_k, !)]$$
$$\Rightarrow_Q x_{k+1} b_1! \qquad\qquad [(a_{k+1}, !, b_1, !)]$$

$$\vdots$$

$$\Rightarrow_Q x_{k+m-1} b_1 \cdots b_{m-1}! \quad [(a_{k+m-1}, !, b_{m-1}, !)]$$
$$\Rightarrow_Q b_1 \cdots b_m f \qquad\qquad [(a_{k+m}, !, b_m, f)]$$

where $k, m \geq 1$, $g = a_0 q_0$, $a_1, \ldots, a_{k+m} \in V - T$, $b_1, \ldots, b_m \in T$, $z_0, \ldots, z_k \in (V-T)^*$, $q_0, \ldots, q_k$, $! \in W - F$, $f \in F$, $x_0, \ldots, x_{k+m-1} \in (V-T)^+$, and $y = b_1 \cdots b_m$.

Proof. Let $Q' = (V', T, W', F', R', g')$ be any queue grammar. Set $\Phi = \{\bar{a} \mid a \in T\}$. Define the homomorphism α from V'^* to $((V' - T) \cup \Phi)^*$ as $\alpha(a) = \bar{a}$, for each $a \in T$ and $\alpha(A) = A$, for each $A \in V' - T$. Set $V = V' \cup \Phi$, $W = W' \cup \{!,f\}$, $F = \{f\}$, and $g = \alpha(a_0)q_0$ for $g' = a_0 q_0$. Define the queue grammar $Q = (V, T, W, F, R, g)$, with R constructed in the following way.

(1) For each $(a, b, x, c) \in R'$, where $c \in W' - F'$, add $(\alpha(a), b, \alpha(x), c)$ to R.
(2) For each $(a, b, x, c) \in R'$

    (2.1) where $x \neq \varepsilon$, $c \in F'$, add $(\alpha(a), b, \alpha(x), !)$ to R.
    (2.2) where $x = \varepsilon$, $c \in F'$, add $(\alpha(a), b, \varepsilon, f)$ to R.

(3) For each $a \in T$,

    (3.1) add $(\bar{a}, !, a, !)$ to R;
    (3.2) add $(\bar{a}, !, a, f)$ to R.

Clearly, each $(a, b, x, c) \in R$ satisfies $a \in V - T$ and either $b \in W'$, $x \in (V - T)^*$, $c \in W' \cup \{!,f\}$ or $b =!$, $x \in T$, $c \in \{!,f\}$.

To see that $L(Q') \subseteq L(Q)$, consider any $v \in L(Q')$. As $v \in L(Q')$, $g' \Rightarrow_{Q'}^* vt$, where $v \in T^*$ and $t \in F'$. Express $g' \Rightarrow_{Q'}^* vt$ as

$$g' \Rightarrow_{Q'}^* axc \Rightarrow_{Q'} vt \; [(a, c, y, t)]$$

where $a \in V'$, $x, y \in T^*$, $xy = v$, and $c \in W' - F'$. This derivation is simulated by Q as follows. First, Q uses rules from (1) to simulate $g' \Rightarrow_{Q'}^* axc$. Then, it uses a rule from (2) to simulate $axc \Rightarrow_{Q'} vt$. For $x = \varepsilon$, a rule from (2.2) can be used to generate $\varepsilon \in L(Q)$ in the case of $\varepsilon \in L(Q')$; otherwise, a rule from (2.1) is used. This part of simulation can be expressed as

$$g \Rightarrow_Q^* \alpha(ax)c \Rightarrow_Q \alpha(v)!$$

At this point, $\alpha(v)$ satisfies $\alpha(v) = \bar{a}_1 \cdots \bar{a}_n$, where $a_i \in T$ for all i, $1 \le i \le n$, for some $n \ge 1$. The rules from (3) of the form $(\bar{a}, !, a, !)$, where $a \in T$, replace every $\bar{a}_j$ with a_j, where $1 \le j \le n - 1$, and, finally, $(\bar{a}, !, a, f)$, where $a \in T$, replaces $\alpha(a_n)$ with a_n. As a result, we obtain the sentence vf, so $L(Q') \subseteq L(Q)$.

To establish $L(Q) \subseteq L(Q')$, observe that the use of a rule from (2.2) in Q before the sentential form is of the form $\alpha(ax)c$, where $a \in V'$, $x \in T^*$, $c \in W' - F'$, leads to an unsuccessful derivation. Similarly, the use of (2.2) if $x \ne \varepsilon$ leads to an unsuccessful derivation as well. The details are left to the reader. As a result, $L(Q) \subseteq L(Q')$.

As $L(Q') \subseteq L(Q)$ and $L(Q) \subseteq L(Q')$, we obtain $L(Q) = L(Q')$. $\square$

Briefly, a queue grammar $Q = (V, T, W, F, R, g)$ in normal form of Definition 3.1.23 generates every string in $L(Q) - \{\varepsilon\}$ so it passes through !. Before it enters !, it generates only strings from $(V - T)^*$; after entering !, it generates only strings from T^*.

Definition 3.1.25. Let $Q = (V, T, W, F, R, g)$ be a left-extended queue grammar. Q satisfies the normal form, if $V = U \cup Z \cup T$ and $W = X \cup Y \cup \{!\}$ such that $U, Z, T, X, Y, \{!\}$ are pairwise disjoint, $F \subseteq Y$, $g \in UW$, and Q derives every $w \in L(Q)$ in this way

$$
\begin{aligned}
\#g &\Rightarrow_Q^m a_0 a_1 \ldots a_m \# b_1 b_2 \cdots b_n! & [t_1 t_2 \ldots t_m] \\
&\Rightarrow_Q a_0 a_1 \ldots a_m b_1 \# b_2 \cdots b_n y_1 p_1 & [r_1] \\
&\Rightarrow_Q a_0 a_1 \ldots a_m b_1 b_2 \# b_3 \cdots b_n y_1 y_2 p_2 & [r_2] \\
&\;\;\vdots \\
&\Rightarrow_Q a_0 a_1 \ldots a_m b_1 b_2 \cdots b_{n-1} \# b_n y_1 y_2 \cdots y_{n-1} p_{n-1} & [r_{n-1}] \\
&\Rightarrow_Q a_0 a_1 \ldots a_m b_1 b_2 \cdots b_{n-1} b_n \# y_1 y_2 \cdots y_n p_n & [r_n]
\end{aligned}
$$

where $m, n \in \mathbb{N}$, $a_j \in U$, $b_i \in Z$, $y_i \in T^*$, $w = y_1 y_2 \cdots y_n$, $p_i \in Y$, $p_n \in F$, $t_j, r_i \in R$, where $t_j = (a_j, b_j, x_j, c_j)$ satisfies $a_j \in U$, $b_j \in X$, $x_j \in (V - T)^*$, $c_j \in X \cup \{!\}$, and $r_i = (a_i, b_i, x_i, c_i)$ satisfies $a_i \in Z$, $b_i \in (Y - F) \cup \{!\}$, $x_i \in T^*$, $c_i \in Y$, for $j = 1, \ldots, m$ and $i = 1, \ldots, n$.

Theorem 3.1.26. *For every left-extended queue grammar K, there is a left-extended queue grammar Q in the normal form from Definition 3.1.25 such that $L(K) = L(Q)$.*

Proof. See Lemma 1 in [Med03c].

Uniform Rewriting

Classical grammars can produce a very broad variety of quite different sentential forms during the generation of their languages. This inconsistent generation represents a highly undesirable grammatical phenomenon. In theory, the demonstration of properties concerning languages generated in this way lead to extremely tedious proofs. In practice, the inconsistent generation of languages is not easy to apply and implement. Therefore, in this section, we explain how to reduce or even overcome this difficulty by making the language generation more uniform. Specifically, phrase-structure grammars are transformed so that they generate only strings that have a uniform permutation-based form.

More precisely, the present section demonstrates that for every phrase-structure grammar G, there exists an equivalent phrase-structure grammar

$$G' = \big(\{S, 0, 1\} \cup T, T, P, S\big)$$

so that every $x \in F(G')$ satisfies

$$x \in T^* \Pi(w)^*$$

where $w \in \{0, 1\}^*$ (recall that $F(G')$ is defined in Definition 2.3.2). Then, it makes this conversion so that for every $x \in F(G)$,

$$x \in \Pi(w)^* T^*$$

Let $G = (V, T, P, S)$ be a phrase-structure grammar. Notice that $\mathrm{alph}(L(G)) \subseteq T$. If $a \in T - \mathrm{alph}(L(G))$, then a actually acts as a pseudoterminal because it appears in no string of $L(G)$. Every transformation described in this section assumes that its input grammar contains no pseudoterminals of this kind, and does not contain any useless nonterminals either.

Let j be a natural number. Set

$$\mathbf{PS}[.j] = \big\{L \mid L = L(G), \text{ where } G = (V, T, P, S) \text{ is a phrase-structure}$$
$$\text{grammar such that } \mathrm{card}(\mathrm{alph}(F(G)) - T) = j \text{ and}$$
$$F(G) \subseteq T^* \Pi(w)^*, \text{ where } w \in (V - T)^*\big\}$$

Analogously, set

$$\mathbf{PS}[j.] = \big\{L \mid L = L(G), \text{ where } G = (V, T, P, S) \text{ is a phrase-structure}$$
$$\text{grammar such that } \mathrm{card}(\mathrm{alph}(F(G)) - T) = j \text{ and}$$
$$F(G) \subseteq \Pi(w)^* T^*, \text{ where } w \in (V - T)^*\big\}$$

Lemma 3.1.27. *Let G be a phrase-structure grammar. Then, there exists a phrase-structure grammar, $G' = (\{S, 0, 1\} \cup T, T, P, S)$, satisfying $L(G') = L(G)$ and $F(G') \subseteq T^* \Pi (1^{n-2} 00)^*$, for some natural number n.*

Proof. Let $G = (V, T, Q, \$)$ be a phrase-structure grammar, where V is the total alphabet of G, T is the terminal alphabet of G, Q is the set of rules of G, and $\$$ is the start symbol of G. Without any loss of generality, assume that $V \cap \{0, 1\} = \emptyset$. The following construction produces an equivalent phrase-structure grammar

$$G' = (\{S, 0, 1\} \cup T, T, P, S)$$

such that $F(G') \subseteq T^* \Pi (1^{n-2} 00)^*$, for some natural number n.

For some integers m, n such that $m \geq 3$ and $2m = n$, introduce an injective homomorphism β from V to

$$(\{1\}^m \{1\}^* \{0\} \{1\}^* \{0\} \cap \{0, 1\}^n) - \{1^{n-2} 00\}$$

Extend the domain of β to V^*. Define the phrase-structure grammar

$$G' = (\{S, 0, 1\} \cup T, T, P, S)$$

with

$$
\begin{aligned}
P = \{ &S \to 1^{n-2} 00 \beta(\$) 1^{n-2} 00 \} \cup \\
&\{\beta(x) \to \beta(y) \mid x \to y \in Q\} \cup \\
&\{1^{n-2} 00 \beta(a) \to a 1^{n-2} 00 \mid a \in T\} \cup \\
&\{1^{n-2} 001^{n-2} 00 \to \varepsilon\}
\end{aligned}
$$

Claim 3.1.28. Let $S \Rightarrow^h_{G'} w$, where $w \in V^*$ and $h \geq 1$. Then,

$$w \in T^* (\{\varepsilon\} \cup \{1^{n-2} 00\} (\beta(V))^* \{1^{n-2} 00\})$$

Proof. The claim is proved by induction on $h \geq 1$.

Basis. Let $h = 1$. That is,

$$S \Rightarrow_{G'} 1^{n-2} 00 \beta(\$) 1^{n-2} 00 \; [S \to 1^{n-2} 00 \beta(\$) 1^{n-2} 00]$$

As

$$1^{n-2} 00 \beta(S) 1^{n-2} 00 \in T^* (\{1^{n-2} 00\} (\beta(V))^* \{1^{n-2} 00\} \cup \{\varepsilon\})$$

the basis holds.

Induction Hypothesis. Suppose that for some $k \geq 0$, if $S \Rightarrow^i_{G'} w$, where $i = 1, \ldots, k$ and $w \in V^*$, then $w \in T^* (\{1^{n-2} 00\} (\beta(V))^* \{1^{n-2} 00\} \cup \{\varepsilon\})$.

Induction Step. Consider any derivation of the form

$$S \Rightarrow_{G'}^{k+1} w$$

where $w \in V^* - T^*$. Express $S \Rightarrow_{G'}^{k+1} w$ as

$$S \Rightarrow_{G'}^{k} u \, \text{lhs}(p)v$$
$$\Rightarrow_{G'} u \, \text{rhs}(p)v \; [p]$$

where $p \in P$ and $w = u \, \text{rhs}(p)v$. Less formally, after k steps, G' derives $u \, \text{lhs}(p)v$. Then, by using p, G' replaces $\text{lhs}(p)$ with $\text{rhs}(p)$ in $u \, \text{lhs}(p)v$, so it obtains $u \, \text{rhs}(p)v$. By the induction hypothesis,

$$u \, \text{lhs}(p)v \in T^*(\{1^{n-2}00\}(\beta(V))^*\{1^{n-2}00\} \cup \{\varepsilon\})$$

As $\text{lhs}(p) \notin T^*$, $u \, \text{lhs}(p)v \notin T^*$. Therefore,

$$u \, \text{lhs}(p)v \in T^*\{1^{n-2}00\}(\beta(V))^*\{1^{n-2}00\}$$

Let

$$u \, \text{lhs}(p)v \in T^*\{1^{n-2}00\}(\beta(V))^j\{1^{n-2}00\}$$

in G', for some $j \geq 1$. By the definition of P, p satisfies one of the following three properties.

(i) Let $\text{lhs}(p) = \beta(x)$ and $\text{rhs}(p) = \beta(y)$, where $x \to y \in Q$, At this point,

$$u \in T^*\{1^{n-2}00\}\{\beta(V)\}^r$$

for some $r \geq 0$, and

$$v \in \{\beta(V)\}^{(j-|\,\text{lhs}(p)|-r)}\{1^{n-2}00\}$$

Distinguish between these two cases: $|x| \leq |y|$ and $|x| > |y|$.

(i.a) Let $|x| \leq |y|$. Set $s = |y| - |x|$. Observe that

$$u \, \text{rhs}(p)v \in T^*\{1^{n-2}00\}(\beta(V))^{(j+s)}\{1^{n-2}00\}$$

As $w = u \, \text{rhs}(p)v$,

$$w \in T^*(\{1^{n-2}00\}(\beta(V))^*\{1^{n-2}00\} \cup \{\varepsilon\})$$

(i.b) Let $|x| > |y|$. By analogy with (i.a), prove that

$$w \in T^*(\{1^{n-2}00\}(\beta(V))^*\{1^{n-2}00\} \cup \{\varepsilon\})$$

(ii) Assume that $\mathrm{lhs}(p) = 1^{n-2}00\beta(a)$ and $\mathrm{rhs}(p) = a1^{n-2}00$, for some $a \in T$. Notice that

$$u\,\mathrm{lhs}(p)v \in T^*\{1^{n-2}00\}(\beta(V))^j\{1^{n-2}00\}$$

implies that $u \in T^*$ and

$$v \in (\beta(V))^{(j-1)}\{1^{n-2}00\}$$

Then,

$$u\,\mathrm{rhs}(p)v \in T^*\{a\}\{1^{n-2}00\}(\beta(V))^{(j-1)}\{1^{n-2}00\}$$

As $w = u\,\mathrm{rhs}(p)v$,

$$w \in T^*(\{1^{n-2}00\}(\beta(V))^*\{1^{n-2}00\} \cup \{\varepsilon\})$$

(iii) Assume that $\mathrm{lhs}(p) = 1^{n-2}001^{n-2}00$ and $\mathrm{rhs}(p) = \varepsilon$. Then, $j = 0$ in

$$T^*\{1^{n-2}00\}(\beta(V))^j\{1^{n-2}00\}$$

so

$$u\,\mathrm{lhs}(p)v \in T^*\{1^{n-2}00\}\{1^{n-2}00\}$$

and $u\,\mathrm{rhs}(p)v \in T^*$. As $w = u\,\mathrm{rhs}(p)v$,

$$w \in T^*(\{1^{n-2}00\}(\beta(V))^*\{1^{n-2}00\} \cup \{\varepsilon\}) \qquad\qquad \square$$

Claim 3.1.29. Let $S \Rightarrow_{G'}^+ u \Rightarrow_{G'}^* z$, where $z \in T^*$. Then, $u \in T^*\Pi(1^{n-2}00)^*$.

Proof. Let $S \Rightarrow_{G'}^+ u \Rightarrow_{G'}^* z$, where $z \in T^*$. By Claim 3.1.28,

$$u \in T^*(\{1^{n-2}00\}(\beta(V))^*\{1^{n-2}00\} \cup \{\varepsilon\})$$

and by the definition of β, $u \in T^*\Pi(1^{n-2}00)^*$. $\qquad\qquad \square$

Claim 3.1.30. Let $\$ \Rightarrow_G^m w$, for some $m \geq 0$. Then, $S \Rightarrow_{G'}^+ 1^{n-2}00\beta(w)1^{n-2}00$.

Proof. The claim is proved by induction on $m \geq 0$.

Basis. Let $m = 0$. That is, $\$ \Rightarrow_G^0 \$$. As

$$S \Rightarrow_{G'} 1^{n-2}00\beta(\$)1^{n-2}00 \ [S \to 1^{n-2}00\beta(\$)1^{n-2}00]$$

the basis holds.

Induction Hypothesis. Suppose that for some $j \geq 1$, if $\$ \Rightarrow^i_G w$, where $i = 1, \ldots, j$ and $w \in V^*$, then $S \Rightarrow^*_{G'} \beta(w)$.

Induction Step. Let $\$ \Rightarrow^{j+1}_G w$. Express $\$ \Rightarrow^{j+1}_G w$ as

$$\$ \Rightarrow^j_G uxv \Rightarrow_G uyv \ [x \to y]$$

where $x \to y \in Q$ and $w = uyv$. By the induction hypothesis,

$$S \Rightarrow^+_{G'} 1^{n-2}00\beta(uxv)1^{n-2}00$$

Express $\beta(uxv)$ as $\beta(uxv) = \beta(u)\beta(x)\beta(v)$. As $x \to y \in P$, $\beta(x) \to \beta(y) \in P$. Therefore,

$$S \Rightarrow^+_{G'} 1^{n-2}00\beta(u)\beta(x)\beta(v)1^{n-2}00$$
$$\Rightarrow_{G'} 1^{n-2}00\beta(u)\beta(y)\beta(v)1^{n-2}00 \ [\beta(x) \to \beta(y)]$$

Because $w = uyv$, $\beta(w) = \beta(u)\beta(y)\beta(v)$, so

$$S \Rightarrow^+_{G'} 1^{n-2}00\beta(w)1^{n-2}00 \qquad\qquad \square$$

Claim 3.1.31. $L(G) \subseteq L(G')$

Proof. Let $w \in L(G)$. Thus, $\$ \Rightarrow^*_G w$ with $w \in T^*$. By Claim 3.1.30,

$$S \Rightarrow^+_{G'} 1^{n-2}00\beta(w)1^{n-2}00$$

Distinguish between these two cases: $w = \varepsilon$ and $w \neq \varepsilon$.

(i) If $w = \varepsilon$, $1^{n-2}00\beta(w)1^{n-2}00 = 1^{n-2}001^{n-2}00$. As $1^{n-2}001^{n-2}00 \to \varepsilon \in P$,

$$S \Rightarrow^*_{G'} 1^{n-2}001^{n-2}00$$
$$\Rightarrow_{G'} \varepsilon \ [1^{n-2}001^{n-2}00 \to \varepsilon]$$

Thus, $w \in L(G')$.

(ii) Assume that $w \neq \varepsilon$. Express w as $w = a_1 a_2 \cdots a_{n-1} a_n$ with $a_i \in T$ for $i = 1, \ldots, n$, $n \geq 0$. Because

$$(\{1^{n-2}00\beta(a) \to a1^{n-2}00 \mid a \in T\} \cup \{1^{n-2}001^{n-2}00 \to \varepsilon\}) \subseteq P$$

there exists

$$S \Rightarrow^*_{G'} 1^{n-2}00\beta(a_1)\beta(a_2)\cdots\beta(a_{n-1})\beta(a_n)1^{n-2}00$$
$$\Rightarrow_{G'} a_1 1^{n-2}00\beta(a_2)\cdots\beta(a_{n-1})\beta(a_n)1^{n-2}00$$
$$[1^{n-2}00\beta(a_1) \to a_1 1^{n-2}00]$$
$$\Rightarrow_{G'} a_1 a_2 1^{n-2}00\beta(a_3)\cdots\beta(a_{n-1})\beta(a_n)1^{n-2}00$$
$$[1^{n-2}00\beta(a_2) \to a_2 1^{n-2}00]$$
$$\vdots$$
$$\Rightarrow_{G'} a_1 a_2 \cdots a_{n-2} 1^{n-2}00\beta(a_{n-1})\beta(a_n)1^{n-2}00$$
$$[1^{n-2}00\beta(a_{n-2}) \to a_{n-2} 1^{n-2}00]$$
$$\Rightarrow_{G'} a_1 a_2 \cdots a_{n-2}a_{n-1} 1^{n-2}00\beta(a_n)1^{n-2}00$$
$$[1^{n-2}00\beta(a_{n-1}) \to a_{n-1} 1^{n-2}00]$$
$$\Rightarrow_{G'} a_1 a_2 \cdots a_{n-2}a_{n-1}a_n 1^{n-2}001^{n-2}00$$
$$[1^{n-2}00\beta(a_n) \to a_n 1^{n-2}00]$$
$$\Rightarrow_{G'} a_1 a_2 \cdots a_{n-2}a_{n-1}a_n$$
$$[1^{n-2}001^{n-2}00 \to \varepsilon]$$

Therefore, $w \in L(G')$. $\qquad\qquad\qquad\qquad\qquad\qquad\qquad\qquad\qquad\qquad$ $\square$

Claim 3.1.32. Let $S \Rightarrow^m_{G'} 1^{n-2}00w1^{n-2}00$, where $w \in \{0,1\}^*$, for some $m \geq 1$. Then, $\$ \Rightarrow^*_G \beta^{-1}(w)$.

Proof. This claim is proved by induction on $m \geq 1$.

Basis. Let $m = 1$. That is,

$$S \Rightarrow_{G'} 1^{n-2}00w1^{n-2}00$$

where $w \in \{0,1\}^*$. Then, $w = \beta(\$)$. As $\$ \Rightarrow^0_G \$$, the basis holds.

Induction Hypothesis. Suppose that for some $j \geq 1$, if $S \Rightarrow^i_{G'} 1^{n-2}00w1^{n-2}00$, where $i = 1,\ldots,j$ and $w \in \{0,1\}^*$, then $\$ \Rightarrow^+_G \beta^{-1}(w)$.

Induction Step. Let

$$S \Rightarrow^{j+1}_{G'} 1^{n-2}00w1^{n-2}00$$

where $w \in \{0,1\}^*$. As $w \in \{0,1\}^*$,

$$S \Rightarrow^{j+1}_{G'} 1^{n-2}00w1^{n-2}00$$

can be expressed as

$$S \Rightarrow^j_{G'} 1^{n-2}00u\beta(x)v1^{n-2}00$$
$$\Rightarrow_{G'} 1^{n-2}00u\beta(y)v1^{n0}200 \quad [\beta(x) \to \beta(y)]$$

where $x,y \in V^*$, $x \to y \in Q$, and $w = u\beta(y)v$. By the induction hypothesis,

$$S \Rightarrow^+_{G'} 1^{n-2}00\beta^{-1}(u\beta(x)v)1^{n-2}00$$

Express $\beta^{-1}(u\beta(x)v)$ as

$$\beta^{-1}(u\beta(x)v) = \beta^{-1}(u)x\beta^{-1}(v)$$

Since $x \to y \in Q$,

$$\begin{aligned}
\$ &\Rightarrow_G^+ \beta^{-1}(u)x\beta^{-1}(v) \\
&\Rightarrow_G \beta^{-1}(u)y\beta^{-1}(v) \ [x \to y]
\end{aligned}$$

Because $w = u\beta(y)v$, $\beta^{-1}(w) = \beta^{-1}(u)y\beta^{-1}(v)$, so

$$\$ \Rightarrow_G^+ \beta^{-1}(w) \qquad\qquad\qquad\qquad \square$$

Claim 3.1.33. $L(G') \subseteq L(G)$

Proof. Let $w \in L(G')$. Distinguish between $w = \varepsilon$ and $w \neq \varepsilon$.

(i) Let $w = \varepsilon$. Observe that G' derives ε as

$$\begin{aligned}
S &\Rightarrow_{G'}^* 1^{n-2}001^{n-2}00 \\
&\Rightarrow_{G'} \varepsilon \ [1^{n-2}001^{n-2}00 \to \varepsilon]
\end{aligned}$$

Because

$$S \Rightarrow_{G'}^* 1^{n-2}001^{n-2}00$$

Claim 3.1.32 implies that $\$ \Rightarrow_G^* \varepsilon$. Therefore, $w \in L(G)$.

(ii) Assume that $w \neq \varepsilon$. Let $w = a_1 a_2 \cdots a_{n-1} a_n$ with $a_i \in T$ for $i = 1, \ldots, n$, where $n \geq 1$. Examine P to see that in G', there exists this derivation

$$\begin{aligned}
S &\Rightarrow_{G'}^* 1^{n-2}00\beta(a_1)\beta(a_2)\cdots\beta(a_{n-1})\beta(a_n)1^{n-2}00 \\
&\Rightarrow_{G'} a_1 1^{n-2}00\beta(a_2)\cdots\beta(a_{n-1})\beta(a_n)1^{n-2}00 \\
&\qquad [1^{n-2}00\beta(a_1) \to a_1 1^{n-2}00] \\
&\Rightarrow_{G'} a_1 a_2 1^{n-2}00\beta(a_3)\cdots\beta(a_{n-1})\beta(a_n)1^{n-2}00 \\
&\qquad [1^{n-2}00\beta(a_2) \to a_2 1^{n-2}00] \\
&\vdots \\
&\Rightarrow_{G'} a_1 a_2 \cdots a_{n-2} 1^{n-2}00\beta(a_{n-1})\beta(a_n)1^{n-2}00 \\
&\qquad [1^{n-2}00\beta(a_{n-2}) \to a_{n-2}1^{n-2}00] \\
&\Rightarrow_{G'} a_1 a_2 \cdots a_{n-2}a_{n-1} 1^{n-2}00\beta(a_n)1^{n-2}00 \\
&\qquad [1^{n-2}00\beta(a_{n-1}) \to a_{n-1}1^{n-2}00] \\
&\Rightarrow_{G'} a_1 a_2 \cdots a_{n-2}a_{n-1}a_n 1^{n-2}001^{n-2}00 \\
&\qquad [1^{n-2}00\beta(a_n) \to a_n 1^{n-2}00] \\
&\Rightarrow_{G'} a_1 a_2 \cdots a_{n-2}a_{n-1}a_n \\
&\qquad [1^{n-2}001^{n-2}00 \to \varepsilon]
\end{aligned}$$

Because

$$S \Rightarrow^*_{G'} 1^{n-2}00\beta(a_1)\beta(a_2) \cdots \beta(a_{n-1})\beta(a_n)1^{n-2}00$$

Claim 3.1.32 implies that

$$\$ \Rightarrow^*_G a_1 a_2 \cdots a_{n-1} a_n$$

Hence, $w \in L(G)$. □

By Claims 3.1.31 and 3.1.33, $L(G) = L(G')$. By Claim 3.1.29, $F(G') \subseteq T^* \Pi(1^{n-2}00)^*$. Thus, Lemma 3.1.27 holds. □

Theorem 3.1.34. PS[.2] = RE

Proof. The inclusion **PS[.2]** $\subseteq$ **RE** follows from Turing-Church thesis (see page 26). By Lemma 3.1.27, **RE** $\subseteq$ **PS[.2]**. Therefore, the theorem holds. □

Lemma 3.1.35. *Let G be a phrase-structure grammar. Then, there exists a phrase-structure grammar* $G' = (\{S, 0, 1\} \cup T, T, P, S)$ *satisfying* $L(G) = L(G')$ *and* $F(G') \subseteq \Pi(1^{n-2}00)^*T^*$, *for some* $n \geq 1$.

Proof. Let $G = (V, T, Q, \$)$ be a phrase-structure grammar, where V is the total alphabet of G, T is the terminal alphabet of G, Q is the set of rules of G, and $\$$ is the start symbol of G. Without any loss of generality, assume that $V \cap \{0, 1\} = \emptyset$. The following construction produces an equivalent phrase-structure grammar

$$G' = (\{S, 0, 1\} \cup T, T, P, S)$$

such that $F(G') \subseteq \Pi(1^{n-2}00)^*T^*$, for some $n \geq 1$.

For some $m \geq 3$ and n such that $2m = n$, introduce an injective homomorphism β from V to

$$(\{1\}^m\{1\}^*\{0\}\{1\}^*\{0\} \cap \{0, 1\}^n) - \{1^{n-2}00\}$$

Extend the domain of β to V^*. Define the phrase-structure grammar

$$G' = (T \cup \{S, 0, 1\}, P, S, T)$$

with

$$\begin{aligned}
P = \{&S \to 1^{n-2}00\beta(\$)1^{n-2}00\} \cup \\
&\{\beta(x) \to \beta(y) \mid x \to y \in Q\} \cup \\
&\{\beta(a)1^{n-2}00 \to 1^{n-2}00a \mid a \in T\} \cup \\
&\{1^{n-2}001^{n-2}00 \to \varepsilon\}
\end{aligned}$$

Complete this proof by analogy with the proof of Lemma 3.1.27. □

Theorem 3.1.36. PS[2.] = RE

Proof. Clearly, **PS**[2.] $\subseteq$ **RE**. By Lemma 3.1.35, **RE** $\subseteq$ **PS**[2.]. Therefore, this theorem holds. $\Box$

Corollary 3.1.37. PS[.2] = PS[2.] = RE $\Box$

There is an open problem area related to the results above.

Open Problem 3.1.38. Recall that in this section we converted any phrase-structure grammar G to an equivalent phrase-structure grammar, $G' = (V, T, P, S)$, so that for every $x \in F(G')$, $x \in T^* \Pi(w)^*$, where w is a string over $V - T$. Then, we made this conversion so that for every $x \in F(G')$, $x \in \Pi(w)^* T^*$. Take into account the length of w. More precisely, for $j, k \geq 1$ set

$$\begin{aligned} \mathbf{PS}[.j, k] = \{L \mid L = L(G), \text{ where } G = (V, T, P, S) \text{ is a phrase-structure} \\ \text{grammar such that } \mathrm{card}(\mathrm{alph}(F(G)) - T) = j \text{ and} \\ F(G) \subseteq T^* \Pi(w)^*, \text{ where } w \in (V - T)^* \text{ and } |w| = k\} \end{aligned}$$

Analogously, set

$$\begin{aligned} \mathbf{PS}[j, k.] = \{L \mid L = L(G), \text{ where } G = (V, T, P, S) \text{ is a phrase-structure} \\ \text{grammar such that } \mathrm{card}(\mathrm{alph}(F(G)) - T) = j \text{ and} \\ F(G) \subseteq \Pi(w)^* T^*, \text{ where } w \in (V - T)^* \text{ and } |w| = k\} \end{aligned}$$

Reconsider this section in terms of these families of languages. $\Box$

3.1.2 Conditional Context Grammars

Context-conditional grammars are based on context-free rules, each of which may be extended by finitely many *permitting* and *forbidding strings*. A rule like this can rewrite a sentential form on the condition that all its permitting strings occur in the current sentential form while all its forbidding strings do not occur there.

This section first defines context-conditional grammars and, after that, it establishes their generative power.

Definitions

Without further ado, we define the basic versions of context-regulated grammars.

Definition 3.1.39. A *context-conditional grammar* is a quadruple

$$G = (V, T, P, S)$$

where V, T, and S are the *total alphabet*, the *terminal alphabet* ($T \subset V$), and the *start symbol* ($S \in V - T$), respectively. P is a finite set of *rules* of the form

$$(A \rightarrow x, Per, For)$$

where $A \in V - T$, $x \in V^*$, and $Per, For \subseteq V^+$ are two finite sets. If $Per \neq \emptyset$ or $For \neq \emptyset$, the rule is said to be *conditional*; otherwise, it is called *context-free*. G has *degree* (r, s), where r and s are natural numbers, if for every $(A \rightarrow x, Per, For) \in P$, max-len($Per$) $\leq r$ and max-len(For) $\leq s$. If $(A \rightarrow x, Per, For) \in P$ implies that $x \neq \varepsilon$, G is said to be *propagating*. Let $u, v \in V^*$ and $(A \rightarrow x, Per, For) \in P$. Then, u *directly derives* v according to $(A \rightarrow x, Per, For)$ in G, denoted by

$$u \Rightarrow_G v \ [(A \rightarrow x, Per, For)]$$

provided that for some $u_1, u_2 \in V^*$, the following conditions hold:

(a) $u = u_1 A u_2$,
(b) $v = u_1 x u_2$,
(c) $Per \subseteq \mathrm{sub}(u)$,
(d) $For \cap \mathrm{sub}(u) = \emptyset$.

When no confusion exists, we simply write $u \Rightarrow_G v$ instead of $u \Rightarrow_G v \ [(A \rightarrow x, Per, For)]$. By analogy with context-free grammars, we extend $\Rightarrow_G$ to $\Rightarrow_G^k$ (where $k \geq 0$), $\Rightarrow_G^+$, and $\Rightarrow_G^*$. The *language* of G, denoted by $L(G)$, is defined as

$$L(G) = \{w \in T^* \mid S \Rightarrow_G^* w\} \qquad \square$$

The families of languages generated by context-conditional grammars and propagating context-conditional grammars of degree (r, s) are denoted by $\mathbf{CG}(r, s)$ and $\mathbf{CG}^{-\varepsilon}(r, s)$, respectively. Furthermore, set

$$\mathbf{CG} = \bigcup_{r=0}^{\infty} \bigcup_{s=0}^{\infty} \mathbf{CG}(r, s)$$

and

$$\mathbf{CG}^{-\varepsilon} = \bigcup_{r=0}^{\infty} \bigcup_{s=0}^{\infty} \mathbf{CG}^{-\varepsilon}(r, s)$$

Generative Power

Next, we prove several theorems concerning the generative power of the general versions of context-conditional grammars. Let us point out, however, that Sects. 3.1.3 through 3.1.6 establish many more results about special cases of these grammars.

Theorem 3.1.40. $\mathbf{CG}^{-\varepsilon}(0,0) = \mathbf{CG}(0,0) = \mathbf{CF}$

Proof. This theorem follows immediately from the definition. Clearly, context-conditional grammars of degree $(0,0)$ are ordinary context-free grammars.   □

Lemma 3.1.41. $\mathbf{CG}^{-\varepsilon} \subseteq \mathbf{CS}$

Proof. Let $r = s = 0$. Then, $\mathbf{CG}^{-\varepsilon}(0,0) = \mathbf{CF} \subset \mathbf{CS}$. The rest of the proof establishes the inclusion for degrees (r,s) such that $r + s > 0$.

Consider a propagating context-conditional grammar

$$G = (V, T, P, S)$$

of degree (r,s), where $r + s > 0$, for some $r, s \geq 0$. Let k be the greater number of r and s. Set

$$M = \{x \in V^+ \mid |x| \leq k\}$$

Next, define

$$\text{cf-rules}(P) = \{A \to x \mid (A \to x, Per, For) \in P,\ A \in (V - T),\ x \in V^+\}$$

Then, set

$$\begin{aligned}
N_F &= \{\lfloor X, x \rfloor \mid X \subseteq M,\ x \in M \cup \{\varepsilon\}\} \\
N_T &= \{\langle X \rangle \mid X \subseteq M\} \\
N_B &= \{\lceil p \rceil \mid p \in \text{cf-rules}(P)\} \cup \{\lceil \emptyset \rceil\} \\
V' &= V \cup N_F \cup N_T \cup N_B \cup \{\triangleright, \triangleleft, \$, S', \#\} \\
T' &= T \cup \{\#\}
\end{aligned}$$

Construct the context-sensitive grammar

$$G' = (V', T', P', S')$$

with the finite set of rules P' defined as follows:

(1) add $S' \to \triangleright \lfloor \emptyset, \varepsilon \rfloor S \triangleleft$ to P';
(2) for all $X \subseteq M$, $x \in (V^k \cup \{\varepsilon\})$ and $y \in V^k$, extend P' by adding

$$\lfloor X, x \rfloor y \to y \lfloor X \cup \text{sub}(xy, k), y \rfloor$$

(3) for all $X \subseteq M$, $x \in (V^k \cup \{\varepsilon\})$ and $y \in V^+$, $|y| \leq k$, extend P' by adding

$$\lfloor X, x \rfloor y \triangleleft \to y \langle X \cup \text{sub}(xy, k) \rangle \triangleleft$$

(4) for all $X \subseteq M$ and every $p = A \to x \in$ cf-rules(P) such that there exists $(A \to x, Per, For) \in P$ satisfying $Per \subseteq X$ and $For \cap X = \emptyset$, extend P' by adding

$$\langle X \rangle \triangleleft \to \lceil p \rceil \triangleleft$$

(5) for every $p \in$ cf-rules(P) and $a \in V$, extend P' by adding

$$a\lceil p \rceil \to \lceil p \rceil a$$

(6) for every $p = A \to x \in$ cf-rules(P), $A \in (V - T)$, $x \in V^+$, extend P' by adding

$$A\lceil p \rceil \to \lceil \emptyset \rceil x$$

(7) for every $a \in V$, extend P' by adding

$$a\lceil \emptyset \rceil \to \lceil \emptyset \rceil a$$

(8) add $\triangleright \lceil \emptyset \rceil \to \triangleright \lfloor \emptyset, \varepsilon \rfloor$ to P';
(9) add $\triangleright \lfloor \emptyset, \varepsilon \rfloor \to \#\$$, $\$\triangleleft \to \#\#$, and $\$a \to a\$$, for all $a \in T$, to P'.

Claim 3.1.42. Every successful derivation in G' has the form

$$\begin{aligned}
S' &\Rightarrow_{G'} \triangleright \lfloor \emptyset, \varepsilon \rfloor S \triangleleft \\
&\Rightarrow_{G'}^+ \triangleright \lfloor \emptyset, \varepsilon \rfloor x \triangleleft \\
&\Rightarrow_{G'} \#\$x\triangleleft \\
&\Rightarrow_{G'}^{|x|} \#x\$\triangleleft \\
&\Rightarrow_{G'} \#x\#\#
\end{aligned}$$

such that $x \in T^+$, and during

$$\triangleright \lfloor \emptyset, \varepsilon \rfloor S \triangleleft \Rightarrow_{G'}^+ \triangleright \lfloor \emptyset, \varepsilon \rfloor x \triangleleft$$

every sentential form w satisfies $w \in \{\triangleright\} H^+ \{\triangleleft\}$, where $H \subseteq V' - \{\triangleright, \triangleleft, \#, \$, S'\}$.

Proof. Observe that the only rule that rewrites S' is $S' \to \triangleright \lfloor \emptyset, \varepsilon \rfloor S \triangleleft$; thus,

$$S' \Rightarrow_{G'} \triangleright \lfloor \emptyset, \varepsilon \rfloor S \triangleleft$$

After that, every sentential form that occurs in

$$\triangleright \lfloor \emptyset, \varepsilon \rfloor S \triangleleft \Rightarrow_{G'}^+ \triangleright \lfloor \emptyset, \varepsilon \rfloor x \triangleleft$$

can be rewritten by using any of the rules (2) through (8) from the construction of P'. By the inspection of these rules, it is obvious that the delimiting symbols $\triangleright$ and $\triangleleft$ remain unchanged and no other occurrences of them appear inside the sentential form. Moreover, there is no rule generating a symbol from $\{\#, \$, S'\}$. Therefore, all these sentential forms belong to $\{\triangleright\}H^+\{\triangleleft\}$.

Next, let us explain how G' generates a string from $L(G')$. Only $\triangleright\lfloor\emptyset, \varepsilon\rfloor \to \#\$$ can rewrite $\triangleright$ to a symbol from T (see (9) in the definition of P'). According to the left-hand side of this rule, we obtain

$$S' \Rightarrow_{G'} \triangleright\lfloor\emptyset, \varepsilon\rfloor S\triangleleft \Rightarrow_{G'}^* \triangleright\lfloor\emptyset, \varepsilon\rfloor x\triangleleft \Rightarrow_{G'} \#\$x\triangleleft$$

where $x \in H^+$. To rewrite $\triangleleft$, G' uses $\$\triangleleft \to \#\#$. Thus, G' needs $\$$ as the left neighbor of $\triangleleft$. Suppose that $x = a_1 a_2 \cdots a_q$, where $q = |x|$ and $a_i \in T$, for all $i \in \{1, \ldots, q\}$. Since for every $a \in T$ there is $\$a \to a\$ \in P'$ (see (9)), we can construct

$$\#\$a_1 a_2 \cdots a_n \triangleleft \Rightarrow_{G'} \quad \#a_1 \$a_2 \cdots a_n \triangleleft$$
$$\Rightarrow_{G'} \quad \#a_1 a_2 \$ \cdots a_n \triangleleft$$
$$\Rightarrow_{G'}^{|x|-2} \#a_1 a_2 \cdots a_n \$\triangleleft$$

Notice that this derivation can be constructed only for x that belong to T^+. Then, $\$\triangleleft$ is rewritten to $\#\#$. As a result,

$$S' \Rightarrow_{G'} \triangleright\lfloor\emptyset, \varepsilon\rfloor S\triangleleft \Rightarrow_{G'}^+ \triangleright\lfloor\emptyset, \varepsilon\rfloor x\triangleleft \Rightarrow_{G'} \#\$x\triangleleft \Rightarrow_{G'}^{|x|} \#x\$\triangleleft \Rightarrow_{G'} \#x\#\#$$

with the required properties. Thus, the claim holds.   □

The following claim demonstrates how G' simulates a direct derivation from G—the heart of the construction.

Let $x \Rightarrow_{G'}^{\oplus} y$ denote the derivation $x \Rightarrow_{G'}^+ y$ such that $x = \triangleright\lfloor\emptyset, \varepsilon\rfloor u\triangleleft$, $y = \triangleright\lfloor\emptyset, \varepsilon\rfloor v\triangleleft$, $u, v \in V^+$, and during $x \Rightarrow_{G'}^+ y$, there is no other occurrence of a string of the form $\triangleright\lfloor\emptyset, \varepsilon\rfloor z\triangleleft$, $z \in V^*$.

Claim 3.1.43. For every $u, v \in V^*$, it holds that

$$\triangleright\lfloor\emptyset, \varepsilon\rfloor u\triangleleft \Rightarrow_{G'}^{\oplus} \triangleright\lfloor\emptyset, \varepsilon\rfloor v\triangleleft \quad \text{if and only if} \quad u \Rightarrow_G v$$

Proof. The proof is divided into the only-if part and the if part.

Only If. Let us show how G' rewrites $\triangleright\lfloor\emptyset, \varepsilon\rfloor u\triangleleft$ to $\triangleright\lfloor\emptyset, \varepsilon\rfloor v\triangleleft$. The simulation consists of two phases.

During the first, forward phase, G' scans u to get all nonempty substrings of length k or less. By repeatedly using rules $\lfloor X, x\rfloor y \to y\lfloor X \cup \mathrm{sub}(xy, k), y\rfloor$, $X \subseteq M$, $x \in (V^k \cup \{\varepsilon\})$, $y \in V^k$ (see (2) in the definition of P'), the occurrence of a symbol of the form $\lfloor X, x\rfloor$ is moved toward the end of the sentential form. Simultaneously, the substrings of u are recorded in X. The forward phase is finished by applying

$\lfloor X, x \rfloor y \triangleleft \rightarrow y \langle X \cup \text{sub}(xy, k) \rangle \triangleleft$, $x \in (V^k \cup \{\varepsilon\})$, $y \in V^+$, $|y| \leq k$ (see (3)); this rule reaches the end of u and completes $X = \text{sub}(u, k)$. Formally,

$$\triangleright \lfloor \emptyset, \varepsilon \rfloor u \triangleleft \Rightarrow_{G'}^+ \triangleright u \langle X \rangle \triangleleft$$

with $X = \text{sub}(u, k)$.

The second, backward phase simulates the application of a conditional rule. Assume that $u = u_1 A u_2$, $u_1, u_2 \in V^*$, $A \in (V - T)$, and there exists a rule $A \rightarrow x \in \text{cf-rules}(P)$ such that $(A \rightarrow x, Per, For) \in P$ for some $Per, For \subseteq M$, where $Per \subseteq X$, $For \cap X = \emptyset$. Let $u_1 x u_2 = v$. Then, G' derives

$$\triangleright u \langle X \rangle \triangleleft \Rightarrow_{G'}^+ \triangleright \lfloor \emptyset, \varepsilon \rfloor v \triangleleft$$

by performing the following five steps

(i) $\langle X \rangle$ is changed to $\lceil p \rceil$, where $p = A \rightarrow x$ satisfies the conditions above (see (4) in the definition of P');
(ii) $\triangleright u_1 A u_2 \lceil p \rceil \triangleleft$ is rewritten to $\triangleright u_1 A \lceil p \rceil u_2 \triangleleft$ by using the rules of the form $a \lceil p \rceil \rightarrow \lceil p \rceil a$, $a \in V$ (see (5));
(iii) $\triangleright u_1 A \lceil p \rceil u_2 \triangleleft$ is rewritten to $\triangleright u_1 \lceil \emptyset \rceil x u_2 \triangleleft$ by using $A \lceil p \rceil \rightarrow \lceil \emptyset \rceil x$ (see (6));
(iv) $\triangleright u_1 \lceil \emptyset \rceil x u_2 \triangleleft$ is rewritten to $\triangleright \lceil \emptyset \rceil u_1 x u_2 \triangleleft$ by using the rules of the form $a \lceil \emptyset \rceil \rightarrow \lceil \emptyset \rceil a$, $a \in V$ (see (7));
(v) finally, $\triangleright \lceil \emptyset \rceil$ is rewritten to $\triangleright \lfloor \emptyset, \varepsilon \rfloor$ by $\triangleright \lceil \emptyset \rceil \rightarrow \triangleright \lfloor \emptyset, \varepsilon \rfloor$.

As a result, we obtain

$$\triangleright \lfloor \emptyset, \varepsilon \rfloor u \triangleleft \Rightarrow_{G'}^+ \triangleright u \langle X \rangle \triangleleft \Rightarrow_{G'} \triangleright u \lceil p \rceil \triangleleft$$
$$\Rightarrow_{G'}^{|u|} \triangleright \lceil \emptyset \rceil v \triangleleft \Rightarrow_{G'} \triangleright \lfloor \emptyset, \varepsilon \rfloor v \triangleleft$$

Observe that this is the only way of deriving

$$\triangleright \lfloor \emptyset, \varepsilon \rfloor u \triangleleft \Rightarrow_{G'}^{\oplus} \triangleright \lfloor \emptyset, \varepsilon \rfloor v \triangleleft$$

Let us show that $u \Rightarrow_G v$. Indeed, the application of $A \lceil p \rceil \rightarrow \lceil \emptyset \rceil x$ implies that there exists $(A \rightarrow x, Per, For) \in P$, where $Per \subseteq \text{sub}(u, k)$ and $For \cap \text{sub}(u, k) = \emptyset$. Hence, there exists a derivation of the form

$$u \Rightarrow_G v \ [p]$$

where $u = u_1 A u_2$, $v = u_1 x u_2$ and $p = (A \rightarrow x, Per, For) \in P$.

If. The converse implication is similar to the only-if part, so we leave it to the reader. $\square$

Claim 3.1.44. $S' \Rightarrow_{G'}^+ \triangleright \lfloor \emptyset, \varepsilon \rfloor x \triangleleft$ if and only if $S \Rightarrow_G^* x$, for all $x \in V^+$.

Proof. The proof is divided into the only-if part and the if part.

Only If. The only-if part is proved by induction on the ith occurrence of the sentential form w satisfying $w = \rhd \lfloor \emptyset, \varepsilon \rfloor u \lhd$, $u \in V^+$ during the derivation in G'.

Basis. Let $i = 1$. Then, $S' \Rightarrow_{G'} \rhd \lfloor \emptyset, \varepsilon \rfloor S \lhd$ and $S \Rightarrow_G^0 S$.

Induction Hypothesis. Suppose that the claim holds for all $i \leq h$, for some $h \geq 1$.

Induction Step. Let $i = h + 1$. Since $h + 1 \geq 2$, we can express

$$S' \Rightarrow_{G'}^+ \rhd \lfloor \emptyset, \varepsilon \rfloor x_i \lhd$$

as

$$S' \Rightarrow_{G'}^+ \rhd \lfloor \emptyset, \varepsilon \rfloor x_{i-1} \lhd \Rightarrow_{G'}^{\oplus} \rhd \lfloor \emptyset, \varepsilon \rfloor x_i \lhd$$

where $x_{i-1}, x_i \in V^+$. By the induction hypothesis,

$$S \Rightarrow_G^* x_{i-1}$$

Claim 3.1.43 says that

$$\rhd \lfloor \emptyset, \varepsilon \rfloor x_{i-1} \lhd \Rightarrow_{G'}^{\oplus} \rhd \lfloor \emptyset, \varepsilon \rfloor x_i \lhd \quad \text{if and only if} \quad x_{i-1} \Rightarrow_G x_i$$

Hence,

$$S \Rightarrow_G^* x_{i-1} \Rightarrow_G x_i$$

and the only-if part holds.

If. By induction on m, we prove that

$$S \Rightarrow_G^m x \text{ implies that } S' \Rightarrow_{G'}^+ \rhd \lfloor \emptyset, \varepsilon \rfloor x \lhd$$

for all $m \geq 0$, $x \in V^+$.

Basis. For $m = 0$, $S \Rightarrow_G^0 S$ and $S' \Rightarrow_{G'} \rhd \lfloor \emptyset, \varepsilon \rfloor S \lhd$.

Induction Hypothesis. Assume that the claim holds for all $m \leq n$, for some $n \geq 0$.

Induction Step. Let

$$S \Rightarrow_G^{n+1} x$$

with $x \in V^+$. Because $n + 1 \geq 1$, there exists $y \in V^+$ such that

$$S \Rightarrow_G^n y \Rightarrow_G x$$

By the induction hypothesis, there is also a derivation

$$S' \Rightarrow_{G'}^{+} \triangleright \lfloor \emptyset, \varepsilon \rfloor y \triangleleft$$

From Claim 3.1.43 it follows that

$$\triangleright \lfloor \emptyset, \varepsilon \rfloor y \triangleleft \Rightarrow_{G'}^{\oplus} \triangleright \lfloor \emptyset, \varepsilon \rfloor x \triangleleft$$

Therefore,

$$S' \Rightarrow_{G'}^{+} \triangleright \lfloor \emptyset, \varepsilon \rfloor x \triangleleft$$

and the converse implication holds as well.   □

From Claims 3.1.42 and 3.1.44, we see that any successful derivation in G' is of the form

$$S' \Rightarrow_{G'}^{+} \triangleright \lfloor \emptyset, \varepsilon \rfloor x \triangleleft \Rightarrow_{G'}^{+} \#x\#\#$$

such that

$$S \Rightarrow_{G}^{*} x, \ x \in T^{+}$$

Therefore, for each $x \in T^{+}$,

$$S' \Rightarrow_{G'}^{+} \#x\#\# \quad \text{if and only if} \quad S \Rightarrow_{G}^{*} x$$

Define the homomorphism h over $(T \cup \{\#\})^{*}$ as $h(\#) = \varepsilon$ and $h(a) = a$ for all $a \in T$. Observe that h is 4-linear erasing with respect to $L(G')$. Furthermore, notice that $h(L(G')) = L(G)$. Because **CS** is closed under linear erasing (see Theorem 10.4 on page 98 in [Sal73]), $L \in$ **CS**. Thus, Lemma 3.1.41 holds.   □

Theorem 3.1.45. CG$^{-\varepsilon}$ = CS

Proof. By Lemma 3.1.41, we have **CG$^{-\varepsilon}$** ⊆ **CS**. **CS** ⊆ **CG$^{-\varepsilon}$** holds as well. In fact, later in this book, we introduce several special cases of propagating context-conditional grammars and prove that even these grammars generate **CS** (see Theorems 3.1.83 and 3.1.89). As a result, **CG$^{-\varepsilon}$** = **CS**.   □

Lemma 3.1.46. CG ⊆ RE

Proof. This lemma follows from Turing-Church thesis (see page 26). To obtain an algorithm converting any context-conditional grammar to an equivalent phrase-structure grammar, use the technique presented in Lemma 3.1.41.   □

Theorem 3.1.47. CG = RE

Proof. By Lemma 3.1.46, **CG** ⊆ **RE**. Later on, we define some special cases of context-conditional grammars and demonstrate that they characterize **RE** (see Theorems 3.1.62, 3.1.86, and 3.1.94). Thus, **RE** ⊆ **CG**.   □

3.1.3 Random-Context Grammars

This section discusses three special cases of context-conditional grammars whose conditions are nonterminal symbols, so their degree is not greater than $(1, 1)$. Specifically, *permitting grammars* are of degree $(1, 0)$. *Forbidding grammars* are of degree $(0, 1)$. Finally, *random context grammars* are of degree $(1, 1)$.

The present section, first, provides definitions and illustrates all the grammars under discussion and, then, it establishes their generative power.

Definitions and Examples

We open this section by defining random context grammars and their two important special cases—permitting and forbidding grammars. Later in this section, we illustrate them.

Definition 3.1.48. Let $G = (V, T, P, S)$ be a context-conditional grammar. G is called a *random context grammar* provided that every $(A \rightarrow x, Per, For) \in P$ satisfies $Per \subseteq N$ and $For \subseteq N$. $\square$

Definition 3.1.49. Let $G = (V, T, P, S)$ be a random context grammar. G is called a *permitting grammar* provided that every $(A \rightarrow x, Per, For) \in P$ satisfies $For = \emptyset$. $\square$

Definition 3.1.50. Let $G = (V, T, P, S)$ be a random context grammar. G is called a *forbidding grammar* provided that every $(A \rightarrow x, Per, For) \in P$ satisfies $Per = \emptyset$. $\square$

The following conventions simplify rules in permitting and forbidding grammars.

Let $G = (V, T, P, S)$ be a permitting grammar, and let $p = (A \rightarrow x, Per, For) \in P$. Since $For = \emptyset$, we usually omit the empty set of forbidding conditions. That is, we write $(A \rightarrow x, Per)$ when no confusion arises.

Let $G = (V, T, P, S)$ be a forbidding grammar, and let $p = (A \rightarrow x, Per, For) \in P$. We write $(A \rightarrow x, For)$ instead of $(A \rightarrow x, Per, For)$ because $Per = \emptyset$ for all $p \in P$.

The families of languages defined by permitting grammars, forbidding grammars, and random context grammars are denoted by **Per**, **For**, and **RC**, respectively. To indicate that only propagating grammars are considered, we use the upper index $-\varepsilon$. That is, $\mathbf{Per}^{-\varepsilon}$, $\mathbf{For}^{-\varepsilon}$, and $\mathbf{RC}^{-\varepsilon}$ denote the families of languages defined by propagating permitting grammars, propagating forbidding grammars, and propagating random context grammars, respectively.

Example 3.1.51 (See [DP89]). Let

$$G = (\{S, A, B, C, D, A', B', C', a, b, c\}, \{a, b, c\}, P, S)$$

be a permitting grammar, where P is defined as follows:

$$P = \{(S \to ABC, \emptyset),$$
$$(A \to aA', \{B\}),$$
$$(B \to bB', \{C\}),$$
$$(C \to cC', \{A'\}),$$
$$(A' \to A, \{B'\}),$$
$$(B' \to B, \{C'\}),$$
$$(C' \to C, \{A\}),$$
$$(A \to a, \{B\}),$$
$$(B \to b, \{C\}),$$
$$(C \to c, \emptyset)\}$$

Consider the string *aabbcc*. G generates this string in the following way

$$S \Rightarrow ABC \Rightarrow aA'BC \Rightarrow aA'bB'C \Rightarrow aA'bB'cC' \Rightarrow$$
$$aAbB'cC' \Rightarrow aAbBcC' \Rightarrow aAbBcC \Rightarrow$$
$$aabBcC \Rightarrow aabbcC \Rightarrow aabbcc$$

Observe that G is propagating and

$$L(G) = \{a^n b^n c^n \mid n \geq 1\}$$

which is a non-context-free language. □

Example 3.1.52 (See [DP89]). Let

$$G = (\{S, A, B, D, a\}, \{a\}, P, S)$$

be a random context grammar. The set of rules P is defined as follows:

$$P = \{(S \to AA, \emptyset, \{B, D\}),$$
$$(A \to B, \emptyset, \{S, D\}),$$
$$(B \to S, \emptyset, \{A, D\}),$$
$$(A \to D, \emptyset, \{S, B\}),$$
$$(D \to a, \emptyset, \{S, A, B\})\}$$

Notice that G is a propagating forbidding grammar. For *aaaaaaaa*, G makes the following derivation

$$S \Rightarrow AA \Rightarrow AB \Rightarrow BB \Rightarrow BS \Rightarrow SS \Rightarrow AAS \Rightarrow AAAA \Rightarrow BAAA \Rightarrow$$
$$BABA \Rightarrow BBBA \Rightarrow BBBB \Rightarrow SBBB \Rightarrow SSBB \Rightarrow SSSB \Rightarrow$$
$$SSSS \Rightarrow AASSS \Rightarrow^3 AAAAAAAA \Rightarrow^8 DDDDDDDD \Rightarrow^8 aaaaaaaa$$

Clearly, G generates this non-context-free language

$$L(G) = \left\{a^{2^n} \mid n \geq 1\right\} \qquad\qquad \square$$

Generative Power

We next establish several theorems concerning the generative power of the grammars defined in the previous section.

Theorem 3.1.53. $\mathbf{CF} \subset \mathbf{Per}^{-\varepsilon} \subset \mathbf{RC}^{-\varepsilon} \subset \mathbf{CS}$

Proof. $\mathbf{CF} \subseteq \mathbf{Per}^{-\varepsilon}$ follows from Example 3.1.51. $\mathbf{Per}^{-\varepsilon} \subset \mathbf{RC}^{-\varepsilon}$ follows from Theorem 2.7 in Chapter 3 in [RS97b]. Finally, $\mathbf{RC}^{-\varepsilon} \subset \mathbf{CS}$ follows from Theorems 1.2.4 and 1.4.5 in [DP89]. $\square$

Theorem 3.1.54. $\mathbf{Per}^{-\varepsilon} = \mathbf{Per} \subset \mathbf{RC} = \mathbf{RE}$

Proof. $\mathbf{Per}^{-\varepsilon} = \mathbf{Per}$ follows from Theorem 1 in [Zet10]. By Theorem 1.2.5 in [DP89], $\mathbf{RC} = \mathbf{RE}$. Furthermore, from Theorem 2.7 in Chapter 3 in [RS97b], it follows that $\mathbf{Per} \subset \mathbf{RC}$; thus, the theorem holds. $\square$

Lemma 3.1.55. $\mathbf{ET0L} \subset \mathbf{For}^{-\varepsilon}$

Proof. (See [Pen75].) Let $L \in \mathbf{ET0L}$, $L = L(G)$ for some ET0L grammar,

$$G = (V, T, P_1, \ldots, P_t, S)$$

Without loss of generality, we assume that G is propagating (see Theorem 2.3.41). We introduce the alphabets

$$\begin{aligned}
V^{(i)} &= \{a^{(i)} \mid a \in V\},\ 1 \leq i \leq t \\
V' &= \{a' \mid a \in V\} \\
V'' &= \{a'' \mid a \in V\} \\
\bar{V} &= \{\bar{a} \mid a \in T\}
\end{aligned}$$

For $w \in V^*$, by $w^{(i)}$, w', w'', and $\bar{w}$, we denote the strings obtained from w by replacing each occurrence of a symbol $a \in V$ by $a^{(i)}$, a', a'', and $\bar{a}$, respectively. Let P' be the set of all random context rules defined as follows:

(1) for every $a \in V$, add $(a' \to a'', \emptyset, \bar{V} \cup V^{(1)} \cup V^{(2)} \cup \cdots \cup V^{(t)})$ to P';

(2) for every $a \in V$ for all $1 \leq i \leq t$, add

$$(a'' \to a^{(i)}, \emptyset, \bar{V} \cup V' \cup V^{(1)} \cup V^{(2)} \cup \cdots \cup V^{(i-1)} \cup V^{(i+1)} \cup \cdots \cup V^{(t)})$$

to P';

(3) for all $i \in \{1, \ldots, t\}$ for every $a \to u \in P_i$, add $(a^{(i)} \to u', \emptyset, V'' \cup \bar{V})$ to P';

(4) for all $a \in T$, add $(a' \rightarrow \bar{a}, \emptyset, V'' \cup V^{(1)} \cup V^{(2)} \cup \cdots \cup V^{(t)})$ to P';

(5) for all $a \in T$, add $(\bar{a} \rightarrow a, \emptyset, V' \cup V'' \cup V^{(1)} \cup V^{(2)} \cup \cdots \cup V^{(t)})$ to P'.

Then, define the random context grammar

$$G' = \left(V' \cup V'' \cup \bar{V} \cup V^{(1)} \cup V^{(2)} \cup \cdots \cup V^{(t)}, T, P', S\right)$$

that has only forbidding context conditions.

Let x' be a string over V'. To x', we can apply only rules whose left-hand side is in V'.

(a) We use $a' \rightarrow a''$ for some $a' \in V'$. The obtained sentential form contains symbols of V' and V''. Hence, we can use only rules of type (1). Continuing in this way, we get $x' \Rightarrow_{G'}^{*} x''$. By analogous arguments, we now have to rewrite all symbols of x'' by rules of (2) with the same index (i). Thus, we obtain $x^{(i)}$. To each symbol $a^{(i)}$ in $x^{(i)}$, we apply a rule $a^{(i)} \rightarrow u'$, where $a \rightarrow u \in P_i$. Since again all symbols in $x^{(i)}$ have to be replaced before starting with rules of another type, we simulate a derivation step in G and get z', where $x \Rightarrow_G z$ in G. Therefore, starting with a rule of (1), we simulate a derivation step in G, and conversely, each derivation step in G can be simulated in this way.

(b) We apply a rule $a' \rightarrow \bar{a}$ to x'. Next, each a' of T' occurring in x' has to be substituted by $\bar{a}$ and then by a by using the rules constructed in (5). Therefore, we obtain a terminal string only if $x' \in T'^{*}$.

By these considerations, any successful derivation in G' is of the form

$$S' \Rightarrow_{G'} S'' \Rightarrow_{G'} S^{(i_0)}$$
$$\Rightarrow_{G'} z_1' \Rightarrow_{G'}^{*} z_1'' \Rightarrow_{G'}^{*} z_1^{(i_1)}$$
$$\vdots$$
$$\Rightarrow_{G'}^{*} z_n' \Rightarrow_{G'}^{*} z_n'' \Rightarrow_{G'}^{*} z_n^{(i_n)}$$
$$\Rightarrow_{G'}^{*} z_{n+1} \Rightarrow_{G'}^{*} \bar{z}_{n+1} \Rightarrow_{G'}^{*} z_{n+1}$$

and such a derivation exists if and only if

$$S \Rightarrow_G z_1 \Rightarrow_G z_2 \Rightarrow_G \cdots \Rightarrow_G z_n \Rightarrow_G z_{n+1}$$

is a successful derivation in G. Thus, $L(G) = L(G')$.

In order to finish the proof, it is sufficient to find a language that is not in **ET0L** and can be generated by a forbidding grammar. A language of this kind is

$$L = \{b(ba^m)^n \mid m \geq n \geq 0\}$$

which can be generated by the grammar

$$G = \left(\{S, A, A', B, B', B'', C, D, E\}, \{a, b\}, P, s\right)$$

with P consisting of the following rules

$$(S \rightarrow SA, \emptyset, \emptyset)$$

$$(S \rightarrow C, \emptyset, \emptyset)$$

$$(C \rightarrow D, \emptyset, \{S, A', B', B'', D, E\})$$

$$(B \rightarrow B'a, \emptyset, \{S, C, E\})$$

$$(A \rightarrow B''a, \emptyset, \{S, C, E, B''\})$$

$$(A \rightarrow A'a, \emptyset, \{S, C, E\})$$

$$(D \rightarrow C, \emptyset, \{A, B\})$$

$$(B' \rightarrow B, \emptyset, \{D\})$$

$$(B'' \rightarrow B, \emptyset, \{D\})$$

$$(A' \rightarrow A, \emptyset, \{D\})$$

$$(D \rightarrow E, \emptyset, \{S, A, A', B', B'', C, E\})$$

$$(B \rightarrow b, \emptyset, \{S, A, A', B', B'', C, D\})$$

$$(E \rightarrow b, \emptyset, \{S, A, A', B, B', B'', C, D\})$$

First, we have the derivation

$$S \Rightarrow_G^* SA^n \Rightarrow_G CA^n \Rightarrow_G DA^n$$

Then, we have to replace all occurrences of A. If we want to replace an occurrence of A by a terminal string in some steps, it is necessary to use $A \rightarrow B''a$. However, this can be done at most once in a phase that replaces all As. Therefore, $m \geq n$. $\square$

Theorem 3.1.56. $\mathbf{CF} \subset \mathbf{ET0L} \subset \mathbf{For}^{-\varepsilon} \subseteq \mathbf{For} \subset \mathbf{CS}$

Proof. According to Example 3.1.52, we already have $\mathbf{CF} \subset \mathbf{For}^{-\varepsilon}$. By [RS80] and Lemma 3.1.55, $\mathbf{CF} \subset \mathbf{ET0L} \subset \mathbf{For}^{-\varepsilon}$. Moreover, in [Pen75], it has been proved that $\mathbf{For}^{-\varepsilon} \subseteq \mathbf{For} \subset \mathbf{CS}$. Therefore, the theorem holds. $\square$

The following corollary summarizes the relations of language families generated by random context grammars.

Corollary 3.1.57.

$$\mathbf{CF} \subset \mathbf{Per}^{-\varepsilon} \subset \mathbf{RC}^{-\varepsilon} \subset \mathbf{CS}$$

$$\mathbf{Per}^{-\varepsilon} = \mathbf{Per} \subset \mathbf{RC} = \mathbf{RE}$$

$$\mathbf{CF} \subset \mathbf{ET0L} \subset \mathbf{For}^{-\varepsilon} \subseteq \mathbf{For} \subset \mathbf{CS}$$

Proof. This corollary follows from Theorems 3.1.53, 3.1.54, and 3.1.56. $\square$

Open Problem 3.1.58. Are **For**$^{-\varepsilon}$ and **For** identical?   □

3.1.4 Forbidding Context Grammars

Generalized forbidding grammars represent a generalized variant of forbidding grammars (see Sect. 3.1.3) in which forbidding context conditions are formed by finite languages.

This section consists of two subsections; Definitions and Generative Power and Reduction. The former defines generalized forbidding grammars, and the latter establishes their power.

Definitions

Next, we define generalized forbidding grammars.

Definition 3.1.59. Let $G = (V, T, P, S)$ be a context-conditional grammar. If every $(A \rightarrow x, Per, For)$ satisfies $Per = \emptyset$, then G is said to be a *generalized forbidding grammar* (a *gf-grammar* for short). □

The following convention simplifies the notation of gf-grammars. Let $G = (V, T, P, S)$ be a gf-grammar of degree (r, s). Since every $(A \rightarrow x, Per, For) \in P$ implies that $Per = \emptyset$, we omit the empty set of permitting conditions. That is, we write $(A \rightarrow x, For)$ instead of $(A \rightarrow x, Per, For)$. For simplicity, we also say that the degree of G is s instead of (r, s).

The families generated by gf-grammars and propagating gf-grammars of degree s are denoted by **GF**(s) and **GF**$^{-\varepsilon}(s)$, respectively. Furthermore, set

$$\mathbf{GF} = \bigcup_{s=0}^{\infty} \mathbf{GF}(s)$$

and

$$\mathbf{GF}^{-\varepsilon} = \bigcup_{s=0}^{\infty} \mathbf{GF}^{-\varepsilon}(s)$$

Generative Power and Reduction

In the present section, we establish the generative power of generalized forbidding grammars, defined in the previous section. In fact, apart from establishing this power, we also give several related results concerning the reduction of these grammars. Indeed, we reduce these grammars with respect to the number of nonterminals, the number of forbidding rules, and the length of forbidding strings.

By analogy with Theorem 3.1.40, it is easy to see that gf-grammars of degree 0 are ordinary context-free grammars.

Theorem 3.1.60. $\mathbf{GF}^{-\varepsilon}(0) = \mathbf{GF}(0) = \mathbf{CF}$ □

Furthermore, gf-grammars of degree 1 are as powerful as forbidding grammars.

Theorem 3.1.61. $\mathbf{GF}(1) = \mathbf{For}$

Proof. This simple proof is left to the reader.   □

Theorem 3.1.62. $\mathbf{GF}(2) = \mathbf{RE}$

Proof. It is straightforward to prove that $\mathbf{GF}(2) \subseteq \mathbf{RE}$; hence it is sufficient to prove the converse inclusion.

Let L be a recursively enumerable language. Without any loss of generality we assume that L is generated by a phrase-structure grammar

$$G = (V, T, P, S)$$

in the Penttonen normal form (see Theorem 3.1.4). Set $N = V - T$.

Let @, \$, S' be new symbols and m be the cardinality of $V \cup \{@\}$. Clearly, $m \geq 1$. Furthermore, let f be an arbitrary bijection from $V \cup \{@\}$ onto $\{1, \ldots, m\}$ and f^{-1} be the inverse of f.

The gf-grammar

$$G' = (V' \cup \{@, \$, S'\}, T, P', S')$$

of degree 2 is defined as follows:

$$V' = W \cup V, \text{ where}$$
$$W = \{[AB \rightarrow AC, j] \mid AB \rightarrow AC \in P, A, B, C \in N, 1 \leq j \leq m + 1\}$$

We assume that W, $\{@, \$, S'\}$, and V are pairwise disjoint alphabets. The set of rules P' is defined in the following way

(1) add $(S' \rightarrow @S, \emptyset)$ to P';
(2) if $A \rightarrow x \in P, A \in N, x \in \{\varepsilon\} \cup T \cup N^2$, then add $(A \rightarrow x, \{\$\})$ to P';
(3) if $AB \rightarrow AC \in P, A, B, C \in N$, then

    (3.a) add $(B \rightarrow \$[AB \rightarrow AC, 1], \{\$\})$ to P';
    (3.b) for all $j = 1, \ldots, m, f(A) \neq j$, extend P' by adding

$$([AB \rightarrow AC, j] \rightarrow [AB \rightarrow AC, j + 1], \{f^{-1}(j)\$\})$$

    (3.c) add $([AB \rightarrow AC, f(A)] \rightarrow [AB \rightarrow AC, f(A) + 1], \emptyset)$ and $([AB \rightarrow AC, m + 1] \rightarrow C, \emptyset)$ to P';

(4) add $(@ \rightarrow \varepsilon, N \cup W \cup \{\$\})$ and $(\$ \rightarrow \varepsilon, W)$ to P'.

Basically, the application of $AB \rightarrow AC$ in G is simulated in G' in the following way. An occurrence of B is rewritten with $\$[AB \rightarrow AC, 1]$. Then, the left adjoining symbol of $\$$ is checked not to be any symbol from $(V \cup \{@\})$ except A. After this, the right adjoining symbol of $\$$ is $[AB \rightarrow AC, m + 1]$. This symbol is rewritten with C. A formal proof is given below.

Immediately from the definition of P' it follows that

$$S' \Rightarrow_{G'}^{+} x$$

where $x \in (V' \cup \{@, S'\})^*$, implies that

1. $S' \notin \mathrm{alph}(x)$;
2. if $\mathrm{alph}(x) \cap W \neq \emptyset$, then $\#_W(x) = 1$ and $\#_{\{\$\}W}(x) = 1$;
3. if $x \notin T^*$, then the leftmost symbol of x is $@$.

Next, we define a finite substitution g from V^* into V'^* such that for all $B \in V$,

$$g(B) = \{B\} \cup \{[AB \rightarrow AC, j] \in W \mid AB \rightarrow AC \in P, \ A, C \in N, \ j = 1, \ldots, m + 1\}$$

Let g^{-1} be the inverse of g.

To show that $L(G) = L(G')$, we first prove that

$$S \Rightarrow_G^n x \quad \text{if and only if} \quad S \Rightarrow_{G'}^{n'} x'$$

where $x' = @v'Xw'$, $X \in \{\$, \varepsilon\}$, $v'w' \in g(x)$, $x \in V^*$, for some $n \geq 0, n' \geq 1$.

Only If. This is established by induction on $n \geq 0$. That is, we have to demonstrate that $S \Rightarrow_G^n x$, $x \in V^*$, $n \geq 0$, implies that $S \Rightarrow_{G'}^{+} x'$ for some x' such that $x' = @v'Xw'$, $X \in \{\$, \varepsilon\}$, $v'w' \in g(x)$.

Basis. Let $n = 0$. The only x is S because $S \Rightarrow_G^0 S$. Clearly, $S' \Rightarrow_{G'} @S$ and $S \in g(S)$.

Induction Hypothesis. Suppose that the claim holds for all derivations of length n or less, for some $n \geq 0$.

Induction Step. Let us consider any derivation of the form

$$S \Rightarrow_G^{n+1} x$$

with $x \in V^*$. Since $n + 1 \geq 1$, there is some $y \in V^+$ and $p \in P$ such that

$$S \Rightarrow_G^n y \Rightarrow_G x \ [p]$$

and by the induction hypothesis, there is also a derivation of the form

$$S \Rightarrow_{G'}^{n'} y'$$

for some $n' \geq 1$, such that $y' = @r'Ys'$, $Y \in \{\$, \varepsilon\}$, and $r's' \in g(y)$.

(i) Let us assume that $p = D \rightarrow y_2 \in P$, $D \in N$, $y_2 \in \{\varepsilon\} \cup T \cup N^2$, $y = y_1 D y_3$, $y_1, y_3 \in V^*$, and $x = y_1 y_2 y_3$. From (2) it is clear that $(D \rightarrow y_2, \{\$\}) \in P'$.

   (a) Let $\$ \notin \text{alph}(y')$. Then, we have $y' = @r's' = @y_1 D y_3$,

$$S' \Rightarrow_{G'}^{n'} @y_1 D y_3 \Rightarrow_{G'} @y_1 y_2 y_3 \ [(D \rightarrow y_2, \{\$\})]$$

   and $y_1 y_2 y_3 \in g(y_1 y_2 y_3) = g(x)$.

   (b) Let $Y = \$ \in \text{sub}(y')$ and $W \cap \text{sub}(y') = \emptyset$. Then, there is the following derivation in G'

$$S' \Rightarrow_{G'}^{n'} @r'\$s' \Rightarrow_{G'} @r's' \ [(\$ \rightarrow \varepsilon, W)]$$

   By analogy with (a) above, we have $@r's' = @y_1 D y_2$, so

$$S' \Rightarrow_{G'}^{n'+1} @y_1 D y_3 \Rightarrow_{G'} @y_1 y_2 y_3 \ [(D \rightarrow y_2, \{\$\})]$$

   where $y_1 y_2 y_3 \in g(x)$.

   (c) Let $\$[AB \rightarrow AC, i] \in \text{sub}(y')$ for some $i \in \{1, \ldots, m + 1\}$, $AB \rightarrow AC \in P$, $A, B, C \in N$. Thus, $y' = @r'\$[AB \rightarrow AC, i]t'$, where $s' = [AB \rightarrow AC, i]t'$. By the inspection of the rules (see (3)) it can be seen (and the reader should be able to produce a formal proof) that we can express the derivation

$$S' \Rightarrow_{G'}^* y'$$

   in the following form

$$\begin{aligned} S' &\Rightarrow_{G'}^* @r'Bt' \\ &\Rightarrow_{G'} @r'\$[AB \rightarrow AC, 1]t' \ [(B \rightarrow \$[AB \rightarrow AC, 1], \{\$\})] \\ &\Rightarrow_{G'}^{i-1} @r'\$[AB \rightarrow AC, i]t' \end{aligned}$$

   Clearly, $r'Bt' \in g(y)$ and $\$ \notin \text{alph}(r'Bt')$. Thus, $r'Bt' = y_1 D y_3$, and there is a derivation

$$S' \Rightarrow_{G'}^* @y_1 D y_3 \Rightarrow_{G'} @y_1 y_2 y_3 \ [(D \rightarrow y_2, \{\$\})]$$

   and $y_1 y_2 y_3 \in g(x)$.

(ii) Let $p = AB \rightarrow AC \in P$, $A, B, C \in N$, $y = y_1 AB y_2$, $y_1, y_2 \in V^*$, and $x = y_1 AC y_2$.

   (a) Let $\$ \notin \text{alph}(y')$. Thus, $r's' = y_1 AB y_2$. By the inspection of the rules introduced in (3) (technical details are left to the reader), there is the following derivation in G'

$$S' \Rightarrow^{n'}_{G'} @y_1ABy_2$$
$$\Rightarrow_{G'} @y_1A\$[AB \to AC, 1]y_2$$
$$[(B \to \$[AB \to AC, 1], \{\$\})]$$
$$\Rightarrow_{G'} @y_1A\$[AB \to AC, 2]y_2$$
$$[([AB \to AC, 1] \to [AB \to AC, 2], \{f^{-1}(1)\$\})]$$

$$\vdots$$

$$\Rightarrow_{G'} @y_1A\$[AB \to AC, f(A)]y_2$$
$$[([AB \to AC, f(A) - 1] \to [AB \to AC, f(A)],$$
$$\{f^{-1}(f(A) - 1)\$\})]$$
$$\Rightarrow_{G'} @y_1A\$[AB \to AC, f(A) + 1]y_2$$
$$[([AB \to AC, f(A)] \to [AB \to AC, f(A) + 1], \emptyset)]$$

$$\vdots$$

$$\Rightarrow_{G'} @y_1A\$[AB \to AC, m + 1]y_2$$
$$[([AB \to AC, m] \to [AB \to AC, m + 1], \{f^{-1}(m)\$\})]$$
$$\Rightarrow_{G'} @y_1A\$Cy_2$$
$$[([AB \to AC, m + 1] \to C, \emptyset)]$$

such that $y_1ACy_2 \in g(y_1ACy_2) = g(x)$.

(b) Let $\$ \in \text{alph}(y')$, $\text{alph}(y') \cap W = \emptyset$. By analogy with (b), the derivation

$$S' \Rightarrow^*_{G'} @r's'$$

with $@r's' = @y_1ABy_2$, can be constructed in G'. Then, by analogy with (a), one can construct the derivation

$$S' \Rightarrow^*_{G'} @y_1ABy_2 \Rightarrow^*_{G'} @y_1A\$Cy_2$$

such that $y_1ACy_2 \in g(x)$.

(c) Let $\#_{\{\$\}W}(y') = 1$. By analogy with (c), one can construct the derivation

$$S' \Rightarrow^*_{G'} @y_1ABy_2$$

Next, by using an analogue from (a), the derivation

$$S' \Rightarrow^*_{G'} @y_1ABy_2 \Rightarrow^*_{G'} @y_1A\$Cy_2$$

can be constructed in G' so $y_1ACy_2 \in g(x)$.

In (i) and (ii) above we have considered all possible forms of p. In cases (a), (b), (c) of (i) and (ii), we have considered all possible forms of y'. In any of these cases, we have constructed the desired derivation of the form

$$S' \Rightarrow^+_{G'} x'$$

such that $x' = @r'Xs'$, $X \in \{\$, \varepsilon\}$, $r's' \in g(x)$. Hence, we have established the only-if part of our claim by the principle of induction.

If. This is also demonstrated by induction on $n' \geq 1$. We have to demonstrate that if $S' \Rightarrow_{G'}^{n'} x'$, $x' = @r'Xs'$, $X \in \{\$, \varepsilon\}$, $r's' \in g(x)$, $x \in V^*$, for some $n' \geq 1$, then $S \Rightarrow_G^* x$.

Basis. For $n' = 1$ the only x' is $@S$ since $S' \Rightarrow_{G'} @S$. Because $S \in g(S)$, we have $x = S$. Clearly, $S \Rightarrow_G^0 S$.

Induction Hypothesis. Assume that the claim holds for all derivations of length at most n' for some $n' \geq 1$. Let us show that it also holds for $n' + 1$.

Induction Step. Consider any derivation of the form

$$S' \Rightarrow_{G'}^{n'+1} x'$$

with $x' = @r'Xs'$, $X \in \{\$, \varepsilon\}$, $r's' \in g(x)$, $x \in V^*$. Since $n' + 1 \geq 2$, we have

$$S' \Rightarrow_{G'}^{n'} y' \Rightarrow_{G'} x' \ [p']$$

for some $p' = (Z' \to w', For) \in P'$, $y' = @q'Yt'$, $Y \in \{\$, \varepsilon\}$, $q't' \in g(y)$, $y \in V^*$, and by the induction hypothesis,

$$S \Rightarrow_G^* y$$

Suppose:

(i) $Z' \in N$, $w' \in \{\varepsilon\} \cup T \cup N^2$. By inspecting P' (see (2)), we have $For = \{\$\}$ and $Z' \to w' \in P$. Thus, $\$ \notin alph(y')$ and so $q't' = y$. Hence, there is the following derivation

$$S \Rightarrow_G^* y \Rightarrow_G x \ [Z' \to w']$$

(ii) $g^{-1}(Z') = g^{-1}(w')$. But then $y = x$, and by the induction hypothesis, we have the derivation

$$S \Rightarrow_G^* y$$

(iii) $p' = (B \to \$[AB \to AC, 1], \{\$\})$; that is, $Z' = B$, $w' = \$[AB \to AC, 1]$, $For = \{\$\}$ and so $w' \in \{\$\}g(Z')$, $Y = \varepsilon$, $X = \$$. By analogy with (ii), we have

$$S \Rightarrow_G^* y$$

and $y = x$.

(iv) $Z' = Y = \$$; that is, $p' = (\$ \to \varepsilon, W)$. Then, $X = \varepsilon$, $r's' = q't' \in g(y)$, and

$$S \Rightarrow_G^* y$$

(v) $p' = ([AB \rightarrow AC, m+1] \rightarrow C, \emptyset)$; that is, $Z' = [AB \rightarrow AC, m+1]$, $w' = C$, $For = \emptyset$. From (3), it follows that there is a rule of the form $AB \rightarrow AC \in P$. Moreover, by inspecting (3), it is not too difficult to see (the technical details are left to the reader) that $Y = \$$, $r' = q'$, $t' = [AB \rightarrow AC, m+1]o'$, $s' = Co'$, and the derivation

$$S' \Rightarrow_{G'}^{n'} y' \Rightarrow_{G'} x' \; [p']$$

can be expressed as

$$
\begin{aligned}
S' &\Rightarrow_{G'}^{*} && @q'Bo' \\
&\Rightarrow_{G'} && @q'\$[AB \rightarrow AC, 1]o' && [(B \rightarrow \$[AB \rightarrow AC, 1], \{\$\})] \\
&\Rightarrow_{G'}^{m+1} && @q'\$[AB \rightarrow AC, m+1]o' && [h] \\
&\Rightarrow_{G'} && @q'\$Co' && [([AB \rightarrow AC, m+1] \rightarrow C, \emptyset)]
\end{aligned}
$$

where

$$
\begin{aligned}
h &= h_1([AB \rightarrow AC, f(A)] \rightarrow [AB \rightarrow AC, f(A)+1], \emptyset)h_2, \\
h_1 &= ([AB \rightarrow AC, 1] \rightarrow [AB \rightarrow AC, 2], \{f^{-1}(1)\$\}) \\
&\quad\;\; ([AB \rightarrow AC, 2] \rightarrow [AB \rightarrow AC, 3], \{f^{-1}(2)\$\}) \\
&\quad\;\; \vdots \\
&\quad\;\; ([AB \rightarrow AC, f(A)-1] \rightarrow [AB \rightarrow AC, f(A)], \{f^{-1}(f(A)-1)\$\})
\end{aligned}
$$

in which $f(A) = 1$ implies that $h_1 = \varepsilon$,

$$
\begin{aligned}
h_2 &= ([AB \rightarrow AC, f(A)+1] \rightarrow [AB \rightarrow AC, f(A)+2], \{f^{-1}(f(A)+1)\$\}) \\
&\quad\;\; \vdots \\
&\quad\;\; ([AB \rightarrow AC, m] \rightarrow [AB \rightarrow AC, m+1], \{f^{-1}(m)\$\})
\end{aligned}
$$

in which $f(A) = m$ implies that $h_2 = \varepsilon$; that is, the rightmost symbol of $q' = r'$ must be A.

Since $q't' \in g(y)$, we have $y = q'Bo'$. Because the rightmost symbol of q' is A and $AB \rightarrow AC \in P$, we have

$$S \Rightarrow_{G}^{*} q'Bo' \Rightarrow_{G} q'Co' \; [AB \rightarrow AC]$$

where $q'Co' = x$.

By inspecting P', we see that (i) through (v) cover all possible derivations of the form

$$S' \Rightarrow_{G'}^{n'} y' \Rightarrow_{G'} x'$$

and thus we have established that

$$S \Rightarrow_{G}^{*} x \quad \text{if and only if} \quad S' \Rightarrow_{G'}^{+} x'$$

where $x' = @r'Xs'$, $r's' \in g(x)$, $X \in \{\$, \varepsilon\}$, $x \in V^*$, by the principle of induction.

A proof of the equivalence of G and G' can easily be derived from above. By the definition of g, we have $g(a) = \{a\}$ for all $a \in T$. Thus, we have for any $x \in T^*$,

$$S \Rightarrow_G^* x \quad \text{if and only if} \quad S' \Rightarrow_{G'}^* @rXs$$

where $X \in \{\$, \varepsilon\}$, $rs = x$. If $X = \varepsilon$, then

$$@x \Rightarrow_{G'} x \left[(@ \rightarrow \varepsilon, N \cup W \cup \{\$\}) \right]$$

If $X = \$$, then

$$@r\$s \Rightarrow_{G'} @x \left[(\$ \rightarrow \varepsilon, W) \right] \Rightarrow_{G'} x \left[(@ \rightarrow \varepsilon, N \cup W \cup \{\$\}) \right]$$

Hence,

$$S \Rightarrow_G^+ x \quad \text{if and only if} \quad S' \Rightarrow_{G'}^+ x$$

for all $x \in T^*$, and so $L(G) = L(G')$. Thus, **RE** = **GF**(2).   □

Theorem 3.1.63. **GF**(2) = **GF** = **RE**

Proof. This theorem follows immediately from the definitions and Theorem 3.1.62.

                                                                                            □

Examine the rules in G' in the proof of Theorem 3.1.62 to establish the following normal form.

Corollary 3.1.64. *Every recursively enumerable language L over some alphabet T can be generated by a gf-grammar $G = (V, T, P \cup \{p_1, p_2\}, S)$ of degree 2 such that*

(i) $(A \rightarrow x, For) \in P$ implies that $|x| = 2$ and the cardinality of For is at most 1;
(ii) $p_i = (A_i \rightarrow \varepsilon, For_i)$, $i = 1, 2$, where $For_i \subseteq V$; that is, max-len(For_i) ≤ 1.   □

In fact, the corollary above represents one of the reduced forms of gf-grammars of degree 2. Perhaps most importantly, it reduces the cardinality of the sets of forbidding conditions so that if a rule contains a condition of length two, this condition is the only context condition attached to the rule. Next, we study another reduced form of gf-grammars of degree 2. We show that we can simultaneously reduce the number of conditional rules and the number of nonterminals in gf-grammars of degree 2 without any decrease of their generative power.

Theorem 3.1.65. *Every recursively enumerable language can be defined by a gf-grammar of degree 2 with no more than 13 forbidding rules and 15 nonterminals.*

Proof. Let $L \in$ **RE**. By Theorem 3.1.11, without any loss of generality, we assume that L is generated by a phrase-structure grammar G of the form

$$G = (V, T, P \cup \{AB \rightarrow \varepsilon, CD \rightarrow \varepsilon\}, S)$$

such that P contains only context-free rules and

$$V - T = \{S, A, B, C, D\}$$

We construct a gf-grammar of degree 2

$$G' = (V', T, P', S')$$

where

$$V' = V \cup W$$
$$W = \{S', @, \tilde{A}, \tilde{B}, \langle \varepsilon_A \rangle, \$, \tilde{C}, \tilde{D}, \langle \varepsilon_C \rangle, \#\}, \ V \cap W = \emptyset$$

in the following way. Let

$$N' = (V' - T) - \{S', @\}$$

Informally, N' denotes the set of all nonterminals in G' except S' and $@$. Then, the set of rules P' is constructed by performing (1) through (4), given next.

(1) If $H \to y \in P$, $H \in V - T$, $y \in V^*$, then add $(H \to y, \emptyset)$ to P';
(2) add $(S' \to @S@, \emptyset)$ and $(@ \to \varepsilon, N')$ to P';
(3) extend P' by adding

$$(A \to \tilde{A}, \{\tilde{A}\})$$
$$(B \to \tilde{B}, \{\tilde{B}\})$$
$$(\tilde{A} \to \langle \varepsilon_A \rangle, \{\tilde{A}a \mid a \in V' - \{\tilde{B}\}\})$$
$$(\tilde{B} \to \$, \{a\tilde{B} \mid a \in V' - \{\langle \varepsilon_A \rangle\}\})$$
$$(\langle \varepsilon_A \rangle \to \varepsilon, \{\tilde{B}\})$$
$$(\$ \to \varepsilon, \{\langle \varepsilon_A \rangle\})$$

(4) extend P' by adding

$$(C \to \tilde{C}, \{\tilde{C}\})$$
$$(D \to \tilde{D}, \{\tilde{D}\})$$
$$(\tilde{C} \to \langle \varepsilon_C \rangle, \{\tilde{C}a \mid a \in V' - \{\tilde{D}\}\})$$
$$(\tilde{D} \to \#, \{a\tilde{D} \mid a \in V' - \{\langle \varepsilon_C \rangle\}\})$$
$$(\langle \varepsilon_C \rangle \to \varepsilon, \{\tilde{D}\})$$
$$(\# \to \varepsilon, \{\langle \varepsilon_C \rangle\})$$

Next, we prove that $L(G') = L(G)$.

Notice that G' has degree 2 and contains only 13 forbidding rules and 15 nonterminals. The rules of (3) simulate the application of $AB \to \varepsilon$ in G' and the rules of (4) simulate the application of $CD \to \varepsilon$ in G'.

Let us describe the simulation of $AB \rightarrow \varepsilon$. First, one occurrence of A and one occurrence of B are rewritten with $\tilde{A}$ and $\tilde{B}$, respectively (no sentential form contains more than one occurrence of $\tilde{A}$ or $\tilde{B}$). The right neighbor of $\tilde{A}$ is checked to be $\tilde{B}$ and $\tilde{A}$ is rewritten with $\langle \varepsilon_A \rangle$. Then, analogously, the left neighbor of $\tilde{B}$ is checked to be $\langle \varepsilon_A \rangle$ and $\tilde{B}$ is rewritten with \$. Finally, $\langle \varepsilon_A \rangle$ and \$ are erased. The simulation of $CD \rightarrow \varepsilon$ is analogical.

To establish $L(G) = L(G')$, we first prove several claims.

Claim 3.1.66. $S' \Rightarrow^+_{G'} w'$ implies that w' has one of the following two forms

(I) $w' = @x'@, x' \in (N' \cup T)^*, \mathrm{alph}(x') \cap N' \neq \emptyset$;
(II) $w' = Xx'Y, x' \in T^*, X, Y \in \{@, \varepsilon\}$.

Proof. The start symbol S' is always rewritten with $@S@$. After this initial step, $@$ can be erased in a sentential form provided that any nonterminal occurring in the sentential form belongs to $\{@, S'\}$ (see N' and (2) in the definition of P'). In addition, notice that only rules of (2) contain $@$ and S'. Thus, any sentential form containing some nonterminals from N' is of the form (I).

Case (II) covers sentential forms containing no nonterminal from N'. At this point, $@$ can be erased, and we obtain a string from $L(G')$.   □

Claim 3.1.67. $S' \Rightarrow^*_{G'} w'$ implies that $\#_{\tilde{X}}(w') \leq 1$ for all $\tilde{X} \in \{\tilde{A}, \tilde{B}, \tilde{C}, \tilde{D}\}$ and some $w' \in V'^*$.

Proof. By the inspection of rules in P', the only rule that can generate $\tilde{X}$ is of the form $(X \rightarrow \tilde{X}, \{\tilde{X}\})$. This rule can be applied only when no $\tilde{X}$ occurs in the rewritten sentential form. Thus, it is impossible to derive w' from S' such that $\#_{\tilde{X}}(w') \geq 2$.   □

Informally, next claim says that every occurrence of $\langle \varepsilon_A \rangle$ in derivations from S' is always followed either by $\tilde{B}$ or \$, and every occurrence of $\langle \varepsilon_C \rangle$ is always followed either by $\tilde{D}$ or #.

Claim 3.1.68. The following two statements hold true.

(I) $S' \Rightarrow^*_{G'} y'_1 \langle \varepsilon_A \rangle y'_2$ implies that $y'_2 \in V'^+$ and $\mathrm{lms}(y'_2) \in \{\tilde{B}, \$\}$ for any $y'_1 \in V'^*$.
(II) $S' \Rightarrow^*_{G'} y'_1 \langle \varepsilon_C \rangle y'_2$ implies that $y'_2 \in V'^+$ and $\mathrm{lms}(y'_2) \in \{\tilde{D}, \#\}$ for any $y'_1 \in V'^*$.

Proof. We establish this claim by examination of all possible forms of derivations that may occur when deriving a sentential form containing $\langle \varepsilon_A \rangle$ or $\langle \varepsilon_C \rangle$.

(I) By the definition of P', the only rule that can generate $\langle \varepsilon_A \rangle$ is $p = (\tilde{A} \rightarrow \langle \varepsilon_A \rangle, \{\tilde{A}a \mid a \in V' - \{\tilde{B}\}\})$. The rule can be applied provided that $\tilde{A}$ occurs in a sentential form. It also holds that $\tilde{A}$ has always a right neighbor (as follows from Claim 3.1.66), and according to the set of forbidding conditions in p, $\tilde{B}$ is the only allowed right neighbor of $\tilde{A}$. Furthermore, by Claim 3.1.67, no other occurrence of $\tilde{A}$ or $\tilde{B}$ can appear in the given sentential form. Consequently, we obtain a derivation

$$S' \Rightarrow^*_{G'} u'_1 \tilde{A} \tilde{B} u'_2 \Rightarrow_{G'} u'_1 \langle \varepsilon_A \rangle \tilde{B} u'_2 \; [p]$$

for some $u'_1, u'_2 \in V'^*$, $\tilde{A}, \tilde{B} \notin \text{alph}(u'_1 u'_2)$. Obviously, $\langle \varepsilon_A \rangle$ is always followed by $\tilde{B}$ in $u'_1 \langle \varepsilon_A \rangle \tilde{B} u'_2$.

Next, we discuss how G' can rewrite the substring $\langle \varepsilon_A \rangle \tilde{B}$ in $u'_1 \langle \varepsilon_A \rangle \tilde{B} u'_2$. There are only two rules having the nonterminals $\langle \varepsilon_A \rangle$ or $\tilde{B}$ on their left-hand side, $p_1 = (\tilde{B} \rightarrow \$, \{a\tilde{B} \mid a \in V' - \{\langle \varepsilon_A \rangle\}\})$ and $p_2 = (\langle \varepsilon_A \rangle \rightarrow \varepsilon, \{\tilde{B}\})$. G' cannot use p_2 to erase $\langle \varepsilon_A \rangle$ in $u'_1 \langle \varepsilon_A \rangle \tilde{B} u'_2$ because p_2 forbids an occurrence of $\tilde{B}$ in the rewritten string. However, we can rewrite $\tilde{B}$ to $\$$ by using p_1 because its set of forbidding conditions defines that the left neighbor of $\tilde{B}$ must be just $\langle \varepsilon_A \rangle$. Hence, we obtain a derivation of the form

$$S' \Rightarrow^*_{G'} u'_1 \tilde{A} \tilde{B} u'_2 \quad \Rightarrow_{G'} u'_1 \langle \varepsilon_A \rangle \tilde{B} u'_2 \; [p]$$
$$\Rightarrow^*_{G'} v'_1 \langle \varepsilon_A \rangle \tilde{B} v'_2 \Rightarrow_{G'} v'_1 \langle \varepsilon_A \rangle \$ v'_2 \; [p_1]$$

Notice that during this derivation, G' may rewrite u'_1 and u'_2 with some v'_1 and v'_2, respectively ($v'_1, v'_2 \in V'^*$); however, $\langle \varepsilon_A \rangle \tilde{B}$ remains unchanged after this rewriting.

In this derivation we obtained the second symbol $\$$, which can appear as the right neighbor of $\langle \varepsilon_A \rangle$. It is sufficient to show that there is no other symbol that can appear immediately after $\langle \varepsilon_A \rangle$. By the inspection of P', only ($\$ \rightarrow \varepsilon, \{\langle \varepsilon_A \rangle\}$) can rewrite $\$$. However, this rule cannot be applied when $\langle \varepsilon_A \rangle$ occurs in the given sentential form. In other words, the occurrence of $\$$ in the substring $\langle \varepsilon_A \rangle \$$ cannot be rewritten before $\langle \varepsilon_A \rangle$ is erased by p_2. Hence, $\langle \varepsilon_A \rangle$ is always followed either by $\tilde{B}$ or $\$$, so the first part of Claim 3.1.68 holds.

(II) By the inspection of rules simulating $AB \rightarrow \varepsilon$ and $CD \rightarrow \varepsilon$ in G' (see (3) and (4) in the definition of P'), these two sets of rules work analogously. Thus, part (II) of Claim 3.1.68 can be proved by analogy with part (I). $\qquad \square$

Let us return to the main part of the proof. Let g be a finite substitution from $(N' \cup T)^*$ to V^* defined as follows:

(a) for all $X \in V$, $g(X) = \{X\}$;
(b) $g(\tilde{A}) = \{A\}$, $g(\tilde{B}) = \{B\}$, $g(\langle \varepsilon_A \rangle) = \{A\}$, $g(\$) = \{B, AB\}$;
(c) $g(\tilde{C}) = \{C\}$, $g(\tilde{D}) = \{D\}$, $g(\langle \varepsilon_C \rangle) = \{C\}$, $g(\#) = \{C, CD\}$.

Having this substitution, we can now prove the following claim.

Claim 3.1.69. $S \Rightarrow^*_G x$ if and only if $S' \Rightarrow^+_{G'} @x'@$ for some $x \in g(x')$, $x \in V^*$, $x' \in (N' \cup T)^*$.

Proof. The claim is proved by induction on the length of derivations.

Only If. We show that

$$S \Rightarrow^m_G x \quad \text{implies} \quad S' \Rightarrow^+_{G'} @x@$$

where $m \geq 0$, $x \in V^*$; clearly $x \in g(x)$. This is established by induction on $m \geq 0$.

Basis. Let $m = 0$. That is, $S \Rightarrow^0_G S$. Clearly, $S' \Rightarrow_{G'} @S@$.

Induction Hypothesis. Suppose that the claim holds for all derivations of length m or less, for some $m \geq 0$.

Induction Step. Let us consider any derivation of the form

$$S \Rightarrow_G^{m+1} x,\ x \in V^*$$

Since $m + 1 \geq 1$, there is some $y \in V^+$ and $p \in P \cup \{AB \to \varepsilon, CD \to \varepsilon\}$ such that

$$S \Rightarrow_G^m y \Rightarrow_G x\ [p]$$

By the induction hypothesis, there is a derivation

$$S' \Rightarrow_{G'}^+ @y@$$

There are the following three cases that cover all possible forms of p.

(i) Let $p = H \to y_2 \in P$, $H \in V - T$, $y_2 \in V^*$. Then, $y = y_1 H y_3$ and $x = y_1 y_2 y_3$, $y_1, y_3 \in V^*$. Because we have $(H \to y_2, \emptyset) \in P'$,

$$S' \Rightarrow_{G'}^+ @y_1 H y_3 @ \Rightarrow_{G'} @y_1 y_2 y_3 @ \ [(H \to y_2, \emptyset)]$$

and $y_1 y_2 y_3 = x$.

(ii) Let $p = AB \to \varepsilon$. Then, $y = y_1 A B y_3$ and $x = y_1 y_3$, $y_1, y_3 \in V^*$. In this case, there is the following derivation

$$
\begin{aligned}
S' &\Rightarrow_{G'}^+ @y_1 A B y_3 @ \\
&\Rightarrow_{G'} @y_1 \tilde{A} B y_3 @ && [(A \to \tilde{A}, \{\tilde{A}\})] \\
&\Rightarrow_{G'} @y_1 \tilde{A} \tilde{B} y_3 @ && [(B \to \tilde{B}, \{\tilde{B}\})] \\
&\Rightarrow_{G'} @y_1 \langle \varepsilon_A \rangle \tilde{B} y_3 @ && [(\tilde{A} \to \langle \varepsilon_A \rangle, \{\tilde{A} a \mid a \in V' - \{\tilde{B}\}\})] \\
&\Rightarrow_{G'} @y_1 \langle \varepsilon_A \rangle \$ y_3 @ && [(\tilde{B} \to \$, \{a\tilde{B} \mid a \in V' - \{\langle \varepsilon_A \rangle\}\})] \\
&\Rightarrow_{G'} @y_1 \$ y_3 @ && [(\langle \varepsilon_A \rangle \to \varepsilon, \{\tilde{B}\})] \\
&\Rightarrow_{G'} @y_1 y_3 @ && [(\$ \to \varepsilon, \{\langle \varepsilon_A \rangle\})]
\end{aligned}
$$

(iii) Let $p = CD \to \varepsilon$. Then, $y = y_1 C D y_3$ and $x = y_1 y_3$, $y_1, y_3 \in V^*$. In this case, there exists the following derivation

$$
\begin{aligned}
S' &\Rightarrow_{G'}^+ @y_1 C D y_3 @ \\
&\Rightarrow_{G'} @y_1 \tilde{C} D y_3 @ && [(C \to \tilde{C}, \{\tilde{C}\})] \\
&\Rightarrow_{G'} @y_1 \tilde{C} \tilde{D} y_3 @ && [(D \to \tilde{D}, \{\tilde{D}\})] \\
&\Rightarrow_{G'} @y_1 \langle \varepsilon_C \rangle \tilde{D} y_3 @ && [(\tilde{C} \to \langle \varepsilon_C \rangle, \{\tilde{C} a \mid a \in V' - \{\tilde{D}\}\})] \\
&\Rightarrow_{G'} @y_1 \langle \varepsilon_C \rangle \# y_3 @ && [(\tilde{D} \to \#, \{a\tilde{D} \mid a \in V' - \{\langle \varepsilon_C \rangle\}\})] \\
&\Rightarrow_{G'} @y_1 \# y_3 @ && [(\langle \varepsilon_C \rangle \to \varepsilon, \{\tilde{D}\})] \\
&\Rightarrow_{G'} @y_1 y_3 @ && [(\# \to \varepsilon, \{\langle \varepsilon_C \rangle\})]
\end{aligned}
$$

If. By induction on the length n of derivations in G', we prove that

$$S' \Rightarrow_{G'}^{n} @x'@ \quad \text{implies} \quad S \Rightarrow_{G}^{*} x$$

for some $x \in g(x')$, $x \in V^*$, $x' \in (N' \cup T)^*$, $n \geq 1$.

Basis. Let $n = 1$. According to the definition of P', the only rule rewriting S' is $(S' \to @S@, \emptyset)$, so $S' \Rightarrow_{G'} @S@$. It is obvious that $S \Rightarrow_{G}^{0} S$ and $S \in g(S)$.

Induction Hypothesis. Assume that the claim holds for all derivations of length n or less, for some $n \geq 1$.

Induction Step. Consider any derivation of the form

$$S' \Rightarrow_{G'}^{n+1} @x'@, \; x' \in (N' \cup T)^*$$

Since $n + 1 \geq 2$, there is some $y' \in (N' \cup T)^+$ and $p' \in P'$ such that

$$S' \Rightarrow_{G'}^{n} @y'@ \Rightarrow_{G'} @x'@ \; [p']$$

and by the induction hypothesis, there is also a derivation

$$S \Rightarrow_{G}^{*} y$$

such that $y \in g(y')$.

By the inspection of P', the following cases (i) through (xiii) cover all possible forms of p'.

(i) Let $p' = (H \to y_2, \emptyset) \in P'$, $H \subset V - T$, $y_2 \in V^*$. Then, $y' = y_1' H y_3'$, $x' = y_1' y_2 y_3'$, $y_1', y_3' \in (N' \cup T)^*$, and y has the form $y = y_1 Z y_3$, where $y_1 \in g(y_1')$, $y_3 \in g(y_3')$, and $Z \in g(H)$. Because for all $X \in V - T$: $g(X) = \{X\}$, the only Z is H; thus, $y = y_1 H y_3$. By the definition of P' (see (1)), there exists a rule $p = H \to y_2$ in P, and we can construct the derivation

$$S \Rightarrow_{G}^{*} y_1 H y_3 \Rightarrow_{G} y_1 y_2 y_3 \; [p]$$

such that $y_1 y_2 y_3 = x$, $x \in g(x')$.

(ii) Let $p' = (A \to \tilde{A}, \{\tilde{A}\})$. Then, $y' = y_1' A y_3'$, $x' = y_1' \tilde{A} y_3'$, $y_1', y_3' \in (N' \cup T)^*$ and $y = y_1 Z y_3$, where $y_1 \in g(y_1')$, $y_3 \in g(y_3')$ and $Z \in g(A)$. Because $g(A) = \{A\}$, the only Z is A, so we can express $y = y_1 A y_3$. Having the derivation $S \Rightarrow_{G}^{*} y$ such that $y \in g(y')$, it is easy to see that also $y \in g(x')$ because $A \in g(\tilde{A})$.

(iii) Let $p' = (B \to \tilde{B}, \{\tilde{B}\})$. By analogy with (ii), $y' = y_1' B y_3'$, $x' = y_1' \tilde{B} y_3'$, $y = y_1 B y_3$, where $y_1', y_3' \in (N' \cup T)^*$, $y_1 \in g(y_1')$, $y_3 \in g(y_3')$; thus, $y \in g(x')$ because $B \in g(\tilde{B})$.

(iv) Let $p' = (\tilde{A} \to \langle \varepsilon_A \rangle, \{\tilde{A}a \mid a \in V' - \{\tilde{B}\}\})$. In this case, it holds that

    (iv.i) application of p' implies that $\tilde{A} \in \mathrm{alph}(y')$, and moreover, by Claim 3.1.67, we have $\#_{\tilde{A}}(y') \leq 1$;

(iv.ii) $\tilde{A}$ has always a right neighbor in @y'@;

(iv.iii) according to the set of forbidding conditions in p', the only allowed right neighbor of $\tilde{A}$ is $\tilde{B}$.

Hence, y' must be of the form $y' = y_1'\tilde{A}\tilde{B}y_3'$, where $y_1', y_3' \in (N' \cup T)^*$ and $\tilde{A} \notin \text{alph}(y_1'y_3')$. Then, $x' = y_1'\langle\varepsilon_A\rangle\tilde{B}y_3'$ and y is of the form $y = y_1Zy_3$, where $y_1 \in g(y_1')$, $y_3 \in g(y_3')$ and $Z \in g(\tilde{A}\tilde{B})$. Because $g(\tilde{A}\tilde{B}) = \{AB\}$, the only Z is AB; thus, we obtain $y = y_1ABy_3$. By the induction hypothesis, we have a derivation $S \Rightarrow_G^* y$ such that $y \in g(y')$. According to the definition of g, $y \in g(x')$ as well because $A \in g(\langle\varepsilon_A\rangle)$ and $B \in g(\tilde{B})$.

(v) Let $p' = (\tilde{B} \to \$, \{a\tilde{B} \mid a \in V' - \{\langle\varepsilon_A\rangle\}\})$. Then, it holds that

(v.i) $\tilde{B} \in \text{alph}(y')$ and, by Claim 3.1.67, $\#_{\tilde{B}}(y') \leq 1$;

(v.ii) $\tilde{B}$ has always a left neighbor in @y'@;

(v.iii) by the set of forbidding conditions in p', the only allowed left neighbor of $\tilde{B}$ is $\langle\varepsilon_A\rangle$.

Therefore, we can express $y' = y_1'\langle\varepsilon_A\rangle\tilde{B}y_3'$, where $y_1', y_3' \in (N' \cup T)^*$ and $\tilde{B} \notin \text{alph}(y_1'y_3')$. Then, $x' = y_1'\langle\varepsilon_A\rangle\$y_3'$ and $y = y_1Zy_3$, where $y_1 \in g(y_1')$, $y_3 \in g(y_3')$, and $Z \in g(\langle\varepsilon_A\rangle\tilde{B})$. By the definition of g, $g(\langle\varepsilon_A\rangle\tilde{B}) = \{AB\}$, so $Z = AB$ and $y = y_1ABy_3$. By the induction hypothesis, we have a derivation $S \Rightarrow_G^* y$ such that $y \in g(y')$. Because $A \in g(\langle\varepsilon_A\rangle)$ and $B \in g(\$)$, $y \in g(x')$ as well.

(vi) Let $p' = (\langle\varepsilon_A\rangle \to \varepsilon, \{\tilde{B}\})$. An application of $(\langle\varepsilon_A\rangle \to \varepsilon, \{\tilde{B}\})$ implies that $\langle\varepsilon_A\rangle$ occurs in y'. Claim 3.1.68 says that $\langle\varepsilon_A\rangle$ has either $\tilde{B}$ or $\$$ as its right neighbor. Since the forbidding condition of p' forbids an occurrence of $\tilde{B}$ in y', the right neighbor of $\langle\varepsilon_A\rangle$ must be $\$$. As a result, we obtain $y' = y_1'\langle\varepsilon_A\rangle\$y_3'$, where $y_1', y_3' \in (N' \cup T)^*$. Then, $x' = y_1'\$y_3'$, and y is of the form $y = y_1Zy_3$, where $y_1 \in g(y_1')$, $y_3 \in g(y_3')$, and $Z \in g(\langle\varepsilon_A\rangle\$)$. By the definition of g, $g(\langle\varepsilon_A\rangle\$) = \{AB, AAB\}$. If $Z = AB$, $y = y_1ABy_3$. Having the derivation $S \Rightarrow_G^* y$, it holds that $y \in g(x')$ because $AB \in g(\$)$.

(vii) Let $p' = (\$ \to \varepsilon, \{\langle\varepsilon_A\rangle\})$. Then, $y' = y_1'\$y_3'$ and $x' = y_1'y_3'$, where $y_1', y_3' \in (N' \cup T)^*$. Express $y = y_1Zy_3$ so that $y_1 \in g(y_1')$, $y_3 \in g(y_3')$, and $Z \in g(\$)$, where $g(\$) = \{B, AB\}$. Let $Z = AB$. Then, $y = y_1ABy_3$, and there exists the derivation

$$S \Rightarrow_G^* y_1ABy_3 \Rightarrow_G y_1y_3 \ [AB \to \varepsilon]$$

where $y_1y_3 = x$, $x \in g(x')$.

In cases (ii) through (vii), we discussed all six rules simulating the application of $AB \to \varepsilon$ in G' (see (3) in the definition of P'). Cases (viii) through (xiii) should cover the rules simulating the application of $CD \to \varepsilon$ in G' (see (4)). However, by the inspection of these two sets of rules, it is easy to see that they work analogously. Therefore, we leave this part of the proof to the reader.

We have completed the proof and established Claim 3.1.69 by the principle of induction.   □

Observe that $L(G) = L(G')$ can be easily derived from the above claim. According to the definition of g, we have $g(a) = \{a\}$ for all $a \in T$. Thus, from Claim 3.1.69, we have for any $x \in T^*$

$$S \Rightarrow_G^* x \quad \text{if and only if} \quad S' \Rightarrow_{G'}^+ @x@$$

Since

$$@x@ \Rightarrow_{G'}^2 x \, [(@ \to \varepsilon, N')(@ \to \varepsilon, N')]$$

we obtain for any $x \in T^*$:

$$S \Rightarrow_G^* x \quad \text{if and only if} \quad S' \Rightarrow_{G'}^+ x$$

Consequently, $L(G) = L(G')$, and the theorem holds. $\qquad \square$

3.1.5 Semi-Conditional Context Grammars

The notion of a semi-conditional grammar, discussed in this section, is defined as a context-conditional grammar in which the cardinality of any context-conditional set is no more than one.

The present section consists of two subsections; Definitions and Examples and Generative Power. Frist defines and illustrates semi-conditional grammars, while the former studies their generative power.

Definitions and Examples

The definition of a semi-conditional grammar opens this section.

Definition 3.1.70. Let $G = (V, T, P, S)$ be a context-conditional grammar. G is called a *semi-conditional grammar* (an *sc-grammar* for short) provided that every $(A \to x, Per, For) \in P$ satisfies card(Per) ≤ 1 and card(For) ≤ 1. $\qquad \square$

Let $G = (V, T, P, S)$ be an sc-grammar, and let $(A \to x, Per, For) \in P$. For brevity, we omit braces in each $(A \to x, Per, For) \in P$, and instead of $\emptyset$, we write 0. For instance, we write $(A \to x, BC, 0)$ instead of $(A \to x, \{BC\}, \emptyset)$.

The families of languages generated by sc-grammars and propagating sc-grammars of degree (r, s) are denoted by $\mathbf{SC}(r, s)$ and $\mathbf{SC}^{-\varepsilon}(r, s)$, respectively. The families of languages generated by sc-grammars and propagating sc-grammars of any degree are defined as

$$\mathbf{SC} = \bigcup_{r=0}^{\infty} \bigcup_{s=0}^{\infty} \mathbf{SC}(r, s)$$

and

$$\mathbf{SC}^{-\varepsilon} = \bigcup_{r=0}^{\infty}\bigcup_{s=0}^{\infty}\mathbf{SC}^{-\varepsilon}(r,s)$$

First, we give examples of sc-grammars with degrees $(1, 0)$, $(0, 1)$, and $(1, 1)$.

Example 3.1.71 (See [Pš5]). Let us consider an sc-grammar

$$G = \left(\{S, A, B, A', B', a, b\}, \{a, b\}, P, S\right)$$

where

$$
\begin{aligned}
P = \{ &(S \to AB, 0, 0), (A \to A'A', B, 0), \\
       &(B \to bB', 0, 0), (A' \to A, B', 0), \\
       &(B' \to B, 0, 0), (B \to b, 0, 0), \\
       &(A' \to a, 0, 0), (A \to a, 0, 0)\}
\end{aligned}
$$

Observe that A can be replaced by $A'A'$ only if B occurs in the rewritten string, and A' can be replaced by A only if B' occurs in the rewritten string. If there is an occurrence of B, the number of occurrences of A and A' can be doubled. However, the application of $(B \to bB', 0, 0)$ implies an introduction of one occurrence of b. As a result,

$$L(G) = \{a^n b^m \mid m \geq 1, \ 1 \leq n \leq 2^m\}$$

which is a non-context-free language.   □

Example 3.1.72 (See [Pš5]). Let

$$G = \left(\{S, A, B, A', A'', B', a, b, c\}, \{a, b, c\}, P, S\right)$$

be an sc-grammar, where

$$
\begin{aligned}
P = \{ &(S \to AB, 0, 0), (A \to A', 0, B'), \\
       &(A' \to A''A'', 0, c), (A'' \to A, 0, B), \\
       &(B \to bB', 0, 0), (B' \to B, 0, 0), \\
       &(B \to c, 0, 0), (A \to a, 0, 0), \\
       &(A'' \to a, 0, 0)\}
\end{aligned}
$$

In this case, we get the non-context-free language

$$L(G) = \{a^n b^m c \mid m \geq 0, \ 1 \leq n \leq 2^{m+1}\}$$ □

Example 3.1.73. Let

$$G = \big(\{S, P, Q, R, X, Y, Z, a, b, c, d, e, f\}, \{a, b, c, d, e, f\}, P, S\big)$$

be an sc-grammar, where

$$
\begin{aligned}
P = \{ & (S \rightarrow PQR, 0, 0), \\
       & (P \rightarrow aXb, Q, Z), \\
       & (Q \rightarrow cYd, X, Z), \\
       & (R \rightarrow eZf, X, Q), \\
       & (X \rightarrow P, Z, Q), \\
       & (Y \rightarrow Q, P, R), \\
       & (Z \rightarrow R, P, Y), \\
       & (P \rightarrow \varepsilon, Q, Z), \\
       & (Q \rightarrow \varepsilon, R, P), \\
       & (R \rightarrow \varepsilon, 0, Y) \}
\end{aligned}
$$

Note that this grammar is an sc-grammar of degree $(1, 1)$. Consider *aabbccddeeff*. For this string, G makes the following derivation

$$
\begin{aligned}
S \Rightarrow\ & PQR \Rightarrow aXbQR \Rightarrow aXbcYdR \Rightarrow aXbcYdeZf \Rightarrow \\
& aPbcYdeZf \Rightarrow aPbcQdeZf \Rightarrow aPbcQdeRf \Rightarrow \\
& aaXbbcQdeRf \Rightarrow aaXbbccYddeRf \Rightarrow aaXbbccYddeeZff \Rightarrow \\
& aaPbbccYddeeZff \Rightarrow aaPbbccQddeeZff \Rightarrow aaPbbccQddeeRff \Rightarrow \\
& aabbccQddeeRff \Rightarrow aabbccddeeRff \Rightarrow aabbccddeeff
\end{aligned}
$$

Clearly, G generates the following language

$$L(G) = \{a^n b^n c^n d^n e^n f^n \mid n \geq 0\}$$

As is obvious, this language is non-context-free.   □

Generative Power

The present section establishes the generative power of sc-grammars.

Theorem 3.1.74. $SC^{-\varepsilon}(0, 0) = SC(0, 0) = CF$

Proof. Follows directly from the definitions.   □

Theorem 3.1.75. $CF \subset SC^{-\varepsilon}(1, 0)$, $CF \subset SC^{-\varepsilon}(0, 1)$

Proof. In Examples 3.1.71 and 3.1.72, we show propagating sc-grammars of degrees $(1, 0)$ and $(0, 1)$ that generate non-context-free languages. Therefore, the theorem holds.   □

Theorem 3.1.76. $SC^{-\varepsilon}(1,1) \subset CS$

Proof. Consider a propagating sc-grammar of degree $(1,1)$

$$G = (V, T, P, S)$$

If $(A \rightarrow x, A, \beta) \in P$, then the permitting condition A does not impose any restriction. Hence, we can replace this rule by $(A \rightarrow x, 0, \beta)$. If $(A \rightarrow x, \alpha, A) \in P$, then this rule cannot ever be applied; thus, we can remove it from P. Let $T' = \{a' \mid a \in T\}$ and $V' = V \cup T' \cup \{S', X, Y\}$. Define a homomorphism τ from V^* to $((V - T) \cup (T'))^*$ as $\tau(a) = a'$ for all $a \in T$ and $\tau(A) = A$ for every $A \in V - T$. Furthermore, introduce a function g from $V \cup \{0\}$ to $2^{((V-T) \cup T')}$ as $g(0) = \emptyset$, $g(a) = \{a'\}$ for all $a \in T$, and $g(A) = \{A\}$ for all $A \in V - T$. Next, construct the propagating random context grammar

$$G' = (V', T \cup \{c\}, P', S')$$

where

$$
\begin{aligned}
P' = \ &\{(S' \rightarrow SX, \emptyset, \emptyset), (X \rightarrow Y, \emptyset, \emptyset), (Y \rightarrow c, \emptyset, \emptyset)\} \cup \\
&\{(A \rightarrow \tau(x), g(\alpha) \cup \{X\}, g(\beta)) \mid (A \rightarrow x, \alpha, \beta) \in P\} \cup \\
&\{(a' \rightarrow a, \{Y\}, \emptyset) \mid a \in T\}
\end{aligned}
$$

It is obvious that $L(G') = L(G)\{c\}$. Therefore, $L(G)\{c\} \in \mathbf{RC}^{-\varepsilon}$. Recall that $\mathbf{RC}^{-\varepsilon}$ is closed under restricted homomorphisms (see page 48 in [DP89]), and by Theorem 3.1.53, it holds that $\mathbf{RC}^{-\varepsilon} \subset \mathbf{CS}$. Thus, we obtain $\mathbf{SC}^{-\varepsilon}(1,1) \subset \mathbf{CS}$. $\square$

The following corollary summarizes the generative power of propagating sc-grammars of degrees $(1,0)$, $(0,1)$, and $(1,1)$—that is, the propagating sc-grammars containing only symbols as their context conditions.

Corollary 3.1.77.

$$
\begin{aligned}
\mathbf{CF} &\subset \mathbf{SC}^{-\varepsilon}(0,1) \subseteq \mathbf{SC}^{-\varepsilon}(1,1) \\
\mathbf{CF} &\subset \mathbf{SC}^{-\varepsilon}(1,0) \subseteq \mathbf{SC}^{-\varepsilon}(1,1) \\
\mathbf{SC}^{-\varepsilon}(1,1) &\subseteq \mathbf{RC}^{-\varepsilon} \subset \mathbf{CS}
\end{aligned}
$$

Proof. This corollary follows from Theorems 3.1.74, 3.1.75, and 3.1.76. $\square$

The next theorem says that propagating sc-grammars of degrees $(1,2)$, $(2,1)$ and propagating sc-grammars of any degree generate exactly the family of context-sensitive languages. Furthermore, if we allow erasing rules, these grammars generate the family of recursively enumerable languages.

Theorem 3.1.78.

$$\text{CF}$$
$$\subset$$
$$\textbf{SC}^{-\varepsilon}(2,1) = \textbf{SC}^{-\varepsilon}(1,2) = \textbf{SC}^{-\varepsilon} = \textbf{CS}$$
$$\subset$$
$$\textbf{SC}(2,1) = \textbf{SC}(1,2) = \textbf{SC} = \textbf{RE}$$

Proof. In the next section, we prove a stronger result in terms of a special variant of sc-grammars—simple semi-conditional grammars (see Theorems 3.1.89 and 3.1.94). Therefore, we omit the proof here. □

In [Oku09], the following theorem is proved. It shows that **RE** can be characterized even by sc-grammars of degree $(2, 1)$ with a reduced number of nonterminals and conditional rules.

Theorem 3.1.79 (See Theorem 1 in [Oku09]). *Every recursively enumerable language can be generated by an sc-grammar of degree $(2, 1)$ having no more than 9 conditional rules and 10 nonterminals.*

3.1.6 Simple Semi-Conditional Context Grammars

The notion of a simple semi-conditional grammar—that is, the subject of this section—is defined as an sc-grammar in which every rule has no more than one condition.

The present section consists of two subsections. First, it defines simple semi-conditional grammars, later, it discusses their generative power and reduction.

Definitions and Examples

First, we define simple semi-conditional grammars. Then, we illustrate them.

Definition 3.1.80. Let $G = (V, T, P, S)$ be a semi-conditional grammar. G is a *simple semi-conditional grammar* (an *ssc-grammar* for short) if $(A \to x, \alpha, \beta) \in P$ implies that $0 \in \{\alpha, \beta\}$. □

The families of languages generated by ssc-grammars and propagating ssc-grammars of degree (r, s) are denoted by $\textbf{SSC}(r, s)$ and $\textbf{SSC}^{-\varepsilon}(r, s)$, respectively. Furthermore, set

$$\textbf{SSC} = \bigcup_{r=0}^{\infty} \bigcup_{s=0}^{\infty} \textbf{SSC}(r, s)$$

and

$$\textbf{SSC}^{-\varepsilon} = \bigcup_{r=0}^{\infty} \bigcup_{s=0}^{\infty} \textbf{SSC}^{-\varepsilon}(r, s)$$

The following proposition provides an alternative definition based on context-conditional grammars. Let $G = (V, T, P, S)$ be a context-conditional grammar. G is an ssc-grammar if and only if every $(A \rightarrow x, Per, For) \in P$ satisfies card(Per) + card(For) ≤ 1.

Example 3.1.81. Let

$$G = (\{S, A, X, C, Y, a, b\}, \{a, b\}, P, S)$$

be an ssc-grammar, where

$$P = \{(S \rightarrow AC, 0, 0),$$
$$(A \rightarrow aXb, Y, 0),$$
$$(C \rightarrow Y, A, 0),$$
$$(Y \rightarrow Cc, 0, A),$$
$$(A \rightarrow ab, Y, 0),$$
$$(Y \rightarrow c, 0, A),$$
$$(X \rightarrow A, C, 0)\}$$

Notice that G is propagating, and it has degree $(1, 1)$. Consider *aabbcc*. G derives this string as follows:

$$S \Rightarrow AC \Rightarrow AY \Rightarrow aXbY \Rightarrow aXbCc \Rightarrow$$
$$aAbCc \Rightarrow aAbYc \Rightarrow aabbYc \Rightarrow aabbcc$$

Obviously,

$$L(G) = \{a^n b^n c^n \mid n \geq 1\}$$ □

Example 3.1.82. Let

$$G = (\{S, A, B, X, Y, a\}, \{a\}, P, S)$$

be an ssc-grammar, where

$$P = \{(S \rightarrow a, 0, 0),$$
$$(S \rightarrow X, 0, 0),$$
$$(X \rightarrow YB, 0, A),$$
$$(X \rightarrow aB, 0, A),$$
$$(Y \rightarrow XA, 0, B),$$
$$(Y \rightarrow aA, 0, B),$$
$$(A \rightarrow BB, XA, 0),$$
$$(B \rightarrow AA, YB, 0),$$
$$(B \rightarrow a, a, 0)\}$$

G is a propagating ssc-grammar of degree $(2, 1)$. Consider the string *aaaaaaaa*. G derives this string as follows:

$$S \Rightarrow X \Rightarrow YB \Rightarrow YAA \Rightarrow XAAA \Rightarrow XBBAA \Rightarrow XBBABB \Rightarrow$$
$$XBBBBBB \Rightarrow aBBBBBBB \Rightarrow aBBaBBBB \Rightarrow^6 aaaaaaaa$$

Observe that G generates the following non-context-free language

$$L(G) = \left\{ a^{2^n} \mid n \geq 0 \right\} \qquad \square$$

Generative Power and Reduction

The power and reduction of ssc-grammars represent the central topic discussed in this section.

Theorem 3.1.83. $\mathbf{SSC}^{-\varepsilon}(2, 1) = \mathbf{CS}$

Proof. Because $\mathbf{SSC}^{-\varepsilon}(2, 1) \subseteq \mathbf{CG}^{-\varepsilon}$ and Lemma 3.1.41 implies that $\mathbf{CG}^{-\varepsilon} \subseteq \mathbf{CS}$, it is sufficient to prove the converse inclusion.

Let $G = (V, T, P, S)$ be a context-sensitive grammar in the Penttonen normal form (see Theorem 3.1.5). We construct an ssc-grammar

$$G' = (V \cup W, T, P', S)$$

that generates $L(G)$. Let

$$W = \left\{ \tilde{B} \mid AB \rightarrow AC \in P, \ A, B, C \in V - T \right\}$$

Define P' in the following way

(1) if $A \rightarrow x \in P, A \in V - T, x \in T \cup (V - T)^2$, then add $(A \rightarrow x, 0, 0)$ to P';
(2) if $AB \rightarrow AC \in P, A, B, C \in V - T$, then add $(B \rightarrow \tilde{B}, 0, \tilde{B})$, $(\tilde{B} \rightarrow C, A\tilde{B}, 0)$, $(\tilde{B} \rightarrow B, 0, 0)$ to P'.

Notice that G' is a propagating ssc-grammar of degree $(2, 1)$. Moreover, from (2), we have for any $\tilde{B} \in W$,

$$S \Rightarrow_{G'}^* w \quad \text{implies} \quad \#_{\tilde{B}}(w) \leq 1$$

for all $w \in V'^*$ because the only rule that can generate $\tilde{B}$ is of the form $(B \rightarrow \tilde{B}, 0, \tilde{B})$.

Let g be a finite substitution from V^* into $(V \cup W)^*$ defined as follows: for all $D \in V$,

(1) if $\tilde{D} \in W$, then $g(D) = \{D, \tilde{D}\}$;
(2) if $\tilde{D} \notin W$, then $g(D) = \{D\}$.

Claim 3.1.84. For any $x \in V^+$, $m, n \geq 0$, $S \Rightarrow_G^m x$ if and only if $S \Rightarrow_{G'}^n x'$ with $x' \in g(x)$.

Proof. The proof is divided into the only-if part and the if part.

Only If. This is proved by induction on $m \geq 0$.

Basis. Let $m = 0$. The only x is S as $S \Rightarrow_G^0 S$. Clearly, $S \Rightarrow_{G'}^n S$ for $n = 0$ and $S \in g(S)$.

Induction Hypothesis. Assume that the claim holds for all derivations of length m or less, for some $m \geq 0$.

Induction Step. Consider any derivation of the form

$$S \Rightarrow_G^{m+1} x$$

where $x \in V^+$. Because $m + 1 \geq 1$, there is some $y \in V^*$ and $p \in P$ such that

$$S \Rightarrow_G^m y \Rightarrow_G x \ [p]$$

By the induction hypothesis,

$$S \Rightarrow_{G'}^n y'$$

for some $y' \in g(y)$ and $n \geq 0$. Next, we distinguish between two cases: case (i) considers p with one nonterminal on its left-hand side, and case considers p with two nonterminals on its left-hand side.

(i) Let $p = D \to y_2 \in P$, $D \in V - T$, $y_2 \in T \cup (V - T)^2$, $y = y_1 D y_3$, $y_1, y_3 \in V^*$, $x = y_1 y_2 y_3$, $y' = y_1' X y_3'$, $y_1' \in g(y_1)$, $y_3' \in g(y_3)$, and $X \in g(D)$. By (1) in the definition of P', $(D \to y_2, 0, 0) \in P$. If $X = D$, then

$$S \Rightarrow_{G'}^n y_1' D y_3' \Rightarrow_{G'} y_1' y_2 y_3' \ [(D \to y_2, 0, 0)]$$

Because $y_1' \in g(y_1)$, $y_3' \in g(y_3)$, and $y_2 \in g(y_2)$, we obtain $y_1' y_2 y_3' \in g(y_1 y_2 y_3) = g(x)$. If $X = \tilde{D}$, we have $(X \to D, 0, 0)$ in P', so

$$S \Rightarrow_{G'}^n y_1' X y_3' \Rightarrow_{G'} y_1' D y_3' \Rightarrow_{G'} y_1' y_2 y_3' \ [(X \to D, 0, 0)(D \to y_2, 0, 0)]$$

and $y_1' y_2 y_3' \in g(x)$.

(ii) Let $p = AB \to AC \in P$, $A, B, C \in V - T$, $y = y_1 A B y_2$, $y_1, y_2 \in V^*$, $x = y_1 A C y_2$, $y' = y_1' X Y y_2'$, $y_1' \in g(y_1)$, $y_2' \in g(y_2)$, $X \in g(A)$, and $Y \in g(B)$. Recall that for any $\tilde{B}$, $\#_{\tilde{B}}(y') \leq 1$ and $(\tilde{B} \to B, 0, 0) \in P'$. Then,

$$y' \Rightarrow_{G'}^i y_1' A B y_2'$$

for some $i \in \{0, 1, 2\}$. At this point, we have

$$S \Rightarrow_{G'}^* y_1' A B y_2'$$
$$\Rightarrow_{G'} y_1' A \tilde{B} y_2' \ [(B \rightarrow \tilde{B}, 0, \tilde{B})]$$
$$\Rightarrow_{G'} y_1' A C y_2' \ [(\tilde{B} \rightarrow C, A\tilde{B}, 0)]$$

where $y_1' A C y_2' \in g(x)$.

If. This is established by induction on $n \geq 0$; in other words, we demonstrate that if $S \Rightarrow_{G'}^n x'$ with $x' \in g(x)$ for some $x \in V^+$, then $S \Rightarrow_G^* x$.

Basis. For $n = 0$, x' surely equals S as $S \Rightarrow_{G'}^0 S$. Because $S \in g(S)$, we have $x = S$. Clearly, $S \Rightarrow_G^0 S$.

Induction Hypothesis. Assume that the claim holds for all derivations of length n or less, for some $n \geq 0$.

Induction Step. Consider any derivation of the form

$$S \Rightarrow_{G'}^{n+1} x'$$

with $x' \in g(x)$, $x \in V^+$. As $n + 1 \geq 1$, there exists some $y \in V^+$ such that

$$S \Rightarrow_{G'}^n y' \Rightarrow_{G'} x' \ [p]$$

where $y' \in g(y)$. By the induction hypothesis,

$$S \Rightarrow_G^* y$$

Let $y' = y_1' B' y_2'$, $y = y_1 B y_2$, $y_1' \in g(y_1)$, $y_2' \in g(y_2)$, $y_1, y_2 \in V^*$, $B' \in g(B)$, $B \in V - T$, $x' = y_1' z' y_2'$, and $p = (B' \rightarrow z', \alpha, \beta) \in P'$. The following three cases cover all possible forms of the derivation step $y' \Rightarrow_{G'} x' \ [p]$.

(i) Let $z' \in g(B)$. Then,

$$S \Rightarrow_G^* y_1 B y_2$$

where $y_1' z' y_2' \in g(y_1 B y_2)$; that is, $x' \in g(y_1 B y_2)$.

(ii) Let $B' = B \in V - T$, $z' \in T \cup (V - T)^2$, $\alpha = \beta = 0$. Then, there exists a rule, $B \rightarrow z' \in P$, so

$$S \Rightarrow_G^* y_1 B y_2 \Rightarrow_G y_1 z' y_2 \ [B \rightarrow z']$$

Since $z' \in g(z')$, we have $x = y_1 z' y_2$ such that $x' \in g(x)$.

(iii) Let $B' = \tilde{B}$, $z' = C$, $\alpha = A\tilde{B}$, $\beta = 0$, $A, B, C \in V - T$. Then, there exists a rule of the form $AB \rightarrow AC \in P$. Since $\#_Z(y') \leq 1$, $Z = \tilde{B}$, and $A\tilde{B} \in \text{sub}(y')$, we have $y_1' = u'A$, $y_1 = uA$, $u' \in g(u)$ for some $u \in V^*$. Thus,

$$S \Rightarrow_G^* uABy_2 \Rightarrow_G uACy_2 \ [AB \to AC]$$

where $uACy_2 = y_1 Cy_2$. Because $C \in g(C)$, we get $x = y_1 Cy_2$ such that $x' \in g(x)$.

As cases (i) through (iii) cover all possible forms of a derivation step in G', we have completed the induction step and established Claim 3.1.84 by the principle of induction.   □

The statement of Theorem 3.1.83 follows immediately from Claim 3.1.84. Because for all $a \in T$, $g(a) = \{a\}$, we have for every $w \in T^+$,

$$S \Rightarrow_G^* w \quad \text{if and only if} \quad S \Rightarrow_{G'}^* w$$

Therefore, $L(G) = L(G')$, so the theorem holds.   □

Corollary 3.1.85. $\mathbf{SSC}^{-\varepsilon}(2,1) = \mathbf{SC}^{-\varepsilon} = \mathbf{SC}^{-\varepsilon}(2,1) = \mathbf{SC}^{-\varepsilon} = \mathbf{CS}$

Proof. This corollary follows from Theorem 3.1.83 and the definitions of propagating ssc-grammars.   □

Next, we turn our investigation to ssc-grammars of degree $(2,1)$ with erasing rules. We prove that these grammars generate precisely the family of recursively enumerable languages.

Theorem 3.1.86. $\mathbf{SSC}(2,1) = \mathbf{RE}$

Proof. Clearly, $\mathbf{SSC}(2,1) \subseteq \mathbf{RE}$; hence, it is sufficient to show that $\mathbf{RE} \subseteq \mathbf{SSC}(2,1)$. Every recursively enumerable language $L \in \mathbf{RE}$ can be generated by a phrase-structure grammar G in the Penttonen normal form (see Theorem 3.1.4). That is, the rules of G are of the form $AB \to AC$ or $A \to x$, where $A, B, C \in V - T$, $x \in \{\varepsilon\} \cup T \cup (V - T)^2$. Thus, the inclusion $\mathbf{RE} \subseteq \mathbf{SSC}(2,1)$ can be proved by analogy with the proof of Theorem 3.1.83. The details are left to the reader.   □

Corollary 3.1.87. $\mathbf{SSC}(2,1) = \mathbf{SSC} = \mathbf{SC}(2,1) = \mathbf{SC} = \mathbf{RE}$ □

To demonstrate that propagating ssc-grammars of degree $(1,2)$ characterize $\mathbf{CS}$, we first establish a normal form for context-sensitive grammars.

Lemma 3.1.88. *Every $L \in \mathbf{CS}$ can be generated by a context-sensitive grammar*

$$G = (\{S\} \cup N_{CF} \cup N_{CS} \cup T, T, P, S)$$

where $\{S\}$, N_{CF}, N_{CS}, and T are pairwise disjoint alphabets, and every rule in P is either of the form $S \to aD$ or $AB \to AC$ or $A \to x$, where $a \in T$, $D \in N_{CF} \cup \{\varepsilon\}$, $B \in N_{CS}$, $A, C \in N_{CF}$, $x \in N_{CS} \cup T \cup (\bigcup_{i=1}^{2} N_{CF}^i)$.

Proof. Let L be a context-sensitive language over an alphabet, T. Without any loss of generality, we can express L as $L = L_1 \cup L_2$, where $L_1 \subseteq T$ and $L_2 \subseteq TT^+$. Thus, by analogy with the proofs of Theorems 1 and 2 in [P85], L_2 can be represented as $L_2 = \bigcup_{a \in T} aL_a$, where each L_a is a context-sensitive language. Let L_a be generated by a context-sensitive grammar

$$G_a = \left(N_{CF_a} \cup N_{CS_a} \cup T, T, P_a, S_a\right)$$

of the form of Theorem 3.1.6. Clearly, we assume that for all as, the nonterminal alphabets N_{CF_a} and N_{CS_a} are pairwise disjoint. Let S be a new start symbol. Consider the context-sensitive grammar

$$G = \left(\{S\} \cup N_{CF} \cup N_{CS} \cup T, T, P, S\right)$$

where

$$N_{CF} = \bigcup_{a \in T} N_{CF_a}$$
$$N_{CS} = \bigcup_{a \in T} N_{CS_a}$$
$$P = \bigcup_{a \in T} P_a \cup \{S \to aS_a \mid a \in T\} \cup \{S \to a \mid a \in L_1\}$$

Obviously, G satisfies the required form, and we have

$$L(G) = L_1 \cup \left(\bigcup_{a \in T} aL(G_a)\right) = L_1 \cup \left(\bigcup_{a \in T} aL_a\right) = L_1 \cup L_2 = L$$

Consequently, the lemma holds. □

We are now ready to characterize **CS** by propagating ssc-grammars of degree $(1, 2)$.

Theorem 3.1.89. $\mathbf{CS} = \mathbf{SSC}^{-\varepsilon}(1, 2)$

Proof. By Lemma 3.1.41, $\mathbf{SSC}^{-\varepsilon}(1, 2) \subseteq \mathbf{CG}^{-\varepsilon} \subseteq \mathbf{CS}$; thus, it is sufficient to prove the converse inclusion.

Let L be a context-sensitive language. Without any loss of generality, we assume that L is generated by a context-sensitive grammar

$$G = \left(\{S\} \cup N_{CF} \cup N_{CS} \cup T, T, P, S\right)$$

of the form of Lemma 3.1.88. Set

$$V = \{S\} \cup N_{CF} \cup N_{CS} \cup T$$

Let q be the cardinality of V; $q \geq 1$. Furthermore, let f be an arbitrary bijection from V onto $\{1, \ldots, q\}$, and let f^{-1} be the inverse of f. Let

$$\tilde{G} = (\tilde{V}, T, \tilde{P}, S)$$

be a propagating ssc-grammar of degree $(1, 2)$, in which

$$\tilde{V} = \left(\bigcup_{i=1}^{4} W_i\right) \cup V$$

where

$$W_1 = \{\langle a, AB \to AC, j \rangle \mid a \in T,\ AB \to AC \in P,\ 1 \le j \le 5\},$$
$$W_2 = \{[a, AB \to AC, j] \mid a \in T,\ AB \to AC \in P,\ 1 \le j \le q + 3\}$$
$$W_3 = \{\widehat{B}, B', B'' \mid B \in N_{CS}\}$$
$$W_4 = \{\bar{a} \mid a \in T\}$$

$\tilde{P}$ is defined as follows:

(1) if $S \to aA \in P, a \in T, A \in (N_{CF} \cup \{\varepsilon\})$, then add $(S \to \bar{a}A, 0, 0)$ to $\tilde{P}$;
(2) if $a \in T, A \to x \in P, A \in N_{CF}, x \in (V - \{S\}) \cup (N_{CF})^2$, then add $(A \to x, \bar{a}, 0)$ to $\tilde{P}$;
(3) if $a \in T, AB \to AC \in P, A, C \in N_{CF}, B \in N_{CS}$, then add the following rules to P' (an informal explanation of these rules can be found below):

    (3.a) $(\bar{a} \to \langle a, AB \to AC, 1 \rangle, 0, 0)$
    (3.b) $(B \to B', \langle a, AB \to AC, 1 \rangle, 0)$
    (3.c) $(B \to \widehat{B}, \langle a, AB \to AC, 1 \rangle, 0)$
    (3.d) $(\langle a, AB \to AC, 1 \rangle \to \langle a, AB \to AC, 2 \rangle, 0, B)$
    (3.e) $(\widehat{B} \to B'', 0, B'')$
    (3.f) $(\langle a, AB \to AC, 2 \rangle \to \langle a, AB \to AC, 3 \rangle, 0, \widehat{B})$
    (3.g) $(B'' \to [a, AB \to AC, 1], \langle a, AB \to AC, 3 \rangle, 0)$
    (3.h) $([a, AB \to AC, j] \to [a, AB \to AC, j + 1], 0, f^{-1}(j)[a, AB \to AC, j])$, for all $j = 1, \ldots, q, f(A) \ne j$
    (3.i) $([a, AB \to AC, f(A)] \to [a, AB \to AC, f(A) + 1], 0, 0)$
    (3.j) $([a, AB \to AC, q + 1] \to [a, AB \to AC, q + 2], 0, B'[a, AB \to AC, q + 1])$
    (3.k) $([a, AB \to AC, q + 2] \to [a, AB \to AC, q + 3], 0, \langle a, AB \to AC, 3 \rangle [a, AB \to AC, q + 2])$
    (3.l) $(\langle a, AB \to AC, 3 \rangle \to \langle a, AB \to AC, 4 \rangle, [a, AB \to AC, q + 3], 0)$
    (3.m) $(B' \to B, \langle a, AB \to AC, 4 \rangle, 0)$
    (3.n) $(\langle a, AB \to AC, 4 \rangle \to \langle a, AB \to AC, 5 \rangle, 0, B')$
    (3.o) $([a, AB \to AC, q + 3] \to C, \langle a, AB \to AC, 5 \rangle, 0)$
    (3.p) $(\langle a, AB \to AC, 5 \rangle \to \bar{a}, 0, [a, AB \to AC, q + 3])$

(4) if $a \in T$, then add $(\bar{a} \to a, 0, 0)$ to $\tilde{P}$.

Let us informally explain the basic idea behind (3)—the heart of the construction. The rules introduced in (3) simulate the application of rules of the form $AB \to AC$ in G as follows: an occurrence of B is chosen, and its left neighbor is checked not to belong to $\tilde{V} - \{A\}$. At this point, the left neighbor necessarily equals A, so B is rewritten with C.

Formally, we define a finite substitution g from V^* into $\tilde{V}^*$ as follows:

(a) if $D \in V$, then add D to $g(D)$;
(b) if $\langle a, AB \to AC, j \rangle \in W_1, a \in T, AB \to AC \in P, B \in N_{CS}, A, C \in N_{CF}, j \in \{1, \ldots, 5\}$, then add $\langle a, AB \to AC, j \rangle$ to $g(a)$;

(c) if $[a, AB \rightarrow AC, j] \in W_2$, $a \in T$, $AB \rightarrow AC \in P$, $B \in N_{CS}$, $A, C \in N_{CF}$, $j \in \{1, \ldots, q + 3\}$, then add $[a, AB \rightarrow AC, j]$ to $g(B)$;
(d) if $\{\widehat{B}, B', B''\} \subseteq W_3$, $B \in N_{CS}$, then include $\{\widehat{B}, B', B''\}$ into $g(B)$;
(e) if $\bar{a} \in W_4$, $a \in T$, then add $\bar{a}$ to $g(a)$.

Let g^{-1} be the inverse of g. To show that $L(G) = L(\tilde{G})$, we first prove three claims.

Claim 3.1.90. $S \Rightarrow_G^+ x$, $x \in V^*$, implies that $x \in T(V - \{S\})^*$.

Proof. Observe that the start symbol S does not appear on the right side of any rule and that $S \rightarrow x \in P$ implies that $x \in T \cup T(V - \{S\})$. Hence, the claim holds. $\quad\square$

Claim 3.1.91. If $S \Rightarrow_{\tilde{G}}^+ x$, $x \in \tilde{V}^*$, then x has one of the following seven forms

(I) $x = ay$, where $a \in T$, $y \in (V - \{S\})^*$;
(II) $x = \bar{a}y$, where $\bar{a} \in W_4$, $y \in (V - \{S\})^*$;
(III) $x = \langle a, AB \rightarrow AC, 1 \rangle y$, where $\langle a, AB \rightarrow AC, 1 \rangle \in W_1$, $y \in ((V - \{S\}) \cup \{B', \widehat{B}, B''\})^*$, $\#_{B''}(y) \le 1$;
(IV) $x = \langle a, AB \rightarrow AC, 2 \rangle y$, where $\langle a, AB \rightarrow AC, 2 \rangle \in W_1$, $y \in ((V - \{S, B\}) \cup \{B', \widehat{B}, B''\})^*$, $\#_{B'}(y) \le 1$;
(V) $x = \langle a, AB \rightarrow AC, 3 \rangle y$, where $\langle a, AB \rightarrow AC, 3 \rangle \in W_1$, $y \in ((V - \{S, B\}) \cup \{B'\})^*(\{[a, AB \rightarrow AC, j] \mid 1 \le j \le q + 3\} \cup \{\varepsilon, B''\})((V - \{S, B\}) \cup \{B'\})^*$;
(VI) $x = \langle a, AB \rightarrow AC, 4 \rangle y$, where $\langle a, AB \rightarrow AC, 4 \rangle \in W_1$, $y \in ((V - \{S\}) \cup \{B'\})^* [a, AB \rightarrow AC, q + 3]((V - \{S\}) \cup \{B'\})^*$;
(VII) $x = \langle a, AB \rightarrow AC, 5 \rangle y$, where $\langle a, AB \rightarrow AC, 5 \rangle \in W_1$, $y \in (V - \{S\})^* \{[a, AB \rightarrow AC, q + 3], \varepsilon\} (V - \{S\})^*$.

Proof. The claim is proved by induction on the length of derivations.

Basis. Consider $S \Rightarrow_{\tilde{G}} x$, $x \in \tilde{V}^*$. By the inspection of the rules, we have

$$S \Rightarrow_{\tilde{G}} \bar{a}A \; [(S \rightarrow \bar{a}A, 0, 0)]$$

for some $\bar{a} \in W_4$, $A \in (\{\varepsilon\} \cup N_{CF})$. Therefore, $x = \bar{a}$ or $x = \bar{a}A$; in either case, x is a string of the required form.

Induction Hypothesis. Assume that the claim holds for all derivations of length n or less, for some $n \ge 1$.

Induction Step. Consider any derivation of the form

$$S \Rightarrow_{\tilde{G}}^{n+1} x$$

where $x \in \tilde{V}^*$. Since $n \ge 1$, we have $n + 1 \ge 2$. Thus, there is some z of the required form, $z \in \tilde{V}^*$, such that

$$S \Rightarrow_{\tilde{G}}^{n} z \Rightarrow_{\tilde{G}} x \; [p]$$

for some $p \in \tilde{P}$.

Let us first prove by contradiction that the first symbol of z does not belong to T. Assume that the first symbol of z belongs to T. As z is of the required form, we have $z = ay$ for some $a \in (V - \{S\})^*$. By the inspection of $\tilde{P}$, there is no $p \in \tilde{P}$ such that $ay \Rightarrow_{\tilde{G}} x$ $[p]$, where $x \in \tilde{V}^*$. We have thus obtained a contradiction, so the first symbol of z is not in T.

Because the first symbol of z does not belong to T, z cannot have form (I); as a result, z has one of forms (II) through (VII). The following cases (i) through (vi) demonstrate that if z has one of these six forms, then x has one of the required forms, too.

(i) Assume that z is of form (II); that is, $z = \bar{a}y$, $\bar{a} \in W_4$, and $y \in (V - \{S\})^*$. By the inspection of the rules in $\tilde{P}$, we see that p has one of the following forms (i.a), (i.b), and (i.c)

   (i.a) $p = (A \to u, \bar{a}, 0)$, where $A \in N_{CF}$ and $u \in (V - \{S\}) \cup N_{CF}^2$;
   (i.b) $p = (\bar{a} \to \langle a, AB \to AC, 1 \rangle, 0, 0)$, where $\langle a, AB \to AC, 1 \rangle \in W_1$;
   (i.c) $p = (\bar{a} \to a, 0, 0)$, where $a \in T$.

Note that rules of forms (i.a), (i.b), and (i.c) are introduced in construction steps (2), (3), and (4), respectively. If p has form (i.a), then x has form (II). If p has form (i.b), then x has form (III). Finally, if p has form (i.c), then x has form (I). In any of these three cases, we obtain x that has one of the required forms.

(ii) Assume that z has form (III); that is, $z = \langle a, AB \to AC, 1 \rangle y$ for some $\langle a, AB \to AC, 1 \rangle \in W_1$, $y \in ((V - \{S\}) \cup \{B', \widehat{B}, B''\})^*$, and $\#_{B''}(y) \le 1$. By the inspection of $\tilde{P}$, we see that z can be rewritten by rules of these four forms

   (ii.a) $(B \to B', \langle a, AB \to AC, 1 \rangle, 0)$.
   (ii.b) $(B \to \widehat{B}, \langle a, AB \to AC, 1 \rangle, 0)$.
   (ii.c) $(\widehat{B} \to B'', 0, B'')$ if $B'' \notin alph(y)$; that is, $\#_{B''}(y) = 0$.
   (ii.d) $(\langle a, AB \to AC, 1 \rangle \to \langle a, AB \to AC, 2 \rangle, 0, B)$ if $B \notin alph(y)$; that is, $\#_B(y) = 0$;

Clearly, in cases (ii.a) and (ii.a), we obtain x of form (III). If $z \Rightarrow_{\tilde{G}} x$ $[p]$, where p is of form (ii.c), then $\#_{B''}(x) = 1$, so we get x of form (III). Finally, if we use the rule of form (ii.d), then we obtain x of form (IV) because $\#_B(z) = 0$.

(iii) Assume that z is of form (IV); that is, $z = \langle a, AB \to AC, 2 \rangle y$, where $\langle a, AB \to AC, 2 \rangle \in W_1$, $y \in ((V - \{S, B\}) \cup \{B', \widehat{B}, B''\})^*$, and $\#_{B''}(y) \le 1$. By the inspection of $\tilde{P}$, we see that the following two rules can be used to rewrite z

   (iii.a) $(\widehat{B} \to B'', 0, B'')$ if $B'' \notin alph(y)$.
   (iii.b) $(\langle a, AB \to AC, 2 \rangle \to \langle a, AB \to AC, 3 \rangle, 0, \widehat{B})$ if $\widehat{B} \notin alph(y)$.

In case (iii.a), we get x of form (IV). In case (iii.b), we have $\#_{\widehat{B}}(y) = 0$, so $\#_{\widehat{B}}(x) = 0$. Moreover, notice that $\#_{B''}(x) \le 1$ in this case. Indeed, the symbol B'' can be generated only if there is no occurrence of B'' in a given rewritten string, so no more than one occurrence of B'' appears in any sentential form.

As a result, we have $\#_{B''}(\langle a, AB \to AC, 3 \rangle y) \leq 1$; that is, $\#_{B''}(x) \leq 1$. In other words, we get x of form (V).

(iv) Assume that z is of form (V); that is, $z = \langle a, AB \to AC, 3 \rangle y$ for some $\langle a, AB \to AC, 3 \rangle \in W_1$, $y \in ((V - \{S, B\}) \cup \{B'\})^*(\{[a, AB \to AC, j] \mid 1 \leq j \leq q + 3\} \cup \{B'', \varepsilon\})((V - \{S, B\}) \cup \{B'\})^*$. Assume that $y = y_1 Y y_2$ with $y_1, y_2 \in ((V - \{S, B\}) \cup \{B'\})^*$. If $Y = \varepsilon$, then we can use no rule from $\tilde{P}$ to rewrite z. Because $z \Rightarrow_{\tilde{G}} x$, we have $Y \neq \varepsilon$. The following cases (iv.a) through (iv.f) cover all possible forms of Y.

(iv.a) Assume $Y = B''$. By the inspection of $\tilde{P}$, we see that the only rule that can rewrite z has the form

$$(B'' \to [a, AB \to AC, 1], \langle a, AB \to AC, 3 \rangle, 0)$$

In this case, we get x of form (V).

(iv.b) Assume $Y = [a, AB \to AC, j], j \in \{1, \dots, q\}$, and $f(A) \neq j$. Then, z can be rewritten only according to the rule

$$([a, AB \to AC, j] \to [a, AB \to AC, j + 1], 0, f^{-1}(j)[a, AB \to AC, j])$$

which can be used if the rightmost symbol of $\langle a, AB \to AC, 3 \rangle y_1$ differs from $f^{-1}(j)$. Clearly, in this case, we again get x of form (V).

(iv.c) Assume $Y = [a, AB \to AC, j], j \in \{1, \dots, q\}, f(A) = j$. This case forms an analogy to case (iv.b) except that the rule of the form

$$([a, AB \to AC, f(A)] \to [a, AB \to AC, f(A) + 1], 0, 0)$$

is now used.

(iv.d) Assume $Y = [a, AB \to AC, q + 1]$. This case forms an analogy to case (iv.b); the only change is the application of the rule

$$([a, AB \to AC, q+1] \to [a, AB \to AC, q+2], 0, B'[a, AB \to AC, q+1])$$

(iv.e) Assume $Y = [a, AB \to AC, q + 2]$. This case forms an analogy to case (iv.b) except that the rule

$$([a, AB \to AC, q + 2] \to [a, AB \to AC, q + 3], 0,$$
$$\langle a, AB \to AC, 3 \rangle[a, AB \to AC, q + 2])$$

is used.

(iv.f) Assume $Y = [a, AB \to AC, q + 3]$. By the inspection of $\tilde{P}$, we see that the only rule that can rewrite z is

$$(\langle a, AB \to AC, 3 \rangle \to \langle a, AB \to AC, 4 \rangle, [a, AB \to AC, q + 3], 0)$$

If this rule is used, we get x of form (VI).

(v) Assume that z is of form (VI); that is, $z = \langle a, AB \to AC, 4 \rangle y$, where $\langle a, AB \to AC, 4 \rangle \in W_1$ and $y \in ((V - \{S\}) \cup \{B'\})^*[a, AB \to AC, q+3]((V - \{S\}) \cup \{B'\})^*$. By the inspection of $\tilde{P}$, these two rules can rewrite z

  (v.a) $(B' \to B, \langle a, AB \to AC, 4 \rangle, 0)$;
  (v.b) $(\langle a, AB \to AC, 4 \rangle \to \langle a, AB \to AC, 5 \rangle, 0, B')$ if $B' \notin \text{alph}(y)$.

  Clearly, in case (v.a), we get x of form (VI). In case (v.b), we get x of form (VII) because $\#_{B'}(y) = 0$, so $y \in (V - \{S\})^*[a, AB \to AC, q+3], \varepsilon\}(V - \{S\})^*$.

(vi) Assume that z is of form (VII); that is, $z = \langle a, AB \to AC, 5 \rangle y$, where $\langle a, AB \to AC, 5 \rangle \in W_1$ and $y \in (V - \{S\})^*\{[a, AB \to AC, q+3], \varepsilon\}(V - \{S\})^*$. By the inspection of $\tilde{P}$, one of the following two rules can be used to rewrite z

  (vi.a) $([a, AB \to AC, q+3] \to C, \langle a, AB \to AC, 5 \rangle, 0)$.
  (vi.b) $(\langle a, AB \to AC, 5 \rangle \to \bar{a}, 0, [a, AB \to AC, q+3])$ if $[a, AB \to AC, q+3] \notin \text{alph}(z)$.

  In case (vi.a), we get x of form (VII). Case (vi.b) implies that $\#_{[a, AB \to AC, q+3]}(y) = 0$; thus, x is of form (II).

This completes the induction step and establishes Claim 3.1.91.   □

Claim 3.1.92. It holds that

$$S \Rightarrow_G^m w \quad \text{if and only if} \quad S \Rightarrow_{\tilde{G}}^n v$$

where $v \in g(w)$ and $w \in V^+$, for some $m, n \geq 0$.

Proof. The proof is divided into the only-if part and the if part.

Only If. The only-if part is established by induction on m; that is, we have to demonstrate that

$$S \Rightarrow_G^m w \quad \text{implies} \quad S \Rightarrow_{\tilde{G}}^* v$$

for some $v \in g(w)$ and $w \in V^+$.

Basis. Let $m = 0$. The only w is S because $S \Rightarrow_G^0 S$. Clearly, $S \Rightarrow_{\tilde{G}}^0 S$, and $S \in g(S)$.

Induction Hypothesis. Suppose that the claim holds for all derivations of length m or less, for some $m \geq 0$.

Induction Step. Let us consider any derivation of the form

$$S \Rightarrow_G^{m+1} x$$

where $x \in V^+$. Because $m + 1 \geq 1$, there are $y \in V^+$ and $p \in P$ such that

$$S \Rightarrow_G^m y \Rightarrow_G x \; [p]$$

and by the induction hypothesis, there is also a derivation

$$S \Rightarrow_{\tilde{G}}^* \tilde{y}$$

for some $\tilde{y} \in g(y)$. The following cases (i) through (iii) cover all possible forms of p.

(i) Let $p = S \to aA \in P$ for some $a \in T$, $A \in N_{CF} \cup \{\varepsilon\}$. Then, by Claim 3.1.90, $m = 0$, so $y = S$ and $x = aA$. By (1) in the construction of $\tilde{G}$, $(S \to \bar{a}A, 0, 0) \in \tilde{P}$. Hence,

$$S \Rightarrow_{\tilde{G}} \bar{a}A$$

where $\tilde{a}A \in g(aA)$.

(ii) Let us assume that $p = D \to y_2 \in P$, $D \in N_{CF}$, $y_2 \in (V - \{S\}) \cup N_{CF}^2$, $y = y_1 D y_3$, $y_1, y_3 \in V^*$, and $x = y_1 y_2 y_3$. From the definition of g, it is clear that $g(Z) = \{Z\}$ for all $Z \in N_{CF}$; therefore, we can express $\tilde{y} = z_1 D z_3$, where $z_1 \in g(y_1)$ and $z_3 \in g(y_3)$. Without any loss of generality, we can also assume that $y_1 = au$, $a \in T$, $u \in (V - \{S\})^*$ (see Claim 3.1.90), so $z_1 = a''u''$, $a'' \in g(a)$, and $u'' \in g(u)$. Moreover, by (2) in the construction, we have $(D \to y_2, \bar{a}, 0) \in \tilde{P}$. The following cases (a) through (e) cover all possible forms of a''.

(a) Let $a'' = \bar{a}$ (see (II) in Claim 3.1.91). Then, we have

$$S \Rightarrow_{\tilde{G}}^n \bar{a}u''Dz_3 \Rightarrow_{\tilde{G}} \bar{a}u''y_2z_3 \ [(D \to y_2, \bar{a}, 0)]$$

and $\bar{a}u''y_2z_3 = z_1 y_2 z_3 \in g(y_1 y_2 y_3) = g(x)$.

(b) Let $a'' = a$ (see (I) in Claim 3.1.91). By (4) in the construction of $\tilde{G}$, we can express the derivation

$$S \Rightarrow_{\tilde{G}}^n au''Dz_3$$

as

$$S \Rightarrow_{\tilde{G}}^{n-1} \bar{a}u''Dz_3 \Rightarrow_{\tilde{G}} au''Dz_3 \ [(\bar{a} \to a, 0, 0)]$$

Thus, there exists the derivation

$$S \Rightarrow_{\tilde{G}}^{n-1} \bar{a}u''Dz_3 \Rightarrow_{\tilde{G}} \bar{a}u''y_2z_3 \ [(D \to y_2, \bar{a}, 0)]$$

with $\bar{a}u''y_2z_3 \in g(x)$.

(c) Let $a'' = \langle a, AB \to AC, 5 \rangle$ for some $AB \to AC \in P$ (see (VII) in Claim 3.1.91), and let $u''Dz_3 \in (V - \{S\})^*$; that is, $[a, AB \to AC, q + 3] \notin$ alph$(u''Dz_3)$. Then, there exists the derivation

$$S \Rightarrow_{\tilde{G}}^{n} \langle a, AB \rightarrow AC, 5 \rangle u'' Dz_3$$
$$\Rightarrow_{\tilde{G}} \bar{a} u'' Dz_3 \ [(\langle a, AB \rightarrow AC, 5 \rangle \rightarrow \bar{a}, 0, [a, AB \rightarrow AC, q+3])]$$
$$\Rightarrow_{\tilde{G}} \bar{a} u'' y_2 z_3 \ [(D \rightarrow y_2, \bar{a}, 0)]$$

and $\bar{a} u'' y_2 z_3 \in g(x)$.

(d) Let $a'' = \langle a, AB \rightarrow AC, 5 \rangle$ (see (VII) in Claim 3.1.91). Let $[a, AB \rightarrow AC, q+3] \in \mathrm{alph}(u'' Dz_3)$. Without any loss of generality, we can assume that $\tilde{y} = \langle a, AB \rightarrow AC, 5 \rangle u'' Do''[a, AB \rightarrow AC, q+3]t''$, where $o''[a, AB \rightarrow AC, q+3]t'' = z_3$, $oBt = y_3$, $o'' \in g(t)$, $o, t \in (V - \{S\})^*$. By the inspection of $\tilde{P}$ (see (3) in the construction of $\tilde{G}$), we can express the derivation

$$S \Rightarrow_{\tilde{G}}^{n} \tilde{y}$$

as

$$S \Rightarrow_{\tilde{G}}^{*} \quad \bar{a} u'' Do'' Bt''$$

$$\Rightarrow_{\tilde{G}} \quad \langle a, AB \rightarrow AC, 1 \rangle u'' Do'' Bt''$$

$$[(\bar{a} \rightarrow \langle a, AB \rightarrow AC, 1 \rangle, 0, 0)]$$

$$\Rightarrow_{\tilde{G}}^{1 + |m_1 m_2|} \langle a, AB \rightarrow AC, 1 \rangle u' Do' \widehat{B} t'$$

$$[m_1 (B \rightarrow \widehat{B}, \langle a, AB \rightarrow AC, 1 \rangle, 0) m_2]$$

$$\Rightarrow_{\tilde{G}} \quad \langle a, AB \rightarrow AC, 2 \rangle u' Do' \widehat{B} t'$$

$$[(\langle a, AB \rightarrow AC, 1 \rangle \rightarrow \langle a, AB \rightarrow AC, 2 \rangle, 0, B)]$$

$$\Rightarrow_{\tilde{G}} \quad \langle a, AB \rightarrow AC, 2 \rangle u' Do' B'' t'$$

$$[\widehat{B} \rightarrow B'', 0, B'']$$

$$\Rightarrow_{\tilde{G}} \quad \langle a, AB \rightarrow AC, 3 \rangle u' Do' B'' t'$$

$$[(\langle a, AB \rightarrow AC, 2 \rangle \rightarrow \langle a, AB \rightarrow AC, 3 \rangle, 0, \widehat{B})]$$

$$\Rightarrow_{\tilde{G}} \quad \langle a, AB \rightarrow AC, 3 \rangle u' Do'[a, AB \rightarrow AC, 1]t'$$

$$[(B'' \rightarrow [a, AB \rightarrow AC, 1], \langle a, AB \rightarrow AC, 3 \rangle, 0)]$$

$$\Rightarrow_{\tilde{G}}^{q+2} \quad \langle a, AB \rightarrow AC, 3 \rangle u' Do'[a, AB \rightarrow AC, q+3]t'$$

$$[\omega]$$

$$\Rightarrow_{\tilde{G}} \quad \langle a, AB \rightarrow AC, 4 \rangle u' Do'[a, AB \rightarrow AC, q+3]t'$$

$$[(\langle a, AB \rightarrow AC, 3 \rangle \rightarrow \langle a, AB \rightarrow AC, 4 \rangle,$$
$$[a, AB \rightarrow AC, q+3], 0)]$$

$$\Rightarrow_{\tilde{G}}^{|m_3|} \quad \langle a, AB \rightarrow AC, 4\rangle u'' Do'' [a, AB \rightarrow AC, q+3]t''$$

$$[m_3]$$

$$\Rightarrow_{\tilde{G}} \quad \langle a, AB \rightarrow AC, 5\rangle u'' Do'' [a, AB \rightarrow AC, q+3]t''$$

$$[(\langle a, AB \rightarrow AC, 4\rangle \rightarrow \langle a, AB \rightarrow AC, 5\rangle, 0, B')]$$

where $m_1, m_2 \in \{(B \rightarrow B', \langle a, AB \rightarrow AC, 1\rangle, 0)\}^*$, $m_3 \in \{(B' \rightarrow B, \langle a, AB \rightarrow AC, 4\rangle, 0)\}^*$, $|m_3| = |m_1 m_2|$,

$$\omega = ([a, AB \rightarrow AC, 1] \rightarrow [a, AB \rightarrow AC, 2], 0,$$
$$f^{-1}(1)[a, AB \rightarrow AC, 1]) \cdots$$
$$([a, AB \rightarrow AC, f(A) - 1] \rightarrow [a, AB \rightarrow AC, f(A)], 0,$$
$$f^{-1}(f(A) - 1)[a, AB \rightarrow AC, f(A) - 1])$$
$$([a, AB \rightarrow AC, f(A)] \rightarrow [a, AB \rightarrow AC, f(A) + 1], 0, 0)$$
$$([a, AB \rightarrow AC, f(A) + 1] \rightarrow [a, AB \rightarrow AC, f(A) + 2], 0,$$
$$f^{-1}(f(A) + 1)[a, AB \rightarrow AC, f(A) + 1]) \cdots$$
$$([a, AB \rightarrow AC, q] \rightarrow [a, AB \rightarrow AC, q + 1], 0,$$
$$f^{-1}(q)[a, AB \rightarrow AC, q])$$
$$([a, AB \rightarrow AC, q + 1] \rightarrow [a, AB \rightarrow AC, q + 2], 0,$$
$$B'[a, AB \rightarrow AC, q + 1])$$
$$([a, AB \rightarrow AC, q + 2] \rightarrow [a, AB \rightarrow AC, q + 3]), 0,$$
$$\langle a, AB \rightarrow AC, 3\rangle [a, AB \rightarrow AC, q + 2])$$

$u' \in ((\text{alph}(u'') - \{B\}) \cup \{B'\})^*$, $g^{-1}(u') = u$, $o' \in ((\text{alph}(o'') - \{B\}) \cup \{B''\})^*$, $g^{-1}(o') = g^{-1}(o'') = o$, $t' \in ((\text{alph}(t'') - \{B\}) \cup \{B'\})^*$, $g^{-1}(t') = g^{-1}(t'') = t$.

Clearly, $\bar{a}u'' Do'' Bt'' \in g(auDoBt) = g(auDy_3) = g(y)$. Thus, there exists the derivation

$$S \Rightarrow_{\tilde{G}}^* \bar{a}u'' Do'' Bt'' \Rightarrow_{\tilde{G}} \bar{a}u'' y_2 o'' Bt'' \quad [(D \rightarrow y_2, \bar{a}, 0)]$$

where $z_1 y_2 z_3 = \bar{a}u'' y_2 o'' Bt'' \in g(auy_2 o Bt) = g(y_1 y_2 y_3) = g(x)$.

(e) Let $a'' = \langle a, AB \rightarrow AC, i\rangle$ for some $AB \rightarrow AC \in P$ and $i \in \{1, \ldots, 4\}$ (see (III)–(VI) in Claim 3.1.91). By analogy with (ii), we can construct the derivation

$$S \Rightarrow_{\tilde{G}}^* \bar{a}u'' Do'' Bt'' \Rightarrow_{\tilde{G}} \bar{a}u'' y_2 o'' Bt'' \quad [(D \rightarrow y_2, \bar{a}, 0)]$$

such that $\bar{a}u'' y_2 o'' Bt'' \in g(y_1 y_2 y_3) = g(x)$. The details are left to the reader.

(iii) Let $p = AB \rightarrow AC \in P$, $A, C \in N_{CF}$, $B \in N_{CS}$, $y = y_1 ABy_3$, $y_1, y_3 \in V^*$, $x = y_1 ACy_3$, $\tilde{y} = z_1 AYz_3$, $Y \in g(B)$, $z_i \in g(y_i)$ where $i \in \{1, 3\}$. Moreover, let $y_1 = au$ (see Claim 3.1.90), $z_1 = a''u''$, $a'' \in g(a)$, and $u'' \in g(u)$. The following cases (a) through (e) cover all possible forms of a''.

(a) Let $a'' = \bar{a}$. Then, by Claim 3.1.91, $Y = B$. By (3) in the construction of $\tilde{G}$, there exists the following derivation

$$S \Rightarrow_{\tilde{G}}^{n} \quad \bar{a}u''ABz_3$$

$$\Rightarrow_{\tilde{G}} \quad \langle a, AB \to AC, 1 \rangle u''ABz_3$$

$$[(\bar{a} \to \langle a, AB \to AC, 1 \rangle, 0, 0)]$$

$$\Rightarrow_{\tilde{G}}^{1+|m_1|} \langle a, AB \to AC, 1 \rangle u'A\widehat{B}u_3$$

$$[m_1(B \to \widehat{B}, \langle a, AB \to AC, 1 \rangle, 0)]$$

$$\Rightarrow_{\tilde{G}} \quad \langle a, AB \to AC, 2 \rangle u'A\widehat{B}u_3$$

$$[(\langle a, AB \to AC, 1 \rangle \to \langle a, AB \to AC, 2 \rangle, 0, B)]$$

$$\Rightarrow_{\tilde{G}} \quad \langle a, AB \to AC, 2 \rangle u'AB''u_3$$

$$[(\widehat{B} \to B'', 0, B'')]$$

$$\Rightarrow_{\tilde{G}} \quad \langle a, AB \to AC, 3 \rangle u'AB''u_3$$

$$[(\langle a, AB \to AC, 2 \rangle \to \langle a, AB \to AC, 3 \rangle, 0, \widehat{B})]$$

$$\Rightarrow_{\tilde{G}} \quad \langle a, AB \to AC, 3 \rangle u'A[a, AB \to AC, 1]u_3$$

$$[(B'' \to [a, AB \to AC, 1], \langle a, AB \to AC, 3 \rangle, 0)]$$

$$\Rightarrow_{\tilde{G}}^{q+2} \langle a, AB \to AC, 3 \rangle u'A[a, AB \to AC, q+3]u_3$$

$$[\omega]$$

$$\Rightarrow_{\tilde{G}} \quad \langle a, AB \to AC, 4 \rangle u'A[a, AB \to AC, q+3]u_3$$

$$[(\langle a, AB \to AC, 3 \rangle \to \langle a, AB \to AC, 4 \rangle,$$

$$[a, AB \to AC, q+3], 0)]$$

$$\Rightarrow_{\tilde{G}}^{|m_2|} \langle a, AB \to AC, 4 \rangle u''A[a, AB \to AC, q+3]z_3$$

$$[m_2]$$

$$\Rightarrow_{\tilde{G}} \quad \langle a, AB \to AC, 5 \rangle u''A[a, AB \to AC, q+3]z_3$$

$$[(\langle a, AB \to AC, 4 \rangle \to \langle a, AB \to AC, 5 \rangle, 0, B')]$$

$$\Rightarrow_{\tilde{G}} \quad \langle a, AB \to AC, 5 \rangle u''ACz_3$$

$$[([a, AB \to AC, q+3] \to C, \langle a, AB \to AC, 5 \rangle, 0)]$$

where $m_1 \in \{(B \to B', \langle a, AB \to AC, 1 \rangle, 0)\}^*$, $m_2 \in \{(B' \to B, \langle a, AB \to AC, 4 \rangle, 0)\}^*$, $|m_1| = |m_2|$,

$$
\begin{aligned}
\omega = &([a, AB \to AC, 1] \to [a, AB \to AC, 2], 0, \\
&\quad f^{-1}(1)[a, AB \to AC, 1]) \cdots \\
&([a, AB \to AC, f(A) - 1] \to [a, AB \to AC, f(A)], 0, \\
&\quad f^{-1}(f(A) - 1)[a, AB \to AC, f(A) - 1]) \\
&([a, AB \to AC, f(A)] \to [a, AB \to AC, f(A) + 1], 0, 0) \\
&([a, AB \to AC, f(A) + 1] \to [a, AB \to AC, f(A) + 2], 0, \\
&\quad f^{-1}(f(A) + 1)[a, AB \to AC, f(A) + 1]) \cdots \\
&([a, AB \to AC, q] \to [a, AB \to AC, q + 1], 0, \\
&\quad f^{-1}(q)[a, AB \to AC, q]) \\
&([a, AB \to AC, q + 1] \to [a, AB \to AC, q + 2], 0, \\
&\quad B'[a, AB \to AC, q + 1]) \\
&([a, AB \to AC, q + 2] \to [a, AB \to AC, q + 3]), 0, \\
&\quad \langle a, AB \to AC, 3 \rangle [a, AB \to AC, q + 2])
\end{aligned}
$$

$u_3 \in ((\mathrm{alph}(z_3) - \{B\}) \cup \{B'\})^*$, $g^{-1}(u_3) = g^{-1}(z_3) = y_3$, $u' \in ((\mathrm{alph}(u'') - \{B\}) \cup \{B'\})^*$, $g^{-1}(u') = g^{-1}(u'') = u$. It is clear that $\langle a, AB \to AC, 5 \rangle \in g(a)$; thus, $\langle a, AB \to AC, 5 \rangle u'' ACz_3 \in g(auACy_3) = g(x)$.

(b) Let $a'' = a$. Then, by Claim 3.1.91, $Y = B$. By analogy with (ii.b) and (iii.a) in the proof of this claim (see above), we obtain

$$
S \Rightarrow_{\tilde{G}}^{n-1} \bar{a} u'' ABz_3 \Rightarrow_{\tilde{G}}^* \langle a, AB \to AC, 5 \rangle u'' ACz_3
$$

so $\langle a, AB \to AC, 5 \rangle u'' ACz_3 \in g(x)$.

(c) Let $a'' = \langle a, AB \to AC, 5 \rangle$ for some $AB \to AC \in P$ (see (VII) in Claim 3.1.91), and let $u'' AYz_3 \in (V - \{S\})^*$. At this point, $Y = B$. By analogy with (ii.c) and (iii.a) in the proof of this claim (see above), we can construct

$$
S \Rightarrow_{\tilde{G}}^{n+1} \bar{a} u'' ABz_3 \Rightarrow_{\tilde{G}}^* \langle a, AB \to AC, 5 \rangle u'' ACz_3
$$

so $\langle a, AB \to AC, 5 \rangle u'' ACz_3 \in g(x)$.

(d) Let $a'' = \langle a, AB \to AC, 5 \rangle$ for some $AB \to AC \in P$ (see (VII) in Claim 3.1.91), and let $[a, AB \to AC, q + 3] \in \mathrm{alph}(u'' AYz_3)$. By analogy with (ii.d) and (iii.a) in the proof of this claim (see above), we can construct

$$
S \Rightarrow_{\tilde{G}}^* \bar{a} u'' ABz_3
$$

and then

$$
S \Rightarrow_{\tilde{G}}^* \bar{a} u'' ABz_3 \Rightarrow_{\tilde{G}}^* \langle a, AB \to AC, 5 \rangle u'' ACz_3
$$

so that $\langle a, AB \to AC, 5 \rangle u'' ACz_3 \in g(auACy_3) = g(x)$.

(e) Let $a'' = \langle a, AB \to AC, i \rangle$ for some $AB \to AC \in P$, $i \in \{1, \ldots, 4\}$, see (III)–(IV) in Claim 3.1.91. By analogy with (ii) and (iii) in the proof of this claim, we can construct

$$S \Rightarrow^*_{\tilde{G}} \bar{a} u'' A C z_3$$

where $\bar{a} u'' A C z_3 \in g(x)$.

If. By induction on n, we next prove that if $S \Rightarrow^n_{\tilde{G}} v$ with $v \in g(w)$ and $w \in V^*$, for some $n \geq 0$, then $S \Rightarrow^*_G w$.

Basis. For $n = 0$, the only v is S as $S \Rightarrow^0_{\tilde{G}} S$. Because $\{S\} = g(S)$, we have $w = S$. Clearly, $S \Rightarrow^0_G S$.

Induction Hypothesis. Assume that the claim holds for all derivations of length n or less, for some $n \geq 0$. Let us show that it also holds true for $n + 1$.

Induction Step. For $n + 1 = 1$, there only exists a direct derivation of the form

$$S \Rightarrow_{\tilde{G}} \bar{a} A \ [(S \to \bar{a} A, 0, 0)]$$

where $A \in N_{CF} \cup \{\varepsilon\}$, $a \in T$, and $\bar{a} A \in g(aA)$. By (1), we have in P a rule of the form $S \to aA$ and, thus, a direct derivation $S \Rightarrow_G aA$.

Suppose that $n + 1 \geq 2$ (i.e. $n \geq 1$). Consider any derivation of the form

$$S \Rightarrow^{n+1}_G x'$$

where $x' \in g(x)$, $x \in V^*$. Because $n + 1 \geq 2$, there exist $\bar{a} \in W_4$, $A \in N_{CF}$, and $y \in V^+$ such that

$$S \Rightarrow_{\tilde{G}} \bar{a} A \Rightarrow^{n-1}_{\tilde{G}} y' \Rightarrow_{\tilde{G}} x' \ [p]$$

where $p \in \tilde{P}$, $y' \in g(y)$, and by the induction hypothesis,

$$S \Rightarrow^*_G y$$

Let us assume that $y' = z_1 Z z_2$, $y = y_1 D y_2$, $z_j \in g(y_j)$, $y_j \in (V - \{S\})^*$, $j = 1, 2$, $Z \in g(D)$, $D \in V - \{S\}$, $p = (Z \to u', \alpha, \beta) \in P'$, $\alpha = 0$ or $\beta = 0$, $x' = z_1 u' z_2$, $u' \in g(u)$ for some $u \in V^*$; that is, $x' \in g(y_1 u y_2)$. The following cases (i) through (iii) cover all possible forms of

$$y' \Rightarrow_{\tilde{G}} x' \ [p]$$

(i) Let $Z \in N_{CF}$. By the inspection of $\tilde{P}$, we see that $Z = D$, $p = (D \to u', \bar{a}, 0) \in \tilde{P}$, $D \to u \in P$ and $u = u'$. Thus,

$$S \Rightarrow^*_G y_1 B y_2 \Rightarrow_G y_1 u y_2 \ [B \to u]$$

(ii) Let $u = D$. Then, by the induction hypothesis, we have the derivation

$$S \Rightarrow_G^* y_1 D y_2$$

and $y_1 D y_2 = y_1 u y_2$ in G.

(iii) Let $p = ([a, AB \rightarrow AC, q + 3] \rightarrow C, \langle a, AB \rightarrow AC, 5 \rangle, 0)$, $Z = [a, AB \rightarrow AC, q + 3]$. Thus, $u' = C$ and $D = B \in N_{CS}$. By case (VI) in Claim 3.1.91 and the form of p, we have $z_1 = \langle a, AB \rightarrow AC, 5 \rangle t$ and $y_1 = ao$, where $t \in g(o)$, $\langle a, AB \rightarrow AC, 5 \rangle \in g(a)$, $o \in (V - \{S\})^*$, and $a \in T$. From (3) in the construction of $\tilde{G}$, it follows that there exists a rule of the form $AB \rightarrow AC \in P$. Moreover, (3) and Claim 3.1.91 imply that the derivation

$$S \Rightarrow_{\tilde{G}} \bar{a}A \Rightarrow_{\tilde{G}}^{n-1} y' \Rightarrow_{\tilde{G}} x' \; [p]$$

can be expressed in the form

$$
\begin{aligned}
S \Rightarrow_{\tilde{G}} \quad & \bar{a}A \\
\Rightarrow_{\tilde{G}}^* \quad & \bar{a}tBz_2 \\
\Rightarrow_{\tilde{G}} \quad & \langle a, AB \rightarrow AC, 1 \rangle tBz_2 \\
& [(\bar{a} \rightarrow \langle a, AB \rightarrow AC, 1 \rangle, 0, 0)] \\
\Rightarrow_{\tilde{G}}^{|\omega'|} \quad & \langle a, AB \rightarrow AC, 1 \rangle v\widehat{B}w_2 \\
& [\omega'] \\
\Rightarrow_{\tilde{G}} \quad & \langle a, AB \rightarrow AC, 1 \rangle vB''w_2 \\
& [(\widehat{B} \rightarrow B'', 0, B'')] \\
\Rightarrow_{\tilde{G}} \quad & \langle a, AB \rightarrow AC, 2 \rangle vB''w_2 \\
& [(\langle a, AB \rightarrow AC, 1 \rangle \rightarrow \langle a, AB \rightarrow AC, 2 \rangle, 0, B)] \\
\Rightarrow_{\tilde{G}} \quad & \langle a, AB \rightarrow AC, 3 \rangle vB''w_2 \\
& [(\langle a, AB \rightarrow AC, 2 \rangle \rightarrow \langle a, AB \rightarrow AC, 3 \rangle, 0, \widehat{B})] \\
\Rightarrow_{\tilde{G}} \quad & \langle a, AB \rightarrow AC, 3 \rangle v[a, AB \rightarrow AC, 1]w_2 \\
& [(B'' \rightarrow [a, AB \rightarrow AC, 1], \langle a, AB \rightarrow AC, 3 \rangle, 0)] \\
\Rightarrow_{\tilde{G}}^{|\omega|} \quad & \langle a, AB \rightarrow AC, 3 \rangle v[a, AB \rightarrow AC, q + 3]w_2 \\
& [\omega] \\
\Rightarrow_{\tilde{G}} \quad & \langle a, AB \rightarrow AC, 4 \rangle v[a, AB \rightarrow AC, q + 3]w_2 \\
& [(\langle a, AB \rightarrow AC, 3 \rangle \rightarrow \langle a, AB \rightarrow AC, 4 \rangle, \\
& \quad [a, AB \rightarrow AC, q + 3], 0)] \\
\Rightarrow_{\tilde{G}}^{|\omega'|-1} \quad & \langle a, AB \rightarrow AC, 4 \rangle t[a, AB \rightarrow AC, q + 3]z_2 \\
& [\omega''] \\
\Rightarrow_{\tilde{G}} \quad & \langle a, AB \rightarrow AC, 5 \rangle t[a, AB \rightarrow AC, q + 3]z_2 \\
& [(\langle a, AB \rightarrow AC, 4 \rangle \rightarrow \langle a, AB \rightarrow AC, 5 \rangle, 0, B')]
\end{aligned}
$$

$$\Rightarrow_{\tilde{G}} \langle a, AB \to AC, 5 \rangle t C z_2$$
$$[([a, AB \to AC, q + 3] \to C, \langle a, AB \to AC, 5 \rangle, 0)]$$

where

$$\omega' \in \{(B \to B', \langle a, AB \to AC, 1 \rangle, 0)\}^*$$
$$\{(B \to \widehat{B}, \langle a, AB \to AC, 1 \rangle, 0)\}$$
$$\{(B \to B', \langle a, AB \to AC, 1 \rangle, 0)\}^*$$

$$g(B) \cap \text{alph}(vw_2) \subseteq \{B'\}, g^{-1}(v) = g^{-1}(t), g^{-1}(w_2) = g^{-1}(z_2),$$

$$\omega = \omega_1$$
$$([a, AB \to AC, f(A)] \to [a, AB \to AC, f(A) + 1], 0, 0)\omega_2$$
$$([a, AB \to AC, q + 1] \to [a, AB \to AC, q + 2], 0$$
$$B'[a, AB \to AC, q + 1])$$
$$([a, AB \to AC, q + 2] \to [a, AB \to AC, q + 3], 0,$$
$$\langle a, AB \to AC, 3 \rangle [a, AB \to AC, q + 2])$$

$$\omega_1 = ([a, AB \to AC, 1] \to [a, AB \to AC, 2], 0,$$
$$f^{-1}(1)[a, AB \to AC, 1]) \cdots$$
$$([a, AB \to AC, f(A) - 1] \to [a, AB \to AC, f(A)], 0,$$
$$f^{-1}(f(A) - 1)[a, AB \to AC, f(A) - 1])$$

where $f(A)$ implies that $q_1 = \varepsilon$,

$$\omega_2 = ([a, AB \to AC, f(A) + 1] \to [a, AB \to AC, f(A) + 2], 0,$$
$$f^{-1}(f(A) + 1)[a, AB \to AC, f(A) + 1]) \cdots$$
$$([a, AB \to AC, q] \to [a, AB \to AC, q + 1], 0,$$
$$f^{-1}(q)[a, AB \to AC, q])$$

where $f(A) = q$ implies that $q_2 = \varepsilon$, $\omega'' \in \{(B' \to B, \langle a, AB \to AC, 4 \rangle, 0)\}^*$. The derivation above implies that the rightmost symbol of t must be A. As $t \in g(o)$, the rightmost symbol of o must be A as well. That is, $t = s'A$, $o = sA$ and $s' \in g(s)$, for some $s \in (V - \{S\})^*$. By the induction hypothesis, there exists a derivation

$$S \Rightarrow_G^* asABy_2$$

Because $AB \to AC \in P$, we get

$$S \Rightarrow_G^* asABy_2 \Rightarrow_G asACy_2 \; [AB \to AC]$$

where $asACy_2 = y_1 u y_2$.

By (i), (ii), and (iii) and the inspection of $\tilde{P}$, we see that we have considered all possible derivations of the form

$$S \Rightarrow_{\tilde{G}}^{n+1} x'$$

so we have established Claim 3.1.92 by the principle of induction. □

The equivalence of G and $\tilde{G}$ can be easily derived from Claim 3.1.92. By the definition of g, we have $g(a) = \{a\}$ for all $a \in T$. Thus, by Claim 3.1.92, we have for all $x \in T^*$,

$$S \Rightarrow_G^* x \quad \text{if and only if} \quad S \Rightarrow_{\tilde{G}}^* x$$

Consequently, $L(G) = L(\tilde{G})$, and the theorem holds. □

Corollary 3.1.93. $\mathbf{SSC}^{-\varepsilon}(1,2) = \mathbf{SSC}^{-\varepsilon} = \mathbf{SC}^{-\varepsilon}(1,2) = \mathbf{SC}^{-\varepsilon} = \mathbf{CS}$ □

We now turn to the investigation of ssc-grammars of degree $(1,2)$ with erasing rules.

Theorem 3.1.94. $\mathbf{SSC}(1,2) = \mathbf{RE}$

Proof. Clearly, we have $\mathbf{SSC}(1,2) \subseteq \mathbf{RE}$. Thus, we only need to show that $\mathbf{RE} \subseteq \mathbf{SSC}(1,2)$. Every language $L \in \mathbf{RE}$ can be generated by a phrase-structure grammar $G = (V, T, P, S)$ in which each rule is of the form $AB \to AC$ or $A \to x$, where $A, B, C \in V - T, x \in \{\varepsilon\} \cup T \cup (V - T)^2$ (see Theorem 3.1.7). Thus, the inclusion can be established by analogy with the proof of Theorem 3.1.89. The details are left to the reader. □

Corollary 3.1.95. $\mathbf{SSC}(1,2) = \mathbf{SSC} = \mathbf{SC}(1,2) = \mathbf{SC} = \mathbf{RE}$ □

The following corollary summarizes the relations of language families generated by ssc-grammars.

Corollary 3.1.96.

$$\mathbf{CF}$$
$$\subset$$
$$\mathbf{SSC}^{-\varepsilon} = \mathbf{SSC}^{-\varepsilon}(2,1) = \mathbf{SSC}^{-\varepsilon}(1,2) =$$
$$= \mathbf{SC}^{-\varepsilon} = \mathbf{SC}^{-\varepsilon}(2,1) = \mathbf{SC}^{-\varepsilon}(1,2) = \mathbf{CS}$$
$$\subset$$
$$\mathbf{SSC} = \mathbf{SSC}(2,1) = \mathbf{SSC}(1,2) = \mathbf{SC} = \mathbf{SC}(2,1) = \mathbf{SC}(1,2) = \mathbf{RE}$$

Proof. This corollary follows from Corollaries 3.1.85, 3.1.87, 3.1.93, and 3.1.95. □

Next, we turn our attention to reduced versions of ssc-grammars. More specifically, we demonstrate that there exist several normal forms of ssc-grammars with a limited number of conditional rules and nonterminals.

Theorem 3.1.97. *Every recursively enumerable language can be defined by an ssc-grammar of degree* $(2, 1)$ *with no more than 12 conditional rules and 13 nonterminals.*

Proof. Let L be a recursively enumerable language. By Theorem 3.1.11, we assume that L is generated by a grammar G of the form

$$G = \left(V, T, P \cup \{AB \to \varepsilon, CD \to \varepsilon\}, S\right)$$

such that P contains only context-free rules and

$$V - T = \{S, A, B, C, D\}$$

Construct an ssc-grammar G' of degree $(2, 1)$,

$$G' = \left(V', T, P', S\right)$$

where

$$V' = V \cup W$$
$$W = \{\tilde{A}, \tilde{B}, \langle \varepsilon_A \rangle, \$, \tilde{C}, \tilde{D}, \langle \varepsilon_C \rangle, \# \}, \ V \cap W = \emptyset$$

The set of rules P' is defined in the following way

1. if $H \to y \in P, H \in V - T, y \in V^*$, then add $(H \to y, 0, 0)$ to P';
2. add the following six rules to P'

$$(A \to \tilde{A}, 0, \tilde{A})$$
$$(B \to \tilde{B}, 0, \tilde{B})$$
$$(\tilde{A} \to \langle \varepsilon_A \rangle, \tilde{A}\tilde{B}, 0)$$
$$(\tilde{B} \to \$, \langle \varepsilon_A \rangle \tilde{B}, 0)$$
$$(\langle \varepsilon_A \rangle \to \varepsilon, 0, \tilde{B})$$
$$(\$ \to \varepsilon, 0, \langle \varepsilon_A \rangle)$$

3. add the following six rules to P'

$$(C \to \tilde{C}, 0, \tilde{C})$$
$$(D \to \tilde{D}, 0, \tilde{D})$$
$$(\tilde{C} \to \langle \varepsilon_C \rangle, \tilde{C}\tilde{D}, 0)$$
$$(\tilde{D} \to \#, \langle \varepsilon_C \rangle \tilde{D}, 0)$$
$$(\langle \varepsilon_C \rangle \to \varepsilon, 0, \tilde{D})$$
$$(\# \to \varepsilon, 0, \langle \varepsilon_C \rangle)$$

Notice that G' has degree $(2, 1)$ and contains only 12 conditional rules and 13 nonterminals. The rules of (2) simulate the application of $AB \to \varepsilon$ in G' and the rules of (3) simulate the application of $CD \to \varepsilon$ in G'.

Let us describe the simulation of $AB \to \varepsilon$. First, one occurrence of A and one occurrence of B are rewritten to $\tilde{A}$ and $\tilde{B}$, respectively (no more than one $\tilde{A}$ and one $\tilde{B}$ appear in any sentential form). The right neighbor of $\tilde{A}$ is checked to be $\tilde{B}$ and $\tilde{A}$ is rewritten to $\langle \varepsilon_A \rangle$. Then, analogously, the left neighbor of $\tilde{B}$ is checked to be $\langle \varepsilon_A \rangle$ and $\tilde{B}$ is rewritten to \$. Finally, $\langle \varepsilon_A \rangle$ and \$ are erased. The simulation of $CD \to \varepsilon$ is analogous.

To establish $L(G) = L(G')$, we first prove two claims.

Claim 3.1.98. $S \Rightarrow_{G'}^* x'$ implies that $\#_{\tilde{X}}(x') \le 1$ for all $\tilde{X} \in \{\tilde{A}, \tilde{B}, \tilde{C}, \tilde{D}\}$ and $x' \in V'^*$.

Proof. By the inspection of rules in P', the only rule that can generate $\tilde{X}$ is of the form $(X \to \tilde{X}, 0, \tilde{X})$. This rule can be applied only when no $\tilde{X}$ occurs in the rewritten sentential form. Thus, it is not possible to derive x' from S such that $\#_{\tilde{X}}(x') \ge 2$. $\square$

Informally, the next claim says that every occurrence of $\langle \varepsilon_A \rangle$ in derivations from S is always followed by either $\tilde{B}$ or \$, and every occurrence of $\langle \varepsilon_C \rangle$ is always followed by either $\tilde{D}$ or #.

Claim 3.1.99. It holds that

(I) $S \Rightarrow_{G'}^* y_1' \langle \varepsilon_A \rangle y_2'$ implies $y_2' \in V'^+$ and $\text{lms}(y_2') \in \{\tilde{B}, \$\}$ for any $y_1' \in V'^*$;
(II) $S \Rightarrow_{G'}^* y_1' \langle \varepsilon_C \rangle y_2'$ implies $y_2' \in V'^+$ and $\text{lms}(y_2') \in \{\tilde{D}, \#\}$ for any $y_1' \in V'^*$.

Proof. We base this proof on the examination of all possible forms of derivations that may occur during a derivation of a sentential form containing $\langle \varepsilon_A \rangle$ or $\langle \varepsilon_C \rangle$.

(I) By the definition of P', the only rule that can generate $\langle \varepsilon_A \rangle$ is $p = (\tilde{A} \to \langle \varepsilon_A \rangle, \tilde{A}\tilde{B}, 0)$. This rule has the permitting condition $\tilde{A}\tilde{B}$, so it can be used provided that $\tilde{A}\tilde{B}$ occurs in a sentential form. Furthermore, by Claim 3.1.98, no other occurrence of $\tilde{A}$ or $\tilde{B}$ can appear in the given sentential form. Consequently, we obtain a derivation

$$S \Rightarrow_{G'}^* u_1' \tilde{A} \tilde{B} u_2' \Rightarrow_{G'} u_1' \langle \varepsilon_A \rangle \tilde{B} u_2' \; [p]$$

for some $u_1', u_2' \in V'^*$, $\tilde{A}, \tilde{B} \notin \text{alph}(u_1' u_2')$, which represents the only way of getting $\langle \varepsilon_A \rangle$. Obviously, $\langle \varepsilon_A \rangle$ is always followed by $\tilde{B}$ in $u_1' \langle \varepsilon_A \rangle \tilde{B} u_2'$.
Next, we discuss how G' can rewrite the substring $\langle \varepsilon_A \rangle \tilde{B}$ in $u_1' \langle \varepsilon_A \rangle \tilde{B} u_2'$. There are only two rules having the nonterminals $\langle \varepsilon_A \rangle$ or $\tilde{B}$ on their left-hand side, $p_1 = (\tilde{B} \to \$, \langle \varepsilon_A \rangle \tilde{B}, 0)$ and $p_2 = (\langle \varepsilon_A \rangle \to \varepsilon, 0, \tilde{B})$. G' cannot use p_2 to erase $\langle \varepsilon_A \rangle$ in $u_1' \langle \varepsilon_A \rangle \tilde{B} u_2'$ because p_2 forbids an occurrence of $\tilde{B}$ in the rewritten string. Rule p_1 has also a context condition, but $\langle \varepsilon_A \rangle \tilde{B} \in \text{sub}(u_1' \langle \varepsilon_A \rangle \tilde{B} u_2')$, and thus p_1 can be used to rewrite $\tilde{B}$ with \$. Hence, we obtain a derivation of the form

$$S \Rightarrow_{G'}^* u_1' \tilde{A} \tilde{B} u_2' \quad \Rightarrow_{G'} u_1' \langle \varepsilon_A \rangle \tilde{B} \tilde{B} u_2' \; [p]$$
$$\Rightarrow_{G'}^* v_1' \langle \varepsilon_A \rangle \tilde{B} v_2' \Rightarrow_{G'} v_1' \langle \varepsilon_A \rangle \$ v_2' \; [p_1]$$

Notice that during this derivation, G' may rewrite u_1' and u_2' to some v_1' and v_2', respectively, where $v_1', v_2' \in V'^*$; however, $\langle \varepsilon_A \rangle \tilde{B}$ remains unchanged after this rewriting.

In this derivation, we obtained the second symbol $ that can appear as the right neighbor of $\langle \varepsilon_A \rangle$. It is sufficient to show that there is no other symbol that can appear immediately after $\langle \varepsilon_A \rangle$. By the inspection of P', only ($ \to \varepsilon, 0, \langle \varepsilon_A \rangle$) can rewrite $. However, this rule cannot be applied when $\langle \varepsilon_A \rangle$ occurs in the given sentential form. In other words, the occurrence of $ in the substring $\langle \varepsilon_A \rangle$$ cannot be rewritten before $\langle \varepsilon_A \rangle$ is erased by rule p_2. Hence, $\langle \varepsilon_A \rangle$ is always followed by either $\tilde{B}$ or $, and thus, the first part of Claim 3.1.99 holds.

(II) By the inspection of rules simulating $AB \to \varepsilon$ and $CD \to \varepsilon$ in G' (see (2) and (3) in the definition of P'), these two sets of rules work analogously. Thus, part (I) of Claim 3.1.99 can be proved by analogy with part (I).   □

Let us return to the main part of the proof. Let g be a finite substitution from V'^* to V^* defined as follows:

1. For all $X \in V$, $g(X) = \{X\}$.
2. $g(\tilde{A}) = \{A\}, g(\tilde{B}) = \{B\}, g(\langle \varepsilon_A \rangle) = \{A\}, g($) = \{B, AB\}$.
3. $g(\tilde{C}) = \{C\}, g(\tilde{D}) = \{D\}, g(\langle \varepsilon_C \rangle) = \{C\}, g(\#) = \{C, CD\}$.

Having this substitution, we can prove the following claim.

Claim 3.1.100. $S \Rightarrow_G^* x$ if and only if $S \Rightarrow_{G'}^* x'$ for some $x \in g(x'), x \in V^*, x' \in V'^*$.

Proof. The claim is proved by induction on the length of derivations.

Only If. We show that

$$S \Rightarrow_G^m x \quad \text{implies} \quad S \Rightarrow_{G'}^* x$$

where $m \geq 0, x \in V^*$; clearly $x \in g(x)$. This is established by induction on $m \geq 0$.

Basis. Let $m = 0$. That is, $S \Rightarrow_G^0 S$. Clearly, $S \Rightarrow_{G'}^0 S$.

Induction Hypothesis. Suppose that the claim holds for all derivations of length m or less, for some $m \geq 0$.

Induction Step. Consider any derivation of the form

$$S \Rightarrow_G^{m+1} x, \ x \in V^*$$

Since $m + 1 \geq 1$, there is some $y \in V^+$ and $p \in P \cup \{AB \to \varepsilon, CD \to \varepsilon\}$ such that

$$S \Rightarrow_G^m y \Rightarrow_G x \ [p]$$

By the induction hypothesis, there is a derivation

$$S \Rightarrow_{G'}^* y$$

The following three cases cover all possible forms of p.

(i) Let $p = H \rightarrow y_2 \in P$, $H \in V - T$, $y_2 \in V^*$. Then, $y = y_1 H y_3$ and $x = y_1 y_2 y_3$, $y_1, y_3 \in V^*$. Because we have $(H \rightarrow y_2, 0, 0) \in P'$,

$$S \Rightarrow^*_{G'} y_1 H y_3 \Rightarrow_{G'} y_1 y_2 y_3 \ [(H \rightarrow y_2, 0, 0)]$$

and $y_1 y_2 y_3 = x$.

(ii) Let $p = AB \rightarrow \varepsilon$. Then, $y = y_1 AB y_3$ and $x = y_1 y_3$, $y_1, y_3 \in V^*$. In this case, there is the derivation

$$
\begin{aligned}
S &\Rightarrow^*_{G'} y_1 AB y_3 \\
&\Rightarrow_{G'} y_1 \tilde{A} B y_3 && [(A \rightarrow \tilde{A}, 0, \tilde{A})] \\
&\Rightarrow_{G'} y_1 \tilde{A} \tilde{B} y_3 && [(B \rightarrow \tilde{B}, 0, \tilde{B})] \\
&\Rightarrow_{G'} y_1 \langle \varepsilon_A \rangle \tilde{B} y_3 && [(\tilde{A} \rightarrow \langle \varepsilon_A \rangle, \tilde{A}\tilde{B}, 0)] \\
&\Rightarrow_{G'} y_1 \langle \varepsilon_A \rangle \$ y_3 && [(\tilde{B} \rightarrow \$, \langle \varepsilon_A \rangle \tilde{B}, 0)] \\
&\Rightarrow_{G'} y_1 \$ y_3 && [(\langle \varepsilon_A \rangle \rightarrow \varepsilon, 0, \tilde{B})] \\
&\Rightarrow_{G'} y_1 y_3 && [(\$ \rightarrow \varepsilon, 0, \langle \varepsilon_A \rangle)]
\end{aligned}
$$

(iii) Let $p = CD \rightarrow \varepsilon$. Then, $y = y_1 CD y_3$ and $x = y_1 y_3$, $y_1, y_3 \in V^*$. By analogy with (ii), there exists the derivation

$$
\begin{aligned}
S &\Rightarrow^*_{G'} y_1 CD y_3 \\
&\Rightarrow_{G'} y_1 \tilde{C} D y_3 && [(C \rightarrow \tilde{C}, 0, \tilde{C})] \\
&\Rightarrow_{G'} y_1 \tilde{C} \tilde{D} y_3 && [(D \rightarrow \tilde{D}, 0, \tilde{D})] \\
&\Rightarrow_{G'} y_1 \langle \varepsilon_C \rangle \tilde{D} y_3 && [(\tilde{C} \rightarrow \langle \varepsilon_C \rangle, \tilde{C}\tilde{D}, 0)] \\
&\Rightarrow_{G'} y_1 \langle \varepsilon_C \rangle \# y_3 && [(\tilde{D} \rightarrow \#, \langle \varepsilon_C \rangle \tilde{D}, 0)] \\
&\Rightarrow_{G'} y_1 \# y_3 && [(\langle \varepsilon_C \rangle \rightarrow \varepsilon, 0, \tilde{D})] \\
&\Rightarrow_{G'} y_1 y_3 && [(\# \rightarrow \varepsilon, 0, \langle \varepsilon_C \rangle)]
\end{aligned}
$$

If. By induction on the length n of derivations in G', we prove that

$$S \Rightarrow^n_{G'} x' \quad \text{implies} \quad S \Rightarrow^*_G x$$

for some $x \in g(x')$, $x \in V^*$, $x' \in V'^*$.

Basis. Let $n = 0$. That is, $S \Rightarrow^0_{G'} S$. It is obvious that $S \Rightarrow^0_G S$ and $S \in g(S)$.

Induction Hypothesis. Assume that the claim holds for all derivations of length n or less, for some $n \geq 0$.

Induction Step. Consider any derivation of the form

$$S \Rightarrow^{n+1}_{G'} x', \ x' \in V'^*$$

Since $n + 1 \geq 1$, there is some $y' \in V'^{+}$ and $p' \in P'$ such that

$$S \Rightarrow_{G'}^{n} y' \Rightarrow_{G'} x' \; [p']$$

and by the induction hypothesis, there is also a derivation

$$S \Rightarrow_{G}^{*} y$$

such that $y \in g(y')$.

By the inspection of P', the following cases (i) through (xiii) cover all possible forms of p'.

(i) Let $p' = (H \rightarrow y_2, 0, 0) \in P'$, $H \in V - T$, $y_2 \in V^{*}$. Then, $y' = y_1' H y_3'$, $x' = y_1' y_2 y_3'$, $y_1', y_3' \in V'^{*}$ and y has the form $y = y_1 Z y_3$, where $y_1 \in g(y_1')$, $y_3 \in g(y_3')$ and $Z \in g(H)$. Because $g(X) = \{X\}$ for all $X \in V - T$, the only Z is H, and thus $y = y_1 H y_3$. By the definition of P' (see (1)), there exists a rule $p = H \rightarrow y_2$ in P, and we can construct the derivation

$$S \Rightarrow_{G}^{*} y_1 H y_3 \Rightarrow_{G} y_1 y_2 y_3 \; [p]$$

such that $y_1 y_2 y_3 = x$, $x \in g(x')$.

(ii) Let $p' = (A \rightarrow \tilde{A}, 0, \tilde{A})$. Then, $y' = y_1' A y_3'$, $x' = y_1' \tilde{A} y_3'$, $y_1', y_3' \in V'^{*}$, and $y = y_1 Z y_3$, where $y_1 \in g(y_1')$, $y_3 \in g(y_3')$ and $Z \in g(A)$. Because $g(A) = \{A\}$, the only Z is A, so we can express $y = y_1 A y_3$. Having the derivation $S \Rightarrow_{G}^{*} y$ such that $y \in g(y')$, it is easy to see that also $y \in g(x')$ because $A \in g(\tilde{A})$.

(iii) Let $p' = (B \rightarrow \tilde{B}, 0, \tilde{B})$. By analogy with (ii), $y' = y_1' B y_3'$, $x' = y_1' \tilde{B} y_3'$, $y = y_1 B y_3$, where $y_1', y_3' \in V'^{*}$, $y_1 \in g(y_1')$, $y_3 \in g(y_3')$, and thus $y \in g(x')$ because $B \in g(\tilde{B})$.

(iv) Let $p' = (\tilde{A} \rightarrow \langle \varepsilon_A \rangle, \tilde{A}\tilde{B}, 0)$. By the permitting condition of this rule, $\tilde{A}\tilde{B}$ surely occurs in y'. By Claim 3.1.98, no more than one $\tilde{A}$ can occur in y'. Therefore, y' must be of the form $y' = y_1' \tilde{A} \tilde{B} y_3'$, where $y_1', y_3' \in V'^{*}$ and $\tilde{A} \notin \text{alph}(y_1' y_3')$. Then, $x' = y_1' \langle \varepsilon_A \rangle \tilde{B} y_3'$ and y is of the form $y = y_1 Z y_3$, where $y_1 \in g(y_1')$, $y_3 \in g(y_3')$ and $Z \in g(\tilde{A}\tilde{B})$. Because $g(\tilde{A}\tilde{B}) = \{AB\}$, the only Z is AB; thus, we obtain $y = y_1 A B y_3$. By the induction hypothesis, we have a derivation $S \Rightarrow_{G}^{*} y$ such that $y \in g(y')$. According to the definition of g, $y \in g(x')$ as well because $A \in g(\langle \varepsilon_A \rangle)$ and $B \in g(\tilde{B})$.

(v) Let $p' = (\tilde{B} \rightarrow \$, \langle \varepsilon_A \rangle \tilde{B}, 0)$. This rule can be applied provided that $\langle \varepsilon_A \rangle \tilde{B} \in \text{sub}(y')$. Moreover, by Claim 3.1.98, $\#_{\tilde{B}}(y') \leq 1$. Hence, we can express $y' = y_1' \langle \varepsilon_A \rangle \tilde{B} y_3'$, where $y_1', y_3' \in V'^{*}$ and $\tilde{B} \notin \text{alph}(y_1' y_3')$. Then, $x' = y_1' \langle \varepsilon_A \rangle \$ y_3'$ and $y = y_1 Z y_3$, where $y_1 \in g(y_1')$, $y_3 \in g(y_3')$ and $Z \in g(\langle \varepsilon_A \rangle \tilde{B})$. By the definition of g, $g(\langle \varepsilon_A \rangle \tilde{B}) = \{AB\}$, so $Z = AB$ and $y = y_1 A B y_3$. By the induction hypothesis, we have a derivation $S \Rightarrow_{G}^{*} y$ such that $y \in g(y')$. Because $A \in g(\langle \varepsilon_A \rangle)$ and $B \in g(\$)$, $y \in g(x')$ as well.

(vi) Let $p' = (\langle \varepsilon_A \rangle \rightarrow \varepsilon, 0, \tilde{B})$. Application of $(\langle \varepsilon_A \rangle \rightarrow \varepsilon, 0, \tilde{B})$ implies that $\langle \varepsilon_A \rangle$ occurs in y'. Claim 3.1.99 says that $\langle \varepsilon_A \rangle$ has either $\tilde{B}$ or $\$$ as its right neighbor. Since the forbidding condition of p' forbids an occurrence of $\tilde{B}$ in y', the right neighbor of $\langle \varepsilon_A \rangle$ must be $\$$. As a result, we obtain $y' = y_1' \langle \varepsilon_A \rangle \$ y_3'$

where $y_1', y_3' \in V'^*$. Then, $x' = y_1'\$y_3'$ and y is of the form $y = y_1 Z y_3$, where $y_1 \in g(y_1')$, $y_3 \in g(y_3')$ and $Z \in g(\langle \varepsilon_A \rangle \$)$. By the definition of g, $g(\langle \varepsilon_A \rangle \$) = \{AB, AAB\}$. If $Z = AB$, $y = y_1 AB y_3$. Having the derivation $S \Rightarrow_G^* y$, it holds that $y \in g(x')$ because $AB \in g(\$)$.

(vii) Let $p' = (\$ \to \varepsilon, 0, \langle \varepsilon_A \rangle)$. Then, $y' = y_1'\$y_3'$ and $x' = y_1'y_3'$, where $y_1', y_3' \in V'^*$. Express $y = y_1 Z y_3$ so that $y_1 \in g(y_1')$, $y_3 \in g(y_3')$ and $Z \in g(\$)$, where $g(\$) = \{B, AB\}$. Let $Z = AB$. Then, $y = y_1 AB y_3$, and there exists the derivation

$$S \Rightarrow_G^* y_1 AB y_3 \Rightarrow_G y_1 y_3 \ [AB \to \varepsilon]$$

where $y_1 y_3 = x, x \in g(x')$.

In cases (ii) through (vii), we discussed all six rules simulating the application of $AB \to \varepsilon$ in G' (see (2) in the definition of P'). Cases (viii) through (xiii) should cover rules simulating the application of $CD \to \varepsilon$ in G' (see (3)). However, by the inspection of these two sets of rules, it is easy to see that they work analogously. Therefore, we leave this part of the proof to the reader.

We have completed the proof and established Claim 3.1.100 by the principle of induction. □

Observe that $L(G) = L(G')$ follows from Claim 3.1.100. Indeed, according to the definition of g, we have $g(a) = \{a\}$ for all $a \in T$. Thus, from Claim 3.1.100, we have for any $x \in T^*$

$$S \Rightarrow_G^* x \quad \text{if and only if} \quad S \Rightarrow_{G'}^* x$$

Consequently, $L(G) = L(G')$, and the theorem holds. □

Let us note that in [Mas06], Theorem 3.1.97 has been improved by demonstrating that even nine conditional rules and ten nonterminals are enough to generate every recursively enumerable language.

Theorem 3.1.101 (See [Mas06]). *Every recursively enumerable language can be generated by an ssc-grammar of degree* $(2, 1)$ *having no more than 9 conditional rules and 10 nonterminals.*

Continuing with the investigation of reduced ssc-grammars, we point out that Vaszil in [Vas05] proved that if we allow permitting conditions of length three— that is, ssc-grammars of degree $(3, 1)$, then the number of conditional rules and nonterminals can be further decreased.

Theorem 3.1.102. *Every recursively enumerable language can be generated by an ssc-grammar of degree* $(3, 1)$ *with no more than 8 conditional rules and 11 nonterminals.*

Proof. (See [Vas05].) Let L be a recursively enumerable language. Without any loss of generality, we assume that L is generated by a phrase-structure grammar

$$G = (V, T, P \cup \{ABC \to \varepsilon\}, S)$$

where

$$V - T = \{S, S', A, B, C\}$$

and P contains only context-free rules of the forms $S \to zSx$, $z \in \{A, B\}^*$, $x \in T$, $S \to S'$, $S' \to uS'v$, $u \in \{A, B\}^*$, $v \in \{B, C\}^*$, $S' \to \varepsilon$ (see Theorem 3.1.9). Every successful derivation in G consists of the following two phases.

1.

$$S \Rightarrow_G^* z_n \cdots z_1 S x_1 \cdots x_n$$
$$\Rightarrow_G z_n \cdots z_1 S' x_1 \cdots x_n \; ; z_i \in \{A, B\}^*, 1 \le i \le n.$$

2.

$$z_n \cdots z_1 S' x_1 \cdots x_n \Rightarrow_G^* z_n \cdots z_1 u_m \cdots u_1 S' v_1 \cdots v_m x_1 \cdots x_n$$
$$\Rightarrow_G z_n \cdots z_1 u_m \cdots u_1 v_1 \cdots v_m x_1 \cdots x_n$$

where $u_j \in \{A, B\}^*$, $v_j \in \{B, C\}^*$, $1 \le j \le m$, and the terminal string $x_1 \cdots x_n$ is generated by G if and only if by using the erasing rule $ABC \to \varepsilon$, the substring $z_n \cdots z_1 u_m \cdots u_1 v_1 \cdots v_m$ can be deleted.

Next, we introduce the ssc-grammar

$$G' = (V', T, P', S)$$

of degree $(3, 1)$, where

$$V' = \{S, S', A, A', A'', B, B', B'', C, C', C''\} \cup T$$

and P' is constructed as follows:

1. for every $H \to y \in P$, add $(H \to y, 0, 0)$ to P';
2. for every $X \in \{A, B, C\}$, add $(X \to X', 0, X')$ to P';
3. add the following six rules to P'

$$(C' \to C'', A'B'C', 0)$$
$$(A' \to A'', A'B'C'', 0)$$
$$(B' \to B'', A''B'C'', 0)$$
$$(A'' \to \varepsilon, 0, C'')$$
$$(C'' \to \varepsilon, 0, B')$$
$$(B'' \to \varepsilon, 0, 0)$$

Observe that G' satisfies all the requirements of this theorem—that is, it contains only 8 conditional rules and 11 nonterminals. G' reproduces the first two phases of generating a terminal string in G by using the rules of the form $(H \to y, 0, 0) \in P'$.

The third phase, during which $ABC \rightarrow \varepsilon$ is applied, is simulated by the additional rules. Examine these rules to see that all strings generated by G can also be generated by G'. Indeed, for every derivation step

$$y_1ABCy_2 \Rightarrow_G y_1y_2 \; [ABC \rightarrow \varepsilon]$$

in G, $y_1, y_2 \in V^*$, there exists the following derivation in G'

$$
\begin{aligned}
y_1ABCy_2 &\Rightarrow_{G'} y_1A'BCy_2 && [(A \rightarrow A', 0, A')] \\
&\Rightarrow_{G'} y_1A'B'Cy_2 && [(B \rightarrow B', 0, B')] \\
&\Rightarrow_{G'} y_1A'B'C'y_2 && [(C \rightarrow C', 0, C')] \\
&\Rightarrow_{G'} y_1A'B'C''y_2 && [(C' \rightarrow C'', A'B'C', 0)] \\
&\Rightarrow_{G'} y_1A''B'C''y_2 && [(A' \rightarrow A'', A'B'C'', 0)] \\
&\Rightarrow_{G'} y_1A''B''C''y_2 && [(B' \rightarrow B'', A''B'C'', 0)] \\
&\Rightarrow_{G'} y_1A''B''y_2 && [(C'' \rightarrow \varepsilon, 0, B')] \\
&\Rightarrow_{G'} y_1B''y_2 && [(A'' \rightarrow \varepsilon, 0, C'')] \\
&\Rightarrow_{G'} y_1y_2 && [(B'' \rightarrow \varepsilon, 0, 0)]
\end{aligned}
$$

As a result, $L(G) \subseteq L(G')$. In the following, we show that G' does not generate strings that cannot be generated by G; thus, $L(G') - L(G) = \emptyset$, so $L(G') = L(G)$.

Let us study how G' can generate a terminal string. All derivations start from S. While the sentential form contains S or S', its form is zSw or $zuS'vw$, $z, u, v \in \{A, B, C, A', B', C'\}^*$, $w \in T^*$, where if $g(X') = X$ for $X \in \{A, B, C\}$ and $g(X) = X$ for all other symbols of V, then $g(zSw)$ or $g(zuS'vw)$ are valid sentential forms of G. Furthermore, zu contains at most one occurrence of A', v contains at most one occurrence of C', and zuv contains at most one occurrence of B' (see 2 in the construction of P'). After $(S' \rightarrow \varepsilon, 0, 0)$ is used, we get a sentential form $zuvw$ with z, u, v, and w as above such that

$$S \Rightarrow_G^* g(zuvw)$$

Next, we demonstrate that

$$zuv \Rightarrow_{G'}^* \varepsilon \quad \text{implies} \quad g(zuv) \Rightarrow_G^* \varepsilon$$

More specifically, we investigate all possible derivations rewriting a sentential form containing a single occurrence of each of the letters A', B', and C'.

Consider a sentential form of the form $zuvw$, where $z, u, v \in \{A, B, C, A', B', C'\}^*$, $w \in T^*$, and $\#_{A'}(zu) = \#_{B'}(zuv) = \#_{C'}(v) = 1$. By the definition of rules rewriting A', B', and C' (see 3 in the construction of P'), we see that these three symbols must form a substring $A'B'C'$; otherwise, no next derivation step can be made. That is, $zuvw = z\bar{u}A'B'C'\bar{v}w$ for some $\bar{u}, \bar{v} \in \{A, B, C\}^*$. Next, observe that the only applicable rule is $(C' \rightarrow C'', A'B'C', 0)$. Thus, we get

$$z\bar{u}A'B'C'\bar{v}w \Rightarrow_{G'} z\bar{u}A'B'C''\bar{v}w$$

This sentential form can be rewritten in two ways. First, we can rewrite A' to A'' by $(A' \to A'', A'B'C'', 0)$. Second, we can replace another occurrence of C with C'. Let us investigate the derivation

$$z\bar{u}A'B'C''\bar{v}w \Rightarrow_{G'} z\bar{u}A''B'C''\bar{v}w \; [(A' \to A'', A'B'C'', 0)]$$

As before, we can either rewrite another occurrence of A to A' or rewrite an occurrence of C to C' or rewrite B' to B'' by using $(B' \to B'', A''B'C'', 0)$. Taking into account all possible combinations of the above-described steps, we see that after the first application of $(B' \to B'', A''B'C'', 0)$, the whole derivation is of the form

$$z\bar{u}A'B'C'\bar{v}w \Rightarrow_{G'}^+ zu_1Xu_2A''B''C''v_1Yv_2w$$

where $X \in \{A', \varepsilon\}$, $Y \in \{C', \varepsilon\}$, $u_1 g(X)u_2 = \bar{u}$, and $v_1 g(Y)v_2 = \bar{v}$. Let $zu_1Xu_2 = x$ and $v_1Yv_2 = y$. The next derivation step can be made in four ways. By an application of $(B \to B', 0, B')$, we can rewrite an occurrence of B in x or y. In both cases, this derivation is blocked in the next step. The remaining two derivations are

$$xA''B''C''yw \Rightarrow_{G'} xA''C''yw \; [(B'' \to \varepsilon, 0, 0)]$$

and

$$xA''B''C''yw \Rightarrow_{G'} xA''B''yw \; [(C'' \to \varepsilon, 0, B')]$$

Let us examine how G' can rewrite $xA''C''yw$. The following three cases cover all possible steps.

(i) If $xA''C''yw \Rightarrow_{G'} x_1B'x_2A''C''yw \; [(B \to B', 0, B')]$, where $x_1Bx_2 = x$, then the derivation is blocked.
(ii) If $xA''C''yw \Rightarrow_{G'} xA''C''y_1B'y_2w \; [(B \to B', 0, B')]$, where $y_1By_2 = y$, then no next derivation step can be made.
(iii) Let $xA''C''yw \Rightarrow_{G'} xA''yw \; [(C'' \to \varepsilon, 0, B')]$. Then, all the following derivations

$$xA''yw \Rightarrow_{G'} xyw$$

and

$$xA''yw \Rightarrow_{G'} x_1B'x_2A''yw \Rightarrow_{G'} x_1B'x_2yw$$

where $x_1Bx_2 = x$, and

$$xA''yw \Rightarrow_{G'} xA''y_1B'y_2w \Rightarrow_{G'} xy_1B'y_2w$$

where $y_1 B y_2 = y$, produce a sentential form in which the substring $A''B''C''$ is erased. This sentential form contains at most one occurrence of A', B', and C'. Return to

$$xA''B''C''yw \Rightarrow_{G'} xA''B''yw$$

Observe that by analogy with case (iii), any rewriting of $xA''B''yw$ removes the substring $A''B''$ and produces a sentential form containing at most one occurrence of A', B', and C'.

To summarize the considerations above, the reader can see that as long as there exists an occurrence of A'', B'', or C'' in the sentential form, only the erasing rules or $(B \rightarrow B', 0, B')$ can be applied. The derivation either enters a sentential form that blocks the derivation or the substring $A'B'C'$ is completely erased, after which new occurrences of A, B, and C can be changed to A', B', and C'. That is,

$$z\bar{u}A'B'C'\bar{v}w \Rightarrow_{G'}^{+} xyw \quad \text{implies} \quad g(z\bar{u}A'B'C'\bar{v}w) \Rightarrow_{G} g(xyw)$$

where $z, \bar{u}, \bar{v} \in \{A, B, C\}^*$, $x, y \in \{A, B, C, A', B', C'\}^*$, $w \in T^*$, and $z\bar{u} = g(x)$, $\bar{v}w = g(yw)$. In other words, the rules constructed in 2 and 3 correctly simulate the application of the only non-context-free rule $ABC \rightarrow \varepsilon$. Recall that $g(a) = a$, for all $a \in T$. Hence, $g(xyw) = g(xy)w$. Thus, $L(G') - L(G) = \emptyset$.

Having $L(G) \subseteq L(G')$ and $L(G') - L(G) = \emptyset$, we get $L(G) = L(G')$, and the theorem holds. $\square$

Theorem 3.1.102 was further slightly improved in [Oku09], where the following result was proved (the number of nonterminals was reduced from 11 to 9).

Theorem 3.1.103 (See [Oku09]). *Every recursively enumerable language can be generated by an ssc-grammar of degree $(3, 1)$ with no more than 8 conditional rules and 9 nonterminals.*

Let us close this section by stating several open problems.

Open Problem 3.1.104. In Theorems 3.1.83, 3.1.86, 3.1.89, and 3.1.94, we proved that ssc-grammars of degrees $(1, 2)$ and $(2, 1)$ generate the family of recursively enumerable languages, and propagating ssc-grammars of degrees $(1, 2)$ and $(2, 1)$ generate the family of context-sensitive languages. However, we discussed no ssc-grammars of degree $(1, 1)$. According to Penttonen (see Theorem 3.1.76), propagating sc-grammars of degree $(1, 1)$ generate a proper subfamily of context-sensitive languages. That is, $\mathbf{SSC}^{-\varepsilon}(1, 1) \subseteq \mathbf{SC}^{-\varepsilon}(1, 1) \subset \mathbf{CS}$. Are propagating ssc-grammars of degree $(1, 1)$ as powerful as propagating sc-grammars of degree $(1, 1)$? Furthermore, consider ssc-grammars of degree $(1, 1)$ with erasing rules. Are they more powerful than propagating ssc-grammars of degree $(1, 1)$? Do they generate the family of all context-sensitive languages or, even more, the family of recursively enumerable languages? $\square$

Open Problem 3.1.105. In Theorems 3.1.97 through 3.1.103, several reduced normal forms of these grammars were presented. These normal forms give rise to

the following questions. Can any of the results be further improved with respect
to the number of conditional rules or nonterminals? Are there analogical reduced
forms of ssc-grammars with degrees $(1, 2)$ and $(1, 3)$? Moreover, reconsider these
results in terms of propagating ssc-grammars. Is it possible to achieve analogical
results if we disallow erasing rules?   □

3.2 Grammars Regulated by States

A *state grammar* G is a context-free grammar extended by an additional state
mechanism that strongly resembles a finite-state control of finite automata. During
every derivation step, G rewrites the leftmost occurrence of a nonterminal that
can be rewritten under the current state; in addition, it moves from a state to
another state, which influences the choice of the rule to be applied in the next
step. If the application of a rule always takes place within the first n occurrences
of nonterminals, G is referred to as *n-limited*.

The present section consists of Sects. 3.2.1 and 3.2.2. The former defines and
illustrates state grammars. The latter describes their generative power.

3.2.1 Definitions and Examples

In this section, we define state grammars and illustrate them by an example.

Definition 3.2.1. A *state grammar* (see [Kas70]) is a quintuple

$$G = (V, W, T, P, S)$$

where

- V is a *total alphabet*;
- W is a finite set of *states*;
- $T \subset V$ is an alphabet of *terminals*;
- $S \in V - T$ is the *start symbol*;
- $P \subseteq (W \times (V - T)) \times (W \times V^+)$ is a finite relation.

Instead of $(q, A, p, v) \in P$, we write $(q, A) \rightarrow (p, v) \in P$. For every $z \in V^*$,
define

$$_G \text{states}(z) = \{q \in W \mid (q, A) \rightarrow (p, v) \in P, A \in \text{alph}(z)\}$$

If $(q, A) \rightarrow (p, v) \in P$, $x, y \in V^*$, and $_G \text{states}(x) = \emptyset$, then G makes a *derivation
step* from (q, xAy) to (p, xvy), symbolically written as

$$(q, xAy) \Rightarrow (p, xvy) \ [(q, A) \rightarrow (p, v)]$$

In addition, if n is a positive integer satisfying that $\#_{V-T}(xA) \leq n$, we say that $(q, xAy) \Rightarrow (p, xvy)$ $[(q, A) \rightarrow (p, v)]$ is *n-limited*, symbolically written as

$$(q, xAy)_n \Rightarrow (p, xvy)\ [(q, A) \rightarrow (p, v)]$$

Whenever there is no danger of confusion, we simplify $(q, xAy) \Rightarrow (p, xvy)$ $[(q, A) \rightarrow (p, v)]$ and $(q, xAy)_n \Rightarrow (p, xvy)\ [(q, A) \rightarrow (p, v)]$ to

$$(q, xAy) \Rightarrow (p, xvy)$$

and

$$(q, xAy)_n \Rightarrow (p, xvy)$$

respectively. In the standard manner, we extend $\Rightarrow$ to $\Rightarrow^m$, where $m \geq 0$; then, based on $\Rightarrow^m$, we define $\Rightarrow^+$ and $\Rightarrow^*$.

Let n be a positive integer, and let $v, \omega \in W \times V^+$. To express that every derivation step in $v \Rightarrow^m \omega$, $v \Rightarrow^+ \omega$, and $v \Rightarrow^* \omega$ is *n*-limited, we write $v_n \Rightarrow^m \omega$, $v_n \Rightarrow^+ \omega$, and $v_n \Rightarrow^* \omega$ instead of $v \Rightarrow^m \omega$, $v \Rightarrow^+ \omega$, and $v \Rightarrow^* \omega$, respectively.

By strings$(v_n \Rightarrow^* \omega)$, we denote the set of all strings occurring in the derivation $v_n \Rightarrow^* \omega$. The *language* of G, denoted by $L(G)$, is defined as

$$L(G) = \left\{ w \in T^* \mid (q, S) \Rightarrow^* (p, w), q, p \in W \right\}$$

Furthermore, for every $n \geq 1$, define

$$L(G, n) = \left\{ w \in T^* \mid (q, S)_n \Rightarrow^* (p, w), q, p \in W \right\}$$

A derivation of the form $(q, S)_n \Rightarrow^* (p, w)$, where $q, p \in W$ and $w \in T^*$, represents a *successful n-limited generation* of w in G. □

Next, we illustrate the previous definition by an example.

Example 3.2.2. Consider the state grammar

$$G = \left(\{S, X, Y, a, b\}, \{p_0, p_1, p_2, p_3, p_4\}, \{a, b\}, P, S\right)$$

with the following nine rules in P

$(p_0, S) \rightarrow (p_0, XY)$	$(p_0, X) \rightarrow (p_3, a)$
$(p_0, X) \rightarrow (p_1, aX)$	$(p_3, Y) \rightarrow (p_0, a)$
$(p_1, Y) \rightarrow (p_0, aY)$	$(p_0, X) \rightarrow (p_4, b)$
$(p_0, X) \rightarrow (p_2, bX)$	$(p_4, Y) \rightarrow (p_0, b)$
$(p_2, Y) \rightarrow (p_0, bY)$	

Observe that G generates the non-context-free language

$$L(G) = \{ww \mid w \in \{a, b\}^+\}$$

Indeed, first, S is rewritten to XY. Then, by using its states, G ensures that whenever X is rewritten to aX, the current state is changed to force the rewrite of Y to aY. Similarly, whenever X is rewritten to bX, the current state is changed to force the rewrite of Y to bY. Every successful derivation is finished by rewriting X to a or b and then Y to a or b, respectively.

For example, *abab* is produced by the following derivation

$$\begin{aligned}
(p_0, S) &\Rightarrow (p_0, XY) & [(p_0, S) \rightarrow (p_0, XY)] \\
&\Rightarrow (p_1, aXY) & [(p_0, X) \rightarrow (p_1, aX)] \\
&\Rightarrow (p_0, aXaY) & [(p_1, Y) \rightarrow (p_0, aY)] \\
&\Rightarrow (p_4, abaY) & [(p_0, X) \rightarrow (p_4, b)] \\
&\Rightarrow (p_0, abab) & [(p_4, Y) \rightarrow (p_0, b)] \qquad\qquad \square
\end{aligned}$$

By **ST**, we denote the family of languages generated by state grammars. For every $n \geq 1$, $\mathbf{ST}_n$ denotes the family of languages generated by n-limited state grammars. Set

$$\mathbf{ST}_\infty = \bigcup_{n \geq 1} \mathbf{ST}_n$$

3.2.2 Generative Power

In this section, which closes the chapter, we give the key result concerning state grammars, originally established in [Kas70].

Theorem 3.2.3. $\mathbf{CF} = \mathbf{ST}_1 \subset \mathbf{ST}_2 \subset \cdots \subset \mathbf{ST}_\infty \subset \mathbf{ST} = \mathbf{CS}$

3.3 Grammars Regulated by Control Languages

In essence, a *grammar with a control language* H is a context-free grammar G extended by a regular *control language* $\varXi$ defined over the set of rules of G. Thus, each control string in $\varXi$ represents, in effect, a sequence of rules in G. A terminal string w is in the language generated by H if and only if $\varXi$ contains a control string according to which G generates w.

The present section is divided into two subsections—Sects. 3.3.1 and 3.3.2. The former defines and illustrates regular-controlled grammars. The latter states their generative power.

3.3.1 Definitions and Examples

In this section, we define the notion of a regular-controlled grammar and illustrate it by examples. Before reading this definition, recall the notion of rule labels, formalized in Definition 2.3.3.

Definition 3.3.1. A *regular-controlled (context-free) grammar* (see [MVMP04]) is a pair

$$H = (G, \Xi)$$

where

- $G = (V, T, \Psi, P, S)$ is a context-free grammar, called *core grammar*;
- $\Xi \subseteq \Psi^*$ is a regular language, called *control language*.

The *language* of H, denoted by $L(H)$, is defined as

$$L(H) = \{w \in T^* \mid S \Rightarrow_G^* w \, [\alpha] \text{ with } \alpha \in \Xi\} \qquad \square$$

In other words, $L(H)$ in the above definition consists of all strings $w \in T^*$ such that there is a derivation in G,

$$S \Rightarrow_G w_1 \, [r_1] \Rightarrow_G w_2 \, [r_2] \Rightarrow_G \cdots \Rightarrow_G w_n \, [r_n]$$

where

$$w = w_n \text{ and } r_1 r_2 \cdots r_n \in \Xi \text{ for some } n \geq 1$$

In what follows, instead of $x \Rightarrow_G y$, we sometimes write $x \Rightarrow_H y$—that is, we use $\Rightarrow_G$ and $\Rightarrow_H$ interchangeably.

Note that if $\Xi = \Psi^*$, then there is no regulation, and thus $L(H) = L(G)$ in this case.

Example 3.3.2. Let $H = (G, \Xi)$ be a regular-controlled grammar, where

$$G = (\{S, A, B, C, a, b, c\}, \{a, b, c\}, \Psi, P, S)$$

is a context-free grammar with P consisting of the following seven rules

$r_1: S \to ABC$	$r_2: A \to aA$	$r_5: A \to \varepsilon$
	$r_3: B \to bB$	$r_6: B \to \varepsilon$
	$r_4: C \to cC$	$r_7: C \to \varepsilon$

and $\Xi = \{r_1\}\{r_2 r_3 r_4\}^* \{r_5 r_6 r_7\}$.

First, r_1 has to be applied. Then, r_2, r_3, and r_4 can be consecutively applied any number of times. The derivation is finished by applying r_5, r_6, and r_7. As a result, this grammar generates the non-context-free language

$$L(H) = \{a^n b^n c^n \mid n \geq 0\}$$

For example, the sentence *aabbcc* is obtained by the following derivation

$$
\begin{aligned}
S &\Rightarrow_H ABC & [r_1] \\
&\Rightarrow_H aABC & [r_2] \\
&\Rightarrow_H aAbBC & [r_3] \\
&\Rightarrow_H aAbBcC & [r_4] \\
&\Rightarrow_H aaAbBcC & [r_2] \\
&\Rightarrow_H aaAbbBcC & [r_3] \\
&\Rightarrow_H aaAbbBccC & [r_4] \\
&\Rightarrow_H aabbBccC & [r_5] \\
&\Rightarrow_H aabbccC & [r_6] \\
&\Rightarrow_H aabbcc & [r_7]
\end{aligned}
$$

As another example, the empty string is derived in this way

$$S \Rightarrow_H ABC\,[r_1] \Rightarrow_H BC\,[r5] \Rightarrow_H C\,[r6] \Rightarrow_H \varepsilon\,[r7] \qquad \square$$

Next, we introduce the concept of appearance checking. Informally, it allows us to skip the application of certain rules if they are not applicable to the current sentential form.

Definition 3.3.3. A *regular-controlled grammar with appearance checking* (see [MVMP04]) is a triple

$$H = (G, \varXi, W)$$

where

- G and $\varXi$ are defined as in a regular-controlled grammar;
- $W \subseteq \Psi$ is the *appearance checking set*.

We say that $x \in V^+$ directly derives $y \in V^*$ in G in the *appearance checking mode* W by application of $r: A \to w \in P$, symbolically written as

$$x \Rightarrow_{(G,W)} y\,[r]$$

if either

$$x = x_1 A x_2 \text{ and } y = x_1 w x_2$$

or

$$A \notin \mathrm{alph}(x), r \in W, \text{ and } x = y$$

Define $\Rightarrow_{(G,W)}^{k}$ for $k \geq 0$, $\Rightarrow_{(G,W)}^{+}$, and $\Rightarrow_{(G,W)}^{*}$ in the standard way. The *language* of H, denoted by $L(H)$, is defined as

$$L(H) = \left\{ w \in T^* \mid S \Rightarrow_{(G,W)}^{*} w \; [\alpha] \text{ with } \alpha \in \Xi \right\} \qquad \square$$

According to Definition 3.3.1, in a regular-controlled grammar without appearance checking, once a control string has been started by G, all its rules have to be applied. G with an appearance checking set somewhat relaxes this necessity, however. Indeed, if the left-hand side of a rule is absent in the sentential form under scan and, simultaneously, this rule is in the appearance checking set, then G skips its application and moves on to the next rule in the control string.

Observe that the only difference between a regular-controlled grammar with and without appearance checking is the derivation mode ($\Rightarrow_{(G,W)}$ instead of $\Rightarrow_G$). Furthermore, note that when $W = \emptyset$, these two modes coincide, so any regular-controlled grammar represents a special case of a regular-controlled grammar with appearance checking.

Example 3.3.4 (From Chapter 3 of [RS97b]). Let $H = (G, \Xi, W)$ be a regular-controlled grammar with appearance checking, where

$$G = \left(\{S, A, X, a\}, \{a\}, \Psi, P, S \right)$$

is a context-free grammar with P consisting of the following rules

$r_1: S \rightarrow AA$ $r_4: A \rightarrow X$
$r_2: S \rightarrow X$ $r_5: S \rightarrow a$
$r_3: A \rightarrow S$

and $\Xi = (\{r_1\}^*\{r_2\}\{r_3\}^*\{r_4\})^*\{r_5\}^*$, $W = \{r_2, r_4\}$.

Assume that we have the sentential form

$$S^{2^m}$$

for some $m \geq 0$, obtained by using a sequence of rules from $(\{r_1\}^*\{r_2\}\{r_3\}^*\{r_4\})^*$. This holds for the start symbol ($m = 0$). We can either repeat this sequence or finish the derivation by using r_5 until we have

$$a^{2^m}$$

In the former case, we might apply r_1 as many times as we wish. However, if we apply it only k many times, where $k < m$, then we have to use r_2, which blocks the derivation. Indeed, there is no rule with X on its left hand side. Thus, this rule guarantees that every S is eventually rewritten to AA. Notice that $r_2 \in W$. As a result, if no S occurs in the sentential form, we can skip it (it is not applicable), so we get

$$S^{2^m} \Rightarrow^*_{(G,W)} (AA)^{2^m} = A^{2^{m+1}}$$

Then, by analogy, we have to rewrite each A to S, so we get

$$A^{2^{m+1}} \Rightarrow^*_{(G,W)} S^{2^{m+1}}$$

which is of the same form as the sentential form from which we started the derivation. Therefore, this grammar generates the non-context-free language

$$L(H) = \left\{ a^{2^n} \mid n \geq 0 \right\}$$

For example, the sentence $aaaa$ is obtained by the following derivation

$$
\begin{aligned}
S &\Rightarrow_{(G,W)} AA & [r_1] \\
&\Rightarrow_{(G,W)} AS & [r_3] \\
&\Rightarrow_{(G,W)} SS & [r_3] \\
&\Rightarrow_{(G,W)} AAS & [r_1] \\
&\Rightarrow_{(G,W)} AAAA & [r_1] \\
&\Rightarrow_{(G,W)} AASA & [r_3] \\
&\Rightarrow_{(G,W)} AASS & [r_3] \\
&\Rightarrow_{(G,W)} SASS & [r_3] \\
&\Rightarrow_{(G,W)} SSSS & [r_3] \\
&\Rightarrow_{(G,W)} SSSa & [r_5] \\
&\Rightarrow_{(G,W)} aSSa & [r_5] \\
&\Rightarrow_{(G,W)} aaSa & [r_5] \\
&\Rightarrow_{(G,W)} aaaa & [r_5]
\end{aligned}
$$

As another example, a single a is generated by

$$S \Rightarrow_{(G,W)} a \; [r5] \qquad\qquad\qquad\qquad \Box$$

We can disallow erasing rules in the underlying core grammar. This is formalized in the following definition.

Definition 3.3.5. Let $H = (G, \Xi)$ ($H = (G, \Xi, W)$) be a regular-controlled grammar (with appearance checking). If G is propagating, then H is said to be a *propagating regular-controlled grammar (with appearance checking)*. $\qquad \Box$

By $\mathbf{rC}_{ac}$, $\mathbf{rC}_{ac}^{-\varepsilon}$, $\mathbf{rC}$, and $\mathbf{rC}^{-\varepsilon}$, we denote the families of languages generated by regular-controlled grammars with appearance checking, propagating regular-controlled grammars with appearance checking, regular-controlled grammars, and propagating regular-controlled grammars, respectively.

3.3.2 Generative Power

The present section concerns the generative power of regular-controlled grammars. More specifically, the next theorem summarizes the relations between the language families defined in the conclusion of the previous section.

Theorem 3.3.6 (See Theorem 1 in [MVMP04]).

(i) All languages in $\mathbf{rC}$ *over a unary alphabet are regular.*

(ii) $\mathbf{CF} \subset \mathbf{rC}^{-\varepsilon} \subset \mathbf{rC}_{ac}^{-\varepsilon} \subset \mathbf{CS}$

(iii) $\mathbf{CF} \subset \mathbf{rC}^{-\varepsilon} \subseteq \mathbf{rC} \subset \mathbf{rC}_{ac} = \mathbf{RE}$

Open Problem 3.3.7. Is $\mathbf{rC} - \mathbf{rC}^{-\varepsilon}$ empty? Put in other words, can any regular-controlled grammar be converted to an equivalent propagating regular-controlled grammar?   □

3.4 Matrix Grammars

As already pointed out in the beginning of this chapter, in essence, any matrix grammar can be viewed as a special regular-controlled grammar with a control language that has the form of the iteration of a finite language. More precisely, a *matrix grammar H* is a context-free grammar *G* extended by a finite set of sequences of its rules, referred to as *matrices*. In essence, *H* makes a derivation so it selects a matrix, and after this selection, it applies all its rules one by one until it reaches the very last rule. Then, it either completes its derivation, or it makes another selection of a matrix and continues the derivation in the same way.

This section is divided into three subsections. Section 3.4.1 defines and illustrates matrix grammars. Section 3.4.2 determines their power. Finally, Sect. 3.4.3 studies even matrix grammars as special cases of matrix grammars, which work in parallel. Consequently, in a very natural way, Sect. 3.4.3 actually introduces the central topic of the next chapter grammatical parallelism (see Chap. 4).

3.4.1 Definitions and Examples

We open this section by giving the rigorous definition of matrix grammars. Then, we illustrate this definition by an example.

Definition 3.4.1. A *matrix grammar with appearance checking* (**M** for short; see [DP89]) is a triple

$$H = (G, M, W)$$

where

- $G = (V, T, \Psi, P, S)$ is a context-free grammar, called *core grammar*;
- $M \subseteq \Psi^+$ is a finite language whose elements are called *matrices*;
- $W \subseteq \Psi$ is the *appearance checking set*.

The *direct derivation relation*, symbolically denoted by $\Rightarrow_H$, is defined over V^* as follows: for $r_1 r_2 \cdots r_n \in M$, for some $n \geq 1$, and $x, y \in V^*$,

$$x \Rightarrow_H y$$

if and only if

$$x = x_0 \Rightarrow_{(G,W)} x_1 \, [r_1] \Rightarrow_{(G,W)} x_2 \, [r_2] \Rightarrow_{(G,W)} \cdots \Rightarrow_{(G,W)} x_n = y \, [r_n]$$

where $x_i \in V^*$, for all i, $1 \leq i \leq n-1$, and the application of rules in the appearance checking mode is defined as in Definition 3.3.3.

Define $\Rightarrow_H^k$ for $k \geq 0$, $\Rightarrow_H^+$, and $\Rightarrow_H^*$ in the standard way. The *language* of H, denoted by $L(H)$, is defined as

$$L(H) = \{ w \in T^* \mid S \Rightarrow_H^* w \} \qquad \square$$

Note that if $M = \Psi$, then there is no regulation, and thus $L(H) = L(G)$ in this case. Sometimes, for brevity, we use rules and rule labels interchangeably.

Definition 3.4.2. Let $H = (G, M, W)$ be a matrix grammar with appearance checking. If $W = \emptyset$, then we say that H is a *matrix grammar without appearance checking* or, simply, a *matrix grammar* and we just write $H = (G, M)$. $\qquad \square$

Without appearance checking, once a matrix has been started, H has to apply all its rules. However, with an appearance checking set of the rules in the matrices, H may sometimes skip the application of a rule within a matrix. More precisely, if the left-hand side of a rule is absent in the current sentential form while the corresponding rule of the applied matrix occurs in the appearance checking set, then H moves on to the next rule in the matrix.

Example 3.4.3 (From [MVMP04]). Let $H = (G, M)$ be a matrix grammar, where $G = (V, T, \Psi, P, S)$ is a context-free grammar with $V = \{S, A, B, a, b\}$, $T = \{a, b\}$, P consists of the following rules

$$r_1: S \to AB \qquad\qquad r_4: A \to bA \qquad\qquad r_7: B \to a$$
$$r_2: A \to aA \qquad\qquad r_5: B \to bB \qquad\qquad r_8: A \to b$$
$$r_3: B \to aB \qquad\qquad r_6: A \to a \qquad\qquad r_9: B \to b$$

and $M = \{r_1, r_2 r_3, r_4 r_5, r_6 r_7, r_8 r_9\}$.

We start with the only applicable matrix r_1 and we get AB. Next, we can either

• terminate the derivation by using the matrix $r_6 r_7$ and obtain aa,
• terminate the derivation by using the matrix $r_8 r_9$ and obtain bb,
• rewrite AB to $aAaB$ by using the matrix $r_2 r_3$, or
• rewrite AB to $bAbB$ by using the matrix $r_4 r_5$.

If the derivation is not terminated, we can continue analogously. For example, the sentence $aabaab$ is obtained by the following derivation

$$S \Rightarrow_H AB$$
$$\Rightarrow_H aAaB$$
$$\Rightarrow_H aaAaaB$$
$$\Rightarrow_H aabaab$$

Clearly, this grammar generates the non-context-free language

$$L(H) = \{ww \mid w \in \{a, b\}^+\} \qquad\qquad \square$$

As with regular-controlled grammars, we can disallow erasing rules in the underlying core grammar.

Definition 3.4.4. Let $H = (G, M, W)$ be a matrix grammar (with appearance checking). If G is propagating, then H is a *propagating matrix grammar (with appearance checking)*. $\square$

The families of languages generated by matrix grammars with appearance checking, propagating matrix grammars with appearance checking, matrix grammars, and propagating matrix grammars are denoted by $\mathbf{M}_{ac}$, $\mathbf{M}_{ac}^{-\varepsilon}$, $\mathbf{M}$, and $\mathbf{M}^{-\varepsilon}$, respectively.

3.4.2 Generative Power

This section states the relations between the language families defined in the conclusion of the previous section.

Theorem 3.4.5 (See Theorem 2 in [MVMP04]).

(i) $\mathbf{M}_{ac} = \mathbf{rC}_{ac}$
(ii) $\mathbf{M}_{ac}^{-\varepsilon} = \mathbf{rC}_{ac}^{-\varepsilon}$
(iii) $\mathbf{M} = \mathbf{rC}$
(iv) $\mathbf{M}^{-\varepsilon} = \mathbf{rC}^{-\varepsilon}$

Notice that the relations between the language families generated by matrix grammars are analogical to the relations between language families generated by regular-controlled grammars (see Theorem 3.3.6).

3.4.3   Even Matrix Grammars

In essence, even matrix grammars can be seen as sequences of context-free grammars, referred to as their components, which work in parallel. More precisely, for a positive integer n, an n-even matrix grammar is an ordered sequence of n context-free grammars with pair-wise disjoint nonterminal alphabets and a shared terminal alphabet, which rules are fixed n-tuples containing one rule of each component—that is, in every derivation step, each of these components rewrites a nonterminal occurring in its current sentential form. A sentential form of an n-even matrix grammar is a concatenation of sentential forms of all of its components from the first to the nth. A derivation is successful if and only if all components generate a terminal string at once. Of course, one-component even matrix grammars are nothing but context-free grammars, so they characterize the family of context-free languages. Surprisingly, two-component even matrix grammars are significantly stronger as the present section demonstrates; indeed, they are as powerful as ordinary matrix grammars (see Sect. 3.4.1). This section also points out that even matrix grammars with more than two components are equivalent with two-component even matrix grammars. Then, it places and studies the following three leftmost derivation restrictions on even matrix grammars. The first restriction requires that every rule is applied so it rewrites the leftmost possible nonterminal in each component. The second restriction parallels the first restriction; in addition, however, in each component, it makes the selection of the rule so the leftmost possible nonterminal is rewritten in the current sentential form. Finally, the third restriction requires that the leftmost nonterminal of each component is rewritten. As the section demonstrates, working under the second and third restriction, even matrix grammars are computational complete—that is, they are equivalent with Turing machines. The section has not precisely determined the generative power of even matrix grammars working under the first restriction, so this determination represents an open problem.

Definitions and Examples

In this section, we define even matrix grammars. Furthermore, we introduce their leftmost variants and illustrate them by an example.

Definition 3.4.6. Let $n \geq 1$. An *even matrix grammar of degree n* ($_nEMG$ for short) is an $(n + 3)$-tuple, $G_n = (N_1, N_2, \ldots, N_n, T, P, S)$, where

(1) $N_1, N_2, \ldots, N_n$ are pairwise disjoint nonterminal alphabets;

(2) T is a terminal alphabet, $T \cap N_i = \emptyset$, for $1 \leq i \leq n$;

(3) S is the *start symbol* such that $S \notin N_1 \cup \cdots \cup N_n \cup T$;

(4) P is a finite set of rewriting rules of the form:

(4.a) $(S) \to (v)$, $v \in T^*$.

(4.b) $(S) \to (v_1 v_2 \ldots v_n)$, $v_i \in (N_i \cup T)^*$, alph$(v_i) \cap N_i \neq \emptyset$, for $1 \leq i \leq n$.

(4.c) $(A_1, A_2, \ldots, A_n) \to (v_1, v_2, \ldots, v_n)$, $A_i \in N_i$, $v_i \in (N_i \cup T)^*$, for $1 \leq i \leq n$. □

Definition 3.4.7. Let $G_n = (N_1, \ldots, N_n, T, P, S)$ be an $_nEMG$, for some $n \geq 1$. Consider any string $u_1 A_1 w_1 \ldots u_n A_n w_n$, where $u_i w_i \in (N_i \cup T)^*$, $A_i \in N_i$, and some rule $(A_1, \ldots, A_n) \to (v_1, \ldots, v_n)$, where $v_i \in (N_i \cup T)^*$, for $1 \leq i \leq n$. Then, G_n makes a *derivation step*

$$u_1 A_1 w_1 \ldots u_n A_n w_n \Rightarrow u_1 v_1 w_1 \ldots u_n v_n w_n$$

Based on additional restrictions, we define the following three *modes of leftmost derivations*:

(1) If $A_i \notin$ alph(u_i), for $1 \leq i \leq n$, then write

$$u_1 A_1 w_1 \ldots u_n A_n w_n \, {}_1{\Rightarrow} u_1 v_1 w_1 \ldots u_n v_n w_n$$

(2) If

$$u_1 A_1 w_1 \ldots u_n A_n w_n = u_1' B_1 w_1' u_2' B_2 w_2' \ldots u_n' B_n w_n'$$

where $u_i', w_i' \in (N_i \cup T)^*$, $B_i \in N_i$, for some $1 \leq j \leq n$: $|u_i'| = |u_i|$, $i < j$, $|u_j'| < |u_j|$, for $1 \leq i \leq n$, and in P, there is no applicable rule

$$(B_1, B_2, \ldots, B_n) \to (x_1, x_2, \ldots, x_n)$$

then write

$$u_1 A_1 w_1 \ldots u_n A_n w_n \, {}_2{\Rightarrow} u_1 v_1 w_1 \ldots u_n v_n w_n$$

(3) If $N_i \cap$ alph$(u_i) = \emptyset$, for $1 \leq i \leq n$, then write

$$u_1 A_1 w_1 \ldots u_n A_n w_n \, {}_3{\Rightarrow} u_1 v_1 w_1 \ldots u_n v_n w_n$$

In what follows, we write $_0{\Rightarrow}$ instead of $\Rightarrow$. We say that $_i{\Rightarrow}$ represents the *direct derivation of mode i*, for $i = 0, 1, 2, 3$. □

For the clarity, let us informally describe the defined modes of leftmost derivations. In the first mode, any applicable rule is chosen and the leftmost possible nonterminals are rewritten. In the second mode, there is a specific rule chosen, which can rewrite nonterminals as to the left as possible—there exists no rule, which could be applied more leftmost—, while the lower components are more prior. In the third mode, always the leftmost nonterminal of each component must be rewritten.

Definition 3.4.8. Let $G_n = (N_1, N_2, \ldots, N_n, T, P, S)$ be an $_nEMG$, and let ϱ be any relation over $(N_1 \cup T)^* (N_2 \cup T)^* \ldots (N_n \cup T)^*$. Set

$$\mathscr{L}(G_n, \varrho) = \{x \mid x \in T^*, S \varrho^* x\}$$

$\mathscr{L}(G_n, \varrho)$ is said to be the *language that G_n generates by ϱ*.

$$_n\mathbf{EM}_\varrho = \{\mathscr{L}(G_n, \varrho) \mid G_n \text{ is a } _nEMG\}$$

is said to be the *family of languages that $_nEMGs$ generate by ϱ*. Let $\mathbf{EM}_\varrho$ denotes the *family of languages generated by all even matrix grammars by ϱ*.   □

We illustrate the previous definitions by the next example.

Example 3.4.9. Let $G = (N_1, N_2, T, P, S)$, where $N_1 = \{A, \overline{A}\}$, $N_2 = \{B, \overline{B}, C, \overline{C}\}$, $T = \{a, b, c, d\}$, be a $_2EMG$ with P containing the following rules:

(1) $(S) \to (AABC)$
(2) $(A, B) \to (aAb, cBd)$ (6) $(A, \overline{B}) \to (aAb, c\overline{B}d)$
(3) $(A, C) \to (aAb, cCd)$ (7) $(A, \overline{C}) \to (aAb, c\overline{C}d)$
(4) $(A, B) \to (\overline{A}, \overline{B})$ (8) $(A, \overline{B}) \to (\overline{A}, \varepsilon)$
(5) $(\overline{A}, C) \to (\varepsilon, \overline{C})$ (9) $(\overline{A}, \overline{C}) \to (\varepsilon, \varepsilon)$

Next, we illustrate $_i\Rightarrow$, $0 \le i \le 3$, in terms of G.

0. Using derivations of mode 0, after applying the starting rule (1), G uses (2) and/or (3). Then, the rule (4) is applied, however, the rule (3) is still applicable, until the rule (5) is used. Next, the derivation proceeds by the rules (6) and/or (7) and eventually finishes with the rules (8) and (9). The derivation may proceed as follows.

$$S_1 \Rightarrow AABC$$
$$_1\Rightarrow^* a^k Ab^k a^l Ab^l c^m Bd^m c^n Cd^n$$
$$_1\Rightarrow a^k Ab^k a^l \overline{A} b^l c^m \overline{B} d^m c^n Cd^n$$
$$_1\Rightarrow^* a^{k+i+j} Ab^{k+i+j} a^l Ab^l c^{m+i} \overline{B} d^{m+i} c^{n+j} Cd^{n+j}$$
$$_1\Rightarrow a^{k+i+j} Ab^{k+i+j} a^l b^l c^{m+i} \overline{B} d^{m+i} c^{n+j} \overline{C} d^{n+j}$$
$$_1\Rightarrow^* a^{k+i+j+o+p} Ab^{k+i+j+o+p} a^l b^l c^{m+i+o} \overline{B} d^{m+i+o} c^{n+j+p} \overline{C} d^{n+j+p}$$
$$_1\Rightarrow a^{k+i+j+o+p} \overline{A} b^{k+i+j+o+p} a^l b^l c^{m+i+o} d^{m+i+o} c^{n+j+p} \overline{C} d^{n+j+p}$$
$$_1\Rightarrow a^{k+i+j+o+p} b^{k+i+j+o+p} a^l b^l c^{m+i+o} d^{m+i+o} c^{n+j+p} d^{n+j+p}$$

for $i, j, k, l, m, n, o, p \geq 0$. Consequently,

$$\mathscr{L}(G,_1\Rightarrow) = \{a^k b^k a^l b^l c^m d^m c^n d^n \mid k+l = m+n, \text{ for } k,l,m,n \geq 0\}$$

which is a non-context-free language.

1. Using mode 1 leftmost derivations, after applying the starting rule (1), G continues using rules (2) and/or (3), until the rule (4) is applied. The rule (3) is still applicable, until the rule (5) is used. Next, the derivation proceeds by the rules (6) and/or (7) and eventually finishes with the rules (8) and (9):

$$S_1 \Rightarrow AABC$$
$$_1\Rightarrow^* a^{i+j} Ab^{i+j} Ac^i Bd^i c^j Cd^j$$
$$_1\Rightarrow a^{i+j} \overline{A} b^{i+j} Ac^i \overline{B} d^i c^j Cd^j$$
$$_1\Rightarrow^* a^{i+j} \overline{A} b^{i+j} a^k Ab^k c^i \overline{B} d^i c^{j+k} Cd^{j+k}$$
$$_1\Rightarrow a^{i+j} b^{i+j} a^k Ab^k c^i \overline{B} d^i c^{j+k} \overline{C} d^{j+k}$$
$$_1\Rightarrow^* a^{i+j} b^{i+j} a^{k+o+p} Ab^{k+o+p} c^{i+o} \overline{B} d^{i+o} c^{j+k+p} \overline{C} d^{j+k+p}$$
$$_1\Rightarrow a^{i+j} b^{i+j} a^{k+o+p} \overline{A} b^{k+o+p} c^{i+o} d^{i+o} c^{j+k+p} \overline{C} d^{j+k+p}$$
$$_1\Rightarrow a^{i+j} b^{i+j} a^{k+o+p} b^{k+o+p} c^{i+o} d^{i+o} c^{j+k+p} d^{j+k+p}$$

for $i, j, k, o, p \geq 0$. Consequently,

$$\mathscr{L}(G,_1\Rightarrow) = \{a^k b^k a^l b^l c^m d^m c^n d^n \mid k+l = m+n, \text{ for } k,l,m,n \geq 0\}$$

2. With derivations of leftmost mode 2, the situation is different. First, the starting rule (1) is applied, however, then only the rule (2) is applicable, until the rule (4) is used. Next, the rule (5) must be applied. The derivation continues with the applications of the rule (6), since the rule (7) is not applicable, until the rules (8) and (9) are used:

$$S_2 \Rightarrow AABC$$
$$_2\Rightarrow^* a^i Ab^i Ac^i Bd^i C$$
$$_2\Rightarrow a^i \overline{A} b^i Ac^i \overline{B} d^i C$$
$$_2\Rightarrow a^i b^i Ac^i \overline{B} d^i \overline{C}$$
$$_2\Rightarrow^* a^i b^i a^j Ab^j c^{i+j} \overline{B} d^{i+j} \overline{C}$$
$$_2\Rightarrow a^i b^i a^j \overline{A} b^j c^{i+j} d^{i+j} \overline{C}$$
$$_2\Rightarrow a^i b^i a^j b^j c^{i+j} d^{i+j}$$

for $i, j \geq 0$. Consequently, $\mathscr{L}(G,_2\Rightarrow) = \{a^i b^i a^j b^j c^{i+j} d^{i+j} \mid \text{ for } i,j \geq 0\}$.

3. The derivation performed by the leftmost mode 3 derivations starts with the rule (1), continues with applying the rule (2) and eventually uses the rule (4). However, then, the derivation is blocked, because there is no rule rewriting $\overline{A}$ and $\overline{B}$:

$$S \, {}_3{\Rightarrow} AABC \, {}_3{\Rightarrow}^* a^i A b^i A c^i B d^i C \, {}_3{\Rightarrow} a^i \overline{A} b^i A c^i \overline{B} d^i C \, {}_3{\Rightarrow} \emptyset$$

Consequently, $\mathscr{L}(G, {}_3{\Rightarrow}) = \emptyset$. $\square$

Generative Power

In this section, we investigate the generative power of even matrix grammars working under the derivation modes introduced in the previous section. We pay a special attention to the number of their components.

Since any context-free grammar is $_1EMG$, we obtain the next theorem.

Theorem 3.4.10. $_1\mathbf{EM}_{i\Rightarrow} = \mathbf{CF}$, *for* $i = 0, 1, 2, 3$. $\square$

Mode 0

From Theorem 3.4.10, every one-component even matrix grammar generates a context-free language. From Example 3.4.9, two-component even matrix grammars are more powerful than one-component ones. Next we prove that additional components do not increase the power of even matrix grammars.

Theorem 3.4.11. $_n\mathbf{EM}_{0\Rightarrow} = {}_2\mathbf{EM}_{0\Rightarrow}$, *for* $n \geq 2$.

Proof. Construction. Let $G = (N_1, N_2, \ldots, N_n, T, P, S)$ be any $_nEMG$, for some $n \geq 2$. Suppose, the rules in P are in the form

$$r : (A_1, A_2, \ldots, A_n) \rightarrow (w_1, w_2, \ldots, w_n)$$

where $1 \leq r \leq \text{card}(P)$ is the unique numeric label. Set $N = N_1 \cup N_2 \cup \cdots \cup N_n$. Define $_2EMG$ $G' = (N, N', T, P', S)$, where $N' = \{Q_j^i \mid 1 \leq i \leq \text{card}(P), 1 \leq j \leq n\}$. Construct P' as follows. Initially, set $P' = \emptyset$. Perform (1) through (3), given next:

(1) for each $r : (S) \rightarrow (w) \in P$, $w \in T^*$, add $(S) \rightarrow (w)$ to P';
(2) for each $r : (S) \rightarrow (w) \in P$, $\text{alph}(w) - T \neq \emptyset$,
    add $(S) \rightarrow (wQ_1^t)$, where $t \in \{1, 2, \ldots, \text{card}(P)\}$, to P';
(3) for each $r : (A_1, A_2, \ldots, A_n) \rightarrow (w_1, w_2, \ldots, w_n) \in P$, add

   (3.a) $(A_i, Q_i^r) \rightarrow (w_i, Q_{i+1}^r)$, for $1 \leq i < n$,
   (3.b) $(A_n, Q_n^r) \rightarrow (w_n, Q_1^t)$, for $t \in \{1, 2, \ldots, \text{card}(P)\}$,
   (3.c) $(A_n, Q_n^r) \rightarrow (w_n, \varepsilon)$, to P'.

Claim 3.4.12. $\mathscr{L}(G, {_0}\Rightarrow) = \mathscr{L}(G', {_0}\Rightarrow)$.

Proof. We establish the proof by proving the following two claims.

Claim 3.4.13. $\mathscr{L}(G, {_0}\Rightarrow) \subseteq \mathscr{L}(G', {_0}\Rightarrow)$.

Proof. To prove the claim, we show that for any sequence of derivation steps of G, generating the sentential form w, there is the corresponding sequence of derivation steps of G' generating the corresponding sentential form or the same terminal string, if it is successful, by induction on m—the number of the derivation steps of G.

Basis. Let $m = 0$. The correspondence of the sentential forms of G and G' is trivial. Let $m = 1$. Then, some starting rule from P of the form $(S) \rightarrow (v)$ is applied. However, in P', there is the corresponding rule $(S) \rightarrow (v)$, in the case the derivation is finished, or $(S) \rightarrow (vQ_1^r)$, for $1 \leq r \leq \text{card}(P)$, otherwise. Without any loss of generality, suppose r is the label of the next rule applied by G. The basis holds.

Induction Hypothesis. Suppose that there exists $k \geq 1$ such that the assumption of correspondence holds for all sequences of the derivation steps of G of the length m, where $0 \leq m \leq k$.

Induction Step. Consider any sequence of moves

$$S_0 \Rightarrow^{k+1} w$$

Since $k + 1 \geq 1$, this sequence can be expressed as

$$S_0 \Rightarrow^k v_1 A_1 u_1 v_2 A_2 u_2 \ldots v_n A_n u_n {_0} \Rightarrow v_1 x_1 u_1 v_2 x_2 u_2 \ldots v_n x_n u_n$$

where $A_i \in N_i$, $v_i, u_i \in (N_i \cup T)^*$ and the last derivation step is performed by some rule

$$r : (A_1, A_2, \ldots, A_n) \rightarrow (x_1, x_2, \ldots, x_n) \in P$$

Then, there exits a sequence of derivation steps of G'

$$S_0 \Rightarrow^* v_1 A_1 u_1 v_2 A_2 u_2 \ldots v_n A_n u_n Q_1^r$$

From the construction of G', there exist rules

$$(A_1, Q_1^r) \rightarrow (x_1, Q_2^r), (A_2, Q_2^r) \rightarrow (x_2, Q_3^r), \ldots, (A_{n-1}, Q_{n-1}^r) \rightarrow (x_{n-1}, Q_n^r),$$
$$(A_n, Q_n^r) \rightarrow (x_n, Q_1^t), (A_n, Q_n^r) \rightarrow (x_n, \varepsilon)$$

for $1 \leq t \leq \text{card}(P)$. Therefore, in G', there is the sequence of derivation steps

$$
\begin{aligned}
S_0 &\Rightarrow^* & v_1 A_1 u_1 v_2 A_2 u_2 \ldots v_n A_n u_n Q_1^r \\
{_0}&\Rightarrow & v_1 x_1 u_1 v_2 A_2 u_2 \ldots v_n A_n u_n Q_2^r \\
{_0}&\Rightarrow & v_1 x_1 u_1 v_2 x_2 u_2 \ldots v_n A_n u_n Q_3^r \\
{_0}&\Rightarrow^{n-3} & v_1 x_1 u_1 v_2 x_2 u_2 \ldots v_{n-1} x_{n-1} u_{n-1} v_n A_n u_n Q_n^r
\end{aligned}
$$

Next, there are two possible situations. First, suppose the derivation of G is not finished and the rule labeled by some t is applied next. Then,

$$v_1 x_1 u_1 v_2 x_2 u_2 \ldots v_n A_n u_n Q_n^r \ {}_0\!\!\Rightarrow v_1 x_1 u_1 v_2 x_2 u_2 \ldots v_n x_n u_n Q_1^t$$

in G', generating the sentential form corresponding to the new sentential form of G, where $v_1 x_1 u_1 v_2 x_2 u_2 \ldots v_n x_n u_n Q_1^t = w Q_1^t$. Second, suppose $w \in T^*$. By the last of the highlighted rules of G'

$$v_1 x_1 u_1 v_2 x_2 u_2 \ldots v_n A_n u_n Q_n^r \ {}_0\!\!\Rightarrow v_1 x_1 u_1 v_2 x_2 u_2 \ldots v_n x_n u_n$$

where $v_1 x_1 u_1 v_2 x_2 u_2 \ldots v_n x_n u_n = w$ and Claim 3.4.13 holds.   □

Claim 3.4.14. $\mathcal{L}(G, {}_0\!\!\Rightarrow) \supseteq \mathcal{L}(G', {}_0\!\!\Rightarrow)$.

Proof. To prove the claim, we show that for any sequence of derivation steps of G', generating the sentential form w, there is the corresponding sequence of derivation steps of G generating the corresponding sentential form or the same terminal string, if it is successful, by induction on m—the number of the derivation steps of G'. First, assume that in G' the symbols Q_y^x of the second component serve as states of computation. With the application of the starting rule, some symbol Q_1^r is inserted. It means, G' is about to simulate the application of the rule r of G. The only applicable rule is then the rule rewriting Q_1^r to Q_2^r. And it holds equally for any Q_y^x, where $y < n$. Therefore, except the starting rules, these rules have always to be applied in sequence. We use this assumption in the next proof.

Basis. Let $m = 0$. The sentential forms of G and G' correspond. Let $m = 1$. Then, some starting rule from P' of the form $(S) \to (v)$ or $(S) \to (v Q_1^r)$, for some $r \in \{1, 2, \ldots, \mathrm{card}(P)\}$ is applied. However, this rule is introduced by the rule $(S) \to (v)$ of P. Therefore, G can make the derivation step corresponding to the one made by G' and the basis holds.

Induction Hypothesis. Suppose that there exists $k \geq 1$ such that the assumption of correspondence holds for all sequences of the derivation steps of G' of the length m, where $0 \leq m \leq k$.

Induction Step. Recall the previous assumption. Let us consider any sequence of moves

$$S \ {}_0\!\!\Rightarrow^{k+n} w$$

This sequence can be written in the form

$$
\begin{aligned}
S \ {}_0\!\!&\Rightarrow^* & v_1 A_1 u_1 v_2 A_2 u_2 \ldots v_n A_n u_n Q_1^r \\
{}_0\!\!&\Rightarrow & v_1 x_1 u_1 v_2 A_2 u_2 \ldots v_n A_n u_n Q_2^r \\
{}_0\!\!&\Rightarrow & v_1 x_1 u_1 v_2 x_2 u_2 \ldots v_n A_n u_n Q_3^r \\
{}_0\!\!&\Rightarrow^{n-2} & v_1 x_1 u_1 v_2 x_2 u_2 \ldots v_n x_n u_n X
\end{aligned}
$$

where $A_i \in N_i$, $v_i, u_i \in (N_i \cup T)^*$ and $X = Q_1^t$, for $1 \le t \le \text{card}(P)$, or $X = \varepsilon$. There are n rules from P' used

$$(A_1, Q_1^r) \to (x_1, Q_2^r), (A_2, Q_2^r) \to (x_2, Q_3^r), \ldots, (A_n, Q_n^r) \to (x_n, X)$$

These rules are introduced by the rule of G

$$r : (A_1, A_2, \ldots, A_n) \to (x_1, x_2, \ldots, x_n) \in P$$

By the induction hypothesis, in the terms of G

$$S_0 \Rightarrow^* v_1 A_1 u_1 v_2 A_2 u_2 \ldots v_n A_n u_n$$

Thus, using the rule r

$$v_1 A_1 u_1 v_2 A_2 u_2 \ldots v_n A_n u_n \,_0\Rightarrow v_1 x_1 u_1 v_2 x_2 u_2 \ldots v_n x_n u_n = w$$

If $X = \varepsilon$, $w \in T^*$ and G' generates the terminal string, G generates the same terminal string and Claim 3.4.14 holds. $\qquad\square$

We have proved $\mathscr{L}(G, _0\Rightarrow) \subseteq \mathscr{L}(G', _0\Rightarrow)$ and $\mathscr{L}(G, _0\Rightarrow) \supseteq \mathscr{L}(G', _0\Rightarrow)$, therefore $\mathscr{L}(G, _0\Rightarrow) = \mathscr{L}(G', _0\Rightarrow)$ and Claim 3.4.12 holds. $\qquad\square$

Since G is any $_nEMG$, for some $n \ge 2$, and G' is $_2EMG$, Theorem 3.4.11 holds. $\qquad\square$

We have proved

$$\mathbf{CF} = {}_1\mathbf{EM}_{0\Rightarrow} \subset {}_2\mathbf{EM}_{0\Rightarrow} = \mathbf{EM}_{0\Rightarrow}$$

however, this classification of even matrix languages is not very precise. Let us recall (see Theorem 3.3.6 and 3.4.5) that

$$\mathbf{CF} \subset \mathbf{M} \subset \mathbf{CS}$$

By the following theorem we prove that even matrix grammars characterize precisely the family of matrix languages.

Theorem 3.4.15. $\mathbf{EM}_{0\Rightarrow} = \mathbf{M}$.

Proof. Without any loss of generality, we consider only $_2EMGs$.

Claim 3.4.16. $_2\mathbf{EM}_{0\Rightarrow} \subseteq \mathbf{M}$.

Proof. Construction. Let $G = (N_1, N_2, T, P, S)$ be any $_2EMG$. Suppose that the rules in P are of the form

$$r : (A_1, A_2) \to (w_1, w_2)$$

where $1 \leq r \leq \text{card}(P)$ is a unique numerical label. Define matrix grammar $H = (G', M)$, where $G' = (N_1 \cup N_2 \cup \{S\} \cup T, T, P', S)$. Construct P' and M as follows. Initially, set $P' = M = \emptyset$. Perform (1) and (2), given next:

(1) for each $r : (S) \to (w) \in P$,   (2) for each $r : (A_1, A_2) \to (w_1, w_2) \in P$,

     (1.a) add $r : S \to w$ to P',   (2.a) add $r_1 : A_1 \to w_1$ and

     (1.b) add r to M;   (2.b) add $r_2 : A_2 \to w_2$ to P',

                                              (2.c) add $r_1 r_2$ to M.

Claim 3.4.17. $\mathscr{L}(G, {}_0\Rightarrow) = \mathscr{L}(H)$.

Proof. Basic Idea. Every ${}_2EMG$ first applies some starting rule. Next, in every derivation step simultaneously in each component there is a nonterminal rewritten to some string. This process can be simulated by matrix grammar as follows. For every starting rule of G, there is the same applicable rule of H. They both generate the equal terminal strings or a sentential forms $w_1 w_2$, where $\text{alph}(w_1) \cap N_2 = \emptyset$ and $\text{alph}(w_2) \cap N_1 = \emptyset$. Then, for every rule $(A_1, A_2) \to (x_1, x_2)$ in G, there are two rules $A_1 \to x_1$ and $A_2 \to x_2$ in H, which must be applied consecutively. Since $N_1 \cap N_2 = \emptyset$, every sentential form remains of the form $w_1 w_2$, where $\text{alph}(w_1) \cap N_2 = \emptyset$ and $\text{alph}(w_2) \cap N_1 = \emptyset$. Thus, H simulates both separated components of G. A detailed version of the proof is left to the reader.

Since for any ${}_2EMG$ we can construct a matrix grammar generating the same language, Claim 3.4.16 holds.   □

Claim 3.4.18. ${}_2\mathbf{EM}_{0\Rightarrow} \supseteq \mathbf{M}$.

Proof. Construction. Let $H = (G, M)$, where $G = (V, T, P, S)$, $V - T = N$, $S \in N$, be any matrix grammar. Without any loss of generality, suppose every $m \in M$ has its unique label. Define ${}_2EMG$ $G' = (N, N', T, P', S')$, where $S' \notin N$,

$$N' = \{Q_i^r \mid r : p_1 p_2 \ldots p_k \in M, i \in \{1, 2, \ldots, k\}, k \geq 1\}$$

Construct P' as follows. Initially set $P' = \emptyset$. Perform (1) given next:

(1) for each $r : p_1 p_2 \ldots p_k \in M$, $k \geq 1$, where $p_i : A_i \to w_i$, $i \in \{1, 2, \ldots, k\}$,

     (1.a) add $(S') \to (S Q_1^r)$,

     (1.b) $(A_i, Q_i^r) \to (w_i, Q_{i+1}^r)$, for $i < k$,

     (1.c) $(A_k, Q_k^r) \to (w_k, Q_1^t)$, for some $t : v \in M$, and

     (1.d) $(A_k, Q_k^r) \to (w_k, \varepsilon)$ to P'.

Claim 3.4.19. $\mathscr{L}(G', {}_0\Rightarrow) = \mathscr{L}(H)$.

Proof. We establish Claim 3.4.19 by proving the following two claims.

Claim 3.4.20. $\mathscr{L}(G', {}_0\Rightarrow) \supseteq \mathscr{L}(H)$.

Proof. We prove the claim by the induction on the number of derivation steps m.

Basis. Let $m = 0$. By the application of the rule $(S') \rightarrow (SQ_1^r)$, G' generates the sentential form corresponding to the starting symbol of H. Without any loss of generality, suppose r is the label of the next matrix applied by H. The basis holds.

Induction Hypothesis. Suppose that there exists $k \geq 0$ such that Claim 3.4.20 holds for all sequences of the derivation steps of the length m, where $0 \leq m \leq k$.

Induction Step. Consider any sequence of moves

$$S \Rightarrow^{k+1} w'$$

where $w' \in (N \cup T)^*$. Since $k + 1 \geq 1$,

$$S \Rightarrow^k w \Rightarrow w'$$

for some $w \in (N \cup T)^*$. Then, the last derivation step is performed by some matrix

$$r : p_1 p_2 \ldots p_n$$

where $n \geq 1$. By the induction hypothesis

$$S'_{\ 0} \Rightarrow SQ_1^s {}_0 \Rightarrow^* wQ_1^r$$

in G', for some $s \in M$. By the construction of G', there exists the sequence of rules corresponding to the matrix r

$$(A_1, Q_1^r) \rightarrow (w_1, Q_2^r), (A_2, Q_2^r) \rightarrow (w_2, Q_3^r) \ldots (A_n, Q_n^r) \rightarrow (w_n, X)$$

where $X = Q_1^t$, where $t \in M$ is the next matrix applied by H, if $\mathrm{alph}(w') \cap N \neq \emptyset$, or $X = \varepsilon$ otherwise. By the application of this sequence of rules

$$wQ_1^r {}_0 \Rightarrow^* w'X$$

which completes the proof.   □

Claim 3.4.21. $\mathscr{L}(G', {}_0\Rightarrow) \subseteq \mathscr{L}(H)$.

Proof. We prove the claim by the induction on the number of derivation steps m.

Basis. Let $m = 1$. Then, G' applied some starting rule $(S') \rightarrow (SQ_1^r)$. The resulting sentential form corresponds to the starting sentential form of H. The basis holds.

Induction Hypothesis. Suppose that there exists $k \geq 1$ such that Claim 3.4.21 holds for all sequences of the derivation steps of the length m, where $1 \leq m \leq k$.

Induction Step. Notice that the rules of G', except the starting ones, form sequences

$$(A_1, Q_1^r) \to (w_1, Q_2^r), (A_2, Q_2^r) \to (w_2, Q_3^r) \dots (A_n, Q_n^r) \to (w_n, X)$$

where $X = Q_1^t, \varepsilon$, for some $t \in M$. These sequences must be always fully applied. Therefore, consider any sequence of moves

$$S'_0 \Rightarrow S_0 \Rightarrow^{k+l} w'$$

where $w' \in (N \cup T)^*$ and $l \geq 1$ is the length of any such sequence of the rules, which is applied last. Then,

$$S'_0 \Rightarrow S_0 \Rightarrow^k w_0 \Rightarrow^l w'$$

The used sequence of the rules is introduced by some matrix $r : v \in M$. By the induction hypothesis

$$S \Rightarrow^* w$$

in H. However, by the matrix r

$$w \Rightarrow w'[r]$$

which completes the proof.   □

Since $\mathscr{L}(G', {}_0\Rightarrow) \supseteq \mathscr{L}(H)$ and $\mathscr{L}(G', {}_0\Rightarrow) \subseteq \mathscr{L}(H)$, $\mathscr{L}(G', {}_0\Rightarrow) = \mathscr{L}(H)$.   □

Since for any matrix grammar we can construct an $_2EMG$ generating the same language, Claim 3.4.18 holds.   □

On the basis of Claims 3.4.16 and 3.4.18 Theorem 3.4.15 holds.   □

Mode 1

The following theorem can be proved by analogy with the proof of Theorem 3.4.11.

Theorem 3.4.22. $_nEM_{1\Rightarrow} = {}_2EM_{1\Rightarrow}$, *for* $n \geq 2$.

Proof. Left to the reader.   □

Open Problem 3.4.23. Does $_2EM_{0\Rightarrow} = {}_2EM_{1\Rightarrow}$ hold?

However, it is still an open problem whether mode 1 leftmost derivations increase the generative power of SMGs.

Mode 2

As we prove next, leftmost derivations of mode 2 significantly increase the generative power of even matrix grammars.

Theorem 3.4.24. $_2EM_{2\Rightarrow} = \mathbf{RE}$.

By the Turing-Church thesis, $_2\mathbf{EM}_{2\Rightarrow} \subseteq \mathbf{RE}$. Thus, we only have to prove the opposite inclusion.

Proof. Construction. Recall Theorem 2.3.18. There exist context-free grammars $G_i = (N_i, T, P_i, S_i)$, where $\mathscr{L}(G_i) = L_i$, for $i = 1, 2$. Without any loss of generality, assume that $N_1 \cap N_2 = \emptyset$ and G_1, G_2 are in Greibach normal form (see Definition 3.1.21). Let $T = \{a_1, \ldots, a_n\}$. Introduce four new symbols—$0, 1, \overline{0}, \overline{1} \notin (N_1 \cup N_2 \cup T)$. Define the following homomorphisms

(1) $c : a_i \mapsto 10^i; \overline{c} : a_i \mapsto \overline{1}\,\overline{0}^i$;
(2) $\pi_1 : N_1 \cup T \mapsto N_1 \cup T \cup \{0, 1\}$,
$$\begin{cases} A \mapsto A, & A \in N_1, \\ a \mapsto h(a)c(a), & a \in T; \end{cases}$$

(3) $o : \overline{a} \mapsto a, a \in \{0, 1\}$;
(4) $\pi_2 : N_2 \cup T \mapsto N_2 \cup \{\overline{0}, \overline{1}\}$,
$$\begin{cases} A \mapsto A, & A \in N_2, \\ a \mapsto \overline{c}(a), & a \in T. \end{cases}$$

Then, let $G = (N_1', N_2', T, P, S)$ be $_2EMG$, where $S \notin N_1' \cup N_2'$,

$$N_1' = N_1 \cup \{0, 1\}, N_2' = N_2 \cup \{\overline{0}, \overline{1}\}$$

Construct P as follows. Initially, set $P = \emptyset$. Perform (1) through (3), given next:

(1) add $(S) \to (S_1 S_2)$ to P;
(2) for each $(A_1) \to (w_1) \in P_1$ and for each $(A_2) \to (w_2) \in P_2$, add
    $(A_1, A_2) \to (\pi_1(w_1), \pi_2(w_2))$ to P;
(3) add

    a. $(0, \overline{0}) \to (\varepsilon, \varepsilon)$,
    b. $(1, \overline{1}) \to (\varepsilon, \varepsilon)$,
    c. $(0, \overline{1}) \to (0, \overline{1})$,
    d. $(1, \overline{0}) \to (1, \overline{0})$ to P.

Claim 3.4.25. $\mathscr{L}(G, _2\Rightarrow) = L$.

Proof. Every derivation of G starts with the application of the rule (1). Next, G simulates the leftmost derivations of G_1 and G_2, respectively, with the rules (2). Without any loss of generality, suppose every $A \in N_1 \cup N_2$

$$A \,_0\Rightarrow^* w, w \in T^*$$

Since G_1 and G_2 are in Greibach normal form, after every application of the rule (2), the leftmost nonterminal symbol in both components of G is the beginning of binary coding of some terminal symbol. Then,

$$S \,_2\Rightarrow S_1 S_2 \,_2\Rightarrow eaw\overline{a}w'$$

where $e \in T$, $a, o(\overline{a}) \in \{0, 1\}^*$, $w \in N_1^*$, $w' \in N_2^*$. If and only if $a = o(\overline{a})$, by the rules (3)a and (3)b

$$eaw\overline{a}w' \,_2\Rightarrow^* eww'$$

and the derivation may possibly continue with an application of another rule (2). Suppose, $a \neq o(\bar{a})$. Then, $|a| \neq |\bar{a}|$ and a or $o(\bar{a})$ contains more 0s, which remain in the sentential form even after erasing the shorter coding. Next, two cases are possible. If the derivation already finished, the remaining 0s or $\bar{0}$s are permanent and the derivation is not terminating. Otherwise, another rule (2) is applied and another codings are inserted. However, then the leftmost nonterminal symbol in one component is 0 or 1 and in the second one is $\bar{1}$ or $\bar{0}$, respectively. The rule (3)c or (3)d must be applied and again infinitely many times, which blocks the derivation. Then, every application of the rule (2) generates the binary codings $a, \bar{a}$, where $a = o(\bar{a})$, $a \in \{0, 1\}^*$, $\bar{a} \in \{\bar{0}, \bar{1}\}^*$, if and only if the derivation is terminating. Thus, if both components of G generate the corresponding encoded symbols in every derivation step of simulated G_1 and G_2, they generate the corresponding encoded strings. Therefore, obviously

$$S_2 \Rightarrow^* x, x \in T^*$$

where $x \in h(L_1 \cap L_2)$. Accordingly, $x \in L$, so $\mathscr{L}(G, {}_2\Rightarrow) = L$.   □

Since L is an arbitrary recursively enumerable language, the proof of Theorem 3.4.24 is completed.   □

Mode 3

By the following theorem we establish the generative power of even matrix grammars working under the leftmost mode 3 derivations.

Theorem 3.4.26. ${}_2\mathbf{EM}_{3\Rightarrow} = \mathbf{RE}$.

We omit the proof, since the proof of Theorem 3.4.24 is fully applicable to prove Theorem 3.4.26. Only notice that the rules (3)c and (3)d are not necessary, the more strict leftmost derivations would block the derivation anyway.

Summary

Let us state all the achieved results.

$$\mathbf{CF} = {}_1\mathbf{EM}_{i\Rightarrow} \subset {}_2\mathbf{EM}_{0\Rightarrow} = \mathbf{EM}_{0\Rightarrow} = \mathbf{M} \subset \mathbf{CS}, i \in \{0, 1, 2, 3\}$$

$$\mathbf{CF} \subset {}_2\mathbf{EM}_{1\Rightarrow} = \mathbf{EM}_{1\Rightarrow} \subset \mathbf{CS}$$

$$_2\mathbf{EM}_{2\Rightarrow} = \mathbf{EM}_{2\Rightarrow} = {}_2\mathbf{EM}_{3\Rightarrow} = \mathbf{EM}_{3\Rightarrow} = \mathbf{RE}$$

We prove that even matrix grammars with two components are exactly as strong as matrix grammars. Additionally, we prove that the presence of more than two components has no influence on their generative power. Three leftmost modes of derivations are introduced and studied. The previous conclusion on the number of components still hold for leftmost derivations. The more strict leftmost derivations increase the generative power significantly—even matrix grammars become Turing complete. However, we are still not sure about the most liberal leftmost derivations, thus we provide Open Problem 3.4.23, which we suggest for the future research.

Of course, even matrix grammars represent variants of ordinary matrix grammars, and as such, from a general viewpoint, they generate their languages in a regulated way. At the same time, however, they can be viewed as language generators working in parallel, which represents the central topic of the next chapter. Before opening this topic, however, we briefly cover one more grammatical regulation—programmed grammars.

3.5 Programmed Grammars

The regulation of a programmed grammar is based upon two binary relations, represented by two sets attached to the grammatical rules. More precisely, a *programmed grammar G* is a context-free grammar, in which two sets, σ_r and φ_r, are attached to each rule r, where σ_r and φ_r are subsets of the entire set of rules in G. G can apply r in the following two ways.

(1) If the left-hand side of r occurs in the sentential form under scan, G rewrites the left-hand side of r to its right-hand side, and during the next derivation step, it has to apply a rule from σ_r.
(2) If the left-hand side of r is absent in the sentential form under scan, then G skips the application of r, and during the next derivation step, it has to apply a rule from φ_r.

This brief section consists of two subsections. Section 3.5.1 defines and illustrates programmed grammars while Sect. 3.5.2 gives their generative power.

3.5.1 Definitions and Examples

In this section, we define programmed grammars and illustrate them by an example.

Definition 3.5.1. A *programmed grammar with appearance checking* (see [DP89]) is a quintuple

$$G = (V, T, \Psi, P, S)$$

where

- V, T, Ψ, and S are defined as in a context-free grammar, $V - T = N$;

- $P \subseteq \Psi \times N \times V^* \times 2^\Psi \times 2^\Psi$ is a finite relation, called the set of *rules*, such that $\text{card}(\Psi) = \text{card}(P)$ and if $(r, A, x, \sigma_r, \varphi_r), (s, A, x, \sigma_s, \varphi_s) \in P$, then $(r, A, x, \sigma_r, \varphi_r) = (s, A, x, \sigma_s, \varphi_s)$.

Instead of $(r, A, x, \sigma_r, \varphi_r) \in P$, we write $(r{:}A \rightarrow x, \sigma_r, \varphi_r) \in P$. For $(r{:}A \rightarrow x, \sigma_r, \varphi_r) \in P$, A is referred to as the *left-hand side* of r, and x is referred to as the *right-hand side* of r.

The *direct derivation relation*, symbolically denoted by $\Rightarrow_G$, is defined over $V^* \times \Psi$ as follows: for $(x_1, r), (x_2, s) \in V^* \times \Psi$,

$$(x_1, r) \Rightarrow_G (x_2, s)$$

if and only if either

$$x_1 = yAz, x_2 = ywz, (r{:}A \rightarrow w, \sigma_r, \varphi_r) \in P, \text{ and } s \in \sigma_r$$

or

$$x_1 = x_2, (r{:}A \rightarrow w, \sigma_r, \varphi_r) \in P, A \notin \text{alph}(x_1), \text{ and } s \in \varphi_r$$

Let $(r{:}A \rightarrow w, \sigma_r, \varphi_r) \in P$. Then, σ_r and φ_r are called the *success field* of r and the *failure field* of r, respectively. Observe that due to our definition of the relation of a direct derivation, if $\sigma_r \cup \varphi_r = \emptyset$, then r is never applicable. Therefore, we assume that $\sigma_r \cup \varphi_r \neq \emptyset$, for all $(r{:}A \rightarrow w, \sigma_r, \varphi_r) \in P$. Define $\Rightarrow_G^k$ for $k \geq 0$, $\Rightarrow_G^*$, and $\Rightarrow_G^+$ in the standard way. Let $(S, r) \Rightarrow_G^* (w, s)$, where $r, s \in \Psi$ and $w \in V^*$. Then, (w, s) is called a *configuration*. The *language* of G is denoted by $L(G)$ and defined as

$$L(G) = \{w \in T^* \mid (S, r) \Rightarrow_G^* (w, s), \text{ for some } r, s \in \Psi\} \qquad \square$$

Definition 3.5.2. Let $G = (V, T, \Psi, P, S)$ be a programmed grammar with appearance checking. G is *propagating* if every $(r{:}A \rightarrow x, \sigma_r, \varphi_r) \in P$ satisfies that $|x| \geq 1$. Rules of the form $(r{:}A \rightarrow \varepsilon, \sigma_r, \varphi_r)$ are called *erasing rules*. If every $(r{:}A \rightarrow x, \sigma_r, \varphi_r) \in P$ satisfies that $\varphi_r = \emptyset$, then G is a *programmed grammar without appearance checking* or, simply, a *programmed grammar*. Then, for the brevity, instead of $(r{:}A \rightarrow x, \sigma_r, \emptyset)$, we write $(r{:}A \rightarrow x, \sigma_r)$.

If G is of index k (see Definition 2.3.13), for some $k \geq 1$, it is *programmed grammar of index k (with appearance checking)*. $\qquad \square$

Example 3.5.3 (From [DP89]). Consider the programmed grammar with appearance checking

$$G = (\{S, A, a\}, \{a\}, \{r_1, r_2, r_3\}, P, S)$$

where P consists of the three rules

$(r_1: S \rightarrow AA, \{r_1\}, \{r_2, r_3\})$
$(r_2: A \rightarrow S, \{r_2\}, \{r_1\})$
$(r_3: A \rightarrow a, \{r_3\}, \emptyset)$

Since the success field of r_i is $\{r_i\}$, for each $i \in \{1, 2, 3\}$, the rules r_1, r_2, and r_3 have to be used as many times as possible. Therefore, starting from S^n, for some $n \geq 1$, the successful derivation has to pass to A^{2n} and then, by using r_2, to S^{2n}, or, by using r_3, to a^{2n}. A cycle like this, consisting of the repeated use of r_1 and r_2, doubles the number of symbols. In conclusion, we obtain the non-context-free language

$$L(G) = \left\{ a^{2^n} \mid n \geq 1 \right\}$$

For example, the sentence $aaaa$ is obtained by the following derivation

$$
\begin{aligned}
(S, r_1) &\Rightarrow_G (AA, r_2) \\
&\Rightarrow_G (AS, r_2) \\
&\Rightarrow_G (SS, r_1) \\
&\Rightarrow_G (AAS, r_1) \\
&\Rightarrow_G (AAAA, r_2) \\
&\Rightarrow_G (AASA, r_2) \\
&\Rightarrow_G (AASS, r_2) \\
&\Rightarrow_G (SASS, r_2) \\
&\Rightarrow_G (SSSS, r_3) \\
&\Rightarrow_G (SSSa, r_3) \\
&\Rightarrow_G (aSSa, r_3) \\
&\Rightarrow_G (aaSa, r_3) \\
&\Rightarrow_G (aaaa, r_3)
\end{aligned}
$$

Notice the similarity between G from this example and H from Example 3.3.4. □

By $\mathbf{P}_{ac}$, $\mathbf{P}_{ac}^{-\varepsilon}$, $\mathbf{P}$, and $\mathbf{P}^{-\varepsilon}$, we denote the families of languages generated by programmed grammars with appearance checking, propagating programmed grammars with appearance checking, programmed grammars, and propagating programmed grammars, respectively.

For a positive integer k, By $_k\mathbf{P}_{ac}$, $_k\mathbf{P}_{ac}^{-\varepsilon}$, $_k\mathbf{P}$, and $_k\mathbf{P}^{-\varepsilon}$, we denote the families of languages generated by programmed grammars of index k with appearance checking, propagating programmed grammars of index k with appearance checking, programmed grammars of index k, and propagating programmed grammars of index k, respectively.

3.5.2 Generative Power

The next theorem states the power of programmed grammars.

Theorem 3.5.4 (See Theorem 5.3.4 in [MZ14]).

  (i) $\mathbf{P}_{ac} = \mathbf{M}_{ac}$
  (ii) $\mathbf{P}_{ac}^{-\varepsilon} = \mathbf{M}_{ac}^{-\varepsilon}$
  (iii) $\mathbf{P} = \mathbf{M}$
  (iv) $\mathbf{P}^{-\varepsilon} = \mathbf{M}^{-\varepsilon}$

Observe that programmed grammars, matrix grammars, and regular-controlled grammars are equally powerful (see Theorems 3.4.5 and 3.5.4).

Theorem 3.5.5 (See Theorems 3.1.2i and 3.1.7 in [DP89]).

$$_k\mathbf{P} \subset {}_{k+1}\mathbf{P}$$

for any $k \geq 1$.

Programmed grammars of finite index establish an infinite hierarchy of language families.

Chapter 4
Parallel Grammars and Computation

Originally, computer programs were always executed strictly sequentially. Indeed, to perform a computational task, an algorithm was written and implemented as an instruction sequence executed on a central processing unit on a single computer. Only one instruction was executed at a time, so after this instruction was completed, the next instruction was executed until all the sequence of instructions was performed in this one-by-one way. In the mid-1980s or so, however, computer programmers introduced the first pioneer programs that performed several parts of a single computational task simultaneously. At that time, parallel computation emerged in computer science.

In general, parallel computation can be thus viewed as a modern type of computation in which many computational processes are carried out simultaneously. From a hardware standpoint, parallel computation is often executed on various computers, such as a single computer with multiple processors or several networked computers with specialized hardware, and it may simultaneously process quite diverse data. It can be performed at various levels, ranging from bit-level through instruction-level up to task-level parallelism. Over the past few years, parallel computing has become the dominant paradigm in computer architecture, mainly in the form of multi-core processors. From a software standpoint, parallel computation is conceptually accomplished by breaking a single computational task into many independent subtasks so that each subtask can be simultaneously executed with the others. All the independent subtasks are processed independently and their results are combined together upon completion.

Making use of mutually cooperating multiprocessor computers, most modern information technologies work in parallel. It thus comes as no surprise that the investigation of parallel computation fulfill a central role within computer science as a whole. In order to build up a systematized body of knowledge about computation in parallel, we need its proper formalization in the first place. The present chapter describes several types of parallel grammars, which can act as a grammatical formalization like this very well.

© Springer International Publishing AG 2017
A. Meduna, O. Soukup, *Modern Language Models and Computation*,
DOI 10.1007/978-3-319-63100-4_4

To give an insight into parallel grammars, recall that up until now, in all grammars under consideration, a single rule was applied during every derivation step. To obtain parallel grammars, this one-rule application is generalized to the application of several rules during a single step. Parallel grammars represent the subject of this chapter. First, it studies partially parallel generation of languages (see Sect. 4.1), after which it investigates the totally parallel generation of languages (see Sect. 4.2).

4.1   Partially Parallel Grammars

Partially parallel language generation is represented by the notion of a scattered context grammar, which is based on finite sequences of context-free rules. According to these sequences, the grammar simultaneously rewrites several nonterminals during a single derivation step while keeping the rest of the rewritten string unchanged.

4.1.1   Definitions and Examples

In this section, we define scattered context grammars and illustrate them by examples.

Definition 4.1.1. A *scattered context grammar* (an SCG for short) is a quadruple

$$G = (V, T, P, S); \quad N = V - T$$

where

- V is a *total alphabet*;
- $T \subset V$ an alphabet of *terminals*;
- $P \subseteq \bigcup_{m=1}^{\infty} N^m \times (V^*)^m$ is a finite set of *rules* of the form

$$(A_1, A_2, \ldots, A_n) \to (x_1, x_2, \ldots, x_n)$$

  where $A_i \in N$, and $x_i \in V^*$, for $1 \leq i \leq n$, for some $n \geq 1$;
- $S \in V - T$ is the *start symbol*;
- N is an alphabet of *nonterminals*.

If

$$u = u_1 A_1 \ldots u_n A_n u_{n+1}$$
$$v = u_1 x_1 \ldots u_n x_n u_{n+1}$$

and $p = (A_1, \ldots, A_n) \to (x_1, \ldots, x_n) \in P$, where $u_i \in V^*$, for all i, $1 \leq i \leq n + 1$, then G makes a *derivation step* from u to v according to p, symbolically written as

$$u \Rightarrow_G v \ [p]$$

or, simply, $u \Rightarrow_G v$. Set

$$\mathrm{lhs}(p) = A_1 \ldots A_n$$
$$\mathrm{rhs}(p) = x_1 \ldots x_n$$

and

$$\mathrm{len}(p) = n$$

If $\mathrm{len}(p) \geq 2$, p is said to be a *context-sensitive rule* while for $\mathrm{len}(p) = 1$, p is said to be *context-free*. Define $\Rightarrow_G^k$, $\Rightarrow_G^*$, and $\Rightarrow_G^+$ in the standard way. The *language* of G is denoted by $L(G)$ and defined as

$$L(G) = \left\{ w \in T^* \mid S \Rightarrow_G^* w \right\}$$

A language L is a *scattered context language* if there exists a scattered context grammar G such that $L = L(G)$. $\qquad\qquad\square$

Definition 4.1.2. A *propagating scattered context grammar* is a scattered context grammar

$$G = (V, T, P, S)$$

in which every $(A_1, \ldots, A_n) \to (x_1, \ldots, x_n) \in P$ satisfies $|x_i| \geq 1$, for all i, $1 \leq i \leq n$. A *propagating scattered context language* is a language generated by a propagating scattered context grammar. $\qquad\qquad\square$

Example 4.1.3. Consider the non-context-free language $L = \{a^n b^n c^n \mid n \geq 1\}$. This language can be generated by the scattered context grammar

$$G = (\{S, A, a, b, c\}, \{a, b, c\}, P, S)$$

where

$$P = \{(S) \to (aAbAcA),$$
$$(A, A, A) \to (aA, bA, cA),$$
$$(A, A, A) \to (\varepsilon, \varepsilon, \varepsilon)\}$$

For example, the sentence *aabbcc* is generated by G as follows:

$$S \Rightarrow_G aAbAcA \Rightarrow_G aaAbbAccA \Rightarrow_G aabbcc$$

Notice, however, that L can be also generated by the propagating scattered context grammar

$$G' = (\{S, A, a, b, c\}, \{a, b, c\}, P', S)$$

where

$$P' = \{(S) \rightarrow (AAA),$$
$$(A, A, A) \rightarrow (aA, bA, cA),$$
$$(A, A, A) \rightarrow (a, b, c)\} \qquad \qquad \square$$

For brevity, we often label rules of scattered context grammars with labels (just like we do in other grammars), as illustrated in the next example.

Example 4.1.4. Consider the non-context-free language

$$L = \{(ab^n)^m \mid m \geq n \geq 2\}$$

This language is generated by the propagating scattered context grammar

$$G = (\{S, S_1, S_2, B, M, X, Y, Z, a, b\}, \{a, b\}, P, S)$$

with P containing the following rules

$$1 : (S) \rightarrow (MS)$$
$$2 : (S) \rightarrow (S_1 S_2)$$
$$3 : (S_1, S_2) \rightarrow (MS_1, BS_2)$$
$$4 : (S_1, S_2) \rightarrow (MX, BY)$$
$$5 : (X, B, Y) \rightarrow (BX, Y, b)$$
$$6 : (M, X, Y) \rightarrow (X, Y, ab)$$
$$7 : (M, X, Y) \rightarrow (Z, Y, ab)$$
$$8 : (Z, B, Y) \rightarrow (Z, b, Y)$$
$$9 : (Z, Y) \rightarrow (a, b)$$

Clearly, by applying rules 1 through 4, G generates a string from

$$\{M\}^+\{X\}\{B\}^+\{Y\}$$

In what follows, we demonstrate that the string is of the form $M^{m-1}XB^{n-1}Y$, where $m, n \geq 2$. Rule 1 allows G to add Ms to the beginning of the sentential form, so $m \geq n$ holds true. Observe that each of the rules 5 through 8 either shifts the last nonterminal Y left or keeps its position unchanged. As a result, always the rightmost nonterminal preceding Y has to be replaced with Y by rules 5 through 7; otherwise, the skipped nonterminals cannot be rewritten during the rest of the derivation. For the same reason, the rightmost nonterminal M preceding X has to be rewritten by the rule 6. Rules 5 and 6 are applied in a cycle consisting of $n-1$ applications of 5 and one application of 6:

$$M^{m-1}XB^{n-1}Y \Rightarrow_G^{n-1} M^{m-1}B^{n-1}XYb^{n-1} \ [5^{n-1}]$$
$$\Rightarrow_G \ M^{m-2}XB^{n-1}Yab^n \ [6]$$

At this point, the substring preceding Y differs from the original string only in the number of Ms decremented by 1, and the cycle can be repeated again. After repeating this cycle $m-2$ times, we obtain $MXB^{n-1}Y(ab^n)^{m-2}$. The derivation is completed as follows:

$$MXB^{n-1}Y(ab^n)^{m-2} \Rightarrow_G^{n-1} MB^{n-1}XYb^{n-1}(ab^n)^{m-2} \ [5^{n-1}]$$
$$\Rightarrow_G ZB^{n-1}Y(ab^n)^{m-1} \ [7]$$
$$\Rightarrow_G^{n-1} Zb^{n-1}Y(ab^n)^{m-1} \ [8^{n-1}]$$
$$\Rightarrow_G (ab^n)^m \ [9] \qquad \square$$

Example 4.1.5 (See [Mas07b, Mas07a]). Consider the non-context-free language

$$L = \left\{ a^{2^n} \mid n \geq 0 \right\}$$

This language is generated by the propagating scattered context grammar

$$G = \big(\{S, W, X, Y, Z, A, a\}, \{a\}, P, S\big)$$

with P containing these rules

$$1 : (S) \to (a)$$
$$2 : (S) \to (aa)$$
$$3 : (S) \to (WAXY)$$
$$4 : (W, A, X, Y) \to (a, W, X, AAY)$$
$$5 : (W, X, Y) \to (a, W, AXY)$$
$$6 : (W, X, Y) \to (Z, Z, a)$$
$$7 : (Z, A, Z) \to (Z, a, Z)$$
$$8 : (Z, Z) \to (a, a)$$

In what follows, we demonstrate that $L(G) = L$. Rules 1 and 2 generate a and aa, respectively. Rule 3 starts off the derivation of longer strings in L. Consider the following derivation of $a^{16} \in L(G)$

$$S \Rightarrow_G WAXY \qquad [3]$$
$$\Rightarrow_G aWXA^2Y \qquad [4]$$
$$\Rightarrow_G a^2WA^3XY \qquad [5]$$
$$\Rightarrow_G a^3WA^2XA^2Y \ [4]$$
$$\Rightarrow_G a^4WAXA^4Y \quad [4]$$
$$\Rightarrow_G a^5WXA^6Y \qquad [4]$$
$$\Rightarrow_G a^6WA^7XY \qquad [5]$$
$$\Rightarrow_G a^6ZA^7Za \qquad [6]$$
$$\Rightarrow_G^7 a^{13}ZZa \qquad [7^7]$$
$$\Rightarrow_G a^{16} \qquad [8]$$

Observe that in any successful derivation, rules 4 and 5 are applied in a cycle, and after the required number of As is obtained, the derivation is finished by rules 6, 7, and 8. In a greater detail, observe that the rule $(W, A, X, Y) \rightarrow (a, W, X, AAY)$ removes one A between W and X, and inserts two As between X and Y. In a successful derivation, this rule has to rewrite the leftmost nonterminal A. After all As are removed between W and X, the rule $(W, X, Y) \rightarrow (a, W, AXY)$ can be used to bring all As occurring between X and Y back between W and X, and the cycle can be repeated again. Alternatively, rule 6 can be used, which initializes the final phase of the derivation in which all As are replaced with as by rules 7 and 8.

By adding one more stage, the above grammar can be extended so that it generates the language

$$\left\{ a^{2^{2^n}} \mid n \geq 0 \right\}$$

The first stage, similar to the above grammar, generates 2^n identical symbols that serve as a counter for the second stage. In the second stage, a string consisting of identical symbols, which are different from those generated during the first stage, is doubled 2^n times, thus obtaining 2^{2^n} identical symbols. This doubling starts from a string consisting of a single symbol. See [Mas07a] for the details.   □

The families of languages generated by scattered context grammars and propagating scattered context grammars are denoted by **SC** and **SC**$^{-\varepsilon}$, respectively.

4.1.2 Generative Power

This brief section establishes the power of scattered context grammars. In addition, it points out a crucially important open problem, referred to as the *PSC = CS problem*, which asks whether **SC**$^{-\varepsilon}$ and **CS** coincide.

Theorem 4.1.6 (See [MT10]). $\mathbf{CF} \subset \mathbf{SC}^{-\varepsilon} \subseteq \mathbf{CS} \subset \mathbf{SC} = \mathbf{RE}$. □

Open Problem 4.1.7. Is the inclusion $\mathbf{SC}^{-\varepsilon} \subseteq \mathbf{CS}$, in fact, an identity? □

4.1.3 Normal Forms

This section demonstrates how to transform any propagating scattered context grammar to an equivalent 2-*limited propagating scattered context grammar*, which represent an important normal form of propagating scattered context grammars. More specifically, in a 2-limited propagating scattered context grammar, each rule consist of no more than two context-free rules, either of which has on their right-hand side no more than two symbols.

Definition 4.1.8. A 2-*limited propagating scattered context grammar* is a propagating scattered context grammar, $G = (V, T, P, S)$, such that

- $(A_1, \ldots, A_n) \to (w_1, \ldots, w_n) \in P$ implies that $n \le 2$, and for every i, $1 \le i \le n$, $1 \le |w_i| \le 2$, and $w_i \in (V - \{S\})^*$;
- $(A) \to (w) \in P$ implies that $A = S$. □

The proof of the transformation is divided into two lemmas.

Lemma 4.1.9. *If $L \subseteq T^*$ is a language generated by a propagating scattered context grammar, $G = (V, T, P, S)$, and if c is a symbol such that $c \notin T$, then there is a 2-limited propagating scattered context grammar, $\bar{G}$, such that $L(\bar{G}) = L\{c\}$.*

Proof. Let $\bar{n}$ be the number of the rules in P. Number the rules of P from 1 to $\bar{n}$. Let $(A_{i1}, \ldots, A_{in_i}) \to (w_{i1}, \ldots, w_{in_i})$ be the ith rule. Let C and $\bar{S}$ be new symbols,

$$W = \{\langle i,j \rangle \mid 1 \le i \le \bar{n}, 1 \le j \le n_i\}$$
$$\bar{V} = V \cup \{C, \bar{S}\} \cup W \cup \{\langle C, i \rangle \mid 1 \le i \le \bar{n}\}$$

Let $G' = (\bar{V}, T \cup \{c\}, P', \bar{S})$ be a propagating scattered context grammar, where P' is defined as follows:

1. for each $1 \le i \le \bar{n}$, add
   $(\bar{S}) \to (S\langle C, i \rangle)$ to P';
2. for each i such that $n_i = 1$ and $1 \le k \le \bar{n}$, add
   $(A_{i1}, \langle C, i \rangle) \to (w_{i1}, \langle C, k \rangle)$ to P';
3. for each i such that $n_i > 1$, $1 \le j \le n_i - 1$, $1 \le k \le \bar{n}$, add

   a. $(A_{i1}, \langle C, i \rangle) \to (\langle i, 1 \rangle, C)$,
   b. $(\langle i, j \rangle, A_{i(j+1)}) \to (w_{ij}, \langle i, j+1 \rangle)$, and
   c. $(\langle i, n_i \rangle, C) \to (w_{in_i}, \langle C, k \rangle)$ to P';

4.

   a. for each i such that $n_i = 1$, add
      $(A_{i1}, \langle C, i \rangle) \to (w_{i1}, c)$ to P';

b. for each i such that $n_i > 1$, add
   $(\langle i, n_i \rangle, C) \rightarrow (w_{in_i}, c)$ to P'.

Clearly, $L(G') = L\{c\}$. Since for some i and j, w_{ij} may satisfy $|w_{ij}| > 2$, G' may not be a 2-limited propagating scattered context grammar. However, by making use of standard techniques, one can obtain a 2-limited propagating scattered context grammar $\bar{G}$ from G' such that $L(\bar{G}) = L(G')$. $\square$

Lemma 4.1.10. *If $L \subseteq T^+$, c is a symbol such that $c \notin T$, and $G = (V, T \cup \{c\}, P, S)$ is a 2-limited propagating scattered context grammar satisfying $L(G) = L\{c\}$, then there is a 2-limited propagating scattered context grammar $\bar{G}$ such that $L(\bar{G}) = L$.*

Proof. For each $a \in T \cup \{S\}$, let $\bar{a}$ be a new symbol. Let

$$L_1 = \{A_1 A_2 A_3 \mid S \Rightarrow_G^* A_1 A_2 A_3, A_i \in V, \text{ for all } i = 1, 2, 3\}$$
$$L_2 = \{A_1 A_2 A_3 A_4 \mid S \Rightarrow_G^* A_1 A_2 A_3 A_4, A_i \in V, \text{ for all } i = 1, 2, 3, 4\}$$

Let h be the homomorphism from V^* to $(\{\bar{a} \mid a \in T\} \cup (V - T))^*$ defined as $h(a) = \bar{a}$, for each $a \in T$, and $h(A) = A$, for each $A \in V - T$. Let

$$V' = h(V) \cup T \cup \{S'\} \cup \{\langle a, b \rangle \mid a, b \in V\}$$

Let $G' = (V', T, P', S')$, where for all $a, b \in T$, $A_1, \ldots, A_6 \in V$, $A \in h(V)$, P' is defined as follows:

1.

   a. for each $a \in T \cap L$, add
      $(S') \rightarrow (a)$ to P';
   b. for each $A_1 A_2 A_3 \in L_1$, add
      $(S') \rightarrow (h(A_1) \langle A_2, A_3 \rangle)$ to P';
   c. for each $A_1 A_2 A_3 A_4 \in L_2$, add
      $(S') \rightarrow (h(A_1 A_2) \langle A_3, A_4 \rangle)$ to P';

2.

   a. for each $(A_1, A_2) \rightarrow (w_1, w_2) \in P$, add
      $(A_1, A_2) \rightarrow (h(w_1), h(w_2))$ to P';
   b. for each $(A_1, A_2) \rightarrow (w_1, w_2) \in P$,
      i. where $|w_2| = 1$, add
         A. $(A_1, \langle A_2, A_3 \rangle) \rightarrow (h(w_1), \langle w_2, A_3 \rangle)$, and
         B. $(A_1, \langle A_3, A_2 \rangle) \rightarrow (h(w_1), \langle A_3, w_2 \rangle)$ to P';
      ii. where $w_2 = A_4 A_5$, add
         A. $(A_1, \langle A_2, A_3 \rangle) \rightarrow (h(w_1), h(A_4) \langle A_5, A_3 \rangle)$, and
         B. $(A_1, \langle A_3, A_2 \rangle) \rightarrow (h(w_1), h(A_3) \langle A_4, A_5 \rangle)$ to P'.
   c. for each $(A_1, A_2) \rightarrow (w_1, w_2) \in P$,
      i. where $|w_1| = |w_2| = 1$, add
         $(A, \langle A_1, A_2 \rangle) \rightarrow (A, \langle w_1, w_2 \rangle)$ to P';

ii. where $w_1 w_2 = A_3 A_4 A_5$, add
$(A, \langle A_1, A_2 \rangle) \rightarrow (A, h(A_3)\langle A_4, A_5 \rangle)$ to P';
iii. where $w_1 w_2 = A_3 A_4 A_5 A_6$, add
$(A, \langle A_1, A_2 \rangle) \rightarrow (A, h(A_3 A_4)\langle A_5, A_6 \rangle)$ to P';

3. for each $a, b \in T$, add

a. $(\bar{a}, \langle b, c \rangle) \rightarrow (a, \langle b, c \rangle)$, and
b. $(\bar{a}, \langle b, c \rangle) \rightarrow (a, b)$ to P'.

Note that the construction simply combines the symbol c with the symbol to its left. The reason for introducing a new symbol $\bar{a}$, for each $a \in T$, is to guarantee that there always exists a nonterminal A whenever a rule from (2.3) is to be applied, and a nonterminal $\bar{a}$ that enables $\langle b, c \rangle$ to be converted to b by a rule from (3). Clearly, $L(G') = L$. G' may not be a 2-limited propagating scattered context grammar since in (2.3), $|h(A_3 A_4)\langle A_5, A_6 \rangle| = 3$. Once again, by standard techniques, we can obtain a 2-limited propagating scattered context grammar $\bar{G}$ from G' such that $L(\bar{G}) = L(G')$. □

By Lemmas 4.1.9 and 4.1.10, any propagating scattered context grammar can be converted to an equivalent 2-limited propagating scattered context grammar as stated in the following theorem.

Theorem 4.1.11. *If G is a propagating scattered context grammar, then there exists a 2-limited propagating scattered context grammar $\bar{G}$ such that $L(\bar{G}) = L(G)$.* □

4.1.4 Reduction

The present section discusses the reduction of scattered context grammars. Perhaps most importantly, it studies how to reduce the size of their components, such as the number of nonterminals or the number of context sensitive rules, without any decrease of their generative power. Indeed, any reduction like this is highly appreciated in both theory and practice because it makes scattered context rewriting more succinct and economical while preserving its power.

Definition 4.1.12. Let $G = (V, T, P, S)$ be a scattered context grammar. Then, its *degree of context sensitivity*, symbolically written as $\operatorname{dcs}(G)$, is defined as

$$\operatorname{dcs}(G) = \operatorname{card}\left(\{p \mid p \in P, \operatorname{lhs}(p) \geq 2\}\right)$$

The *maximum context sensitivity* of G, denoted by $\operatorname{mcs}(G)$, is defined as

$$\operatorname{mcs}(G) = \max\left(\{\operatorname{len}(p) - 1 \mid p \in P\}\right)$$

The *overall context sensitivity* of G, denoted by $\text{ocs}(G)$, is defined as

$$\text{ocs}(G) = \text{len}(p_1) + \cdots + \text{len}(p_n) - n$$

where $P = \{p_1, \ldots, p_n\}$.   □

We present several results that reduce one of these measures while completely ignoring the other measures. Frequently, however, results of this kind are achieved at the cost of an enormous increase of the other measures. Therefore, we also undertake a finer approach to this descriptional complexity by simultaneously reducing several of these measures while keeping the generative power unchanged.

We start by pointing out a result regarding scattered context grammars with a single nonterminal.

Theorem 4.1.13 (See Theorem 5 in [Med00c]). *One-nonterminal scattered context grammars cannot generate all recursively enumerable languages.*

For scattered context grammars containing only one context-sensitive rule (see Definition 4.1.1), the following theorem holds.

Theorem 4.1.14. *There exists a scattered context grammar G such that G defines a non-context-free language, and*

$$\text{dcs}(G) = \text{mcs}(G) = \text{ocs}(G) = 1$$

Proof. Consider the scattered context grammar

$$G = \left(\{S, A, C, a, b, c\}, \{a, b, c\}, P, S\right)$$

where the set of rules P is defined as

$$P = \{(S) \rightarrow (AC),$$
$$(A) \rightarrow (aAb),$$
$$(C) \rightarrow (cC),$$
$$(A, C) \rightarrow (\varepsilon, \varepsilon)\}$$

It is easy to verify that $L(G) = \{a^n b^n c^n \mid n \geq 0\}$ and $\text{dcs}(G) = \text{mcs}(G) = \text{ocs}(G) = 1$.   □

Next, we concentrate our attention on reducing the number of nonterminals in scattered context grammars. We first demonstrate how the number of nonterminals can be reduced to three.

Theorem 4.1.15. *For every recursively enumerable language L, there is a scattered context grammar $G = (V, T, P, S)$ such that $L(G) = L$, and*

$$\text{card}\left(V - T\right) = 3$$

Proof. Let L be a recursively enumerable language. By Theorem 2.3.43, there exists a queue grammar $Q = (\bar{V}, T, W, F, R, g)$ such that $L = L(Q)$. Without any loss of generality, assume that Q satisfies the normal form of Definition 3.1.23. Set $n =$ card$(\bar{V} \cup W)$. Introduce a bijective homomorphism β from $\bar{V} \cup W$ to $\{B\}^*\{A\}\{B\}^* \cap \{A, B\}^n$. Without any loss of generality, assume that $(\bar{V} \cup W) \cap \{A, B, S\} = \emptyset$. Define the scattered context grammar

$$G = (T \cup \{A, B, S\}, T, P, S)$$

where P is constructed in the following way

(1) for $g = ab$, where $a \in \bar{V} - T$ and $b \in W - F$, add $(S) \rightarrow (\beta(b)SS\beta(a)SA)$ to P;
(2) for each $a \in \{A, B\}$, add $(S, S, a, S) \rightarrow (S, \varepsilon, aS, S)$ to P;
(3) for each $(a, b, x, c) \in R$, where $a \in \bar{V} - T$, $x \in (\bar{V} - T)^*$, and $b, c \in W - F - \{1\}$, extend P by adding

$$(b_1, \ldots, b_n, S, a_1, \ldots, a_n, S, S)$$
$$\rightarrow (c_1, \ldots, c_n, \varepsilon, e_1, \ldots, e_n, SS, \beta(x)S)$$

where $b_1 \cdots b_n = \beta(b)$, $a_1 \cdots a_n = \beta(a)$, $c_1 \cdots c_n = \beta(c)$, and $e_1 \cdots e_n = \varepsilon$;
(4) for each $(a, b, x, c) \in R$, where $a \in \bar{V} - T$, $b \in W - F - \{1\}$, $x \in (\bar{V} - T)^*$, and

(4.1) $c = 1$, extend P by adding

$$(b_1, \ldots, b_n, S, a_1, \ldots, a_n, S, S)$$
$$\rightarrow (c_1, \ldots, c_n, \varepsilon, e_1, \ldots, e_n, SS, \beta(x)S)$$

(4.2) $c \in F$ and $x = \varepsilon$, extend P by adding

$$(b_1, \ldots, b_n, S, a_1, \ldots, a_n, S, S, A)$$
$$\rightarrow (e_1, \ldots, e_n, \varepsilon, e_{n+1}, \ldots, e_{2n}, \varepsilon, \varepsilon, \varepsilon)$$

where $b_1 \cdots b_n = \beta(b)$, $a_1 \cdots a_n = \beta(a)$, $c_1 \cdots c_n = \beta(c)$, and $e_1 \cdots e_{2n} = \varepsilon$;
(5) for each $(a, 1, x, c) \in R$, where $a \in \bar{V} - T$, $x \in T$, and

(5.1) $c = 1$, extend P by adding

$$(b_1, \ldots, b_n, S, a_1, \ldots, a_n, S, S)$$
$$\rightarrow (c_1, \ldots, c_n, \varepsilon, e_1, \ldots, e_n, SS, xS)$$

(5.2) $c \in F$, extend P by adding

$$(b_1, \ldots, b_n, S, a_1, \ldots, a_n, S, S, A)$$
$$\rightarrow (e_1, \ldots, e_n, \varepsilon, e_{n+1}, \ldots, e_{2n}, \varepsilon, \varepsilon, x)$$

where $b_1 \cdots b_n = \beta(1)$, $a_1 \cdots a_n = \beta(a)$, $c_1 \cdots c_n = \beta(c)$, and $e_1 \cdots e_{2n} = \varepsilon$.

The constructed scattered context grammar G simulates the queue grammar Q that satisfies the normal form of Definition 3.1.23. The rule from (1), applied only once, initializes the derivation. One of the rules from (4.2) and (5.2) terminates the derivation. In a greater detail, a rule from (4.2) is used in the derivation of $\varepsilon \in L(Q)$; in a derivation of every other string, a rule from (5.2) is used in the last step of the derivation.

Every sentential form of G can be divided into two parts. The first n nonterminals encode the state of Q. The second part represents the queue, where the first symbol S always occurs at the beginning of the queue and the third S always occurs at the end of the queue, followed by the ultimate nonterminal A.

During any successful derivation of G, a rule introduced in (2) is always applied after the application of a rule introduced in (1), (3), (4.1), and (5.1). More precisely, to go on performing the successful derivation, after applying rules from (1), (3), (4.1), and (5.1), G shifts the second occurrence of S right in the current sentential form. G makes this shift by using rules introduced in (2) to obtain a sentential form having precisely n occurrences of $d \in \{A, B\}$ between the first occurrence of S and the second occurrence of S.

The following claims demonstrate that the rule from (1) can be used only once during a successful derivation.

Claim 4.1.16. Let $S \Rightarrow_G^* x$ be a derivation during which G uses the rules introduced in (1) i times, for some $i \geq 1$. Then, $\#_S(x) = 1 + 2i - 3j$, $\#_B(x) = (n - 1)k$, and $\#_A(x) = k+i-j$, where k is a non-negative integer and j is the number of applications of rules introduced in (4.2) and (5.2) such that $j \geq 1$ and $1 + 2i \geq 3j$.

Proof. Notice that the rules introduced in (2), (3), (4.1), and (5.1) preserve the number of As, Bs, and Ss present in the sentential form. Next, observe that every application of the rule from (1) adds 2 Ss to the sentential form and every application of a rule from (4.2) or (5.2) removes 3 Ss from the sentential form. Finally, notice the last A on the right-hand side of the rule from (1) and on the left-hand sides of the rules from (4.2) and (5.2). Based on these observations, it is easy to see that Claim 4.1.16 holds.   □

Claim 4.1.17. Let $S \Rightarrow_G^* x$ be a derivation during which G applies the rule introduced in (1) two or more times. Then, $x \notin T^*$.

Proof. Let $S \Rightarrow_G^* x$, where $x \in T^*$. Because $x \in T^*$, $\#_S(x) = \#_B(x) = \#_A(x) = 0$. As a result, we get $k = 0$, and $i = j = 1$ from the equations introduced in Claim 4.1.16. Thus, for $i \geq 2$, $x \notin T^*$.   □

Next, we demonstrate that rules from (4.2) and (5.2) can only be used during the last derivation step of a successful derivation.

Claim 4.1.18. G generates every $w \in L(G)$ as follows:

$$S \Rightarrow_G u \ [p] \Rightarrow_G^* v \Rightarrow_G w \ [q]$$

where p is the rule introduced in (1), q is a rule introduced in (4.2) or (5.2), and during $u \Rightarrow_G^* v$, G makes every derivation step by a rule introduced in (2), (3), (4.1), or (5.1).

Proof. Let $w \in L(G)$. By Claim 4.1.17, as $w \in T^*$, G uses the rule introduced in (1) only once. Because $S \Rightarrow_G^* w$ begins from S, we can express $S \Rightarrow_G^* w$ as

$$S \Rightarrow_G u \ [p] \Rightarrow_G^* w$$

where p is the rule introduced in (1), and G never uses this rule during $u \Rightarrow_G^* w$. Observe that every rule r introduced in (2), (3), (4.1), and (5.1) satisfies $\#_S(\mathrm{lhs}(r)) = 3$ and $\#_S(\mathrm{rhs}(r)) = 3$. Furthermore, notice that every rule q introduced in (4.2) and (5.2) satisfies $\#_S(\mathrm{lhs}(q)) = 3$ and $\#_S(\mathrm{rhs}(q)) = 0$. These observations imply

$$S \Rightarrow_G u \ [p] \Rightarrow_G^* v \Rightarrow_G w \ [q]$$

where p is the rule introduced in (1), q is a rule introduced in (4.2) or (5.2), and during $u \Rightarrow_G^* v$, G makes every step by a rule introduced in (2), (3), (4.1), or (5.1). $\qquad\square$

In what follows, we demonstrate that in order to apply a rule from (3) through (5), there have to be exactly n nonterminals between the first and the second occurrence of S. This can be accomplished by one or more applications of a rule from (2).

Claim 4.1.19. If $x \Rightarrow_G y \ [p]$ is a derivation step in a successful derivation of G, where p is a rule from (3) through (5), then $x = x_1 S x_2 S x_3 S A$, where $x_1, x_2, x_3 \in (T \cup \{A, B\})^+$, $\#_{\{A,B\}}(x_1) = k$, $\#_{\{A,B\}}(x_2) = m$, and $k = m = n$.

Proof. If $k < n$ or $m < n$, no rule introduced in (3) through (5) can be used. Therefore, $k \geq n$ and $m \geq n$.

Assume that $k > n$. The only rules that remove the symbols from $\{A, B\}$ in front of the first symbol S are those introduced in (4.2) and (5.2), and these rules remove precisely n nonterminals preceding the first symbol S. For $k > n$, $k - n$ nonterminals remain in the sentential form after the last derivation step so the derivation is unsuccessful. Therefore, $k = n$.

Assume that $m > n$. Then, after the application of a rule introduced in (3) through (5), m symbols from $\{A, B\}$ appear in front of the first S. Therefore, the number of nonterminals appearing in front of the first occurrence of S is greater than n, which contradicts the argument given in the previous paragraph. As a result, $m = n$. $\qquad\square$

Based on Claims 4.1.16 through 4.1.19 and the properties of Q, we can express every successful derivation of G as

- either $S \Rightarrow_G \mathrm{rhs}(p_1) \ [p_1] \Rightarrow_G^* u \ [\Xi] \Rightarrow_G v \ [p_{4a}] \Rightarrow_G^* w \ [\Psi] \Rightarrow_G z \ [p_{5b}]$ for $z \neq \varepsilon$;
- or $S \Rightarrow_G \mathrm{rhs}(p_1) \ [p_1] \Rightarrow_G^* u \ [\Xi] \Rightarrow_G \varepsilon \ [p_{4b}]$;

where p_1, p_{4a}, p_{4b}, and p_{5b}, are rules introduced in (1), (4.1), (4.2), and (5.2), respectively, $\varXi$ is a sequence of rules from (2) and (3), $\varPsi$ is a sequence of rules from (2) and (5.1), and the derivation satisfies the following properties.

- Every derivation step in $\mathrm{rhs}(p_1) \Rightarrow_G^* u \, [\varXi]$ has one of these forms:

$$\beta(b_1)Sa_1'Sa_1''d_1y_1'SA \Rightarrow_G \beta(b_1)Sa_1'a_1''d_1Sy_1'SA \, [p_2], \text{ or}$$
$$\beta(b_1)S\beta(a_1)S\beta(y_1)SA \Rightarrow_G \beta(c_1)SS\beta(y_1x_1)SA \, [p_3]$$

  where a_1', a_1'', $y_1' \in \{A, B\}^*$, $d_1 \in \{A, B\}$, $(a_1, b_1, x_1, c_1) \in R$, $b_1 \neq 1$, $c_1 \neq 1$, $y_1 \in (\bar{V} - T)^*$, and p_2, p_3 are rules introduced in (2), (3), respectively.
- The derivation step $u \Rightarrow_G v \, [p_{4a}]$ has this form

$$\beta(b_2)S\beta(a_2)S\beta(y_2)SA \Rightarrow_G \beta(1)SS\beta(y_2x_2)SA \, [p_{4a}]$$

  where $(a_2, b_2, x_2, 1) \in R$, $b_2 \neq 1$, and $y_2 \in (\bar{V} - T)^+$. Observe that if $y_2x_2 = \varepsilon$, no rule is applicable after this step and the derivation is blocked.
- The derivation step $u \Rightarrow_G \varepsilon \, [p_{4b}]$ has this form

$$\beta(b_3)S\beta(a_3)S\beta(y_3)SA \Rightarrow_G \varepsilon \, [p_{4b}]$$

  where $(a_3, b_3, \varepsilon, c_3) \in R$, $b_3 \neq 1$, $c_3 \in F$, and $y_3 = \varepsilon$. As no rule can be applied after a rule from (4.2) is used, if $y_3 \neq \varepsilon$, there remain some nonterminals in the sentential form so the derivation is unsuccessful.
- Every derivation step in $v \Rightarrow_G^* w \, [\varPsi]$ has one of these forms

$$\beta(1)Sa_4'Sa_4''d_4y_4't_4SA \Rightarrow_G \beta(1)Sa_4'a_4''d_4Sy_4t_4SA \, [p_2], \text{ or}$$
$$\beta(1)S\beta(a_4)S\beta(y_4)t_4SA \Rightarrow_G \beta(1)SS\beta(y_4)t_4x_4SA \, [p_{5a}]$$

  where a_4', a_4'', $y_4' \in \{A, B\}^*$, $d_4 \in \{A, B\}$, $(a_4, 1, x_4, 1) \in R$, $y_4 \in (\bar{V} - T)^*$, $t_4 \in T^*$, and p_2, p_{5a} are rules introduced in (2), (5.1), respectively.
- The derivation step $w \Rightarrow_G z \, [p_{5b}]$ has this form

$$\beta(1)S\beta(a_5)St_5SA \Rightarrow_G t_5x_5 \, [p_{5b}]$$

  where $(a_5, 1, x_5, c_5) \in R$, $c_5 \in F$, and $t_5 \in T^*$.

Observe that

$$S \Rightarrow_G \mathrm{rhs}(p_1) \, [p_1] \Rightarrow_G^* u \, [\varXi] \Rightarrow_G v \, [p_{4a}] \Rightarrow_G^* w \, [\varPsi] \Rightarrow_G z \, [p_{5b}], \text{ for } z \neq \varepsilon$$

if and only if

$$g \Rightarrow_Q^* a_2y_2b_2 \Rightarrow_Q y_2x_21 \, [(a_2, b_2, x_2, 1)]$$
$$\Rightarrow_Q^* a_5t_51 \Rightarrow_Q zc_5 \, [(a_5, 1, x_5, c_5)]$$

or

$$S \Rightarrow_G \text{rhs}(p_1) \; [p_1] \Rightarrow_G^* u \; [\Xi] \Rightarrow_G \varepsilon \; [p_{4b}]$$

if and only if

$$g \Rightarrow_Q^* a_3 y_3 b_3 \Rightarrow_Q c_3 \; [(a_3, b_3, \varepsilon, c_3)]$$

As a result, $L(Q) = L(G)$, so the theorem holds. □

Recall that one-nonterminal scattered context grammars are incapable of generating all recursively enumerable languages (see Theorem 4.1.13). By Theorem 4.1.15, three-nonterminal scattered context grammars characterize **RE**. As stated in the following theorem, the optimal bound for the needed number of nonterminals is, in fact, two. This very recent result is proved in [CVV10].

Theorem 4.1.20 (See [CVV10]). *For every recursively enumerable language L, there is a scattered context grammar* $G = (V, T, P, S)$ *such that* $L(G) = L$, *and*

$$\text{card}\,(V - T) = 2$$

Up until now, we have reduced only one measure of descriptional complexity regardless of all the other measures. We next reconsider this topic in a finer way by simultaneously reducing several measures. It turns out that this simultaneous reduction results in an increase of all the measures involved. In addition, reducing the number of nonterminals necessarily leads to an increase of the number of context-sensitive rules and vice versa.

Theorem 4.1.21. *For every recursively enumerable language L, there is a scattered context grammar* $G = (V, T, P, S)$ *such that* $L(G) = L$, *and*

$$\text{card}(V - T) = 5$$
$$\text{dcs}(G) = 2$$
$$\text{mcs}(G) = 3$$
$$\text{ocs}(G) = 6$$

Proof. (See [Vas05].) Let

$$G' = (\{S', A, B, C, D\} \cup T, T, P' \cup \{AB \to \varepsilon, CD \to \varepsilon\}, S')$$

be a phrase-structure grammar in the Geffert normal form, where P' is a set of context-free rules, and $L(G') = L$ (see Theorem 3.1.11). Define the homomorphism h from $\{A, B, C, D\}^*$ to $\{0, 1\}^*$ so that $h(A) = h(B) = 00$, $h(C) = 10$, and $h(D) = 01$. Define the scattered context grammar

$$G = (\{S, \bar{S}, 0, 1, \$\} \cup T, T, P, S)$$

with P constructed as follows:

(1) for each $S' \to zS'a \in P'$, where $z \in \{A, C\}^*$, $a \in T$, extend P by adding

$$(S) \to (h(z)Sa)$$

(2) add $(S) \to (\bar{S})$ to P;

(3) for each $S' \to uS'v \in P'$, where $u \in \{A, C\}^*$, $v \in \{B, D\}^*$, extend P by adding

$$(\bar{S}) \to (h(u)\bar{S}h(v))$$

(4) extend P by adding

  (4.a) $(\bar{S}) \to (\$\$)$,
  (4.b) $(0, \$, \$, 0) \to (\$, \varepsilon, \varepsilon, \$)$,
  (4.c) $(1, \$, \$, 1) \to (\$, \varepsilon, \varepsilon, \$)$,
  (4.d) $(\$) \to (\varepsilon)$.

Observe that G' generates every $a_1 \cdots a_k \in L(G')$ in the following way

$$
\begin{aligned}
S' &\Rightarrow_{G'} z_{a_k} S' a_k \\
&\Rightarrow_{G'} z_{a_k} z_{a_{k-1}} S' a_{k-1} a_k \\
&\ \ \vdots \\
&\Rightarrow_{G'} z_{a_k} \cdots z_{a_2} S' a_2 \cdots a_k \\
&\Rightarrow_{G'} z_{a_k} \cdots z_{a_2} z_{a_1} S' a_1 a_2 \cdots a_k \\
&\Rightarrow_{G'} z_{a_k} \cdots z_{a_2} z_{a_1} u_l S' v_l a_1 a_2 \cdots a_k \\
&\ \ \vdots \\
&\Rightarrow_{G'} z_{a_k} \cdots z_{a_1} u_l \cdots u_2 S' v_2 \cdots v_l a_1 \cdots a_k \\
&\Rightarrow_{G'} z_{a_k} \cdots z_{a_1} u_l \cdots u_2 u_1 v_1 v_2 \cdots v_l a_1 \cdots a_k \\
&= \quad\ d_m \cdots d_2 d_1 e_1 e_2 \cdots e_n a_1 \cdots a_k \\
&\Rightarrow_{G'} d_m \cdots d_2 e_2 \cdots e_n a_1 \cdots a_k \\
&\ \ \vdots \\
&\Rightarrow_{G'} d_m e_n a_1 \cdots a_k \\
&\Rightarrow_{G'} a_1 \cdots a_k
\end{aligned}
$$

where $a_1, \ldots, a_k \in T$, $z_{a_1}, \ldots, z_{a_k}$, $u_1, \ldots, u_l \in \{A, C\}^*$, $v_1, \ldots, v_l \in \{B, D\}^*$, $d_1, \ldots, d_m \in \{A, C\}$, and $e_1, \ldots, e_n \in \{B, D\}$. After erasing S' from the sentential form, G' verifies that the generated strings $z_{a_k} \cdots z_{a_1} u_l \cdots u_1$ and $v_1 \cdots v_l$ are identical. If $m \neq n$, or $d_i e_i \notin \{AB, CD\}$, for some $i \geq 1$, the generated strings do not coincide, and the derivation is blocked, so $a_1 \cdots a_k$ does not belong to the generated language.

The above derivation can be straightforwardly simulated by G as follows:

$$S \Rightarrow_G h(z_{a_k})Sa_k$$
$$\Rightarrow_G h(z_{a_k})h(z_{a_{k-1}})Sa_{k-1}a_k$$

$$\vdots$$

$$\Rightarrow_G h(z_{a_k})\cdots h(z_{a_2})Sa_2\cdots a_k$$
$$\Rightarrow_G h(z_{a_k})\cdots h(z_{a_2})h(z_{a_1})Sa_1a_2\cdots a_k$$
$$\Rightarrow_G h(z_{a_k}\cdots z_{a_2}z_{a_1})\bar{S}a_1a_2\cdots a_k \qquad [p_2]$$
$$\Rightarrow_G h(z_{a_k}\cdots z_{a_2}z_{a_1})h(u_l)\bar{S}h(v_l)a_1a_2\cdots a_k$$

$$\vdots$$

$$\Rightarrow_G h(z_{a_k}\cdots z_{a_1})h(u_l)\cdots h(u_2)\bar{S}h(v_2)\cdots h(v_l)a_1\cdots a_k$$
$$\Rightarrow_G h(z_{a_k}\cdots z_{a_1})h(u_l)\cdots h(u_2)h(u_1)\bar{S}h(v_1)h(v_2)\cdots h(v_l)a_1\cdots a_k$$
$$\Rightarrow_G h(z_{a_k}\cdots z_{a_1})h(u_l\cdots u_2u_1)\$\$h(v_1v_2\cdots v_l)a_1\cdots a_k \qquad [p_{4a}]$$
$$= \quad f_r\cdots f_2f_1\$\$g_1g_2\cdots g_sa_1\cdots a_k$$
$$\Rightarrow_G f_r\cdots f_2\$\$g_2\cdots g_sa_1\cdots a_k$$

$$\vdots$$

$$\Rightarrow_G f_r\$\$g_sa_1\cdots a_k$$
$$\Rightarrow_G \$\$a_1\cdots a_k$$
$$\Rightarrow_G \$a_1\cdots a_k \qquad [p_{4d}]$$
$$\Rightarrow_G a_1\cdots a_k \qquad [p_{4d}]$$

where $f_1,\ldots,f_r,g_1,\ldots,g_s \in \{0,1\}$, and p_2, p_{4a}, and p_{4d} are rules introduced in (2), (4.a), and (4.d), respectively. In this derivation, the context-free rules of G' are simulated by the rules introduced in (1) through (3), and the context-sensitive rules of G' are simulated by the rules introduced in (4.b) and (4.c). There are the following differences between the derivations in G' and G.

- Instead of verifying the identity of $z_{a_k}\cdots z_{a_1}u_l\cdots u_1$ and $v_1\cdots v_l$, G verifies that $h(z_{a_k}\cdots z_{a_1}u_l\cdots u_1)$ and $h(v_1\cdots v_l)$ coincide. This means that instead of comparing strings over $\{A,B,C,D\}$, G compares the strings $f_r\cdots f_1$ and $g_1\cdots g_s$ over $\{0,1\}$.
- The rule introduced in (2) guarantees that no rule from (1) can be used after its application. Similarly, the rule introduced in (4.a) prevents the rules of (1) through (3) from being applied.
- When applying the rules from (4.b) and (4.c), some symbols f_i and g_j, where i, $j \geq 1$, can be skipped. However, if some 0s and 1s that do not directly neighbor with the $\$$s are rewritten, the form of these rules guarantees that the skipped nonterminals can never be rewritten later in the derivation, so the derivation is necessarily unsuccessful in this case.

- The rule from (4.d) can be used anytime the symbol $ appears in the sentential form. However, when this rule is used and some nonterminals from $\{0, 1\}$ occur in the sentential form, these nonterminals can never be removed from the sentential form, so the derivation is blocked. As a result, the rule from (4.d) has to be applied at the very end of the derivation.

These observations imply that $L = L(G) = L(G')$. As obvious, $\text{card}(V - T) = 5$, $\text{dcs}(G) = 2$, $\text{mcs}(G) = 3$, $\text{ocs}(G) = 6$. Thus, the theorem holds.   □

Theorem 4.1.22. *For every recursively enumerable language L, there is a scattered context grammar* $\bar{G} = (V, T, \bar{P}, S)$ *such that* $L(\bar{G}) = L$, *and*

$$\text{card}(V - T) = 8$$
$$\text{dcs}(\bar{G}) = 6$$
$$\text{mcs}(\bar{G}) = 1$$
$$\text{ocs}(\bar{G}) = 6$$

Proof. We slightly modify the construction given in the proof of Theorem 4.1.21. Define the scattered context grammar

$$\bar{G} = \left(\{S, \bar{S}, 0, 1, \$_L, \$_R, \$_0, \$_1\} \cup T, T, \bar{P}, S\right)$$

and initialize $\bar{P}$ with the set of all rules introduced in steps (1) through (3) of the construction given in the proof of Theorem 4.1.21. Then, add the following rules to $\bar{P}$

(4)

    (4.a) $(\bar{S}) \rightarrow (\$_L \$_R)$,
    (4.b) $(0, \$_L) \rightarrow (\$_0, \varepsilon)$, $(\$_R, 0) \rightarrow (\varepsilon, \$_0)$, $(\$_0, \$_0) \rightarrow (\$_L, \$_R)$,
    (4.c) $(1, \$_L) \rightarrow (\$_1, \varepsilon)$, $(\$_R, 1) \rightarrow (\varepsilon, \$_1)$, $(\$_1, \$_1) \rightarrow (\$_L, \$_R)$,
    (4.d) $(\$_L) \rightarrow (\varepsilon)$, and $(\$_R) \rightarrow (\varepsilon)$.

Observe that a single derivation step made by a rule introduced in step (4.b) or (4.c) of the construction of G is simulated in $\bar{G}$ by the above rules from (4.b) or (4.c) in three derivation steps. In a greater detail, a derivation of the form

$$x0\$\$0yz \Rightarrow_G x\$\$yz \ [(0, \$, \$, 0) \rightarrow (\$, \varepsilon, \varepsilon, \$)]$$

is simulated by $\bar{G}$ as follows:

$$x0\$_L\$_R0yz \Rightarrow_{\bar{G}} x\$_0\$_R0yz \ [(0, \$_L) \rightarrow (\$_0, \varepsilon)]$$
$$\Rightarrow_{\bar{G}} x\$_0\$_0yz \ [(\$_R, 0) \rightarrow (\varepsilon, \$_0)]$$
$$\Rightarrow_{\bar{G}} x\$_L\$_Ryz \ [(\$_0, \$_0) \rightarrow (\$_L, \$_R)]$$

where $x, y \in \{0, 1\}^*$, and $z \in T^*$. The rest of the proof resembles the proof of Theorem 4.1.21 and is, therefore, left to the reader. □

Theorem 4.1.23. *For every recursively enumerable language L, there is a scattered context grammar $G = (V, T, P, S)$ such that $L(G) = L$, and*

$$\text{card}(V - T) = 4$$
$$\text{dcs}(G) = 4$$
$$\text{mcs}(G) = 5$$
$$\text{ocs}(G) = 20$$

Proof. Let

$$G' = (\{S', A, B, C, D\} \cup T, T, P' \cup \{AB \to \varepsilon, CD \to \varepsilon\}, S')$$

be a phrase-structure grammar in the Geffert normal form, where P' is a set of context-free rules, and $L(G') = L$ (see Theorem 3.1.11). Define the homomorphism h from $\{A, B, C, D\}^*$ to $\{0, 1\}^*$ so that $h(A) = h(B) = 00$, $h(C) = 10$, and $h(D) = 01$. Define the scattered context grammar

$$G = (\{S, 0, 1, \$\} \cup T, T, P, S)$$

with P constructed as follows:

(1) add $(S) \to (11S11)$ to P;
(2) for each $S' \to zS'a \in P'$, add $(S) \to (h(z)S1a1)$ to P;
(3) for each $S' \to uS'v \in P'$, add $(S) \to (h(u)Sh(v))$ to P;
(4) for each $S' \to uv \in P'$, add $(S) \to (h(u)\$\$h(v))$ to P.
(5) add

    (5.a) $(0, 0, \$, \$, 0, 0) \to (\$, \varepsilon, \varepsilon, \varepsilon, \varepsilon, \$)$,
    (5.b) $(1, 0, \$, \$, 0, 1) \to (\$, \varepsilon, \varepsilon, \varepsilon, \varepsilon, \$)$,
    (5.c) $(1, 1, \$, \$, 1, 1) \to (11\$, \varepsilon, \varepsilon, \varepsilon, \varepsilon, \$)$, and
    (5.d) $(1, 1, \$, \$, 1, 1) \to (\varepsilon, \varepsilon, \varepsilon, \varepsilon, \varepsilon, \varepsilon)$ to P.

Every successful derivation starts by an application of the rule introduced in (1), and this rule is not used during the rest of the derivation. Rules from (2) through (4) simulate the context-free rules of G'. After the rule from (4) is used, only rules from (5) are applicable. The rules from (5.a) and (5.b) verify that the strings over $\{0, 1\}$, generated by the rules from (2) through (4), coincide. The rule from (5.c) removes the 1s between the terminal symbols and, in addition, makes sure that rules from (2) can never be used in a successful derivation after a rule from (3) is applied. Finally, the rule from (5.d) completes the derivation.

The proof of the theorem is based on five claims, established next.

Claim 4.1.24. Every successful derivation in G can be expressed as

$$
\begin{aligned}
S &\Rightarrow_G^* v \quad [\varXi] \\
&\Rightarrow_G w \quad [p_4] \\
&\Rightarrow_G^* y \quad [\Psi] \\
&\Rightarrow_G z \quad [p_{5d}]
\end{aligned}
$$

where

$$
\begin{aligned}
v &\in \{0, 1\}^* \{S\} \big(\{0, 1\} \cup T\big)^* \\
w &\in \{0, 1\}^+ \{\$\}\{\$\} \big(\{0, 1\} \cup T\big)^* \\
y &\in \{1\}\{1\}\{\$\}T^*\{\$\}T^*\{1\}T^*\{1\}T^*
\end{aligned}
$$

$z \in T^*$, p_4 and p_{5d} are rules introduced in (4) and (5.d), respectively, and $\varXi$ and Ψ are sequences of rules introduced in (1) through (3) and (5.a) through (5.c), respectively.

Proof. As S appears on the left-hand side of every rule introduced in (1) through (4), all of them are applicable while S occurs in the sentential form. On the other hand, no rule from (5) can be used at this point. After p_4 is used, it replaces S with \$\$, so rules from (1) through (4) are not applicable and only rules from (5) can be used. Therefore, the beginning of the derivation can be expressed as

$$
\begin{aligned}
S &\Rightarrow_G^* v \quad [\varXi] \\
&\Rightarrow_G w \quad [p_4]
\end{aligned}
$$

Because all rules, except for p_{5d}, contain nonterminals on their right-hand sides, p_{5d} has to be applied in the last derivation step and no other rule can be applied after its use. Applications of rules from (2) through (4) may introduce some nonterminals 0 and 1 to the sentential form, so in this case, the rules from (5.a) and (5.b) are applied to remove them. As a result,

$$
\begin{aligned}
w &\Rightarrow_G^* y \quad [\Psi] \\
&\Rightarrow_G z \quad [p_{5d}]
\end{aligned}
$$

and the sentential forms satisfy the conditions given in the claim.   □

Claim 4.1.25. In $w \Rightarrow_G^+ z$ from Claim 4.1.24, every sentential form s satisfies

$$
s \in \{0, 1\}^* \{\$\}T^*\{\$\} \big(\{0, 1\} \cup T\big)^*
$$

Proof. The form of the rules introduced in (5) implies that whenever a nonterminal appears between the two occurrences of \$, it can never be removed during the rest of the derivation. Therefore, the claim holds.   □

Claim 4.1.26. In $w \Rightarrow_G^+ z$ from Claim 4.1.24, every sentential form s satisfies

$$s \in \{1\}\{1\}\left(\{1\}\{0\} \cup \{0\}\{0\}\right)^* \{\$\} T^* \{\$\}\left(\{0, 1\} \cup T\right)^*$$

Proof. Claim 4.1.25 implies that whenever rules from (5) are used, each of these rules is applied to the nonterminals from $\{0, 1\}$ immediately preceding the first occurrence of $ and immediately following the second occurrence of $; otherwise, the derivation is unsuccessful. As a result, the only rule that removes the substring 11 preceding the first occurrence of $ is (5.d). However, by Claim 4.1.24, (5.d) is used during the very last derivation step, so the substring 11 has to appear at the beginning of the sentential form in order to generate a string over T.   □

Claim 4.1.27. The derivation

$$S \Rightarrow_G^* v \ [\Xi]$$

from Claim 4.1.24 can be expressed, in a greater detail, as

$$\begin{aligned} S &\Rightarrow_G 11S11 \ [p_1] \\ &\Rightarrow_G^* v \end{aligned}$$

where p_1 is the rule introduced in (1), and this rule is not used during the rest of the derivation.

Proof. The rule introduced in (1) is the only rule that introduces the substring 11 in front of the first occurrence of $. By Claim 4.1.26, in front of the first $, this substring appears only at the beginning of every sentential form in $w \Rightarrow_G^+ z$, so p_1 has to be applied at the beginning of the derivation and cannot be used later in the derivation.   □

Claim 4.1.28. The derivation

$$\begin{aligned} w &\Rightarrow_G^* y \ [\Psi] \\ &\Rightarrow_G z \ [p_{5d}] \end{aligned}$$

from Claim 4.1.24 can be expressed, in a greater detail, as

$$\begin{aligned} w &\Rightarrow_G^* x \ [\Psi_1] \\ &\Rightarrow_G^* y \ [\Psi_2] \\ &\Rightarrow_G z \ [p_{5d}] \end{aligned}$$

where

$$x \in \{1\}\{1\}\left(\{1\}\{0\} \cup \{0\}\{0\}\right)^* \{\$\} T^* \{\$\}\left(\{0, 1\} \cup T\right)^*$$

Ψ_1 is a sequence of rules introduced in (5.a) and (5.b), and Ψ_2 is a sequence of rules introduced in (5.c).

Proof. The proof of this claim follows immediately from Claims 4.1.25 and 4.1.26.

$\square$

Claim 4.1.29. The derivation

$$11S11 \Rightarrow_G^* v$$

from Claims 4.1.24 and 4.1.27 can be expressed in a greater detail as

$$11S11 \Rightarrow_G^* u \; [\varXi_1]$$
$$\Rightarrow_G^* v \; [\varXi_2]$$

where

$$u \in \{1\}\{1\}\big(\{1\}\{0\} \cup \{0\}\{0\}\big)^* S\big(\{0\}\{1\} \cup \{0\}\{0\}\big)^* \big(\{1\}T\{1\}\big)^* \{1\}\{1\}$$

and $\varXi_1$, $\varXi_2$ are sequences of rules introduced in (2), (3), respectively.

Proof. By Claim 4.1.27, every derivation starts by an application of the rule from (1). Therefore, u ends with 11. Next, notice that the two nonterminals 1 surrounding a, where $a \in T$, introduced by every application of a rule from (2) can only be removed by the rule from (5.c). Indeed, by Claim 4.1.25, any other rule leaves a nonterminal between the two symbols \$, so the derivation is unsuccessful. By Claim 4.1.28, rules from (5.a) and (5.b) cannot be applied after the rule from (5.c) is used. As a result, the generation of the strings over $\{0, 1\}$ by rules from (2) and (3) has to correspond to their removal by (5.a), (5.b), and (5.c). This implies that rules from (2) have to be applied before rules from (3). $\square$

Based upon Claims 4.1.24 through 4.1.29, we see that every successful derivation is of this form

$$S \Rightarrow_G 11S11 \; [p_1]$$
$$\Rightarrow_G^* u \qquad [\varXi_1]$$
$$\Rightarrow_G^* v \qquad [\varXi_2]$$
$$\Rightarrow_G w \qquad [p_4]$$
$$\Rightarrow_G^* x \qquad [\Psi_1]$$
$$\Rightarrow_G^* y \qquad [\Psi_2]$$
$$\Rightarrow_G z \qquad [p_{5d}]$$

As the rest of this proof can be made by analogy with the proof of Theorem 4.1.21, we leave it to the reader. $\square$

4.1.5 Economical Transformations

The generation of languages is frequently performed in a specifically required way and based upon a prescribed set of grammatical components, such as a certain collection of nonterminals or rules. On the other hand, if these requirements are met, the generation can be based upon grammars of various types. For this purpose, we often make use of transformations that convert grammars of some type to equivalent grammars of another type so the transformed grammars strongly resemble the original grammars regarding the way they work as well as the components they consist of. In other words, we want the output grammars resulting from these transformations to work similarly to the way the given original grammars work and, perhaps even more importantly, to contain the same set of grammatical components possibly extended by very few additional components. Transformations that produce scattered context grammars in this economical way are discussed throughout the rest of this section. Because phrase-structure grammars represent one of the very basic grammatical models in formal language theory (see Sect. 2.3.1), this section pays a special attention to the economical transformations that convert these fundamental grammars to equivalent scattered context grammars.

To compare the measures of scattered context and phrase-structure grammars, we first define the degree of context-sensitivity of phrase-structure grammars analogously to the degree of context sensitivity of scattered context grammars (see Definition 4.1.12).

Definition 4.1.30. Let $G = (V, T, P, S)$ be a phrase-structure grammar. Its *degree of context sensitivity*, symbolically written as $\mathrm{dcs}(G)$, is defined as

$$\mathrm{dcs}(G) = \mathrm{card}\left(\{x \to y \mid x \to y \in P, |x| \geq 2\}\right) \qquad \square$$

Theorem 4.1.31. *For every phrase-structure grammar* $G = (V, T, P, S)$ *in the Kuroda normal form, there is a scattered context grammar* $\bar{G} = (\bar{V}, T, \bar{P}, \bar{S})$ *such that* $L(\bar{G}) = L(G)$, *and*

$$\mathrm{card}(\bar{V}) = \mathrm{card}(V) + 5$$
$$\mathrm{card}(\bar{P}) = \mathrm{card}(P) + 4$$
$$\mathrm{dcs}(\bar{G}) = \mathrm{dcs}(G) + 2$$

Proof. Let $G = (V, T, P, S)$ be a phrase-structure grammar in the Kuroda normal form. Without any loss of generality, assume that $V \cap \{\bar{S}, F, 0, 1, \$\} = \emptyset$. Set $\bar{V} = V \cup \{\bar{S}, F, 0, 1, \$\}$. Define the scattered context grammar

$$\bar{G} = (\bar{V}, T, \bar{P}, \bar{S})$$

where $\bar{P}$ is constructed as follows:

(1) add $(\bar{S}) \to (FFFS)$ to $\bar{P}$;
(2) for each $AB \to CD \in P$, add $(A, B) \to (C0, 1D)$ to $\bar{P}$;
(3) for each $A \to BC \in P$, add $(A) \to (BC)$ to $\bar{P}$;
(4) for each $A \to a \in P$, where $a \in T \cup \{\varepsilon\}$, add $(A) \to (\$a)$ to $\bar{P}$;
(5) add

    (5.a) $(F, 0, 1, F, F) \to (\varepsilon, F, F, \varepsilon, F)$,
    (5.b) $(F, F, F, \$) \to (\varepsilon, \varepsilon, F, FF)$,
    (5.c) $(F) \to (\varepsilon)$ to $\bar{P}$.

The rule from (1) starts a derivation and introduces three occurrences of the nonterminal F, which are present in every sentential form until three applications of the rule from (5.c) complete the derivation. Rules from (2), (3), and (4) simulate the corresponding rules of the Kuroda normal form behind the last occurrence of F. The rules from (5.a) and (5.b) guarantee that before (5.c) is applied for the first time, every sentential form in a successful derivation belongs to

$$T^*\{F\}(T \cup \{\varepsilon\})\{0^i 1^i \mid i \geq 0\}\{F\}\{F\}(V \cup \{0, 1, \$\})^*$$

and, thereby, the simulation of every derivation of G is performed properly. Notice that there are only terminals in front of the first nonterminal F. Moreover, the only nonterminals appearing between the first occurrence and the second occurrence of F are from $\{0, 1\}$, and there is no symbol between the second and the third occurrence of F in a successful derivation.

Next, we establish several claims to demonstrate that $L(G) = L(\bar{G})$ in a rigorous way.

Claim 4.1.32. Every successful derivation of $\bar{G}$ can be expressed as

$$\begin{aligned}
\bar{S} &\Rightarrow_{\bar{G}} FFFS \quad [p_1] \\
&\Rightarrow_{\bar{G}}^* uFvFxFy \ [\Psi] \\
&\Rightarrow_{\bar{G}}^* w \\
&\Rightarrow_{\bar{G}}^3 z \quad [p_{5c}p_{5c}p_{5c}]
\end{aligned}$$

where $u, z \in T^*$, $v, x, y \in (\bar{V} - \{\bar{S}, F\})^*$, $w \in (\bar{V} - \{\bar{S}\})^*$, p_1 and p_{5c} are rules introduced in (1) and (5.3), respectively, and Ψ is a sequence of rules introduced in (2) through (5.2).

Proof. The only rule with $\bar{S}$ on its left-hand side is the rule introduced in (1), and because no rule contains $\bar{S}$ on its right-hand side, this rule is not used during the rest of the derivation process. As a result,

$$\bar{S} \Rightarrow_{\bar{G}} FFFS \ [p_1]$$

Observe that no rule from (2) through (4) contains the nonterminal F and rules from (5.a) and (5.b) contain three nonterminals F on their left-hand sides as well as their right-hand sides. The rule from (5.c), which is the only rule with its right-hand side over T, removes F from the sentential form, so no rule from (5.a) and (5.b) can be used once it is applied. Notice that rules from (4) simulate $A \rightarrow a$, where $A \in V - T$, $a \in T \cup \{\varepsilon\}$, and these rules introduce $\$$ to the sentential form. In addition, observe that only the rule from (5.b) rewrites $\$$. Consequently, to generate a string over T, rules from (2) through (4) cannot be used after the rule from (5.c) is applied. Therefore,

$$w \Rightarrow_G^3 z \; [p_{5c}p_{5c}p_{5c}]$$

Notice that rules from (5.a) and (5.b) cannot rewrite any symbol in u. If $\mathrm{alph}(u) \cap (\bar{V}-T) \neq \emptyset$, then a nonterminal from $\{0, 1, \$\}$ remains in front of the first F because rules from (2) through (4) cannot rewrite u to a string over T, so the derivation would be unsuccessful in this case. Therefore, $u \in T^*$, and the claim holds. □

Claim 4.1.33. Let

$$\bar{S} \Rightarrow_G^+ uFvFxFy \Rightarrow_G^* w \Rightarrow_G^3 z$$

where $u, z \in T^*$, $v, x, y \in \left(\bar{V} - \{\bar{S}, F\}\right)^*$, and $w \in \left(\bar{V} - \{\bar{S}\}\right)^*$. Then, $x \in T^*$.

Proof. First, notice that if $(\bar{V} - T) \cap \mathrm{alph}(x) \neq \emptyset$, x cannot be rewritten to a string over T by using only rules from (2) through (4). Next, examine the rules from (5.a) and (5.b) to see that these rules cannot rewrite any symbol from x, and the rule from (5.b) moves x in front of the first occurrence of F. However, by Claim 4.1.32, no nonterminal can appear in front of the first F. As a result, $(\bar{V}-T) \cap \mathrm{alph}(x) = \emptyset$, so $x \in T^*$. □

Claim 4.1.34. Let

$$\bar{S} \Rightarrow_G^+ uFvFxFy \Rightarrow_G^* w \Rightarrow_G^3 z$$

where $u, z \in T^*$, $v, x, y \in \left(\bar{V} - \{\bar{S}, F\}\right)^*$, and $w \in \left(\bar{V} - \{\bar{S}\}\right)^*$. Then, $v = v'v''$, where $v' \in \left(\{0\} \cup T\right)^*$, $v'' \in \left(\{1\} \cup T\right)^*$, and $\#_0(v') = \#_1(v'')$.

Proof. First, notice that if $(\bar{V} - T) \cap \mathrm{alph}(v) \neq \emptyset$, v cannot be rewritten to a string over T by using only rules from (2) through (4). Next, examine the rules from (5.a) and (5.b).

First, observe that the rule from (5.b) can only be applied if $v \in T^*$. Indeed, (5.b) moves v in front of the first F, and if $(\bar{V} - T) \cap \mathrm{alph}(v) \neq \emptyset$, then Claim 4.1.32 implies that the derivation is unsuccessful. Therefore, $(\bar{V}-T) \cap \mathrm{alph}(v) = \emptyset$ before the rule from (5.b) is applied. Second, observe that because the rule from (5.a) rewrites only nonterminals over $\{0, 1\}$ in v, $\left((V - T) \cup \{\$\}\right) \cap \mathrm{alph}(v) = \emptyset$. Finally, observe that the rule from (5.a) has to be applied so that the first 0 following the first F and the first 1 preceding the second F is rewritten by (5.a). If this property is not satisfied, the form of (5.a) implies that 0 appears in front of the first F or 1 appears in between the second F and the third F. However, by Claims 4.1.32 and 4.1.33, this results into an unsuccessful derivation.

Based on these observations, we see that in order to generate $z \in T^*$, v has to satisfy $v = v'v''$, where $v' \in (\{0\} \cup T)^*$, $v'' \in (\{1\} \cup T)^*$, and $\#_0(v') = \#_1(v'')$. $\quad\square$

Claim 4.1.35. Let

$$\bar{S} \Rightarrow^+_G uFvFxFy \Rightarrow^*_G w \Rightarrow^3_G z$$

where $u, z \in T^*$, $v, x, y \in (\bar{V} - \{\bar{S}, F\})^*$, and $w \in (\bar{V} - \{\bar{S}\})^*$. Then,

$$y \in (T \cup \{\varepsilon\})(\{0^i 1^i \mid i \geq 0\}K)^*$$

with $K = (V - T) \cup \{\$\}(T \cup \{\varepsilon\})$, $v \in (T \cup \{\varepsilon\})\{0^i 1^i \mid i \geq 0\}$, and $x = \varepsilon$.

Proof. First, consider the rule introduced in (5.b). This rule rewrites \$ to FF in its last component. Because the nonterminal \$ is introduced by rules from (4), \$ may be followed by $a \in T$. Therefore, after (5.b) is applied, the last nonterminal F may be followed by a. As a result, the prefix of y is always over $T \cup \{\varepsilon\}$.

Second, notice that when the rule (5.b) is used, the first nonterminal \$ following the third nonterminal F has to be rewritten. In addition, the substring appearing between these symbols has to be in $\{0^i 1^i \mid i \geq 0\}$. The form of the rule introduced in (5.b) implies that after its application, this substring is moved in between the first occurrence of F and the second occurrence of F, so the conditions given by Claim 4.1.34 are satisfied. Therefore,

$$v \in (T \cup \{\varepsilon\})\{0^i 1^i \mid i \geq 0\}$$

and because no terminal appears in the suffix of v, the proof of Claim 4.1.34 implies that $x = \varepsilon$. By induction, prove that

$$y \in (T \cup \{\varepsilon\})(\{0^i 1^i \mid i \geq 0\}K)^*$$

The induction part is left to the reader. $\quad\square$

Next, we define the homomorphism α from $\bar{V}^*$ to V^* as $\alpha(\bar{A}) = \varepsilon$, for all $\bar{A} \in \bar{V} - V$, and $\alpha(A) = A$, for all $A \in V$, and use this homomorphism in the following claims.

Claim 4.1.36. Let $\bar{S} \Rightarrow^m_G w \Rightarrow^*_G z$, where $m \geq 1$, $z \in T^*$, and $w \in \bar{V}^*$. Then, $S \Rightarrow^*_G \alpha(w)$.

Proof. This claim is established by induction on $m \geq 1$.

Basis. Let $m = 1$. Then, $\bar{S} \Rightarrow_{\bar{G}} FFFS$. Because $\alpha(FFFS) = S$, $S \Rightarrow^0_G S$, so the basis holds.

Induction Hypothesis. Suppose that the claim holds for every $m \leq j$, for some $j \geq 1$.

Induction Step. Let $\bar{S} \Rightarrow_{\bar{G}}^{j+1} w \Rightarrow_{\bar{G}}^* z$, where $z \in T^*$ and $w \in \bar{V}^*$. Based on Claims 4.1.32 and 4.1.35, express this derivation as

$$\bar{S} \Rightarrow_{\bar{G}}^{j} uFvFxFy$$
$$\Rightarrow_{\bar{G}} w \; [p]$$
$$\Rightarrow_{\bar{G}}^* z$$

where $u \in T^*$, $x = \varepsilon$,

$$y \in \big(T \cup \{\varepsilon\}\big)\big(\{0^i 1^i \mid i \geq 0\}K\big)^*$$

with $K = (V - T) \cup \{\$\}\big(T \cup \{\varepsilon\}\big)$, and

$$v \in \big(T \cup \{\varepsilon\}\big)\{0^i 1^i \mid i \geq 0\}$$

By the induction hypothesis, $S \Rightarrow_G^* \alpha(uFvFxFy)$. Next, this proof considers all possible forms of p.

- Assume that $p = (A, B) \rightarrow (C0, 1D) \in \bar{P}$, where $A, B, C, D \in V - T$. Claim 4.1.35 and its proof imply $y = y'Ay''By'''$, where $y'' \in \{0^i 1^i \mid i \geq 0\}$, and

$$w = uFvFxFy'C0y''1Dy'''$$

As $(A, B) \rightarrow (C0, 1D) \in \bar{P}$, $AB \rightarrow CD \in P$ holds true. Because $\alpha(y'') = \varepsilon$,

$$\alpha(uFvFxFy'Ay''By''') \Rightarrow_G \alpha(uFvFxFy'C0y''1Dy''')$$

Therefore, $S \Rightarrow_G^* \alpha(w)$.
- Assume that $p = (A) \rightarrow (BC) \in \bar{P}$, where $A, B, C \in V - T$. Claim 4.1.35 implies that $y = y'Ay''$, and

$$w = uFvFxFy'BCy''$$

As $(A) \rightarrow (BC) \in \bar{P}$, $A \rightarrow BC \in P$ holds true. Notice that

$$\alpha(uFvFxFy'Ay'') \Rightarrow_G \alpha(uFvFxFy'BCy'')$$

Therefore, $S \Rightarrow_G^* \alpha(w)$.
- Assume that $p = (A) \rightarrow (\$a) \in \bar{P}$, where $A \in V - T$ and $a \in T \cup \{\varepsilon\}$. Claim 4.1.35 implies that $y = y'Ay''$, and

$$w = uFvFxFy'\$ay''$$

As $(A) \to (\$a) \in \bar{P}$, $A \to a \in P$ holds true. Notice that

$$\alpha(uFvFxFy'Ay'') \Rightarrow_G \alpha(uFvFxFy'\$ay'')$$

Therefore, $S \Rightarrow_G^* \alpha(w)$.
- Assume that p is a rule from (5). Notice that these rules rewrite only nonterminals over $\{0, 1, F, \$\}$. Therefore, $\alpha(w) = \alpha(uFvFxFy)$, so $S \Rightarrow_G^* \alpha(w)$.

Based on the arguments above, $\bar{S} \Rightarrow_{\bar{G}}^j uFvFxFy \Rightarrow_{\bar{G}} w$ $[p]$, for any $p \in \bar{P}$, implies that $S \Rightarrow_G^* \alpha(w)$. Thus, the claim holds.   □

Claim 4.1.37. $L(\bar{G}) \subseteq L(G)$

Proof. By Claim 4.1.36, if $\bar{S} \Rightarrow_{\bar{G}}^+ z$ with $z \in T^*$, then $S \Rightarrow_G^* z$. Therefore, the claim holds.   □

Claim 4.1.38. Let $S \Rightarrow_G^m w \Rightarrow_G^* z$, where $m \geq 0$, $w \in V^*$, and $z \in T^*$. Then, $\bar{S} \Rightarrow_{\bar{G}}^+ uFvFxFy$, where $u \in T^*$,

$$y \in (T \cup \{\varepsilon\})(\{0^i 1^i \mid i \geq 0\}K)^*$$

with $K = (V - T) \cup \{\$\}(T \cup \{\varepsilon\})$,

$$v \in (T \cup \{\varepsilon\})\{0^i 1^i \mid i \geq 0\}$$

and $x = \varepsilon$, so that $w = \alpha(uFvFxFy)$.

Proof. This claim is established by induction on $m \geq 0$.

Basis. Let $m = 0$. Then, $S \Rightarrow_G^0 S \Rightarrow_G^* z$. Notice that $\bar{S} \Rightarrow_{\bar{G}} FFFS$ by using the rule introduced in (1), and $S = \alpha(FFFS)$. Thus, the basis holds.

Induction Hypothesis. Suppose that the claim holds for every $m \leq j$, where $j \geq 1$.

Induction Step. Let $S \Rightarrow_G^{j+1} w \Rightarrow_G^* z$, where $w \in V^*$, and $z \in T^*$. Express this derivation as

$$\begin{aligned} S &\Rightarrow_G^j t \\ &\Rightarrow_G w \ [p] \\ &\Rightarrow_G^* z \end{aligned}$$

where $w \in V^*$ and $p \in P$. By the induction hypothesis, $\bar{S} \Rightarrow_{\bar{G}}^+ uFvFxFy$, where $u \in T^*$,

$$y \in (T \cup \{\varepsilon\})(\{0^i 1^i \mid i \geq 0\}K)^*$$

with $K = (V - T) \cup \{\$\}(T \cup \{\varepsilon\})$,

$$v \in (T \cup \{\varepsilon\})\{0^i 1^i \mid i \geq 0\}$$

and $x = \varepsilon$ so that $t = \alpha(uFvFxFy)$. Next, this proof considers all possible forms of p:

- Assume that $p = AB \to CD \in P$, where $A, B, C, D \in V - T$. Express $t \Rightarrow_G w$ as $t'ABt'' \Rightarrow_G t'CDt''$, where $t'ABt'' = t$ and $t'CDt'' = w$. Claim 4.1.35 implies that $y = y'A0^k1^kBy''$, where $k \geq 0$, $\alpha(uFvFxFy') = t'$, and $\alpha(y'') = t''$. As $AB \to CD \in P$, $(A, B) \to (C0, 1D) \in \bar{P}$ holds true. Then,

$$uFvFxFy'A0^k1^kBy'' \Rightarrow_{\bar{G}} uFvFxFy'C0^{k+1}1^{k+1}Dy''$$

  Therefore,

$$\bar{S} \Rightarrow^+_{\bar{G}} uFvFxFy'C0^{k+1}1^{k+1}Dy''$$

  and $w = \alpha(uFvFxFy'C0^{k+1}1^{k+1}Dy'')$.

- Assume that $p = A \to BC \in P$, where $A, B, C \in V - T$. Express $t \Rightarrow_G w$ as $t'At'' \Rightarrow_G t'BCt''$, where $t'At'' = t$ and $t'BCt'' = w$. Claim 4.1.35 implies that $y = y'Ay''$, where $\alpha(uFvFxFy') = t'$ and $\alpha(y'') = t''$. As $A \to BC \in P$, $(A) \to (BC) \in \bar{P}$ holds true. Then,

$$uFvFxFy'Ay'' \Rightarrow_{\bar{G}} uFvFxFy'BCy''$$

  Therefore,

$$\bar{S} \Rightarrow^+_{\bar{G}} uFvFxFy'BCy''$$

  and $w = \alpha(uFvFxFy'BCy'')$.

- Assume that $p = A \to a \in P$, where $A \in V - T$ and $a \in T \cup \{\varepsilon\}$. Express $t \Rightarrow_G w$ as $t'At'' \Rightarrow_G t'at''$, where $t'At'' = t$ and $t'at'' = w$. Claim 4.1.35 implies that $y = y'Ay''$, where $\alpha(uFvFxFy') = t'$ and $\alpha(y'') = t''$. As $A \to a \in P$, $(A) \to (\$a) \in \bar{P}$ holds true. Then,

$$uFvFxFy'Ay'' \Rightarrow_{\bar{G}} uFvFxFy'\$ay''$$

  Therefore,

$$\bar{S} \Rightarrow^+_{\bar{G}} uFvFxFy'\$ay''$$

  and $w = \alpha(uFvFxFy'\$ay'')$.

Consider the arguments above to see that $S \Rightarrow^j_G t \Rightarrow_G w$ $[p]$, for any $p \in P$, implies that $\bar{S} \Rightarrow^+_{\bar{G}} s$, where $w = \alpha(s)$. Thus, the claim holds. □

Claim 4.1.39. $L(G) \subseteq L(\bar{G})$

Proof. By Claims 4.1.32, 4.1.35, and 4.1.38, if $S \Rightarrow^*_G z$, where $z \in T^*$, then $\bar{S} \Rightarrow^+_{\bar{G}} z$. Therefore, Claim 4.1.39 holds. □

By Claims 4.1.37 and 4.1.39, $L(\bar{G}) = L(G)$. Observe that $\text{card}(\bar{V}) = \text{card}(V)+5$, $\text{card}(\bar{P}) = \text{card}(P) + 4$, and $\text{dcs}(\bar{G}) = \text{dcs}(G) + 2$. Thus, the theorem holds.   □

In the conclusion of this section, we point out several open problem areas.

Open Problem 4.1.40. By Theorem 4.1.21, scattered context grammars with two context-sensitive rules characterize **RE**. What is the generative power of scattered context grammars with one context-sensitive rule?   □

Open Problem 4.1.41. Revert the transformation under discussion and study economical transformations of scattered context grammars to phrase-structure grammars.   □

Open Problem 4.1.42. From a much broader perspective, apart from the transformations between scattered context grammars and phrase-structure grammars, study economical transformations between other types of grammars.   □

4.2 Totally Parallel Grammars

The totally parallel generation of languages works so that all symbols of the current sentential form are simultaneously rewritten during every single derivation step. The present section discusses this rewriting performed by *Extended tabled zero-sided Lindenmayer grammars* or, more briefly, ET0L grammars (see Section 2.3.4). Recall that these grammars can be understood as generalized parallel versions of context-free grammars. More precisely, there exist three main conceptual differences between them and context-free grammars. First, instead of a single set of rules, they have finitely many sets of rules. Second, the left-hand side of a rule may be formed by any grammatical symbol, including a terminal. Third, all symbols of a string are simultaneously rewritten during a single derivation step. The present section restricts its attention to ET0L grammars that work in a context-conditional way. Specifically, by analogy with context-conditional grammars that work in a sequential way (see Sect. 3.1), the section discusses *context-conditional ET0L grammars* that capture this dependency so each of their rules may be associated with finitely many strings representing *permitting conditions* and, in addition, finitely many strings representing *forbidding conditions*. A rule like this can rewrite a symbol if all its permitting conditions occur in the current rewritten sentential form and, simultaneously, all its forbidding conditions do not. Otherwise, these grammars work just like ordinary ET0L grammars. The section consists of four subsections. Section 4.2.1 defines the basic version of context-conditional ET0L grammars. The other sections investigate three variants of the basic version—*forbidding ET0L grammars* (Sect. 4.2.2), *simple semi-conditional ET0L grammars* (Sect. 4.2.3), and *left random context ET0L grammars* (Sect. 4.2.4). All these sections concentrate their attention on establishing the generative power of the ET0L grammars under investigation.

4.2.1 Context-Conditional ET0L Grammars

In the present subsection, we demonstrate that context-conditional ET0L grammars characterize the family of recursively enumerable languages (see Theorem 4.2.11), and, without erasing rules, they characterize the family of context-sensitive languages (see Theorem 4.2.9).

Definitions

In this section, we define context-conditional ET0L grammars.

Definition 4.2.1. A *context-conditional ET0L grammar* (a *C-ET0L grammar* for short) is a $(t + 3)$-tuple

$$G = (V, T, P_1, \ldots, P_t, S)$$

where $t \geq 1$, and V, T, and S are the *total alphabet*, the *terminal alphabet* ($T \subset V$), and the *start symbol* ($S \in V - T$), respectively. Every P_i, where $1 \leq i \leq t$, is a finite set of rules of the form

$$(a \to x, Per, For)$$

with $a \in V$, $x \in V^*$, and $Per, For \subseteq V^+$ are finite languages. If every $(a \to x, Per, For) \in P_i$ for $i = 1, 2, \ldots, t$ satisfies that $|x| \geq 1$, then G is said to be *propagating* (a *C-EPT0L grammar* for short). G has *degree* (r, s), where r and s are natural numbers, if for every $i = 1, \ldots, t$ and $(a \to x, Per, For) \in P_i$, max-len($Per$) $\leq r$ and max-len(For) $\leq s$.

Let $u, v \in V^*$, $u = a_1 a_2 \cdots a_q$, $v = v_1 v_2 \cdots v_q$, $q = |u|$, $a_j \in V$, $v_j \in V^*$, and $p_1, p_2, \ldots, p_q$ be a sequence of rules $p_j: (a_j \to v_j, Per_j, For_j) \in P_i$ for all $j = 1, \ldots, q$ and some $i \in \{1, \ldots, t\}$. If for every p_j, $Per_j \subseteq \text{sub}(u)$ and $For_j \cap \text{sub}(u) = \emptyset$, then u *directly derives* v according to $p_1, p_2, \ldots, p_q$ in G, denoted by

$$u \Rightarrow_G v \, [p_1, p_2, \ldots, p_q]$$

In the standard way, define $\Rightarrow_G^k$ for $k \geq 0$, $\Rightarrow_G^*$, and $\Rightarrow_G^+$. The *language* of G is denoted by $L(G)$ and defined as

$$L(G) = \{x \in T^* \mid S \Rightarrow_G^* x\} \qquad \square$$

Definition 4.2.2. Let $G = (V, T, P_1, \ldots, P_t, S)$ be a C-ET0L grammar, for some $t \geq 1$. If $t = 1$, then G is called a *context-conditional E0L grammar* (a *C-E0L grammar* for short). If G is a propagating C-E0L grammar, then G is said to be a *C-EP0L grammar*. $\qquad \square$

The language families defined by C-EPT0L, C-ET0L, C-EP0L, and C-E0L grammars of degree (r, s) are denoted by $\mathbf{C\text{-}EPT0L}(r, s)$, $\mathbf{C\text{-}ET0L}(r, s)$, $\mathbf{C\text{-}EP0L}(r, s)$, and $\mathbf{C\text{-}E0L}(r, s)$, respectively. Set

$$\mathbf{C\text{-}EPT0L} = \bigcup_{r=0}^{\infty} \bigcup_{s=0}^{\infty} \mathbf{C\text{-}EPT0L}(r, s) \qquad \mathbf{C\text{-}ET0L} = \bigcup_{r=0}^{\infty} \bigcup_{s=0}^{\infty} \mathbf{C\text{-}ET0L}(r, s)$$

$$\mathbf{C\text{-}EP0L} = \bigcup_{r=0}^{\infty} \bigcup_{s=0}^{\infty} \mathbf{C\text{-}EP0L}(r, s) \qquad \mathbf{C\text{-}E0L} = \bigcup_{r=0}^{\infty} \bigcup_{s=0}^{\infty} \mathbf{C\text{-}E0L}(r, s)$$

Generative Power

In this section, we discuss the generative power of context-conditional grammars.

Lemma 4.2.3. $\mathbf{C\text{-}EP0L} \subseteq \mathbf{C\text{-}EPT0L} \subseteq \mathbf{C\text{-}ET0L}$ *and* $\mathbf{C\text{-}EP0L} \subseteq \mathbf{C\text{-}E0L} \subseteq$ $\mathbf{C\text{-}ET0L}$. *For any* $r, s \geq 0$, $\mathbf{C\text{-}EP0L}(r, s) \subseteq \mathbf{C\text{-}EPT0L}(r, s) \subseteq \mathbf{C\text{-}ET0L}(r, s)$, *and* $\mathbf{C\text{-}EP0L}(r, s) \subseteq \mathbf{C\text{-}E0L}(r, s) \subseteq \mathbf{C\text{-}ET0L}(r, s)$.

Proof. This lemma follows from Definitions 4.2.1 and 4.2.2.   □

Theorem 4.2.4.

$$\mathbf{CF}$$
$$\subset$$
$$\mathbf{C\text{-}E0L}(0, 0) = \mathbf{C\text{-}EP0L}(0, 0) = \mathbf{E0L} = \mathbf{EP0L}$$
$$\subset$$
$$\mathbf{C\text{-}ET0L}(0, 0) = \mathbf{C\text{-}EPT0L}(0, 0) = \mathbf{ET0L} = \mathbf{EPT0L}$$
$$\subset$$
$$\mathbf{CS}$$

Proof. Clearly, C-EP0L and C-E0L grammars of degree $(0, 0)$ are ordinary EP0L and E0L grammars, respectively. Analogously, C-EPT0L and C-ET0L grammars of degree $(0, 0)$ are EPT0L and ET0L grammars, respectively. Since $\mathbf{CF} \subset \mathbf{E0L} = \mathbf{EP0L} \subset \mathbf{ET0L} = \mathbf{EPT0L} \subset \mathbf{CS}$ (see Theorem 2.3.41), we get $\mathbf{CF} \subset \mathbf{C\text{-}E0L}(0, 0) = \mathbf{C\text{-}EP0L}(0, 0) = \mathbf{E0L} \subset \mathbf{C\text{-}ET0L}(0, 0) = \mathbf{C\text{-}EPT0L}(0, 0) = \mathbf{ET0L} \subset \mathbf{CS}$; therefore, the theorem holds.   □

Lemma 4.2.5. $\mathbf{C\text{-}EPT0L}(r, s) \subseteq \mathbf{CS}$, *for any* $r \geq 0$, $s \geq 0$.

Proof. For $r = 0$ and $s = 0$, we have

$$\mathbf{C\text{-}EPT0L}(0, 0) = \mathbf{EPT0L} \subset \mathbf{CS}$$

The following proof demonstrates that the inclusion holds for any r and s such that $r + s \geq 1$.

Let L be a language generated by a C-EPT0L grammar

$$G = (V, T, P_1, \ldots, P_t, S)$$

of degree (r, s), for some $r, s \geq 0$, $r + s \geq 1$, $t \geq 1$. Let k be the greater number of r and s. Set

$$M = \{x \in V^+ \mid |x| \leq k\}$$

For every P_i, where $1 \leq i \leq t$, define

$$\text{cf-rules}(P_i) = \{a \to z \mid (a \to z, Per, For) \in P_i, \ a \in V, \ z \in V^+\}$$

Then, set

$$
\begin{aligned}
N_F &= \{\lfloor X, x \rfloor \mid X \subseteq M, \ x \in M \cup \{\varepsilon\}\} \\
N_T &= \{\langle X \rangle \mid X \subseteq M\} \\
N_B &= \{\lceil Q \rceil \mid Q \subseteq \text{cf-rules}(P_i) \ 1 \leq i \leq t\} \\
V' &= V \cup N_F \cup N_T \cup N_B \cup \{\triangleright, \triangleleft, \$, S', \#\} \\
T' &= T \cup \{\#\}
\end{aligned}
$$

Construct the context-sensitive grammar

$$G' = (V', T', P', S')$$

with the finite set of rules P' constructed by performing (1) through (7), given next.

(1) Add $S' \to \triangleright \lfloor \emptyset, \varepsilon \rfloor S \triangleleft$ to P'.

(2) For all $X \subseteq M$, $x \in (V^k \cup \{\varepsilon\})$ and $y \in V^k$, extend P' by adding

$$\lfloor X, x \rfloor y \to y \lfloor X \cup \text{sub}(xy, k), y \rfloor$$

(3) For all $X \subseteq M$, $x \in (V^k \cup \{\varepsilon\})$ and $y \in V^+$, $|y| \leq k$, extend P' by adding

$$\lfloor X, x \rfloor y \triangleleft \to y \langle X \cup \text{sub}(xy, k) \rangle \triangleleft$$

(4) For all $X \subseteq M$ and $Q \subseteq \text{cf-rules}(P_i)$, where $i \in \{1, \ldots, t\}$, such that for every $a \to z \in Q$, there exists $(a \to z, Per, For) \in P_i$ satisfying $Per \subseteq X$ and $For \cap X = \emptyset$, extend P' by adding

$$\langle X \rangle \triangleleft \to \lceil Q \rceil \triangleleft$$

(5) For every $Q \subseteq$ cf-rules(P_i) for some $i \in \{1, \ldots, t\}$, $a \in V$ and $z \in V^+$ such that $a \to z \in Q$, extend P' by adding

$$a\lceil Q \rceil \to \lceil Q \rceil z$$

(6) For all $Q \subseteq$ cf-rules(P_i) for some $i = \{1, \ldots, t\}$, extend P' by adding

$$\triangleright \lceil Q \rceil \to \triangleright \lfloor \emptyset, \varepsilon \rfloor$$

(7) Add $\triangleright \lfloor \emptyset, \varepsilon \rfloor \to \#\$$, $\$ \triangleleft \to \#\#$, and $\$a \to a\$$, for all $a \in T$, to P'.

To prove that $L(G) = L(G')$, we first establish Claims 4.2.6 through 4.2.8.

Claim 4.2.6. Every successful derivation in G' has the form

$$
\begin{aligned}
S' &\Rightarrow_{G'} \triangleright \lfloor \emptyset, \varepsilon \rfloor S \triangleleft \\
&\Rightarrow_{G'}^+ \triangleright \lfloor \emptyset, \varepsilon \rfloor x \triangleleft \\
&\Rightarrow_{G'} \#\$x \triangleleft \\
&\Rightarrow_{G'}^{|x|} \#x\$ \triangleleft \\
&\Rightarrow_{G'} \#x\#\#
\end{aligned}
$$

such that $x \in T^+$ and during $\triangleright \lfloor \emptyset, \varepsilon \rfloor S \triangleleft \Rightarrow_{G'}^+ \triangleright \lfloor \emptyset, \varepsilon \rfloor x \triangleleft$, every sentential form w satisfies $w \in \{\triangleright\}H^+\{\triangleleft\}$, where $H \subseteq V' - \{\triangleright, \triangleleft, \#, \$, S'\}$.

Proof. The only rule that can rewrite the start symbol is $S' \to \triangleright \lfloor \emptyset, \varepsilon \rfloor S \triangleleft$; thus,

$$S' \Rightarrow_{G'} \triangleright \lfloor \emptyset, \varepsilon \rfloor S \triangleleft$$

After that, every sentential form that occurs in

$$\triangleright \lfloor \emptyset, \varepsilon \rfloor S \triangleleft \Rightarrow_{G'}^+ \triangleright \lfloor \emptyset, \varepsilon \rfloor x \triangleleft$$

can be rewritten by using any of the rules introduced in (2) through (6) from the construction of P'. By the inspection of these rules, it is obvious that the edge symbols $\triangleright$ and $\triangleleft$ remain unchanged and no other occurrences of them appear inside the sentential form. Moreover, there is no rule generating a symbol from $\{\#, \$, S'\}$. Therefore, all these sentential forms belong to $\{\triangleright\}H^+\{\triangleleft\}$.

Next, let us explain how G' generates a string from $L(G')$. Only $\triangleright \lfloor \emptyset, \varepsilon \rfloor \to \#\$$ can rewrite $\triangleright$ to a symbol from T (see (7) in the definition of P'). According to the left-hand side of this rule, we obtain

$$S' \Rightarrow_{G'} \triangleright \lfloor \emptyset, \varepsilon \rfloor S \triangleleft \Rightarrow_{G'}^* \triangleright \lfloor \emptyset, \varepsilon \rfloor x \triangleleft \Rightarrow_{G'} \#\$x \triangleleft$$

where $x \in H^+$. To rewrite $\triangleleft$, G' uses $\$ \triangleleft \to \#\#$. Thus, G' needs $\$$ as the left neighbor of $\triangleleft$. Suppose that $x = a_1 a_2 \cdots a_q$, where $q = |x|$ and $a_i \in T$, for all

$i \in \{1, \ldots, q\}$. Since for every $a \in T$ there is $\$a \to a\$ \in P'$ (see (7)), we can construct

$$
\begin{aligned}
\#\$a_1 a_2 \cdots a_n \triangleleft &\Rightarrow_{G'} \quad \#a_1 \$a_2 \cdots a_n \triangleleft \\
&\Rightarrow_{G'} \quad \#a_1 a_2 \$ \cdots a_n \triangleleft \\
&\Rightarrow_{G'}^{|x|-2} \#a_1 a_2 \cdots a_n \$ \triangleleft
\end{aligned}
$$

Notice that this derivation can be constructed only for x that belong to T^+. Then, $\$\triangleleft$ is rewritten to $\#\#$. As a result,

$$
S' \Rightarrow_{G'} \triangleright \lfloor \emptyset, \varepsilon \rfloor S \triangleleft \Rightarrow_{G'}^+ \triangleright \lfloor \emptyset, \varepsilon \rfloor x \triangleleft \Rightarrow_{G'} \#\$x \triangleleft \Rightarrow_{G'}^{|x|} \#x\$ \triangleleft \Rightarrow_{G'} \#x\#\#
$$

with the required properties. Thus, the claim holds. $\qquad\square$

The following claim demonstrates how G' simulates a direct derivation from G—the heart of the construction.

Let $x \Rightarrow_{G'}^{\oplus} y$ denote the derivation $x \Rightarrow_{G'}^+ y$ such that $x = \triangleright \lfloor \emptyset, \varepsilon \rfloor u \triangleleft$, $y = \triangleright \lfloor \emptyset, \varepsilon \rfloor v \triangleleft$, $u, v \in V^+$, and during $x \Rightarrow_{G'}^+ y$, there is no other occurrence of a string of the form $\triangleright \lfloor \emptyset, \varepsilon \rfloor z \triangleleft$, $z \in V^*$.

Claim 4.2.7. For every $u, v \in V^*$,

$$
\triangleright \lfloor \emptyset, \varepsilon \rfloor u \triangleleft \Rightarrow_{G'}^{\oplus} \triangleright \lfloor \emptyset, \varepsilon \rfloor v \triangleleft \quad \text{if and only if} \quad u \Rightarrow_G v
$$

Proof. The proof is divided into the only-if part and the if part.

Only If. Let us show how G' rewrites $\triangleright \lfloor \emptyset, \varepsilon \rfloor u \triangleleft$ to $\triangleright \lfloor \emptyset, \varepsilon \rfloor v \triangleleft$ by performing a derivation consisting of a forward phase and a backward phase.

During the first, forward phase, G' scans u to obtain all nonempty substrings of length k or less. By repeatedly using rules

$$
\lfloor X, x \rfloor y \to y \lfloor X \cup \mathrm{sub}(xy, k), y \rfloor
$$

where $X \subseteq M$, $x \in (V^k \cup \{\varepsilon\})$, $y \in V^k$ (see (2) in the definition of P'), the occurrence of a symbol with form $\lfloor X, x \rfloor$ is moved toward the end of the sentential form. Simultaneously, the substrings of u are collected in X. The forward phase is finished by

$$
\lfloor X, x \rfloor y \triangleleft \to y \langle X \cup \mathrm{sub}(xy, k) \rangle \triangleleft
$$

where $x \in (V^k \cup \{\varepsilon\})$, $y \in V^+$, $|y| \le k$ (see (3)); the rule reaches the end of u and completes $X = \mathrm{sub}(u, k)$. Formally,

$$
\triangleright \lfloor \emptyset, \varepsilon \rfloor u \triangleleft \Rightarrow_{G'}^+ \triangleright u \langle X \rangle \triangleleft
$$

such that $X = \text{sub}(u, k)$. Then, $\langle X \rangle$ is changed to $\lceil Q \rceil$, where

$$Q = \{a \to z \mid (a \to z, Per, For) \in P_i, \ a \in V, \ z \in V^+,$$
$$Per, For \subseteq M, \ Per \subseteq X, \ For \cap X = \emptyset\}$$

for some $i \in \{1, \ldots, t\}$, by

$$\langle X \rangle \lhd \to \lceil Q \rceil \lhd$$

(see (4)). In other words, G' selects a subset of rules from P_i that could be used to rewrite u in G.

The second, backward phase simulates rewriting of all symbols in u in parallel. Since

$$a\lceil Q \rceil \to \lceil Q \rceil z \in P'$$

for all $a \to z \in Q, a \in V, z \in V^+$ (see (5)),

$$\rhd u \lceil Q \rceil \lhd \Rightarrow_{G'}^{|u|} \rhd \lceil Q \rceil v \lhd$$

such that $\lceil Q \rceil$ moves left and every symbol $a \in V$ in u is rewritten to some z provided that $a \to z \in Q$. Finally, $\lceil Q \rceil$ is rewritten to $\lfloor \emptyset, \varepsilon \rfloor$ by

$$\rhd \lceil Q \rceil \to \rhd \lfloor \emptyset, \varepsilon \rfloor$$

As a result, we obtain

$$\rhd \lfloor \emptyset, \varepsilon \rfloor u \lhd \Rightarrow_{G'}^+ \rhd u \langle X \rangle \lhd \Rightarrow_{G'} \rhd u \lceil Q \rceil \lhd$$
$$\Rightarrow_{G'}^{|u|} \rhd \lceil Q \rceil v \lhd \Rightarrow_{G'} \rhd \lfloor \emptyset, \varepsilon \rfloor v \lhd$$

Observe that this is the only way of deriving

$$\rhd \lfloor \emptyset, \varepsilon \rfloor u \lhd \Rightarrow_{G'}^{\oplus} \rhd \lfloor \emptyset, \varepsilon \rfloor v \lhd$$

Let us show that $u \Rightarrow_G v$. Indeed, because we have $(a \to z, Per, For) \in P_i$ for every $a\lceil Q \rceil \to \lceil Q \rceil z \in P$ used in the backward phase, where $Per \subseteq \text{sub}(u, k)$ and $For \cap \text{sub}(u, k) = \emptyset$ (see the construction of Q), there exists a derivation

$$u \Rightarrow_G v \ [p_1 \cdots p_q]$$

where $|u| = q$, and $p_j: (a \to z, Per, For) \in P_i$ such that $a\lceil Q \rceil \to \lceil Q \rceil z$ has been applied in the $(q - j + 1)$th derivation step in

$$\rhd u \lceil Q \rceil \lhd \Rightarrow_{G'}^{|u|} \rhd \lceil Q \rceil v \lhd$$

where $a \in V, z \in V^+, 1 \leq j \leq q$.

If. The converse implication can be proved similarly to the only-if part, so we leave it to the reader. □

Claim 4.2.8. $S' \Rightarrow_{G'}^{+} \triangleright \lfloor \emptyset, \varepsilon \rfloor x \triangleleft$ if and only if $S \Rightarrow_{G}^{*} x$, for all $x \in V^{+}$.

Proof. The proof is divided into the only-if part and the if part.

Only If. The only-if part is proved by induction on the ith occurrence of the sentential form w satisfying $w = \triangleright \lfloor \emptyset, \varepsilon \rfloor u \triangleleft$, $u \in V^{+}$, during the derivation in G'.

Basis. Let $i = 1$. Then, $S' \Rightarrow_{G'} \triangleright \lfloor \emptyset, \varepsilon \rfloor S \triangleleft$ and $S \Rightarrow_{G}^{0} S$.

Induction Hypothesis. Suppose that the claim holds for all $1 \leq i \leq h$, for some $h \geq 1$.

Induction Step. Let $i = h + 1$. Since $h + 1 \geq 2$, we can express

$$S' \Rightarrow_{G'}^{+} \triangleright \lfloor \emptyset, \varepsilon \rfloor x_i \triangleleft$$

as

$$S' \Rightarrow_{G'}^{+} \triangleright \lfloor \emptyset, \varepsilon \rfloor x_{i-1} \triangleleft \Rightarrow_{G'}^{\oplus} \triangleright \lfloor \emptyset, \varepsilon \rfloor x_i \triangleleft$$

where $x_{i-1}, x_i \in V^{+}$. By the induction hypothesis,

$$S \Rightarrow_{G}^{*} x_{i-1}$$

Claim 4.2.7 says that

$$\triangleright \lfloor \emptyset, \varepsilon \rfloor x_{i-1} \triangleleft \Rightarrow_{G'}^{\oplus} \triangleright \lfloor \emptyset, \varepsilon \rfloor x_i \triangleleft \quad \text{if and only if} \quad x_{i-1} \Rightarrow_{G} x_i$$

Hence,

$$S \Rightarrow_{G}^{*} x_{i-1} \Rightarrow_{G} x_i$$

and the only-if part holds.

If. By induction on h, we prove that

$$S \Rightarrow_{G}^{h} x \quad \text{implies that} \quad S' \Rightarrow_{G'}^{+} \triangleright \lfloor \emptyset, \varepsilon \rfloor x \triangleleft$$

for all $h \geq 0$, $x \in V^{+}$.

Basis. For $h = 0$, $S \Rightarrow_{G}^{0} S$ and $S' \Rightarrow_{G'} \triangleright \lfloor \emptyset, \varepsilon \rfloor S \triangleleft$.

Induction Hypothesis. Assume that the claim holds for all $0 \leq h \leq n$, for some $n \geq 0$.

Induction Step. Consider any derivation of the form

$$S \Rightarrow_{G}^{n+1} x$$

where $x \in V^+$. Since $n + 1 \geq 1$, there exists $y \in V^+$ such that

$$S \Rightarrow_G^n y \Rightarrow_G x$$

and by the induction hypothesis, there is also a derivation

$$S' \Rightarrow_{G'}^+ \triangleright \lfloor \emptyset, \varepsilon \rfloor y \triangleleft$$

From Claim 4.2.7, we have

$$\triangleright \lfloor \emptyset, \varepsilon \rfloor y \triangleleft \Rightarrow_{G'}^\oplus \triangleright \lfloor \emptyset, \varepsilon \rfloor x \triangleleft$$

Therefore,

$$S' \Rightarrow_{G'}^+ \triangleright \lfloor \emptyset, \varepsilon \rfloor y \triangleleft \Rightarrow_{G'}^\oplus \triangleright \lfloor \emptyset, \varepsilon \rfloor x \triangleleft$$

and the converse implication holds as well.   □

From Claims 4.2.6 and 4.2.8, we see that any successful derivation in G' is of the form

$$S' \Rightarrow_{G'}^+ \triangleright \lfloor \emptyset, \varepsilon \rfloor x \triangleleft \Rightarrow_{G'}^+ \#x\#\#$$

such that

$$S \Rightarrow_G^* x, \ x \in T^+$$

Therefore, for each $x \in T^+$, we have

$$S' \Rightarrow_{G'}^+ \#x\#\# \quad \text{if and only if} \quad S \Rightarrow_G^* x$$

Define the homomorphism h over $(T \cup \{\#\})^*$ as $h(\#) = \varepsilon$ and $h(a) = a$ for all $a \in T$. Observe that h is 4-linear erasing with respect to $L(G')$. Furthermore, notice that $h(L(G')) = L(G)$. Since **CS** is closed under linear erasing (see Theorem 10.4 on page 98 in [Sal73]), $L \in$ **CS**. Thus, Lemma 4.2.5 holds.   □

Theorem 4.2.9. **C - EPT0L** = **CS**

Proof. By Lemma 4.2.5, **C - EPT0L** $\subseteq$ **CS**. Later in this section, we define two special cases of C-EPT0L grammars and prove that they generate all the family of context-sensitive languages (see Theorems 4.2.30 and 4.2.47). Therefore, **CS** $\subseteq$ **C - EPT0L**, and hence **C - EPT0L** = **CS**.   □

Lemma 4.2.10. **C - ET0L** $\subseteq$ **RE**

Proof. This lemma follows from Turing-Church thesis. To obtain an algorithm converting any C-ET0L grammar to an equivalent phrase-structure grammar, use the technique presented in Lemma 4.2.5.   □

Theorem 4.2.11. C-$ET0L = RE$

Proof. By Lemma 4.2.10, C-$ET0L \subseteq RE$. In Sects. 4.2.2 and 4.2.3, we introduce two special cases of C-ET0L grammars and demonstrate that even these grammars generate RE (see Theorems 4.2.33 and 4.2.44); therefore, $RE \subseteq C$-$ET0L$. As a result, C-$ET0L = RE$. □

4.2.2 Forbidding ET0L Grammars

Forbidding ET0L grammars, discussed in the present section, represent context-conditional ET0L grammars in which no rule has any permitting condition. First, this section defines and illustrates them. Then, it establishes their generative power and reduces their degree without affecting the power.

Definitions and Examples

In this section, we define forbidding ET0L grammars.

Definition 4.2.12. Let $G = (V, T, P_1, \ldots, P_t, S)$ be a C-ET0L grammar. If every $p: (a \rightarrow x, Per, For) \in P_i$, where $i = 1, \ldots, t$, satisfies $Per = \emptyset$, then G is said to be *forbidding ET0L grammar* (an *F-ET0L grammar* for short). If G is a propagating F-ET0L grammar, then G is said to be an *F-EPT0L grammar*. If $t = 1$, G is called an *F-E0L grammar*. If G is a propagating F-E0L grammar, G is called an *F-EP0L grammar*. □

Let $G = (V, T, P_1, \ldots, P_t, S)$ be an F-ET0L grammar of degree (r, s). From the above definition, $(a \rightarrow x, Per, For) \in P_i$ implies that $Per = \emptyset$ for all $i = 1, \ldots, t$. By analogy with sequential forbidding grammars, we thus omit the empty set in the rules. For simplicity, we also say that the degree of G is s instead of (r, s).

The families of languages generated by F-E0L grammars, F-EP0L grammars, F-ET0L grammars, and F-EPT0L grammars of degree s are denoted by F-$E0L(s)$, F-$EP0L(s)$, F-$ET0L(s)$, and F-$EPT0L(s)$, respectively. Moreover, set

$$F\text{-}EPT0L = \bigcup_{s=0}^{\infty} F\text{-}EPT0L(s) \qquad F\text{-}ET0L = \bigcup_{s=0}^{\infty} F\text{-}ET0L(s)$$

$$F\text{-}EP0L = \bigcup_{s=0}^{\infty} F\text{-}EP0L(s) \qquad F\text{-}E0L = \bigcup_{s=0}^{\infty} F\text{-}E0L(s)$$

Example 4.2.13. Let

$$G = \big(\{S, A, B, C, a, \bar{a}, b\}, \{a, b\}, P, S\big)$$

be an F-EP0L grammar, where

$$P = \{(S \rightarrow ABA, \emptyset),$$
$$(A \rightarrow aA, \{\bar{a}\}),$$
$$(B \rightarrow bB, \emptyset),$$
$$(A \rightarrow \bar{a}, \{\bar{a}\}),$$
$$(\bar{a} \rightarrow a, \emptyset),$$
$$(B \rightarrow C, \emptyset),$$
$$(C \rightarrow bC, \{A\}),$$
$$(C \rightarrow b, \{A\}),$$
$$(a \rightarrow a, \emptyset),$$
$$(b \rightarrow b, \emptyset)\}$$

Obviously, G is an F-EP0L grammar of degree 1. Observe that for every string from $L(G)$, there exists a derivation of the form

$$S \Rightarrow_G ABA$$
$$\Rightarrow_G aAbBaA$$
$$\Rightarrow_G^+ a^{m-1}Ab^{m-1}Ba^{m-1}A$$
$$\Rightarrow_G a^{m-1}\bar{a}b^{m-1}Ca^{m-1}\bar{a}$$
$$\Rightarrow_G a^m b^m Ca^m$$
$$\Rightarrow_G^+ a^m b^{n-1} Ca^m$$
$$\Rightarrow_G a^m b^n a^m$$

with $1 \leq m \leq n$. Hence,

$$L(G) = \{a^m b^n a^m \mid 1 \leq m \leq n\}$$

Note that $L(G) \notin \mathbf{E0L}$ (see page 268 in [RS97a]); however, $L(G) \in \mathbf{F\text{-}EP0L}(1)$. As a result, F-EP0L grammars of degree 1 are more powerful than ordinary E0L grammars.   □

Generative Power and Reduction

Next, we investigate the generative power of F-ET0L grammars of all degrees.

Theorem 4.2.14. $\mathbf{F\text{-}EPT0L}(0) = \mathbf{EPT0L}$, $\mathbf{F\text{-}ET0L}(0) = \mathbf{ET0L}$, $\mathbf{F\text{-}EP0L}(0) = \mathbf{EP0L}$, *and* $\mathbf{F\text{-}E0L}(0) = \mathbf{E0L}$

Proof. This theorem follows from Definition 4.2.12.   □

Lemmas 4.2.15, 4.2.18, 4.2.20, and 4.2.21, given next, inspect the generative power of forbidding ET0L grammars of degree 1. As a conclusion, in

Theorem 4.2.22, we demonstrate that both F-EPT0L(1) and F-ET0L(1) grammars generate precisely the family of ET0L languages.

Lemma 4.2.15. EPT0L $\subseteq$ F-EP0L(1)

Proof. Let

$$G = (V, T, P_1, \ldots, P_t, S)$$

be an EPT0L grammar, where $t \geq 1$. Set

$$W = \{\langle a, i \rangle \mid a \in V, \, i = 1, \ldots, t\}$$

and

$$F_i = \{\langle a, j \rangle \in W \mid j \neq i\}$$

Then, construct an F-EP0L grammar of degree 1

$$G' = (V', T, P', S)$$

where $V' = V \cup W$, $(V \cap W = \emptyset)$ and the set of rules P' is defined as follows:

(1) for each $a \in V$ and $i = 1, \ldots, t$, add $(a \to \langle a, i \rangle, \emptyset)$ to P';
(2) if $a \to z \in P_i$ for some $i \in \{1, \ldots, t\}$, $a \in V$, $z \in V^+$, add $(\langle a, i \rangle \to z, F_i)$ to P'.

Next, to demonstrate that $L(G) = L(G')$, we prove Claims 4.2.16 and 4.2.17.

Claim 4.2.16. For each derivation $S \Rightarrow_{G'}^n x$, $n \geq 0$,

 (I) if $n = 2k + 1$ for some $k \geq 0$, $x \in W^+$;
(II) if $n = 2k$ for some $k \geq 0$, $x \in V^+$.

Proof. The claim follows from the definition of P'. Indeed, every rule in P' is either of the form $(a \to \langle a, i \rangle, \emptyset)$ or $(\langle a, i \rangle \to z, F_i)$, where $a \in V$, $\langle a, i \rangle \in W$, $z \in V^+$, $i \in \{1, \ldots, t\}$. Since $S \in V$,

$$S \Rightarrow_{G'}^{2k+1} x \quad \text{implies} \quad x \in W^+$$

and

$$S \Rightarrow_{G'}^{2k} x \quad \text{implies} \quad x \in V^+$$

Thus, the claim holds. $\quad\square$

Define the finite substitution γ from V^* to V'^* such that for every $a \in V$,

$$\gamma(a) = \{a\} \cup \{\langle a, i \rangle \in W \mid i = 1, \ldots, t\}$$

Claim 4.2.17. $S \Rightarrow^*_G x$ if and only if $S \Rightarrow^*_{G'} x'$ for some $x' \in \gamma(x)$, $x \in V^+$, $x' \in V'^+$.

Proof. The proof is divided into the only-if part and the if part.

Only If. By induction on $h \geq 0$, we show that for all $x \in V^+$,

$$S \Rightarrow^h_G x \quad \text{implies} \quad S \Rightarrow^{2h}_{G'} x$$

Basis. Let $h = 0$. Then, the only x is S; therefore, $S \Rightarrow^0_G S$ and also $S \Rightarrow^0_{G'} S$.

Induction Hypothesis. Suppose that

$$S \Rightarrow^h_G x \quad \text{implies} \quad S \Rightarrow^{2h}_{G'} x$$

for all derivations of length $0 \leq h \leq n$, for some $n \geq 0$.

Induction Step. Consider any derivation of the form

$$S \Rightarrow^{n+1}_G x$$

Since $n + 1 \geq 1$, this derivation can be expressed as

$$S \Rightarrow^n_G y \Rightarrow_G x \ [p_1, p_2, \ldots, p_q]$$

such that $y \in V^+$, $q = |y|$, and $p_j \in P_i$ for all $j = 1, \ldots, q$ and some $i \in \{1, \ldots, t\}$. By the induction hypothesis,

$$S \Rightarrow^{2n}_{G'} y$$

Suppose that $y = a_1 a_2 \cdots a_q$, $a_j \in V$. Let

$$
\begin{aligned}
S &\Rightarrow^{2n}_{G'} a_1 a_2 \cdots a_q \\
&\Rightarrow_{G'} \langle a_1, i\rangle \langle a_2, i\rangle \cdots \langle a_q, i\rangle \quad [p'_1, p'_2, \cdots, p'_q] \\
&\Rightarrow_{G'} z_1 z_2 \cdots z_q \qquad\qquad\quad [p''_1, p''_2, \cdots, p''_q]
\end{aligned}
$$

where $p'_j : (a_j \rightarrow \langle a_j, i\rangle, \emptyset)$ and $p''_j : (\langle a_j, i\rangle \rightarrow z_j, F_i)$ such that $p_j : a_j \rightarrow z_j$, $z_j \in V^+$, for all $j = 1, \ldots, q$. Then, $z_1 z_2 \cdots z_q = x$; therefore,

$$S \Rightarrow^{2(n+1)}_{G'} x$$

If. The converse implication is established by induction on $h \geq 0$. That is, we prove that

$$S \Rightarrow^h_{G'} x' \quad \text{implies} \quad S \Rightarrow^*_G x$$

for some $x' \in \gamma(x)$, $h \geq 0$.

Basis. For $h = 0$, $S \Rightarrow^0_{G'} S$ and $S \Rightarrow^0_G S$; clearly, $S \in \gamma(S)$.

Induction Hypothesis. Assume that there exists a natural number n such that the claim holds for every h, where $0 \le h \le n$.

Induction Step. Consider any derivation of the form

$$S \Rightarrow^{n+1}_{G'} x'$$

Express this derivation as

$$S \Rightarrow^n_{G'} y' \Rightarrow_{G'} x' \; [p'_1, p'_2, \ldots, p'_q]$$

where $y' \in V'^+$, $q = |y'|$, and $p'_1, p'_2, \ldots, p'_q$ is a sequence of rules from P'. By the induction hypothesis,

$$S \Rightarrow^*_G y$$

where $y \in V^+$, $y' \in \gamma(y)$. Claim 4.2.16 says that there exist the following two cases—(i) and (ii).

(i) Let $n = 2k$ for some $k \ge 0$. Then, $y' \in V^+$, $x' \in W^+$, and every rule

$$p'_j : (a_j \rightarrow \langle a_j, i \rangle, \emptyset)$$

where $a_j \in V$, $\langle a_j, i \rangle \in W$, $i \in \{1, \ldots, t\}$, $1 \le j \le q$. In this case, $\langle a_j, i \rangle \in \gamma(a_j)$ for every a_j and any i (see the definition of g); hence, $x' \in \gamma(y)$ as well.

(ii) Let $n = 2k + 1$. Then, $y' \in W^+$, $x' \in V^+$, and each p'_j is of the form

$$p'_j : (\langle a_j, i \rangle \rightarrow z_j, F_i)$$

where $\langle a_j, i \rangle \in W$, $z_j \in V^+$, $i \in \{1, \ldots, t\}$, $1 \le j \le q$. Moreover, according to the forbidding conditions of p'_j, all $\langle a_j, i \rangle$ in y' have the same i. Thus, $y' = \langle a_1, i \rangle \langle a_2, i \rangle \cdots \langle a_q, i \rangle$, $y = \gamma^{-1}(y') = a_1 a_2 \cdots a_q$, and $x' = z_1 z_2 \cdots z_q$. By the definition of P',

$$(\langle a_j, i \rangle \rightarrow z_j, F_i) \in P' \quad \text{implies} \quad a_j \rightarrow z_j \in P_i$$

Therefore,

$$S \Rightarrow^*_G a_1 a_2 \cdots a_q \Rightarrow_G z_1 z_2 \cdots z_q \; [p_1, p_2, \ldots, p_q]$$

where $p_j : a_j \rightarrow z_j \in P_i$ such that $p'_j : (\langle a_j, i \rangle \rightarrow z_j, F_i)$. Obviously, $x' = x = z_1 z_2 \cdots z_q$.

This completes the induction and establishes Claim 4.2.17. □

By Claim 4.2.17, for any $x \in T^+$,

$$S \Rightarrow_G^* x \quad \text{if and only if} \quad S \Rightarrow_{G'}^* x$$

Therefore, $L(G) = L(G')$, so the lemma holds. $\qquad\qquad\qquad\qquad\qquad$ □

In order to simplify the notation in the proof of the following lemma, for every subset of rules

$$P \subseteq \{(a \to z, F) \mid a \in V,\ z \in V^*,\ F \subseteq V\}$$

define

$$\text{left}(P) = \{a \mid (a \to z, F) \in P\}$$

Informally, $\text{left}(P)$ denotes the set of the left-hand sides of all rules in P.

Lemma 4.2.18. $\text{F-EPT0L}(1) \subseteq \text{EPT0L}$

Proof. Let

$$G = (V, T, P_1, \dots, P_t, S)$$

be an F-EPT0L grammar of degree 1, $t \geq 1$. Let Q be the set of all subsets $O \subseteq P_i$, $1 \leq i \leq t$, such that every $(a \to z, F) \in O$, $a \in V$, $z \in V^+$, $F \subseteq V$, satisfies $F \cap \text{left}(O) = \emptyset$. Introduce a new set Q' so that for each $O \in Q$, add

$$\{a \to z \mid (a \to z, F) \in O\}$$

to Q'. Express

$$Q' = \{Q'_1, \dots, Q'_m\}$$

where m is the cardinality of Q'. Then, construct the EPT0L grammar

$$G' = (V, T, Q'_1, \dots, Q'_m, S)$$

To see the basic idea behind the construction of G', consider a pair of rules $p_1 : (a_1 \to z_1, F_1)$ and $p_2 : (a_2 \to z_2, F_2)$ from P_i, for some $i \in \{1, \dots, t\}$. During a single derivation step, p_1 and p_2 can concurrently rewrite a_1 and a_2 provided that $a_2 \notin F_1$ and $a_1 \notin F_2$, respectively. Consider any $O \subseteq P_i$ containing no pair of rules $(a_1 \to z_1, F_1)$ and $(a_2 \to z_2, F_2)$ such that $a_1 \in F_2$ or $a_2 \in F_1$. Observe that for any derivation step based on O, no rule from O is blocked by its forbidding conditions; thus, the conditions can be omitted. A formal proof is given next.

Claim 4.2.19. $S \Rightarrow_G^h x$ if and only if $S \Rightarrow_{G'}^h x$, $x \in V^*$, $m \geq 0$.

Proof. The claim is proved by induction on $h \geq 0$.

Only If. By induction $h \geq 0$, we prove that

$$S \Rightarrow_G^h x \quad \text{implies} \quad S \Rightarrow_{G'}^h x$$

for all $x \in V^*$.

Basis. Let $h = 0$. As obvious, $S \Rightarrow_G^0 S$ and $S \Rightarrow_{G'}^0 S$.

Induction Hypothesis. Suppose that the claim holds for all derivations of length $0 \leq h \leq n$, for some $n \geq 0$.

Induction Step. Consider any derivation of the form

$$S \Rightarrow_G^{n+1} x$$

Since $n + 1 \geq 1$, there exists $y \in V^+$, $q = |y|$, and a sequence $p_1, \ldots, p_q$, where $p_j \in P_i$ for all $j = 1, \ldots, q$ and some $i \in \{1, \ldots, t\}$, such that

$$S \Rightarrow_G^n y \Rightarrow_G x \, [p_1, \ldots, p_q]$$

By the induction hypothesis,

$$S \Rightarrow_{G'}^n y$$

Set

$$O = \{p_j \mid 1 \leq j \leq q\}$$

Observe that

$$y \Rightarrow_G x \, [p_1, \ldots, p_q]$$

implies that $\text{alph}(y) = \text{left}(O)$. Moreover, every p_j: $(a \rightarrow z, F) \in O$, $a \in V$, $z \in V^+$, $F \subseteq V$, satisfies $F \cap \text{alph}(y) = \emptyset$. Hence, $(a \rightarrow z, F) \in O$ implies $F \cap \text{left}(O) = \emptyset$. Inspect the definition of G' to see that there exists

$$Q'_r = \{a \rightarrow z \mid (a \rightarrow z, F) \in O\}$$

for some r, $1 \leq r \leq m$. Therefore,

$$S \Rightarrow_{G'}^n y \Rightarrow_{G'} x \, [p'_1, \ldots, p'_q]$$

where p'_j: $a \rightarrow z \in Q'_r$ such that p_j: $(a \rightarrow z, F) \in O$, for all $j = 1, \ldots, q$.

If. The if part demonstrates for every $h \geq 0$,

$$S \Rightarrow_{G'}^h x \text{ implies that } S \Rightarrow_G^h x$$

where $x \in V^*$.

Basis. Suppose that $h = 0$. As obvious, $S \Rightarrow_{G'}^0 S$ and $S \Rightarrow_G^0 S$.

Induction Hypothesis. Assume that the claim holds for all derivations of length $0 \leq h \leq n$, for some $n \geq 0$.

Induction Step. Consider any derivation of the form

$$S \Rightarrow_{G'}^{n+1} x$$

As $n + 1 \geq 1$, there exists a derivation

$$S \Rightarrow_{G'}^n y \Rightarrow_{G'} x [p_1', \ldots, p_q']$$

such that $y \in V^+$, $q = |y|$, each $p_i' \in Q_r'$ for some $r \in \{1, \ldots, m\}$, and by the induction hypothesis,

$$S \Rightarrow_G^n y$$

Then, by the definition of Q_r', there exists P_i and $O \subseteq P_i$ such that every $(a \rightarrow z, F) \in O$, $a \in V$, $z \in V^+$, $F \subseteq V$, satisfies $a \rightarrow z \in Q_r'$ and $F \cap \text{left}(O) = \emptyset$. Since $\text{alph}(y) \subseteq \text{left}(O)$, $(a \rightarrow z, F) \in O$ implies that $F \cap \text{alph}(y) = \emptyset$. Hence,

$$S \Rightarrow_G^n y \Rightarrow_G x [p_1, \ldots, p_q]$$

where $p_j: (a \rightarrow z, F) \in O$ for all $j = 1, \ldots, q$.   □

From the claim above,

$$S \Rightarrow_G^* x \quad \text{if and only if} \quad S \Rightarrow_{G'}^* x$$

for all $x \in T^*$. Consequently, $L(G) = L(G')$, and the lemma holds.   □

The following two lemmas can be proved by analogy with Lemmas 4.2.15 and 4.2.18. The details are left to the reader.

Lemma 4.2.20. ET0L $\subseteq$ F-E0L(1)   □

Lemma 4.2.21. F-ET0L(1) $\subseteq$ ET0L   □

Theorem 4.2.22.

$$\text{F-EP0L}(1) = \text{F-EPT0L}(1) = \text{F-E0L}(1) = \text{F-ET0L}(1) = \text{ET0L} = \text{EPT0L}$$

Proof. By Lemmas 4.2.15 and 4.2.18, **EPT0L** $\subseteq$ **F - EP0L**(1) and **F - EPT0L**(1) $\subseteq$ **EPT0L**, respectively. Since **F - EP0L**(1) $\subseteq$ **F - EPT0L**(1), we get **F - EP0L**(1) = **F - EPT0L**(1) = **EPT0L**. Analogously, from Lemmas 4.2.20 and 4.2.21, we have **F - E0L**(1) = **F - ET0L**(1) = **ET0L**. Theorem 2.3.41 implies that **EPT0L** = **ET0L**. Therefore,

$$\mathbf{F\text{-}EP0L}(1) = \mathbf{F\text{-}EPT0L}(1) = \mathbf{F\text{-}E0L}(1) = \mathbf{F\text{-}ET0L}(1) = \mathbf{EPT0L} = \mathbf{ET0L}$$

Thus, the theorem holds. □

Next, we investigate the generative power of F-EPT0L grammars of degree 2. The following lemma establishes a normal form for context-sensitive grammars so that the grammars satisfying this form generate only sentential forms containing no nonterminal from N_{CS} as the leftmost symbol of the string. We make use of this normal form in Lemma 4.2.24.

Lemma 4.2.23. *Every context-sensitive language $L \in$ **CS** can be generated by a context-sensitive grammar, $G = (N_1 \cup N_{CF} \cup N_{CS} \cup T, T, P, S_1)$, where N_1, N_{CF}, N_{CS}, and T are pairwise disjoint alphabets, $S_1 \in N_1$, and in P, every rule has one of the following forms*

  (i) $AB \to AC$, where $A \in (N_1 \cup N_{CF})$, $B \in N_{CS}$, $C \in N_{CF}$;
 (ii) $A \to B$, where $A \in N_{CF}$, $B \in N_{CS}$;
(iii) $A \to a$, where $A \in (N_1 \cup N_{CF})$, $a \in T$;
 (iv) $A \to C$, where $A, C \in N_{CF}$;
  (v) $A_1 \to C_1$, where $A_1, C_1 \in N_1$;
 (vi) $A \to DE$, where $A, D, E \in N_{CF}$;
(vii) $A_1 \to D_1E$, where $A_1, D_1 \in N_1$, $E \in N_{CF}$.

Proof. Let

$$G' = (N_{CF} \cup N_{CS} \cup T, T, P', S)$$

be a context-sensitive grammar of the form defined in Theorem 3.1.6. From this grammar, we construct a grammar

$$G = (N_1 \cup N_{CF} \cup N_{CS} \cup T, T, P, S_1)$$

where

$$N_1 = \{X_1 \mid X \in N_{CF}\}$$
$$P = P' \cup \{A_1B \to A_1C \mid AB \to AC \in P', A, C \in N_{CF}, B \in N_{CS}, A_1 \in N_1\}$$
$$\cup \{A_1 \to a \mid A \to a \in P', A \in N_{CF}, A_1 \in N_1, a \in T\}$$
$$\cup \{A_1 \to C_1 \mid A \to C \in P', A, C \in N_{CF}, A_1, C_1 \in N_1\}$$
$$\cup \{A_1 \to D_1E \mid A \to DE \in P', A, D, E \in N_{CF}, A_1, D_1 \in N_1\}$$

G works by analogy with G' except that in G every sentential form starts with a symbol from $N_1 \cup T$ followed by symbols that are not in N_1. Notice, however, that by $AB \to AC$, G' can never rewrite the leftmost symbol of any sentential form. Based on these observations, it is rather easy to see that $L(G) = L(G')$; a formal proof of this identity is left to the reader. As G is of the required form, Lemma 4.2.23 holds.   □

Lemma 4.2.24. $CS \subseteq F\text{-}EP0L(2)$

Proof. Let L be a context-sensitive language generated by a grammar

$$G = \left(N_1 \cup N_{CF} \cup N_{CS} \cup T, T, P, S_1\right)$$

of the form of Lemma 4.2.23. Set

$$
\begin{aligned}
V &= N_1 \cup N_{CF} \cup N_{CS} \cup T \\
P_{CS} &= \{AB \to AC \mid AB \to AC \in P, A \in (N_1 \cup N_{CF}), B \in N_{CS}, C \in N_{CF}\} \\
P_{CF} &= P - P_{CS}
\end{aligned}
$$

Informally, P_{CS} and P_{CF} are the sets of context-sensitive and context-free rules in P, respectively, and V denotes the total alphabet of G.

Let f be an arbitrary bijection from V to $\{1, \dots, m\}$, where m is the cardinality of V, and let f^{-1} be the inverse of f.

Construct an F-EP0L grammar of degree 2,

$$G' = \left(V', T, P', S_1\right)$$

with V' defined as

$$
\begin{aligned}
W_0 &= \{\langle A, B, C\rangle \mid AB \to AC \in P_{CS}\} \\
W_S &= \{\langle A, B, C, j\rangle \mid AB \to AC \in P_{CS}, 1 \leq j \leq m+1\} \\
W &= W_0 \cup W_S \\
V' &= V \cup W
\end{aligned}
$$

where V, W_0, and W_S are pairwise disjoint alphabets. The set of rules P' is constructed by performing (1) through (3), given next.

(1) For every $X \in V$, add $(X \to X, \emptyset)$ to P'.
(2) For every $A \to u \in P_{CF}$, add $(A \to u, W)$ to P'.
(3) For every $AB \to AC \in P_{CS}$, extend P' by adding

    (3.a) $(B \to \langle A, B, C\rangle, W)$;
    (3.b) $(\langle A, B, C\rangle \to \langle A, B, C, 1\rangle, W - \{\langle A, B, C\rangle\})$;
    (3.c) $(\langle A, B, C, j\rangle \to \langle A, B, C, j+1\rangle, \{f^{-1}(j)\langle A, B, C, j\rangle\})$ for all $1 \leq j \leq m$
        such that $f(A) \neq j$;
    (3.d) $(\langle A, B, C, f(A)\rangle \to \langle A, B, C, f(A)+1\rangle, \emptyset)$;
    (3.e) $(\langle A, B, C, m+1\rangle \to C, \{\langle A, B, C, m+1\rangle^2\})$.

Let us informally explain how G' simulates the non-context-free rules of the form $AB \to AC$ (see rules of (3) in the construction of P'). First, chosen occurrences of B are rewritten with $\langle A, B, C \rangle$ by $(B \to \langle A, B, C \rangle, W)$. The forbidding condition of this rule guarantees that there is no simulation already in process. After that, left neighbors of all occurrences of $\langle A, B, C \rangle$ are checked not to be any symbols from $V - \{A\}$. In a greater detail, G' rewrites $\langle A, B, C \rangle$ with $\langle A, B, C, i \rangle$ for $i = 1$. Then, in every $\langle A, B, C, i \rangle$, G' increments i by one as long as i is less or equal to the cardinality of V; simultaneously, it verifies that the left neighbor of every $\langle A, B, C, i \rangle$ differs from the symbol that f maps to i except for the case when $f(A) = i$. Finally, G' checks that there are no two adjoining symbols $\langle A, B, C, m + 1 \rangle$. At this point, the left neighbors of $\langle A, B, C, m + 1 \rangle$ are necessarily equal to A, so every occurrence of $\langle A, B, C, m + 1 \rangle$ is rewritten to C.

Observe that the other symbols remain unchanged during the simulation. Indeed, by the forbidding conditions, the only rules that can rewrite symbols $X \notin W$ are of the form $(X \to X, \emptyset)$. Moreover, the forbidding condition of $(\langle A, B, C \rangle \to \langle A, B, C, 1 \rangle, W - \{\langle A, B, C \rangle\})$ implies that it is not possible to simulate two different non-context-free rules at the same time.

To establish that $L(G) = L(G')$, we first prove Claims 4.2.25 through 4.2.29.

Claim 4.2.25. $S_1 \Rightarrow_{G'}^h x'$ implies that $\mathrm{lms}(x') \in (N_1 \cup T)$ for every $h \geq 0$, $x' \in V'^*$.

Proof. The claim is proved by induction on $h \geq 0$.

Basis. Let $h = 0$. Then, $S_1 \Rightarrow_{G'}^0 S_1$ and $S_1 \in N_1$.

Induction Hypothesis. Assume that the claim holds for all derivations of length $h \leq n$, for some $n \geq 0$.

Induction Step. Consider any derivation of the form

$$S_1 \Rightarrow_{G'}^{n+1} x'$$

where $x' \in V'^*$. Since $n + 1 \geq 1$, there is a derivation

$$S_1 \Rightarrow_{G'}^n y' \Rightarrow_{G'} x' \ [p_1, \ldots, p_q]$$

$y' \in V'^*$, $q = |y'|$, and by the induction hypothesis, $\mathrm{lms}(y') \in (N_1 \cup T)$. Inspect P' to see that the rule p_1 that rewrites the leftmost symbol of y' is one of the following forms $(A_1 \to A_1, \emptyset)$, $(a \to a, \emptyset)$, $(A_1 \to a, W)$, $(A_1 \to C_1, W)$, or $(A_1 \to D_1 E, W)$, where $A_1, C_1, D_1 \in N_1$, $a \in T$, $E \in N_{CF}$ (see (1) and (2) in the definition of P' and Lemma 4.2.23). It is obvious that the leftmost symbols of the right-hand sides of these rules belong to $(N_1 \cup T)$. Hence, $\mathrm{lms}(x') \in (N_1 \cup T)$, so the claim holds. $\square$

Claim 4.2.26. $S_1 \Rightarrow_{G'}^n y_1' X y_3'$, where $X \in W_S$, implies that $y_1' \in V'^+$ and $y_3' \in V'^*$, for any $n \geq 0$.

Proof. Informally, the claim says that every occurrence of a symbol from W_S has always a left neighbor. Clearly, this claim follows from the statement of Claim 4.2.25. Since $W_S \cap (N_1 \cup T) = \emptyset$, X cannot be the leftmost symbol in a sentential form and the claim holds.   □

Claim 4.2.27. $S_1 \Rightarrow^h_{G'} x'$, $h \geq 0$, implies that x' has one of the following three forms

(I) $x' \in V^*$;

(II) $x' \in (V \cup W_0)^*$ and $\#_{W_0}(x') > 0$;

(III) $x' \in (V \cup \{\langle A, B, C, j \rangle\})^*$, $\#_{\{\langle A,B,C,j\rangle\}}(x') > 0$, and $\{f^{-1}(k)\langle A, B, C, j \rangle \mid 1 \leq k < j, k \neq f(A)\} \cap \text{sub}(x') = \emptyset$, where $\langle A, B, C, j \rangle \in W_S$, $A \in (N_1 \cup N_{CF})$, $B \in N_{CS}$, $C \in N_{CF}$, $1 \leq j \leq m + 1$.

Proof. We prove the claim by induction on $h \geq 0$.

Basis. Let $h = 0$. Clearly, $S_1 \Rightarrow^0_{G'} S_1$ and S_1 is of type (I).

Induction Hypothesis. Suppose that the claim holds for all derivations of length $h \leq n$, for some $n \geq 0$.

Induction Step. Consider any derivation of the form

$$S_1 \Rightarrow^{n+1}_{G'} x'$$

Since $n + 1 \geq 1$, there exists $y' \in V'^*$ and a sequence of rules $p_1, \ldots, p_q$, where $p_i \in P'$, $1 \leq i \leq q$, $q = |y'|$, such that

$$S_1 \Rightarrow^n_{G'} y' \Rightarrow_{G'} x' \; [p_1, \ldots, p_q]$$

Let $y' = a_1 a_2 \ldots a_q$, $a_i \in V'$.

By the induction hypothesis, y' can only be of forms (I) through (III). Thus, the following three cases cover all possible forms of y'.

(i) Let $y' \in V^*$ (form (I)). In this case, every rule p_i can be either of the form $(a_i \rightarrow a_i, \emptyset)$, $a_i \in V$, or $(a_i \rightarrow u, W)$ such that $a_i \rightarrow u \in P_{CF}$, or $(a_i \rightarrow \langle A, a_i, C \rangle, W)$, $a_i \in N_{CS}$, $\langle A, a_i, C \rangle \in W_0$ (see the definition of P').

  Suppose that for every $i \in \{1, \ldots, q\}$, p_i has one of the first two listed forms. According to the right-hand sides of these rules, we obtain $x' \in V^*$; that is, x' is of form (I).

  If there exists i such that $p_i: (a_i \rightarrow \langle A, a_i, C \rangle, W)$ for some $A \in (N_1 \cup N_{CF})$, $a_i \in N_{CS}$, $C \in N_{CF}$, $\langle A, a_i, C \rangle \in W_0$, we get $x' \in (V \cup W_0)^*$ with $\#_{W_0}(x') > 0$. Thus, x' belongs to (II).

(ii) Let $y' \in (V \cup W_0)^*$ and $\#_{W_0}(y') > 0$ (form (II)). At this point, p_i is either $(a_i \rightarrow a_i, \emptyset)$ (rewriting $a_i \in V$ to itself) or $(\langle A, B, C \rangle \rightarrow \langle A, B, C, 1 \rangle, W - \{\langle A, B, C \rangle\})$ rewriting $a_i = \langle A, B, C \rangle \in W_0$ to $\langle A, B, C, 1 \rangle \in W_S$, where $A \in (N_1 \cup N_{CF})$, $B \in N_{CS}$, $C \in N_{CF}$. Since $\#_{W_0}(y') > 0$, there exists at least one i such that $a_i = \langle A, B, C \rangle \in W_0$. The corresponding rule p_i can be used provided that $\#_{W - \{\langle A,B,C\rangle\}}(y') = 0$. Therefore, $y' \in (V \cup \{\langle A, B, C \rangle\})^*$, so $x' \in (V \cup \{\langle A, B, C, 1 \rangle\})^*$, $\#_{\{\langle A,B,C,1\rangle\}}(x') > 0$. That is, x' is of type (III).

(iii) Assume that $y' \in (V \cup \{\langle A, B, C, j \rangle\})^*$, $\#_{\{\langle A,B,C,j \rangle\}}(y') > 0$, and

$$\text{sub}(y') \cap \{f^{-1}(k)\langle A, B, C, j \rangle \mid 1 \leq k < j, k \neq f(A)\} = \emptyset$$

where $\langle A, B, C, j \rangle \in W_S$, $A \in (N_1 \cup N_{CF})$, $B \in N_{CS}$, $C \in N_{CF}$, $1 \leq j \leq m + 1$ (form (III)). By the inspection of P', we see that the following four forms of rules can be used to rewrite y' to x'

(iii.a) $(a_i \rightarrow a_i, \emptyset)$, $a_i \in V$;
(iii.b) $(\langle A, B, C, j \rangle \rightarrow \langle A, B, C, j + 1 \rangle, \{f^{-1}(j)\langle A, B, C, j \rangle\})$, $1 \leq j \leq m$, $j \neq f(A)$;
(iii.c) $(\langle A, B, C, f(A) \rangle \rightarrow \langle A, B, C, f(A) + 1 \rangle, \emptyset)$;
(iii.d) $(\langle A, B, C, m + 1 \rangle \rightarrow C, \{\langle A, B, C, m + 1 \rangle^2\})$.

Let $1 \leq j \leq m$, $j \neq f(A)$. Then, symbols from V are rewritten to themselves (case (iii.a)) and every occurrence of $\langle A, B, C, j \rangle$ is rewritten to $\langle A, B, C, j + 1 \rangle$ by (iii.b). Clearly, we obtain $x' \in (V \cup \{\langle A, B, C, j + 1 \rangle\})^*$ such that $\#_{\{\langle A,B,C,j+1 \rangle\}}(x') > 0$. Furthermore, (iii.b) can be used only when $f^{-1}(j)\langle A, B, C, j \rangle \notin \text{sub}(y')$. As

$$\text{sub}(y') \cap \{f^{-1}(k)\langle A, B, C, j \rangle \mid 1 \leq k < j, k \neq f(A)\} = \emptyset$$

it holds that

$$\text{sub}(y') \cap \{f^{-1}(k)\langle A, B, C, j \rangle \mid 1 \leq k \leq j, k \neq f(A)\} = \emptyset$$

Since every occurrence of $\langle A, B, C, j \rangle$ is rewritten to $\langle A, B, C, j + 1 \rangle$ and other symbols are unchanged,

$$\text{sub}(x') \cap \{f^{-1}(k)\langle A, B, C, j + 1 \rangle \mid 1 \leq k < j + 1, k \neq f(A)\} = \emptyset$$

Therefore, x' is of form (III).

Next, assume that $j = f(A)$. Then, all occurrences of $\langle A, B, C, j \rangle$ are rewritten to $\langle A, B, C, j + 1 \rangle$ by (iii.c), and symbols from V are rewritten to themselves. As before, we obtain $x' \in (V \cup \{\langle A, B, C, j + 1 \rangle\})^*$ and $\#_{\{\langle A,B,C,j+1 \rangle\}}(x') > 0$. Moreover, because

$$\text{sub}(y') \cap \{f^{-1}(k)\langle A, B, C, j \rangle \mid 1 \leq k < j, k \neq f(A)\} = \emptyset$$

and j is $f(A)$,

$$\text{sub}(x') \cap \{f^{-1}(k)\langle A, B, C, j + 1 \rangle \mid 1 \leq k < j + 1, k \neq f(A)\} = \emptyset$$

and x' belongs to (III) as well.

Finally, let $j = m + 1$. Then, every occurrence of $\langle A, B, C, j \rangle$ is rewritten to C (case (iii.d)). Therefore, $x' \in V^*$, so x' has form (I).

In (i), (ii), and (iii), we have considered all derivations that rewrite y' to x', and in each of these cases, we have shown that x' has one of the requested forms. Therefore, Claim 4.2.27 holds.   □

To prove the following claims, we need a finite symbol-to-symbols substitution γ from V^* into V'^* defined as

$$\gamma(X) = \{X\} \cup \{\langle A, X, C \rangle \mid \langle A, X, C \rangle \in W_0\}$$
$$\cup \{\langle A, X, C, j \rangle \mid \langle A, X, C, j \rangle \in W_S, 1 \leq j \leq m + 1\}$$

for all $X \in V$, $A \in (N_1 \cup N_{CF})$, $C \in N_{CF}$. Let γ^{-1} be the inverse of γ.

Claim 4.2.28. Let $y' = a_1 a_2 \cdots a_q$, $a_i \in V'$, $q = |y'|$, and $\gamma^{-1}(a_i) \Rightarrow_G^{h_i} \gamma^{-1}(u_i)$ for all $i \in \{1, \ldots, q\}$ and some $h_i \in \{0, 1\}$, $u_i \in V'^+$. Then, $\gamma^{-1}(y') \Rightarrow_G^r \gamma^{-1}(x')$ such that $x' = u_1 u_2 \cdots u_q$, $r = \sum_{i=1}^q h_i$, $r \leq q$.

Proof. First, consider any derivation of the form

$$\gamma^{-1}(X) \Rightarrow_G^h \gamma^{-1}(u)$$

where $X \in V'$, $u \in V'^+$, $h \in \{0, 1\}$. If $h = 0$, then $\gamma^{-1}(X) = \gamma^{-1}(u)$. Let $h = 1$. Then, there surely exists a rule $p: \gamma^{-1}(X) \to \gamma^{-1}(u) \in P$ such that

$$\gamma^{-1}(X) \Rightarrow_G \gamma^{-1}(u) \ [p]$$

Return to the statement of this claim. We can construct

$$\gamma^{-1}(a_1)\gamma^{-1}(a_2) \cdots \gamma^{-1}(a_q) \Rightarrow_G^{h_1} \gamma^{-1}(u_1)\gamma^{-1}(a_2) \cdots \gamma^{-1}(a_q)$$
$$\Rightarrow_G^{h_2} \gamma^{-1}(u_1)\gamma^{-1}(u_2) \cdots \gamma^{-1}(a_q)$$
$$\vdots$$
$$\Rightarrow_G^{h_q} \gamma^{-1}(u_1)\gamma^{-1}(u_2) \cdots \gamma^{-1}(u_q)$$

where

$$\gamma^{-1}(y') = \gamma^{-1}(a_1) \cdots \gamma^{-1}(a_q)$$

and

$$\gamma^{-1}(u_1) \cdots \gamma^{-1}(u_q) = \gamma^{-1}(u_1 \cdots u_q) = \gamma^{-1}(x')$$

In such a derivation, each $\gamma^{-1}(a_i)$ is either left unchanged (if $h_i = 0$) or rewritten to $\gamma^{-1}(u_i)$ by the corresponding rule $\gamma^{-1}(a_i) \to \gamma^{-1}(u_i)$. Obviously, the length of this derivation is $\sum_{i=1}^q h_i$.   □

Claim 4.2.29. $S_1 \Rightarrow_G^* x$ if and only if $S_1 \Rightarrow_{G'}^* x'$, where $x \in V^*$, $x' \in V'^*$, $x' \in \gamma(x)$.

Proof. The proof is divided into the only-if part and the if part.

Only If. The only-if part is established by induction on $h \geq 0$. That is, we show that

$$S_1 \Rightarrow_G^h x \quad \text{implies} \quad S_1 \Rightarrow_{G'}^* x$$

where $x \in V^*$, for $h \geq 0$.

Basis. Let $h = 0$. Then, $S_1 \Rightarrow_G^0 S_1$ and $S_1 \Rightarrow_{G'}^0 S_1$ as well.

Induction Hypothesis. Assume that the claim holds for all derivations of length $0 \leq h \leq n$, for some $n \geq 0$.

Induction Step. Consider any derivation of the form

$$S_1 \Rightarrow_G^{n+1} x$$

Since $n + 1 > 0$, there exists $y \in V^*$ and $p \in P$ such that

$$S_1 \Rightarrow_G^n y \Rightarrow_G x \; [p]$$

and by the induction hypothesis, there is also a derivation

$$S_1 \Rightarrow_{G'}^* y$$

Let $y = a_1 a_2 \cdots a_q$, $a_i \in V$, $1 \leq i \leq q$, $q = |y|$. The following cases (i) and (ii) cover all possible forms of p.

(i) Let $p\colon A \to u \in P_{CF}$, $A \in (N_1 \cup N_{CF})$, $u \in V^*$. Then, $y = y_1 A y_3$ and $x = y_1 u y_3$, $y_1, y_3 \in V^*$. Let $s = |y_1| + 1$. Since we have $(A \to u, W) \in P'$, we can construct a derivation

$$S_1 \Rightarrow_{G'}^* y \Rightarrow_{G'} x \; [p_1, \cdots, p_q]$$

such that $p_s\colon (A \to u, W)$ and $p_i\colon (a_i \to a_i, \emptyset)$ for all $i \in \{1, \cdots, q\}$, $i \neq s$.

(ii) Let $p\colon AB \to AC \in P_{CS}$, $A \in (N_1 \cup N_{CF})$, $B \in N_{CS}$, $C \in N_{CF}$. Then, $y = y_1 A B y_3$ and $x = y_1 A C y_3$, $y_1, y_3 \in V^*$. Let $s = |y_1| + 2$. In this case, there is the following derivation

$$
\begin{aligned}
S_1 \Rightarrow_{G'}^* & y_1 A B y_3 \\
\Rightarrow_{G'} & y_1 A \langle A, B, C \rangle y_3 && [p_s\colon (B \to \langle A, B, C \rangle, W)] \\
\Rightarrow_{G'} & y_1 A \langle A, B, C, 1 \rangle y_3 && [p_s\colon (\langle A, B, C \rangle \to \langle A, B, C, 1 \rangle, \\
& && \quad W - \{\langle A, B, C \rangle\})] \\
\Rightarrow_{G'} & y_1 A \langle A, B, C, 2 \rangle y_3 && [p_s\colon (\langle A, B, C, 1 \rangle \to \langle A, B, C, 2 \rangle, \\
& && \quad \{f^{-1}(1)\langle A, B, C, j \rangle\})] \\
\vdots \\
\Rightarrow_{G'} & y_1 A \langle A, B, C, f(A) \rangle y_3 && [p_s\colon (\langle A, B, C, f(A) - 1 \rangle \to \\
& && \quad \langle A, B, C, f(A) \rangle, \{f^{-1}(f(A) - 1) \\
& && \quad \langle A, B, C, f(A) - 1 \rangle\})]
\end{aligned}
$$

$$\Rightarrow_{G'} y_1 A \langle A, B, C, f(A) + 1 \rangle y_3 \ [p_s \colon (\langle A, B, C, f(A) \rangle \to$$
$$\langle A, B, C, f(A) + 1 \rangle, \emptyset)]$$
$$\Rightarrow_{G'} y_1 A \langle A, B, C, f(A) + 2 \rangle y_3 \ [p_s \colon (\langle A, B, C, f(A) + 1 \rangle \to$$
$$\langle A, B, C, f(A) + 2 \rangle, \{f^{-1}(f(A) + 1)$$
$$\langle A, B, C, f(A) + 1 \rangle\})]$$

$$\vdots$$

$$\Rightarrow_{G'} y_1 A \langle A, B, C, m + 1 \rangle y_3 \quad [p_s \colon (\langle A, B, C, m \rangle \to \langle A, B, C, m + 1 \rangle,$$
$$\{f^{-1}(m) \langle A, B, C, m \rangle\})]$$
$$\Rightarrow_{G'} y_1 A C y_3 \qquad\qquad [p_s \colon (\langle A, B, C, m + 1 \rangle \to C,$$
$$\{\langle A, B, C, m + 1 \rangle^2\})]$$

such that $p_i \colon (a_i \to a_i, \emptyset)$ for all $i \in \{1, \dots, q\}$, $i \neq s$.

If. By induction on $h \geq 0$, we prove that

$$S_1 \Rightarrow_{G'}^h x' \quad \text{implies} \quad S_1 \Rightarrow_G^* x$$

where $x' \in V'^*$, $x \in V^*$ and $x' \in \gamma(x)$.

Basis. Let $h = 0$. The only x' is S_1 because $S_1 \Rightarrow_{G'}^0 S_1$. Obviously, $S_1 \Rightarrow_G^0 S_1$ and $S_1 \in \gamma(S_1)$.

Induction Hypothesis. Suppose that the claim holds for any derivation of length $0 \leq h \leq n$, for some $n \geq 0$.

Induction Hypothesis. Consider any derivation of the form

$$S_1 \Rightarrow_{G'}^{n+1} x'$$

Since $n + 1 \geq 1$, there exists $y' \in V'^*$ and a sequence of rules $p_1, \dots, p_q$ from P', $q = |x'|$, such that

$$S_1 \Rightarrow_{G'}^n y' \Rightarrow_{G'} x' \ [p_1, \dots, p_q]$$

Let $y' = a_1 a_2 \cdots a_q$, $a_i \in V'$, $1 \leq i \leq q$. By the induction hypothesis, we have

$$S_1 \Rightarrow_G^* y$$

where $y \in V^*$ such that $y' \in \gamma(y)$.

From Claim 4.2.27, y' has one of the following forms (i), (ii), or (iii), described next.

(i) Let $y' \in V'^*$ (see (I) in Claim 4.2.27). Inspect P' to see that there are three forms of rules rewriting symbols a_i in y':

(i.a) $p_i: (a_i \to a_i, \emptyset) \in P', a_i \in V$. In this case,

$$\gamma^{-1}(a_i) \Rightarrow_G^0 \gamma^{-1}(a_i)$$

(i.b) $p_i: (a_i \to u_i, W) \in P'$ such that $a_i \to u_i \in P_{CF}$. Since $a_i = \gamma^{-1}(a_i)$, $u_i = \gamma^{-1}(u_i)$ and $a_i \to u_i \in P$,

$$\gamma^{-1}(a_i) \Rightarrow_G \gamma^{-1}(u_i) \ [a_i \to u_i]$$

(i.c) $p_i: (a_i \to \langle A, a_i, C \rangle, W) \in P', a_i \in N_{CS}, A \in (N_1 \cup N_{CF}), C \in N_{CF}$. Since $\gamma^{-1}(a_i) = \gamma^{-1}(\langle A, a_i, C \rangle)$, we have

$$\gamma^{-1}(a_i) \Rightarrow_G^0 \gamma^{-1}(\langle A, a_i, C \rangle)$$

We see that for all a_i, there exists a derivation

$$\gamma^{-1}(a_i) \Rightarrow_G^{h_i} \gamma^{-1}(z_i)$$

for some $h_i \in \{0, 1\}$, where $z_i \in V'^+$, $x' = z_1 z_2 \cdots z_q$. Therefore, by Claim 4.2.28, we can construct

$$S_1 \Rightarrow_G^* y \Rightarrow_G^r x$$

where $0 \le r \le q, x = \gamma^{-1}(x')$.

(ii) Let $y' \in (V \cup W_0)^*$ and $\#_{W_0}(y') > 0$ (see (II)). At this point, the following two forms of rules can be used to rewrite a_i in y'—(ii.a) or (ii.b).

(ii.a) $p_i: (a_i \to a_i, \emptyset) \in P', a_i \in V$. As in case (i.a),

$$\gamma^{-1}(a_i) \Rightarrow_G^0 \gamma^{-1}(a_i)$$

(ii.b) $p_i: (\langle A, B, C \rangle \to \langle A, B, C, 1 \rangle, W - \{\langle A, B, C \rangle\}), a_i = \langle A, B, C \rangle \in W_0, A \in (N_1 \cup N_{CF}), B \in N_{CS}, C \in N_{CF}$. Since $\gamma^{-1}(\langle A, B, C \rangle) = \gamma^{-1}(\langle A, B, C, 1 \rangle)$,

$$\gamma^{-1}(\langle A, B, C \rangle) \Rightarrow_G^0 \gamma^{-1}(\langle A, B, C, 1 \rangle)$$

Thus, there exists a derivation

$$S_1 \Rightarrow_G^* y \Rightarrow_G^0 x$$

where $x = \gamma^{-1}(x')$.

(iii) Let $y' \in (V \cup \{\langle A, B, C, j \rangle\})^*$, $\#_{\{\langle A,B,C,j \rangle\}}(y') > 0$, and

$$sub(y') \cap \{f^{-1}(k)\langle A, B, C, j \rangle \mid 1 \le k < j, \ k \ne f(A)\} = \emptyset$$

where $\langle A, B, C, j \rangle \in W_S$, $A \in (N_1 \cup N_{CF})$, $B \in N_{CS}$, $C \in N_{CF}$, $1 \le j \le m + 1$ (see (III)). By the inspection of P', the following four forms of rules can be used to rewrite y' to x':

(iii.a) p_i: $(a_i \to a_i, \emptyset)$, $a_i \in V$;
(iii.b) p_i: $(\langle A, B, C, j \rangle \to \langle A, B, C, j + 1 \rangle, \{f^{-1}(j)\langle A, B, C, j \rangle\})$, $1 \le j \le m$, $j \ne f(A)$;
(iii.c) p_i: $(\langle A, B, C, f(A) \rangle \to \langle A, B, C, f(A) + 1 \rangle, \emptyset)$;
(iii.d) p_i: $(\langle A, B, C, m + 1 \rangle \to C, \{\langle A, B, C, m + 1 \rangle^2\})$.

Let $1 \le j \le m$. G' can rewrite such y' using only the rules (iii.a) through (iii.c). Since $\gamma^{-1}(\langle A, B, C, j \rangle) = \gamma^{-1}(\langle A, B, C, j + 1 \rangle)$ and $\gamma^{-1}(a_i) = \gamma^{-1}(a_i)$, by analogy with (ii), we obtain

$$S_1 \Rightarrow^*_G y \Rightarrow^0_G x$$

such that $x = \gamma^{-1}(x')$.

Let $j = m + 1$. In this case, only the rules (iii.a) and (iii.d) can be used. Since $\#_{\{\langle A,B,C,j \rangle\}}(y') > 0$, there is at least one occurrence of $\langle A, B, C, m + 1 \rangle$ in y', and by the forbidding condition of the rule (iii.d), $\langle A, B, C, m + 1 \rangle^2 \notin \mathrm{sub}(y')$. Observe that for $j = m + 1$,

$$\{f^{-1}(k)\langle A, B, C, m + 1 \rangle \mid 1 \le k < j, \ k \ne f(A)\}$$
$$= \{X\langle A, B, C, m + 1 \rangle \mid X \in V, \ X \ne A\}$$

and thus

$$\mathrm{sub}(y') \cap \{X\langle A, B, C, m + 1 \rangle \mid X \in V, \ X \ne A\} = \emptyset$$

According to Claim 4.2.26, $\langle A, B, C, m + 1 \rangle$ has always a left neighbor in y'. As a result, the left neighbor of every occurrence of $\langle A, B, C, m + 1 \rangle$ is A. Therefore, we can express y', y, and x' as follows:

$$y' = y_1 A\langle A, B, C, m + 1 \rangle y_2 A\langle A, B, C, m + 1 \rangle y_3 \cdots y_r A\langle A, B, C, m + 1 \rangle y_{r+1}$$
$$y = \gamma^{-1}(y_1)AB\gamma^{-1}(y_2)AB\gamma^{-1}(y_3) \cdots \gamma^{-1}(y_r)AB\gamma^{-1}(y_{r+1})$$
$$x' = y_1 A C y_2 A C y_3 \cdots y_r A C y_{r+1}$$

where $r \ge 1$, $y_s \in V^*$, $1 \le s \le r + 1$. Since we have p: $AB \to AC \in P$, there is a derivation

$$S_1 \Rightarrow^*_G \gamma^{-1}(y_1)AB\gamma^{-1}(y_2)AB\gamma^{-1}(y_3) \cdots \gamma^{-1}(y_r)AB\gamma^{-1}(y_{r+1})$$
$$\Rightarrow_G \gamma^{-1}(y_1)AC\gamma^{-1}(y_2)AB\gamma^{-1}(y_3) \cdots \gamma^{-1}(y_r)AB\gamma^{-1}(y_{r+1}) \ [p]$$
$$\Rightarrow_G \gamma^{-1}(y_1)AC\gamma^{-1}(y_2)AC\gamma^{-1}(y_3) \cdots \gamma^{-1}(y_r)AB\gamma^{-1}(y_{r+1}) \ [p]$$
$$\vdots$$
$$\Rightarrow_G \gamma^{-1}(y_1)AC\gamma^{-1}(y_2)AC\gamma^{-1}(y_3) \cdots \gamma^{-1}(y_r)AC\gamma^{-1}(y_{r+1}) \ [p]$$

where

$$\gamma^{-1}(y_1)AC\gamma^{-1}(y_2)AC\gamma^{-1}(y_3)\cdots\gamma^{-1}(y_r)AC\gamma^{-1}(y_{r+1}) = \gamma^{-1}(x') = x$$

Since cases (i), (ii), and (iii) cover all possible forms of y', we have completed the induction and established Claim 4.2.29. □

The equivalence of G and G' follows from Claim 4.2.29. Indeed, observe that by the definition of γ, we have $\gamma(a) = \{a\}$ for all $a \in T$. Therefore, by Claim 4.2.29, we have for any $x \in T^*$,

$$S_1 \Rightarrow_G^* x \quad \text{if and only if} \quad S_1 \Rightarrow_{G'}^* x$$

Thus, $L(G) = L(G')$, and the lemma holds. □

Theorem 4.2.30. $\mathbf{CS} = \mathbf{F\text{-}EP0L}(2) = \mathbf{F\text{-}EPT0L}(2) = \mathbf{F\text{-}EP0L} = \mathbf{F\text{-}EPT0L}$

Proof. By Lemma 4.2.24, $\mathbf{CS} \subseteq \mathbf{F\text{-}EP0L}(2) \subseteq \mathbf{F\text{-}EPT0L}(2) \subseteq \mathbf{F\text{-}EPT0L}$. From Lemma 4.2.5 and the definition of F-ET0L grammars, $\mathbf{F\text{-}EPT0L}(s) \subseteq \mathbf{F\text{-}EPT0L} \subseteq \mathbf{C\text{-}EPT0L} \subseteq \mathbf{CS}$ for any $s \geq 0$. Moreover, $\mathbf{F\text{-}EP0L}(s) \subseteq \mathbf{F\text{-}EP0L} \subseteq \mathbf{F\text{-}EPT0L}$. Thus, $\mathbf{CS} = \mathbf{F\text{-}EP0L}(2) = \mathbf{F\text{-}EPT0L}(2) = \mathbf{F\text{-}EP0L} = \mathbf{F\text{-}EPT0L}$, and the theorem holds. □

Return to the proof of Lemma 4.2.24. Observe the form of the rules in the F-EP0L grammar G'. This observation gives rise to the next corollary.

Corollary 4.2.31. *Every context-sensitive language can be generated by an F-EP0L grammar $G = (V, T, P, S)$ of degree 2 such that every rule from P has one of the following forms*
  (i) $(a \rightarrow a, \emptyset)$, $a \in V$;
  (ii) $(X \rightarrow x, F)$, $X \in V - T$, $|x| \in \{1, 2\}$, max-len$(F) = 1$;
  (iii) $(X \rightarrow Y, \{z\})$, $X, Y \in V - T$, $z \in V^2$. □

Next, we demonstrate that the family of recursively enumerable languages is generated by the forbidding E0L grammars of degree 2.

Lemma 4.2.32. $\mathbf{RE} \subseteq \mathbf{F\text{-}E0L}(2)$

Proof. Let L be a recursively enumerable language generated by a phrase-structure grammar

$$G = (V, T, P, S)$$

having the form defined in Theorem 3.1.7, where

$$\begin{aligned}
V &= N_{CF} \cup N_{CS} \cup T \\
P_{CS} &= \{AB \rightarrow AC \in P \mid A, C \in N_{CF}, B \in N_{CS}\} \\
P_{CF} &= P - P_{CS}
\end{aligned}$$

Let \$ be a new symbol and m be the cardinality of $V \cup \{\$\}$. Furthermore, let f be an arbitrary bijection from $V \cup \{\$\}$ onto $\{1, \ldots, m\}$, and let f^{-1} be the inverse of f.

Define the F-E0L grammar

$$G' = (V', T, P', S')$$

of degree 2 as follows:

$$W_0 = \{\langle A, B, C \rangle \mid AB \to AC \in P\}$$
$$W_S = \{\langle A, B, C, j \rangle \mid AB \to AC \in P, 1 \le j \le m\}$$
$$W = W_0 \cup W_S$$
$$V' = V \cup W \cup \{S', \$\}$$

where $A, C \in N_{CF}, B \in N_{CS}$, and V, W_0, W_S, and $\{S', \$\}$ are pairwise disjoint alphabets. The set of rules P' is constructed by performing (1) through (4), given next.

(1) Add $(S' \to \$S, \emptyset)$, $(\$ \to \$, \emptyset)$ and $(\$ \to \varepsilon, V' - T - \{\$\})$ to P'.
(2) For all $X \in V$, add $(X \to X, \emptyset)$ to P'.
(3) For all $A \to u \in P_{CF}, A \in N_{CF}, u \in \{\varepsilon\} \cup N_{CS} \cup T \cup (\bigcup_{i=1}^{2} N_{CF}^i)$, add $(A \to u, W)$ to P'.
(4) If $AB \to AC \in P_{CS}, A, C \in N_{CF}, B \in N_{CS}$, then add the following rules into P'.

   (4.a) $(B \to \langle A, B, C \rangle, W)$;
   (4.b) $(\langle A, B, C \rangle \to \langle A, B, C, 1 \rangle, W - \{\langle A, B, C \rangle\})$;
   (4.c) $(\langle A, B, C, j \rangle \to \langle A, B, C, j + 1 \rangle, \{f^{-1}(j)\langle A, B, C, j \rangle\})$ for all $1 \le j \le m$ such that $f(A) \ne j$;
   (4.d) $(\langle A, B, C, f(A) \rangle \to \langle A, B, C, f(A) + 1 \rangle, \emptyset)$;
   (4.e) $(\langle A, B, C, m + 1 \rangle \to C, \{\langle A, B, C, m + 1 \rangle^2\})$.

Let us only give a gist of the reason why $L(G) = L(G')$. The construction above resembles the construction in Lemma 4.2.24 very much. Indeed, to simulate the non-context-free rules $AB \to AC$ in F-E0L grammars, we use the same technique as in F-EP0L grammars from Lemma 4.2.24. We only need to guarantee that no sentential form begins with a symbol from N_{CS}. This is solved by an auxiliary nonterminal $\$$ in the definition of G'. The symbol is always generated in the first derivation step by $(S' \to \$S, \emptyset)$ (see (1) in the definition of P'). After that, it appears as the leftmost symbol of all sentential forms containing some nonterminals. The only rule that can erase it is $(\$ \to \varepsilon, V' - T - \{\$\})$.

Therefore, by analogy with the technique used in Lemma 4.2.24, we can establish

$$S \Rightarrow_G^* x \quad \text{if and only if} \quad S' \Rightarrow_{G'}^+ \$x'$$

such that $x \in V^*$, $x' \in (V' - \{S', \$\})^*$, $x' \in \gamma(x)$, where γ is a finite substitution from V^* into $(V' - \{S', \$\})^*$ defined as

$$\gamma(X) = \{X\} \cup \{\langle A, X, C \rangle \mid \langle A, X, C \rangle \in W_0\}$$
$$\cup \{\langle A, X, C, j \rangle \mid \langle A, X, C, j \rangle \in W_S, 1 \le j \le m + 1\}$$

for all $X \in V, A, C \in N_{CF}$. The details are left to the reader.

As in Lemma 4.2.24, we have $\gamma(a) = \{a\}$ for all $a \in T$; hence, for all $x \in T^*$,

$$S \Rightarrow^*_G x \quad \text{if and only if} \quad S' \Rightarrow^+_{G'} \$x$$

Since

$$\$x \Rightarrow_{G'} x \; [(\$ \to \varepsilon, V' - T - \{\$\})]$$

we obtain

$$S \Rightarrow^*_G x \quad \text{if and only if} \quad S' \Rightarrow^+_{G'} x$$

Consequently, $L(G) = L(G')$; thus, $\mathbf{RE} \subseteq \mathbf{F \text{-} E0L}(2)$. $\qquad \square$

Theorem 4.2.33. $\mathbf{RE} = \mathbf{F \text{-} E0L}(2) = \mathbf{F \text{-} ET0L}(2) = \mathbf{F \text{-} E0L} = \mathbf{F \text{-} ET0L}$

Proof. By Lemma 4.2.32, we have $\mathbf{RE} \subseteq \mathbf{F \text{-} E0L}(2) \subseteq \mathbf{F \text{-} ET0L}(2) \subseteq \mathbf{F \text{-} ET0L}$. From Lemma 4.2.10, it follows that $\mathbf{F \text{-} ET0L}(s) \subseteq \mathbf{F \text{-} ET0L} \subseteq \mathbf{C \text{-} ET0L} \subseteq \mathbf{RE}$, for any $s \geq 0$. Thus, $\mathbf{RE} = \mathbf{F \text{-} E0L}(2) = \mathbf{F \text{-} ET0L}(2) = \mathbf{F \text{-} E0L} = \mathbf{F \text{-} ET0L}$, so the theorem holds. $\qquad \square$

By analogy with Corollary 4.2.31, we obtain the following normal form.

Corollary 4.2.34. *Every recursively enumerable language can be generated by an F-E0L grammar $G = (V, T, P, S)$ of degree 2 such that every rule from P has one of the following forms*

(i) $(a \to a, \emptyset)$, $a \in V$;
(ii) $(X \to x, F)$, $X \in V - T$, $|x| \leq 2$, $F \neq \emptyset$, and $\text{max-len}(F) = 1$;
(iii) $(X \to Y, \{z\})$, $X, Y \in V - T$, $z \in V^2$. $\qquad \square$

Moreover, we obtain the following relations between F-ET0L language families.

Corollary 4.2.35.

$$\mathbf{CF}$$
$$\subset$$
$$\mathbf{F \text{-} EP0L}(0) = \mathbf{F \text{-} E0L}(0) = \mathbf{EP0L} = \mathbf{E0L}$$
$$\subset$$
$$\mathbf{F \text{-} EP0L}(1) = \mathbf{F \text{-} EPT0L}(1) = \mathbf{F \text{-} E0L}(1) = \mathbf{F \text{-} ET0L}(1)$$
$$= \mathbf{F \text{-} EPT0L}(0) = \mathbf{F \text{-} ET0L}(0) = \mathbf{EPT0L} = \mathbf{ET0L}$$
$$\subset$$
$$\mathbf{F \text{-} EP0L}(2) = \mathbf{F \text{-} EPT0L}(2) = \mathbf{F \text{-} EP0L} = \mathbf{F \text{-} EPT0L} = \mathbf{CS}$$
$$\subset$$
$$\mathbf{F \text{-} E0L}(2) = \mathbf{F \text{-} ET0L}(2) = \mathbf{F \text{-} E0L} = \mathbf{F \text{-} ET0L} = \mathbf{RE}$$

Proof. This corollary follows from Theorems 4.2.14, 4.2.22, 4.2.30, and 4.2.33. $\qquad \square$

4.2.3 Simple Semi-Conditional ET0L Grammars

Simple semi-conditional ET0L grammars represent another variant of context-
conditional ET0L grammars with restricted sets of context conditions. By analogy
with sequential simple semi-conditional grammars (see Sect. 3.1.6), these grammars
are context-conditional ET0L grammars in which every rule contains no more
than one context condition. This section defines them, establishes their power and
reduces their degree.

Definitions

In this section, we define simple semi-conditional ET0L grammars.

Definition 4.2.36. Let $G = (V, T, P_1, \ldots, P_t, S)$ be a context-conditional ET0L
grammar, for some $t \geq 1$. If for all $p: (a \to x, Per, For) \in P_i$ for every $i = 1, \ldots, t$
holds that $\text{card}(Per) + \text{card}(For) \leq 1$, G is said to be a *simple semi-conditional
ET0L grammar* (*SSC-ET0L grammar* for short). If G is a propagating SSC-ET0L
grammar, then G is called an *SSC-EPT0L grammar*. If $t = 1$, then G is called an
SSC-E0L grammar; if, in addition, G is a propagating SSC-E0L grammar, G is said
to be an *SSC-EP0L grammar*.   □

Let $G = (V, T, P_1, \ldots, P_t, S)$ be an SSC-ET0L grammar of degree (r, s). By
analogy with ssc-grammars (see Sect. 3.1.6), in each rule $(a \to x, Per, For) \in P_i$,
$i = 1, \ldots, t$, we omit braces and instead of $\emptyset$, we write 0. For example, we write
$(a \to x, EF, 0)$ instead of $(a \to x, \{EF\}, \emptyset)$.

Let **SSC-EPT0L**(r, s), **SSC-ET0L**(r, s), **SSC-EP0L**(r, s), and **SSC-E0L**(r, s)
denote the families of languages generated by SSC-EPT0L, SSC-ET0L, SSC-EP0L,
and SSC-E0L grammars of degree (r, s), respectively. Furthermore, the families
of languages generated by SSC-EPT0L, SSC-ET0L, SSC-EP0L, and SSC-E0L
grammars of any degree are denoted by **SSC-EPT0L**, **SSC-ET0L**, **SSC-EP0L**,
and **SSC-E0L**, respectively. Moreover, set

$$\textbf{SSC-EPT0L} = \bigcup_{r=0}^{\infty} \bigcup_{s=0}^{\infty} \textbf{SSC-EPT0L}(r, s)$$

$$\textbf{SSC-ET0L} = \bigcup_{r=0}^{\infty} \bigcup_{s=0}^{\infty} \textbf{SSC-ET0L}(r, s)$$

$$\textbf{SSC-EP0L} = \bigcup_{r=0}^{\infty} \bigcup_{s=0}^{\infty} \textbf{SSC-EP0L}(r, s)$$

$$\textbf{SSC-E0L} = \bigcup_{r=0}^{\infty} \bigcup_{s=0}^{\infty} \textbf{SSC-E0L}(r, s)$$

Generative Power and Reduction

Next, let us investigate the generative power of SSC-ET0L grammars. The following lemma proves that every recursively enumerable language can be defined by an SSC-E0L grammar of degree $(1, 2)$.

Lemma 4.2.37. RE $\subseteq$ SSC-E0L$(1, 2)$

Proof. Let

$$G = \left(N_{CF} \cup N_{CS} \cup T, T, P, S \right)$$

be a phrase-structure grammar of the form of Theorem 3.1.7. Then, let $V = N_{CF} \cup N_{CS} \cup T$ and m be the cardinality of V. Let f be an arbitrary bijection from V to $\{1, \ldots, m\}$, and f^{-1} be the inverse of f. Set

$$
\begin{aligned}
M = \{\#\} \cup \\
\{\langle A, B, C \rangle \mid AB \to AC \in P, A, C \in N_{CF}, B \in N_{CS}\} \cup \\
\{\langle A, B, C, i \rangle \mid AB \to AC \in P, A, C \in N_{CF}, B \in N_{CS}, 1 \le i \le m + 2\}
\end{aligned}
$$

and

$$W = \{[A, B, C] \mid AB \to AC \in P, A, C \in N_{CF}, B \in N_{CS}\}$$

Next, construct an SSC-E0L grammar of degree $(1, 2)$

$$G' = \left(V', T, P', S' \right)$$

where

$$V' = V \cup M \cup W \cup \{S'\}$$

Without any loss of generality, we assume that V, M, W, and $\{S'\}$ are pairwise disjoint. The set of rules P' is constructed by performing (1) through (5), given next.

(1) Add $(S' \to \#S, 0, 0)$ to P'.
(2) For all $A \to x \in P, A \in N_{CF}, x \in \{\varepsilon\} \cup N_{CS} \cup T \cup N_{CF}^2$, add $(A \to x, \#, 0)$ to P'.
(3) For every $AB \to AC \in P, A, C \in N_{CF}, B \in N_{CS}$, add the following rules to P'

    (3.a) $(\# \to \langle A, B, C \rangle, 0, 0)$;
    (3.b) $(B \to [A, B, C], \langle A, B, C \rangle, 0)$;
    (3.c) $(\langle A, B, C \rangle \to \langle A, B, C, 1 \rangle, 0, 0)$;
    (3.d) $([A, B, C] \to [A, B, C], 0, \langle A, B, C, m + 2 \rangle)$;
    (3.e) $(\langle A, B, C, i \rangle \to \langle A, B, C, i + 1 \rangle, 0, f^{-1}(i)[A, B, C])$ for all $1 \le i \le m$, $i \ne f(A)$;
    (3.f) $(\langle A, B, C, f(A) \rangle \to \langle A, B, C, f(A) + 1 \rangle, 0, 0)$;
    (3.g) $(\langle A, B, C, m + 1 \rangle \to \langle A, B, C, m + 2 \rangle, 0, [A, B, C]^2)$;

(3.h) $(\langle A, B, C, m + 2 \rangle \to \#, 0, \langle A, B, C, m + 2 \rangle [A, B, C])$;
(3.i) $([A, B, C] \to C, \langle A, B, C, m + 2 \rangle, 0)$.

(4) For all $X \in V$, add $(X \to X, 0, 0)$ to P'.
(5) Add $(\# \to \#, 0, 0)$ and $(\# \to \varepsilon, 0, 0)$ to P'.

Let us explain how G' works. During the simulation of a derivation in G, every sentential form starts with an auxiliary symbol from M, called the master. This symbol determines the current simulation mode and controls the next derivation step. Initially, the master is set to $\#$ (see (1) in the definition of P'). In this mode, G' simulates context-free rules (see (2)); notice that symbols from V can always be rewritten to themselves by (4). To start the simulation of a non-context-free rule of the form $AB \to AC$, G' rewrites the master to $\langle A, B, C \rangle$. In the following step, chosen occurrences of B are rewritten to $[A, B, C]$; no other rules can be used except rules introduced in (4). At the same time, the master is rewritten to $\langle A, B, C, i \rangle$ with $i = 1$ (see (3.c)). Then, i is repeatedly incremented by one until i is greater than the cardinality of V (see rules (3.e) and (3.f)). Simultaneously, the master's conditions make sure that for every i such that $f^{-1}(i) \neq A$, no $f^{-1}(i)$ appears as the left neighbor of any occurrence of $[A, B, C]$. Finally, G' checks that there are no two adjoining $[A, B, C]$ (see (3.g)) and that $[A, B, C]$ does not appear as the right neighbor of the master (see (3.h)). At this point, the left neighbors of $[A, B, C]$ are necessarily equal to A and every occurrence of $[A, B, C]$ is rewritten to C. In the same derivation step, the master is rewritten to $\#$.

Observe that in every derivation step, the master allows G' to use only a subset of rules according to the current mode. Indeed, it is not possible to combine context-free and non-context-free simulation modes. Furthermore, no two different non-context-free rules can be simulated at the same time. The simulation ends when $\#$ is erased by $(\# \to \varepsilon, 0, 0)$. After this erasure, no other rule can be used.

The following three claims demonstrate some important properties of derivations in G' to establish $L(G) = L(G')$.

Claim 4.2.38. $S' \Rightarrow_{G'}^{+} w'$ implies that $w' \in M(V \cup W)^*$ or $w' \in (V \cup W)^*$. Furthermore, if $w' \in M(V \cup W)^*$, every v' such that $S' \Rightarrow_{G'}^{+} v' \Rightarrow_{G'}^{*} w'$ belongs to $M(V \cup W)^*$ as well.

Proof. When deriving w', G' first rewrites S' to $\#S$ by using $(S' \to \#S, 0, 0)$, where $\# \in M$ and $S \in V$. Next, inspect P' to see that every symbol from M is always rewritten to a symbol belonging to M or, in the case of $\#$, erased by $(\# \to \varepsilon, 0, 0)$. Moreover, there are no rules generating new occurrences of symbols from $(M \cup \{S'\})$. Thus, all sentential forms derived from S' belong either to $M(V \cup W)^*$ or to $(V \cup W)^*$. In addition, if a sentential form belongs to $M(V \cup W)^*$, all previous sentential forms (except for S') are also from $M(V \cup W)^*$. □

Claim 4.2.39. Every successful derivation in G' is of the form

$$S' \Rightarrow_{G'} \#S \Rightarrow_{G'}^{+} \#u' \Rightarrow_{G'} w' \Rightarrow_{G'}^{*} w'$$

where $u' \in V^*$, $w' \in T^*$.

Proof. From Claim 4.2.38 and its proof, every successful derivation has the form

$$S' \Rightarrow_{G'} \#S \Rightarrow_{G'}^+ \#u' \Rightarrow_{G'} v' \Rightarrow_{G'}^* w'$$

where $u', v' \in (V \cup W)^*, w' \in T^*$. This claim shows that

$$\#u' \Rightarrow_{G'} v' \Rightarrow_{G'}^* w'$$

implies that $u' \in V^*$ and $v' = w'$. Consider

$$\#u' \Rightarrow_{G'} v' \Rightarrow_{G'}^* w'$$

where $u', v' \in (V \cup W)^*, w' \in T^*$. Assume that u' contains a nonterminal $[A, B, C] \in W$. There are two rules rewriting $[A, B, C]$:

$$p_1: ([A, B, C] \to [A, B, C], 0, \langle A, B, C, m+2 \rangle)$$

and

$$p_2: ([A, B, C] \to C, \langle A, B, C, m+2 \rangle, 0)$$

Because of its permitting condition, p_2 cannot be applied during $\#u' \Rightarrow_{G'} v'$. If $[A, B, C]$ is rewritten by p_1—that is, $[A, B, C] \in \text{alph}(v')$—$[A, B, C]$ necessarily occurs in all sentential forms derived from v'. Thus, no u' containing a nonterminal from W results in a terminal string; hence, $u' \in V^*$. By analogical considerations, establish that also $v' \in V^*$. Next, assume that v' contains some $A \in N_{CF}$ or $B \in N_{CS}$. The first one can be rewritten by $(A \to z, \#, 0), z \in V^*$, and the second one by $(B \to [A, B, C], \langle A, B, C \rangle, 0), [A, B, C] \in W, \langle A, B, C \rangle \in M$. In both cases, the permitting condition forbids an application of the rule. Consequently, $v' \in T^*$. It is sufficient to show that $v' = w'$. Indeed, every rule rewriting a terminal is of the form $(a \to a, 0, 0), a \in T$. $\qquad\square$

Claim 4.2.40. Let $S' \Rightarrow_{G'}^n Zx', Z \in M, x' \in (V \cup W)^*, n \geq 1$. Then, Zx' has one of the following forms

(I) $Z = \#, x' \in V^*$;

(II) $Z = \langle A, B, C \rangle, x' \in V^*$, for some $A, C \in N_{CF}, B \in N_{CS}$;

(III) $Z = \langle A, B, C, i \rangle, x' \in (V \cup \{[A, B, C]\})^*, 1 \leq i \leq m+1$, and $\{f^{-1}(j)[A, B, C] \mid 1 \leq j < i, j \neq f(A)\} \cap \text{sub}(x') = \emptyset$ for some $A, C \in N_{CF}, B \in N_{CS}$;

(IV) $Z = \langle A, B, C, m+2 \rangle, x' \in (V \cup \{[A, B, C]\})^*, \{X[A, B, C] \mid X \in V, X \neq A\} \cap \text{sub}(x') = \emptyset$, and $[A, B, C]^2 \notin \text{sub}(x')$ for some $A, C \in N_{CF}, B \in N_{CS}$.

Proof. This claim is proved by induction on $h \geq 1$.

Basis. Let $h = 1$. Then, $S' \Rightarrow_{G'} \#S$, where $\#S$ is of type (I).

Induction Hypothesis. Suppose that the claim holds for all derivations of length $1 \leq h \leq n$, for some $n \geq 1$.

Induction Step. Consider any derivation of the form

$$S' \Rightarrow_{G'}^{n+1} Qx'$$

where $Q \in M$, $x' \in (V \cup W)^*$. Since $n + 1 \geq 2$, by Claim 4.2.38, there exists $Zy' \in M(V \cup W)^*$ and a sequence of rules $p_0, p_1, \ldots, p_q$, where $p_i \in P'$, $0 \leq i \leq q$, $q = |y'|$, such that

$$S' \Rightarrow_{G'}^{n} Zy' \Rightarrow_{G'} Qx' \, [p_0, p_1, \ldots, p_q]$$

Let $y' = a_1 a_2 \cdots a_q$, where $a_i \in (V \cup W)$ for all $i = 1, \ldots, q$. By the induction hypothesis, the following cases (i) through (iv) cover all possible forms of Zy'.

(i) Let $Z = \#$ and $y' \in V^*$ (form (I)). According to the definition of P', p_0 is either $(\# \to \langle A, B, C \rangle, 0, 0)$, $A, C \in N_{CF}$, $B \in N_{CS}$, or $(\# \to \#, 0, 0)$, or $(\# \to \varepsilon, 0, 0)$, and every p_i is either of the form $(a_i \to z, \#, 0)$, $z \in \{\varepsilon\} \cup N_{CS} \cup T \cup N_{CF}^2$, or $(a_i \to a_i, 0, 0)$. Obviously, y' is always rewritten to a string $x' \in V^*$. If $\#$ is rewritten to $\langle A, B, C \rangle$, we get $\langle A, B, C \rangle x'$ that is of form (II). If $\#$ remains unchanged, $\#x'$ is of type (I). In case that $\#$ is erased, the resulting sentential form does not belong to $M(V \cup W)^*$ required by this claim (which also holds for all strings derived from x' (see Claim 4.2.38)).

(ii) Let $Z = \langle A, B, C \rangle$, $y' \in V^*$, for some $A, C \in N_{CF}$, $B \in N_{CS}$ (form (II)). In this case, $p_0 : (\langle A, B, C \rangle \to \langle A, B, C, 1 \rangle, 0, 0)$ and every p_i is either $(a_i \to [A, B, C], \langle A, B, C \rangle, 0)$ or $(a_i \to a_i, 0, 0)$ (see the definition of P'). It is easy to see that $\langle A, B, C, 1 \rangle x'$ belongs to (III).

(iii) Let $Z = \langle A, B, C, j \rangle$, $y' \in (V \cup \{[A, B, C]\})^*$, and y' satisfies

$$\{f^{-1}(k)[A, B, C] \mid 1 \leq k < j, \ k \neq f(A)\} \cap \mathrm{sub}()(y') = \emptyset$$

$1 \leq j \leq m + 1$, for some $A, C \in N_{CF}$, $B \in N_{CS}$ (form (III)). The only rules rewriting symbols from y' are $(a_i \to a_i, 0, 0)$, $a_i \in V$, and $([A, B, C] \to [A, B, C], 0, \langle A, B, C, m + 2 \rangle)$; thus, y' is rewritten to itself. By the inspection of P', p_0 can be of the following three forms.

(a) If $j \neq f(A)$ and $j < m + 1$,

$$p_0 = (\langle A, B, C, j \rangle \to \langle A, B, C, j + 1 \rangle, 0, f^{-1}(j)[A, B, C])$$

Clearly, p_0 can be used only when $f^{-1}(j)[A, B, C] \notin \mathrm{sub}(Zy')$. As

$$\{f^{-1}(k)[A, B, C] \mid 1 \leq k < j, \ k \neq f(A)\} \cap \mathrm{sub}()(y') = \emptyset$$

it also

$$\{f^{-1}(k)[A,B,C] \mid 1 \le k \le j,\ k \ne f(A)\} \cap \mathrm{sub}()(y') = \emptyset$$

Since $\langle A,B,C,j \rangle$ is rewritten to $\langle A,B,C,j+1 \rangle$ and y' is unchanged, we get $\langle A,B,C,j+1 \rangle y'$ with

$$\{f^{-1}(k)[A,B,C] \mid 1 \le k < j+1,\ k \ne f(A)\} \cap \mathrm{sub}()(y') = \emptyset$$

which is of form (III).

(b) If $j = f(A)$,

$$p_0 = (\langle A,B,C,f(A) \rangle \to \langle A,B,C,f(A)+1 \rangle, 0, 0)$$

As before, $Qx' = \langle A,B,C,j+1 \rangle y'$. Moreover, because

$$\{f^{-1}(k)[A,B,C] \mid 1 \le k < j,\ k \ne f(A)\} \cap \mathrm{sub}()(y') = \emptyset$$

and $j = f(A)$,

$$\{f^{-1}(k)[A,B,C] \mid 1 \le k < j+1,\ k \ne f(A)\} \cap \mathrm{sub}()(x') = \emptyset$$

Consequently, Qx' belongs to (III) as well.

(c) If $j = m+1$,

$$p_0 = (\langle A,B,C,m+1 \rangle \to \langle A,B,C,m+2 \rangle, 0, [A,B,C]^2)$$

Then, $Qx' = \langle A,B,C,m+2 \rangle y'$. The application of p_0 implies that $[A,B,C]^2 \notin \mathrm{sub}(x')$. In addition, observe that for $j = m+1$,

$$\{f^{-1}(k)[A,B,C] \mid 1 \le k < j,\ k \ne f(A)\}$$
$$= \{X[A,B,C] \mid X \in V,\ X \ne A\}$$

Hence,

$$\{X[A,B,C] \mid X \in V,\ X \ne A\} \cap \mathrm{sub}()(x') = \emptyset$$

As a result, Qx' is of form (IV).

(iv) Let $Z = \langle A,B,C,m+2 \rangle$, $y' \in (V \cup \{[A,B,C]\})^*$, $[A,B,C]^2 \notin \mathrm{sub}(y')$, and

$$\{X[A,B,C] \mid X \in V,\ X \ne A\} \cap \mathrm{sub}()(y') = \emptyset$$

for some $A,C \in N_{CF}$, $B \in N_{CS}$ (form (IV)). Inspect P' to see that

$$p_0 = (\langle A,B,C,m+2 \rangle \to \#, 0, \langle A,B,C,m+2 \rangle [A,B,C])$$

and p_i is either

$$(a_i \to a_i, 0, 0), \ a_i \in V$$

or

$$([A, B, C] \to C, \langle A, B, C, m + 2 \rangle, 0)$$

where $1 \leq i \leq q$. According to the right-hand sides of these rules, $Qx' \in \{\#\}V^*$; that is, Qx' belongs to (I).

In cases (i) through (iv), we have demonstrated that every sentential form obtained in $n + 1$ derivation steps satisfies the statement of this claim. Therefore, we have finished the induction step and established Claim 4.2.40.   □

To prove the following claims, define the finite substitution γ from V^* into $(V \cup W)^*$ as

$$\gamma(X) = \{X\} \cup \{[A, B, C] \in W \mid A, C \in N_{CF}, B \in N_{CS}\}$$

for all $X \in V$. Let γ^{-1} be the inverse of γ.

Claim 4.2.41. Let $y' = a_1 a_2 \cdots a_q$, $a_i \in (V \cup W)^*$, $q = |y'|$, and $\gamma^{-1}(a_i) \Rightarrow_G^{h_i} \gamma^{-1}(x_i')$ for all $i \in \{1, \ldots, q\}$ and some $h_i \in \{0, 1\}$, $x_i' \in (V \cup W)^*$. Then, $\gamma^{-1}(y') \Rightarrow_G^{h} \gamma^{-1}(x')$ such that $x' = x_1' x_2' \cdots x_q'$, $h = \sum_{i=1}^{q} h_i$, $h \leq q$.

Proof. Consider any derivation of the form

$$\gamma^{-1}(X) \Rightarrow_G^{l} \gamma^{-1}(u)$$

$X \in (V \cup W)$, $u \in (V \cup W)^*$, $l \in \{0, 1\}$. If $l = 0$, $\gamma^{-1}(X) = \gamma^{-1}(u)$. Let $l = 1$. Then, there surely exists a rule $p: \gamma^{-1}(X) \to \gamma^{-1}(u) \in P$ such that

$$\gamma^{-1}(X) \Rightarrow_G \gamma^{-1}(u) \ [p]$$

Return to the statement of this claim. We can construct this derivation

$$\begin{aligned}
\gamma^{-1}(a_1)\gamma^{-1}(a_2) \cdots \gamma^{-1}(a_q) &\Rightarrow_G^{h_1} \gamma^{-1}(x_1')\gamma^{-1}(a_2) \cdots \gamma^{-1}(a_q) \\
&\Rightarrow_G^{h_2} \gamma^{-1}(x_1')\gamma^{-1}(x_2') \cdots \gamma^{-1}(a_q) \\
&\ \ \vdots \\
&\Rightarrow_G^{h_q} \gamma^{-1}(x_1')\gamma^{-1}(x_2') \cdots \gamma^{-1}(x_q')
\end{aligned}$$

where

$$\gamma^{-1}(y') = \gamma^{-1}(a_1) \cdots \gamma^{-1}(a_q)$$

and

$$\gamma^{-1}(x_1') \cdots \gamma^{-1}(x_q') = \gamma^{-1}(x_1' \cdots x_q') = \gamma^{-1}(x')$$

In such a derivation, each $\gamma^{-1}(a_i)$ is either left unchanged (if $h_i = 0$) or rewritten to $\gamma^{-1}(x_i')$ by the corresponding rule $\gamma^{-1}(a_i) \to \gamma^{-1}(x_i')$. Obviously, the length of this derivation is $\sum_{i=1}^{q} h_i$. □

Claim 4.2.42. $S \Rightarrow_G^* x$ if and only if $S' \Rightarrow_{G'}^+ Qx'$, where $\gamma^{-1}(x') = x$, $Q \in M$, $x \in V^*$, $x' \in (V \cup W)^*$.

Proof. The proof is divided into the only-if part and the if part.

Only If. By induction on $h \geq 0$, we show that

$$S \Rightarrow_G^h x \quad \text{implies} \quad S' \Rightarrow_{G'}^+ \#x$$

where $x \in V^*$, $h \geq 0$. Clearly, $\gamma^{-1}(x) = x$.

Basis. Let $h = 0$. Then, $S \Rightarrow_G^0 S$. In G', $S' \Rightarrow_{G'} \#S$ by using $(S' \to \#S, 0, 0)$.

Induction Hypothesis. Assume that the claim holds for all derivations of length $0 \leq h \leq n$, for some $n \geq 0$.

Induction Step. Consider any derivation of the form

$$S \Rightarrow_G^{n+1} x$$

As $n + 1 \geq 1$, there exists $y \in V^*$ and $p \in P$ such that

$$S \Rightarrow_G^n y \Rightarrow_G x \ [p]$$

Let $y = a_1 a_2 \cdots a_q$, $a_i \in V$ for all $1 \leq i \leq q$, where $q = |y|$. By the induction hypothesis,

$$S' \Rightarrow_{G'}^+ \#y$$

The following cases investigate all possible forms of p.

(i) Let $p: A \to z$, $A \in N_{CF}$, $z \in \{\varepsilon\} \cup N_{CS} \cup T \cup N_{CF}^2$. Then, $y = y_1 A y_3$ and $x = y_1 z y_3$, $y_1, y_3 \in V^*$. Let $l = |y_1| + 1$. In this case, we can construct

$$S' \Rightarrow_{G'}^+ \#y \Rightarrow_{G'} \#x \ [p_0, p_1, \ldots, p_q]$$

such that $p_0: (\# \to \#, 0, 0)$, $p_l: (A \to z, \#, 0)$, and $p_i: (a_i \to a_i, 0, 0)$ for all $1 \leq i \leq q$, $i \neq l$.

(ii) Let $p: AB \to AC$, $A, C \in N_{CF}$, $B \in N_{CS}$. Then, $y = y_1 ABy_3$ and $x = y_1 ACy_3$, $y_1, y_3 \in V^*$. Let $l = |y_1| + 2$. At this point, there exists the following derivation

$$
\begin{aligned}
S' &\Rightarrow^+_{G'} \#y_1 ABy_3 \\
&\Rightarrow_{G'} \langle A, B, C \rangle y_1 ABy_3 \\
&\Rightarrow_{G'} \langle A, B, C, 1 \rangle y_1 A[A, B, C]y_3 \\
&\Rightarrow_{G'} \langle A, B, C, 2 \rangle y_1 A[A, B, C]y_3 \\
&\;\;\vdots \\
&\Rightarrow_{G'} \langle A, B, C, f(A) \rangle y_1 A[A, B, C]y_3 \\
&\Rightarrow_{G'} \langle A, B, C, f(A) + 1 \rangle y_1 A[A, B, C]y_3 \\
&\;\;\vdots \\
&\Rightarrow_{G'} \langle A, B, C, m + 1 \rangle y_1 A[A, B, C]y_3 \\
&\Rightarrow_{G'} \langle A, B, C, m + 2 \rangle y_1 A[A, B, C]y_3 \\
&\Rightarrow_{G'} \#y_1 ACy_3
\end{aligned}
$$

If. The if part establishes that

$$
S' \Rightarrow^h_{G'} Qx' \quad \text{implies} \quad S \Rightarrow^*_{G'} x
$$

where $\gamma^{-1}(x') = x$, $Q \in M$, $x' \in (V \cup W)^*$, $x \in V^*$, $h \geq 1$. This claim is proved by induction on $h \geq 1$.

Basis. Assume that $h = 1$. Since the only rule that can rewrite S' is $(S' \to \#S, 0, 0)$, $S' \Rightarrow_{G'} \#S$. Clearly, $S \Rightarrow^0_G S$ and $\gamma^{-1}(S) = S$.

Induction Hypothesis. Suppose that the claim holds for any derivation of length $1 \leq h \leq n$, for some $n \geq 1$.

Induction Step. Consider any derivation of the form

$$
S' \Rightarrow^{n+1}_{G'} Qx'
$$

where $Qx' \in M(V \cup W)^*$. Since $n + 1 \geq 2$, by Claim 4.2.38, there exists a derivation

$$
S' \Rightarrow^+_{G'} Zy' \Rightarrow_{G'} Qx' \; [p_0, p_1, \ldots, p_q]
$$

where $Zy' \in M(V \cup W)^*$, and $p_i \in P'$ for all $i \in \{0, 1, \ldots, q\}$, $q = |y'|$. By the induction hypothesis, there is also a derivation

$$
S \Rightarrow^*_{G'} y
$$

where $y \in V^*$, $\gamma^{-1}(y') = y$. Let $y' = a_1 a_2 \cdots a_q$. Claim 4.2.40 says that Zy' has one of the following forms.

(i) Let $Z = \#$ and $y' \in V^*$. Then, there are the following two forms of rules rewriting a_i in y'.

(i.a) Let $(a_i \to a_i, 0, 0)$, $a_i \in V$. In this case,

$$\gamma^{-1}(a_i) \Rightarrow_G^0 \gamma^{-1}(a_i)$$

(i.b) Let $(a_i \to x_i, \#, 0)$, $x_i \in \{\varepsilon\} \cup N_{CS} \cup T \cup N_{CF}^2$. Since $a_i = \gamma^{-1}(a_i)$, $x_i = \gamma^{-1}(x_i)$ and $a_i \to x_i \in P$,

$$\gamma^{-1}(a_i) \Rightarrow_G \gamma^{-1}(x_i) \, [a_i \to x_i]$$

We see that for all a_i, there exists a derivation

$$\gamma^{-1}(a_i) \Rightarrow_G^{h_i} \gamma^{-1}(x_i)$$

for some $h_i \in \{0, 1\}$, where $x_i \in V^*$, $x' = x_1 x_2 \cdots x_q$. Therefore, by Claim 4.2.41, we can construct

$$S' \Rightarrow_G^* y \Rightarrow_G^{\bar{h}} x$$

where $0 \leq \bar{h} \leq q$, $x = \gamma^{-1}(x')$.

(ii) Let $Z = \langle A, B, C \rangle$, $y' \in V^*$, for some $A, C \in N_{CF}$, $B \in N_{CS}$. At this point, the following two forms of rules can be used to rewrite a_i in y'.

(ii.a) Let $(a_i \to a_i, 0, 0)$, $a_i \in V$. As in case (i.a),

$$\gamma^{-1}(a_i) \Rightarrow_G^0 \gamma^{-1}(a_i)$$

(ii.b) Let $(a_i \to [A, B, C], \langle A, B, C \rangle, 0)$, $a_i = B$. Since $\gamma^{-1}([A, B, C]) = \gamma^{-1}(B)$, we have

$$\gamma^{-1}(a_i) \Rightarrow_G^0 \gamma^{-1}([A, B, C])$$

Thus, there exists the derivation

$$S \Rightarrow_G^* y \Rightarrow_G^0 x, \ x = \gamma^{-1}(x')$$

(iii) Let $Z = \langle A, B, C, j \rangle$, $y' \in (V \cup \{[A, B, C]\})^*$, and

$$\{f^{-1}(k)[A, B, C] \mid 1 \leq k < j, \ k \neq f(A)\} \cap \mathrm{sub}()(y') = \emptyset$$

$1 \leq j \leq m + 1$, for some $A, C \in N_{CF}$, $B \in N_{CS}$. Then, the only rules rewriting symbols from y' are

$$(a_i \to a_i, 0, 0), \ a_i \in V$$

and

$$([A, B, C] \rightarrow [A, B, C], 0, \langle A, B, C, m + 2 \rangle)$$

Hence, $x' = y'$. Since we have

$$S \Rightarrow_G^* y, \ \gamma^{-1}(y') = y$$

it also holds that $\gamma^{-1}(x') = y$.

(iv) Let $Z = \langle A, B, C, m + 2 \rangle$, $y' \in (V \cup \{[A, B, C]\})^*$, $[A, B, C]^2 \notin \mathrm{sub}(y')$,

$$\{X[A, B, C] \mid X \in V, \ X \neq A\} \cap \mathrm{sub}()(y') = \emptyset$$

for some $A, C \in N_{CF}$, $B \in N_{CS}$. G' rewrites $\langle A, B, C, m + 2 \rangle$ by using

$$(\langle A, B, C, m + 2 \rangle \rightarrow \#, 0, \langle A, B, C, m + 2 \rangle [A, B, C])$$

which forbids $\langle A, B, C, m+2 \rangle [A, B, C]$ as a substring of Zy'. As a result, the left neighbor of every occurrence of $[A, B, C]$ in $\langle A, B, C, m+2 \rangle y'$ is A. Inspect P' to see that a_i can be rewritten either by $(a_i \rightarrow a_i, 0, 0)$, $a_i \in V$, or by $([A, B, C] \rightarrow C, \langle A, B, C, m + 2 \rangle, 0)$. Therefore, we can express

$$y' = y_1 A[A, B, C] y_2 A[A, B, C] y_3 \cdots y_l A[A, B, C] y_{l+1}$$
$$y = y_1 A B y_2 A B y_3 \cdots y_l A B y_{l+1}$$
$$x' = y_1 A C y_2 A C y_3 \cdots y_l A C y_{l+1}$$

where $l \geq 0$, $y_k \in V^*$, $1 \leq k \leq l + 1$. Since we have $p : AB \rightarrow AC \in P$, there is a derivation

$$\begin{aligned}
S &\Rightarrow_G^* y_1 A B y_2 A B y_3 \cdots y_l A B y_{l+1} \\
&\Rightarrow_G y_1 A C y_2 A B y_3 \cdots y_l A B y_{l+1} \ [p] \\
&\Rightarrow_G y_1 A C y_2 A C y_3 \cdots y_l A B y_{l+1} \ [p] \\
&\vdots \\
&\Rightarrow_G y_1 A C y_2 A C y_3 \cdots y_l A C y_{l+1} \ [p]
\end{aligned}$$

Since cases (i) through (iv) cover all possible forms of y', we have completed the induction and established Claim 4.2.42. $\square$

Let us finish the proof of Lemma 4.2.37. Consider any derivation of the form

$$S \Rightarrow_G^* w, \ w \in T^*$$

From Claim 4.2.42, it follows that

$$S' \Rightarrow_{G'}^+ \#w$$

because $\gamma(a) = \{a\}$ for every $a \in T$. Then, as shown in Claim 4.2.39,

$$S' \Rightarrow^+_{G'} \#w \Rightarrow_{G'} w$$

and hence,

$$S \Rightarrow^*_G w \quad \text{implies} \quad S' \Rightarrow^+_{G'} w$$

for all $w \in T^*$. To prove the converse implication, consider a successful derivation of the form

$$S' \Rightarrow^+_{G'} \#u \Rightarrow_{G'} w \Rightarrow^*_{G'} w$$

$u \in V^*$, $w \in T^*$ (see Claim 4.2.39). Observe that by the definition of P', for every

$$S' \Rightarrow^+_{G'} \#u \Rightarrow_{G'} w$$

there also exists a derivation

$$S' \Rightarrow^+_{G'} \#u \Rightarrow^*_{G'} \#w \Rightarrow_{G'} w$$

Then, according to Claim 4.2.42, $S \Rightarrow^*_G w$. Consequently, we get for every $w \in T^*$,

$$S \Rightarrow^*_G w \quad \text{if and only if} \quad S' \Rightarrow^*_{G'} w$$

Therefore, $L(G) = L(G')$. □

Lemma 4.2.43. $\mathbf{SSC\text{-}ET0L}(r, s) \subseteq \mathbf{RE}$ *for any* $r, s \geq 0$.

Proof. By Lemma 4.2.10, $\mathbf{C\text{-}ET0L} \subseteq \mathbf{RE}$. Since $\mathbf{SSC\text{-}ET0L}(r, s) \subseteq \mathbf{C\text{-}ET0L}$ for all $r, s \geq 0$ (see Definition 4.2.36), $\mathbf{SSC\text{-}ET0L}(r, s) \subseteq \mathbf{RE}$ for all $r, s \geq 0$ as well. □

Inclusions established in Lemmas 4.2.37 and 4.2.43 imply the following theorem.

Theorem 4.2.44.

$$\mathbf{SSC\text{-}E0L}(1, 2) = \mathbf{SSC\text{-}ET0L}(1, 2) = \mathbf{SSC\text{-}E0L} = \mathbf{SSC\text{-}ET0L} = \mathbf{RE}$$

Proof. From Lemmas 4.2.37 and 4.2.43, we have that $\mathbf{RE} \subseteq \mathbf{SSC\text{-}E0L}(1, 2)$ and $\mathbf{SSC\text{-}ET0L}(r, s) \subseteq \mathbf{RE}$ for any $r, s \geq 0$. By the definitions it holds that $\mathbf{SSC\text{-}E0L}(1, 2) \subseteq \mathbf{SSC\text{-}ET0L}(1, 2) \subseteq \mathbf{SSC\text{-}ET0L}$ and $\mathbf{SSC\text{-}E0L}(1, 2) \subseteq \mathbf{SSC\text{-}E0L} \subseteq \mathbf{SSC\text{-}ET0L}$. Hence, $\mathbf{SSC\text{-}E0L}(1, 2) = \mathbf{SSC\text{-}ET0L}(1, 2) = \mathbf{SSC\text{-}E0L} = \mathbf{SSC\text{-}ET0L} = \mathbf{RE}$. □

Next, let us investigate the generative power of propagating SSC-ET0L grammars.

Lemma 4.2.45. $CS \subseteq SSC\text{-}EP0L(1,2)$

Proof. We can base this proof on the same technique as in Lemma 4.2.37. However, we have to make sure that the construction produces no erasing rules. This requires some modifications of the original algorithm; in particular, we have to eliminate the rule $(\# \to \varepsilon, 0, 0)$.

Let L be a context-sensitive language generated by a context-sensitive grammar

$$G = (V, T, P, S)$$

of the normal form of Theorem 3.1.6, where

$$V = N_{CF} \cup N_{CS} \cup T$$

Let m be the cardinality of V. Define a bijection f from V to $\{1, \ldots, m\}$. Let f^{-1} be the inverse of f. Set

$$
\begin{aligned}
M = \;& \{\langle \# \mid X \rangle \mid X \in V\} \cup \\
& \{\langle A, B, C \mid X \rangle \mid AB \to AC \in P, \; X \in V\} \cup \\
& \{\langle A, B, C, i \mid X \rangle \mid AB \to AC \in P, \; 1 \leq i \leq m+2, \; X \in V\} \\
W = \;& \{[A, B, C, X] \mid AB \to AC \in P, \; X \in V\}, \text{ and} \\
V' = \;& V \cup M \cup W
\end{aligned}
$$

where V, M, and W are pairwise disjoint. Then, construct the SSC-EP0L grammar of degree $(1, 2)$,

$$G' = (V', T, P', \langle \# \mid S \rangle)$$

with the set of rules P' constructed by performing (1) through (4), given next.

(1) For all $A \to x \in P$, $A \in N_{CF}$, $x \in T \cup N_{CS} \cup N_{CF}^2$,

    (1.a) for all $X \in V$, add $(A \to x, \langle \# \mid X \rangle, 0)$ to P';

    (1.b) if $x \in T \cup N_{CS}$, add $(\langle \# \mid A \rangle \to \langle \# \mid x \rangle, 0, 0)$ to P';

    (1.c) if $x = YZ$, $YZ \in N_{CF}^2$, add $(\langle \# \mid A \rangle \to \langle \# \mid Y \rangle Z, 0, 0)$ to P'.

(2) For all $X \in V$ and for every $AB \to AC \in P$, $A, C \in N_{CF}$, $B \in N_{CS}$, extend P' by adding

    (2.a) $(\langle \# \mid X \rangle \to \langle A, B, C \mid X \rangle, 0, 0)$;

    (2.b) $(B \to [A, B, C, X], \langle A, B, C \mid X \rangle, 0)$;

    (2.c) $(\langle A, B, C \mid X \rangle \to \langle A, B, C, 1 \mid X \rangle, 0, 0)$;

    (2.d) $([A, B, C, X] \to [A, B, C, X], 0, \langle A, B, C, m+2 \rangle X)$;

    (2.e) $(\langle A, B, C, i \mid X \rangle \to \langle A, B, C, i+1 \mid X \rangle, 0, f^{-1}(i)[A, B, C, X])$ for all $1 \leq i \leq m$, $i \neq f(A)$;

    (2.f) $(\langle A, B, C, f(A) \mid X \rangle \to \langle A, B, C, f(A)+1 \mid X \rangle, 0, 0)$

    (2.g) $(\langle A, B, C, m+1 \mid X \rangle \to \langle A, B, C, m+2 \mid X \rangle, 0, [A, B, C, X]^2)$;

(2.h) $(\langle A, B, C, m + 2 \mid X \rangle \rightarrow \langle \# \mid X \rangle, 0, 0)$ for $X = A$,
$(\langle A, B, C, m + 2 \mid X \rangle \rightarrow \langle \# \mid X \rangle, 0, \langle A, B, C, m + 2 \mid X \rangle [A, B, C, X])$
otherwise;
(2.i) $([A, B, C, X] \rightarrow C, \langle A, B, C, m + 2 \mid X \rangle, 0)$.

(3) For all $X \in V$, add $(X \rightarrow X, 0, 0)$ to P'.
(4) For all $X \in V$, add $(\langle \# \mid X \rangle \rightarrow \langle \# \mid X \rangle, 0, 0)$ and $(\langle \# \mid X \rangle \rightarrow X, 0, 0)$ to P'.

Consider the construction above and the construction used in the proof of Lemma 4.2.37. Observe that the present construction does not attach the master as an extra symbol before sentential forms. Instead, the master is incorporated with its right neighbor into one composite symbol. For example, if G generates $AabCadd$, the corresponding sentential form in G' is $\langle \# \mid A \rangle abCadd$, where $\langle \# \mid A \rangle$ is one symbol. At this point, we need no rule erasing #; the master is simply rewritten to the symbol with which it is incorporated (see rules of (4)). In addition, this modification involves some changes to the algorithm: First, G' can rewrite symbols incorporated with the master (see rules of (1.b) and (1.c)). Second, conditions of the rules depending on the master refer to the composite symbols. Finally, G' can make context-sensitive rewriting of the composite master's right neighbor (see rules of (2.h)). For instance, if

$$ABadC \Rightarrow_G ACadC \ [AB \rightarrow AC]$$

in G, G' derives

$$\langle \# \mid A \rangle BadC \Rightarrow_{G'}^+ \langle \# \mid A \rangle CadC$$

Based on the observations above, the reader can surely establish $L(G) = L(G')$ by analogy with the proof of Lemma 4.2.37. Thus, the fully rigorous version of this proof is omitted. □

Lemma 4.2.46. $\mathbf{SSC\text{-}EPT0L}(r, s) \subseteq \mathbf{CS}$, *for all* $r, s \geq 0$.

Proof. By Lemma 4.2.5, $\mathbf{C\text{-}EPT0L}(r, s) \subseteq \mathbf{CS}$, for any $r \geq 0, s \geq 0$. Since every SSC-EPT0L grammar is a special case of a C-EPT0L grammar (see Definition 4.2.36), we obtain $\mathbf{SSC\text{-}EPT0L}(r, s) \subseteq \mathbf{CS}$, for all $r, s \geq 0$. □

Theorem 4.2.47.

$$\mathbf{CS} = \mathbf{SSC\text{-}EP0L}(1, 2) = \mathbf{SSC\text{-}EPT0L}(1, 2) = \mathbf{SSC\text{-}EP0L} = \mathbf{SSC\text{-}EPT0L}$$

Proof. By Lemma 4.2.45, we have $\mathbf{CS} \subseteq \mathbf{SSC\text{-}EP0L}(1, 2)$. Lemma 4.2.46 says that $\mathbf{SSC\text{-}EPT0L}(r, s) \subseteq \mathbf{CS}$ for all $r, s \geq 0$. From the definitions, it follows that $\mathbf{SSC\text{-}EP0L}(1, 2) \subseteq \mathbf{SSC\text{-}EPT0L}(1, 2) \subseteq \mathbf{SSC\text{-}EPT0L}$ and $\mathbf{SSC\text{-}EP0L}(1, 2) \subseteq \mathbf{SSC\text{-}EP0L} \subseteq \mathbf{SSC\text{-}EPT0L}$. Hence, we have the identity $\mathbf{SSC\text{-}EP0L}(1, 2) = \mathbf{SSC\text{-}EPT0L}(1, 2) = \mathbf{SSC\text{-}EP0L} = \mathbf{SSC\text{-}EPT0L} = \mathbf{CS}$, so the theorem holds. □

The following corollary summarizes the established relations between the language families generated by SSC-ET0L grammars.

Corollary 4.2.48.

$$\mathbf{CF}$$
$$\subset$$
$$\mathbf{SSC\text{-}EP0L}(0,0) = \mathbf{SSC\text{-}E0L}(0,0) = \mathbf{EP0L} = \mathbf{E0L}$$
$$\subset$$
$$\mathbf{SSC\text{-}EPT0L}(0,0) = \mathbf{SSC\text{-}ET0L}(0,0) = \mathbf{EPT0L} = \mathbf{ET0L}$$
$$\subset$$
$$\mathbf{SSC\text{-}EP0L}(1,2) = \mathbf{SSC\text{-}EPT0L}(1,2) = \mathbf{SSC\text{-}EP0L} = \mathbf{SSC\text{-}EPT0L} = \mathbf{CS}$$
$$\subset$$
$$\mathbf{SSC\text{-}E0L}(1,2) = \mathbf{SSC\text{-}ET0L}(1,2) = \mathbf{SSC\text{-}E0L} = \mathbf{SSC\text{-}ET0L} = \mathbf{RE} \quad \square$$

Open Problem 4.2.49. Notice that Corollary 4.2.48 does not include some related language families. For instance, it contains no language families generated by SSC-ET0L grammars with degrees $(1, 1)$, $(1, 0)$, and $(0, 1)$. What is their generative power? What is the generative power of SSC-ET0L grammars of degree $(2, 1)$? Are they as powerful as SSC-ET0L grammars of degree $(1, 2)$?

4.2.4   Left Random Context ET0L Grammars

As their name indicates, *left random context ET0L grammars* (*LRC-ET0L grammars* for short) represent another variant of context-conditional ET0L grammars. In this variant, a set of *permitting symbols* and a set of *forbidding symbols* are attached to each of their rules, just like in random context grammars (see Sect. 3.1.3). A rule like this can rewrite a symbol if each of its permitting symbols occurs to the left of the rewritten symbol in the current sentential form while each of its forbidding symbols does not occur there. LRC-ET0L grammars represent the principal subject of this section.

In the present section, we demonstrate that LRC-ET0L grammars are computationally complete—that is, they characterize the family of recursively enumerable languages (see Theorem 4.2.60). In fact, we prove that the family of recursively enumerable languages is characterized even by LRC-ET0L grammars with a limited number of nonterminals (see Theorem 4.2.62). We also demonstrate how to characterize the family of context-sensitive languages by these grammars without erasing rules (see Theorem 4.2.59).

In addition, we study a variety of special cases of LRC-ET0L grammars. First, we introduce *left random context E0L grammars* (*LRC-E0L grammars* for short), which represent LRC-ET0L grammars with a single set of rules. We prove that the above characterizations hold in terms of LRC-E0L grammars as well. Second, we introduce *left permitting E0L grammars* (*LP-E0L grammars* for short), which represent LRC-E0L grammars where each rule has only a set of permitting symbols. Analogously, we define *left forbidding E0L grammars* (*LF-E0L grammars* for short) as LRC-E0L grammars where each rule has only a set of forbidding symbols. We demonstrate that LP-E0L grammars are more powerful than ordinary E0L grammars and that LF-E0L grammars are at least as powerful as ordinary ET0L grammars.

Definitions and Examples

In this section, we define LRC-ET0L grammars and their variants. In addition, we illustrate them by examples.

Definition 4.2.50. A *left random context ET0L grammar* (a *LRC-ET0L grammar* for short) is an $(n + 3)$-tuple

$$G = (V, T, P_1, P_2, \ldots, P_n, w)$$

where V, T, and w are defined as in an ET0L grammar, $N = V - T$ is the alphabet of *nonterminals*, and $P_i \subseteq V \times V^* \times 2^N \times 2^N$ is a finite relation, for all i, $1 \le i \le n$, for some $n \ge 1$. By analogy with phrase-structure grammars, elements of P_i are called *rules* and instead of $(X, y, U, W) \in P_i$, we write $(X \to y, U, W)$ throughout this section. The *direct derivation relation* over V^*, symbolically denoted by $\Rightarrow_G$, is defined as follows:

$$u \Rightarrow_G v$$

if and only if

- $u = X_1 X_2 \cdots X_k$,
- $v = y_1 y_2 \cdots y_k$,
- $(X_i \to y_i, U_i, W_i) \in P_h$,

- $U_i \subseteq \mathrm{alph}(X_1 X_2 \cdots X_{i-1})$, and
- $\mathrm{alph}(X_1 X_2 \cdots X_{i-1}) \cap W_i = \emptyset$,

for all i, $1 \le i \le k$, for some $k \ge 1$ and $h \le n$. For $(X \to y, U, W) \in P_i$, U and W are called the *left permitting context* and the *left forbidding context*, respectively. Let $\Rightarrow_G^m$, $\Rightarrow_G^*$, and $\Rightarrow_G^+$ denote the mth power of $\Rightarrow_G$, for $m \ge 0$, the reflexive-transitive closure of $\Rightarrow_G$, and the transitive closure of $\Rightarrow_G$, respectively. The *language* of G is denoted by $L(G)$ and defined as

$$L(G) = \{x \in T^* \mid w \Rightarrow_G^* x\} \qquad \square$$

Definition 4.2.51. Let $G = (V, T, P_1, P_2, \ldots, P_n, w)$ be an LRC-ET0L grammar, for some $n \ge 1$. If every $(X \to y, U, W) \in P_i$ satisfies that $W = \emptyset$, for all i, $1 \le i \le n$, then G is a *left permitting ET0L grammar* (an *LP-ET0L grammar* for short). If every $(X \to y, U, W) \in P_i$ satisfies that $U = \emptyset$, for all i, $1 \le i \le n$, then G is a *left forbidding ET0L grammar* (an *LF-ET0L grammar* for short). $\qquad \square$

By analogy with ET0L grammars (see their definition in Sect. 2.3.4), we define *LRC-EPT0L, LP-EPT0L, LF-EPT0L, LRC-E0L, LP-E0L, LF-E0L, LRC-EP0L, LP-EP0L,* and *LF-EP0L grammars.*

The language families that are generated by LRC-ET0L, LP-ET0L, LF-ET0L, LRC-EPT0L, LP-EPT0L and LF-EPT0L grammars are denoted by **LRC - ET0L, LP - ET0L, LF - ET0L, LRC - EPT0L, LP - EPT0L,** and **LF - EPT0L,** respectively. The language families generated by LRC-E0L, LP-E0L, LF-E0L, LRC-EP0L, LP-EP0L, and LF-EP0L grammars are denoted by **LRC - E0L, LP - E0L, LF - E0L, LRC - EP0L, LP - EP0L,** and **LF - EP0L,** respectively.

Next, we illustrate the above-introduced notions by two examples.

Example 4.2.52. Consider $K = \{a^m b^n a^m \mid 1 \leq m \leq n\}$. This language is generated by the LF-EP0L grammar

$$G = \big(\{A, B, B', \bar{a}, a, b\}, \{a, b\}, P, ABA\big)$$

with P containing the following nine rules

$(A \rightarrow aA, \emptyset, \{\bar{a}\})$	$(a \rightarrow a, \emptyset, \emptyset)$	$(B' \rightarrow bB', \emptyset, \{A\})$
$(A \rightarrow \bar{a}, \emptyset, \{\bar{a}\})$	$(B \rightarrow bB, \emptyset, \emptyset)$	$(B' \rightarrow b, \emptyset, \{A\})$
$(\bar{a} \rightarrow a, \emptyset, \{A\})$	$(B \rightarrow B', \emptyset, \emptyset)$	$(b \rightarrow b, \emptyset, \emptyset)$

To rewrite A to a string not containing A, $(A \rightarrow \bar{a}, \emptyset, \{\bar{a}\})$ has to be used. Since the only rule which can rewrite $\bar{a}$ is $(\bar{a} \rightarrow a, \emptyset, \{A\})$, and the rules that can rewrite A have $\bar{a}$ in their forbidding contexts, it is guaranteed that both As are rewritten to $\bar{a}$ simultaneously; otherwise, the derivation is blocked. The rules $(B' \rightarrow bB', \emptyset, \{A\})$ and $(B' \rightarrow b, \emptyset, \{A\})$ are applicable only if there is no A to the left of B'. Therefore, after these rules are applied, no more as can be generated. Consequently, we see that for every string from $L(G)$, there exists a derivation of the form

$$\begin{aligned} ABA &\Rightarrow_G^* a^{m-1}Ab^{m-1}Ba^{m-1}A \\ &\Rightarrow_G a^{m-1}\bar{a}b^{m-1}B'a^{m-1}\bar{a} \\ &\Rightarrow_G^+ a^m b^n a^m \end{aligned}$$

with $1 \leq m \leq n$. Hence, $L(G) = K$.   □

Recall that $K \notin$ **E0L** (see page 268 in [RS97a]); however, $K \in$ **LF - EP0L**. As a result, LF-EP0L grammars are more powerful than ordinary E0L grammars.

The next example shows how to generate K by an LP-EP0L grammar, which implies that LP-E0L grammars have greater expressive power than E0L grammars.

Example 4.2.53. Consider the LRC-EP0L grammar

$$H = \big(\{S, A, A', B, B', \bar{a}, a, b\}, \{a, b\}, P, S\big)$$

with P containing the following fourteen rules

$(S \rightarrow ABA', \emptyset, \emptyset)$ $(A' \rightarrow aA', \{A\}, \emptyset)$ $(B \rightarrow bB, \emptyset, \emptyset)$

$(S \rightarrow \bar{a}B'\bar{a}, \emptyset, \emptyset)$ $(A' \rightarrow \bar{a}, \{\bar{a}\}, \emptyset)$ $(B \rightarrow bB', \emptyset, \emptyset)$

$(A \rightarrow aA, \emptyset, \emptyset)$ $(\bar{a} \rightarrow \bar{a}, \emptyset, \emptyset)$ $(B' \rightarrow bB', \{\bar{a}\}, \emptyset)$

$(A \rightarrow a\bar{a}, \emptyset, \emptyset)$ $(\bar{a} \rightarrow a, \emptyset, \emptyset)$ $(B' \rightarrow b, \emptyset, \{\bar{a}\}, \emptyset)$

$(a \rightarrow a, \emptyset, \emptyset)$ $(b \rightarrow b, \emptyset, \emptyset)$

If the first applied rule is $(S \rightarrow \bar{a}B'\bar{a}, \emptyset, \emptyset)$, then the generated string of terminals clearly belongs to K from Example 4.2.52. By using this rule, we can obtain a string with only two as, which is impossible if $(S \rightarrow ABA', \emptyset, \emptyset)$ is used instead. Therefore, we assume that $(S \rightarrow ABA', \emptyset, \emptyset)$ is applied as the first rule. Observe that $(A' \rightarrow aA', \{A\}, \emptyset)$ can be used only when there is A present to the left of A' in the current sentential form. Also, $(A' \rightarrow a, \{\bar{a}\}, \emptyset)$ can be applied only after $(A \rightarrow a\bar{a}, \emptyset, \emptyset)$ is used. Finally, note that $(B' \rightarrow bB', \{\bar{a}\}, \emptyset)$ and $(B' \rightarrow b, \emptyset, \{\bar{a}\}, \emptyset)$ can be applied only if there is $\bar{a}$ to the left of B'. Therefore, after these rules are used, no more as can be generated. Consequently, we see that for every string from $L(G)$ with more than two as, there exists a derivation of the form

$$\begin{aligned}
S &\Rightarrow_H ABA' \\
&\Rightarrow_H^* a^{m-2}Ab^{m-2}Ba^{m-2}A' \\
&\Rightarrow_H a^{m-1}\bar{a}b^{m-1}B'a^{m-1}A' \\
&\Rightarrow_H^+ a^{m-1}\bar{a}b^{n-1}B'a^{m-1}\bar{a} \\
&\Rightarrow_H a^m b^n a^m
\end{aligned}$$

with $2 \le m \le n$. Hence, $L(H) = K$. $\square$

Generative Power and Reduction

In this section, we establish the generative power of LRC-ET0L grammars and their special variants. More specifically, we prove that **LRC-EPT0L** = **LRC-EP0L** = **CS** (Theorem 4.2.59), **LRC-ET0L** = **LRC-E0L** = **RE** (Theorem 4.2.60), **ET0L** ⊆ **LF-EP0L** (Theorem 4.2.69), and **E0L** ⊂ **LP-EP0L** (Theorem 4.2.70).

First, we consider LRC-EPT0L and LRC-EP0L grammars.

Lemma 4.2.54. CS ⊆ LRC-EP0L

Proof. Let $G = (V, T, P, S)$ be a context-sensitive grammar and let $N = V - T$. Without any loss of generality, making use of Theorem 3.1.5, we assume that G is in the Penttonen normal form. Next, we construct a LRC-EP0L grammar H such that $L(H) = L(G)$. Set

$$\bar{N} = \{\bar{A} \mid A \in N\}$$
$$\hat{N} = \{\hat{A} \mid A \in N\}$$
$$N' = N \cup \bar{N} \cup \hat{N}$$

Without any loss of generality, we assume that $\bar{N}, \hat{N}, N$, and T are pairwise disjoint. Construct

$$H = (V', T, P', S)$$

as follows. Initially, set $V' = N' \cup T$ and $P' = \emptyset$. Perform (1) through (5), given next.

(1) For each $A \to a \in P$, where $A \in N$ and $a \in T$, add $(A \to a, \emptyset, N')$ to P'.
(2) For each $A \to BC \in P$, where $A, B, C \in N$, add $(A \to BC, \emptyset, \bar{N} \cup \hat{N})$ to P'.
(3) For each $AB \to AC \in P$, where $A, B, C \in N$,

   (3.1) add $(B \to C, \{\hat{A}\}, N \cup (\hat{N} - \{\hat{A}\}))$ to P';
   (3.2) for each $D \in N$, add $(D \to D, \{\hat{A}, B\}, \hat{N} - \{\hat{A}\})$ to P'.

(4) For each $D \in N$, add $(D \to \bar{D}, \emptyset, \bar{N} \cup \hat{N})$, $(D \to \hat{D}, \emptyset, \bar{N} \cup \hat{N})$, $(\bar{D} \to D, \emptyset, N \cup \hat{N})$, and $(\hat{D} \to D, \emptyset, N \cup \hat{N})$ to P'.
(5) For each $a \in T$ and each $D \in N$, add $(a \to a, \emptyset, N')$ and $(D \to D, \emptyset, \bar{N} \cup \hat{N})$ to P'.

Before proving that $L(H) = L(G)$, let us give an insight into the construction. The simulation of context-free rules of the form $A \to BC$, where $A, B, C \in N$, is done by rules introduced in (2). Rules from (5) are used to rewrite all the remaining symbols.

H simulates context-sensitive rules—that is, rules of the form $AB \to AC$, where $A, B, C \in N$—as follows. First, it rewrites all nonterminals to the left of A to their barred versions by rules from (4), A to $\hat{A}$ by $(A \to \hat{A}, \emptyset, \bar{N} \cup \hat{N})$ from (4), and all the remaining symbols by passive rules from (5). Then, it rewrites B to C by $(B \to C, \{\hat{A}\}, N \cup (\hat{N} - \{\hat{A}\}))$ from (3.1), barred nonterminals to non-barred nonterminals by rules from (4), $\hat{A}$ back to A by $(\hat{A} \to A, \emptyset, N \cup \hat{N})$ from (4), all other nonterminals by passive rules from (3.2), and all terminals by passive rules from (5). For example, for

$$abXYABZ \Rightarrow_G abXYACZ$$

there is

$$abXYABZ \Rightarrow_H ab\bar{X}\bar{Y}\hat{A}BZ \Rightarrow_H abXYACZ$$

Observe that if H makes an improper selection of the symbols rewritten to their barred and hatted versions, like in $AXB \Rightarrow_H \hat{A}\bar{X}B$, then the derivation is blocked because every rule of the form $(\bar{D} \to D, \emptyset, N \cup \hat{N})$, where $D \in D$, requires that there are no hatted nonterminals to the left of $\bar{D}$.

To prevent $AAB \Rightarrow_H AaB \Rightarrow_H \hat{A}aB \Rightarrow_H AaC$, rules simulating $A \to a$, where $A \in N$ and $a \in T$, introduced in (1), can be used only if there are no nonterminals to the left of A. Therefore, a terminal can never appear between two nonterminals, and so every sentential form generated by H is of the form $x_1 x_2$, where $x_1 \in T^*$ and $x_2 \in N'^*$.

To establish $L(H) = L(G)$, we prove three claims. Claim 4.2.55 demonstrates that every $y \in L(G)$ can be generated in two stages; first, only nonterminals are generated, and then, all nonterminals are rewritten to terminals. Claim 4.2.56 shows how such derivations of every $y \in L(G)$ in G are simulated by H. Finally, Claim 4.2.57 shows how derivations of H are simulated by G.

Claim 4.2.55. Let $y \in L(G)$. Then, in G, there exists a derivation $S \Rightarrow_G^* x \Rightarrow_G^* y$, where $x \in N^+$, and during $x \Rightarrow_G^* y$, only rules of the form $A \rightarrow a$, where $A \in N$ and $a \in T$, are applied.

Proof. Let $y \in L(G)$. Since there are no rules in P with symbols from T on their left-hand sides, we can always rearrange all the applications of the rules occurring in $S \Rightarrow_G^* y$ so the claim holds. □

Claim 4.2.56. If $S \Rightarrow_G^h x$, where $x \in N^+$, for some $h \geq 0$, then $S \Rightarrow_H^* x$.

Proof. This claim is established by induction on $h \geq 0$.

Basis. For $h = 0$, this claim obviously holds.

Induction Hypothesis. Suppose that there exists $n \geq 0$ such that the claim holds for all derivations of length h, where $0 \leq h \leq n$.

Induction Step. Consider any derivation of the form

$$S \Rightarrow_G^{n+1} w$$

where $w \in N^+$. Since $n + 1 \geq 1$, this derivation can be expressed as

$$S \Rightarrow_G^n x \Rightarrow_G w$$

for some $x \in N^+$. By the induction hypothesis, $S \Rightarrow_H^* x$.

Next, we consider all possible forms of $x \Rightarrow_G w$, covered by the following two cases—(i) and (ii).

(i) Let $A \rightarrow BC \in P$ and $x = x_1 A x_2$, where $A, B, C \in N$ and $x_1, x_2 \in N^*$. Then, $x_1 A x_2 \Rightarrow_G x_1 BC x_2$. By (2), $(A \rightarrow BC, \emptyset, \bar{N} \cup \hat{N}) \in P'$, and by (5), $(D \rightarrow D, \emptyset, \bar{N} \cup \hat{N}) \in P'$, for each $D \in N$. Since $\mathrm{alph}(x_1 A x_2) \cap (\bar{N} \cup \hat{N}) = \emptyset$,

$$x_1 A x_2 \Rightarrow_H x_1 BC x_2$$

which completes the induction step for (i).

(ii) Let $AB \rightarrow AC \in P$ and $x = x_1 AB x_2$, where $A, B, C \in N$ and $x_1, x_2 \in N^*$. Then, $x_1 AB x_2 \Rightarrow_G x_1 BC x_2$. Let $x_1 = X_1 X_2 \cdots X_k$, where $X_i \in N$, for all i, $1 \leq i \leq k$, for some $k \geq 1$. By (4), $(X_i \rightarrow \bar{X}_i, \emptyset, \bar{N} \cup \hat{N}) \in P'$, for all i, $1 \leq i \leq k$, and $(A \rightarrow \hat{A}, \emptyset, \bar{N} \cup \hat{N}) \in P'$. By (5), $(D \rightarrow D, \emptyset, \bar{N} \cup \hat{N}) \in P'$, for all $D \in \mathrm{alph}(B x_2)$. Since $\mathrm{alph}(x_1 AB x_2) \cap (\bar{N} \cup \hat{N}) = \emptyset$,

$$X_1 X_2 \cdots X_k AB x_2 \Rightarrow_H \bar{X}_1 \bar{X}_2 \cdots \bar{X}_k \hat{A} B x_2$$

By (3.1), $(B \rightarrow C, \{\hat{A}\}, N \cup (\hat{N} - \{\hat{A}\})) \in P'$. By (4), $(\bar{X}_i \rightarrow X_i, \emptyset, N \cup \hat{N}) \in P'$, for all i, $1 \leq i \leq k$, and $(\hat{A} \rightarrow A, \emptyset, N \cup \hat{N}) \in P'$. By (3.2), $(D \rightarrow D, \{\hat{A}, B\}, \hat{N} - \{\hat{A}\}) \in P'$, for all $D \in \mathrm{alph}(x_2)$. Since $\mathrm{alph}(\bar{X}_1\bar{X}_2 \cdots \bar{X}_k) \cap (N \cup \hat{N}) = \emptyset$,

$$\bar{X}_1\bar{X}_2 \cdots \bar{X}_k\hat{A}Bx_2 \Rightarrow_H X_1X_2 \cdots X_kACx_2$$

which completes the induction step for (ii).

Observe that cases (i) and (ii) cover all possible forms of $x \Rightarrow_G w$. Thus, the claim holds. $\square$

Define the homomorphism τ from V'^* to V^* as $\tau(\bar{A}) = \tau(\hat{A}) = \tau(A) = A$, for all $A \in N$, and $\tau(a) = a$, for all $a \in T$.

Claim 4.2.57. If $S \Rightarrow_H^h x$, where $x \in V'^+$, for some $h \geq 0$, then $S \Rightarrow_G^* \tau(x)$, and x is of the form x_1x_2, where $x_1 \in T^*$ and $x_2 \in N'^*$.

Proof. This claim is established by induction on $h \geq 0$.

Basis. For $h = 0$, this claim obviously holds.

Induction Hypothesis. Suppose that there exists $n \geq 0$ such that the claim holds for all derivations of length h, where $0 \leq h \leq n$.

Induction Step. Consider any derivation of the form

$$S \Rightarrow_H^{n+1} w$$

Since $n + 1 \geq 1$, this derivation can be expressed as

$$S \Rightarrow_H^n x \Rightarrow_H w$$

for some $x \in V'^+$. By the induction hypothesis, $S \Rightarrow_G^* \tau(x)$, and x is of the form x_1x_2, where $x_1 \in T^*$ and $x_2 \in N'^*$.

Next, we make the following four observations regarding the possible forms of $x \Rightarrow_H w$.

(i) A rule from (1) can be applied only to the leftmost occurrence of a nonterminal in x_2. Therefore, w is always of the required form.

(ii) Rules from (1) and (2) can be applied only if $\mathrm{alph}(x) \cap (\bar{N} \cup \hat{N}) = \emptyset$. Furthermore, every rule from (1) and (2) is constructed from some $A \rightarrow a \in P$ and $A \rightarrow BC \in P$, respectively, where $A, B, C \in N$ and $a \in T$. If two or more rules are applied at once, G can apply them sequentially.

(iii) When a rule from (3.1)—that is, $(B \rightarrow C, \{\hat{A}\}, N \cup (\hat{N} - \{\hat{A}\}))$—is applied, $\hat{A}$ has to be right before the occurrence of B that is rewritten to C. Otherwise, the symbols between $\hat{A}$ and that occurrence of B cannot be rewritten by any rule and, therefore, the derivation is blocked. Furthermore, H can apply only a single such rule. Since every rule in (3.1) is constructed from some $AB \rightarrow AC \in P$, where $A, B, C \in N$, G applies $AB \rightarrow AC$ to simulate this rewrite.

(iv) If rules introduced in (3.2), (4), or (5) are applied, the induction step follows directly from the induction hypothesis.

Based on these observations, we see that the claim holds. □

Next, we establish $L(H) = L(G)$. Let $y \in L(G)$. Then, by Claim 4.2.55, in G, there exists a derivation $S \Rightarrow_G^* x \Rightarrow_G^* y$ such that $x \in N^+$ and during $x \Rightarrow_G^* y$, G uses only rules of the form $A \rightarrow a$, where $A \in N$ and $a \in T$. By Claim 4.2.56, $S \Rightarrow_H^* x$. Let $x = X_1 X_2 \cdots X_k$ and $y = a_1 a_2 \cdots a_k$, where $X_i \in N$, $a_i \in T$, $X_i \rightarrow a_i \in P$, for all i, $1 \leq i \leq k$, for some $k \geq 1$. By (1), $(X_i \rightarrow a_i, \emptyset, N') \in P'$, for all i. By (5), $(a_i \rightarrow a_i, \emptyset, N') \in P'$ and $(X_i \rightarrow X_i, \emptyset, \bar{N} \cup \hat{N}) \in P'$, for all i. Therefore,

$$
\begin{aligned}
X_1 X_2 \cdots X_k &\Rightarrow_H a_1 X_2 \cdots X_k \\
&\Rightarrow_H a_1 a_2 \cdots X_k \\
&\;\;\vdots \\
&\Rightarrow_H a_1 a_2 \cdots a_k
\end{aligned}
$$

Consequently, $y \in L(G)$ implies that $y \in L(H)$, so $L(G) \subseteq L(H)$.

Consider Claim 4.2.57 with $x \in T^+$. Then, $x \in L(H)$ implies that $\tau(x) = x \in L(G)$, so $L(H) \subseteq L(G)$. As $L(G) \subseteq L(H)$ and $L(H) \subseteq L(G)$, $L(H) = L(G)$, so the lemma holds. □

Lemma 4.2.58. LRC-EPT0L $\subseteq$ CS

Proof. Let $G = (V, T, P_1, P_2, \ldots, P_n, w)$ be an LRC-EPT0L grammar, for some $n \geq 1$. From G, we can construct a phrase-structure grammar, $H = (N', T, P', S)$, such that $L(G) = L(H)$ and if $S \Rightarrow_H^* x \Rightarrow_H^* z$, where $x \in (N' \cup T)^+$ and $z \in T^+$, then $|x| \leq 4|z|$. Consequently, by the workspace theorem (see Theorem 2.3.19), $L(H) \in$ **CS**. Since $L(G) = L(H)$, $L(G) \in$ **CS**, so the lemma holds. □

Theorem 4.2.59. LRC-EPT0L = LRC-EP0L = CS

Proof. **LRC-EP0L $\subseteq$ LRC-EPT0L** follows from the definition of an LRC-EP0L grammar. By Lemma 4.2.54, we have **CS $\subseteq$ LRC-EP0L**, which implies that **CS $\subseteq$ LRC-EP0L $\subseteq$ LRC-EPT0L**. Since **LRC-EPT0L $\subseteq$ CS** by Lemma 4.2.58, **LRC-EP0L $\subseteq$ LRC-EPT0L $\subseteq$ CS**. Hence, **LRC-EPT0L = LRC-EP0L = CS**, so the theorem holds. □

Hence, LRC-EP0L grammars characterize **CS**. Next, we focus on LRC-ET0L and LRC-E0L grammars.

Theorem 4.2.60. LRC-ET0L = LRC-E0L = RE

Proof. The inclusion **LRC-E0L $\subseteq$ LRC-ET0L** follows from the definition of a LRC-E0L grammar. The inclusion **LRC-ET0L $\subseteq$ RE** follows from Turing-Church thesis. The inclusion **RE $\subseteq$ LRC-E0L** can be proved by analogy with the proof of Lemma 4.2.54. Observe that by Theorem 3.1.4, G can additionally contain rules of the form $A \rightarrow \varepsilon$, where $A \in N$. We can simulate these context-free rules in the same way we simulate $A \rightarrow BC$, where $A, B, C \in N$—that is, for each $A \rightarrow \varepsilon \in P$, we introduce $(A \rightarrow \varepsilon, \emptyset, \bar{N} \cup \hat{N})$ to P'. As **LRC-E0L** $\subseteq$ **LRC-ET0L** $\subseteq$ **RE** and **RE** $\subseteq$ **LRC-E0L** $\subseteq$ **LRC-ET0L**, **LRC-ET0L** = **LRC-E0L** = **RE**, so the theorem holds.   □

The following corollary compares the generative power of LRC-E0L and LRC-ET0L grammars to the power of E0L and ET0L grammars.

Corollary 4.2.61.

$$\mathbf{CF} \subset \mathbf{E0L} = \mathbf{EP0L} \subset \mathbf{ET0L} = \mathbf{EPT0L}$$
$$\subset$$
$$\mathbf{LRC\text{-}EPT0L} = \mathbf{LRC\text{-}EP0L} = \mathbf{CS}$$
$$\subset$$
$$\mathbf{LRC\text{-}ET0L} = \mathbf{LRC\text{-}E0L} = \mathbf{RE}$$

Proof. This corollary follows from Theorem 2.3.41 in Sect. 2.3.4 and from Theorems 4.2.59 and 4.2.60 above.   □

Next, we show that the family of recursively enumerable languages is characterized even by LRC-E0L grammars with a limited number of nonterminals. Indeed, we prove that every recursively enumerable language can be generated by a LRC-E0L grammar with seven nonterminals.

Theorem 4.2.62. *Let K be a recursively enumerable language. Then, there is an LRC-E0L grammar, $H = (V, T, P, w)$, such that $L(H) = K$ and $\mathrm{card}(V - T) = 7$.*

Proof. Let K be a recursively enumerable language. By Theorem 3.1.9, there is a phrase-structure grammar in the Geffert normal form

$$G = \big(\{S, A, B, C\}, T, P \cup \{ABC \rightarrow \varepsilon\}, S\big)$$

satisfying $L(G) = K$. Next, we construct an LRC-E0L grammar H such that $L(H) = L(G)$. Set $N = \{S, A, B, C\}$, $V = N \cup T$, and $N' = N \cup \{\bar{A}, \bar{B}, \#\}$ (without any loss of generality, we assume that $V \cap \{\bar{A}, \bar{B}, \#\} = \emptyset$). Construct

$$H = \big(V', T, P', S\#\big)$$

as follows. Initially, set $V' = N' \cup T$ and $P' = \emptyset$. Perform (1) through (8), given next.

(1) For each $a \in T$,
    add $(a \rightarrow a, \emptyset, \emptyset)$ to P'.
(2) For each $X \in N$,
    add $(X \rightarrow X, \emptyset, \{\bar{A}, \bar{B}, \#\})$ and $(X \rightarrow X, \{\bar{A}, \bar{B}, C\}, \{S, \#\})$ to P'.
(3) Add $(\# \rightarrow \#, \emptyset, \{\bar{A}, \bar{B}\})$, $(\# \rightarrow \#, \{\bar{A}, \bar{B}, C\}, \{S\}\})$, and $(\# \rightarrow \varepsilon, \emptyset, N' - \{\#\})$ to P'.
(4) For each $S \rightarrow uSa \in P$, where $u \in \{A, AB\}^*$ and $a \in T$,
    add $(S \rightarrow uS\#a, \emptyset, \{\bar{A}, \bar{B}, \#\})$ to P'.

(5) For each $S \to uSv \in P$, where $u \in \{A, AB\}^*$ and $v \in \{BC, C\}^*$,
   add $(S \to uSv, \emptyset, \{\bar{A}, \bar{B}, \#\})$ to P'.
(6) For each $S \to uv \in P$, where $u \in \{A, AB\}^*$ and $v \in \{BC, C\}^*$,
   add $(S \to uv, \emptyset, \{\bar{A}, \bar{B}, \#\})$ to P'.
(7) Add $(A \to \bar{A}, \emptyset, \{S, \bar{A}, \bar{B}, \#\})$ and $(B \to \bar{B}, \emptyset, \{S, \bar{A}, \bar{B}, \#\})$ to P'.
(8) Add $(\bar{A} \to \varepsilon, \emptyset, \{S, \bar{A}, \bar{B}, C, \#\})$, $(\bar{B} \to \varepsilon, \{\bar{A}\}, \{S, \bar{B}, C, \#\})$, and $(C \to \varepsilon, \{\bar{A}, \bar{B}\},$
   $\{S, C, \#\})$ to P'.

Before proving that $L(H) = L(G)$, let us informally explain (1) through (8). H simulates the derivations of G that satisfy the form described in Theorem 3.1.9. Since H works in a parallel way, rules from (1) through (3) are used to rewrite symbols that are not actively rewritten. The context-free rules in P are simulated by rules from (4) through (6). The context-sensitive rule $ABC \to \varepsilon$ is simulated in a two-step way. First, rules introduced in (7) rewrite A and B to $\bar{A}$ and $\bar{B}$, respectively. Then, rules from (8) erase $\bar{A}$, $\bar{B}$, and C; for example,

$$AABCBC\#a\# \Rightarrow_H A\bar{A}\bar{B}CBC\#a\# \Rightarrow_H ABC\#a\#$$

The role of # is twofold. First, it ensures that every sentential form of H is of the form $w_1 w_2$, where $w_1 \in (N' - \{\#\})^*$ and $w_2 \in (T \cup \{\#\})^*$. Since left permitting and left forbidding contexts cannot contain terminals, a mixture of symbols from T and N in H could produce a terminal string out of $L(G)$. For example, observe that $AaBC \Rightarrow_H^* a$, but such a derivation does not exist in G. Second, if any of $\bar{A}$ and $\bar{B}$ are present, $ABC \to \varepsilon$ has to be simulated. Therefore, it prevents derivations of the form $Aa \Rightarrow_H \bar{A}a \Rightarrow_H a$ (notice that the start string of H is $S\#$). Since H works in a parallel way, if rules from (7) are used improperly, the derivation is blocked, so no partial erasures are possible.

Observe that every sentential form of G and H contains at most one occurrence of S. In every derivation step of H, only a single rule from $P \cup \{ABC \to \varepsilon\}$ can be simulated at once. $ABC \to \varepsilon$ can be simulated only if there is no S. #s can be eliminated by an application of rules from (7); however, only if no nonterminals occur to the left of # in the current sentential form. Consequently, all #s are erased at the end of every successful derivation. Based on these observations and on Theorem 3.1.9, we see that every successful derivation in H is of the form

$$S\# \Rightarrow_H^* w_1 w_2 \# a_1 \# a_2 \cdots \# a_n \#$$
$$\Rightarrow_H^* \# a_1 \# a_2 \cdots \# a_n \#$$
$$\Rightarrow_H^* a_1 a_2 \cdots a_n$$

where $w_1 \in \{A, AB\}^*$, $w_2 \in \{BC, C\}^*$, and $a_i \in T$ for all $i = 1, \ldots, n$, for some $n \geq 0$.

To establish $L(H) = L(G)$, we prove two claims. First, Claim 4.2.63 shows how derivations of G are simulated by H. Then, Claim 4.2.64 demonstrates the converse—that is, it shows how derivations of H are simulated by G.

Define the homomorphism φ from V^* to V'^* as $\varphi(X) = X$ for all $X \in N$, and $\varphi(a) = \#a$ for all $a \in T$.

Claim 4.2.63. If $S \Rightarrow_G^h x \Rightarrow_G^* z$, for some $h \geq 0$, where $x \in V^*$ and $z \in T^*$, then $S\# \Rightarrow_H^* \varphi(x)\#$.

Proof. This claim is established by induction on $h \geq 0$.

Basis. For $h = 0$, this claim obviously holds.

Induction Hypothesis. Suppose that there exists $n \geq 0$ such that the claim holds for all derivations of length h, where $0 \leq h \leq n$.

Induction Step. Consider any derivation of the form

$$S \Rightarrow_G^{n+1} w \Rightarrow_G^* z$$

where $w \in V^*$ and $z \in T^*$. Since $n + 1 \geq 1$, this derivation can be expressed as

$$S \Rightarrow_G^n x \Rightarrow_G w \Rightarrow_G^* z$$

for some $x \in V^+$. Without any loss of generality, we assume that $x = x_1 x_2 x_3 x_4$, where $x_1 \in \{A, AB\}^*$, $x_2 \in \{S, \varepsilon\}$, $x_3 \in \{BC, C\}^*$, and $x_4 \in T^*$ (see Theorem 3.1.9 and the form of rules in P). Next, we consider all possible forms of $x \Rightarrow_G w$, covered by the following four cases—(i) through (iv).

(i) *Application of* $S \to uSa \in P$. Let $x = x_1 S x_4$, $w = x_1 uSax_4$, and $S \to uSa \in P$, where $u \in \{A, AB\}^*$ and $a \in T$. Then, by the induction hypothesis,

$$S\# \Rightarrow_H^* \varphi(x_1 S x_4)\#$$

By (4), $r: (S \to uS\#a, \emptyset, \{\bar{A}, \bar{B}, \#\}) \in P'$. Since $\varphi(x_1 S x_4)\# = x_1 S \varphi(x_4)\#$ and $\text{alph}(x_1 S) \cap \{\bar{A}, \bar{B}, \#\} = \emptyset$, by (1), (2), (3), and by r,

$$x_1 S \varphi(x_4)\# \Rightarrow_H x_1 uS\#a\varphi(x_4)\#$$

As $\varphi(x_1 uSax_4)\# = x_1 uS\#a\varphi(x_4)\#$, the induction step is completed for (i).

(ii) *Application of* $S \to uSv \in P$. Let $x = x_1 Sx_3 x_4$, $w = x_1 uSvx_3 x_4$, and $S \to uSv \in P$, where $u \in \{A, AB\}^*$ and $v \in \{BC, C\}^*$. To complete the induction step for (ii), proceed by analogy with (i), but use a rule from (5) instead of a rule from (4).

(iii) *Application of* $S \to uv \in P$. Let $x = x_1 Sx_3 x_4$, $w = x_1 uvx_3 x_4$, and $S \to uv \in P$, where $u \in \{A, AB\}^*$ and $v \in \{BC, C\}^*$. To complete the induction step for (iii), proceed by analogy with (i), but use a rule from (6) instead of a rule from (4).

(iv) *Application of* $ABC \to \varepsilon$. Let $x = x_1' ABCx_3' x_4$, $w = x_1' x_3' x_4$, where $x_1 x_2 x_3 = x_1' ABCx_3'$, so $x \Rightarrow_G w$ by $ABC \to \varepsilon$. Then, by the induction hypothesis,

$$S\# \Rightarrow_H^* \varphi(x_1' ABCx_3' x_4)\#$$

Since $\varphi(x_1'ABCx_3'x_4)\# = x_1'ABCx_3'\varphi(x_4)\#$ and $\mathrm{alph}(x_1'ABCx_3') \cap \{\bar{A}, \bar{B}, \#\} = \emptyset$,

$$x_1'ABCx_3'\varphi(x_4)\# \Rightarrow_H x_1'\bar{A}BCx_3'\varphi(x_4)\#$$

by rules from (1), (2), (3), and (7). Since $\mathrm{alph}(x_1') \cap \{S, \bar{A}, \bar{B}, C, \#\} = \emptyset$, $\{\bar{A}\} \subseteq \mathrm{alph}(x_1'\bar{A})$, $\mathrm{alph}(x_1'\bar{A}) \cap \{S, \bar{B}, C, \#\} = \emptyset$, $\{\bar{A}, \bar{B}\} \subseteq \mathrm{alph}(x_1'\bar{A}\bar{B})$, and $\{S, C, \#\} \cap \mathrm{alph}(x_1'\bar{A}\bar{B}) = \emptyset$,

$$x_1'\bar{A}\bar{B}Cx_3'\varphi(x_4)\# \Rightarrow_H x_1'x_3'\varphi(x_4)\#$$

by rules from (1), (2), (3), and (8). As $\varphi(x_1'x_3'x_4)\# = x_1'x_3'\varphi(x_4)\#$, the induction step is completed for (iv).

Observe that cases (i) through (iv) cover all possible forms of $x \Rightarrow_G w$, so the claim holds. $\square$

Define the homomorphism τ from V'^* to V^* as $\tau(X) = X$ for all $X \in N$, $\tau(a) = a$ for all $a \in T$, and $\tau(\bar{A}) = A$, $\tau(\bar{B}) = B$, $\tau(\#) = \varepsilon$.

Claim 4.2.64. If $S\# \Rightarrow_H^h x \Rightarrow_H^* z$, for some $h \geq 0$, where $x \in V'^*$ and $z \in T^*$, then $S \Rightarrow_G^* \tau(x)$.

Proof. This claim is established by induction on $h \geq 0$.

Basis. For $h = 0$, this claim obviously holds.

Induction Hypothesis. Suppose that there exists $n \geq 0$ such that the claim holds for all derivations of length h, where $0 \leq h \leq n$.

Induction Step. Consider any derivation of the form

$$S\# \Rightarrow_H^{n+1} w \Rightarrow_H^* z$$

where $w \in V'^*$ and $z \in T^*$. Since $n + 1 \geq 1$, this derivation can be expressed as

$$S\# \Rightarrow_H^n x \Rightarrow_H w \Rightarrow_H^* z$$

for some $x \in V'^+$. By the induction hypothesis, $S \Rightarrow_G^* \tau(x)$. Next, we consider all possible forms of $x \Rightarrow_H w$, covered by the following five cases—(i) through (v).

(i) Let $x = x_1 S x_2$ and $w = x_1 uS\#ax_2$, where $x_1, x_2, \in V'^*$, such that $x_1 S x_2 \Rightarrow_H$ $x_1 uS\#ax_2$ by $(S \rightarrow uS\#a, \emptyset, \{\bar{A}, \bar{B}, \#\})$—introduced in (4) from $S \rightarrow uSa \in P$, where $u \in \{A, AB\}^*$, $a \in T$—and by the rules introduced in (1), (2), and (3). Since $\tau(x_1 S x_2) = \tau(x_1)S\tau(x_2)$,

$$\tau(x_1)S\tau(x_2) \Rightarrow_G \tau(x_1)uSa\tau(x_2)$$

As $\tau(x_1)uSa\tau(x_2) = \tau(x_1 uS\#ax_2)$, the induction step is completed for (i).

(ii) Let $x = x_1 S x_2$ and $w = x_1 u S v x_2$, where $x_1, x_2, \in V'^*$, such that $x_1 S x_2 \Rightarrow_H$
$x_1 u S v x_2$ by $(S \to u S v, \emptyset, \{\bar{A}, \bar{B}, \#\})$—introduced in (5) from $S \to u S v \in P$,
where $u \in \{A, AB\}^*$, $v \in \{BC, C\}^*$—and by the rules introduced in (1), (2),
and (3). Proceed by analogy with (i).

(iii) Let $x = x_1 S x_2$ and $w = x_1 u v x_2$, where $x_1, x_2, \in V'^*$, such that $x_1 S x_2 \Rightarrow_H$
$x_1 u v x_2$ by $(S \to u v, \emptyset, \{\bar{A}, \bar{B}, \#\})$—introduced in (6) from $S \to u v \in P$, where
$u \in \{A, AB\}^*$, $v \in \{BC, C\}^*$—and by the rules introduced in (1), (2), and (3).
Proceed by analogy with (i).

(iv) Let $x = x_1 \bar{A} \bar{B} C x_2$ and $w = x_1' x_2$, where $x_1, x_2 \in V'^*$ and $\tau(x_1') = x_1$, such
that $x_1 \bar{A} \bar{B} C x_2 \Rightarrow_H x_1' x_2$ by rules introduced in (1), (2), (3), (7), and (8). Since
$\tau(x_1 \bar{A} \bar{B} C x_2) = \tau(x_1) A B C \tau(x_2)$,

$$\tau(x_1) A B C \tau(x_2) \Rightarrow_G \tau(x_1) \tau(x_2)$$

by $ABC \to \varepsilon$. As $\tau(x_1) \tau(x_2) = \tau(x_1' x_2)$, the induction step is completed
for (iv).

(v) Let $x \Rightarrow_H w$ only by rules from (1), (2), (3), and from (7). As $\tau(x) = \tau(w)$, the
induction step is completed for (v).

Observe that cases (i) through (v) cover all possible forms of $x \Rightarrow_H w$, so the
claim holds. $\square$

Next, we prove that $L(H) = L(G)$. Consider Claim 4.2.63 with $x \in T^*$. Then,
$S \Rightarrow_G^* x$ implies that $S\# \Rightarrow_H^* \varphi(x)\#$. By (3), $(\# \to \varepsilon, \emptyset, N' - \{\#\}) \in P'$, and by (1),
$(a \to a, \emptyset, \emptyset) \in P'$ for all $a \in T$. Since $\mathrm{alph}(\varphi(x)\#) \cap (N' - \{\#\}) = \emptyset$, $\varphi(x)\# \Rightarrow_H$
x. Hence, $L(G) \subseteq L(H)$. Consider Claim 4.2.64 with $x \in T^*$. Then, $S\# \Rightarrow_H^* x$
implies that $S \Rightarrow_G^* x$. Hence, $L(H) \subseteq L(G)$. Since $\mathrm{card}(N') = 7$, the theorem
holds. $\square$

We turn our attention to LRC-E0L grammars containing only forbidding conditions.

Lemma 4.2.65. EPT0L $\subseteq$ LF - EP0L

Proof. Let $G = (V, T, P_1, P_2, \ldots, P_t, w)$ be an EPT0L grammar, for some $t \geq 1$.
Set

$$R = \{\langle X, i \rangle \mid X \in V, 1 \leq i \leq t\}$$

and

$$F_i = \{\langle X, j \rangle \in R \mid j \neq i\} \text{ for } i = 1, 2, \ldots, t$$

Without any loss of generality, we assume that $V \cap R = \emptyset$. Define the LF-EP0L
grammar

$$H = (V', T, P', w)$$

where $V' = V \cup R$, and P' is constructed by performing the following two steps:

(1) for each $X \in V$ and each $i \in \{1, 2, \ldots, t\}$, add $(X \rightarrow \langle X, i \rangle, \emptyset, \emptyset)$ to P';
(2) for each $X \rightarrow y \in P_i$, where $1 \le i \le t$, add $(\langle X, i \rangle \rightarrow y, \emptyset, F_i)$ to P'.

To establish $L(H) = L(G)$, we prove three claims. Claim 4.2.66 points out that the every sentential form in H is formed either by symbols from R or from V, depending on whether the length of the derivation is even or odd. Claim 4.2.67 shows how derivations of G are simulated by H. Finally, Claim 4.2.68 demonstrates the converse—that is, it shows how derivations of H are simulated by G.

Claim 4.2.66. For every derivation $w \Rightarrow_H^n x$, where $n \ge 0$,

 (i) if $n = 2k + 1$, for some $k \ge 0$, then $x \in R^+$;
(ii) if $n = 2k$, for some $k \ge 0$, then $x \in V^+$.

Proof. The claim follows from the construction of P'. Indeed, every rule in P' is either of the form $(X \rightarrow \langle X, i \rangle, \emptyset, \emptyset)$ or $(\langle X, i \rangle \rightarrow y, \emptyset, F_i)$, where $X \in V, 1 \le i \le t$, and $y \in V^+$. Since $w \in V^+$, $w \Rightarrow_H^{2k+1} x$ implies that $x \in R^+$, and $w \Rightarrow_H^{2k} x$ implies that $x \in V^+$. Thus, the claim holds. □

Claim 4.2.67. If $w \Rightarrow_G^h x$, where $x \in V^+$, for some $h \ge 0$, then $w \Rightarrow_H^* x$.

Proof. This claim is established by induction on $h \ge 0$.

Basis. For $h = 0$, this claim obviously holds.

Induction Hypothesis. Suppose that there exists $n \ge 0$ such that the claim holds for all derivations of length h, where $0 \le h \le n$.

Induction Step. Consider any derivation of the form

$$w \Rightarrow_G^{n+1} y$$

where $y \in V^+$. Since $n + 1 \ge 1$, this derivation can be expressed as

$$w \Rightarrow_G^n x \Rightarrow_G y$$

for some $x \in V^+$. Let $x = X_1 X_2 \cdots X_h$ and $y = y_1 y_2 \cdots y_h$, where $h = |x|$. As $x \Rightarrow_G y$, $X_i \rightarrow y_i \in P_m$, for all i, $1 \le i \le h$, for some $m \le t$.

By the induction hypothesis, $w \Rightarrow_H^* x$. By (1), $(X_i \rightarrow \langle X_i, m \rangle, \emptyset, \emptyset) \in P'$, for all i, $1 \le i \le h$. Therefore,

$$X_1 X_2 \cdots X_h \Rightarrow_H \langle X_1, m \rangle \langle X_2, m \rangle \cdots \langle X_h, m \rangle$$

By (2), $(\langle X_i, m \rangle \rightarrow y_i, \emptyset, F_m) \in P'$, for all i, $1 \le i \le h$. Since $\text{alph}(\langle X_1, m \rangle \langle X_2, m \rangle \cdots \langle X_h, m \rangle) \cap F_m = \emptyset$,

$$\langle X_1, m \rangle \langle X_2, m \rangle \cdots \langle X_h, m \rangle \Rightarrow_H y_1 y_2 \cdots y_h$$

which proves the induction step.   □

Define the homomorphism ψ from V'^* to V^* as $\psi(X) = \psi(\langle X, i \rangle) = X$, for all $X \in V$ and all i, $1 \le i \le t$.

Claim 4.2.68. If $w \Rightarrow_H^h x$, where $x \in V'^+$, for some $h \ge 0$, then $w \Rightarrow_G^* \psi(x)$.

Proof. This claim is established by induction on $h \ge 0$.

Basis. For $h = 0$, this claim obviously holds.

Induction Hypothesis. Suppose that there exists $n \ge 0$ such that the claim holds for all derivations of length h, where $0 \le h \le n$.

Induction Step. Consider any derivation of the form

$$w \Rightarrow_H^{n+1} y$$

where $y \in V'^+$. Since $n + 1 \ge 1$, this derivation can be expressed as

$$w \Rightarrow_H^n x \Rightarrow_H y$$

for some $x \in V'^+$. By the induction hypothesis, $w \Rightarrow_G^* \psi(x)$. By Claim 4.2.66, there exist the following two cases—(i) and (ii).

(i) Let $n = 2k + 1$, for some $k \ge 0$. Then, $x \in R^+$, so let $x = \langle X_1, m_1 \rangle \langle X_2, m_2 \rangle \cdots \langle X_h, m_h \rangle$, where $h = |x|$, $X_i \in V$, for all i, $1 \le i \le h$, and $m_j \in \{1, 2, \ldots, t\}$, for all j, $1 \le j \le h$. The only possible derivation in H is

$$\langle X_1, m_1 \rangle \langle X_2, m_2 \rangle \cdots \langle X_h, m_h \rangle \Rightarrow_H y_1 y_2 \cdots y_h$$

by rules introduced in (2), where $y_i \in V^*$, for all i, $1 \le i \le h$. Observe that $m_1 = m_2 = \cdots = m_h$; otherwise, $\langle X_h, m_h \rangle$ cannot be rewritten (see the form of left forbidding contexts of the rules introduced to P' in (2)). By (2), $X_j \to y_j \in P_{m_h}$, for all j, $1 \le j \le h$. Since $\psi(x) = X_1 X_2 \cdots X_h$,

$$X_1 X_2 \cdots X_h \Rightarrow_G y_1 y_2 \cdots y_h$$

which proves the induction step for (i).

(ii) Let $n = 2k$, for some $k \ge 0$. Then, $x \in V^+$, so let $x = X_1 X_2 \cdots X_h$, where $h = |x|$. The only possible derivation in H is

$$X_1 X_2 \cdots X_h \Rightarrow_H \langle X_1, m_1 \rangle \langle X_2, m_2 \rangle \cdots \langle X_h, m_h \rangle$$

by rules introduced in (1), where $m_j \in \{1, 2, \ldots, t\}$, for all j, $1 \le j \le h$. Since $\psi(y) = \psi(x)$, where $y = \langle X_1, m_1 \rangle \langle X_2, m_2 \rangle \cdots \langle X_h, m_h \rangle$, the induction step for (ii) follows directly from the induction hypothesis.

Hence, the claim holds. $\square$

Next, we establish $L(H) = L(G)$. Consider Claim 4.2.67 with $x \in T^+$. Then, $w \Rightarrow_G^* x$ implies that $w \Rightarrow_H^* x$, so $L(G) \subseteq L(H)$. Consider Claim 4.2.68 with $x \in T^+$. Then, $w \Rightarrow_H^* x$ implies that $w \Rightarrow_G^* \psi(x) = x$, so $L(H) \subseteq L(G)$. Hence, $L(H) = L(G)$, so the lemma holds. □

Theorem 4.2.69.

$$\mathbf{E0L} = \mathbf{EP0L} \subset \mathbf{ET0L} = \mathbf{EPT0L} \subseteq \mathbf{LF\text{-}EP0L} \subseteq \mathbf{LF\text{-}E0L}$$

Proof. The inclusions **E0L** = **EP0L**, **ET0L** = **EPT0L**, and **E0L** ⊂ **ET0L** follow from Theorem 2.3.41. From Lemma 4.2.65, we have **EPT0L** ⊆ **LF-EP0L**. The inclusion **LF-EP0L** ⊆ **LF-E0L** follows directly from the definition of an LF-E0L grammar. □

Next, we briefly discuss LRC-E0L grammars containing only permitting conditions.

Theorem 4.2.70. $\mathbf{E0L} = \mathbf{EP0L} \subset \mathbf{LP\text{-}EP0L} \subseteq \mathbf{LP\text{-}E0L}$

Proof. The identity **E0L** = **EP0L** follows from Theorem 2.3.41. The inclusions **EP0L** ⊆ **LP-EP0L** ⊆ **LP-E0L** follow directly from the definition of an LP-E0L grammar. The properness of the inclusion **EP0L** ⊂ **LP-EP0L** follows from Example 4.2.53. □

To conclude this section, we compare LRC-ET0L grammars and their special variants to a variety of conditional ET0L grammars with respect to their generative power. Then, we formulate some open problem areas.

Consider *random context ET0L grammars* (abbreviated *RC-ET0L grammars*), see [RS78, Sol76] and Section 8 in [DP89]. These grammars have been recently discussed in connection to various grammar systems (see [BCVHV05, BCVHV07, BH00, BH08, CVDV08, CVPS95, Das07, FHF01]) and membrane systems (P systems, see [Sos03]). Recall that as a generalization of LRC-ET0L grammars, they check the occurrence of symbols in the entire sequential form. Notice, however, that contrary to our definition of LRC-ET0L grammars, in [RS78, Sol76] and in other works, RC-ET0L grammars are defined so that they have permitting and forbidding conditions attached to whole sets of rules rather than to each single rule. Since we also study LRC-E0L grammars, which contain just a single set of rules, attachment to rules is more appropriate in our case, just like in terms of other types of regulated ET0L grammars discussed in this section.

The language families generated by RC-ET0L grammars and propagating RC-ET0L grammars are denoted by **RC-ET0L** and **RC-EPT0L**, respectively (for the definitions of these families, see [Das07]).

Theorem 4.2.71 (See [BH00]).

$$\mathbf{RC\text{-}EPT0L} \subset \mathbf{CS} \text{ and } \mathbf{RC\text{-}ET0L} \subseteq \mathbf{RE}$$

Let us point out that it is not known whether the inclusion $\mathbf{RC\text{-}ET0L} \subseteq \mathbf{RE}$ is, in fact, proper (see [BCVHV05, Das07, Sos03]).

Corollary 4.2.72.

$$\mathbf{RC\text{-}EPT0L} \subset \mathbf{LRC\text{-}EP0L} \text{ and } \mathbf{RC\text{-}ET0L} \subseteq \mathbf{LRC\text{-}E0L}$$

Proof. This corollary follows from Theorems 4.2.59, 4.2.60, and 4.2.71.   □

Corollary 4.2.72 is of some interest because LRC-E0L grammars (i) have only a single set of rules and (ii) they check only prefixes of sentential forms.

A generalization of LF-ET0L grammars, called forbidding ET0L grammars (abbreviated F-ET0L grammars), is introduced and discussed in Sect. 4.2.2. Recall that as opposed to LF-ET0L grammars, these grammars check the absence of forbidding symbols in the entire sentential form. Furthermore, recall that $\mathbf{F\text{-}ET0L}(1)$ denotes the family of languages generated by F-ET0L grammars whose forbidding strings are of length one.

Corollary 4.2.73. $\mathbf{F\text{-}ET0L}(1) \subseteq \mathbf{LF\text{-}EP0L}$

Proof. This corollary follows from Lemma 4.2.65 and from Theorem 4.2.22, which says that $\mathbf{F\text{-}ET0L}(1) = \mathbf{ET0L}$.   □

This result is also of some interest because LF-EP0L grammars (i) have only a single set of rules, (ii) have no rules of the form $(A \rightarrow \varepsilon, \emptyset, W)$, and (iii) they check only prefixes of sentential forms.

Furthermore, consider conditional ET0L grammars (*C-ET0L* grammars for short) and simple semi-conditional ET0L grammars (*SSC-ET0L grammars* for short) from Sects. 4.2.1 and 4.2.3, respectively. Recall that these grammars differ from RC-ET0L grammars by the form of their permitting and forbidding sets. In C-ET0L grammars, these sets contain strings rather than single symbols. SSC-ET0L grammars are C-ET0L grammars in which every rule can either forbid or permit the occurrence of a single string.

Recall that $\mathbf{C\text{-}ET0L}$ and $\mathbf{C\text{-}EPT0L}$ denote the language families generated by C-ET0L grammars and propagating C-ET0L grammars, respectively. The language families generated by SSC-ET0L grammars and propagating SSC-ET0L grammars are denoted by $\mathbf{SSC\text{-}ET0L}$ and $\mathbf{SSC\text{-}EPT0L}$, respectively.

Corollary 4.2.74.

$$\mathbf{C\text{-}EPT0L} = \mathbf{SSC\text{-}EPT0L} = \mathbf{LRC\text{-}EP0L}$$
$$\subset$$
$$\mathbf{C\text{-}ET0L} = \mathbf{SSC\text{-}ET0L} = \mathbf{LRC\text{-}E0L}$$

Proof. This corollary follows from Theorems 4.2.59 and 4.2.60 and from Theorems 4.2.9, 4.2.11, 4.2.47, and 4.2.44, which say that $\mathbf{C\text{-}EPT0L} = \mathbf{SSC\text{-}EPT0L} = \mathbf{CS}$ and $\mathbf{C\text{-}ET0L} = \mathbf{SSC\text{-}ET0L} = \mathbf{RE}$.   □

We close this section by formulating several open problem areas suggested as topics of future investigation related to the present study.

Open Problem 4.2.75. By Theorem 4.2.69, **ET0L** $\subseteq$ **LF - E0L**. Is this inclusion, in fact, an identity?

Open Problem 4.2.76. ET0L and EPT0L grammars have the same generative power (see Theorem 2.3.41). Are LF-E0L and LF-EP0L grammars equally powerful? Are LP-E0L and LP-EP0L grammars equally powerful?

Open Problem 4.2.77. What is the relation between the language families generated by ET0L grammars and by LP-E0L grammars?

Open Problem 4.2.78. Establish the generative power of LP-ET0L and LF-ET0L grammars.

Open Problem 4.2.79. Theorem 4.2.62 has proved that every recursively enumerable language can be generated by a LRC-E0L grammar with seven nonterminals. Can this result be improved?

Open Problem 4.2.80. Recall that LRC-E0L grammars without erasing rules characterize the family of context-sensitive languages (see Theorem 4.2.59). Can we establish this characterization based upon these grammars with a limited number of nonterminals?

Chapter 5
Jumping Grammars and Discontinuous Computation

Indisputably, processing information in a largely discontinuous way has become a quite common computational phenomenon [BYRN11, BCC10, MRS08]. Indeed, consider a process p that deals with information i. During a single computational step, p can read a piece of information x in i, erase it, generate a new piece of information y, and insert y into i possibly far away from the original occurrence of x, which was erased. Therefore, intuitively speaking, during its computation, p keeps jumping across i as a whole. To explore computation like this systematically and rigorously, the language theory should provide computer science with language-generating models to explore various information processors mathematically, so it should do so for the purpose sketched above, too.

However, the classical versions of grammars (see Sect. 2.3) work on words strictly continuously, and as such, they can hardly serve as appropriate models of this kind. Therefore, a proper formalization of processors that work in the way described above necessities an adaptation of classical grammars so they work on words discontinuously. At the same time, any adaptation of this kind should conceptually maintain the original structure of these models as much as possible so computer science can quite naturally base its investigation upon these newly adapted grammatical models by analogy with the standard approach based upon their classical versions. Simply put, these new models should work on words in a discontinuous way while keeping their structural conceptualization unchanged. This chapter introduces and studies grammars that work in this discontinuous way. Indeed, the grammars discussed in this section are conceptualized just like classical grammars except that during the applications of their rules, they can jump over symbols in either direction within the rewritten strings, and in this jumping way, they generate their languages.

The present chapter consists of two sections. Section 5.1 studies the jumping generation of language by classical phrase-structure grammars, which work in a sequential way (see Sect. 2.3). Then, Sect. 5.2 discusses the same topic in terms of scattered context grammars, which work in a parallel way (see Sect. 4.1).

© Springer International Publishing AG 2017
A. Meduna, O. Soukup, *Modern Language Models and Computation*,
DOI 10.1007/978-3-319-63100-4_5

5.1   Jumping Grammars: Sequential Versions

Consider a classical phrase-structure grammar G (see Sect. 2.3.1). Recall that
G represents a language-generating rewriting system based upon an alphabet of
symbols and a finite set of rules. The alphabet of symbols is divided into two disjoint
subalphabets—the alphabet of terminal symbols and the alphabet of nonterminal
symbols. Each rule is of the form $x \to y$, where x and y are strings over the alphabet
of G, where x and y are referred to as the left-hand side and the right-hand side of
$x \to y$. G applies $x \to y$ strictly sequentially so it rewrites a string z according to
$x \to y$ so it

(1) selects an occurrence of x in z,
(2) erases it, and
(3) inserts y precisely at the position of this erasure.

More formally, let $z = uxv$, where u and v are strings. By using $x \to y$, G rewrites
uxv as uyv. Starting from a special start nonterminal symbol, G repeatedly rewrites
strings according to its rules in this sequential way until it obtains a sentence—that
is, a string that solely consists of terminal symbols; the set of all sentences represents
the language generated by the grammar.

The notion of a *jumping grammar*, discussed in this chapter, is conceptualized
just like that of a classical grammar; however, it rewrites strings in a jumping way.
Consider G, described above, as a grammar that works in a jumping way. Let z and
$x \to y$ have the same meaning as above. G rewrites a string z according to a rule
$x \to y$ in such a way that it selects an occurrence of x in z, erases it, and inserts y
anywhere in the rewritten string, so this insertion may occur at a different position
than the erasure of x. In other words, G rewrites a string z according to $x \to y$ so it
performs (1) and (2) as described above, but during (3), G can jump over a portion of
the rewritten string in either direction and insert y there. Formally, by using $x \to y$,
G rewrites ucv as udv, where u, v, w, c, d are strings such that either (i) $c = xw$ and
$d = wy$ or (ii) $c = wx$ and $d = yw$.

The present section narrows its investigation to the study of the generative power
of jumping grammars. First, it compares the generative power of jumping grammars
with the accepting power of jumping finite automata. More specifically, it demonstrates that regular jumping grammars are as powerful as jumping finite automata.
Regarding grammars, the general versions of jumping grammars are as powerful as
classical phrase-structure grammars. As there exist many important special versions
of these classical grammars, we discuss their jumping counterparts in the present
section as well. We study the jumping versions of context-free grammars and their
special cases, including regular grammars, right-linear grammars, linear grammars,
and context-free grammars of finite index (see Sect. 2.3.1). Surprisingly, all of
them have a different power than their classical counterparts. In the conclusion
of this section, the section formulates several open problems and suggests future
investigation areas.

Next, we define four modes of derivation relations, three of which represent jumping derivation steps. For the sake of convenience, we also recall some terminology, such as the notion of a phrase-structure grammar, introduced earlier in this book (see Sect. 2.3.1).

Definition 5.1.1. Let $G = (V, T, P, S)$ be a phrase-structure grammar. We introduce four *modes of derivation steps* as derivation relations over V^*—namely, $_s\Rightarrow$, $_{lj}\Rightarrow$, $_{rj}\Rightarrow$, and $_j\Rightarrow$.

Let $u, v \in V^*$. We define the four derivation relations as follows

(i) $u \;_s\Rightarrow v$ in G iff there exist $x \to y \in P$ and $w, z \in V^*$ such that $u = wxz$ and $v = wyz$;

(ii) $u \;_{lj}\Rightarrow v$ in G iff there exist $x \to y \in P$ and $w, t, z \in V^*$ such that $u = wtxz$ and $v = wytz$;

(iii) $u \;_{rj}\Rightarrow v$ in G iff there exist $x \to y \in P$ and $w, t, z \in V^*$ such that $u = wxtz$ and $v = wtyz$;

(iv) $u \;_j\Rightarrow v$ in G iff $u \;_{lj}\Rightarrow v$ or $u \;_{rj}\Rightarrow v$ in G.   □

Let $_h\Rightarrow$ be one of the four derivation relations (i) through (iv) over V^*; in other words, h equals s, lj, rj, or j. As usual, for every $n \geq 0$, the nth power of $_h\Rightarrow$ is denoted by $_h\Rightarrow^n$. The transitive-reflexive closure and the transitive closure of $_h\Rightarrow$ are denoted by $_h\Rightarrow^*$ and $_h\Rightarrow^+$, respectively.

Example 5.1.2. Consider the following RG

$$G = (\{A, B, C, a, b, c\}, \Sigma = \{a, b, c\}, P, A)$$

where $P = \{A \to aB, B \to bC, C \to cA, C \to c\}$. Observe that

$$L(G, \;_s\Rightarrow) = \{abc\}\{abc\}^*, \text{ but}$$

$$L(G, \;_j\Rightarrow) = \{w \in \Sigma^* \mid \#_{\{a\}}(w) = \#_{\{b\}}(w) = \#_{\{c\}}(w)\}$$

Notice that although $L(G, \;_s\Rightarrow)$ is regular, $L(G, \;_j\Rightarrow) \in \textbf{CS}$ is a well-known non-context-free language.   □

Example 5.1.3. Consider the following CSG $G = (\{S, A, B, a, b\}, \{a, b\}, P, S)$ containing the following rules

$$
\begin{aligned}
S &\to aABb \\
S &\to ab \\
AB &\to AABB \\
aA &\to aa \\
Bb &\to bb
\end{aligned}
$$

Trivially, $L(G, \;_s\Rightarrow) = \{a^n b^n \mid n \geq 1\}$. Using $_j\Rightarrow$, we can make the following derivation sequence (the rewritten substring is underlined):

$$\underline{S}\ _j\!\Rightarrow aA\underline{B}b\ _j\!\Rightarrow aAAB\underline{B}b\ _j\!\Rightarrow \underline{aA}ABbb\ _j\!\Rightarrow a\underline{a}ABbb\ _j\!\Rightarrow a\underline{B}bbaa\ _j\!\Rightarrow abbbaa$$

Notice that $L(G,\ _s\!\Rightarrow)$ is context-free, but we cannot generate this language by any CFG, CSG or even MONG in jumping derivation mode.   □

Lemma 5.1.4. $\{a\}^*\{b\}^* \notin \mathscr{L}(\Gamma_{MONG},\ _j\!\Rightarrow)$.

Proof. Assume that there exists a MONG $G = (V, T, P, S)$ such that $L(G,\ _j\!\Rightarrow) = \{a\}^*\{b\}^*$. Let $p\colon x \to y \in P$ be the last applied rule during a derivation $S\ _j\!\Rightarrow^+ w$, where $w \in L(G,\ _j\!\Rightarrow)$; that is, $S\ _j\!\Rightarrow^* uxv\ _j\!\Rightarrow w\ [p]$, where $u, v, w \in T^*$ and $y \in \{a\}^+ \cup \{b\}^+ \cup \{a\}^+\{b\}^+$. In addition, assume that the sentential form uxv is longer than x such that $uv \in \{a\}^+\{b\}^+$.

(i) If y contains at least one symbol b, the last jumping derivation step can place y at the beginning of the sentence and create a string from $\{a, b\}^*\{b\}\{a, b\}^*\{a\}\{a, b\}^*$ that does not belong to $\{a\}^*\{b\}^*$.

(ii) By analogy, if y contains at least one symbol a, the last jumping derivation step can place y at the end of the sentence and therefore, place at least one a behind some bs.

This is a contradiction, so there is no MONG that generates regular language $\{a\}^*\{b\}^*$ using $_j\!\Rightarrow$.   □

We re-open a discussion related to Lemma 5.1.4 at the end of this section.

Corollary 5.1.5. *The following pairs of language families are incomparable, but not disjoint:*

(i) **REG** *and* **JMON***;*
(ii) **CF** *and* **JMON***;*
(iii) **REG** *and* **JREG***;*
(iv) **CF** *and* **JREG***.*

Proof. Since **REG** ⊂ **CF**, it is sufficient to prove that **REG** − **JMON**, **JREG** − **CF**, and **REG** ∩ **JREG** are non-empty. By Lemma 5.1.4, $\{a\}^*\{b\}^* \in$ **REG** − **JMON**. In Example 5.1.2, we define a jumping RG that generates a non-context-free language that belongs to **JREG** − **CF**. Observe that regular language $\{a\}^*$ belongs to **JREG**, so **REG** ∩ **JREG** is non-empty.   □

As even some very simple regular language such as $\{a\}^+\{b\}^+$ cannot be generated by jumping derivation in CSGs or even MONGs, we pinpoint the following open problem and state a theorem comparing these families with context-sensitive languages.

Open Problem 5.1.6. Is **JCS** ⊆ **JMON** proper?

Theorem 5.1.7. **JMON** ⊂ **CS**.

Proof. To see that **JMON** ⊆ **CS**, we demonstrate how to transform any jumping MONG, $G = (V_G, T, P_G, S)$, to a MONG, $H = (V_H, T, P_H, S)$, such that $L(G,\ _j\!\Rightarrow) = L(H,\ _s\!\Rightarrow)$. Set $V_H = N_H \cup T$ and $N_H = N_G \cup \{\bar{X} \mid X \in V_G\}$. Let π

be the homomorphism from V_G^* to V_H^* defined by $\pi(X) = \bar{X}$ for all $X \in V_G$. Set $P_H = P_1 \cup P_2$, where

$$P_1 = \bigcup_{\alpha \to \beta \in P_G} \{\alpha \to \pi(\beta),\, \pi(\beta) \to \beta\}$$

and

$$P_2 = \bigcup_{\alpha \to \beta \in P_G} \{X\pi(\beta) \to \pi(\beta)X,\, \pi(\beta)X \to X\pi(\beta) \mid X \in V_G\}$$

As obvious, $L(G, {}_j\Rightarrow) = L(H, {}_s\Rightarrow)$. Clearly, $\{a\}^*\{b\}^* \in \mathbf{CS}$. Thus, by Lemma 5.1.4, $\mathbf{CS} - \mathbf{JMON} \neq \emptyset$, so this theorem holds. $\qquad\square$

Example 5.1.8. Consider the language of all well-written arithmetic expressions with parentheses $(,\,)$ and $[,\,]$. Eliminate everything but the parentheses in this language to obtain the language $L(G, {}_s\Rightarrow)$ defined by the CFG $G = (V = \{E, (,), [,]\}, T = \{(,), [,]\}, \{E \to (E)E,\, E \to [E]E,\, E \to \varepsilon\}, E)$. G is not of a finite index (see Example 10.1 on page 210 in [Sal73]). Consider the jumping RLG $H = (V, T, P_H, E)$, where P_H contains

$$
\begin{aligned}
E &\;\to\; ()E \\
E &\;\to\; []E \\
E &\;\to\; \varepsilon
\end{aligned}
$$

Since H is a RLG, there is at most one occurrence of E in any sentential form derived from E in H, so H is of index 1. Next, we sketch a proof that $L(G, {}_s\Rightarrow) = L(H, {}_j\Rightarrow)$. As obvious, $\{\varepsilon, (), []\} \subseteq L(G, {}_s\Rightarrow) \cap L(H, {}_j\Rightarrow)$. Consider $\alpha E\beta \,{}_s\Rightarrow\, \alpha(E)E\beta$ $[E \to (E)E]\,{}_s\Rightarrow^*\, \alpha(\gamma)\delta\beta$ in G with $\gamma \neq \varepsilon$. H can simulate this derivation as follows

$$\alpha E\beta \,{}_j\Rightarrow\, \alpha()E\beta \,{}_j\Rightarrow^*\, \alpha()\delta'E\delta''\beta \,{}_j\Rightarrow\, \alpha(xE)\delta\beta \,{}_j\Rightarrow^*\, \alpha(\gamma)\delta\beta$$

where $\delta = \delta'\delta''$, $x \in \{(), []\}$, and $\alpha, \beta, \gamma, \delta \in V^*$. For $\gamma = \varepsilon$, we modify the previous jumping derivation so we make a jumping derivation step from $\alpha()\delta'E\delta''\beta$ to $\alpha()\delta\beta$ by $E \to \varepsilon$ in H. We deal with $E \to [E]E$ analogically, so $L(G, {}_s\Rightarrow) \subseteq L(H, {}_j\Rightarrow)$. Since $L(G, {}_s\Rightarrow)$ contains all proper strings with the three types of parentheses, to prove $L(H, {}_j\Rightarrow) \subseteq L(G, {}_s\Rightarrow)$, we have to show that H cannot generate an improper string of parentheses. As each non-erasing rule of H inserts both left and right parenthesis in the sentential form at once, the numbers of parentheses are always well-balanced. In addition, in H we cannot generate an improper mixture of two kinds of parentheses, such as $([])$, or an improper parenthesis order, such as $)($, so $L(G, {}_s\Rightarrow) = L(H, {}_j\Rightarrow)$. $\qquad\square$

5.1.1 Results

Relations Between the Language Families Resulting from Various Jumping Grammars

We establish several relations between the language families generated by jumping versions of grammars defined earlier in this section.

Theorem 5.1.9. $\mathbf{JRLIN} = \mathbf{JLIN} = \mathbf{JCF}_{fin}$.

Proof. Since $\mathbf{JRLIN} \subseteq \mathbf{JLIN} \subseteq \mathbf{JCF}_{fin}$ follows from the definitions, it suffices to proof that $\mathbf{JCF}_{fin} \subseteq \mathbf{JRLIN}$.

Construction. Let V and T be an alphabet and an alphabet of terminals, respectively. Set $N = V - T$. Let $\eta: V \rightarrow N \cup \{\varepsilon\}$ be the homomorphism such that $\eta(X) = X$ if $X \in N$; otherwise $\eta(X) = \varepsilon$. Let $\tau: V \rightarrow T \cup \{\varepsilon\}$ be the homomorphism such that $\tau(X) = X$ if $X \in T$; otherwise $\eta(X) = \varepsilon$. As usual, extend η and τ to strings of symbols.

For every CFG $G = (V_G, T, P_G, S)$ and index $k \geq 1$, we construct a RLG $H = (V_H, T, P_H, \langle S \rangle)$ such that $L(G, {}_j\Rightarrow_k) = L(H, {}_j\Rightarrow)$. Set

$$V_H = \{\langle x \rangle \mid x \in \bigcup_{i=1}^{k}(V_G - T)^i\} \cup T$$

and set

$$P_H = \{\langle \alpha A \beta \rangle \rightarrow \tau(x)\langle \gamma \rangle \mid A \rightarrow x \in P_G, \alpha, \beta \in N^*, \gamma = \alpha\beta\eta(x), 1 \leq |\gamma| \leq k\}$$

$$\cup \{\langle A \rangle \rightarrow x \mid A \rightarrow x \in P_G, x \in T^*\}$$

Basic Idea. CFG G working with index k means that every sentential form contains at most k nonterminal symbols. In jumping derivation mode, the position of nonterminal symbol does not matter for context-free rewriting. Together with the finiteness of N, we can store the list of nonterminals using just one nonterminal from constructed $V_H - T$ in the simulating RLG.

For every jumping derivation step $\gamma A \delta \, {}_j\Rightarrow_k \gamma' x \delta'$ by $A \rightarrow x$ in G, there is a simulating jumping derivation step $\tau(\bar{\gamma})\langle \eta(\gamma A \delta) \rangle \tau(\bar{\delta}) \, {}_j\Rightarrow \tau(\bar{\gamma'})\tau(x)\langle \eta(\gamma \delta x) \rangle \tau(\bar{\delta'})$ in H, where $\gamma \delta = \gamma' \delta' = \bar{\gamma}\bar{\delta} = \bar{\gamma'}\bar{\delta'}$. The last simulating step of jumping application of $A \rightarrow w$ with $w \in T^*$ replaces the only nonterminal of the form $\langle A \rangle$ by w that can be placed anywhere in the string. $\square$

Consider the finite index restriction in the family $\mathbf{JCF}_{fin}$ in Theorem 5.1.9. Dropping this restriction gives rise to the question, whether the inclusion $\mathbf{JCF}_{fin} \subseteq \mathbf{JCF}$ is proper or not. The next theorem was recently proved.

Theorem 5.1.10 (See Theorem 3.10 in [Mad16]).

$$\mathbf{JCF}_{fin} \subset \mathbf{JCF}.$$

Indeed, from a broader perspective, an investigation of finite-index-based restrictions placed upon various jumping grammars and their effect on the resulting generative power represents a challenging open problem area as illustrated by Example 5.1.8.

Theorem 5.1.11. $\mathbf{JCF}^{-\varepsilon} = \mathbf{JCF}$.

Proof. It is straightforward to establish this theorem by analogy with the same statement reformulated in terms of ordinary CFGs, which work based on $_s\Rightarrow$ (see Theorem 5.1.3.2.4 on page 328 in [Med00a]). $\square$

Lemma 5.1.12. $\mathbf{RE} \subseteq \mathbf{JRE}$.

Proof. Construction. For every PSG $G = (V_G, T, P_G, S_G)$, we construct another PSG $H = (V_H = V_G \cup \{S_H, \$, \#, \lfloor, \rfloor\}, T, P_H, S_H)$ such that $L(G, {}_s\Rightarrow) = L(H, {}_j\Rightarrow)$. $S_H, \$, \#, \lfloor$, and $\rfloor$ are new nonterminal symbols in H. Set

$$P_H = \{S_H \rightarrow \#S_G, \# \rightarrow \lfloor\$, \lfloor\rfloor \rightarrow \#, \# \rightarrow \varepsilon\}$$

$$\cup\{\$\alpha \rightarrow \rfloor\beta \mid \alpha \rightarrow \beta \in P_G\}$$

Basic Idea. Nonterminal $\#$ has at most one occurrence in the sentential form. $\#$ is generated by the initial rule $S_H \rightarrow \#S_G$. This symbol participates in the beginning and end of every simulation of the application of a rule from P_G. Each simulation consists of several jumping derivation steps:

(i) $\#$ is expanded to a string of two nonterminals—marker of a position ($\lfloor$), where the rule is applied in the sentential form, and auxiliary symbol ($\$$) presented as a left context symbol in the left-hand side of every simulated rule from P_G.

(ii) For each $x \rightarrow y$ from P_G, $\$x \rightarrow \rfloor y$ is applied in H. To be able to finish the simulation properly, the right-hand side ($\rfloor y$) of applied rule has to be placed right next to the marker symbol $\lfloor$; otherwise, we cannot generate a sentence.

(iii) The end of the simulation (rule $\lfloor\rfloor \rightarrow \#$) checks that the jumping derivation was applied like in terms of $_s\Rightarrow$.

(iv) In the end, $\#$ is removed to finish the generation of a string of terminal symbols.

Claim 5.1.13. Let y be a sentential form of H; that is, $S_H {}_j\Rightarrow^* y$. For every $X \in \{\#, \$, \lfloor, \rfloor, S_H\}$, $\#_{\{X\}}(y) \leq 1$.

Proof. The claim follows from the rules in P_H (see the construction in the proof of Lemma 5.1.12). Note that $\#_{\{\#, \$, \lfloor, \rfloor, S_H\}}(y) \leq 2$ and in addition, if symbol $\#$ occurs in y then $\#_{\{\$, \lfloor, \rfloor, S_H\}}(y) = 0$. $\square$

Define the homomorphism $h: V_H^* \rightarrow V_G^*$ as $h(X) = X$ for all $X \in V_G$, $h(S_H) = S_G$, and $h(Y) = \varepsilon$ for all $Y \in \{\$, \#, \lfloor, \rfloor\}$.

Claim 5.1.14. If $S_{G\,s}\Rightarrow^m w$ in G, where $w \in T^*$ and $m \geq 0$, then $S_{H\,j}\Rightarrow^* w$ in H.

Proof. First, we prove by induction on $m \geq 0$ that for every $S_{G\,s}\Rightarrow^m x$ in G with $x \in V_G^*$, there is $S_{H\,j}\Rightarrow^* x'$ in H such that $h(x') = x$.

Basis. For $S_{G\,s}\Rightarrow^0 S_G$ in G, there is $S_{H\,j}\Rightarrow \#S_G$ in H.

Induction Hypothesis. Suppose there exists $k \geq 0$ such that $S_{G\,s}\Rightarrow^m x$ in G implies that $S_{H\,j}\Rightarrow^* x'$ in H, where $h(x') = x$, for all $0 \leq m \leq k$.

Induction Step. Assume that $S_{G\,s}\Rightarrow^k y_{\,s}\Rightarrow x$ in G. By the induction hypothesis, $S_{H\,j}\Rightarrow^* y'$ in H with $h(y') = y$.

The derivation step $y_{\,s}\Rightarrow x$ in G is simulated by an application of three jumping rules from P_H in H to get $y'_{\,j}\Rightarrow^3 x'$ with $h(x') = x$ as follows.

$$
\begin{aligned}
y' = u'\#v' \quad _j\Rightarrow & \quad u''\lfloor\$\alpha v'' \quad && [\# \to \lfloor\$] \\
_j\Rightarrow & \quad u''\lfloor\rfloor\beta v'' \quad && [\$ \to \rfloor] \\
_j\Rightarrow & \quad u'''\#v''' \quad && [\lfloor\rfloor\beta \to \#] \quad = x'
\end{aligned}
$$

where $u'v' = u''\alpha v''$ and $u''\beta v'' = u'''v'''$.

In case $x \in T^*$, there is one additional jumping derivation step during the simulation that erases the only occurrence of #-symbol (see Claim 5.1.13) by rule $\# \to \varepsilon$.

Note that $h(x)$ for $x \in T^*$ is the identity. Therefore, in case $x \in T^*$ the induction proves the claim.   □

Claim 5.1.15. If $S_{H\,j}\Rightarrow^m w$ in H, for some $m \geq 0$, where $w \in T^*$, then $S_{G\,s}\Rightarrow^* w$ in G.

Proof. To prove this claim, first, we prove by induction on $m \geq 0$ that for every $S_{H\,j}\Rightarrow^m x$ in H with $x \in V_H^*$ such that there exists a jumping derivation $x_{\,j}\Rightarrow^* w$, where $w \in T^*$, then $S_{G\,s}\Rightarrow^* x'$ in G such that $h(x) = x'$.

Basis. For $m = 0$, when we have $S_{H\,j}\Rightarrow^0 S_{H\,j}\Rightarrow^* w$ in H, then there is $S_{G\,s}\Rightarrow^0 S_G$ in G such that $h(S_H) = S_G$. Furthermore, for $m = 1$, we have $S_{H\,j}\Rightarrow^1 \#S_{G\,j}\Rightarrow^* w$ in H, then again there is $S_{G\,s}\Rightarrow^0 S_G$ in G such that $h(\#S_G) = S_G$, so the basis holds.

Induction Hypothesis. Suppose there exists $k \geq 1$ such that $S_{H\,j}\Rightarrow^m x_{\,j}\Rightarrow^* w$ in H implies that $S_{G\,s}\Rightarrow^* x'$ in G, where $h(x) = x'$, for all $1 \leq m \leq k$.

Induction Step. Assume that $S_{H\,j}\Rightarrow^k y_{\,j}\Rightarrow x_{\,j}\Rightarrow^* w$ in H with $w \in T^*$. By the induction hypothesis, $S_{G\,s}\Rightarrow^* y'$ in G such that $h(y) = y'$. Let $u, v \in V_G^*$ and $\bar{u}, \bar{v} \in V_H^*$. Let us examine the following possibilities of $y_{\,j}\Rightarrow x$ in H:

(i) $y = u\#v_{\,j}\Rightarrow \bar{u}\lfloor\$\bar{v} = x$ in H such that $uv = \bar{u}\bar{v}$: Simply, $y' = uv_{\,s}\Rightarrow^0 uv$ in G and by Claim 5.1.13 $h(\bar{u}\lfloor\$\bar{v}) = h(\bar{u}\bar{v}) = h(uv) = uv$.

(ii) $u\lfloor\$\alpha v_{\,j}\Rightarrow \bar{u}\lfloor\rfloor\beta\bar{v}$ in H by rule $\$\alpha \to \rfloor\beta$ such that $uv = \bar{u}\bar{v}$: In fact, to be able to rewrite $\lfloor$, the symbol $\lfloor$ needs $\rfloor$ as its right neighbor, so $u = \bar{u}$ and $v = \bar{v}$ in this jumping derivation step; otherwise the jumping derivation is prevent from generating a string of terminals. According to rule $\alpha \rightarrow \beta$, $u\alpha v_{\,s}\Rightarrow u\beta v$ in G and $h(\bar{u}\rfloor\beta\bar{v}) = u\beta v$.

(iii) $u\lfloor\,\rfloor v_{\,j}\Rightarrow \bar{u}\#\bar{v}$ in H such that $uv = \bar{u}\bar{v}$: In G, $uv_{\,s}\Rightarrow^0 uv$ and $h(\bar{u}\#\bar{v}) = h(\bar{u}\bar{v}) = h(uv) = uv$.

(iv) $u\#v_{\,j}\Rightarrow uv$ in H by $\# \rightarrow \varepsilon$: Trivially, $uv_{\,s}\Rightarrow^0 uv$ in G and $h(uv) = uv$.

If $x \in T^*$, then the induction proves the claim. □

This closes the proof of Lemma 5.1.12. □

Theorem 5.1.16. $\mathbf{JRE} = \mathbf{RE}$.

Proof. By Turing-Church thesis, $\mathbf{JRE} \subseteq \mathbf{RE}$. The opposite inclusion holds by Lemma 5.1.12 that is proved in details by Claims 5.1.14 and 5.1.15. □

Properties of Jumping Derivations

We demonstrate that the order of nonterminals in a sentential form of jumping CFGs is irrelevant. Then, in this section, we study the semilinearity of language families generated by various jumping grammars.

As a generalization of the proof of Theorem 5.1.9, we give the following lemma demonstrating that the order in which nonterminals occur in sentential forms is irrelevant in jumping derivation mode based on context-free rules in terms of generative power.

Lemma 5.1.17. *Let η and τ be the homomorphisms from the proof of Theorem 5.1.9. For every $G \in \Gamma_X$ with $X \in \{RG, RLG, LG, CFG\}$ and $G = (V, T, P, S)$ with $N = V - T$, if $S_{\,j}\Rightarrow^* \gamma_{\,j}\Rightarrow^m w$ in G, $m \geq 0$, $\gamma \in V^*$, $w \in T^*$, then for every $\delta \in V^*$ such that $\tau(\gamma) = \tau(\delta)$ and $\eta(\delta) \in \mathrm{perm}(\eta(\gamma))$, there is $\delta_{\,j}\Rightarrow^* w$ in G.*

Proof. We prove this lemma by induction on $m \geq 0$.

Basis. Let $m = 0$. That is, $S_{\,j}\Rightarrow^* \gamma_{\,j}\Rightarrow^0 w$ in G, so $\gamma = w$. By $\tau(\delta) = \tau(\gamma)$, we have $\gamma = w = \delta$, so $\delta_{\,j}\Rightarrow^0 w$ in G.

Induction Hypothesis. Assume that there exists $k \geq 0$ such that the lemma holds for all $0 \leq m \leq k$.

Induction Step. Assume that $S_{\,j}\Rightarrow^* \gamma_{\,j}\Rightarrow \gamma'\;[A \rightarrow x]_{\,j}\Rightarrow^k w$ in G with $k \geq 0$. Observe that $\tau(\delta) = \tau(\gamma)$ and $\eta(\delta) \in \mathrm{perm}(\eta(\gamma))$. By the above-mentioned assumption, $|\eta(\gamma)| \geq 1$—that is $|\eta(\delta)| \geq 1$. Thus, the jumping derivation $\delta_{\,j}\Rightarrow^* w$ in G can be written as $\delta_{\,j}\Rightarrow \delta'\;[A \rightarrow x]_{\,j}\Rightarrow^* w$. Since all the rules in G are context-free, the position of A in δ and its context is irrelevant, and the occurrence of A in δ is guaranteed by the lemma precondition. During the application of $A \rightarrow x$, (1) an occurrence of A is found in δ, (2) removed, and (3) the right-hand side of the rule, x, is inserted anywhere in δ instead of A without preserving the position of the rewritten A. Assume x is inserted into δ' so that $\tau(\delta') = \tau(\gamma')$. We also preserve that $\eta(\delta') \in \mathrm{perm}(\eta(\gamma'))$; therefore, the lemma holds. □

Notice that even if there is no derivation $S_j \Rightarrow^* \delta$ in G, the lemma holds.

Note that based on the proof of Lemma 5.1.17, we can turn any jumping version of a CFG to an equivalent jumping CFG satisfying a modified Greibach normal form, in which each rule is of the form $A \rightarrow \alpha\beta$, where $\alpha \in T^*, \beta \in N^*$. Observe that $\alpha \notin T$. Consider, for instance, a context-free rule p with $\alpha = a_1 \cdots a_n$. By an application of p during a derivation of a string of terminals w, we arrange that a_1 appears somewhere in front of a_n in w. In other words, from Theorem 13 and Corollary 14 in [MZ12a] together with Theorem 8.2.68, it follows that for any language L, $L \in$ **JREG** implies $L = \text{perm}(L)$, which means that the order of all terminals in $w \in L$ is utterly irrelevant.

Corollary 5.1.18. *For every* $G \in \Gamma_X$ *with* $X \in \{RG, RLG, LG, CFG\}$, $S_j \Rightarrow^* \gamma_j \Rightarrow^* w$ *in* G *implies an existence of a derivation of the following form*

$$S_j \Rightarrow^* \alpha\beta_j \Rightarrow^* w \text{ in } G$$

where $\alpha = \tau(\gamma)$, $\beta \in \text{perm}(\eta(\gamma))$, S *is the start nonterminal, and* w *is a string of terminals.*

Definition 5.1.19 ([Gin66]). Let $w \in V^*$ with $V = \{a_1, \ldots, a_n\}$. We define *Parikh vector* of w by $\psi_V(w) = (\#_{a_1}(w), \#_{a_2}(w), \ldots, \#_{a_n}(w))$. A set of vectors is called *semilinear* if it can be represented as a union of a finite number of sets of the form $\{v_0 + \sum_{i=1}^m \alpha_i v_i \mid \alpha_i \in \mathbb{N}, 1 \leq i \leq m\}$, where v_i for $0 \leq i \leq m$ is an n-dimensional vector. A language $L \subseteq V^*$ is called *semilinear* if the set $\psi_V(L) = \{\psi_V(w) \mid w \in L\}$ is a semilinear set. A language family is *semilinear* if all its languages are semilinear. □

Lemma 5.1.20. *For* $X \in \{RG, RLG, LG, CFG\}$, $\mathscr{L}(\Gamma_X, _j\Rightarrow)$ *is semilinear.*

Proof. By Parikh's Theorem (see Theorem 6.9.2 on page 228 in [Har78]), for each context-free language $L \subseteq V^*$, $\psi_V(L)$ is semilinear. Let G be a CFG such that $L(G, _s\Rightarrow) = L$. From the definition of $_j\Rightarrow$ and CFG it follows that $\psi(L(G, _s\Rightarrow)) = \psi(L(G, _j\Rightarrow))$ therefore $\psi(L(G, _j\Rightarrow))$ is semilinear as well. □

Recall that the family of context-sensitive languages is not semilinear (for instance, Example 2.3.1 and Theorem 2.3.1 in [DP89] implies that $\{a^{2^n} \mid n \geq 0\} \in$ **CS**, but is not semilinear language). By no means, this result rules out that $\mathscr{L}(\Gamma_{CSG}, _j\Rightarrow)$ or $\mathscr{L}(\Gamma_{MONG}, _j\Rightarrow)$ are semilinear. There is, however, another kind of results concerning multiset grammars (see [KMVP00]) saying that a context-sensitive multiset grammar generates a non-semilinear language. The multiset grammars work with Parikh vector of a sentential form so the order of symbols in the sentential form is irrelevant. Then, all permutations of terminal strings generated by the grammar belong to the generated language.

Instead of the full definition of multiset grammars (see [KMVP00]), based on notions from the theory of macrosets, we introduce *multiset derivation mode* concerning the classical string formal language theory.

Definition 5.1.21. Let $G = (V, T, P, S) \in \Gamma_{PSG}$ be a grammar and $u, v \in V^*$; then, $u \, _m\!\Rightarrow v \, [x \rightarrow y]$ in G iff there exist $x \rightarrow y \in P$ and $t, t', z, z' \in V^*$ such that $txt' \in \mathrm{perm}(u)$, $zyz' \in \mathrm{perm}(v)$, and $tt' \in \mathrm{perm}(zz')$. $\qquad\square$

Lemma 5.1.22. Let $G \in \Gamma_{PSG}$; then, $w \in L(G, \, _m\!\Rightarrow)$ implies that $\mathrm{perm}(w) \subseteq L(G, \, _m\!\Rightarrow)$.

Proof. Consider Definition 5.1.21 with v representing every permutation of v in every $u \, _m\!\Rightarrow v$ in G to see that this lemma holds true. $\qquad\square$

Recall that $\mathscr{L}(\Gamma_{MONG}, \, _m\!\Rightarrow)$ is not semilinear (see [KMVP00]). As every context-sensitive multiset grammar can be transformed into a CSG that generates the same language under jumping derivation mode, we establish the following theorem.

Theorem 5.1.23. $\mathscr{L}(\Gamma_{CSG}, \, _j\!\Rightarrow)$ is not semilinear. Neither is $\mathscr{L}(\Gamma_{MONG}, \, _j\!\Rightarrow)$.

Proof. Recall that $\mathscr{L}(\Gamma_{MONG}, \, _m\!\Rightarrow)$ contains non-semilinear languages (see Theorem 1 in [KMVP00]). Thus, to prove Theorem 5.1.23, we only need to prove that $\mathscr{L}(\Gamma_{MONG}, \, _m\!\Rightarrow) \subseteq \mathscr{L}(\Gamma_{CSG}, \, _j\!\Rightarrow)$ because $\mathscr{L}(\Gamma_{CSG}, \, _j\!\Rightarrow) \subseteq \mathscr{L}(\Gamma_{MONG}, \, _j\!\Rightarrow)$ follows from Definition 5.1.1.

Construction. For every MONG $G = (V_G, T, P_G, S)$, we next construct a CSG $H = (V_H, T, P_H, S)$ such that $L(G, \, _m\!\Rightarrow) = L(H, \, _j\!\Rightarrow)$. Let $N_G = V_G - T$ and h be the homomorphism $h \colon V_G^* \rightarrow V_H^*$ defined as $h(X) = X$ for all $X \in N_G$ and $h(a) = \langle a \rangle$ for all $a \in T$. First, set $V_H = V_G \cup N_t \cup N_{cs}$, where $N_t = \{\langle a \rangle \mid a \in T\}$ and $N_{cs} = \{_pX \mid X \in N_G \cup N_t, p \in P_G \text{ with } |\mathrm{lhs}(p)| > 1\}$. For every $p \in P_G$ with $|\mathrm{lhs}(p)| > 1$, let $g_p \colon (N_G \cup N_t)^* \rightarrow N_{cs}^*$ be the homomorphism defined as $g_p(X) = \, _pX$ for all $X \in N_G \cup N_t$. Set $P_t = \{\langle a \rangle \rightarrow a \mid a \in T\}$, $P_{cf} = \{A \rightarrow h(x) \mid A \rightarrow x \in P_G, A \in V_G - T \text{ and } x \in V_G^*\}$, and $P_{cs} = \emptyset$. For every rule $p \colon X_1 X_2 \cdots X_n \rightarrow Y_1 Y_2 \cdots Y_m \in P_G$ with $2 \leq n \leq m$, where $X_i, Y_{i'} \in V_G$, $1 \leq i \leq n$, and $1 \leq i' \leq m$, add these $2n$ new rules with labels $p_1, p_2, \ldots, p_{2n}$

$$
\begin{aligned}
p_1: & & h(X_1 X_2 \cdots X_n) & \rightarrow g_p(h(X_1)) h(X_2 \cdots X_n) \\
p_2: & & g_p(h(X_1)) h(X_2 \cdots X_n) & \rightarrow g_p(h(X_1 X_2)) h(X_3 \cdots X_n) \\
& & & \vdots \\
p_n: & & g_p(h(X_1 X_2 \cdots X_{n-1})) h(X_n) & \rightarrow g_p(h(X_1 X_2 \cdots X_{n-1} X_n)) \\
p_{n+1}: & & g_p(h(X_1 X_2 \cdots X_n)) & \rightarrow h(Y_1) g_p(h(X_2 \cdots X_n)) \\
p_{n+2}: & & h(Y_1) g_p(h(X_2 \cdots X_n)) & \rightarrow h(Y_1 Y_2) g_p(h(X_3 \cdots X_n)) \\
& & & \vdots \\
p_{2n}: & & h(Y_1 Y_2 \cdots Y_{n-1}) g_p(h(X_n)) & \rightarrow h(Y_1 Y_2 \cdots Y_{n-1} Y_n Y_{n+1} \cdots Y_m)
\end{aligned}
$$

into P_{cs}. Set $P_c = \{A \rightarrow A \mid A \in V_H - T\}$. Finally, set $P_H = P_{cf} \cup P_t \cup P_c \cup P_{cs}$.

Basic Idea. There are two essential differences between multiset derivation mode of a MONG and jumping derivation mode of a CSG.

(I) While a MONG rewrites a string at once in a single derivation step, a CSG rewrites only a single nonterminal that occurs within a given context during a single derivation step.

(II) In the multiset derivation mode, the mutual neighborhood of the rewritten symbols is completely irrelevant—that is, G applies any rule without any restriction placed upon the mutual adjacency of the rewritten symbols in the multiset derivation mode (see Definition 5.1.21). To put this in a different way, G rewrites any permutation of the required context in this way.

In the construction of the jumping CSG H, which simulates the multiset MONG G, we arrange (II) as follows.

(I.a) In H, the only rules generating terminals belong to P_t. By using homomorphism h, in every other rule, each terminal a is changed to the corresponding nonterminal $\langle a \rangle$.

(II.b) In P_c, there are rules that can rearrange the order of all nonterminals arbitrarily in any sentential form of H. Thus, considering (I.a), just like in G, no context restriction placed upon the mutual adjacency of rewritten symbols occurs in H. Indeed, H only requires the occurrence of the symbols from $h(\mathrm{lhs}(p))$ during the simulation of an application of $p \in P_G$.

In order to arrange (I), an application of a monotone context-sensitive rule $p: X_1 X_2 \cdots X_n \to Y_1 Y_2 \cdots Y_m \in P_G$, $2 \leq n \leq m$ in $u \ {}_m\!\!\Rightarrow v \ [p]$ in G is simulated in H by the following two phases.

(i) First, H verifies that a sentential form u contains all symbols from $h(\mathrm{lhs}(p))$ and marks them by subscript p for the consecutive rewriting. Therefore, to finish the simulation of the application of p, H has to use rules created based on p during the construction of P_{cs} since no other rules from P_H rewrite symbols ${}_p X$, $X \in N_G \cup N_t$.

$$
\begin{array}{lllll}
u & {}_j\!\!\Rightarrow^* & \alpha_0 X_1' X_2' \cdots X_n' \beta_0 \ [\rho_0] & {}_j\!\!\Rightarrow & u_1 \ [p_1] \\
& {}_j\!\!\Rightarrow^* & \alpha_1 {}_p X_1' X_2' \cdots X_n' \beta_1 \ [\rho_1] & {}_j\!\!\Rightarrow & u_2 \ [p_2] \\
& {}_j\!\!\Rightarrow^* & \alpha_2 {}_p X_1' {}_p X_2' \cdots X_n' \beta_2 \ [\rho_2] & {}_j\!\!\Rightarrow & u_3 \ [p_3] \\
& \vdots & & & \\
& {}_j\!\!\Rightarrow^* & \alpha_{n-1} {}_p X_1' {}_p X_2' \cdots {}_p X_{n-1}' X_n' \beta_{n-1} \ [\rho_{n-1}] & {}_j\!\!\Rightarrow & u_n \ [p_n]
\end{array}
$$

where $\rho_i \in P_c^*$ for $0 \leq i < n$ and $X_\ell' = h(X_\ell)$ for $1 \leq \ell \leq n$.

(ii) Then, by performing $u_n \ {}_j\!\!\Rightarrow^* v$, H simulates the application of p in G.

$$
\begin{array}{lllll}
u_n & {}_j\!\!\Rightarrow^* & \alpha_n {}_p X_1' {}_p X_2' \cdots {}_p X_n' \beta_n \ [\rho_n] & {}_j\!\!\Rightarrow & u_{n+1} \ [p_{n+1}] \\
& {}_j\!\!\Rightarrow^* & \alpha_{n+1} Y_1' {}_p X_2' \cdots {}_p X_n' \beta_{n+1} \ [\rho_{n+1}] & {}_j\!\!\Rightarrow & u_{n+2} \ [p_{n+2}] \\
& \vdots & & & \\
& {}_j\!\!\Rightarrow^* & \alpha_{2n-1} Y_1' Y_2' \cdots Y_{n-1}' {}_p X_n' \beta_{2n-1} \ [\rho_{2n-1}] & {}_j\!\!\Rightarrow & u_{2n} \ [p_{2n}] \\
& = & \alpha_{2n} Y_1' Y_2' \cdots Y_m' \beta_{2n} & {}_j\!\!\Rightarrow^* & v \ [\rho_{2n}]
\end{array}
$$

where $\rho_i \in P_c^*$ for $n \leq i \leq 2n$, $X_\ell' = h(X_\ell)$ for $1 \leq \ell \leq n$, and $Y_k' = h(Y_k)$ for $1 \leq k \leq m$.

The simulation of application of rules of P_G is repeated using rules from $P_c \cup P_{cf} \cup P_{cs}$ in H until a multiset derivation of a string of terminals in G is simulated. (In fact, we can simultaneously simulate more than one application of a rule from P_G if there is no interference in H.)

Then, in the final phase of the entire simulation, each nonterminal $\langle a \rangle$ is replaced with terminal a by using rules from P_t. To be precise, the rules of P_t can be applied even sooner, but symbols rewritten by these rules can be no longer rewritten by rules from $P_c \cup P_{cf} \cup P_{cs}$ in H.

To formally prove that $L(G, {}_m\Rightarrow) = L(H, {}_j\Rightarrow)$, we establish the following claims.

Claim 5.1.24. Every $w \in L(H, {}_j\Rightarrow)$ can be generated by a derivation of the form

$$S {}_j\Rightarrow^* w' {}_j\Rightarrow^* w \text{ in } H \text{ such that } w' = h(w) \text{ and } w \in T^*$$

Proof. In the construction given in the proof of Theorem 5.1.23, we introduce P_{cf} and P_{cs} such that for every $p \in P_H - P_t$, $\mathrm{rhs}(p) \in (V_H - T)^*$. In $S {}_j\Rightarrow^* w'$, we apply rules only from $P_H - P_t$ so $w' \in N_t^*$, and no terminal symbol occurs in any sentential form in $S {}_j\Rightarrow^* w'$. Then, by rules from P_t, we generate w such that $w = h(w')$. □

Claim 5.1.25. If $w \in L(H, {}_j\Rightarrow)$, then $\mathrm{perm}(w) \subseteq L(H, {}_j\Rightarrow)$.

Proof. Let $w \in T^*$. Assume that w is generated in H as described in Claim 5.1.24—that is, $S {}_j\Rightarrow^* w' {}_j\Rightarrow^* w$ such that $w' = h(w)$. Since rules from P_t rewrite nonterminals in w' one by one in the jumping derivation mode, we have $w' {}_j\Rightarrow^* w''$ in H for every $w'' \in \mathrm{perm}(w)$. □

Claim 5.1.26. If $S {}_m\Rightarrow^\ell v$ in G for some $\ell \geq 0$, then $S {}_j\Rightarrow^* v'$ in H such that $v' \in \mathrm{perm}(h(v))$.

Proof. We prove this claim by induction on $\ell \geq 0$.

Basis. Let $\ell = 0$. That is, $S {}_m\Rightarrow^0 S$ in G, so $S {}_j\Rightarrow^0 S$ in G. By $h(S) = S$, $S \in \mathrm{perm}(h(S))$.

Induction Hypothesis. Assume that the claim holds for all $0 \leq \ell \leq k$, for some $k \geq 0$.

Induction Step. Take any $S {}_m\Rightarrow^{k+1} v$. Express $S {}_m\Rightarrow^{k+1} v$ as

$$S {}_m\Rightarrow^k u {}_m\Rightarrow v \ [p : x \to y]$$

in G. By the induction hypothesis, $S {}_j\Rightarrow^* u'$ in H such that $u' \in \mathrm{perm}(h(u))$. According to the form of monotone rule $p : x \to y \in P_G$, there are the following two cases, (i) and (ii), concerning $u {}_m\Rightarrow v$ in G to examine.

(i) $|x| = 1$: Let $x = A$. By the induction hypothesis, $\#_{\{A\}}(u) \geq 1$ implies $\#_{\{A\}}(u') \geq 1$. By the construction according to p, we have $p' : A \to h(y) \in P_{cf}$. Assume $u = u_1 A u_2 \; {}_m\!\!\Rightarrow v$ in G with $u_1 y u_2 \in \mathrm{perm}(v)$. Then, $u' = u'_1 A u'_2 \; {}_j\!\!\Rightarrow u'_3 h(y) u'_4 \; [p'] = v'$ in H, where $u'_1 u'_2 = u'_3 u'_4$, so $v' \in \mathrm{perm}(h(v))$.

(ii) $|x| \geq 2$: Let $x = X_1 X_2 \cdots X_n$, $y = Y_1 Y_2 \cdots Y_m$, where $|x| = n \leq m = |y|$, $X_i \in V_G$, $1 \leq i \leq n$, but $x \notin T^*$, $Y_{i'} \in V_G$, $1 \leq i' \leq m$. By construction of P_{cs}, we have $p_1, p_2, \ldots, p_{2n} \in P_H$. If p can be applied in G, then, by the induction hypothesis, $\#_{\{X_i\}}(u) = \#_{\{h(X_i)\}}(u')$ for $1 \leq i \leq n$. To simulate the application of p in H, first, apply rules from P_c to yield $u' \; {}_j\!\!\Rightarrow^* u'_1 h(X_1 X_2 \cdots X_n) u'_2$. Next, consecutively apply $p_1, p_2, \ldots, p_{2n}$ so $u'_1 h(X_1 X_2 \cdots X_n) u'_2 \; {}_j\!\!\Rightarrow^* u'_3 h(Y_1 Y_2 \cdots Y_m) u'_4 = v'$ with $u'_1 u'_2 = u'_3 u'_4$ and $v' \in \mathrm{perm}(h(v))$. □

By Claim 5.1.26 with $v = w$ and $w \in T^*$, for every $S \; {}_m\!\!\Rightarrow^* w$ in G, there is a derivation $S \; {}_j\!\!\Rightarrow^* w''$ in H such that $w'' \in \mathrm{perm}(h(w))$. By Claim 5.1.24, there is a jumping derivation in H from w'' to w' such that $w' \in T^*$ and $w' \in \mathrm{perm}(w)$. Therefore, by Lemma 5.1.22 and Claim 5.1.25, if $w \in L(G, \; {}_m\!\!\Rightarrow)$, then $\mathrm{perm}(w) \subseteq L(H, \; {}_j\!\!\Rightarrow)$, so $L(G, \; {}_m\!\!\Rightarrow) \subseteq L(H, \; {}_j\!\!\Rightarrow)$.

Claim 5.1.27. If $S \; {}_j\!\!\Rightarrow^\ell v \; {}_j\!\!\Rightarrow^* \bar{v}$ in H for some $\ell \geq 0$, then $S \; {}_m\!\!\Rightarrow^* v'$ in G such that $\bar{v} \in \mathrm{perm}(h(v'))$.

Proof. We prove this claim by induction on $\ell \geq 0$.

Basis. Let $\ell = 0$. Express $S \; {}_j\!\!\Rightarrow^0 S \; {}_j\!\!\Rightarrow^* S$ as $S \; {}_j\!\!\Rightarrow^0 S \; {}_j\!\!\Rightarrow^0 S$ in H, therefore $S \; {}_m\!\!\Rightarrow^0 S$ in G. By $h(S) = S$, $S \in \mathrm{perm}(h(S))$.

Induction Hypothesis. Assume that the claim holds for all $0 \leq \ell \leq k$, for some $k \geq 0$.

Induction Step. Take any $S \; {}_j\!\!\Rightarrow^{k+1} v \; {}_j\!\!\Rightarrow^* \bar{v}$. Express $S \; {}_j\!\!\Rightarrow^{k+1} v \; {}_j\!\!\Rightarrow^* \bar{v}$ as

$$S \; {}_j\!\!\Rightarrow^k u \; {}_j\!\!\Rightarrow v \; [q : x \to y] \; {}_j\!\!\Rightarrow^* \bar{v}$$

in H. Without any loss of generality, assume that $q \in P_H - P_t$ so $u, v \in (V_H - T)^*$ (see Claim 5.1.24). If $q \in P_{cf} \cup P_{cs}$, then p denotes the rule from P_G that implied the addition of q into P_{cf} or P_{cs} during the construction in the proof of Theorem 5.1.23. Without any loss of generality and with respect to p from P_G, assume that there is no simulation of another context-sensitive rule from P_G in progress in H so $\#_{N_{cs}}(u_1 u_2) = \#_{N_{cs}}(v_1 v_2) = 0$, where $u = u_1 x u_2$ and $v = v_1 y v_2$. By the induction hypothesis, $S \; {}_j\!\!\Rightarrow^* u \; {}_j\!\!\Rightarrow^* \bar{u}$ in H implies $S \; {}_m\!\!\Rightarrow^* u'$ in G such that $\bar{u} \in \mathrm{perm}(h(u'))$. Now, we study several cases based on the form of q:

(i) $q \in P_c$ and $x = y = A$: Then, in a jumping derivation $u \; {}_j\!\!\Rightarrow v \; [q] \; {}_j\!\!\Rightarrow^0 \bar{v}$ in H, $u = u_1 A u_2$ and $v = v_1 A v_2$, where $u_1 u_2 = v_1 v_2$, so $v = \bar{v} \in \mathrm{perm}(u)$. By the induction hypothesis, with $u \; {}_j\!\!\Rightarrow^0 \bar{u}$ in H so $u = \bar{u}$, there is a derivation $S \; {}_m\!\!\Rightarrow^* u'$ in G such that $u \in \mathrm{perm}(h(u'))$. Together with $\bar{v} \in \mathrm{perm}(u)$, there is also a derivation $S \; {}_m\!\!\Rightarrow^* u' \; {}_m\!\!\Rightarrow^0 v'$ in G with $\bar{v} \in \mathrm{perm}(h(v'))$.

(ii) $q \in P_{cf}$ and $x = A$: Then, $u = u_1 A u_2$ and $v = v_1 y v_2$ with $u_1 u_2 = v_1 v_2$ and $v_j \Rightarrow^0 \bar{v}$ in H, so $v = \bar{v}$. By the induction hypothesis, with $u_j \Rightarrow^0 \bar{u}$ in H so $u = \bar{u}$, there is $S_m \Rightarrow^* u'$ in G with $u \in \text{perm}(h(u'))$ and we can write $u' = u'_1 A u'_2$. By the construction, $p: A \rightarrow y \in P_G$, so together with the induction hypothesis we have $S_m \Rightarrow^* u'_1 A u'_2 \, _m \Rightarrow v'$ $[p]$ in G, where $v' \in \text{perm}(u'_1 y u'_2)$, so $\bar{v} \in \text{perm}(h(v'))$.

(iii) $q = p_i \in P_{cs}$, where $1 \leq i \leq 2n$ and $n = |\text{lhs}(p)|$: Express $S_j \Rightarrow^k u_j \Rightarrow v_j \Rightarrow^* \bar{v}$ in H as

$$S_j \Rightarrow^{k-i+1} \tilde{u}_j \Rightarrow^{i-1} u \, [\bar{\rho}]_j \Rightarrow v \, [p_i]_j \Rightarrow^* \alpha_{2n} h(Y_1 Y_2 \cdots Y_m) \beta_{2n} \, [\bar{\rho}] = \bar{v}$$

in H. By the construction of P_{cs} according to p and by the induction hypothesis, $\tilde{\rho} = p_1 \cdots p_{i-1}$ and $\bar{\rho} = p_{i+1} \cdots p_{2n}$. By the induction hypothesis, $S_m \Rightarrow^* \tilde{u}'$ in G such that $\tilde{u} \in \text{perm}(h(\tilde{u}'))$. Then, by the application of $p \in P_G$, we have $S_m \Rightarrow^* \tilde{u}' \, _m \Rightarrow v'$ such that $\bar{v} \in \text{perm}(h(v'))$.

In (iii), there are three subcases of $u_j \Rightarrow v$ with $u_1 u_2 = v_1 v_2$ in H:

(iii.a) $1 \leq i \leq n$: Then, $u = u_1 g_p(h(X_1 \cdots X_{i-1})) h(X_i X_{i+1} \cdots X_n) u_2$ and $v = v_1 g_p(h(X_1 \cdots X_{i-1} X_i)) h(X_{i+1} \cdots X_n) v_2$.

(iii.b) $n < i < 2n$ and $i' = i - n$: Then, $u = u_1 h(Y_1 \cdots Y_{i'-1}) g_p(h(X_{i'} X_{i'+1} \cdots X_n)) u_2$ and $v = v_1 h(Y_1 \cdots Y_{i'}) g_p(h(X_{i'+1} \cdots X_n)) v_2$.

(iii.c) $i = 2n$: Then, $u = u_1 h(Y_1 \cdots Y_{n-1}) g_p(h(X_n)) u_2$ and $v = v_1 h(Y_1 \cdots Y_{n-1} Y_n \cdots Y_m) v_2$.

Therefore, the claim holds for $k + 1$ as well. $\square$

Assume $v \in N_t^*$ in Claim 5.1.27 so $v' \in T^*$. Based on Claim 5.1.24, without any loss of generality, we can assume that all rules from P_t are applied in the end of a derivation of $w \in T^*$ in H. Specifically, $S_j \Rightarrow^* v \, [\rho_v]_j \Rightarrow^* w \, [\rho_w]$ in H, where $\rho_v \in (P_H - P_t)^*$, $\rho_w \in P_t^*$, and $v = h(w)$. By Claim 5.1.27, we have $S_m \Rightarrow^* v'$ in G with $v \in \text{perm}(h(v'))$. Recall that $v \in N_t^*$ and $v' \in T^*$. Therefore, $S_m \Rightarrow^* v'$ in G and $w \in \text{perm}(v')$.

Next, by Claim 5.1.25, $w \in L(H, _j\Rightarrow)$ implies $\text{perm}(w) \subseteq L(H, _j\Rightarrow)$. By the previous paragraph and Lemma 5.1.22, for w, we generate $\text{perm}(w)$ in G included in $L(G, _m\Rightarrow)$, that is, $L(H, _j\Rightarrow) \subseteq L(G, _m\Rightarrow)$.

This closes the proof of Theorem 5.1.23. $\square$

Concerning the semilinearity of language families defined by jumping grammars under investigation, the following corollary sums up all important properties established in this section.

Corollary 5.1.28. $\mathscr{L}(\Gamma_X, _j\Rightarrow)$ *is semilinear for* $X \in \{RG, RLG, LG, CFG\}$ *and not semilinear for* $X \in \{CSG, MONG, PSG\}$.

Proof. For **JRE**, the non-semilinearity follows from the well-known facts that **CS** is not semilinear (see Example 2.3.1 and Theorem 2.3.1 in [DP89]) and **CS** $\subset$ **RE** and from Theorem 5.1.16. The rest follows from Lemma 5.1.20 and Theorem 5.1.23. $\square$

Corollary 5.1.29. $\mathbf{JCF} \subset \mathbf{JCS}$.

Proof. Obviously, by Definition 5.1.1, $\mathbf{JCF} \subseteq \mathbf{JCS}$. By Corollary 5.1.28, $\mathbf{JCS}$ contains a non-semilinear language that does not belong to $\mathbf{JCF}$.   □

We close this section by proposing several future investigation areas concerning jumping grammars. Some of them relate to specific open questions pointed out earlier in the section; the present section, however, formulates them more generally and broadly.

I *Other Types of Grammars.* The present section has concentrated its attention to the language families resulting from classical grammars, such as the grammars classified by Chomsky (see [Cho59]). Apart from them, however, the formal language theory has introduced many other types of grammars, ranging from regulated grammars through parallel grammars up to grammar systems. Reconsider the present study in their terms.

II *Left and Right Jumping Mode.* Considering the left and right jumps introduced in Definition 5.1.1, study them in terms of classical types of grammars. Later in Sect. 8.2.5, this book gives an introduction to discussion of left and right jumping derivation modes in terms of automata.

III *Closure Properties.* Several results and some open problems concerning closure properties follows from Sect. 8.2.7. Additionally, study closure properties of language families generated in a jumping way. Specifically, investigate these properties in terms of CFGs, CSGs, and MONGs.

IV *Alternative Definition of Jumping Mode with Context.* Assume context-sensitive rules (CSG) of the following form

$$\alpha A \beta \to \alpha \gamma \beta, \text{ where } A \in N, \alpha, \beta, \gamma \in V^*, \gamma \neq \varepsilon.$$

There are three interesting ways of defining a jumping derivation step:

IV.a Using the previous definition (see Definition 5.1.1) of jumping derivation; that is, find $\alpha A \beta$ in the current sentential form $u\alpha A \beta v$, remove $\alpha A \beta$, and place $\alpha \gamma \beta$ anywhere in uv. For instance,

$$aAbc \ _j\!\!\Rightarrow caxb \ [aAb \to axb]$$

IV.b Do not move the context of the rewritten nonterminal; that is, find A with left context α and right context β, remove this A from the current sentential form, and place γ in the new sentential form, such that string γ will be again in the context of both α and β (but it can be different occurrence of α and β). For instance,

$$aAbab \ _j\!\!\Rightarrow abaxb \ [aAb \to axb]$$

IV.c Similarly to (b), in the third variant we do not move the context of the rewritten nonterminal either and, in addition, γ has to be placed between the same occurrence of α and β. As a consequence, context-sensitive rules are applied sequentially even in this jumping derivation mode. For instance,

$$aAbab \;_{j''}\Rightarrow axbab \; [aAb \rightarrow axb]$$

Notice that this derivation mode influences only the application of context-free rules (i.e. $\alpha = \beta = \varepsilon$).

Example 5.1.30. Example 5.1.3 shows a CSG that generates $\{a^n b^n \mid n \geq 1\}$ when the alternative jumping derivation mode $_j\Rightarrow$ for CSGs is used. In context of Lemma 5.1.4, the alternative jumping derivation mode (b) can increase the generative power of jumping CSGs (a). In fact, it is an open question whether $\mathscr{L}(\Gamma_{CSG}, \;_j\Rightarrow) \subseteq \mathscr{L}(\Gamma_{MONG}, \;_j\Rightarrow)$. □

V *Relationship with Formal Macroset Theory.* Recently, formal language theory has introduced various rewriting devices that generate different objects than classical formal languages. Specifically, in this way, Formal Macroset Theory has investigated the generation of macrosets—that is, sets of multisets over alphabets. Notice that some of its results resemble results achieved in the present study (c.f., for instance, Theorem 1 in [KMVP00] and Theorems 5.1.9 and 5.1.10 above). Explain this resemblance mathematically.

5.2 Jumping Grammars: Semi-Parallel Versions

This section introduces and studies jumping versions of scattered context grammars (see Sect. 4.1). To give an insight into the key motivation and reason for this study, let us take a closer look at a more specific kind of information processing in a discontinuous way. Consider a process p that deals with information i. Typically, during a single computational step, p (1) reads n pieces of information, x_1 through x_n, in i, (2) erases them, (3) generate n new pieces of information, y_1 through y_n, and (4) inserts them into i possibly at different positions than the original occurrence of x_1 through x_n, which was erased. To explore computation like this systematically and rigorously, the present section introduces and discusses jumping versions of scattered context grammars (see [GH69]), which represent suitable grammatical models of computation like this.

To see this suitability, recall that the notion of a scattered context grammar G represents a language-generating rewriting system based upon an alphabet of symbols and a finite set of rules. The alphabet of symbols is divided into two disjoint subalphabets—the alphabet of terminal symbols and the alphabet of nonterminal symbols. In G, a rule r is of the form

$$(A_1, A_2, \ldots, A_n) \rightarrow (x_1, x_2, \ldots, x_n)$$

for some positive integer n. On the left-hand side of r, the As are nonterminals. On the right-hand side, the xs are strings. G can apply r to any string u of the form

$$u = u_0 A_1 u_1 \ldots u_{n-1} A_n u_n$$

where us are any strings. Notice that A_1 through A_n are scattered throughout u, but they occur in the order prescribed by the left-hand side of r. In essence, G applies r to u so

(1) it deletes $A_1, A_2, \ldots, A_n$ in u, after which
(2) it inserts $x_1, x_2, \ldots, x_n$ into the string resulting from the deletion (1).

By this application, G makes a derivation step from u to a string v of the form

$$v = v_0 x_1 v_1 \ldots v_{n-1} x_n v_n$$

Notice that $x_1, x_2, \ldots, x_n$ are inserted in the order prescribed by the right-hand side of r. However, they are inserted in a scattered way—that is, in between the inserted xs, some substrings vs occur.

To formalize the above-described computation, consisting of phases (1) through (4), the present section introduces and studies the following nine jumping derivation modes of the standard application.

(1) Mode 1 requires that $u_i = v_i$ for all $i = 0, \ldots, n$ in the above described derivation step.
(2) Mode 2 obtains v from u as follows:

  (2.a) $A_1, A_2, \ldots, A_n$ are deleted;
  (2.b) x_1 through x_n are inserted in between u_0 and u_n.

(3) Mode 3 obtains v from u so it changes u by performing (3.a) through (3.c), described next:

  (3.a) $A_1, A_2, \ldots, A_n$ are deleted;
  (3.b) x_1 and x_n are inserted into u_0 and u_n, respectively;
  (3.c) x_2 through x_{n-1} are inserted in between the newly inserted x_1 and x_n.

(4) In mode 4, the derivation from u to v is performed by the following steps:

  (4.a) $A_1, A_2, \ldots, A_n$ are deleted;
  (4.b) a central u_i is nondeterministically chosen, for some $0 \le i \le n$;
  (4.c) x_i and x_{i+1} are inserted into u_i;
  (4.d) x_j is inserted between u_j and u_{j+1}, for all $j < i$;
  (4.e) x_k is inserted between u_{k-2} and u_{k-1}, for all $k > i + 1$.

(5) In mode 5, v is obtained from u by (5.a) through (5.e), given next:

  (5.a) $A_1, A_2, \ldots, A_n$ are deleted;
  (5.b) a central u_i is nondeterministically chosen, for some $0 \le i \le n$;
  (5.c) x_1 and x_n are inserted into u_0 and u_n, respectively;

(5.d) x_j is inserted between u_{j-2} and u_{j-1}, for all $1 < j \le i$;

(5.e) x_k is inserted between u_k and u_{k+1}, for all $i + 1 \le k < n$.

(6) Mode 6 derives v from u applying the next steps:

(6.a) $A_1, A_2, \ldots, A_n$ are deleted;

(6.b) a central u_i is nondeterministically chosen, for some $0 \le i \le n$;

(6.c) x_j is inserted between u_j and u_{j+1}, for all $j < i$;

(6.d) x_k is inserted between u_{k-2} and u_{k-1}, for all $k > i + 1$.

(7) Mode 7 obtains v from u performing the steps stated below:

(7.a) $A_1, A_2, \ldots, A_n$ are deleted;

(7.b) a central u_i is nondeterministically chosen, for some $0 \le i \le n$;

(7.c) x_j is inserted between u_{j-2} and u_{j-1}, for all $1 < j \le i$;

(7.d) x_k is inserted between u_k and u_{k+1}, for all $i + 1 \le k < n$.

(8) In mode 8, v is produced from u by following the given steps:

(8.a) $A_1, A_2, \ldots, A_n$ are deleted;

(8.b) x_1 and x_n are inserted into u_1 and u_{n-1}, respectively;

(8.c) x_i is inserted into $u_{i-1}u_i$, for all $1 < i < n$, to the right of x_{i-1} and to the left of x_{i+1}.

(9) Mode 9 derives v from u by the next procedure:

(9.a) $A_1, A_2, \ldots, A_n$ are deleted;

(9.b) x_1 and x_n are inserted into u_0 and u_n, respectively;

(9.c) x_i is inserted into $u_{i-1}u_i$, for all $1 < i < n$, to the right of x_{i-1} and to the left of x_{i+1}.

As obvious, all these jumping derivation modes reflect and formalize the above-described four-phase computation performed in a discontinuous way more adequately than their standard counterpart. Consequently, applications of these grammars are expected in any scientific area involving this kind of computation, ranging from applied mathematics through computational linguistics and compiler writing up to data mining and bioinformatics.

This section is organized as follows. It formally introduces all the new jumping derivation modes in scattered context grammars. After that, each of them is illustrated and investigated in a separate subsection. Most importantly, it is demonstrated that scattered context grammars working under any of the newly introduced derivation modes are computationally complete—that is, they characterize the family of recursively enumerable languages. Finally, it suggests four open problem areas to be discussed in the future.

5.2.1 Definitions

Let us recall notation concerning scattered context grammars (see Sect. 4.1). In this section, we formally define nine derivation modes (1) through (9), sketched in the previous introductory section.

Definition 5.2.1. Let $G = (V, T, P, S)$ be an SCG, and let ϱ be a relation over V^*. Set

$$\mathcal{L}(G, \varrho) = \{x \mid x \in T^*, S \, \varrho^* \, x\}$$

$\mathcal{L}(G, \varrho)$ is said to be the *language that G generates by* ϱ. Set

$$\mathbf{JSC}_\varrho = \{\mathcal{L}(G, \varrho) \mid G \text{ is an SCG}\}$$

$\mathbf{JSC}_\varrho$ is said to be *the language family that SCGs generate by* ϱ.   □

Definition 5.2.2. Let $G = (V, T, P, S)$ be an SCG. Next, we rigorously define the following direct derivation relations $_1\!\Rightarrow$ through $_9\!\Rightarrow$ over V^*, intuitively sketched in the previous introductory section.

First, let $(A) \to (x) \in P$ and $u = w_1 A w_2 \in V^*$. Then,

$$w_1 A w_2 \;_i\!\Rightarrow w_1 x w_2, \text{ for } i = 1, \ldots, 9$$

Second, let $(A_1, A_2, \ldots, A_n) \to (x_1, x_2, \ldots, x_n) \in P$, $u = u_0 A_1 u_1 \ldots A_n u_n$, and $u_0 u_1 \ldots u_n = v_0 v_1 \ldots v_n$, where $u_i, v_i \in V^*$, $0 \le i \le n$, for some $n \ge 2$. Then,

(1) $u_0 A_1 u_1 A_2 u_2 \ldots A_n u_n \;_1\!\Rightarrow u_0 x_1 u_1 x_2 v_2 \ldots x_n u_n$;

(2) $u_0 A_1 u_1 A_2 u_2 \ldots A_n u_n \;_2\!\Rightarrow v_0 x_1 v_1 x_2 v_2 \ldots x_n v_n$, where $u_0 z_1 = v_0$, $z_2 u_n = v_n$;

(3) $u_0 A_1 u_1 A_2 u_2 \ldots A_n u_n \;_3\!\Rightarrow v_0 x_1 v_1 x_2 v_2 \ldots x_n v_n$, where $u_0 = v_0 z_1$, $u_n = z_2 v_n$;

(4) $u_0 A_1 u_1 A_2 u_2 \ldots u_{i-1} A_i u_i A_{i+1} u_{i+1} \ldots u_{n-1} A_n u_n \;_4\!\Rightarrow$
    $u_0 u_1 x_1 u_2 x_2 \ldots u_{i-1} x_{i-1} u_{i_1} x_i u_{i_2} x_{i+1} u_{i_3} x_{i+2} u_{i+1} \ldots x_n u_{n-1} u_n$,   where $u_i = u_{i_1} u_{i_2} u_{i_3}$;

(5) $u_0 A_1 u_1 A_2 \ldots u_{i-1} A_{i-1} u_i A_i u_{i+1} \ldots A_n u_n \;_5\!\Rightarrow$
    $u_{0_1} x_1 u_{0_2} x_2 u_1 \ldots x_{i-1} u_{i-1} u_i u_{i+1} x_i \ldots u_{n_1} x_n u_{n_2}$,
    where $u_0 = u_{0_1} u_{0_2}$, $u_n = u_{n_1} u_{n_2}$;

(6) $u_0 A_1 u_1 A_2 u_2 \ldots u_{i-1} A_i u_i A_{i+1} u_{i+1} \ldots u_{n-1} A_n u_n \;_6\!\Rightarrow$
    $u_0 u_1 x_1 u_2 x_2 \ldots u_{i-1} x_{i-1} u_i x_{i+2} u_{i+1} \ldots x_n u_{n-1} u_n$;

(7) $u_0 A_1 u_1 A_2 \ldots u_{i-1} A_i u_i A_{i+1} u_{i+1} \ldots A_n u_n \;_7\!\Rightarrow$
    $u_0 x_2 u_1 \ldots x_i u_{i-1} u_i u_{i+1} x_{i+1} \ldots u_n$;

(8) $u_0 A_1 u_1 A_2 u_2 \ldots A_n u_n \;_8\!\Rightarrow v_0 x_1 v_1 x_2 v_2 \ldots x_n v_n$, where $u_0 z_1 = v_0$, $z_2 u_n = v_n$,
    $|u_0 u_1 \ldots u_{j-1}| \le |v_0 v_1 \ldots v_j|$, $|u_{j+1} \ldots u_n| \le |v_j v_{j+1} \ldots v_n|$, $0 < j < n$;

(9) $u_0 A_1 u_1 A_2 u_2 \ldots A_n u_n \;_9\!\Rightarrow v_0 x_1 v_1 x_2 v_2 \ldots x_n v_n$, where $u_0 = v_0 z_1$, $u_n = z_2 v_n$,
    $|u_0 u_1 \ldots u_{j-1}| \le |v_0 v_1 \ldots v_j|$, $|u_{j+1} \ldots u_n| \le |v_j v_{j+1} \ldots v_n|$,
    $0 < j < n$.   □

We close this section by illustrating the above-introduced notation in Definition 5.2.1. Let $G = (V, T, P, S)$ be an SCG; then, $\mathcal{L}(G, {}_5\Rightarrow) = \{x \mid x \in T^*, S \, {}_5\Rightarrow^* x\}$ and $\mathbf{JSC}_{{}_5\Rightarrow} = \{\mathcal{L}(G, {}_5\Rightarrow) \mid G$ is an SCG$\}$. To give another example, $\mathbf{JSC}_{{}_1\Rightarrow}$ denotes the family of all scattered context languages.

5.2.2 Results

This section is divided into nine subsections, each of which is dedicated to the discussion of one of the nine jumping derivation modes introduced in the previous section. More specifically, the section (1) repeats the definition of the mode in question, (2) illustrates it by an example, and (3) determines the generative power of SCGs using this mode. Most importantly, the section demonstrates that scattered context grammars working under any of these newly introduced derivation modes are computationally complete–that is, they characterize the family of recursively enumerable languages.

Let us recall Theorem 2.3.18 in Sect. 2.3.1, which fulfills an important role in the proofs throughout this section.

Jumping Derivation Mode 1

${}_1\Rightarrow$ represents, in fact, the ordinary scattered context derivation mode.

Definition 5.2.3. Let $G = (V, T, P, S)$ be an SCG. Let $u_0 A_1 u_1 \ldots A_n u_n \in V^*$ and $(A_1, A_2, \ldots, A_n) \to (x_1, x_2, \ldots, x_n) \in P$, for $n \geq 1$. Then,

$$u_0 A_1 u_1 A_2 u_2 \ldots A_n u_n \, {}_1\Rightarrow u_0 x_1 u_1 x_2 v_2 \ldots x_n u_n \qquad \square$$

Example 5.2.4. Let $G = (V, T, P, S)$ be an SCG, where $V = \{S, S', S'', S''', A, B, C \, A', B', C', a, b, c\}$, $T = \{a, b, c\}$, and P contains the following rules:

(i)	$(S) \to (aSA)$	(vii)	$(S', C) \to (cS', C')$
(ii)	$(S) \to (bSB)$	(viii)	$(S', S'') \to (\varepsilon, S''')$
(iii)	$(S) \to (cSC)$	(ix)	$(S''', A') \to (S''', a)$
(iv)	$(S) \to (S'S'')$	(x)	$(S''', B') \to (S''', b)$
(v)	$(S', A) \to (aS', A')$	(xi)	$(S''', C') \to (S''', c)$
(vi)	$(S', B) \to (bS', B')$	(xii)	$(S''' \to \varepsilon)$

Consider ${}_1\Rightarrow$. Then, the derivation of G is as follows.

First, G generates any string $w \in T^*$ to the left of S and its reversal in capital letters to the right of S with linear rules. Then, it replaces S with $S'S''$. Next, while nondeterministically rewriting nonterminal symbols to the right of S'' to their prime versions, it generates the sequence of terminals in the same order to the left of S', which we denote w'. Since all the symbols to the right of S' must be rewritten, the sequence of symbols generated to the left of S' must have the same composition of symbols. Otherwise, no terminal string can be generated, so the derivation is blocked. Thereafter, S' is erased, and S'' is rewritten to S'''. Finally, the prime versions of symbols to the right of S''' are rewritten to the terminal string denoted w''. Consequently,

$$\mathcal{L}(G, {}_1\Rightarrow) = \{x \in T^* \mid x = ww'w'', w = \text{reversal}(w''),$$

$$w' \text{ is any permutation of } w\}$$

For instance, the string $abccabcba$ is generated by G in the following way:

$$S \;{}_1\Rightarrow aSA \;{}_1\Rightarrow abSBA \;{}_1\Rightarrow abcSCBA \;{}_1\Rightarrow abcS'S''CBA \;{}_1\Rightarrow abccS'S''C'BA$$

$${}_1\Rightarrow abccaS'S''C'BA' \;{}_1\Rightarrow abccabS'S''C'B'A' \;{}_1\Rightarrow abccabS'''C'B'A'$$

$${}_1\Rightarrow abccabS'''cB'A' \;{}_1\Rightarrow abccabS'''cbA' \;{}_1\Rightarrow abccabS'''cba \;{}_1\Rightarrow abccabcba \qquad \square$$

Next, we prove that SCGs working under ${}_1\Rightarrow$ characterize **RE**.

Theorem 5.2.5 (See [FM03b]). $\mathbf{JSC}_{1\Rightarrow} = \mathbf{RE}$.

Proof. As obvious, any SCG G can be turned to a Turing machine M so M accepts $\mathcal{L}(G, {}_1\Rightarrow)$. Thus, $\mathbf{JSC}_{1\Rightarrow} \subseteq \mathbf{RE}$. Therefore, we only need to prove $\mathbf{RE} \subseteq \mathbf{JSC}_{1\Rightarrow}$.

Let $L \in \mathbf{RE}$. Express $L = h(L_1 \cap L_2)$, where h, L_1, and L_2 have the same meaning as in Theorem 2.3.18. Since L_2 is context-free, so is reversal(L_2) (see page 419 in [Woo87]). Thus, there are context-free grammars G_1 and G_2 that generate L_1 and reversal(L_2), respectively. More precisely, let $G_i = (V_i, T, P_i, S_i)$ for $i = 1, 2$. Let $T = \{a_1, \ldots, a_n\}$ and $0, 1, \$, S \notin V_1 \cup V_2$ be the new symbols. Without any loss of generality, assume that $V_1 \cap V_2 = \emptyset$. Define the new morphisms

(I) $c : a_i \mapsto 10^i 1$;

(II) $C_1 : V_1 \cup T \to V_1 \cup \Sigma \cup \{0, 1\}^*$,
$\begin{cases} A \mapsto A, & A \in V_1, \\ a \mapsto f(a), & a \in T; \end{cases}$

(III) $C_2 : V_2 \cup T \to V_2 \cup \{0, 1\}^*$,
$\begin{cases} A \mapsto A, & A \in V_2, \\ a \mapsto c(a), & a \in T; \end{cases}$

(IV) $f : a_i \mapsto h(a_i)c(a_i)$;

(V) $t : \Sigma \cup \{0, 1, \$\} \to \Sigma$,
$\begin{cases} a \mapsto a, & a \in \Sigma, \\ A \mapsto \varepsilon, & A \notin \Sigma; \end{cases}$

(VI) $t' : \Sigma \cup \{0, 1, \$\} \to \{0, 1\}$,
$\begin{cases} a \mapsto a, & a \in \{0, 1\}, \\ A \mapsto \varepsilon, & A \notin \{0, 1\}. \end{cases}$

Finally, let $G = (V, \Sigma, P, S)$ be SCG, with $V = V_1 \cup V_2 \cup \{S, 0, 1, \$\}$ and P containing the rules

(1) $(S) \to (\$S_1 1111 S_2 \$)$;

(2) $(A) \to (C_i(w))$, for all $A \to w \in P_i$, where $i = 1, 2$;

(3) $(\$, a, a, \$) \to (\varepsilon, \$, \$, \varepsilon)$, for $a = 0, 1$;

(4) $(\$) \to (\varepsilon)$.

Claim 5.2.6. $\mathcal{L}(G, {}_1\Rightarrow) = L$.

Proof. Basic Idea. First, the starting rule (1) is applied. The starting nonterminals S_1 and S_2 are inserted into the current sentential form. Then, by using the rules (2) G simulates derivations of both G_1 and G_2 and generates the sentential form $w = \$w_1 1111 w_2\$$.

Suppose $S_1 \Rightarrow^* w$, where alph$(w) \cap (N_1 \cup N_2) = \emptyset$. Recall, N_1 and N_2 denote the nonterminal alphabets of G_1 and G_2, respectively. If $t'(w_1) = $ reversal(w_2), then $t(w_1) = h(v)$, where $v \in L_1 \cap L_2$ and $h(v) \in L$. In other words, w represents a successful derivation of both G_1 and G_2, where both grammars have generated the same sentence v; therefore G must generate the sentence $h(v)$.

The rules (3) serve to check, whether the simulated grammars have generated the identical words. Binary codings of the generated words are erased while checking the equality. Always the leftmost and the rightmost symbols are erased, otherwise some symbol is skipped. If the codings do not match, some 0 or 1 cannot be erased and no terminal string can be generated.

Finally, the symbols $ are erased with the rule (4). If G_1 and G_2 generated the same sentence and both codings were successfully erased, G has generated the terminal sentence $h(v) \in L$.   □

Claim 5.2.6 implies $\mathbf{RE} \subseteq \mathbf{JSC}_{1\Rightarrow}$. Thus, Theorem 5.2.5 holds.   □

Jumping Derivation Mode 2

Definition 5.2.7. Let $G = (V, T, P, S)$ be an SCG. Let $u = u_0 A_1 u_1 \ldots A_n u_n \in V^*$ and $(A_1, A_2, \ldots, A_n) \to (x_1, x_2, \ldots, x_n) \in P$, for $n \geq 1$. Then,

$$u_0 A_1 u_1 A_2 u_2 \ldots A_n u_n \,_2\!\!\Rightarrow v_0 x_1 v_1 x_2 v_2 \ldots x_n v_n$$

where $u_0 u_1 \ldots u_n = v_0 v_1 \ldots v_n$, $u_0 z_1 = v_0$ and $z_2 u_n = v_n$, $z_1, z_2 \in V^*$.   □

Informally, by using $(A_1, A_2, \ldots, A_n) \to (x_1, x_2, \ldots, x_n) \in P$ G obtains $v = v_0 x_1 v_1 x_2 v_2 \ldots x_n v_n$ from $u = u_0 A_1 u_1 A_2 u_2 \ldots A_n u_n$ in $_2\!\!\Rightarrow$ as follows:

(1) $A_1, A_2, \ldots, A_n$ are deleted;
(2) x_1 through x_n are inserted in between u_0 and u_n.

Notice, the mutual order of inserted right-hand-side strings must be always preserved.

Example 5.2.8. Consider SCG defined in Example 5.2.4 and $_2\!\!\Rightarrow$. Context-free rules act in the same way as in $_1\!\!\Rightarrow$ unlike context-sensitive rules. Let us focus on the differences.

First, G generates the sentential form $wS'S''\overline{w}$, where $w \in T^*$ and $\overline{w}$ is the reversal of w in capital letters, with context-free derivations. Then, the nonterminals to the right of S' are rewritten to their prime versions and possibly randomly shifted closer to S', which may arbitrarily change their order. Additionally, the sequence of terminals in the same order is generated to the left of S', which we denote w'. S'

may be also shifted, however, in such case it appears to the right of S'' and future application of the rule (viii) is excluded and no terminal string can be generated. Since all the symbols to the right of S' must be rewritten, the sequence generated to the left of S' must have the same composition of symbols. Next, S' is erased and S'' is rewritten to S'''' at once, which ensures their mutual order is preserved. If any prime symbol occurs to the left of S''', it cannot be erased and the derivation is blocked. Finally, the prime versions of symbols to the right of S''' are rewritten to the terminal string denoted w'', which also enables random disordering. Consequently,

$$\mathcal{L}(G, {}_2\Rightarrow) = \{x \in T^* \mid x = ww'w'', w', w'' \text{ are any permutations of } w\}$$

For example, the string $abcacbbac$ is generated by G in the following way:

$$S \; {}_2\Rightarrow aSA \; {}_2\Rightarrow abSBA \; {}_2\Rightarrow abcSCBA \; {}_2\Rightarrow abcS'S''CBA \; {}_2\Rightarrow abcaS'S''A'CB$$

$${}_2\Rightarrow abcacS'S''A'C'B \; {}_2\Rightarrow abcacbS'S''B'A'C' \; {}_2\Rightarrow abcacbS'''B'A'C'$$

$${}_2\Rightarrow abcacbS'''B'A'c \; {}_2\Rightarrow abcacbS'''bA'c \; {}_2\Rightarrow abcacbS'''bac \; {}_2\Rightarrow abcacbbac \quad \square$$

Theorem 5.2.9. $\mathbf{JSC}_{2\Rightarrow} = \mathbf{RE}$.

Proof. Clearly $\mathbf{JSC}_{2\Rightarrow} \subseteq \mathbf{RE}$, so we only need to prove $\mathbf{RE} \subseteq \mathbf{JSC}_{2\Rightarrow}$.

Let $G = (V, \Sigma, P, S)$ be the SCG constructed in the proof of Theorem 5.2.5. First, we modify G to a new SCG G' so $\mathcal{L}(G, {}_1\Rightarrow) = \mathcal{L}(G', {}_1\Rightarrow)$. Then, we prove $\mathcal{L}(G', {}_2\Rightarrow) = \mathcal{L}(G', {}_1\Rightarrow)$.

Construction. Set

$$N = \{\lceil, \rceil, \lfloor, \rfloor, \mid, X, \underline{X}, \overline{X}, \overline{\underline{X}}, Y, \underline{Y}, \overline{Y}, \overline{\underline{Y}}\}$$

where $V \cap N = \emptyset$. Define the new morphisms

(I) $\overline{C}_1 : V_1 \cup T$,
$$\begin{cases} A \mapsto A, & A \in V_1, \\ a \mapsto \lceil f(a) \rceil \mid, & a \in T; \end{cases}$$

(II) $\overline{C}_2 : V_2 \cup T$,
$$\begin{cases} A \mapsto A & A \in V_2, \\ a \mapsto \mid \lceil c(a) \rceil, & a \in T; \end{cases}$$

(III) $b : \Sigma \cup \{0, 1, \$\} \cup N \rightarrow \{0, 1\}$,
$$\begin{cases} A \mapsto A, & A \in \{0, 1\}, \\ A \mapsto \varepsilon, & A \notin \{0, 1\}. \end{cases}$$

(IV) $\overline{t}' : \Sigma \cup \{0, 1, \$\} \cup N \rightarrow \{0, 1, \$\} \cup N$,
$$\begin{cases} A \mapsto A, & A \in \{\$\} \cup N, \\ A \mapsto t'(A), & A \notin \{\$\} \cup N. \end{cases}$$

Let $G' = (V', \Sigma, P', S)$ be SCG, with $V' = V \cup N$ and P' containing

(1) $(S) \rightarrow (\lceil X\$S_1 \lceil 11 \parallel 11 \rceil S_2 \$ Y \rceil)$;
(2) $(A) \rightarrow (\overline{C}_i(w))$ for $A \rightarrow w \in P_i$, where $i = 1, 2$;
(3) $(\lceil, X, \lceil) \rightarrow (\lfloor, \overline{\underline{X}}, \rfloor), (\lceil, Y, \lceil) \rightarrow (\lfloor, \overline{\underline{Y}}, \rfloor)$;
(4) $(\lfloor, \overline{\underline{X}}, \rfloor) \rightarrow (\lfloor, \overline{X}, \rfloor), (\lfloor, \overline{\underline{Y}}, \rfloor) \rightarrow (\lfloor, \overline{Y}, \rfloor)$;
(5) $(\$, 0, \overline{X}, \overline{Y}, 0, \$) \rightarrow (\varepsilon, \$, \underline{X}, \underline{Y}, \$, \varepsilon)$;

(6) $(\$, \overline{X}, \overline{Y}, \$) \rightarrow (\varepsilon, \overline{X}\$, \$\overline{Y}, \varepsilon)$;
(7) $(\rfloor, \overline{X}, \$, \rfloor, \lfloor, \$, \overline{Y}, \lfloor) \rightarrow (\varepsilon, \varepsilon, \varepsilon, \underline{X}\$, \$\underline{Y}, \varepsilon, \varepsilon, \varepsilon)$;
(8) $(\underline{X}, 1, 1, \rfloor, \lfloor, 1, 1, \underline{Y}) \rightarrow (\varepsilon, \varepsilon, \varepsilon, X, Y, \varepsilon, \varepsilon, \varepsilon)$;
(9) $(\$) \rightarrow (\varepsilon), (X) \rightarrow (\varepsilon), (Y) \rightarrow (\varepsilon)$.

Notice that X and Y hold the current state of computation and force the context-sensitive rules to be used in the following order:

(a) after applying the rule (3), only the rule (4) may be applied;
(b) after applying the rule (4), only the rule (5) or (6) may be applied;
(c) after applying the rule (5), only the rule (4) may be applied;
(d) after applying the rule (6), only the rule (7) may be applied;
(e) after applying the rule (7), only the rule (8) may be applied;
(f) after applying the rule (8), only the rule (3) may be applied.

Claim 5.2.10. $\mathcal{L}(G', {}_1\Rightarrow) = \mathcal{L}(G, {}_1\Rightarrow)$.

Proof. The context-free rules (1) and (2) of G' correspond one to one to the rules (1) and (2) of G, only the codings of terminals contain additional symbols. Thus, for every derivation in G

$$S\,{}_1\Rightarrow^* \$v_1 1111 v_2 \$ = v$$

where v is generated by using the rules (1) and (2) and $\mathrm{alph}(v) \cap (N_1 \cup N_2) = \emptyset$, there is

$$S\,{}_1\Rightarrow^* \rceil X\$w_1 \lceil 11 \;||\; 11 \rceil w_2 \$ Y \lceil = w$$

in G' generated by the rules (1) and (2), where $b(w_1) = t'(v_1)$, $b(w_2) = v_2$. This also holds vice versa. Since such a sentential form represents a successful derivations of both G_1 and G_2, without any loss of generality, we can consider it in every successful derivation of either G, or G'. Additionally, in G

$$v\,{}_1\Rightarrow^* v', v' \in \Sigma^*$$

if and only if $t'(v_1) = \mathrm{reversal}(v_2)$. Note, $v' = t(v)$. Therefore, we have to prove

$$w\,{}_1\Rightarrow^* w', w' \in \Sigma^*$$

if and only if $\overline{t'}(w_1) = \mathrm{reversal}(w_2)$. Then, $v' = w'$.

Claim 5.2.11. In G', for

$$S\,{}_1\Rightarrow^* \rceil X\$w_1 \lceil 11 \;||\; 11 \rceil w_2 \$ Y \lceil = w, \mathrm{alph}(w) \cap (N_1 \cup N_2) = \emptyset$$

where w is generated by using the rules (1) and (2),

$$w_1 \Rightarrow^* w'$$

where $w' \in \Sigma^*$ if and only if $\bar{t}'(w_1) = \text{reversal}(w_2)$.

For the sake of readability, in the next proof we omit all symbols from Σ in w_1—that is, we consider only nonterminal symbols, which are to be erased.

Proof. If. Suppose $w_1 = \text{reversal}(w_2)$, then $w_1 \Rightarrow^* \varepsilon$. From the construction of G', $w_1 = (\lceil 10^{i_1} 1 \rceil \;|)(\lceil 10^{i_2} 1 \rceil \;|) \ldots (\lceil 10^{i_n} 1 \rceil \;|)$, where $i_j \in \{1, \ldots, |\Sigma|\}$, $1 \le j \le n$, $n \ge 0$. Consider two cases—(I) $n = 0$ and (II) $n \ge 1$.

(I) If $n = 0$, $w = \rceil X\$ \lceil 11 \;||\; 11 \rceil \$ Y \lceil$. Then, by using the rules (3) and (4), the rules (7) and (8), and four times the rules (9), we obtain

$$
\begin{aligned}
&\rceil X\$\lceil 11 \;||\; 11\rceil\$Y\lceil_1 \Rightarrow \rfloor\overline{X}\$\rfloor 11 \;||\; 11\rceil\$Y\lceil_1 \Rightarrow \\
&\rfloor\overline{X}\$\rfloor 11 \;||\; 11\lfloor\$\overline{Y}\lfloor_1 \Rightarrow \rfloor\overline{X}\$\rfloor 11 \;||\; 11\lfloor\$\overline{Y}\lfloor_1 \Rightarrow \\
&\rfloor\overline{X}\$\rfloor 11 \;||\; 11\lfloor\$\overline{Y}\lfloor_1 \Rightarrow \underline{X}\$ 11 \;||\; 11\$\underline{Y} \quad_1 \Rightarrow \\
&\$XY\$_1 \Rightarrow XY\$_1 \Rightarrow Y\$_1 \Rightarrow \$_1 \Rightarrow \varepsilon
\end{aligned}
$$

and the claim holds.

(II) Let $n \ge 1$,

$$
\begin{aligned}
w &= \rceil X\$\lceil 10^{i'} 1 \rceil \;|\; (\lceil 10^{i_m} 1 \rceil \;|)^k \lceil 11 \;||\; 11 \rceil (|\; \lceil 10^{j_{m'}} 1 \rceil)^k \;|\; \lceil 10^{j'} 1 \rceil \$ Y \lceil \\
&= \rceil X\$\lceil 10^{i'} 1 \rceil \;|\; u \;|\; \lceil 10^{j'} 1 \rceil \$ Y \lceil
\end{aligned}
$$

where $k \ge 0$, $m, m' \in \{1, \ldots, k\}$, $i', i_m, j', j_{m'} \in \{1, \ldots, |\Sigma|\}$. Sequentially using both rules (3) and (4) and the rule (7) we obtain the derivation

$$
\begin{aligned}
&\rceil X\$\lceil 10^{i'} 1 \rceil \;|\; u \;|\; \lceil 10^{j'} 1 \rceil \$Y\lceil_1 \Rightarrow \rfloor\overline{X}\$\rfloor 10^{i'} 1 \rceil \;|\; u \;|\; \lceil 10^{j'} 1 \rceil \$Y\lceil_1 \Rightarrow \\
&\rfloor\overline{X}\$\rfloor 10^{i'} 1 \rceil \;|\; u \;|\; \lceil 10^{j'} 1\lfloor\$\overline{Y}\lfloor_1 \Rightarrow \rfloor\overline{X}\$\rfloor 10^{i'} 1 \rceil \;|\; u \;|\; \lceil 10^{j'} 1\lfloor\$\overline{Y}\lfloor_1 \Rightarrow \\
&\rfloor\overline{X}\$\rfloor 10^{i'} 1 \rceil \;|\; u \;|\; \lceil 10^{j'} 1\lfloor\$\overline{Y}\lfloor_1 \Rightarrow \underline{X}\$ 10^{i'} 1 \rceil \;|\; u \;|\; \lceil 10^{j'} 1\$\underline{Y}
\end{aligned}
$$

Next, we prove

$$w' = \underline{X}\$ 10^{i'} 1\rceil \;|\; (\lceil 10^{i_m} 1 \rceil \;|)^k \lceil 11 \;||\; 11 \rceil (|\; \lceil 10^{j_{m'}} 1 \rceil)^k \;|\; \lceil 10^{j'} 1\$\underline{Y}_1 \Rightarrow^* \varepsilon$$

by induction on $k \ge 0$.

Basis. Let $k = 0$. Then,

$$w' = \underline{X}\$ 10^{i'} 1\rceil \;|\; \lceil 11 \;||\; 11 \rceil \;|\; \lceil 10^{j'} 1\$\underline{Y}$$

By using a rule (8) and twice a rule (3) G' performs

$$\underline{X}\$10^{i'}1\rceil \mid \lceil 11 \parallel 11 \rceil \mid \lceil 10^{j'}1\$\underline{Y}_{\,1}\Rightarrow \$0^{i'}\rceil X\lceil 11 \parallel 11 \rceil Y\lceil 0'\$$$
$$_{1}\Rightarrow \$0^{i'}\rfloor \overline{X}\rfloor 11 \parallel 11\rceil Y\lceil 0'\$ \qquad _{1}\Rightarrow \$0^{i'}\rfloor \overline{X}\rfloor 11 \parallel 11\lfloor \overline{Y}\lfloor 0'\$$$

Since $i' = j'$, both sequences of 0s are simultaneously erased by repeatedly using both rules (4) and the rule (5). Observe that

$$\$0^{i'}\rfloor \overline{X}\rfloor 11 \parallel 11\lfloor \overline{Y}\lfloor 0'\$ \,_{1}\Rightarrow^{*}\$\rfloor \overline{X}\rfloor 11 \parallel 11\lfloor \overline{Y}\lfloor \$$$

Finally, by applying the rules (4), (6), (7), (8), and (9), we finish the derivation as

$$\$\rfloor \overline{X}\rfloor 11 \parallel 11\lfloor \overline{Y}\lfloor \$ \,_{1}\Rightarrow \rfloor \overline{X}\$\rfloor 11 \parallel 11\lfloor \$\overline{Y}\lfloor \,_{1}\Rightarrow$$
$$\underline{X}\$11 \parallel 11\$\underline{Y}\,_{1}\Rightarrow \$XY\$\,_{1}\Rightarrow^{*}\varepsilon$$

and the basis holds.

Induction Hypothesis. Suppose there exists $k \geq 0$ such that

$$w' = \underline{X}\$10^{i'}1\rceil \mid (\lceil 10^{i_m}1\rceil \mid)^{l}\lceil 11 \parallel 11\rceil(\mid \lceil 10^{j_{m'}}1\rceil)^{l} \mid \lceil 10^{j'}1\$\underline{Y}\,_{1}\Rightarrow^{*}\varepsilon$$

where $m, m' \in \{1, \ldots, l\}$, $i', i_m, j', j_{m'} \in \{1, \ldots, |\Sigma|\}$, for all $0 \leq l \leq k$.

Induction Step. Consider any

$$w' = \underline{X}\$10^{i'}1\rceil \mid (\lceil 10^{i_m}1\rceil \mid)^{k+1}\lceil 11 \parallel 11\rceil(\mid \lceil 10^{j_{m'}}1\rceil)^{k+1} \mid \lceil 10^{j'}1\$\underline{Y}$$

where $m, m' \in \{1, \ldots, k+1\}$, $i', i_m, j', j_{m'} \in \{1, \ldots, |\Sigma|\}$. Since $k+1 \geq 1$

$$w' = \underline{X}\$10^{i'}1\rceil \mid \lceil 10^{i''}1\rceil \mid u \mid \lceil 10^{j''}1\rceil \mid \lceil 10^{j'}1\$\underline{Y}$$
$$u = (\lceil 10^{i_m}1\rceil \mid)^{k}\lceil 11 \parallel 11\rceil(\mid \lceil 10^{j_{m'}}1\rceil)^{k}$$

By using the rule (8) and both rules (3) G' performs

$$\underline{X}\$10^{i'}1\rceil \mid \lceil 10^{i''}1\rceil \mid u \mid \lceil 10^{j''}1\rceil \mid \lceil 10^{j'}1\$\underline{Y}\,_{1}\Rightarrow$$
$$\$0^{i'}\rceil X\lceil 10^{i''}1\rceil \mid u \mid \lceil 10^{j''}1\rceil Y\lceil 0'\$ \qquad _{1}\Rightarrow$$
$$\$0^{i'}\rfloor \overline{X}\rfloor 10^{i''}1\rceil \mid u \mid \lceil 10^{j''}1\rceil Y\lceil 0'\$ \qquad _{1}\Rightarrow$$
$$\$0^{i'}\rfloor \overline{X}\rfloor 10^{i''}1\rceil \mid u \mid \lceil 10^{j''}1\lfloor \overline{Y}\lfloor 0'\$$$

Since $i' = j'$, the prefix of 0s and the suffix of 0s are simultaneously erased by repeatedly using the rules (4) and the rule (5).

$$\$0^{i'}\rfloor \overline{X}\rfloor 10^{i''}1\rceil \mid u \mid \lceil 10^{j''}1\lfloor \overline{Y}\lfloor 0'\$ \,_{1}\Rightarrow^{*}\$\rfloor \overline{X}\rfloor 10^{i''}1\rceil \mid u \mid \lceil 10^{j''}1\lfloor \overline{Y}\lfloor \$$$

Finally, G' uses the rule (6) and the rule (7)

$$\$\lfloor\overline{X}\rfloor 10^{i''}1\rceil \mid u \mid \lceil 10^{i''}1\lfloor\overline{Y}\lfloor\$ \,_1 \Rightarrow \lfloor\overline{X}\$\rfloor 10^{i''}1\rceil \mid u \mid \lceil 10^{i''}1\lfloor\$\overline{Y}\lfloor\,_1 \Rightarrow$$

$$\underline{X}\$10^{i''}1\rceil \mid u \mid \lceil 10^{i''}1\$\underline{Y} = w''$$

where

$$w'' = \underline{X}\$10^{i''}1\rceil \mid (\lceil 10^{im}1\rceil \mid)^k \lceil 11 \mid\mid 11\rceil(\mid \lceil 10^{im'}1\rceil)^k \mid \lceil 10^{i''}1\$\underline{Y}$$

By the induction hypothesis, $w''_1 \Rightarrow^* \varepsilon$, which completes the proof.

Only If. Suppose that $w_1 \neq \text{reversal}(w_2)$, then there is no w' satisfying $w_1 \Rightarrow^* w'$ and $w' = \varepsilon$.

From the construction of G', there is no rule shifting the left $\$$ to the left and no rule shifting the right $\$$ to the right. Since the rule (5) is the only one erasing 0s and these 0s must occur between two $\$$s, if there is any 0, which is not between the two $\$$s, it is unable to be erased. Moreover, an application of the rule (5) moves the left $\$$ on the previous position of erased left 0; if it is not the leftmost, the derivation is blocked. It is symmetric on the right. A similar situation is regarding 1s, X, and Y. Thus, for the sentential form w, if 0 or 1 is the rightmost or the leftmost symbol of w, no terminal string can be generated.

Since $w_1 \neq \text{reversal}(w_2)$, the codings of terminal strings generated by G_1 and G_2 are different. Then, there is a and a', where $w_1 = vau$, $w_2 = u'a'\,\text{reversal}(v)$, and $a \neq a'$. For always the outermost 0 or 1 is erased, otherwise the derivation is blocked, suppose the derivation correctly erases both strings v, so a and a' are the outermost symbols. The derivation can continue in the following two ways.

(I) Suppose the outermost 0s are erased before the outermost 1s. Then, the rule (5) is used, which requires $\overline{X}$ and $\overline{Y}$ between the previous positions of 0s. However, there is 1, a or a', which is not between X and Y.

(II) Suppose the outermost 1s are erased before the outermost 0s. Then, the rule (8) is used, which requires $\underline{X}$ and $\underline{Y}$ in the current sentential form. The symbols $\underline{X}$ and $\underline{Y}$ are produced by the rule (7), which requires X and $\$$ between two symbols $\rfloor$ and Y and $\$$ between two symbols $\lfloor$. Suppose w' is the current sentential form. Since w_1 or $\text{reversal}(w_2)$ is of the form

$$\ldots \lceil 10^{i_0}1\rceil \mid \lceil 10^{i_1}1\rceil \mid \lceil 10^{i_2}1\rceil \mid \ldots$$

where $i_0, i_1, i_2 \in \{1, \ldots, |\Sigma|\}$, there is 0 as the leftmost or rightmost symbol of w' and $X\$$ and $\$Y$ occurs between $\rfloor$s and $\lfloor$s, respectively. However, this 0 is obviously not between the two $\$$ and remains permanently in the sentential form.

We showed that G' can generate the terminal string from the sentential form w if and only if $\bar{t}'(w_1) = \text{reversal}(w_2)$, so the claim holds. $\qquad\square$

We proved that for any $w \in \Sigma^*$, $S_1 \Rightarrow^* w$ in G if and only if $S_1 \Rightarrow^* w$ in G', and Claim 5.2.10 holds.   □

Let us turn to $_2\Rightarrow$.

Claim 5.2.12. $\mathcal{L}(G', _2\Rightarrow) = \mathcal{L}(G', _1\Rightarrow)$.

Proof. In $_2\Rightarrow$, applications of context-free rules progress in the same way as in $_1\Rightarrow$. While using context-sensitive rules inserted right-hand-side strings can be nondeterministically scattered between the previous positions of the leftmost and rightmost affected nonterminals, only their order is preserved. We show, we can control this by the construction of G'.

Recall the observations made at the beginning of the proof of Claim 5.2.10. Since the behaviour of context-free rules remains unchanged in terms of $_2\Rightarrow$, these still hold true. It remains to prove that Claim 5.2.11 also holds in $_2\Rightarrow$.

In a special case, $_2\Rightarrow$ behave exactly as $_1\Rightarrow$, hence definitely $\mathcal{L}(G', _1\Rightarrow) \subseteq \mathcal{L}(G', _2\Rightarrow)$. We prove

$$w \notin \mathcal{L}(G', _1\Rightarrow) \Rightarrow w \notin \mathcal{L}(G', _2\Rightarrow)$$

Therefore, to complete the proof of Claim 5.2.12, we establish the following claim.

Claim 5.2.13. In G', for

$$S_1 \Rightarrow^* \rceil X\$w_1 \lceil 11 \;\|\; 11 \rceil w_2 \$Y \lceil = w, \text{alph}(w) \cap (N_1 \cup N_2) = \emptyset$$

where w is generated only by using the rules (1) and (2), and $\overline{t'}(w_1) \neq \text{reversal}(w_2)$, there is no w', where

$$w_1 \Rightarrow^* w', w' \in \Sigma^*$$

For the sake of readability, in the next proof we omit all symbols from Σ in w_1—we consider only nonterminal symbols, which are to be erased.

Proof. Suppose any w, where

$$S_1 \Rightarrow^* w = \rceil X\$w_1 \lceil 11 \;\|\; 11 \rceil w_2 \$Y \lceil$$

in G' and w is generated by using the rules (1) and (2), $\text{alph}(w) \cap (N_1 \cup N_2) = \emptyset$, and $w_1 \neq \text{reversal}(w_2)$.

From the construction of G', there is no rule shifting the left \$ to the left and no rule shifting the right \$ to the right. Neither $_2\Rightarrow$ can do this. Since the rule (5) is the only one erasing 0s and these 0s must be between two \$s, if there is any 0, which is not between the two \$s, it cannot be erased. A similar situation is regarding 1s, X, and Y. Thus, for the sentential form w, if 0 or 1 is the outermost symbol of w, no terminal string can be generated.

Consider two cases (I) $w_1 = \varepsilon$ or $w_2 = \varepsilon$ and (II) $w_1 \neq \varepsilon$ and $w_2 \neq \varepsilon$.

(I) Suppose the condition does not apply. Without any loss of generality, suppose $w_1 = \varepsilon$. Since $w_1 \neq \text{reversal}(w_2)$, $w_2 \neq \varepsilon$. Then,

$$w = \,]X\$\lceil 11 \parallel 11\rceil (\mid \lceil 10^{i_m} 1 \rceil)^k \mid \lceil 10^{i'} 1 \rceil \$ Y\lceil$$

where $k \geq 0$, $m \in \{1, \ldots, k\}$, $i_m, i' \in \{1, \ldots, |\Sigma|\}$.

First, the rules (3) and (9) are the only applicable, however, application of the rule (9) would block the derivation, so we do not consider it. While rewriting X, the leftmost $]$ is rewritten. Unless the leftmost $\lceil$ is chosen, it becomes unpaired and, thus, cannot be erased. It is symmetric with Y. After the application of the rules (3), the rules (4) becomes applicable. The positions of the symbols $\$$ must be preserved for future usage of the rule (7). Then, the only way of continuing a successful derivation is

$$\begin{aligned}
&]X\$\lceil 11 \parallel 11\rceil (\mid \lceil 10^{i_m} 1 \rceil)^k \mid \lceil 10^{i'} 1 \rceil \$ Y\lceil \, {}_2\Rightarrow \\
&]\underline{X}\$]11 \parallel 11\rceil (\mid \lceil 10^{i_m} 1 \rceil)^k \mid \lceil 10^{i'} 1 \rceil \$ Y\lceil \, {}_2\Rightarrow \\
&]\overline{\underline{X}}\$]11 \parallel 11\rceil (\mid \lceil 10^{i_m} 1 \rceil)^k \mid \lceil 10^{i'} 1\lfloor\$\overline{\underline{Y}}\lfloor \, {}_2\Rightarrow \\
&]\overline{X}\$]11 \parallel 11\rceil (\mid \lceil 10^{i_m} 1 \rceil)^k \mid \lceil 10^{i'} 1\lfloor\$\overline{\underline{Y}}\lfloor \, {}_2\Rightarrow \\
&]\overline{X}\$]11 \parallel 11\rceil (\mid \lceil 10^{i_m} 1 \rceil)^k \mid \lceil 10^{i'} 1\lfloor\$\overline{Y}\lfloor
\end{aligned}$$

Notice that if neighboring nonterminals are rewritten, ${}_2\Rightarrow$ do not shift any symbol.

Next, the rule (7) is the only applicable possibly shifting $\underline{X}$, $\underline{Y}$, and $\$$s anywhere into the current sentential form. However, if any shift is performed, there is a symbol 1 as the outer most symbol, which is obviously unable to be erased.

Thus,

$$]\overline{X}\$]11 \parallel 11\rceil (\mid \lceil 10^{i_m} 1 \rceil)^k \mid \lceil 10^{i'} 1\lfloor\$\overline{Y}\lfloor {}_2\Rightarrow$$

$$\underline{X}\$11 \parallel 11\rceil (\mid \lceil 10^{i_m} 1 \rceil)^k \mid \lceil 10^{i'} 1\$\underline{Y} = w'$$

Next, consider two cases depending on k.

(I.i) Suppose $k = 0$. Then,

$$w' = \underline{X}\$11 \parallel 11\rceil \mid \lceil 10^{i'} 1\$\underline{Y}$$

Since $i' > 0$, the rule (5) must be used. It requires presence of $\overline{X}$ and $\overline{Y}$ in the current sentential form. These can be obtained only by application of the rule (8) and both rules from (3) and (4). However, it must rewrite two pairs of $]$, $\lceil$, but there is only one remaining. Therefore, there are i' symbols 0, which cannot be erased, and no terminal string can be generated.

(I.ii) Suppose $k > 0$. Then, w' is of the form

$$\underline{X}\$11 \parallel 11\rceil \mid \lceil u\rceil \mid \lceil 10^{i'} 1\$\underline{Y}$$

The rule (8) is the only applicable. It rewrites $\underline{X}$ to X, $\underline{Y}$ to Y and put them potentially anywhere into the current sentential form. However, the rules (3), which are the only containing X and Y on the left-hand side, require X and Y situated between $\rceil$ and $\lceil$.

$$\underline{X}\$11 \parallel 11\rceil \mid \lceil u\rceil \mid \lceil 10^{i'} 1\$\underline{Y} \;_2\Rightarrow \$11 \parallel 11\rceil X\lceil u\rceil Y\lceil 0^{i'}\$$$

Without any loss of generality, we omit other possibilities of erasing the symbols $\mid$ or 1, because the derivation would be blocked in the same way. Since there is no 0 to the left of X, the future application of the rule (5) is excluded and the rightmost sequence of 0s is obviously skipped and cannot be erased any more.

(II) Suppose the condition applies. Then,

$$
\begin{aligned}
w &= \rceil X\$\lceil 10^i 1\rceil \mid (\lceil 10^{j_m} 1\rceil \mid)^k\lceil 11 \parallel 11\rceil(\mid \lceil 10^{j_{m'}} 1\rceil)^{k'} \mid \lceil 10^{i'} 1\rceil\$Y\lceil \\
&= \rceil X\$\lceil 10^i 1\rceil \mid \lceil u\rceil \mid \lceil 10^{i'} 1\rceil\$Y\lceil
\end{aligned}
$$

where $k, k' \geq 0$, $m \in \{1, \ldots, k\}$, $m' \in \{1, \ldots, k'\}$, $i_m, i'_m, j, j' \in \{1, \ldots, |\Sigma|\}$.

First, the situation is completely the same as in (I), the only possibly non-blocking derivation consists of application of both rules (3) and (4) followed by application of the rule (7). No left-hand-side string may be shifted during the application of these rules or the derivation is blocked.

$$
\begin{aligned}
&\rceil X\$\lceil 10^i 1\rceil \mid \lceil u\rceil \mid \lceil 10^{i'} 1\rceil\$Y\lceil \;_2\Rightarrow \rceil\underline{X}\$\rfloor 10^i 1\rceil \mid \lceil u\rceil \mid \lceil 10^{i'} 1\rceil\$Y\lceil \;_2\Rightarrow \\
&\rceil\underline{X}\$\rfloor 10^i 1\rceil \mid \lceil u\rceil \mid \lceil 10^{i'} 1\lfloor\$\underline{Y}\lfloor \;_2\Rightarrow \rceil\underline{X}\$\rfloor 10^i 1\rceil \mid \lceil u\rceil \mid \lceil 10^{i'} 1\lfloor\$\underline{Y}\lfloor \;_2\Rightarrow \\
&\rceil\underline{X}\$\rfloor 10^i 1\rceil \mid \lceil u\rceil \mid \lceil 10^{i'} 1\lfloor\$\overline{Y}\lfloor \;_2\Rightarrow \underline{X}\$10^i 1\rceil \mid \lceil u\rceil \mid \lceil 10^{i'} 1\$\underline{Y}
\end{aligned}
$$

Next, the rule (8) is the only applicable rule, which erases four symbols 1, two $\mid$, rewrites $\underline{X}$ to X and $\underline{Y}$ to Y, and inserts them possibly anywhere into the current sentential form. However, X must be inserted between $\rceil$ and $\lceil$, otherwise the rule (3) is not applicable and X remains permanently in the sentential form. Unless the leftmost pair of $\rceil$ and $\lceil$ is chosen, there are skipped symbols 1 remaining to the left of X. The rules (6) and (7) ensures the derivation is blocked, if X is shifted to the right. Additionally, the only way to erase 1s is the rule (8), but these 1s must be to the right of X. Thus, the skipped symbols 1 cannot be erased. Therefore, the pair of $\rceil$ and $\lceil$ is the leftmost or the derivation is blocked. Moreover, the two erased 1s are also the leftmost or they cannot be erased in the future and the same holds for the left erased symbol $\mid$. A similar situation is regarding Y. Then,

$$\underline{X}\$10^i1\rceil \mid \lceil u\rceil \mid \lceil 10^{i'}1\$\underline{Y}\ _2 \Rightarrow \$0^i\rceil X\lceil u\rceil Y\lceil 0^{i'}\$$$

and by using the rules (3) and repeatedly the rules (4) and (5) both outer most sequences of 0s can be erased, if $i = i'$. Additionally, the rules (4) ensure, X and Y are never shifted. If there is any 0 skipped, it cannot be erased and the derivation is blocked.

$$\$0^i\rceil X\lceil u\rceil Y\lceil 0^{i'}\$\ _2\Rightarrow^* \$0^i\rfloor\overline{X}\rfloor u\lfloor\overline{Y}\lfloor 0^{i'}\$\ _2\Rightarrow^* \$\rfloor\overline{X}\rfloor u\lfloor\overline{Y}\lfloor\$$$

Finally, by the rules (6) and (7) both terminal codings can be completely erased and $\underline{X}$, $\underline{Y}$, and two \$ are the outermost symbols, if no symbol is skipped.

$$\$\rfloor\overline{X}\rfloor u\lfloor\overline{Y}\lfloor\$\ _2\Rightarrow \rfloor X\$\rfloor u\lfloor\$Y\lfloor\ _2\Rightarrow \underline{X}\$u\$\underline{Y}$$

Since $w_1 \neq w_2$, $w_1 = vau$ and $w_2 = u'a'reversall(v)$, where $a \neq a'$ are the outermost non-identical terminal codings. Derivation can always erase vs, as it was described, or be blocked before. Without any loss of generality, we have to consider two cases.

(II.i) Suppose $au = \varepsilon$. Then, $u'a' \neq \varepsilon$ and the situation is the same as in (I), no terminal string can be generated and the derivation is blocked.

(II.ii) Suppose $au \neq \varepsilon$, $u'a' \neq \varepsilon$. If the derivation is not blocked before, it may generate the sentential form

$$\$0^i\rceil X\lceil u\rceil Y\lceil 0^{i'}\$$$

where $10^i1 = a$, $10^{i'}1 = a'$. Then, $i \neq i'$ and while simultaneously erasing the sequences of 0s of both codings, one is erased before the second one. The rule (5) becomes inapplicable and there is no way not to skip the remaining part of the second sequence of 0s. The derivation is blocked.

We covered all possibilities and showed, there is no way to generate terminal string $w' \notin \mathcal{L}(G',\ _1\Rightarrow)$, and the claim holds. □

Since $S\ _1\Rightarrow^*w$, $w \in \Sigma^*$ if and only if $S\ _2\Rightarrow^*w$, Claim 5.2.12 holds. □

We proved that $\mathcal{L}(G',\ _2\Rightarrow)=\mathcal{L}(G',\ _1\Rightarrow)$, $\mathcal{L}(G',\ _1\Rightarrow)=\mathcal{L}(G,\ _1\Rightarrow)$, and $\mathcal{L}(G,\ _1\Rightarrow) = L$, then $\mathcal{L}(G',\ _2\Rightarrow) = L$, so the proof of Theorem 5.2.9 is completed. □

Jumping Derivation Mode 3

Definition 5.2.14. Let $G = (V, T, P, S)$ be an SCG. Let $u = u_0A_1u_1 \ldots A_nu_n \in V^*$ and $(A_1,A_2,\ldots,A_n) \to (x_1,x_2,\ldots,x_n) \in P$, for $n \geq 1$. Then,

$$u_0 A_1 u_1 A_2 u_2 \ldots A_n u_n \; {}_3\!\Rightarrow v_0 x_1 v_1 x_2 v_2 \ldots x_n v_n$$

where $u_0 u_1 \ldots u_n = v_0 v_1 \ldots v_n$, $u_0 = v_0 z_1$ and $u_n = z_2 v_n$, $z_1, z_2 \in V^*$. □

Informally, G obtains $v = v_0 x_1 v_1 x_2 v_2 \ldots x_n v_n$ from $u = u_0 A_1 u_1 A_2 u_2 \ldots A_n u_n$ by $(A_1, A_2, \ldots, A_n) \to (x_1, x_2, \ldots, x_n) \in P$ in terms of ${}_3\!\Rightarrow$ as follows:

(1) $A_1, A_2, \ldots, A_n$ are deleted;
(2) x_1 and x_n are inserted into u_0 and u_n, respectively;
(3) x_2 through x_{n-1} are inserted in between the newly inserted x_1 and x_n.

Example 5.2.15. Let $G = (V, T, P, S)$, where $V = \{S, A, \$, a, b\}$, $T = \{a, b\}$, be an SCG with P containing the following rules:

(i)	$(S) \to (A\$)$	(iv) $(A) \to (\varepsilon)$
(ii)	$(A) \to (aAb)$	(v) $(\$) \to (\varepsilon)$
(iii)	$(A, \$) \to (A, \$)$	

Consider G uses ${}_3\!\Rightarrow$. Notice that context-free rules are not influenced by ${}_3\!\Rightarrow$.

After applying starting rule (i), G generates $a^n b^n$, where $n \geq 0$, by using the rule (ii) or finishes the derivation with rules (iv) and (v). However, at any time during the derivation the rule (iii) can be applied. It inserts or erases nothing, but it potentially shifts A to the left. Notice, the symbol $\$$ is always the rightmost and, thus, cannot be shifted. Then,

$$\mathcal{L}(G, {}_3\!\Rightarrow) = \{x \in T^* \mid x = \varepsilon \text{ or } x = uvwb^n, uw = a^n, n \geq 0,$$

and v is defined recursively as $x\}$

For example, the string $aaaababbabbb$ is generated by G in the following way:

$$S \; {}_3\!\Rightarrow A\$ \; {}_3\!\Rightarrow aAb\$ \; {}_3\!\Rightarrow aaAbb\$ \; {}_3\!\Rightarrow aaaAbbb\$ \; {}_3\!\Rightarrow aaAabbb\$$$

$${}_3\!\Rightarrow aaaAbabbb\$ \; {}_3\!\Rightarrow aaaaAbbabbb\$ \; {}_3\!\Rightarrow aaaAabbabbb\$$$

$${}_3\!\Rightarrow aaaaAbabbabbb\$ \; {}_3\!\Rightarrow aaaababbabbb\$ \; {}_3\!\Rightarrow aaaababbabbb$$ □

Theorem 5.2.16. $\mathbf{JSC}_{3\Rightarrow} = \mathbf{RE}$.

Proof. Clearly $\mathbf{JSC}_{3\Rightarrow} \subseteq \mathbf{RE}$, so we only need to prove $\mathbf{RE} \subseteq \mathbf{JSC}_{3\Rightarrow}$.

Let $G = (V, \Sigma, P, S)$ be the SCG constructed in the proof of Theorem 5.2.5. Next, we modify G to a new SCG G' satisfying $\mathcal{L}(G, {}_1\!\Rightarrow) = \mathcal{L}(G', {}_1\!\Rightarrow)$. Finally, we prove $\mathcal{L}(G', {}_3\!\Rightarrow) = \mathcal{L}(G', {}_1\!\Rightarrow)$.

Construction. Let $G' = (V, \Sigma, P', S)$ be SCG with P' containing

(1) $(S) \to (S_1 11\$\$11 S_2)$;
(2) $(A) \to (C_i(w))$ for $A \to w \in P_i$, where $i = 1, 2$;

(3) $(a, \$, \$, a) \rightarrow (\$, \varepsilon, \varepsilon, \$)$, for $a = 0, 1$;
(4) $(\$) \rightarrow (\varepsilon)$.

We establish the proof of Theorem 5.2.16 by demonstrating the following two claims.

Claim 5.2.17. $\mathcal{L}(G', {}_1\Rightarrow) = \mathcal{L}(G, {}_1\Rightarrow)$.

Proof. G' is closely related to G, only the rules (1) and (3) are slightly modified. As a result the correspondence of the sentences generated by the simulated G_1, G_2, respectively, is not checked in the direction from the outermost to the central symbols but from the central to the outermost symbols. Again, if the current two symbols do not match, they cannot be erased both and the derivation blocks.   □

Claim 5.2.18. $\mathcal{L}(G', {}_3\Rightarrow) = \mathcal{L}(G', {}_1\Rightarrow)$.

Proof. Without any loss of generality, we can suppose the rules (1) and (2) are used only before the first usage of the rule (3). The context-free rules work unchanged with ${}_3\Rightarrow$. Then, for every derivation

$$S\ {}_1\Rightarrow^* w = w_1 11\$\$11 w_2$$

generated only by the rules (1) and (2), where $\mathrm{alph}(w) \cap (N_1 \cup N_2) = \emptyset$, there is the identical derivation

$$S\ {}_3\Rightarrow^* w$$

and vice versa. Since

$$w\ {}_1\Rightarrow^* w', w' \in \Sigma^*$$

if and only if $t'(w_1) = \mathrm{reversal}(w_2)$, we can complete the proof of the previous claim by the following one.

Claim 5.2.19. Let the sentential form w be generated only by the rules (1) and (2). Without any loss of generality, suppose $\mathrm{alph}(w) \cap (N_1 \cup N_2) = \emptyset$. Consider

$$S\ {}_3\Rightarrow^* w = w_1 11\$\$11 w_2$$

Then, $w\ {}_3\Rightarrow^* w'$, where $w' \in \Sigma^*$ if and only if $t'(w_1) = \mathrm{reversal}(w_2)$.

For better readability, in the next proof we omit all symbols of w_1 from Σ—we consider only nonterminal symbols, which are to be erased.

Basic Idea. The rules (3) are the only with 0s and 1s on their left-hand sides. These symbols are simultaneously erasing to the left and to the right of $\$$s checking the equality. While proceeding from the center to the edges, when there is any symbol skipped, which is remaining between $\$$s, there is no way, how to erase it, and no terminal string can be generated.

Consider $_3\Rightarrow$. Even when the symbols are erasing one after another from the center to the left and right, $_3\Rightarrow$ can potentially shift the left \$ to the left and the right \$ to the right skipping some symbols. Also in this case the symbols between \$s cannot be erased anymore.

Proof. If. Recall

$$w = 10^{m_1}110^{m_2}1\ldots10^{m_l}111\$\$1110^{m_l}1\ldots10^{m_2}110^{m_1}1$$

Suppose the check works properly not skipping any symbol. Then,

$$w\ _3\Rightarrow^* w' = \$\$$$

and twice applying the rule (4) the derivation finishes.   □

Proof. Only If. If $w_1 \neq$ reversal(w_2), though the check works properly,

$$w\ _1\Rightarrow^* w' = w_1'x\$\$x'w_2'$$

and $x, x' \in \{0, 1\}$, $x \neq x'$. Continuing the check with application of the rules (3) will definitely skip x or x'. Consequently, no terminal string can be generated.

We showed that G' can generate the terminal string from the sentential form w if and only if $t'(w_1) =$ reversal(w_2), and the claim holds.   □

Since $S\ _1\Rightarrow^* w$, $w \in \Sigma^*$ if and only if $S\ _3\Rightarrow^* w$, Claim 5.2.18 holds.   □

We proved that $\mathcal{L}(G, _1\Rightarrow) = L$, $\mathcal{L}(G', _1\Rightarrow) = \mathcal{L}(G, _1\Rightarrow)$, $\mathcal{L}(G', _3\Rightarrow) = \mathcal{L}(G', _1\Rightarrow)$; therefore, $\mathcal{L}(G', _3\Rightarrow) = L$ holds. Thus, the proof of Theorem 5.2.16 is completed.   □

Jumping Derivation Mode 4

Definition 5.2.20. Let $G = (V, T, P, S)$ be an SCG. Let $uAv \in V^*$ and $(A) \rightarrow (x) \in P$. Then, $uAv\ _4\Rightarrow uxv$. Let $u = u_0A_1u_1\ldots A_nu_n \in V^*$ and $(A_1, A_2, \ldots, A_n) \rightarrow (x_1, x_2, \ldots, x_n) \in P$, for $n \geq 2$. Then,

$$u_0A_1u_1A_2u_2\ldots u_{i-1}A_iu_iA_{i+1}u_{i+1}\ldots u_{n-1}A_nu_n\ _4\Rightarrow$$

$$u_0u_1x_1u_2x_2\ldots u_{i-1}x_{i-1}u_{i_1}x_iu_{i_2}x_{i+1}u_{i_3}x_{i+2}u_{i+1}\ldots x_nu_{n-1}u_n$$

where $u_i = u_{i_1}u_{i_2}u_{i_3}$.   □

Informally, $v = u_0u_1x_1u_2x_2\ldots u_{i-1}x_{i-1}u_{i_1}x_iu_{i_2}x_{i+1}u_{i_3}x_{i+2}u_{i+1}\ldots x_nu_{n-1}u_n$ is obtained from $u = u_0A_1u_1A_2u_2\ldots u_{i-1}A_iu_iA_{i+1}u_{i+1}\ldots u_{n-1}A_nu_n$ in G by $(A_1, A_2, \ldots, A_n) \rightarrow (x_1, x_2, \ldots, x_n) \in P$ in $_4\Rightarrow$ as follows:

(1) $A_1, A_2, \ldots, A_n$ are deleted;
(2) a central u_i is nondeterministically chosen, for some $i \in \{0, \ldots, n\}$;
(3) x_i and x_{i+1} are inserted into u_i;
(4) x_j is inserted between u_j and u_{j+1}, for all $j < i$;
(5) x_k is inserted between u_{k-2} and u_{k-1}, for all $k > i + 1$.

Example 5.2.21. Let $G = (V, T, P, S)$, where $V = \{S, A, B, C, \$, a, b, c, d\}$, $T = \{a, b, c, d\}$, be an SCG with P containing the following rules:

   (i) $(S) \rightarrow (AB\$\$BA)$ (iv) $(A, B, B, A) \rightarrow (A, C, C, A)$
   (ii) $(A) \rightarrow (aAb)$ (v) $(\$, C, C, \$) \rightarrow (\varepsilon, \varepsilon, \varepsilon, \varepsilon)$
   (iii) $(B) \rightarrow (cBd)$ (vi) $(A) \rightarrow (\varepsilon)$

Consider G uses $_4\Rightarrow$. Then, every context-sensitive rule is applied in the following way. First, all affected nonterminals are erased. Next, some position of the current sentential form called center is nondeterministically chosen. Finally, the corresponding right-hand sides of the selected rule are inserted each at the original place of the neighboring erased nonterminal closer to the center. The central right-hand-side strings are randomly put closer to the chosen central position. In this example, we show how to control the choice.

First the rule (i) rewrites S to $AB\$\$BA$. Then, G uses the rules (ii) and (iii) generating a sentential form

$$a^{n_1} A b^{n_1} c^{n_2} B d^{n_2} \$\$ c^{n_3} B d^{n_3} a^{n_4} A b^{n_4}$$

where $n_i \geq 0$, for $i \in \{1, 2, 3, 4\}$. If the rule (vi) is used, derivation is blocked. Next, G uses the context-sensitive rule (iv), which may act in several different ways. In any case, it inserts two Cs into the current sentential form and the only possibility to erase them is the rule (v). However, thereby we force the rule (iv) to choose the center for interchanging nonterminals between Bs and moreover to insert Cs between the two symbols $\$$. Finally, G continues by using the rule (ii) and eventually finishes twice using the rule (vi). Consequently,

$$\mathcal{L}(G, {}_4\Rightarrow) = \{x \in T^* \mid x = a^{n_1} b^{n_1} c^{n_2} a^{n_3} b^{n_3} d^{n_2} c^{n_4} a^{n_5} b^{n_5} d^{n_4} a^{n_6} b^{n_6},$$

$$n_i \geq 0, i \in \{1, 2, 3, 4, 5, 6\}\}$$

Then, the string $aabbcabdccddab$ is generated by G in the following way:

$$S \, _4\Rightarrow AB\$\$BA \, _4\Rightarrow aAbB\$\$BA \, _4\Rightarrow aaAbbB\$\$BA \, _4\Rightarrow aaAbbcBd\$\$BA$$

$$_4\Rightarrow aaAbbcBd\$\$cBdA \, _4\Rightarrow aaAbbcBd\$\$ccBddA \, _4\Rightarrow aaAbbcBd\$\$ccBddaAb$$

$$_4\Rightarrow aabbcAd\$CC\$ccAddab \, _4\Rightarrow aabbcAdccAddab \, _4\Rightarrow aabbcaAbdccAddab$$

$$_4\Rightarrow aabbcabdccAddab \, _4\Rightarrow aabbcabdccddab \qquad\qquad \square$$

Theorem 5.2.22. JSC$_{4\Rightarrow}$ = RE.

Proof. As obvious, **JSC$_{4\Rightarrow}$** $\subseteq$ **RE**, so we only prove **RE** $\subseteq$ **JSC$_{4\Rightarrow}$**.

Let $G = (V, \Sigma, P, S)$ be the SCG constructed in the proof of Theorem 5.2.5. Next, we modify G to a new SCG G' so $\mathcal{L}(G, _1\Rightarrow) = \mathcal{L}(G', _4\Rightarrow)$.

Construction. Introduce five new symbols—$D, E, F, |$, and T. Set $N = \{D, E, F, |, \mathsf{T}\}$. Let $G' = (V', \Sigma, P', S)$ be SCG, with $V' = V \cup N$ and P' containing the rules

(1) $(S) \to (F\$S_1 11|E|11S_2\$F)$;
(2) $(A) \to (C_i(w))$ for $A \to w \in P_i$, where $i = 1, 2$;
(3) $(F) \to (FF)$;
(4) $(\$, a, a, \$) \to (\varepsilon, D, D, \varepsilon)$, for $a = 0, 1$;
(5) $(F, D, |, |, D, F) \to (\$, \varepsilon, \mathsf{T}, \mathsf{T}, \varepsilon, \$)$;
(6) $(\mathsf{T}, E, \mathsf{T}) \to (\varepsilon, |E|, \varepsilon)$;
(7) $(\$) \to (\varepsilon), (E) \to (\varepsilon), (|) \to (\varepsilon)$.

Claim 5.2.23. $\mathcal{L}(G, _1\Rightarrow) = \mathcal{L}(G', _4\Rightarrow)$.

Proof. The behaviour of context-free rules remains unchanged under $_4\Rightarrow$. Since the rules of G' simulating the derivations of G_1 and G_2 are identical to the ones of G simulating both grammars, for every derivation of G

$$S_1 \Rightarrow^* \$w_1 1111w_2\$ = w$$

where w is generated only by using the rules (1) and(2) and alph$(w) \cap (N_1 \cup N_2) = \emptyset$, there is

$$S_4 \Rightarrow^* F\$w_1 11|E|11w_2\$\#\#F = w'$$

in G', generated by the corresponding rules (1) and (2), and vice versa. Without any loss of generality, we can consider such a sentential form in every successful derivation. Additionally, in G

$$w_1 \Rightarrow^* v, v \in \Sigma^*$$

if and only if $t'(w_1) = $ reversal(w_2); then $v = t(w)$. Therefore, we have to prove

$$w'_4 \Rightarrow^* v', v' \in \Sigma^*$$

if and only if $t'(w_1) = $ reversal(w_2). Then, obviously $v' = v$ and we can complete the proof by the following claim.

Claim 5.2.24. In G', for

$$S_4 \Rightarrow^* w = F^{i_1}\$w_1 11|E|11w_2\$\#\#F^{i_2}, \text{ alph}(w) \cap (N_1 \cup N_2) = \emptyset$$

where w is generated only by using the rules (1) and (2),

$$w \,_4{\Rightarrow}^* w'$$

where $w' \in \Sigma^*$ if and only if $t'(w_1) = \text{reversal}(w_2)$, for some $i_1, i_2 \geq 0$.

The new rule (3) potentially arbitrarily multiplies the number of Fs to the left and right. Then, Fs from both sequences are simultaneously erasing by using the rule (5). Thus, without any loss of generality, suppose $i_1 = i_2$ equal the number of future usages of the rule (5).

For the sake of readability, in the next proof, in w_1, we omit all symbols from Σ—we consider only nonterminal symbols, which are to be erased.

Proof. If. Suppose $w_1 = \text{reversal}(w_2)$, then $w \,_4{\Rightarrow}^* \varepsilon$. We prove this by the induction on the length of w_1, w_2, where $|w_1| = |w_2| = k$.

Basis. Let $k = 0$. Then, $w = FF\$11|E|11\$FF$. Except the rules (7), the rule (4) is the only applicable. The center for interchanging the right-hand-side strings must be chosen between the two rewritten 1s and additionally inserted Ds must remain on the different sides of the central string $|E|$. Moreover, if any 1 stays outside the two Ds, it cannot be erased, so

$$FF\$11|E|11\$FF \,_4{\Rightarrow} FFD1|E|1DFF$$

Next, the rule (5) rewrites Ds back to $\$$s, erases Fs, and changes $|$s to Ts. The center must be chosen between the two $|$s and inserted Ts may not be shifted, otherwise they appear on the same side of E and the rule (6) is inapplicable. It secures the former usage of the rule (4) was as expected, so

$$FFD1|E|1DFF \,_4{\Rightarrow} F\$1\mathsf{T}E\mathsf{T}1\$F$$

By the rule (6) the symbols T may be rewritten back to $|$s. No left-hand-side string may be shifted during the application of the rule and the choice of the central position has no influence, because the neighboring symbols are rewritten. It secures the former usage of the rule (5) was as expected; therefore,

$$F\$1\mathsf{T}E\mathsf{T}1\$F \,_4{\Rightarrow} F\$1|E|1\$F$$

Then, the same sequence of rules with the same restrictions can be used again to erase remaining 1s and the check is finished by the rules (7) as

$$F\$1|E|1\$F \,_4{\Rightarrow} FD|E|DF \,_4{\Rightarrow} \$\mathsf{T}E\mathsf{T}\$ \,_4{\Rightarrow} \$|E|\$ \,_4{\Rightarrow}^* \varepsilon$$

and the basis holds.

Induction Hypothesis. Suppose there exists $k \geq 0$ such that the claim holds for all $0 \leq m \leq k$, where

$$w = F^{i_1}\$w_1 11|E|11w_2\$F^{i_2}, |w_1| = |w_2| = m$$

Induction Step. Consider G' generating w with

$$w = F^{i_1}\$w_1 11|E|11w_2\$F^{i_2}$$

where $|w_1| = |w_2| = k + 1$, $w_1 = \text{reversal}(w_2) = aw_1'$, and $a \in \{0, 1\}$. Except the rules (7), the rule (4) is the only applicable. The center for interchanging of the right-hand-side strings must be chosen between the two rewritten 0s or 1s and additionally inserted Ds must remain on the different sides of the central string $|E|$. Moreover, the outermost 0s or 1s must be rewritten and Ds may not be shifted between the new outermost ones, otherwise they cannot be erased.

$$F^{i_1}\$w_1 11|E|11w_2\$F^{i_2} {}_4\Rightarrow F^{i_1}Dw_1' 11|E|11w_2'DF^{i_2}$$

Next, the rule (5) rewrites Ds back to \$s, erases Fs, and changes $|$s to Ts. The center must be chosen between the two $|$s and inserted Ts may not be shifted, otherwise they appear on the same side of E and the rule (6) is inapplicable. It secures the former usage of the rule (4) was as expected.

$$F^{i_1}Dw_1' 11|E|11w_2'DF^{i_2} {}_4\Rightarrow F^{i_1'}\$w_1' 11\mathsf{T}E\mathsf{T}11w_2'\$F^{i_2'}$$

By the rule (6) the symbols T may be rewritten back to $|$s. No left-hand-side string may be shifted during the application of the rule and the position of the chosen center has no influence, because the neighboring symbols are rewritten. It secures the former usage of the rule (5) was as expected.

$$F^{i_1'}\$w_1' 11\mathsf{T}E\mathsf{T}11w_2'\$F^{i_2'} {}_4\Rightarrow F^{i_1'}\$w_1' 11|E|11w_2'\$F^{i_2'} = w'$$

By the induction hypothesis, $w'{}_4\Rightarrow^*\varepsilon$, which completes the proof.

Only If. Suppose $w_1 \neq \text{reversal}(w_2)$; there is no w', where $w\, _4\Rightarrow^* w'$ and $w' = \varepsilon$.
    Since $w_1 \neq \text{reversal}(w_2)$, $w_1 = vau$, $w_2 = u'a'\,\text{reversal}(v)$, and $a \neq a'$. Suppose both vs are correctly erased and no symbol is skipped producing the sentential form

$$F^{i_1}\$au11|E|11u'a'\$F^{i_2}$$

Next, the rule (4) can be applied to erase outermost 0s or 1s. However, then, there is 0 or 1 outside inserted Ds and, thus, unable to be erased, which completes the proof.
    We showed that G' can generate the terminal string from the sentential form w if and only if $t'(w_1) = \text{reversal}(w_2)$, and the claim holds. $\square$

We proved that for some $w \in \Sigma^*$, $S_1 \Rightarrow^* w$ in G if and only if $S_4 \Rightarrow^* w$ in G', and the claim holds.   □

Since $\mathcal{L}(G, {}_1\Rightarrow) = \mathcal{L}(G', {}_4\Rightarrow) = L$, the proof of Theorem 5.2.22 is completed.   □

Jumping Derivation Mode 5

Definition 5.2.25. Let $G = (V, T, P, S)$ be an SCG. Let $uAv \in V^*$ and $(A) \rightarrow (x) \in P$. Then, $uAv\ {}_5\Rightarrow uxv$. Let $u = u_0 A_1 u_1 \ldots A_n u_n \in V^*$ and $(A_1, A_2, \ldots, A_n) \rightarrow (x_1, x_2, \ldots, x_n) \in P$, for $n \geq 2$. Then,

$$u_0 A_1 u_1 A_2 \ldots u_{i-1} A_{i-1} u_i A_i u_{i+1} \ldots A_n u_n\ {}_5\Rightarrow$$

$$u_{0_1} x_1 u_{0_2} x_2 u_1 \ldots x_{i-1} u_{i-1} u_i u_{i+1} x_i \ldots u_{n_1} x_n u_{n_2}$$

where $u_0 = u_{0_1} u_{0_2}$, $u_n = u_{n_1} u_{n_2}$.   □

Informally, G obtains $u_{0_1} x_1 u_{0_2} x_2 u_1 \ldots x_{i-1} u_{i-1} u_i u_{i+1} x_i \ldots u_{n_1} x_n u_{n_2}$ from $u_0 A_1 u_1 A_2 \ldots u_{i-1} A_{i-1} u_i A_i u_{i+1} \ldots A_n u_n$ by $(A_1, A_2, \ldots, A_n) \rightarrow (x_1, x_2, \ldots, x_n) \in P$ in ${}_5\Rightarrow$ as follows:

(1) $A_1, A_2, \ldots, A_n$ are deleted;
(2) a central u_i is nondeterministically chosen, for some $i \in \{0, \ldots, n\}$;
(3) x_1 and x_n are inserted into u_0 and u_n, respectively;
(4) x_j is inserted between u_{j-2} and u_{j-1}, for all $1 < j \leq i$;
(5) x_k is inserted between u_k and u_{k+1}, for all $i + 1 \leq k < n$.

Example 5.2.26. Let $G = (V, T, P, S)$, where $V = \{S, A, B, \$, a, b\}$, $T = \{a, b\}$, be an SCG with P containing the following rules:

(i)	$(S) \rightarrow (\$AA\$)$	(iv) $(B, \$, \$, B) \rightarrow (A, \varepsilon, \varepsilon, A)$
(ii)	$(A) \rightarrow (aAb)$	(v) $(A) \rightarrow (\varepsilon)$
(iii)	$(A, A) \rightarrow (B, B)$	

Recall Example 5.2.21. ${}_4\Rightarrow$ interchanges the positions of nonterminals influenced by context-sensitive rules in the direction from the outer ones to the central ones. Opposed to ${}_4\Rightarrow$, ${}_5\Rightarrow$ interchanges nonterminals in the direction from a nondeterministically chosen center. In the present example, we show one possibility to control the choice.

Consider G uses ${}_5\Rightarrow$. First the rule (i) rewrites S to $\$AA\$$. Then, G uses the rule (ii) generating the sentential form

$$\$a^m A b^m a^n A b^n \$$$

where $m, n \geq 0$. If the rule (v) is used, derivation is blocked, because there is no way to erase the symbols \$. Next, G uses the context-sensitive rule (iii), which nondeterministically chooses a center and nondeterministically shifts Bs from the previous positions of As in the direction from this center. However, for the future application of the rule (iv) the chosen center must lie between As and moreover Bs must be inserted as the leftmost and the rightmost symbols of the current sentential form. The subsequent usage of the rule (iv) preserves As as the leftmost and the rightmost symbols independently of the effect of $_5\Rightarrow$. Finally, G continues by using the rule (ii) and eventually finishes twice using the rule (v). If the rule (iii) is used again, there is no possibility to erase inserted Bs. Consequently,

$$\mathcal{L}(G, _5\Rightarrow) = \{x \in T^* \mid x = a^k b^k a^l b^l a^m b^m a^n b^n, k, l, m, n \geq 0\}$$

Then, the string $aabbabaaabbb$ is generated by G in the following way:

$$S \ _5\Rightarrow \$AA\$ \ _5\Rightarrow \$aAbA\$ \ _5\Rightarrow \$aaAbbA\$ \ _5\Rightarrow \$aaAbbaAb\$$$

$$_5\Rightarrow B\$aabbab\$B \ _5\Rightarrow AaabbabA \ _5\Rightarrow AaabbabaaAb \ _5\Rightarrow AaabbabaaAbb$$

$$_5\Rightarrow AaabbabaaaAbbb \ _5\Rightarrow aabbabaaaAbbb \ _5\Rightarrow aabbabaaabbb \qquad \square$$

Theorem 5.2.27. $\mathbf{JSC}_{_5\Rightarrow} = \mathbf{RE}$.

Proof. As obvious, $\mathbf{JSC}_{_5\Rightarrow} \subseteq \mathbf{RE}$, so we only prove $\mathbf{RE} \subseteq \mathbf{JSC}_{_5\Rightarrow}$.

Let $G = (V, \Sigma, P, S)$ be the SCG constructed in the proof of Theorem 5.2.5. Next, we modify G to a new SCG G' so $\mathcal{L}(G, _1\Rightarrow) = \mathcal{L}(G', _5\Rightarrow)$.

Construction. Introduce four new symbols—D,E,F, and $\circ$. Set $N = \{D,E,F,\circ\}$. Let $G' = (V', \Sigma, P', S)$ be SCG, with $V' = V \cup N$ and P' containing the rules

(1) $(S) \rightarrow (\$S_1 1111 S_2\$ \circ E \circ F)$;
(2) $(A) \rightarrow (C_i(w))$ for $A \rightarrow w \in P_i$, where $i = 1, 2$;
(3) $(F) \rightarrow (FF)$;
(4) $(\$, a, a, \$, E, F) \rightarrow (\varepsilon, \varepsilon, \$, \$, \varepsilon, D)$, for $a = 0, 1$;
(5) $(\circ, D, \circ) \rightarrow (\varepsilon, \circ E\circ, \varepsilon)$;
(6) $(\$) \rightarrow (\varepsilon), (E) \rightarrow (\varepsilon), (\circ) \rightarrow (\varepsilon)$.

Claim 5.2.28. $\mathcal{L}(G, _1\Rightarrow) = \mathcal{L}(G', _5\Rightarrow)$.

Proof. Context-free rules are not influenced by $_5\Rightarrow$. The rule (3) must generate precisely as many Fs as the number of applications of the rule (4). Context-sensitive rules of G' correspond to context-sensitive rules of G, except the special rule (5). We show, the construction of G' forces context-sensitive rules to work exactly in the same way as the rules of G do.

Every application of the rule (4) must be followed by the application of the rule (5) to rewrite D back to E, which requires the symbol D between two $\circ$s. It ensures the previous usage of context-sensitive rule selected the center to the right

of the rightmost affected nonterminal and all right-hand-side strings changed their
positions with the more left ones. The leftmost right-hand-side string is then shifted
randomly to the left, but it is always ε. $_5\Rightarrow$ has no influence on the rule (5).

From the construction of G', it works exactly in the same way as G does.   □

$\mathcal{L}(G, {}_1\Rightarrow) = \mathcal{L}(G', {}_5\Rightarrow)$ and $\mathcal{L}(G, {}_1\Rightarrow) = L$; therefore $\mathcal{L}(G', {}_5\Rightarrow) = L$. Thus,
the proof of Theorem 5.2.27 is completed.   □

Jumping Derivation Mode 6

Definition 5.2.29. Let $G = (V, T, P, S)$ be an SCG. Let $uAv \in V^*$ and $(A) \rightarrow$
$(x) \in P$. Then, $uAv \ {}_6\Rightarrow uxv$. Let $u = u_0A_1u_1 \ldots A_nu_n \in V^*$ and $(A_1, A_2, \ldots, A_n) \rightarrow$
$(x_1, x_2, \ldots, x_n) \in P$, for $n \geq 2$. Then,

$$u_0A_1u_1A_2u_2 \ldots u_{i-1}A_iu_iA_{i+1}u_{i+1} \ldots u_{n-1}A_nu_n \ {}_6\Rightarrow$$

$$u_0u_1x_1u_2x_2 \ldots u_{i-1}x_{i-1}u_ix_{i+2}u_{i+1} \ldots x_nu_{n-1}u_n \qquad \qquad □$$

Informally, G obtains $u_0u_1x_1u_2x_2 \ldots u_{i-1}x_{i-1}u_ix_{i+2}u_{i+1} \ldots x_nu_{n-1}u_n$ from
$u_0A_1u_1A_2u_2 \ldots u_{i-1}A_iu_iA_{i+1}u_{i+1} \ldots u_{n-1}A_nu_n$ by using $(A_1, A_2, \ldots, A_n) \rightarrow (x_1,$
$x_2, \ldots, x_n) \in P$ in $_6\Rightarrow$ as follows:

(1) $A_1, A_2, \ldots, A_n$ are deleted;
(2) a central u_i is nondeterministically chosen, for some $i \in \{0, \ldots, n\}$;
(3) x_j is inserted between u_j and u_{j+1}, for all $j < i$;
(4) x_k is inserted between u_{k-2} and u_{k-1}, for all $k > i + 1$.

Example 5.2.30. Let $G = (V, T, P, S)$, where $V = \{S, A, B, a, b\}$, $T = \{a, b\}$, be
an SCG with P containing the following rules:

     (i) $(S) \rightarrow (ABBA)$ (iii) $(A, B, B, A) \rightarrow (AB, B, B, BA)$
     (ii) $(A) \rightarrow (aAb)$ (iv) $(A, B, B, A) \rightarrow (\varepsilon, B, B, \varepsilon)$

Consider G uses $_6\Rightarrow$. $_6\Rightarrow$ interchanges nonterminals similarly as $_4\Rightarrow$ does in
Example 5.2.21, however, the central nonterminals are removed. This property
can be used to eliminate nondeterminism of choosing of the center, which we
demonstrate next.

The rules (i) and (ii) are context-free, not affected by $_6\Rightarrow$. First the starting rule (i)
rewrites S to $ABBA$. Then, G uses the rule (ii) generating the sentential form

$$a^mAb^mBBa^nAb^n$$

where $m, n \geq 0$. Next, G uses the context-sensitive rule (iii) or (iv). Notice, there
is no rule erasing Bs, thus in both cases the center of interchanging of nonterminals
must be chosen between the two Bs. Otherwise, in both cases there is exactly one

A remaining, thus the only applicable rule is the rule (ii), which is context-free and not erasing. Therefore, G uses the rule (iii) generating the sentential form

$$a^m b^m ABBAa^n b^n$$

and continues by using the rule (ii) or it uses the rule (iv) and finishes the derivation.
Subsequently, the language G generates is

$$\mathcal{L}(G, {}_6\Rightarrow) = \left\{ x \in T^* \mid x = a^{n_1} b^{n_1} a^{n_2} b^{n_2} \ldots a^{n_{2k}} b^{n_{2k}}, k, n_i \geq 0, 1 \leq i \leq 2k \right\}$$

Then, the string $aabbabaabbab$ is generated by G in the following way:

$$S \ {}_6{\Rightarrow} ABBA \ {}_6{\Rightarrow} aAbBBA \ {}_6{\Rightarrow} aaAbbBBA \ {}_6{\Rightarrow} aaAbbBBaAb$$

$${}_6{\Rightarrow} aabbABBAab \ {}_6{\Rightarrow} aabbaAbBBAab \ {}_6{\Rightarrow} aabbaAbBBaAbab$$

$${}_6{\Rightarrow} aabbaAbBBaaAbbab \ {}_6{\Rightarrow} aabbabaabbab \qquad \qquad \square$$

Theorem 5.2.31. $\mathbf{JSC}_{6\Rightarrow} = \mathbf{RE}$.

Proof. Clearly, $\mathbf{JSC}_{6\Rightarrow} \subseteq \mathbf{RE}$. Next, we prove $\mathbf{RE} \subseteq \mathbf{JSC}_{6\Rightarrow}$.
  Let $G = (V, \Sigma, P, S)$ be the SCG constructed in the proof of Theorem 5.2.5.
Next, we modify G to a new SCG G' so $\mathcal{L}(G, {}_1\Rightarrow) = \mathcal{L}(G', {}_6\Rightarrow)$.

Construction. Introduce two new symbols—E and F. Let $G' = (V', \Sigma, P', S)$ be SCG, with $V' = V \cup \{E, F\}$ and P' containing the rules

(1) $(S) \to (F\$S_1 1111 S_2 \$)$;
(2) $(A) \to (C_i(w))$ for $A \to w \in P_i$, where $i = 1, 2$;
(3) $(F) \to (FF)$;
(4) $(F, \$, a, a, \$) \to (E, E, \varepsilon, \$, \$)$, for $a = 0, 1$;
(5) $(\$) \to (\varepsilon)$.

Claim 5.2.32. $\mathcal{L}(G, {}_1\Rightarrow) = \mathcal{L}(G', {}_6\Rightarrow)$.

Proof. Context-free rules are not influenced by ${}_6\Rightarrow$. Context-sensitive rules of G' closely correspond to context-sensitive rules of G. The new symbols are used to force modified rules to act in the same way as sample ones do. The symbols F are first multiplied and then consumed by context-sensitive rules, so their number must equal the number of usages of these rules. The new symbols E are essential. E never appears on the left-hand side of any rule, thus whenever it is inserted into the sentential form, no terminal string can be generated. Therefore, the center is always chosen between two Es, which are basically never inserted, and other right-hand-side strings are then inserted deterministically.
  G' with ${}_6\Rightarrow$ works in the same way as G with ${}_1\Rightarrow$ does. $\qquad \square$

  $\mathcal{L}(G, {}_1\Rightarrow) = \mathcal{L}(G', {}_6\Rightarrow)$, hence $\mathcal{L}(G', {}_6\Rightarrow) = L$. Thus, the proof of Theorem 5.2.31 is completed. $\qquad \square$

Jumping Derivation Mode 7

Definition 5.2.33. Let $G = (V, T, P, S)$ be an SCG. Let $(A) \to (x) \in P$ and $uAv \in V^*$. Then, $uAv \ _7{\Rightarrow} uxv$. Let $u = u_0A_1u_1 \ldots A_nu_n \in V^*$ and $(A_1,A_2,\ldots,A_n) \to (x_1,x_2,\ldots,x_n) \in P$, for $n \geq 2$. Then,

$$u_0A_1u_1A_2 \ldots u_{i-1}A_iu_iA_{i+1}u_{i+1} \ldots A_nu_n \ _7{\Rightarrow}$$

$$u_0x_2u_1 \ldots x_iu_{i-1}u_iu_{i+1}x_{i+1} \ldots u_n \qquad\qquad \square$$

Informally, by using the rule $(A_1,A_2,\ldots,A_n) \to (x_1,x_2,\ldots,x_n) \in P$, G obtains $u_0x_2u_1 \ldots x_iu_{i-1}u_iu_{i+1}x_{i+1} \ldots u_n$ from $u_0A_1u_1A_2 \ldots u_{i-1}A_iu_iA_{i+1}u_{i+1} \ldots A_nu_n$ in $_7{\Rightarrow}$ as follows:

(1) $A_1, A_2, \ldots, A_n$ are deleted;
(2) a central u_i is nondeterministically chosen, for some $i \in \{0 \ldots, n\}$;
(3) x_j is inserted between u_{j-2} and u_{j-1}, for all $1 < j \leq i$;
(4) x_k is inserted between u_k and u_{k+1}, for all $i + 1 \leq k < n$.

Example 5.2.34. Let $G = (V, T, P, S)$, where $V = \{S, A, B, C, \$, a, b, c\}$, $T = \{a, b, c\}$, be an SCG with P containing the following rules:

  (i) $(S) \to (ABC\$)$ (v) $(A, B, C) \to (A, B, C)$
  (ii) $(A) \to (aAa)$ (vi) $(A, B) \to (A, B)$
  (iii) $(B) \to (bBb)$ (vii) $(A, \$) \to (\varepsilon, \varepsilon)$
  (iv) $(C) \to (cCc)$

Consider G uses $_7{\Rightarrow}$. $_7{\Rightarrow}$ interchanges nonterminals in the direction from the nondeterministically chosen center and erases the outermost nonterminals. In this example, we show that we may force the center to lie outside the part of a sentential form between the affected nonterminals.

The derivation starts by using the starting rule (i) and continues by using the rules (ii) through (iv) generating the sentential form

$$a^mAa^mb^nBb^nc^lCc^l\$$$

where $m, n, l \geq 0$. Next, G uses the context-sensitive rule (v) choosing the center to the left of A erasing C. If a different central position is chosen, the symbol A is erased while B or C cannot be erased in the future and the derivation is blocked. There is the same situation, if one of the rules (vi) or (vii) is used instead. Notice, no rule erases B or C. Then, the derivation continues by using the rules (ii) and (iii) and eventually the rule (vi) rewriting B to A and erasing B. Otherwise, A is erased and the symbol $\$ cannot be erased any more. G continues by using the rule (ii) and finally finishes the derivation with the rule (vii). Subsequently,

$$\mathcal{L}(G, _7{\Rightarrow}) = \{x \in T^* \mid x = a^{2m_1}b^{n_1}a^{2m_2}b^{n_1}c^lb^{n_2}a^{2m_3}b^{n_2}c^l,$$

$$m_1, m_2, m_3, n_1, n_2, l \geq 0\}$$

Then, the string $aabaabccbaabcc$ is generated by G in the following way:

$$S {_7}{\Rightarrow} ABC\$ {_7}{\Rightarrow} aAaBC\$ {_7}{\Rightarrow} aAabBbC\$ {_7}{\Rightarrow} aAabBbcCc\$$$

$$_7{\Rightarrow} aAabBbccCcc\$ {_7}{\Rightarrow} aabAbccBcc\$ {_7}{\Rightarrow} aabaAabccBcc\$ {_7}{\Rightarrow}$$

$$aabaAabccbBbcc\$$$

$$_7{\Rightarrow} aabaabccbAbcc\$ {_7}{\Rightarrow} aabaabccbaAabcc\$ {_7}{\Rightarrow} aabaabccbaabcc \qquad \Box$$

Theorem 5.2.35. $\mathbf{JSC}_{7\Rightarrow} = \mathbf{RE}$.

Proof. Clearly, $\mathbf{JSC}_{7\Rightarrow} \subseteq \mathbf{RE}$. We prove $\mathbf{RE} \subseteq \mathbf{JSC}_{7\Rightarrow}$.

Let $G = (V, \Sigma, P, S)$ be the SCG constructed in the proof of Theorem 5.2.5. Next, we modify G to a new SCG G' so $\mathcal{L}(G, {_1}{\Rightarrow}) = \mathcal{L}(G', {_7}{\Rightarrow})$.

Construction. Introduce four new symbols—E, F, H, and $|$. Set $N = \{E, F, H, |\}$. Let $G' = (V', \Sigma, P', S)$ be SCG, with $V' = V \cup N$ and P' containing the rules

(1) $(S) \rightarrow (FHS_1 11\$|\$11S_2)$;
(2) $(A) \rightarrow (C_i(w))$ for $A \rightarrow w \in P_i$, where $i = 1, 2$;
(3) $(F) \rightarrow (FF)$;
(4) $(a, \$, \$, a) \rightarrow (\varepsilon, E, E, \varepsilon)$, for $a = 0, 1$;
(5) $(F, H, E, |, E) \rightarrow (H, \$, |, \$, \varepsilon)$;
(6) $(\$) \rightarrow (\varepsilon)$, $(H) \rightarrow (\varepsilon)$, $(|) \rightarrow (\varepsilon)$.

Claim 5.2.36. $\mathcal{L}(G, {_1}{\Rightarrow}) = \mathcal{L}(G', {_7}{\Rightarrow})$.

Proof. The behaviour of context-free rules remains unchanged under $_7{\Rightarrow}$. Since the rules of G' simulating the derivations of G_1, G_2, respectively, are identical to the ones of G simulating both grammars, for every derivation of G

$$S {_1}{\Rightarrow}^* \$w_1 1111w_2\$ = w$$

where w is generated only by using the rules (1) and (2) and $\mathrm{alph}(w) \cap (N_1 \cup N_2) = \emptyset$, there is

$$S {_7}{\Rightarrow}^* FHw_1 11\$|\$11w_2 = w'$$

in G', generated by the corresponding rules (1) and (2), and vice versa. Without any loss of generality, we can consider such a sentential form in every successful derivation. Additionally, in G

$$w {_1}{\Rightarrow}^* v, v \in \Sigma^*$$

if and only if $t'(w_1) = \mathrm{reversal}(w_2)$; then $v = t(w)$. Therefore, we have to prove

$$w' {_4}{\Rightarrow}^* v', v' \in \Sigma^*$$

if and only if $t'(w_1) = \text{reversal}(w_2)$. Then, obviously $v' = v$ and we can complete the proof by the following claim.

Claim 5.2.37. In G', for some $i \geq 1$,

$$S \, _7\Rightarrow^* w = F^i H w_1 \$|\$ w_2 E$$

where w is generated only by using the rules (1) through (3) and $\text{alph}(w) \cap (N_1 \cup N_2) = \emptyset$. Then,

$$w \, _7\Rightarrow^* w'$$

where $w' \in \Sigma^*$ if and only if $t'(w_1) = \text{reversal}(w_2)$.

The new rule (3) may potentially arbitrarily multiply the number of Fs to the left. Then, Fs are erasing by using the rule (5). Thus, without any loss of generality, suppose i equals the number of the future usages of the rule (5).

For the sake of readability, in the next proof we omit all symbols in w_1 from Σ—we consider only nonterminal symbols, which are to be erased.

Proof. If. Suppose $w_1 = \text{reversal}(w_2)$, then $w \, _7\Rightarrow^* \varepsilon$. We prove this by the induction on the length of w_1, w_2, where $|w_1| = |w_2| = k$. Then, obviously $i = k$. By the construction of G', the least k equals 2, but we prove the claim for all $k \geq 0$.

Basis. Let $k = 0$. Then,

$$w = H\$|\$$$

By the rules (6)

$$H\$|\$ \, _7\Rightarrow^* \varepsilon$$

and the basis holds.

Induction Hypothesis. Suppose there exists $k \geq 0$ such that the claim holds for all m, where

$$w = F^m H w_1 \$|\$ w_2, |w_1| = |w_2| = m, 0 \leq m \leq k$$

Induction Step. Consider G' generates w, where

$$w = F^{k+1} H w_1 \$|\$ w_2, |w_1| = |w_2| = k + 1$$

Since $w_1 = \text{reversal}(w_2)$ and $|w_1| = |w_2| = k + 1$, $w_1 = w_1'a$, $w_2 = aw_2'$. The symbols a can be erased by application of the rules (4) and (5) under several conditions. First, when the rule (4) is applied, the center for interchanging right-hand-side strings must be chosen between the two $\$$s, otherwise both Es appear on the same side of the symbol | and the rule (5) is not applicable. Next, no 0 or 1 may be skipped, while proceeding in the direction from the center to the edges. Finally, when the rule (5) is applied, a center must be chosen to the left of F, otherwise H is erased and the future application of this rule is excluded.

$$F^{k+1}Hw_1'a\$|\$aw_2'\ _7\Rightarrow F^{k+1}Hw_1'D|Dw_2'\ _7\Rightarrow F^kHw_1'\$|\$w_2' = w'$$

By the induction hypothesis, $w'_7\Rightarrow^*\varepsilon$, which completes the proof.

Only If. Suppose $w_1 \neq$ reversal(w_2), then, there is no w', where $w\ _7\Rightarrow^* w'$ and $w' = \varepsilon$.

Since $w_1 \neq$ reversal(w_2), $w_1 = uav$, $w_2 =$ reversal(v)$a'u'$, and $a \neq a'$. Suppose both vs are correctly erased and no symbol is skipped producing the sentential form

$$F^iHua\$|\$a'u'$$

Next the rule (4) can be applied to erase innermost 0s or 1s. However, since $a \neq a'$, even if the center is chosen properly between the two \$s, there is 0 or 1 between inserted Es and, thus, unable to be erased, which completes the proof.

We showed that G' can generate the terminal string from the sentential form w if and only if $t'(w_1) =$ reversal(w_2), and the claim holds.   □

We proved $S\ _1\Rightarrow^*w$, $w \in \Sigma^*$, in G if and only if $S\ _7\Rightarrow^*w$ in G', hence $\mathcal{L}(G, _1\Rightarrow) = \mathcal{L}(G', _7\Rightarrow)$ and the claim holds.   □

Since $\mathcal{L}(G, _1\Rightarrow) = \mathcal{L}(G', _7\Rightarrow)$ and $\mathcal{L}(G, _1\Rightarrow) = L$, the proof of Theorem 5.2.35 is completed.   □

Jumping Derivation Mode 8

Definition 5.2.38. Let $G = (V, T, P, S)$ be an SCG. Let $u = u_0A_1u_1\ldots A_nu_n \in V^*$ and $(A_1,A_2,\ldots,A_n) \to (x_1,x_2,\ldots,x_n) \in P$, for $n \geq 1$. Then,

$$u_0A_1u_1A_2u_2\ldots A_nu_n\ _8\Rightarrow v_0x_1v_1x_2v_2\ldots x_nv_n$$

where $u_0z_1 = v_0$, $z_2u_n = v_n$, $|u_0u_1\ldots u_{j-1}| \leq |v_0v_1\ldots v_j|$, $|u_{j+1}\ldots u_n| \leq |v_jv_{j+1}\ldots v_n|$, $0 < j < n$, and $z_1, z_2 \in V^*$.   □

Informally, G obtains $v_0x_1v_1x_2v_2\ldots x_nv_n$ from $u_0A_1u_1A_2u_2\ldots A_nu_n$ by using $(A_1,A_2,\ldots,A_n) \to (x_1,x_2,\ldots,x_n) \in P$ in $_8\Rightarrow$ as follows:

(1) $A_1, A_2, \ldots, A_n$ are deleted;
(2) x_1 and x_n are inserted into u_1 and u_{n-1}, respectively;
(3) x_i is inserted into $u_{i-1}u_i$, for all $1 < i < n$, to the right of x_{i-1} and to the left of x_{i+1}.

Example 5.2.39. Let $G = (V, T, P, S)$, where $V = \{S, \overline{S}, A, B, C, a, b, c\}$, $T = \{a, b, c\}$, be an SCG with P containing the following rules:

(i) $(S) \to (AS)$ (iv) $(\overline{S}) \to (B)$
(ii) $(S) \to (\overline{S})$ (v) $(B) \to (BB)$
(iii) $(\overline{S}) \to (b\overline{S}cC)$ (vi) $(A, B, C) \to (a, \varepsilon, \varepsilon)$

Consider G uses $_8\Rightarrow$. $_8\Rightarrow$ acts in a similar way as $_2\Rightarrow$ does. When a rule is to be applied, there is a nondeterministically chosen center in between the affected nonterminals and rule right-hand-side strings can be shifted in the direction to this center, but not farther than the neighboring affected nonterminal was.

The rules (i) through (v) are context-free. Without any loss of generality, we suppose these rules are used only before the first application of the rule (vi) producing the string

$$A^m b^n B^l (cC)^n$$

The derivation finishes with the sequence of applications of the rule (vi). For As, Bs, and Cs are being rewritten together, $m = n = l$. Moreover, inserted a is always between the rewritten A and B. Subsequently,

$$\mathcal{L}(G, {_8\Rightarrow}) = \{x \in T^* \mid x = wc^n, w \in \{a, b\}^*, \#_a(w) = \#_b(w) = n, n \geq 1\}$$

For example, the string *baabbaccc* is generated by G in the following way:

$$S \,{_8\Rightarrow}\, AS \,{_8\Rightarrow}\, AAS \,{_8\Rightarrow}\, AAAS \,{_8\Rightarrow}\, AAA\overline{S} \,{_8\Rightarrow}\, AAAb\overline{S}cC \,{_8\Rightarrow}\, AAAbb\overline{S}cCcC$$

$${_8\Rightarrow}\, AAAbbb\overline{S}cCcCcC \,{_8\Rightarrow}\, AAAbbbBcCcCcC \,{_8\Rightarrow}\, AAAbbbBBcCcCcC$$

$${_8\Rightarrow}\, AAAbbbBBBcCcCcC \,{_8\Rightarrow}\, AAbbbaBBccCcC \,{_8\Rightarrow}\, AbabbaBcccC \,{_8\Rightarrow}\,$$

$$baabbaccc \qquad\qquad\qquad \square$$

Theorem 5.2.40. $\mathbf{JSC}_{8\Rightarrow} = \mathbf{RE}$.

Proof. Prove this theorem by analogy with the proof of Theorem 5.2.9. $\qquad \square$

Jumping Derivation Mode 9

Definition 5.2.41. Let $G = (V, T, P, S)$ be an SCG. Let $u = u_0 A_1 u_1 \ldots A_n u_n \in V^*$ and $(A_1, A_2, \ldots, A_n) \to (x_1, x_2, \ldots, x_n) \in P$, for $n \geq 1$. Then,

$$u_0 A_1 u_1 A_2 u_2 \ldots A_n u_n \,{_9\Rightarrow}\, v_0 x_1 v_1 x_2 v_2 \ldots x_n v_n$$

where $u_0 = v_0 z_1$, $u_n = z_2 v_n$, $|u_0 u_1 \ldots u_{j-1}| \leq |v_0 v_1 \ldots v_j|$, $|u_{j+1} \ldots u_n| \leq |v_j v_{j+1} \ldots v_n|$, $0 < j < n$, and $z_1, z_2 \in V^*$. $\qquad \square$

Informally, G obtains $v_0 x_1 v_1 x_2 v_2 \ldots x_n v_n$ from $u_0 A_1 u_1 A_2 u_2 \ldots A_n u_n$ by using $(A_1, A_2, \ldots, A_n) \to (x_1, x_2, \ldots, x_n) \in P$ in $_9\Rightarrow$ as follows:

(1) $A_1, A_2, \ldots, A_n$ are deleted;

(2) x_1 and x_n are inserted into u_0 and u_n, respectively;

(3) x_i is inserted into $u_{i-1} u_i$, for all $1 < i < n$, to the right of x_{i-1} and to the left of x_{i+1}.

Example 5.2.42. Let $G = (V, T, P, S)$, where $V = \{S, \bar{S}, A, B, C, \$, a, b, c\}$, $T = \{a, b, c\}$, be an SCG with P containing the following rules:

   (i) $(S) \to (aSa)$ (v) $(C) \to (cBC\$)$

   (ii) $(S) \to (A)$ (vi) $(C) \to (\varepsilon)$

   (iii) $(A) \to (\$A)$ (vii) $(\$, B, \$) \to (\varepsilon, b, \varepsilon)$

   (iv) $(A) \to (C)$

Consider G uses $_9\Rightarrow$. $_9\Rightarrow$ acts similarly to $_3\Rightarrow$ with respect to the direction of shift of the rule right-hand sides, but with limitation as in $_8\Rightarrow$. When a rule is to be applied, there is a nondeterministically chosen center in between the affected nonterminals and rule right-hand-side strings can be shifted in the direction from this center, but not farther than the neighboring affected nonterminal was.

The rules (i) through (vi) are context-free. Without any loss of generality, we can suppose these rules are used only before the first application of the rule (vii), which produce the sentential form

$$a^m \$^n (cB)^l \$^l a^m$$

The derivation finishes with the sequence of applications of the rule (vii). The symbols \$ and Bs are being rewritten together, thus $n = l$ must hold. Additionally, $_9\Rightarrow$ ensures, b is always inserted between the rewritten \$s. Subsequently,

$$\mathcal{L}(G, _9\Rightarrow) = \{x \in T^* \mid x = a^m w a^m, w \in \{b, c\}^*, \#_b(w) = \#_c(w), m \geq 0\}$$

For example, the string $aabcbcaa$ is generated by G in the following way:

$$S \; _9\Rightarrow aSa \; _9\Rightarrow aaSaa \; _9\Rightarrow aaAaa \; _9\Rightarrow aa\$Aaa \; _9\Rightarrow aa\$\$Aaa$$

$$_9\Rightarrow aa\$\$Caa \; _9\Rightarrow aa\$\$cBC\$aa \; _9\Rightarrow aa\$\$cBcBC\$\$aa$$

$$_9\Rightarrow aa\$\$cBcB\$\$aa \; _9\Rightarrow aa\$bccB\$aa \; _9\Rightarrow aabcbcaa \qquad \square$$

Theorem 5.2.43. $\mathbf{JSC}_{9\Rightarrow} = \mathbf{RE}$.

Proof. Prove this theorem by analogy with the proof of Theorem 5.2.16. $\qquad \square$

Open Problem Areas

Finally, let us suggest some open problem areas concerning the subject of this section.

Open Problem 5.2.44. Return to derivation modes (1) through (9) in Sect. 5.2.2. Introduce and study further modes. For instance, in a more general way, discuss a jumping derivation mode, in which the only restriction is to preserve a mutual order of the inserted right-hand-side strings, which can be nondeterministically spread across the whole sentential form regardless of the positions of the rewritten nonterminals. In a more restrictive way, study a jumping derivation mode over words satisfying some prescribed requirements, such as a membership in a regular language.

Open Problem 5.2.45. Consider propagating versions of jumping scattered context grammars. In other words, rule out erasing rules in them. Reconsider the investigation of the present section in its terms.

Open Problem 5.2.46. The present section has often demonstrated that some jumping derivation modes work just like ordinary derivation modes in scattered context grammars. State general combinatorial properties that guarantee this behaviour.

Open Problem 5.2.47. Establish normal forms of scattered context grammars working in jumping ways.

Chapter 6
Algebra, Grammars, and Computation

In terms of algebra, the context-free and E0L grammatical derivations are traditionally defined over the free monoids generated by total alphabets of these grammars under the operation of concatenation. The present chapter, however, introduces and investigates these derivations over different algebraic structures in order to increase the generative power of these grammars. (see [Med90a, Med95b, MK02]).

Specifically, in this chapter, we define the context-free and E0L derivations over free groups, which represent a fundamental algebraic structure in mathematics. By this natural modification of the grammatical derivations, we significantly increase the generative power of context-free and E0L grammars. As a matter of fact, we show that they characterize the family of recursively enumerable languages under this alternative definition. In addition, this characterization is very economical because it is based upon the context-free grammars with no more than eight nonterminals and the E0L grammars with no more than six nonterminals.

From a broader perspective, the results achieved in this chapter demonstrate that algebraically alternative definitions of some fundamental concepts, such as derivations, in grammatically based models may result into a significant increase of their generative power. As a result, the study of these alternative grammatical concepts definitely represents an important investigation area in the formal language theory today.

6.1 Sequential and Parallel Generation over Free Groups: Conceptualization

The classical theory of formal languages defines grammatical derivations over the free monoids generated by total alphabets of these grammars under the operation of concatenation. The present section, however, defines derivations over different

© Springer International Publishing AG 2017
A. Meduna, O. Soukup, *Modern Language Models and Computation*,
DOI 10.1007/978-3-319-63100-4_6

algebraic structures. Specifically, it introduces sequential derivations made by context-free grammars and, in addition, parallel derivations made by EOL grammars over free groups.

6.1.1 Definitions

Let us recall basic notions concerning languages (see Sect. 2.1) and phrase-structure grammars in Kuroda normal form (see Definition 3.1.1). Next, this section mentions only the new notions used in this chapter.

For an alphabet, V, V^* represents the set of all strings over V, and is, thus, a free monoid generated by V under the operation of concatenation. Furthermore, V° represents the free group generated by V under the operation of concatenation. The unit of V° is denoted by ε. For every string, $w \in V^\circ$, there is the *inverse string of w*, denoted by $\overline{w}$, with the property that $w\overline{w} = \overline{w}w = \varepsilon$.

The inverse string of $w = a_1 a_2 \ldots a_n$, where $a_i \in V$, $i = 1, 2, \ldots, n$, $n \geq 0$, is defined as $\overline{w} = \overline{a_n a_{n-1}} \ldots \overline{a_1}$. The string is said to be *reduced*, if it contains no pairs of the form $x\overline{x}$ or $\overline{x}x$, where $x, \overline{x} \in V^\circ$. Let $w = uxyv \in V^\circ$ be a string, where $x, y, u, v \in V^\circ$ and $x = \overline{y}$. To express that x and y are mutually inverse and can be distracted, we underline xy in $ux\underline{xy}v$. As well as for any string $w \in V^*$, for any reduced string $w \in V^\circ$, $|w|$ denotes the length of w.

For example, if $V = \{a, b, c, \overline{a}, \overline{b}, \overline{c}\}$, then the inverse string of $bcaa \in V^\circ$ is $\overline{aacb} \in V^\circ$. Because $\overline{a}a = a\overline{a} = \varepsilon$, $\overline{b}b = b\overline{b} = \varepsilon$ and $\overline{c}c = c\overline{c} = \varepsilon$, it is obvious that $bcaaaacb = \overline{aacbbcaa} = \varepsilon$.

Definition 6.1.1. A context-free grammar (see Sect. 2.3.1) over a free group (a CFG$^\circ$ for short) is a quadruple, $G = (V, T, P, S)$, where V, S, and T have the same meaning as for classical context-free grammar and P is a finite set of rules of the form $A \to x$, where $A \in V - T$ and $x \in V^\circ$. If $A \to x \in P$, $u = u_1 A u_2$ and $v = u_1 x u_2$, where $u, v \in V^\circ$, then u directly derives v over V° by using $A \to x$ in G, symbolically written as $u \circ\!\Rightarrow v \, [A \to x]$ in G or, simply, $u \circ\!\Rightarrow v$. Let $\circ\!\Rightarrow^n$ denotes the n–fold product of $\circ\!\Rightarrow$, where $n \geq 0$. Beyond, let $\circ\!\Rightarrow^+$ and $\circ\!\Rightarrow^*$ denote transitive closure of $\circ\!\Rightarrow$ and the transitive-reflexive closure of $\circ\!\Rightarrow$, respectively. The language generated by G over V°, $L(G)^\circ$, is defined as $L(G)^\circ = \{y \in T^* : S \circ\!\Rightarrow^* y$ in $G\}$.   □

Definition 6.1.2. An EOL grammar (see Sect. 2.3.4) over a free group (an EOL$^\circ$ grammar for short) is a quadruple, $G = (V, T, P, w)$, where V and T have the same meaning as for classical EOL grammars, P is a finite set of rules of the form $X \to x$, where $X \in V$, $x \in V^\circ$, and $w \in V^\circ$ is the axiom. If $A_1 \to x_1, \ldots, A_n \to x_n \in P$, $u = A_1 \ldots A_n$ and $v = x_1 \ldots x_n$, where $u, v \in V^\circ$, then u directly derives v over V° by using $A_i \to x_i$ in G for $i = 1, 2, \ldots, n$, symbolically written as $u \circ\!\Rightarrow v \, [A_1 \to x_1, \ldots, A_n \to x_n]$ or, simply, $u \circ\!\Rightarrow v$ in G. Let $\circ\!\Rightarrow^n$ denotes the n–fold product of $\circ\!\Rightarrow$, where $n \geq 0$. Furthermore, let $\circ\!\Rightarrow^+$ and $\circ\!\Rightarrow^*$ denote the transitive closure of $\circ\!\Rightarrow$ and the transitive-reflexive closure of $\circ\!\Rightarrow$, respectively. The language generated by G over V°, $L(G)^\circ$, is defined as $L(G)^\circ = \{y \in T^* : w \circ\!\Rightarrow^* y$ in EOL grammar $G\}$.   □

The families of languages generated by context-free grammars over free groups and languages generated by E0L grammars over free groups are denoted by $\mathbf{CF}^\circ$ and $\mathbf{E0L}^\circ$, respectively.

6.2 Results: Computational Completeness

In the present section, we show that by the modification of the grammatical derivations described in the previous section, we significantly increase the generative power of context-free and E0L grammars. In fact, we demonstrate that they characterize the family of recursively enumerable languages under this alternative definition.

Lemma 6.2.1. *For every phrase-structure grammar, $H = (V, T, P, S)$, there exists an equivalent phrase-structure grammar, $G = (V_G, T, P_G, S)$, so that each rule in P_G has one of these forms:*

(i) $AB \to CD$, where $A \neq C$;
(ii) $A \to BC$, where $A \neq B$;
(iii) $A \to x$,

where $V_G = N_G \cup T$, $A, B, C, D \in N_G$, and $x \in T \cup \{\varepsilon\}$.

Proof. Let $H = (V, T, P, S)$ be a grammar, $N = V - T$. Without any loss of generality, assume that H satisfies the Kuroda normal form. Define the grammar $G = (V_G, T, P_G, S)$, $V_G = N_G \cup T$, where N_G and P_G are constructed as follows:

   I. set $N_G = N$ and add P's rules that satisfy (i) through (iii) to P_G;
   II. for every $AB \to AD \in P$, add $AB \to A'D'$, $A'D' \to AD$ to P_G and A', D' to N_G, where A' and D' are two new nonterminals;
   III. for every $A \to AB \in P$, add $A \to A'B'$, $A'B' \to AB$ to P_G and A', B' to N_G, where A' and B' are two new nonterminals.

A formal proof that H and G are equivalent is left for the reader.   □

Theorem 6.2.2. $\mathbf{CF}^\circ = \mathbf{RE}$.

Proof. Construction. Consider that $G = (V, T, P, S)$ is a phrase-structure grammar, $N = V - T$. Without any loss of generality, assume that G satisfies the properties described in Lemma 6.2.1.

We construct the $\mathbf{CF}^\circ$ grammar, $\Gamma = (V_\Gamma, T, P_\Gamma, S_\Gamma)$, where $N_\Gamma = V_\Gamma - T = \{0, \bar{0}, 1, \bar{1}, 2, \bar{2}, S_\Gamma, \overline{S_\Gamma}\}$. Define the injections, $g : N \to \{0, 1\}^n$ and $h : N \to \{0, 1\}^{2n}$, such that $h(A) = g(A)\,\mathrm{reversal}(g(A))$, where $A \in N$ and $n = \lceil \log_2(\mathrm{card}(N)) \rceil$. Note that the inverses of $0, 1, 2 \in N_\Gamma$ are $\bar{0}, \bar{1}, \bar{2} \in N_\Gamma$, respectively. Furthermore, if $h(A) = a_1 \ldots a_n a_n \ldots a_1$, $a_i \in \{0, 1\}$ for $i = 1, 2, \ldots, n$, $n \geq 0$, then $\bar{h}(A) = \overline{a_1} \ldots \overline{a_n a_n} \ldots \overline{a_1}$. The set of rules, P_Γ, is constructed as follows:

I. add $S_\Gamma \rightarrow h(S)2$ to P_Γ;

II. for every $AB \rightarrow CD \in P$, add $2 \rightarrow \overline{h}(B)\overline{2h}(A)\overline{2}2h(C)2h(D)2$ to P_Γ;

III. for every $A \rightarrow BC \in P$, add $2 \rightarrow \overline{h}(A)\overline{2}2h(B)2h(C)2$ to P_Γ;

IV. for every $A \rightarrow x \in P$, add $2 \rightarrow \overline{h}(A)x$ to P_Γ,

where $A, B, C, D \in N$ and $x \in T \cup \{\varepsilon\}$. The construction of Γ is completed.

Basic Idea. In essence, Γ codes G's nonterminals in a binary way. Γ simulates G's application of a context-free rule of the form $A \rightarrow BC$ (see III.) so it rewrites 2 that follows $h(A)$ to $\overline{h}(A)\overline{2}2h(B)2h(C)2$; as a result, a substring $h(A)\overline{h}(A)$ is produced. This proper form is eliminated by a group reduction; any improper form—that is, any form different from $h(A)\overline{h}(A)$—cannot be eliminated in this way, so a terminal string is not generated at this point. Γ simulates the application of a rule of the form $A \rightarrow x$ from IV. analogously. In a similar way, Γ simulates the application of $AB \rightarrow CD$ (see II.). In that case, however, it rewrites 2 that follows $h(A)2h(B)$ to a string that starts with the inverse binary code of two nonterminals, $\overline{h}(B)\overline{2h}(A)$, rather than a single nonterminal.

Proof. First, we prove $L(G) \subseteq L(\Gamma)^\circ$. By induction, we demonstrate Claims 6.2.3 and 6.2.4. Without any loss of generality, assume that every $w \in L(G)$ can be generated by derivation of the form $S \Rightarrow^* w' \Rightarrow^* w$, where $w' \in N^*$ is generated from S using only rules of the form $AB \rightarrow CD$, $A \rightarrow BC$, and $A \rightarrow \varepsilon$, while w is obtained from w' using only rules of type $A \rightarrow a$, where $A, B, C, D \in N$ and $a \in T$.

Claim 6.2.3. If $S \Rightarrow^i y_1 y_2 \ldots y_m$ in G, then $S_\Gamma {}^\circ \Rightarrow^{i+1} h(y_1)2h(y_2)2 \ldots h(y_m)2$ in Γ, where $y_1, \ldots, y_m \in N$, $m \geq 0$.

Proof. Basis. Let $i = 0$. Then, $S \Rightarrow^0 S$ in G. By construction, $S_\Gamma \rightarrow h(S)2 \in P_\Gamma$, so $S_\Gamma \Rightarrow^1 h(S)2$ in Γ.

Induction Hypothesis. Assume that the implication of Claim 6.2.3 holds for all $0 \leq i \leq l$, where l is a non-negative integer.

Induction Step. Consider any derivation of the form $S \Rightarrow^{l+1} \beta$, where $\beta \in N^*$. Express this derivation as $S \Rightarrow^l \alpha \Rightarrow \beta$, where $\alpha \in N^*$. Express $\alpha = y_1 y_2 \ldots y_k$, where $y_1, \ldots, y_k \in N$, $k \geq 0$. By the induction hypothesis, $S_\Gamma {}^\circ \Rightarrow^{l+1} h(y_1)2h(y_2)2 \ldots h(y_k)2 = \alpha_\Gamma$ in Γ. Note that in the proof of this claim, we express the prefix and suffix of the current sentential form as $u = p_1 \ldots p_r$ and $v = q_1 \ldots q_s$, respectively, where $p_j, q_k \in N$, for $j = 1, \ldots, r$, $k = 1, \ldots, s$, $r, s \geq 0$. The following cases 1. through 3. cover all possibilities how G can make $\alpha \Rightarrow \beta$.

1. Let $AB \rightarrow CD \in P$, where $A, B, C, D \in N$. Then, $\alpha = uABv \Rightarrow uCDv = \beta$ in G. By II., $2 \rightarrow \overline{h}(B)\overline{2h}(A)\overline{2}2h(C)2h(D)2 \in P_\Gamma$ and $\alpha_\Gamma = h(p_1)2 \ldots$
$\ldots h(p_r)2h(A)2h(B)2h(q_1)2 \ldots h(q_s)2 {}^\circ \Rightarrow h(p_1)2 \ldots$
$\ldots h(p_r)2h(A)2h(B)\overline{h}(B)\overline{2h}(A)\overline{2}2h(C)2h(D)2h(q_1)2 \ldots h(q_s)2 =$
$= h(p_1)2 \ldots h(p_r)2h(C)2h(D)2h(q_1)2 \ldots h(q_s)2 = \beta_\Gamma$ in Γ. Thus $S \Rightarrow^{l+1} \beta$ in G and $S_\Gamma {}^\circ \Rightarrow^{l+2} \beta_\Gamma$ in Γ.

2. Let $A \rightarrow BC \in P$, where $A, B, C \in N$. Then, $\alpha = uAv \Rightarrow uBCv = \beta$ in G. By III., $2 \rightarrow \overline{h}(A)\overline{2}2h(B)2h(C)2 \in P_\Gamma$ and $\alpha_\Gamma = h(p_1)2 \ldots$
$\ldots h(p_r)2h(A)2h(q_1)2 \ldots h(q_s)2° \Rightarrow h(p_1)2 \ldots$
$\ldots h(p_r)2h(A)\overline{h}(A)\overline{2}2h(B)2h(C)2h(q_1)2 \ldots h(q_s)2 = h(p_1)2 \ldots$
$\ldots h(p_r)2h(B)2h(C)2h(q_1)2 \ldots h(q_s)2 = \beta_\Gamma$ in Γ. Clearly, $S \Rightarrow^{l+1} \beta$ in G and $S_\Gamma° \Rightarrow^{l+2} \beta_\Gamma$ in Γ as well.

3. Let $A \rightarrow \varepsilon \in P$, where $A \in N$. Then, $\alpha = uAv \Rightarrow uv = \beta$ in G. By IV., $2 \rightarrow \overline{h}(A) \in P_\Gamma$ and $\alpha_\Gamma = h(p_1)2 \ldots$
$\ldots h(p_r)2h(A)2h(q_1)2 \ldots h(q_s)2° \Rightarrow h(p_1)2 \ldots$
$\ldots h(p_r)2h(A)\overline{h}(A)\overline{2}2h(q_1)2 \ldots h(q_s)2 = h(p_1)2 \ldots$
$\ldots h(p_r)2h(q_1)2 \ldots h(q_s)2 = \beta_\Gamma$ in Γ. Clearly, $S \Rightarrow^{l+1} \beta$ in G and $S_\Gamma° \Rightarrow^{l+2} \beta_\Gamma$ in Γ.

The induction step is completed, so Claim 6.2.3 holds. □

Now, we establish Claim 6.2.4.

Claim 6.2.4. If $y_1 \ldots y_k \Rightarrow^k w$ in G using only rules of the form $A \rightarrow a \in P$, then $h(y_1)2 \ldots h(y_k)2° \Rightarrow^k w$ in Γ, where $A, y_i \in N$, for $i = 1, 2, \ldots, k$, $k \geq 0$, $a \in T$, and $w \in T^*$.

Without any loss of generality, assume that w is produced by the leftmost derivations.

Proof. Basis. Let $k = 0$. Then, $\varepsilon \Rightarrow^0 \varepsilon$ in G. Clearly, $\varepsilon° \Rightarrow^0 \varepsilon$ in Γ.

Induction Hypothesis. Assume that the implication of Claim 6.2.4 holds for every $0 \leq k \leq l$, where l is a non-negative integer.

Induction Step. Consider any sentential form $y_1y_2 \ldots y_ly_{l+1}$, where $y_i \in N$, for $i = 1, 2, \ldots, l + 1$. Express the derivation from Claim 6.2.4 as $y_1y_2 \ldots y_ly_{l+1} \Rightarrow^l wy_{l+1} \Rightarrow wa$, where $w \in T^*$, $a \in T$.

Let $y_{l+1} \rightarrow a \in P$, where $y_{l+1} \in N$ and $a \in T$. Then, $wy_{l+1} \Rightarrow wa$ in G. By IV., $2 \rightarrow \overline{h}(y_{l+1})a \in P_\Gamma$, so $wh(y_{l+1})2° \Rightarrow wh(y_{l+1})\overline{h}(y_{l+1})a = wa$ in Γ. Clearly, $h(y_1)2h(y_2)2 \ldots h(y_l)2h(y_{l+1})2° \Rightarrow^{l+1} wa$ in Γ.

The induction step is completed, so Claim 6.2.4 holds. □

By Claims 6.2.3 and 6.2.4, $L(G) \subseteq L(\Gamma)°$.

Next, we prove $L(\Gamma)° \subseteq L(G)$. Again, assume that every $w \in L(\Gamma)°$ can be generated by a derivation of the form $S_\Gamma° \Rightarrow^* w'° \Rightarrow^* w$, where w' is generated from S_Γ using only rules of form $S_\Gamma \rightarrow h(S)2$, $2 \rightarrow \overline{h}(B)\overline{2}h(A)\overline{2}2h(C)2h(D)2$, $2 \rightarrow \overline{h}(A)\overline{2}2h(B)2h(C)2$, and $2 \rightarrow \overline{h}(A)$, while w is obtained from w' using only rules of type $2 \rightarrow \overline{h}(A)a$, where $A, B, C, D \in N$ and $a \in T$. Note that w' is of the form $h(y_1)2 \ldots h(y_m)2$, where $y_j \in N$, for $j = 1, \ldots, m$, $m \geq 0$. By induction on any $i \geq 0$, we first establish Claim 6.2.5.

Claim 6.2.5. If $S_\Gamma° \Rightarrow h(S)2° \Rightarrow^i h(y_1)2h(y_2)2 \ldots h(y_m)2$ in Γ, then $S \Rightarrow^i y_1y_2 \ldots y_m$ in G, where $y_j \in N$, for $j = 1, 2, \ldots, m$.

Proof. Basis. Let $i = 0$. Then, $m = 1$, $h(y_1) = h(S)$ and $S_\Gamma \circ \Rightarrow h(S)2 \circ \Rightarrow^0 h(S)2$ in Γ. In G, $y_1 = S$ and $S \Rightarrow^0 S$.

Induction Hypothesis. Assume that Claim 6.2.5 holds for all $0 \leq i \leq l$, where l is a non-negative integer.

Induction Step. Consider any derivation of the form $S_\Gamma \circ \Rightarrow h(S)2 \circ \Rightarrow^{l+1} \beta$. More precisely, express this derivation as $S_\Gamma \circ \Rightarrow h(S)2 \circ \Rightarrow^l \alpha \circ \Rightarrow \beta$, where $\alpha, \beta \in V_\Gamma^\circ$. To be accurate, α and β are of the form $h(y_1)2 \ldots h(y_k)2$, where $y_j \in N$, for $j = 1, \ldots, k, k \geq 0$.

Assume that $\alpha = h(y_1)2h(y_2)2 \ldots h(y_k)2$, where $y_j \in N$, for $j = 1, \ldots, k, k \geq 0$. By the induction hypothesis, $S \Rightarrow^l y_1 y_2 \ldots y_k$. There are the following possibilities how Γ can make the derivation $\alpha \circ \Rightarrow \beta$. Note that in what follows, $u, v \in V_\Gamma^\circ$ and they are of the same form as α and β; in the case of G, $u_G, v_G \in N^*$.

1. Let $2 \to \overline{h}(A) \in P_\Gamma$, where $A \in N$. Then,

$$\alpha = uh(X)2h(Y)2v \Rightarrow uh(X)\overline{h}(A)h(Y)2v = \beta$$

where $X, Y \in N$. Express the substrings $h(X)$, $h(A)$, and $h(Y)$ as $h(X) = X_1 \ldots X_n X_n \ldots X_1$, $h(A) = A_1 \ldots A_n A_n \ldots A_1$, and $h(Y) = Y_1 \ldots Y_n Y_n \ldots Y_1$, respectively, where $X_i, A_i, Y_i \in N_\Gamma$, for $i = 1, 2, \ldots n$. Recall that

$$n = \lceil \log_2(\text{card}(N)) \rceil$$

and more precisely, $X_i, A_i, Y_i \in \{0, 1\}$. In 1.a. through 1.d, given next, we cover all possible cases concerning X, Y, and A.

1.a Consider that $X \neq A$ and $A \neq Y$. Let $X_i = A_i$ and $A_j = Y_j$, for every $i = 1, 2, \ldots, r$ and $j = 1, 2, \ldots, s$, where $0 \leq r < n, 0 \leq s < n$. Then, $\beta = uX_1 \ldots X_n X_n \ldots X_1 \overline{A_1} \ldots \overline{A_n A_n} \ldots \overline{A_1} Y_1 \ldots Y_n Y_n \ldots Y_1 2v$ is reduced to $\beta = uX_1 \ldots X_n X_n \ldots X_{r+1} \overline{A_{r+1}} \ldots \overline{A_n A_n} \ldots \overline{A_{s+1}} Y_{s+1} \ldots Y_n Y_n \ldots Y_1 2v$. Since the reduced sentential form contains the substring of inverses, $\overline{A_{r+1}} \ldots \overline{A_n A_n} \ldots \overline{A_{s+1}}$, and, moreover,

$$|X_1 \ldots X_n X_n \ldots X_{r+1} \overline{A_{r+1}} \ldots \overline{A_n A_n} \ldots \overline{A_{s+1}} Y_{s+1} \ldots Y_n Y_n \ldots Y_1| \neq 2n$$

and P_Γ contains no rule rewriting $\overline{0}$s and $\overline{1}$s, there is no way of removing them. Thus, we are not able to produce a valid terminal string.

1.b Now, consider that $X = A$ and $A \neq Y$. Then, β is reduced to $\beta = uY_1 \ldots Y_n Y_n \ldots Y_1 2v = uh(Y)2v$. Observe that $h(X)$ is removed. By IV., $A \to \varepsilon \in P$. Since $A = X$, then $\alpha_G = u_G XY v_G \Rightarrow u_G Y v_G = \beta_G$ in G. Clearly, $S \Rightarrow^{l+1} \beta_G$ in G and Claim 6.2.5 holds.

1.c Next, consider that $X \neq A$ and $A = Y$. Then, β is reduced to $\beta = uX_1 \ldots X_n X_n \ldots X_1 2v = uh(X)2v$. Observe that, in this case, $h(Y)$ is removed. By IV., $A \to \varepsilon \in P$. Since $A = Y$, then $\alpha_G = u_G XY v_G \Rightarrow u_G X v_G = \beta_G$ in G. Therefore, $S \Rightarrow^{l+1} \beta_G$ in G and Claim 6.2.5 holds too.

1.d Finally, consider that $X = A$ and $A = Y$. Then, $\beta = uX_1 \ldots X_n X_n \ldots$ $\ldots X_1 2v = uh(X)2v = uY_1 \ldots Y_n Y_n \ldots Y_1 2v = uh(Y)2v$. In that case, $h(X)$ or $h(Y)$ is removed. By IV., $A \rightarrow \varepsilon \in P$. Since $A = X = Y$, then $\alpha_G = u_G XY v_G \Rightarrow u_G X v_G = u_G Y v_G = \beta_G$ and, thus, $S \Rightarrow^{l+1} \beta_G$ in G. You can see that Claim 6.2.5 holds as well.

2. Let $2 \rightarrow \overline{h}(A)\overline{2}2h(B)2h(C)2 \in P_\Gamma$, where $A, B, C \in N$. Then,

$$\alpha = uh(X)2h(Y)2v \Rightarrow uh(X)\overline{h}(A)\overline{2}2h(B)2h(C)2h(Y)2v = \beta$$

where $X, Y \in N$. By Lemma 6.2.1, $A \neq B$. In 2.a. and 2.b, given next, we cover all possible cases concerning X and A.

2.a Consider that $X \neq A$. Observe that the situation is analogical to 1.a, so the derivation of a terminal string is blocked.

2.b Now, consider that $X = A$. In that case, the situation is analogical to 1.b. The resulting sentential form is $\beta = uh(B)2h(C)2h(Y)2v$. By III., $A \rightarrow BC \in P$, so $\alpha_G = u_G XY v_G = u_G AY v_G \Rightarrow u_G BCY v_g = \beta_G$. Obviously, $S \Rightarrow^{l+1} \beta_G$ in G and Claim 6.2.5 holds.

3. Let $2 \rightarrow \overline{h}(B)\overline{2h}(A)\overline{2}2h(C)2h(D)2 \in P_\Gamma$, where $A, B, C, D \in N$. Then, $\alpha = uh(X)2h(Y)2h(Z)2v \, ° \Rightarrow uh(X)2h(Y)\overline{h}(B)\overline{2h}(A)\overline{2}2h(C)2h(D)2h(Z)2v = \beta$. By Lemma 6.2.1, $A \neq C$. In 3.a. through 3.c, given next, we cover all possible cases concerning X, Y, A, and B.

3.a Consider that $Y = B$ and $X \neq A$. Then, the subsequence $2h(Y)\overline{h}(B)\overline{2}$ is erased and the resulting, still non-completely-reduced, sentential form is $\beta = uh(X)\overline{h}(A)h(C)2h(D)2h(Z)2v$, so the situation is analogical to 1.a. and the derivation of a valid terminal string is blocked.

3.b Now, consider that $Y \neq B$. Then, the resulting non-completely-reduced sentential form is $\beta = uh(X)2h(Y)\overline{h}(B)\overline{2h}(A)h(C)2h(D)2h(Z)2v$. Express the subsequences $h(X)$, $h(Y)$, $h(B)$, $h(A)$, and $h(C)$ as $h(X) = X_1 \ldots X_n X_n \ldots X_1$, $h(Y) = Y_1 \ldots Y_n Y_n \ldots Y_1$, $h(B) = B_1 \ldots B_n B_n \ldots$ $\ldots B_1$, $h(A) = A_1 \ldots A_n A_n \ldots A_1$, and $h(C) = C_1 \ldots C_n C_n \ldots C_1$, respectively, where $B_i, C_i, X_i, Y_i \in \{0, 1\}$, for $i = 1, 2, \ldots n$. Recall that by Lemma 6.2.1, $A \neq C$. Assume that $Y_i = B_i$, for every $i = 1, 2, \ldots, r$ and $A_j = C_j$, for every $j = 1, 2, \ldots, s$, where $0 \le r < n$ and $0 \le s < n$. Then, β is reduced to $uX_1 \ldots X_n X_n \ldots X_1 2Y_1 \ldots Y_n Y_n \ldots Y_{r+1}\overline{B_{r+1}} \ldots$ $\ldots \overline{B_n B_n} \ldots \overline{B_1} 2\overline{A_1} \ldots \overline{A_n A_n} \ldots \overline{A_{s+1}} C_{s+1} \ldots C_n C_n \ldots C_1 2h(D)2h(z)2v$. Observe that the possible equality or inequality of X and A is not important in this case. Since the reduced sentential form contains the substring $\overline{B_{r+1}} \ldots \overline{B_n B_n} \ldots \overline{B_1} 2\overline{A_1} \ldots \overline{A_n A_n} \ldots \overline{A_{s+1}}$ and moreover, the constant length $2n$ of encoded nonterminals is violated, there is no way of deriving a valid terminal string, so the derivation is blocked.

3.c Finally, consider that $Y = B$ and $X = A$. Then, β is reduced to $\beta = uh(C)2h(D)2h(Z)2v$. This is the correct sentential form. By II., $AB \rightarrow CD \in P$, so $\alpha_G = u_G XYZ v_G = u_G ABZ v_G \Rightarrow u_G CDZ v_G = \beta_G$. Indeed, $S \Rightarrow^{l+1} \beta_G$ in G and Claim 6.2.5 holds.

The induction step is completed, so Claim 6.2.5 holds.   □

Now, we establish Claim 6.2.6.

Claim 6.2.6. If $h(y_1)2\ldots h(y_k)2\,^{\circ}\Rightarrow^k w$ in Γ using only rules of the form $2 \to \overline{h}(A)a$, then $y_1\ldots y_k \Rightarrow^k w$ in G, where $A, y_i \in N$, for $i = 1, 2,\ldots k$, $k \geq 0$, $a \in T$, and $w \in T^*$. Without any loss of generality, assume that the terminal string is produced by the leftmost derivations.

Proof. Basis. Let $k = 0$. Then, $\varepsilon\,^{\circ}\Rightarrow^0 \varepsilon$ in Γ. Clearly, $\varepsilon \Rightarrow^0 \varepsilon$ in G.

Induction Hypothesis. Assume that the implication of Claim 6.2.6 holds for every $0 \leq k \leq l$, where l is a non-negative integer.

Induction Step. Consider any sentential form of the form

$$h(y_1)2h(y_2)2\ldots h(y_l)2h(y_{l+1})2$$

where $y_i \in N$ for $i = 1, 2,\ldots l + 1$. Express the derivation from Claim 6.2.6 as $h(y_1)2h(y_2)2\ldots h(l_k)2h(y_{l+1})2\,^{\circ}\Rightarrow^k wh(y_{l+1})2\,^{\circ}\Rightarrow wa$, where $w \in T^*$, $a \in T$.

Let $2 \to \overline{h}(y_{l+1})a \in P_\Gamma$, where $y_{l+1} \in N$ and $a \in T$. Then, $wh(y_{l+1})2\,^{\circ}\Rightarrow wh(y_{l+1})\overline{h}(y_{l+1})a = wa$ in Γ. By IV., $y_{l+1} \to a \in P$, so $wy_{l+1} \Rightarrow wa$ in G. Clearly, $y_1y_2\ldots y_ky_{l+1} \Rightarrow^{l+1} wa$ in G too.

The induction step is completed, so Claim 6.2.6 holds.   □

By Claims 6.2.5 and 6.2.6, $L(\Gamma)^{\circ} \subseteq L(G)$.

All together, Claims 6.2.3, 6.2.4, 6.2.5, and 6.2.6 imply $L(G) = L(\Gamma)^{\circ}$, so **CF$^{\circ}$ = RE**.   □

Theorem 6.2.7. E0L$^{\circ}$ = RE.

Proof. Consider a phrase-structure grammar, $G = (V, T, P, S)$, satisfying the properties described in Lemma 6.2.1. Define the injections, $g : N \to \{0, 1\}^n$ and $h : N \to \{0, 1\}^{2n}$, such that $h(A) = g(A)\,\text{reversal}(g(A))$, where $A \in N$ and $n = \lceil \log_2(\text{card}(N)) \rceil$. If $h(A) = a_1\ldots a_n a_n\ldots a_1$, $a_i \in \{0, 1\}$, where $i = 1, 2,\ldots, n$, $n \geq 0$, then $\overline{h}(A) = \overline{a_1}\ldots\overline{a_n}\overline{a_n}\ldots\overline{a_1}$.

We construct the **E0L$^{\circ}$** grammar, $\Gamma = (V_\Gamma, T, P_\Gamma, w_\Gamma)$, where $V_\Gamma - T = \{0, \overline{0}, 1, \overline{1}, 2, \overline{2}\}$ and $w_\Gamma = h(S)2$ as follows:

   I.   add $0 \to 0$, $1 \to 1$, and $2 \to 2$ to P_Γ;
   II. for every $AB \to CD \in P$, add $2 \to \overline{h}(B)\overline{2}h(A)\overline{2}2h(C)2h(D)2$ to P_Γ;
   III. for every $A \to BC \in P$, add $2 \to \overline{h}(A)\overline{2}2h(B)2h(C)2$ to P_Γ;
   IV. for every $A \to x \in P$, add $2 \to \overline{h}(A)x$ to P_Γ,

where $A, B, C, D \in N$ and $x \in T \cup \{\varepsilon\}$.

A formal verification of this construction is analogical to the proof of Theorem 6.2.7, so it is left to the reader.   □

Corollary 6.2.8. CF$^{\circ}$ = E0L$^{\circ}$.

6.2.1   Conclusion

As the main result of this chapter, we have demonstrated that the context-free and E0L grammars defined over free groups define the entire family of recursively enumerable languages. It is noteworthy, however, that the proofs of Theorems 6.2.2 and 6.2.7 are based on a proof technique that can be also used to achieve some well known results of the formal language theory in an alternative way.

Specifically, consider the characterization of recursively enumerable languages by phrase-structure grammars (defined standardly over free monoids) with only two context rules—$AB \rightarrow \varepsilon$ and $CD \rightarrow \varepsilon$ (see Theorem 3.1.11). By an easy modification of the proof technique mentioned above, we can achieve this result in a completely new, alternative way. We leave the details of this modification to the reader because it is simple and, as already noted, out of the main topic of the present chapter.

Part III
Modern Automata

To a large extent, in terms of automata, this four-chapter part parallels what Part II covers in terms of grammars. Indeed, Chap. 7 gives the fundamentals of automata that formalize regulated computation. Similarly to grammars discussed in Chap. 5, Chap. 8 discusses automata that formalize a discontinuous way of computation so they jump across the words they work on discontinuously. Chapter 9 discusses language models for computation based upon new data structures. More specifically, it studies deep pushdown automata, underlined by stacks that can be modified deeper than on their top. Finally, Chap. 10 studies automata that work over free groups, so it parallels Chap. 6, which studies the same topic in grammatical terms.

Part III
Modern Automata

Chapter 7
Regulated Automata and Computation

Just like there exist regulated grammars, which formalize regulated computation (see Chap. 3), there also exist their automata-based counterparts for this purpose. Basically, in a very natural and simple way, these automata regulate the selection of rules according to which their sequences of moves are made. These regulated automata represent the principle subject of the present chapter, which covers their most essential types.

The chapter is divided into three sections. Section 7.1 discusses self-regulating versions of finite and pushdown automata. Section 7.2 studies these automata with control languages by which they regulate their computation. Section 7.3 investigates self-reproducing pushdown transducers.

7.1 Self-Regulating Automata

This chapter studies finite and pushdown automata that regulate the selection of a rule according to which the current move is made by a rule according to which a previous move was made, hence their name—self-regulating automata. To give a more precise insight into self-regulating automata, consider a finite automaton M with a finite binary relation R over the set of rules in M. Furthermore, suppose that M makes a sequence of moves ρ that leads to the acceptance of a string, so ρ can be expressed as a concatenation of $n + 1$ consecutive subsequences, $\rho = \rho_0 \rho_1 \cdots \rho_n$, where $|\rho_k| = |\rho_j|$, $0 \leq k, j \leq n$, in which r_i^j denotes the rule according to which the ith move in ρ_j is made, for all $0 \leq j \leq n$ and $1 \leq i \leq |\rho_j|$ (as usual, $|\rho_j|$ denotes the length of ρ_j). If for all $0 \leq j < n$, $(r_1^j, r_1^{j+1}) \in R$, then M represents an n-turn

© Springer International Publishing AG 2017
A. Meduna, O. Soukup, *Modern Language Models and Computation*,
DOI 10.1007/978-3-319-63100-4_7

first-move self-regulating finite automaton with respect to R. If for all $0 \leq j < n$ and all $1 \leq i \leq |\rho_i|$, $(r_i^j, r_i^{j+1}) \in R$, then M represents an *n-turn all-move self-regulating finite automaton with respect to R.*

Section 7.1 is divided into two sections. In Sect. 7.1.1, based on the number of turns, we establish two infinite hierarchies of language families that lie between the families of regular and context-sensitive languages. First, we demonstrate that n-turn first-move self-regulating finite automata give rise to an infinite hierarchy of language families coinciding with the hierarchy resulting from $(n + 1)$-parallel right-linear grammars (see [RW73, RW75, Woo73, Woo75]). Recall that n-parallel right-linear grammars generate a proper language subfamily of the language family generated by $(n + 1)$-parallel right-linear grammars (see Theorem 5 in [RW75]). As a result, n-turn first-move self-regulating finite automata accept a proper language subfamily of the language family accepted by $(n+1)$-turn first-move self-regulating finite automata, for all $n \geq 0$. Similarly, we prove that n-turn all-move self-regulating finite automata give rise to an infinite hierarchy of language families coinciding with the hierarchy resulting from $(n + 1)$-right-linear simple matrix grammars (see [DP89, Iba70, Woo75]). As n-right-linear simple matrix grammars generate a proper subfamily of the language family generated by $(n + 1)$-right-linear simple matrix grammars (see Theorem 1.5.4 in [DP89]), n-turn all-move self-regulating finite automata accept a proper language subfamily of the language family accepted by $(n + 1)$-turn all-move self-regulating finite automata. Furthermore, since the families of right-linear simple matrix languages coincide with the language families accepted by multi-tape non-writing automata (see [FR68]) and by finite-turn checking automata (see [Sir71]), all-move self-regulating finite automata characterize these families, too. Finally, we summarize the results about both infinite hierarchies.

In Sect. 7.1.2, by analogy with self-regulating finite automata, we introduce and discuss *self-regulating pushdown automata*. Regarding self-regulating all-move pushdown automata, we prove that they do not give rise to any infinite hierarchy analogical to the achieved hierarchies resulting from the self-regulating finite automata. Indeed, zero-turn all-move self-regulating pushdown automata define the family of context-free languages while one-turn all-move self-regulating pushdown automata define the family of recursively enumerable languages. On the other hand, as far as self-regulating first-move pushdown automata are concerned, the question whether they define an infinite hierarchy is open.

7.1.1 Self-Regulating Finite Automata

First, the present section defines n-turn first-move self-regulating finite automata and n-turn all-move self-regulating finite automata. Then, it determines the accepting power of these automata.

Definitions and Examples

In this section, we define and illustrate n-turn first-move self-regulating finite automata and n-turn all-move self-regulating finite automata. Recall the formalization of rule labels from Definition 2.4.3 because we make use of it in this section frequently.

Definition 7.1.1. A *self-regulating finite automaton* (an *SFA* for short) is a septuple

$$M = (Q, \Sigma, \delta, q_0, q_t, F, R)$$

where

1. $(Q, \Sigma, \delta, q_0, F)$ is a finite automaton,
2. $q_t \in Q$ is a *turn state*, and
3. $R \subseteq \Psi \times \Psi$ is a finite relation on the alphabet of rule labels. □

In this chapter, we consider two ways of self-regulation—first-move and all-move. According to these two types of self-regulation, two types of n-turn self-regulating finite automata are defined.

Definition 7.1.2. Let $n \geq 0$ and $M = (Q, \Sigma, \delta, q_0, q_t, F, R)$ be a self-regulating finite automaton. M is said to be an *n-turn first-move self-regulating finite automaton* (an *n-first-SFA* for short) if every $w \in L(M)$ is accepted by M in the following way

$$q_0 w \vdash_M^* f [\mu]$$

such that

$$\mu = r_1^0 \cdots r_k^0 r_1^1 \cdots r_k^1 \cdots r_1^n \cdots r_k^n$$

where $k \geq 1$, r_k^0 is the first rule of the form $qx \rightarrow q_t$, for some $q \in Q$, $x \in \Sigma^*$, and

$$(r_1^j, r_1^{j+1}) \in R$$

for all $j = 0, 1, \ldots, n$. □

The family of languages accepted by n-first-SFAs is denoted by **FSFA**$_n$.

Example 7.1.3. Consider a 1-first-SFA

$$M = (\{s, t, f\}, \{a, b\}, \delta, s, t, \{f\}, \{(1, 3)\})$$

with δ containing rules (see Fig. 7.1)

$$
\begin{aligned}
&1\colon sa \rightarrow s \\
&2\colon sa \rightarrow t \\
&3\colon tb \rightarrow f \\
&4\colon fb \rightarrow f
\end{aligned}
$$

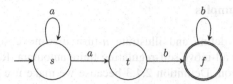

Fig. 7.1 1-turn first-move self-regulating finite automaton M

With *aabb*, M makes

$$saabb \vdash_M sabb\,[1] \vdash_M tbb\,[2] \vdash_M fb\,[3] \vdash_M f\,[4]$$

In brief, $saabb \vdash_M^* f\,[1234]$. Observe that $L(M) = \{a^n b^n \mid n \geq 1\}$, which belongs to $\mathbf{CF} - \mathbf{REG}$. □

Definition 7.1.4. Let $n \geq 0$ and $M = (Q, \Sigma, \delta, q_0, q_t, F, R)$ be a self-regulating finite automaton. M is said to be an *n-turn all-move self-regulating finite automaton* (an *n-all-SFA* for short) if every $w \in L(M)$ is accepted by M in the following way

$$q_0 w \vdash_M^* f\,[\mu]$$

such that

$$\mu = r_1^0 \cdots r_k^0 r_1^1 \cdots r_k^1 \cdots r_1^n \cdots r_k^n$$

where $k \geq 1$, r_k^0 is the first rule of the form $qx \to q_t$, for some $q \in Q, x \in \Sigma^*$, and

$$(r_i^j, r_i^{j+1}) \in R$$

for all $i = 1, 2, \ldots, k$ and $j = 0, 1, \ldots, n-1$. □

The family of languages accepted by n-all-SFAs is denoted by $\mathbf{ASFA}_n$.

Example 7.1.5. Consider a 1-all-SFA

$$M = \big(\{s, t, f\}, \{a, b\}, \delta, s, t, \{f\}, \{(1, 4), (2, 5), (3, 6)\}\big)$$

with δ containing the following rules (see Fig. 7.2)

$$
\begin{aligned}
&1\!: sa \to s\\
&2\!: sb \to s\\
&3\!: s \to t\\
&4\!: ta \to t\\
&5\!: tb \to t\\
&6\!: t \to f
\end{aligned}
$$

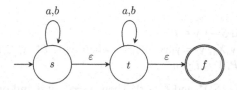

Fig. 7.2 1-turn all-move self-regulating finite automaton M

With *abab*, M makes

$$sabab \vdash_M sbab\,[1] \vdash_M sab\,[2] \vdash_M tab\,[3] \vdash_M tb\,[4] \vdash_M t\,[5] \vdash_M f\,[6]$$

In brief, $sabab \vdash_M^* f$ [123456]. Observe that $L(M) = \{ww \mid w \in \{a, b\}^*\}$, which belongs to **CS − CF**. □

Accepting Power

In this section, we discuss the accepting power of n-first-SFAs and n-all-SFAs.

n-Turn First-Move Self-Regulating Finite Automata

We prove that the family of languages accepted by n-first-SFAs coincides with the family of languages generated by so-called $(n + 1)$-parallel right-linear grammars (see [RW73, RW75, Woo73, Woo75]). First, however, we define these grammars formally.

Definition 7.1.6. For $n \geq 1$, an *n-parallel right-linear grammar* (see [RW73, RW75, Woo73, Woo75]) (an *n-PRLG* for short) is an $(n + 3)$-tuple

$$G = (N_1, \ldots, N_n, T, S, P)$$

where N_i, $1 \leq i \leq n$, are pairwise disjoint *nonterminal alphabets*, T is a *terminal alphabet*, $S \notin N$ is an *initial symbol*, where $N = N_1 \cup \cdots \cup N_n$, and P is a finite set of *rules* that contains these three kinds of rules

1. $S \rightarrow X_1 \cdots X_n$, $X_i \in N_i, 1 \leq i \leq n$;
2. $X \rightarrow wY$, $X, Y \in N_i$ for some i, $1 \leq i \leq n, w \in T^*$;
3. $X \rightarrow w$, $X \in N, w \in T^*$.

For $x, y \in (N \cup T \cup \{S\})^*$,

$$x \Rightarrow_G y$$

if and only if

1. either $x = S$ and $S \to y \in P$,
2. or $x = y_1 X_1 \cdots y_n X_n$, $y = y_1 x_1 \cdots y_n x_n$, where $y_i \in T^*$, $x_i \in T^* N \cup T^*$, $X_i \in N_i$, and $X_i \to x_i \in P$, $1 \leq i \leq n$.

Let $x, y \in (N \cup T \cup \{S\})^*$ and $\ell > 0$. Then, $x \Rightarrow_G^\ell y$ if and only if there exists a sequence

$$x_0 \Rightarrow_G x_1 \Rightarrow_G \cdots \Rightarrow_G x_\ell$$

where $x_0 = x$, $x_\ell = y$. As usual, $x \Rightarrow_G^+ y$ if and only if there exists $\ell > 0$ such that $x \Rightarrow_G^\ell y$, and $x \Rightarrow_G^* y$ if and only if $x = y$ or $x \Rightarrow_G^+ y$.

The *language* of G is defined as

$$L(G) = \{ w \in T^* \mid S \Rightarrow_G^+ w \}$$

A language $K \subseteq T^*$ is an *n-parallel right-linear language* (*n*-PRLL for short) if there is an *n*-PRLG G such that $K = L(G)$. □

The family of *n*-PRLLs is denoted by $\mathbf{PRL}_n$.

Definition 7.1.7. Let $G = (N_1, \ldots, N_n, T, S, P)$ be an *n*-PRLG, for some $n \geq 1$, and $1 \leq i \leq n$. By the *ith component* of G, we understand the 1-PRLG

$$G = (N_i, T, S', P')$$

where P' contains rules of the following forms:

1. $S' \to X_i$ if $S \to X_1 \cdots X_n \in P$, $X_i \in N_i$;
2. $X \to wY$ if $X \to wY \in P$ and $X, Y \in N_i$;
3. $X \to w$ if $X \to w \in P$ and $X \in N_i$. □

To prove that the family of languages accepted by *n*-first-SFAs coincides with the family of languages generated by $(n+1)$-PRLGs, we need the following normal form of PRLGs.

Lemma 7.1.8. *For every n-PRLG $G = (N_1, \ldots, N_n, T, S, P)$, there is an equivalent n-PRLG $G' = (N_1', \ldots, N_n', T, S, P')$ that satisfies:*

(i) *if $S \to X_1 \cdots X_n \in P'$, then X_i does not occur on the right-hand side of any rule, for $i = 1, 2, \ldots, n$;*
(ii) *if $S \to \alpha$, $S \to \beta \in P'$ and $\alpha \neq \beta$, then $\mathrm{alph}(\alpha) \cap \mathrm{alph}(\beta) = \emptyset$.*

Proof. If G does not satisfy the conditions from the lemma, then we construct a new *n*-PRLG

$$G' = (N_1', \ldots, N_n', T, S, P')$$

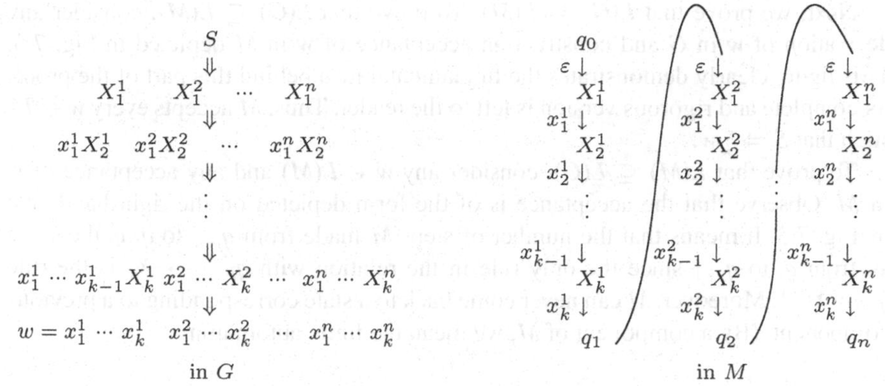

$$
\begin{array}{ccc}
S \\
\Downarrow \\
X_1^1 \quad X_1^2 \quad \cdots \quad X_1^n \\
\Downarrow \\
x_1^1 X_2^1 \quad x_1^2 X_2^2 \quad \cdots \quad x_1^n X_2^n \\
\Downarrow \\
\vdots \\
\Downarrow \\
x_1^1 \cdots x_{k-1}^1 X_k^1 \; x_1^2 \cdots X_k^2 \; \cdots \; x_1^n \cdots X_k^n \\
\Downarrow \\
w = x_1^1 \cdots x_k^1 \quad x_1^2 \cdots x_k^2 \quad \cdots \quad x_1^n \cdots x_k^n \\
\text{in } G
\end{array}
$$

Fig. 7.3 A derivation of w in G and the corresponding acceptance of w in M

where P' contains all rules of the form $X \to \beta \in P$, $X \neq S$, and $N_j \subseteq N'_j$, $1 \le j \le n$. For each rule $S \to X_1 \cdots X_n \in P$, we add new nonterminals $Y_j \notin N'_j$ into N'_j, and rules include $S \to Y_1 \cdots Y_n$ and $Y_j \to X_j$ in P', $1 \le j \le n$. Clearly,

$$S \Rightarrow_G X_1 \cdots X_n \text{ if and only if } S \Rightarrow_{G'} Y_1 \cdots Y_n \Rightarrow_{G'} X_1 \cdots X_n$$

Thus, $L(G) = L(G')$. $\qquad\square$

Lemma 7.1.9. *Let G be an n-PRLG. Then, there is an $(n-1)$-first-SFA M such that $L(G) = L(M)$.*

Proof. Informally, M is divided into n parts (see Fig. 7.3). The ith part represents a finite automaton accepting the language of the ith component of G, and R also connects the ith part to the $(i+1)$st part as depicted in Fig. 7.3.

Formally, without loss of generality, we assume $G = (N_1, \ldots, N_n, T, S, P)$ to be in the form from Lemma 7.1.8. We construct an $(n-1)$-first-SFA

$$M = (Q, T, \delta, q_0, q_t, F, R)$$

where

$Q = \{q_0, \ldots, q_n\} \cup N, N = N_1 \cup \cdots \cup N_n, \{q_0, q_1, \ldots, q_n\} \cap N = \emptyset, F = \{q_n\}$

$\delta = \{q_i \to X_{i+1} \mid S \to X_1 \cdots X_n \in P, 0 \le i < n\} \cup$
$\quad \{Xw \to Y \mid X \to wY \in P\} \cup$
$\quad \{Xw \to q_i \mid X \to w \in P, w \in T^*, X \in N_i, i \in \{1, \ldots, n\}\}$

$q_t = q_1$

$\Psi = \delta$

$R = \{(q_i \to X_{i+1}, q_{i+1} \to X_{i+2}) \mid S \to X_1 \cdots X_n \in P, 0 \le i \le n-2\}$

Next, we prove that $L(G) = L(M)$. To prove that $L(G) \subseteq L(M)$, consider any derivation of w in G and construct an acceptance of w in M depicted in Fig. 7.3. This figure clearly demonstrates the fundamental idea behind this part of the proof; its complete and rigorous version is left to the reader. Thus, M accepts every $w \in T^*$ such that $S \Rightarrow_G^* w$.

To prove that $L(M) \subseteq L(G)$, consider any $w \in L(M)$ and any acceptance of w in M. Observe that the acceptance is of the form depicted on the right-hand side of Fig. 7.3. It means that the number of steps M made from q_{i-1} to q_i is the same as from q_i to q_{i+1} since the only rule in the relation with $q_{i-1} \rightarrow X_1^i$ is the rule $q_i \rightarrow X_1^{i+1}$. Moreover, M can never come back to a state corresponding to a previous component. (By a component of M, we mean the finite automaton

$$M_i = (Q, \Sigma, \delta, q_{i-1}, \{q_i\})$$

for $1 \le i \le n$.) Next, construct a derivation of w in G. By Lemma 7.1.8, we have

$$\text{card}\left(\{X \mid (q_i \rightarrow X_1^{i+1}, q_{i+1} \rightarrow X) \in R\}\right) = 1$$

for all $0 \le i < n - 1$. Thus, $S \rightarrow X_1^1 X_1^2 \cdots X_1^n \in P$. Moreover, if $X_j^i x_j^i \rightarrow X_{j+1}^i$, we apply $X_j^i \rightarrow x_j^i X_{j+1}^i \in P$, and if $X_k^i x_k^i \rightarrow q_i$, we apply $X_k^i \rightarrow x_k^i \in P$, $1 \le i \le n$, $1 \le j < k$.

Hence, Lemma 7.1.9 holds. $\square$

Lemma 7.1.10. *Let M be an n-first-SFA. Then, there is an $(n+1)$-PRLG G such that $L(G) = L(M)$.*

Proof. Let $M = (Q, \Sigma, \delta, q_0, q_t, F, R)$. Consider

$$G = (N_0, \dots, N_n, \Sigma, S, P)$$

where

$N_i = (Q\Sigma^l \times Q \times \{i\} \times Q) \cup (Q \times \{i\} \times Q)$
$l = \max(\{|w| \mid qw \rightarrow p \in \delta\}), 0 \le i \le n$
$P = \{S \rightarrow [q_0 x_0, q^0, 0, q_t][q_t x_1, q^1, 1, q_{i_1}][q_{i_1} x_2, q^2, 2, q_{i_2}] \cdots [q_{i_{n-1}} x_n, q^n, n, q_{i_n}] \mid$
$\qquad r_0: q_0 x_0 \rightarrow q^0, r_1: q_t x_1 \rightarrow q^1, r_2: q_{i_1} x_2 \rightarrow q^2, \dots, r_n: q_{i_{n-1}} x_n \rightarrow q^n \in \delta,$
$\qquad (r_0, r_1), (r_1, r_2), \dots, (r_{n-1}, r_n) \in R, q_{i_n} \in F\} \cup$
$\qquad \{[px, q, i, r] \rightarrow x[q, i, r]\} \cup$
$\qquad \{[q, i, q] \rightarrow \varepsilon \mid q \in Q\} \cup$
$\qquad \{[q, i, p] \rightarrow w[q', i, p] \mid qw \rightarrow q' \in \delta\}$

Next, we prove that $L(G) = L(M)$. To prove that $L(G) \subseteq L(M)$, observe that we make $n + 1$ copies of M and go through them similarly to Fig. 7.3. Consider a derivation of w in G. Then, in a greater detail, this derivation is of the form

$$S \Rightarrow_G [q_0x_0^0, q_1^0, 0, q_t][q_t x_0^1, q_1^1, 1, q_{i_1}] \cdots [q_{i_{n-1}} x_0^n, q_1^n, n, q_{i_n}]$$

$$\Rightarrow_G x_0^0[q_1^0, 0, q_t] x_0^1[q_1^1, 1, q_{i_1}] \cdots x_0^n[q_1^n, n, q_{i_n}]$$

$$\Rightarrow_G x_0^0 x_1^0[q_2^0, 0, q_t] x_0^1 x_1^1[q_2^1, 1, q_{i_1}] \cdots x_0^n x_1^n[q_2^n, n, q_{i_n}] \tag{7.1}$$

$$\vdots$$

$$\Rightarrow_G x_0^0 x_1^0 \cdots x_k^0[q_t, 0, q_t] x_0^1 x_1^1 \cdots x_k^1[q_{i_1}, 1, q_{i_1}] \cdots x_0^n x_1^n \cdots x_k^n[q_{i_n}, n, q_{i_n}]$$

$$\Rightarrow_G x_0^0 x_1^0 \cdots x_k^0 x_0^1 x_1^1 \cdots x_k^1 \cdots x_0^n x_1^n \cdots x_k^n$$

and

$$r_0 : q_0 x_0^0 \to q_1^0, r_1 : q_t x_0^1 \to q_1^1, r_2 : q_{i_1} x_0^2 \to q_1^2, \ldots, r_n : q_{i_{n-1}} x_0^n \to q_1^n \in \delta$$

$$(r_0, r_1), (r_1, r_2), \ldots, (r_{n-1}, r_n) \in R$$

and $q_{i_n} \in F$.

Thus, the sequence of rules used in the acceptance of w in M is

$$\mu = (q_0 x_0^0 \to q_1^0)(q_1^0 x_1^0 \to q_2^0) \cdots (q_k^0 x_k^0 \to q_t)$$

$$(q_t x_0^1 \to q_1^1)(q_1^1 x_1^1 \to q_2^1) \cdots (q_k^1 x_k^1 \to q_{i_1})$$

$$(q_{i_1} x_0^2 \to q_1^2)(q_1^2 x_1^2 \to q_2^2) \cdots (q_k^2 x_k^2 \to q_{i_2}) \tag{7.2}$$

$$\vdots$$

$$(q_{i_{n-1}} x_0^n \to q_1^n)(q_1^n x_1^n \to q_2^n) \cdots (q_k^n x_k^n \to q_{i_n}).$$

Next, we prove that $L(M) \subseteq L(G)$. Informally, the acceptance is divided into $n+1$ parts of the same length. Grammar G generates the ith part by the ith component and records the state from which the next component starts.

Let μ be a sequence of rules used in an acceptance of

$$w = x_0^0 x_1^0 \cdots x_k^0 x_0^1 x_1^1 \cdots x_k^1 \cdots x_0^n x_1^n \cdots x_k^n$$

in M of the form (7.2). Then, the derivation of the form (7.1) is the corresponding derivation of w in G since $[q_j^i, i, p] \to x_j^i[q_{j+1}^i, i, p] \in P$ and $[q, i, q] \to \varepsilon$, for all $0 \le i \le n$, $1 \le j < k$.

Hence, Lemma 7.1.10 holds. $\square$

The first main result of this chapter follows next.

Theorem 7.1.11. *For all* $n \ge 0$, $\mathbf{FSFA}_n = \mathbf{PRL}_{n+1}$.

Proof. This proof follows from Lemmas 7.1.9 and 7.1.10. $\square$

Corollary 7.1.12. *The following statements hold true.*

(i) $\mathbf{REG} = \mathbf{FSFA}_0 \subset \mathbf{FSFA}_1 \subset \mathbf{FSFA}_2 \subset \cdots \subset \mathbf{CS}$
(ii) $\mathbf{FSFA}_1 \subset \mathbf{CF}$
(iii) $\mathbf{FSFA}_2 \nsubseteq \mathbf{CF}$
(iv) $\mathbf{CF} \nsubseteq \mathbf{FSFA}_n$ *for any* $n \geq 0$.
(v) *For all* $n \geq 0$, $\mathbf{FSFA}_n$ *is closed under union, finite substitution, homomorphism, intersection with a regular language, and right quotient with a regular language.*
(vi) *For all* $n \geq 1$, $\mathbf{FSFA}_n$ *is not closed under intersection and complement.*

Proof. Recall the following statements that are proved in [RW75].

- $\mathbf{REG} = \mathbf{PRL}_1 \subset \mathbf{PRL}_2 \subset \mathbf{PRL}_3 \subset \cdots \subset \mathbf{CS}$
- $\mathbf{PRL}_2 \subset \mathbf{CF}$
- $\mathbf{CF} \nsubseteq \mathbf{PRL}_n$, $n \geq 1$.
- For all $n \geq 1$, $\mathbf{PRL}_n$ is closed under union, finite substitution, homomorphism, intersection with a regular language, and right quotient with a regular language.
- For all $n \geq 2$, $\mathbf{PRL}_n$ is not closed under intersection and complement.

These statements and Theorem 7.1.11 imply statements (i), (ii), (iv), (v), and (vi) in Corollary 7.1.12. Moreover, observe that

$$\{a^n b^n c^{2n} \mid n \geq 0\} \in \mathbf{FSFA}_2 - \mathbf{CF}$$

which proves (iii).   □

Theorem 7.1.13. *For all* $n \geq 1$, $\mathbf{FSFA}_n$ *is not closed under inverse homomorphism.*

Proof. For $n = 1$, let $L = \{a^k b^k \mid k \geq 1\}$, and let the homomorphism $h : \{a, b, c\}^* \rightarrow \{a, b\}^*$ be defined as $h(a) = a$, $h(b) = b$, and $h(c) = \varepsilon$. Then, $L \in \mathbf{FSFA}_1$, but

$$L' = h^{-1}(L) \cap c^* a^* b^* = \{c^* a^k b^k \mid k \geq 1\} \notin \mathbf{FSFA}_1$$

Assume that L' is in $\mathbf{FSFA}_1$. Then, by Theorem 7.1.11, there is a 2-PRLG

$$G = (N_1, N_2, T, S, P)$$

such that $L(G) = L'$. Let

$$k > \mathrm{card}(P) \cdot \max\left(\{|w| \mid X \rightarrow wY \in P\}\right)$$

Consider a derivation of $c^k a^k b^k \in L'$. The second component can generate only finitely many as; otherwise, it derives $\{a^k b^n \mid k < n\}$, which is not regular. Analogously, the first component generates only finitely many bs. Therefore, the first component generates any number of as, and the second component generates any number of bs. Moreover, there is a derivation of the form $X \Rightarrow_G^m X$, for some $X \in N_2$, and $m \geq 1$, used in the derivation in the second component. In the first component, there is a derivation $A \Rightarrow_G^l a^s A$, for some $A \in N_1$, and $s, l \geq 1$. Then, we can modify the derivation of $c^k a^k b^k$ so that in the first component, we repeat the cycle $A \Rightarrow_G^l a^s A$ $(m + 1)$-times, and in the second component, we repeat the cycle $X \Rightarrow_G^m X$ $(l + 1)$-times. The derivations of both components have the same length—the added cycles are of length ml, and the rest is of the same length as in the derivation of $c^k a^k b^k$. Therefore, we have derived $c^k a^r b^k$, where $r > k$, which is not in L'—a contradiction.

For $n > 1$, the proof is analogous and left to the reader. $\square$

Corollary 7.1.14. *For all* $n \geq 1$, *FSFA_n is not closed under concatenation. Therefore, it is not closed under Kleene closure either.*

Proof. For $n = 1$, let $L_1 = \{c\}^*$ and $L_2 = \{a^k b^k \mid k \geq 1\}$. Then,

$$L_1 L_2 = \{c^j a^k b^k \mid k \geq 1, j \geq 0\}$$

Analogously, prove this corollary for $n > 1$. $\square$

n-Turn All-Move Self-Regulating Finite Automata

We next turn our attention to n-all-SFAs. We prove that the family of languages accepted by n-all-SFAs coincides with the family of languages generated by so-called n-right-linear simple matrix grammars (see [DP89, Iba70, Woo75]). First, however, we define these grammars formally.

Definition 7.1.15. For $n \geq 1$, an *n-right-linear simple matrix grammar* (see [DP89, Iba70, Woo75]), an *n-RLSMG* for short, is an $(n + 3)$-tuple

$$G = \left(N_1, \ldots, N_n, T, S, P\right)$$

where N_i, $1 \leq i \leq n$, are pairwise disjoint *nonterminal alphabets*, T is *a terminal alphabet*, $S \notin N$ is an *initial symbol*, where $N = N_1 \cup \cdots \cup N_n$, and P is a finite set of *matrix rules*. A matrix rule can be in one of the following three forms

1. $(S \rightarrow X_1 \cdots X_n)$, $X_i \in N_i, 1 \leq i \leq n$;
2. $(X_1 \rightarrow w_1 Y_1, \cdots, X_n \rightarrow w_n Y_n)$, $w_i \in T^*, X_i, Y_i \in N_i, 1 \leq i \leq n$;
3. $(X_1 \rightarrow w_1, \cdots, X_n \rightarrow w_n)$, $X_i \in N_i, w_i \in T^*, 1 \leq i \leq n$.

Let m be a matrix. Then, $m[i]$ denotes the ith rule of m. For $x, y \in (N \cup T \cup \{S\})^*$,

$$x \Rightarrow_G y$$

if and only if

1. either $x = S$ and $(S \rightarrow y) \in P$,
2. or $x = y_1 X_1 \cdots y_n X_n$, $y = y_1 x_1 \cdots y_n x_n$, where $y_i \in T^*$, $x_i \in T^* N \cup T^*$, $X_i \in N_i$, $1 \le i \le n$, and $(X_1 \rightarrow x_1, \cdots, X_n \rightarrow x_n) \in P$.

We define $x \Rightarrow_G^+ y$ and $x \Rightarrow_G^* y$ as in Definition 7.1.6.

The *language* of G is defined as

$$L(G) = \{w \in T^* \mid S \Rightarrow_G^* w\}$$

A language $K \subseteq T^*$ is an *n-right linear simple matrix language* (an *n*-RLSML for short) if there is an *n*-RLSMG G such that $K = L(G)$.   □

The family of *n*-RLSMLs is denoted by **RLSM**$_n$. Furthermore, the *i*th component of an *n*-RLSMG is defined analogously to the *i*th component of an *n*-PRLG (see Definition 7.1.7).

To prove that the family of languages accepted by *n*-all-SFAs coincides with the family of languages generated by *n*-RLSMGs, the following lemma is needed.

Lemma 7.1.16. *For every n-RLSMG, $G = (N_1, \ldots, N_n, T, S, P)$, there is an equivalent n-RLSMG G' that satisfies (i) through (iii), given next.*

(i) *If $(S \rightarrow X_1 \cdots X_n)$, then X_i does not occur on the right-hand side of any rule, $1 \le i \le n$.*
(ii) *If $(S \rightarrow \alpha)$, $(S \rightarrow \beta) \in P$ and $\alpha \neq \beta$, then $\mathrm{alph}(\alpha) \cap \mathrm{alph}(\beta) = \emptyset$.*
(iii) *For any two matrices $m_1, m_2 \in P$, if $m_1[i] = m_2[i]$, for some $1 \le i \le n$, then $m_1 = m_2$.*

Proof. The first two conditions can be proved analogously to Lemma 7.1.8. Suppose that there are matrices m and m' such that $m[i] = m'[i]$, for some $1 \le i \le n$. Let

$$m = (X_1 \rightarrow x_1, \ldots, X_n \rightarrow x_n)$$
$$m' = (Y_1 \rightarrow y_1, \ldots, Y_n \rightarrow y_n)$$

Replace these matrices with matrices

$$m_1 = (X_1 \rightarrow X_1', \ldots, X_n \rightarrow X_n')$$
$$m_2 = (X_1' \rightarrow x_1, \ldots, X_n' \rightarrow x_n)$$
$$m_1' = (Y_1 \rightarrow Y_1'', \ldots, Y_n \rightarrow Y_n'')$$
$$m_2' = (Y_1'' \rightarrow y_1, \ldots, Y_n'' \rightarrow y_n)$$

where X_i', Y_i'' are new nonterminals for all i. These new matrices satisfy condition (iii). Repeat this replacement until the resulting grammar satisfies the properties of G' given in this lemma.   □

Lemma 7.1.17. *Let G be an n-RLSMG. There is an $(n-1)$-all-SFA M such that $L(G) = L(M)$.*

Proof. Without loss of generality, we assume that $G = (N_1, \ldots, N_n, T, S, P)$ is in the form described in Lemma 7.1.16. We construct an $(n-1)$-all-SFA

$$M = (Q, T, \delta, q_0, q_t, F, R)$$

where

$$Q = \{q_0, \ldots, q_n\} \cup N, N = N_1 \cup \cdots \cup N_n, \{q_0, q_1, \ldots, q_n\} \cap N = \emptyset$$
$$F = \{q_n\}$$
$$\delta = \{q_i \rightarrow X_{i+1} \mid (S \rightarrow X_1 \cdots X_n) \in P, \ 0 \le i < n\} \cup$$
$$\{X_i w_i \rightarrow Y_i \mid (X_1 \rightarrow w_1 Y_1, \ldots, X_n \rightarrow w_n Y_n) \in P, \ 1 \le i \le n\} \cup$$
$$\{X_i w_i \rightarrow q_i \mid (X_1 \rightarrow w_1, \ldots, X_n \rightarrow w_n) \in P, \ w_i \in T^*, \ 1 \le i \le n\}$$
$$q_t = q_1$$
$$\Psi = \delta$$
$$R = \{(q_i \rightarrow X_{i+1}, q_{i+1} \rightarrow X_{i+2}) \mid$$
$$(S \rightarrow X_1 \cdots X_n) \in P, \ 0 \le i \le n-2\} \cup$$
$$\{(X_i w_i \rightarrow Y_i, X_{i+1} w_{i+1} \rightarrow Y_{i+1}) \mid$$
$$(X_1 \rightarrow w_1 Y_1, \ldots, X_n \rightarrow w_n Y_n) \in P, \ 1 \le i < n\} \cup$$
$$\{(X_i w_i \rightarrow q_i, X_{i+1} w_{i+1} \rightarrow q_{i+1}) \mid$$
$$(X_1 \rightarrow w_1, \ldots, X_n \rightarrow w_n) \in P, \ w_i \in T^*, \ 1 \le i < n\}$$

Next, we prove that $L(G) = L(M)$. A proof of $L(G) \subseteq L(M)$ can be made by analogy with the proof of the same inclusion of Lemma 7.1.9, which is left to the reader.

To prove that $L(M) \subseteq L(G)$, consider $w \in L(M)$ and an acceptance of w in M. As in Lemma 7.1.9, the derivation looks like the one depicted on the right-hand side of Fig. 7.3. Next, we describe how G generates w. By Lemma 7.1.16, there is matrix

$$(S \rightarrow X_1^1 X_1^2 \cdots X_1^n) \in P$$

Moreover, if $X_j^i x_j^i \rightarrow X_{j+1}^i, 1 \le i \le n$, then

$$(X_j^i \rightarrow x_j^i X_{j+1}^i, X_j^{i+1} \rightarrow x_j^{i+1} X_{j+1}^{i+1}) \in R$$

for $1 \le i < n$, $1 \le j < k$. We apply

$$(X_j^1 \rightarrow x_j^1 X_{j+1}^1, \ldots, X_j^n \rightarrow x_j^n X_{j+1}^n) \in P$$

If $X_k^i x_k^i \rightarrow q_i, 1 \le i \le n$, then

$$(X_k^i \rightarrow x_k^i, X_k^{i+1} \rightarrow x_k^{i+1}) \in R$$

for $1 \leq i < n$, and we apply

$$(X_k^1 \to x_k^1, \ldots, X_k^n \to x_k^n) \in P$$

Thus, $w \in L(G)$.

Hence, Lemma 7.1.17 holds.   □

Lemma 7.1.18. *Let M be an n-all-SFA. There is an $(n+1)$-RLSMG G such that $L(G) = L(M)$.*

Proof. Let $M = (Q, \Sigma, \delta, q_0, q_t, F, R)$. Consider

$$G = \big(N_0, \ldots, N_n, \Sigma, S, P\big)$$

where

$$
\begin{aligned}
N_i &= (Q\Sigma^l \times Q \times \{i\} \times Q) \cup (Q \times \{i\} \times Q) \\
l &= \max(\{|w| \mid qw \to p \in \delta\}), 0 \leq i \leq n \\
P &= \{(S \to [q_0 x_0, q^0, 0, q_t][q_t x_1, q^1, 1, q_{i_1}] \cdots [q_{i_{n-1}} x_n, q^n, n, q_{i_n}]) \mid \\
&\qquad r_0 \colon q_0 x_0 \to q^0, r_1 \colon q_t x_1 \to q^1, \ldots, r_n \colon q_{i_{n-1}} x_n \to q^n \in \delta \\
&\qquad (r_0, r_1), \ldots, (r_{n-1}, r_n) \in R, q_{i_n} \in F\} \cup \\
&\quad \{([p_0 x_0, q_0, 0, r_0] \to x_0[q_0, 0, r_0], \ldots, [p_n x_n, q_n, n, r_n] \to x_n[q_n, n, r_n])\} \cup \\
&\quad \{([q_0, 0, q_0] \to \varepsilon, \ldots, [q_n, n, q_n] \to \varepsilon) : q_i \in Q, 0 \leq i \leq n\} \cup \\
&\quad \{([q_0, 0, p_0] \to w_0[q_0', 0, p_0], \ldots, [q_n, n, p_n] \to w_n[q_n', n, p_n]) \mid \\
&\qquad r_j \colon q_j w_j \to q_j' \in \delta, 0 \leq j \leq n, (r_i, r_{i+1}) \in R, 0 \leq i < n\}
\end{aligned}
$$

Next, we prove that $L(G) = L(M)$. To prove that $L(G) \subseteq L(M)$, consider a derivation of w in G. Then, the derivation is of the form (7.1) and there are rules

$$r_0 \colon q_0 x_0^0 \to q_1^0, r_1 \colon q_t x_0^1 \to q_1^1, \ldots, r_n \colon q_{i_{n-1}} x_0^n \to q_1^n \in \delta$$

such that $(r_0, r_1), \ldots, (r_{n-1}, r_n) \in R$. Moreover, $(r_j^l, r_j^{l+1}) \in R$, where $r_j^l \colon q_j^l x_j^l \to q_{j+1}^l \in \delta$, and $(r_k^l, r_k^{l+1}) \in R$, where $r_k^l \colon q_k^l x_k^l \to q_{i_l} \in \delta, 0 \leq l < n, 1 \leq j < k$, q_{i_0} denotes q_t, and $q_{i_n} \in F$. Thus, M accepts w with the sequence of rules μ of the form (7.2).

To prove that $L(M) \subseteq L(G)$, let μ be a sequence of rules used in an acceptance of

$$w = x_0^0 x_1^0 \cdots x_k^0 x_0^1 x_1^1 \cdots x_k^1 \cdots x_0^n x_1^n \cdots x_k^n$$

in M of the form (7.2). Then, the derivation is of the form (7.1) because

$$([q_j^0, 0, q_t] \to x_j^0[q_{j+1}^0, 0, q_t], \ldots, [q_j^n, n, q_{i_n}] \to x_j^n[q_{j+1}^n, n, q_{i_n}]) \in P$$

for all $q_j^i \in Q, 1 \leq i \leq n, 1 \leq j < k$, and $([q_t, 0, q_t] \to \varepsilon, \ldots, [q_{i_n}, n, q_{i_n}] \to \varepsilon) \in P$.

Hence, Lemma 7.1.18 holds. □

Next, we establish another important result of this chapter.

Theorem 7.1.19. *For all $n \geq 0$, $\mathrm{ASFA}_n = \mathrm{RLSM}_{n+1}$.*

Proof. This proof follows from Lemmas 7.1.17 and 7.1.18. □

Corollary 7.1.20. *The following statements hold true.*

(i) $\mathrm{REG} = \mathrm{ASFA}_0 \subset \mathrm{ASFA}_1 \subset \mathrm{ASFA}_2 \subset \cdots \subset \mathrm{CS}$
(ii) $\mathrm{ASFA}_1 \nsubseteq \mathrm{CF}$
(iii) $\mathrm{CF} \nsubseteq \mathrm{ASFA}_n$, *for every $n \geq 0$.*
(iv) *For all $n \geq 0$, ASFA_n is closed under union, concatenation, finite substitution, homomorphism, intersection with a regular language, and right quotient with a regular language.*
(v) *For all $n \geq 1$, ASFA_n is not closed under intersection, complement, and Kleene closure.*

Proof. Recall the following statements that are proved in [Woo75].

- $\mathrm{REG} = \mathrm{RLSM}_1 \subset \mathrm{RLSM}_2 \subset \mathrm{RLSM}_3 \subset \cdots \subset \mathrm{CS}$
- For all $n \geq 1$, RLSM_n is closed under union, finite substitution, homomorphism, intersection with a regular language, and right quotient with a regular language.
- For all $n \geq 2$, RLSM_n is not closed under intersection and complement.

Furthermore, recall these statements proved in [Sir69] and [Sir71].

- For all $n \geq 1$, RLSM_n is closed under concatenation.
- For all $n \geq 2$, RLSM_n is not closed under Kleene closure.

These statements and Theorem 7.1.19 imply statements (i), (iv), and (v) of Corollary 7.1.20. Moreover, observe that

$$\{ww \mid w \in \{a, b\}^*\} \in \mathrm{ASFA}_1 - \mathrm{CF}$$

(see Example 7.1.5), which proves (ii). Finally, let

$$L = \{wcw^R \mid w \in \{a, b\}^*\}$$

By Theorem 1.5.2 in [DP89], $L \notin \mathrm{RLSM}_n$, for any $n \geq 1$. Thus, (iii) follows from Theorem 7.1.19. □

Theorem 7.1.21, given next, follows from Theorem 7.1.19 and from Corollary 3.3.3 in [Sir71]. However, Corollary 3.3.3 in [Sir71] is not proved effectively. We next prove Theorem 7.1.21 effectively.

Theorem 7.1.21. *ASFA_n is closed under inverse homomorphism, for all $n \geq 0$.*

Proof. For $n = 1$, let $M = (Q, \Sigma, \delta, q_0, q_t, F, R)$ be a 1-all-SFA, and let $h : \Delta^* \rightarrow \Sigma^*$ be a homomorphism. Next, we construct a 1-all-SFA

$$M' = (Q', \Delta, \delta', q'_0, q'_t, \{q'_f\}, R')$$

accepting $h^{-1}(L(M))$ as follows. Set

$$k = \max\left(\{|w| \mid qw \rightarrow p \in \delta\}\right) + \max\left(\{|h(a)| \mid a \in \Delta\}\right)$$

and

$$Q' = \{q'_0\} \cup \{[x, q, y] \mid x, y \in \Sigma^*, |x|, |y| \leq k, q \in Q\}$$

Initially, set δ' and R' to $\emptyset$. Then, extend δ' and R' by performing (1) through (5), given next.

(1) For $y \in \Sigma^*$, $|y| \leq k$, add
    $(q'_0 \rightarrow [\varepsilon, q_0, y], q'_t \rightarrow [y, q_t, \varepsilon])$ to R'.
(2) For $A \in Q'$, $q \neq q_t$, add
    $([x, q, y]a \rightarrow [xh(a), q, y], A \rightarrow A)$ to R'.
(3) For $A \in Q'$, add
    $(A \rightarrow A, [x, q, \varepsilon]a \rightarrow [xh(a), q, \varepsilon])$ to R'.
(4) For $(qx \rightarrow p, q'x' \rightarrow p') \in R$, $q \neq q_t$, add
    $([xw, q, y] \rightarrow [w, p, y], [x'w', q', \varepsilon] \rightarrow [w', p', \varepsilon])$ to R'.
(5) For $q_f \in F$, add
    $([y, q_t, y] \rightarrow q'_t, [\varepsilon, q_f, \varepsilon] \rightarrow q'_f)$ to R'.

In essence, M' simulates M in the following way. In a state of the form $[x, q, y]$, the three components have the following meaning

- $x = h(a_1 \cdots a_n)$, where $a_1 \cdots a_n$ is the input string that M' has already read;
- q is the current state of M;
- y is the suffix remaining as the first component of the state that M' enters during a turn; y is thus obtained when M' reads the last symbol right before the turn occurs in M; M reads y after the turn.

More precisely, $h(w) = w_1 y w_2$, where w is an input string, w_1 is accepted by M before making the turn—that is, from q_0 to q_t, and $y w_2$ is accepted by M after making the turn—that is, from q_t to $q_f \in F$. A rigorous version of this proof is left to the reader.

For $n > 1$, the proof is analogous and left to the reader. $\square$

Language Families Accepted by n-First-SFAs and n-All-SFAs

Next, we compare the family of languages accepted by n-first-SFAs with the family of languages accepted by n-all-SFAs.

Theorem 7.1.22. *For all* $n \geq 1$, $\mathbf{FSFA}_n \subset \mathbf{ASFA}_n$.

Proof. In [RW75] and [Woo75], it is proved that for all $n > 1$, $\mathbf{PRL}_n \subset \mathbf{RLSM}_n$. The proof of Theorem 7.1.22 thus follows from Theorems 7.1.11 and 7.1.19. ☐

Theorem 7.1.23. $\mathbf{FSFA}_n \not\subseteq \mathbf{ASFA}_{n-1}$, $n \geq 1$.

Proof. Recall that $\mathbf{FSFA}_n = \mathbf{PRL}_{n+1}$ (see Theorem 7.1.11) and $\mathbf{ASFA}_{n-1} = \mathbf{RLSM}_n$ (see Theorem 7.1.19). It is easy to see that

$$L = \{a_1^k a_2^k \cdots a_{n+1}^k \mid k \geq 1\} \in \mathbf{PRL}_{n+1}$$

However, Lemma 1.5.6 in [DP89] implies that

$$L \notin \mathbf{RLSM}_n$$

Hence, the theorem holds. ☐

Lemma 7.1.24. *For each regular language* L, $\{w^n \mid w \in L\} \in \mathbf{ASFA}_{n-1}$.

Proof. Let $L = L(M)$, where M is a finite automaton. Make n copies of M. Rename their states so all the sets of states are pairwise disjoint. In this way, also rename the states in the rules of each of these n automata; however, keep the labels of the rules unchanged. For each rule label r, include (r, r) into R. As a result, we obtain an n-all-SFA that accepts $\{w^n \mid w \in L\}$. A rigorous version of this proof is left to the reader. ☐

Theorem 7.1.25. $\mathbf{ASFA}_n - \mathbf{FSFA} \neq \emptyset$, *for all* $n \geq 1$, *where* $\mathbf{FSFA} = \bigcup_{m=1}^{\infty} \mathbf{FSFA}_m$.

Proof. By induction on $n \geq 1$, we prove that

$$L = \{(cw)^{n+1} \mid w \in \{a, b\}^*\} \notin \mathbf{FSFA}$$

From Lemma 7.1.24, it follows that $L \in \mathbf{ASFA}_n$.

Basis. For $n = 1$, let G be an m-PRLG generating L, for some positive integer m. Consider a sufficiently large string $cw_1 cw_2 \in L$ such that $w_1 = w_2 = a^{n_1} b^{n_2}$, $n_2 > n_1 > 1$. Then, there is a derivation of the form

$$S \Rightarrow_G^p x_1 A_1 x_2 A_2 \cdots x_m A_m$$
$$\Rightarrow_G^k x_1 y_1 A_1 x_2 y_2 A_2 \cdots x_m y_m A_m \tag{7.3}$$

in G, where cycle (7.3) generates more than one a in w_1. The derivation continues as

$$x_1 y_1 A_1 \cdots x_m y_m A_m \Rightarrow_G^r$$
$$x_1 y_1 z_1 B_1 \cdots x_m y_m z_m B_m \Rightarrow_G^l x_1 y_1 z_1 u_1 B_1 \cdots x_m y_m z_m u_m B_m \tag{7.4}$$

(cycle (7.4) generates no as) $\Rightarrow_G^s cw_1 cw_2$

Next, modify the left derivation, the derivation in components generating cw_1, so that the a-generating cycle (7.3) is repeated $(l + 1)$-times. Similarly, modify the right derivation, the derivation in the other components, so that the no-a-generating cycle (7.4) is repeated $(k + 1)$-times. Thus, the modified left derivation is of length

$$p + k(l + 1) + r + l + s = p + k + r + l(k + 1) + s$$

which is the length of the modified right derivation. Moreover, the modified left derivation generates more as in w_1 than the right derivation in w_2—a contradiction.

Induction Hypothesis. Suppose that the theorem holds for all $k \leq n$, for some $n \geq 1$.

Induction Step. Consider $n + 1$ and let

$$\{(cw)^{n+1} \mid w \in \{a, b\}^*\} \in \textbf{FSFA}_l$$

for some $l \geq 1$. As $\textbf{FSFA}_l$ is closed under the right quotient with a regular language, and language $\{cw \mid w \in \{a, b\}^*\}$ is regular, we obtain

$$\{(cw)^n \mid w \in \{a, b\}^*\} \in \textbf{FSFA}_l \subseteq \textbf{FSFA}$$

which is a contradiction.   □

7.1.2 Self-Regulating Pushdown Automata

The present section consists of two subsections. Section Definitions defines n-turn first-move self-regulating pushdown automata and n-turn all-move self-regulating pushdown automata. Section Accepting Power determines the accepting power of n-turn all-move self-regulating pushdown automata.

Definitions

Before defining self-regulating pushdown automata, recall the formalization of rule labels from Definition 2.4.10 because this formalization is often used throughout this section.

Definition 7.1.26. A *self-regulating pushdown automaton* (an *SPDA* for short) M is a 9-tuple

$$M = (Q, \Sigma, \Gamma, \delta, q_0, q_t, Z_0, F, R)$$

where

1. $(Q, \Sigma, \Gamma, \delta, q_0, Z_0, F)$ is a pushdown automaton entering a final state and emptying its pushdown,
2. $q_t \in Q$ is a *turn state*, and
3. $R \subseteq \Psi \times \Psi$ is a finite relation, where Ψ is an alphabet of rule labels. □

Definition 7.1.27. Let $n \geq 0$ and

$$M = (Q, \Sigma, \Gamma, \delta, q_0, q_t, Z_0, F, R)$$

be a self-regulating pushdown automaton. M is said to be an *n-turn first-move self-regulating pushdown automaton*, *n-first-SPDA*, if every $w \in L(M)$ is accepted by M in the following way

$$Z_0 q_0 w \vdash_M^* f [\mu]$$

such that

$$\mu = r_1^0 \cdots r_k^0 r_1^1 \cdots r_k^1 \cdots r_1^n \cdots r_k^n$$

where $k \geq 1$, r_k^0 is the first rule of the form $Zqx \rightarrow \gamma q_t$, for some $Z \in \Gamma$, $q \in Q$, $x \in \Sigma^*$, $\gamma \in \Gamma^*$, and

$$(r_1^j, r_1^{j+1}) \in R$$

for all $0 \leq j < n$. □

The family of languages accepted by *n*-first-SPDAs is denoted by **FSPDA$_n$**.

Definition 7.1.28. Let $n \geq 0$ and

$$M = (Q, \Sigma, \Gamma, \delta, q_0, q_t, Z_0, F, R)$$

be a self-regulating pushdown automaton. M is said to be an *n-turn all-move self-regulating pushdown automaton* (an *n-all-SPDA* for short) if every $w \in L(M)$ is accepted by M in the following way

$$Z_0 q_0 w \vdash_M^* f [\mu]$$

such that

$$\mu = r_1^0 \cdots r_k^0 r_1^1 \cdots r_k^1 \cdots r_1^n \cdots r_k^n$$

where $k \geq 1$, r_k^0 is the first rule of the form $Zqx \rightarrow \gamma q_t$, for some $Z \in \Gamma$, $q \in Q$, $x \in \Sigma^*$, $\gamma \in \Gamma^*$, and

$$(r_i^j, r_i^{j+1}) \in R$$

for all $1 \leq i \leq k$, $0 \leq j < n$.   □

The family of languages accepted by n-all-SPDAs is denoted by $\mathbf{ASPDA}_n$.

Accepting Power

In this section, we investigate the accepting power of self-regulating pushdown automata.

As every n-first-SPDA and every n-all-SPDA without any turn state represents, in effect, an ordinary pushdown automaton, we obtain the following theorem.

Theorem 7.1.29. $\mathbf{FSPDA}_0 = \mathbf{ASPDA}_0 = \mathbf{CF}$ □

However, if we consider 1-all-SPDAs, their power is that of phrase-structure grammars.

Theorem 7.1.30. $\mathbf{ASPDA}_1 = \mathbf{RE}$

Proof. For any $L \in RE$, $L \subseteq \Delta^*$, there are context-free languages $L(G)$ and $L(H)$ and a homomorphism $h : \Sigma^* \rightarrow \Delta^*$ such that

$$L = h\big(L(G) \cap L(H)\big)$$

(see Theorem 2.3.18). Suppose that $G = (N_G, \Sigma, P_G, S_G)$ and $H = (N_H, \Sigma, P_H, S_H)$ are in the Greibach normal form (see Definition 3.1.21)—that is, all rules are of the form $A \rightarrow a\alpha$, where A is a nonterminal, a is a terminal, and α is a (possibly empty) string of nonterminals. Let us construct a 1-all-SPDA

$$M = \big(\{q_0, q, q_t, p, f\}, \Delta, \Sigma \cup N_G \cup N_H \cup \{Z\}, \delta, q_0, Z, \{f\}, R\big)$$

where $Z \notin \Sigma \cup N_G \cup N_H$, with R constructed by performing (1) through (4), stated next.

(1) Add $(Zq_0 \rightarrow ZS_G q, Zq_t \rightarrow ZS_H p)$ to R.
(2) Add $(Aq \rightarrow B_n \cdots B_1 aq, Cp \rightarrow D_m \cdots D_1 ap)$ to R if
$\quad A \rightarrow aB_1 \cdots B_n \in P_G$ and
$\quad C \rightarrow aD_1 \cdots D_m \in P_H$.
(3) Add $(aqh(a) \rightarrow q, ap \rightarrow p)$ to R.
(4) Add $(Zq \rightarrow Zq_t, Zp \rightarrow f)$ to R.

Moreover, δ contains only the rules from the definition of R.

Next, we prove that $w \in h(L(G) \cap L(H))$ if and only if $w \in L(M)$.

Only If. Let $w \in h(L(G) \cap L(H))$. There are $a_1, a_2, \ldots, a_n \in \Sigma$ such that

$$a_1 a_2 \cdots a_n \in L(G) \cap L(H)$$

and $w = h(a_1 a_2 \cdots a_n)$, for some $n \geq 0$. There are leftmost derivations

$$S_G \Rightarrow_G^n a_1 a_2 \cdots a_n$$

and

$$S_H \Rightarrow_H^n a_1 a_2 \cdots a_n$$

of length n in G and H, respectively, because in every derivation step exactly one terminal element is derived. Thus, M accepts $h(a_1)h(a_2) \cdots h(a_n)$ as

$$Zq_0 h(a_1)h(a_2) \cdots h(a_n)$$
$$\vdash_M ZS_G q h(a_1)h(a_2) \cdots h(a_n)$$
$$\vdots$$
$$\vdash_M Za_n q h(a_n)$$
$$\vdash_M Zq$$
$$\vdash_M Zq_t$$
$$\vdash_M ZS_H p$$
$$\vdots$$
$$\vdash_M Za_n p$$
$$\vdash_M Zp$$
$$\vdash_M f$$

In state q, by using its pushdown, M simulates a derivation of $a_1 \cdots a_n$ in G but reads $h(a_1) \cdots h(a_n)$ as the input. In p, M simulates a derivation of $a_1 a_2 \cdots a_n$ in H but reads no input. As $a_1 a_2 \cdots a_n$ can be derived in both G and H by making the same number of steps, the automaton can successfully complete the acceptance of w.

If. Notice that in one step, M can read only $h(a) \in \Delta^*$, for some $a \in \Sigma$. Let $w \in L(M)$, then $w = h(a_1)h(a_2) \cdots h(a_n)$, for some $a_1, a_2, \ldots, a_n \in \Sigma$. Consider the following acceptance of w in M

$$Zq_0 h(a_1)h(a_2) \cdots h(a_n)$$
$$\vdash_M ZS_G q h(a_1)h(a_2) \cdots h(a_n)$$
$$\vdots$$
$$\vdash_M Za_n q h(a_n)$$
$$\vdash_M Zq$$
$$\vdash_M Zq_t$$
$$\vdash_M ZS_H p$$
$$\vdots$$
$$\vdash_M Za_n p$$
$$\vdash_M Zp$$
$$\vdash_M f$$

As stated above, in q, M simulates a derivation of $a_1 a_2 \cdots a_n$ in G, and then in p, M simulates a derivation of $a_1 a_2 \cdots a_n$ in H. It successfully completes the acceptance of w only if $a_1 a_2 \cdots a_n$ can be derived in both G and H. Hence, the if part holds, too.   □

7.1.3   Open Problems

Although the fundamental results about self-regulating automata have been achieved in this chapter, there still remain several open problems concerning them.

Open Problem 7.1.31. What is the language family accepted by n-turn first-move self-regulating pushdown automata, when $n \geq 1$ (see Definition 7.1.27)?   □

Open Problem 7.1.32. By analogy with the standard deterministic finite and pushdown automata (see Sect. 2.4), introduce the deterministic versions of self-regulating automata. What is their power?   □

Open Problem 7.1.33. Discuss the closure properties of other language operations, such as the reversal.   □

7.2   Regulated Acceptance with Control Languages

This section discusses automata in which the application of rules is regulated by control languages by analogy with context-free grammars regulated by control languages (see Sect. 3.3). Section 7.2.1 studies this topic in terms of finite automata while Sect. 7.2.2 investigates pushdown automata regulated in this way. More precisely, Sect. 7.2.1 discusses finite automata working under two kinds of regulation—*state-controlled regulation* and *transition-controlled regulation*. It establishes conditions under which any state-controlled finite automaton can be turned to an equivalent transition-controlled finite automaton and vice versa. Then, it proves that under either of the two regulations, finite automata controlled by regular languages characterize the family of regular languages, and an analogical result is then reformulated in terms of context-free languages. However, Sect. 7.2.1 also demonstrates that finite automata controlled by languages generated by propagating programmed grammars with appearance checking increase their power significantly; in fact, they are computationally complete. Section 7.2.2 first shows that pushdown automata regulated by regular languages are as powerful as ordinary pushdown automata. Then, however, it proves that pushdown automata regulated by linear languages characterize the family of recursively enumerable languages; in fact, this characterization holds even in terms of one-turn pushdown automata.

7.2.1   Finite Automata Regulated by Control Languages

The present section studies finite automata regulated by control languages. In fact, it studies two kinds of this regulation—*state-controlled regulation* and *transition-controlled regulation*. To give an insight into these two types of regulation, consider a finite automaton M controlled by a language C, and a sequence $\tau \in C$ that resulted into the acceptance of an input word w. Working under the former regulation, M has C defined over the set of states, and it accepts w by going through all the states in τ and ending up in a final state. Working under the latter regulation, M has C defined over the set of transitions, and it accepts w by using all the transitions in τ and ending up in a final state.

First, we define these two types of controlled finite automata formally. After that, we establish conditions under which it is possible to convert any state-controlled finite automaton to an equivalent transition-controlled finite automaton and vice versa (Theorem 7.2.5). Then, we prove that under both regulations, finite automata controlled by regular languages characterize the family of regular languages (Theorem 7.2.7 and Corollary 7.2.8). Finally, we show that finite automata controlled by context-free languages characterize the family of context-free languages (Theorem 7.2.10 and Corollary 7.2.11).

After that, we demonstrate a close relation of controlled finite automata to programmed grammars with appearance checking (see Sect. 3.5). Recall that programmed grammars with appearance checking are computationally complete—that is, they are as powerful as phrase-structure grammars; the language family generated by propagating programmed grammars with appearance checking is properly included in the family of context-sensitive languages (see Theorems 3.3.6, 3.4.5, and 3.5.4). This section proves that finite automata that are controlled by languages generated by propagating programmed grammars with appearance checking are computationally complete (Theorem 7.2.17 and Corollary 7.2.18). More precisely, state-controlled finite automata are computationally complete with $n + 1$ states, where n is the number of symbols in the accepted language (Corollary 7.2.19). Transition-controlled finite automata are computationally complete with a single state (Theorem 7.2.20).

Definitions

We begin by defining state-controlled and transition-controlled finite automata formally.

Definition 7.2.1. Let $M = (Q, \Sigma, R, s, F)$ be a finite automaton. Based on $\vdash_M$, we define a relation $\rhd_M$ over $Q\Sigma^* \times Q^*$ as follows: if $\alpha \in Q^*$ and $pax \vdash_M qx$, where $p, q \in Q, x \in \Sigma^*$, and $a \in \Sigma \cup \{\varepsilon\}$, then

$$(pax, \alpha) \rhd_M (qx, \alpha p)$$

Let $\rhd_M^n$, $\rhd_M^*$, and $\rhd_M^+$ denote the nth power of $\rhd_M$, for some $n \geq 0$, the reflexive-transitive closure of $\rhd_M$, and the transitive closure of $\rhd_M$, respectively.

Let $C \subseteq Q^*$ be a *control language*. The *state-controlled language of M with respect to C* is denoted by $_\triangleright L(M, C)$ and defined as

$$_\triangleright L(M, C) = \{w \in \Sigma^* \mid (sw, \varepsilon) \triangleright_M^* (f, \alpha), f \in F, \alpha \in C\}$$

The pair (M, C) is called a *state-controlled finite automaton.*   □

Before defining transition-controlled finite automata, recall the formalization of rule labels from Definition 2.4.3.

Definition 7.2.2. Let $M = (Q, \Sigma, \Psi, R, s, F)$ be a finite automaton. Based on $\vdash_M$, we define a relation $\blacktriangleright_M$ over $Q\Sigma^* \times \Psi^*$ as follows: if $\beta \in \Psi^*$ and $pax \vdash_M qx\ [r]$, where $r: pa \to q \in R$ and $x \in \Sigma^*$, then

$$(pax, \beta)\ \blacktriangleright_M\ (qx, \beta r)$$

Let $\blacktriangleright_M^n$, $\blacktriangleright_M^*$, and $\blacktriangleright_M^+$ denote the nth power of $\blacktriangleright_M$, for some $n \geq 0$, the reflexive-transitive closure of $\blacktriangleright_M$, and the transitive closure of $\blacktriangleright_M$, respectively.

Let $C \subseteq \Psi^*$ be a *control language*. The *transition-controlled language of M with respect to C* is denoted by $_\blacktriangleright L(M, C)$ and defined as

$$_\blacktriangleright L(M, C) = \{w \in \Sigma^* \mid (sw, \varepsilon) \blacktriangleright_M^* (f, \beta), f \in F, \beta \in C\}$$

The pair (M, C) is called a *transition-controlled finite automaton.*   □

For any family of languages $\mathscr{L}$, **SCFA**$(\mathscr{L})$ and **TCFA**$(\mathscr{L})$ denote the language families defined by state-controlled finite automata controlled by languages from $\mathscr{L}$ and transition-controlled finite automata controlled by languages from $\mathscr{L}$, respectively.

Conversions

First, we show that under certain circumstances, it is possible to convert any state-controlled finite automaton to an equivalent transition-controlled finite automaton and vice versa. These conversions will be helpful to prove that **SCFA**$(\mathscr{L})$ = **TCFA**$(\mathscr{L}) = \mathscr{J}$, where $\mathscr{L}$ satisfies the required conditions, we only have to prove that either **SCFA**$(\mathscr{L}) = \mathscr{J}$ or **TCFA**$(\mathscr{L}) = \mathscr{J}$.

Lemma 7.2.3. *Let $\mathscr{L}$ be a language family that is closed under finite ε-free substitution. Then,* **SCFA**$(\mathscr{L}) \subseteq$ **TCFA**$(\mathscr{L})$.

Proof. Let $\mathscr{L}$ be a language family that is closed under finite ε-free substitution, $M = (Q, \Sigma, R, s, F)$ be a finite automaton, and $C \in \mathscr{L}$ be a control language. Without any loss of generality, assume that $C \subseteq Q^*$. We next construct a finite automaton M' and a language $C' \in \mathscr{L}$ such that $_\triangleright L(M, C) = {}_\blacktriangleright L(M', C')$. Define

$$M' = (Q, \Sigma, \Psi, R', s, F)$$

where

$$\Psi = \{\langle p, a, q \rangle \mid pa \to q \in R\}$$
$$R' = \{\langle p, a, q \rangle : pa \to q \mid pa \to q \in R\}$$

Define the finite ε-free substitution π from Q^* to Ψ^* as

$$\pi(p) = \{\langle p, a, q \rangle \mid pa \to q \in R\}$$

Let $C' = \pi(C)$. Since $\mathscr{L}$ is closed under finite ε-free substitution, $C' \in \mathscr{L}$. Observe that $(sw, \varepsilon) \triangleright_M^n (f, \alpha)$, where $w \in \Sigma^*$, $f \in F$, $\alpha \in C$, and $n \geq 0$, if and only if $(sw, \varepsilon) \blacktriangleright_{M'}^n (f, \beta)$, where $\beta \in \pi(\alpha)$. Hence, $_\triangleright L(M, C) = {}_\blacktriangleright L(M', C')$, so the lemma holds.   □

Lemma 7.2.4. *Let $\mathscr{L}$ be a language family that contains all finite languages and is closed under concatenation. Then,* $\mathbf{TCFA}(\mathscr{L}) \subseteq \mathbf{SCFA}(\mathscr{L})$.

Proof. Let $\mathscr{L}$ be a language family that contains all finite languages and is closed under concatenation, $M = (Q, \Sigma, \Psi, R, s, F)$ be a finite automaton, and $C \in \mathscr{L}$ be a control language. Without any loss of generality, assume that $C \subseteq \Psi^*$. We next construct a finite automaton M' and a language $C' \in \mathscr{L}$ such that $_\blacktriangleright L(M, C) = {}_\triangleright L(M', C')$. Define

$$M' = (Q', \Sigma, R', s', F')$$

where

$$Q' = \Psi \cup \{s', \ell\} \quad (s', \ell \notin \Psi)$$
$$R' = \{s' \to r \mid r : sa \to q \in R\} \cup$$
$$\qquad \{ra \to t \mid r : pa \to q, t : qb \to m \in R\} \cup$$
$$\qquad \{ra \to \ell \mid r : pa \to q \in R, q \in F\}$$
$$F' = \{r \mid r : pa \to q \in R, p \in F\} \cup \{\ell\}$$

Finally, if $s \in F$, then add s' to F'. Set $C' = \{s', \varepsilon\}C$. Since $\mathscr{L}$ is closed under concatenation and contains all finite languages, $C' \in \mathscr{L}$. Next, we prove that $_\blacktriangleright L(M, C) = {}_\triangleright L(M', C')$. First, notice that $s \in F$ if and only if $s' \in F$. Hence, by the definition of C', it is sufficient to consider nonempty sequences of moves of both M and M'. Indeed, $(s, \varepsilon) \blacktriangleright_M^0 (s, \varepsilon)$ with $s \in F$ and $\varepsilon \in C$ if and only if $(s', \varepsilon) \triangleright_{M'}^0 (s', \varepsilon)$ with $s' \in F$ and $\varepsilon \in C'$. Observe that

$$(sw, \varepsilon) \blacktriangleright_M (p_1 w_1, r_1) \blacktriangleright_M (p_2 w_2, r_1 r_2) \blacktriangleright_M \cdots \blacktriangleright_M (p_n w_n, r_1 r_2 \cdots r_n)$$

by

$$r_1: \quad p_0 a_1 \rightarrow p_1$$
$$r_2: \quad p_1 a_2 \rightarrow p_2$$
$$\vdots$$
$$r_n: p_{n-1} a_n \rightarrow p_n$$

where $w \in \Sigma^*$, $p_i \in Q$ for $i = 1, 2, \ldots, n$, $p_n \in F$, $w_i \in \Sigma^*$ for $i = 1, 2, \ldots, n$, $a_i \in \Sigma \cup \{\varepsilon\}$ for $i = 1, 2, \ldots n$, and $n \geq 1$ if and only if

$$(s'w, \varepsilon) \rhd_{M'} (r_1 w, s') \rhd_{M'} (r_2 w_1, s' r_1) \rhd_{M'} \cdots \rhd_{M'} (r_{n+1} w_n, s' r_1 r_2 \cdots r_n)$$

by

$$s' \rightarrow r_1$$
$$r_1 a_1 \rightarrow r_2$$
$$r_2 a_2 \rightarrow r_3$$
$$\vdots$$
$$r_n a_n \rightarrow r_{n+1}$$

with $r_{n+1} \in F'$ (recall that $p_n \in F$). Hence, $_{\blacktriangleright}L(M, C) = _{\rhd}L(M', C')$ and the lemma holds. □

Theorem 7.2.5. *Let $\mathscr{L}$ be a language family that is closed under finite ε-free substitution, contains all finite languages, and is closed under concatenation. Then,* **SCFA($\mathscr{L}$) = TCFA($\mathscr{L}$).**

Proof. This theorem follows directly from Lemmas 7.2.3 and 7.2.4. □

Regular-Controlled Finite Automata

Initially, we consider finite automata controlled by regular control languages.

Lemma 7.2.6. SCFA(REG) ⊆ REG

Proof. Let $M = (Q, \Sigma, R, s, F)$ be a finite automaton and $C \subseteq Q^*$ be a regular control language. Since C is regular, there is a complete finite automaton $H = (\hat{Q}, Q, \hat{R}, \hat{s}, \hat{F})$ such that $L(H) = C$. We next construct a finite automaton M' such that $L(M') = _{\rhd}L(M, L(H))$. Define

$$M' = \left(Q', \Sigma, R', s', F'\right)$$

where

$$Q' = \{\langle p, q \rangle \mid p \in Q, q \in \hat{Q}\}$$
$$R' = \{\langle p, r \rangle a \to \langle q, t \rangle \mid pa \to q \in R, rp \to t \in \hat{R}\}$$
$$s' = \langle s, \hat{s} \rangle$$
$$F' = \{\langle p, q \rangle \mid p \in F, q \in \hat{F}\}$$

Observe that a move in M' by $\langle p, r \rangle a \to \langle q, t \rangle \in R'$ simultaneously simulates a move in M by $pa \to q \in R$ and a move in H by $rp \to t \in \hat{R}$. Based on this observation, it is rather easy to see that M' accepts an input string $w \in \Sigma^*$ if and only if M reads w and enters a final state after going through a sequence of states from $L(H)$. Therefore, $L(M') = {}_\triangleright L(M, L(H))$. A rigorous proof of the identity $L(M') = {}_\triangleright L(M, L(H))$ is left to the reader. □

The following theorem shows that finite automata controlled by regular languages are of little or no interest because they are as powerful as ordinary finite automata.

Theorem 7.2.7. SCFA(REG) = REG

Proof. The inclusion **REG** $\subseteq$ **SCFA(REG)** is obvious. The converse inclusion follows from Lemma 7.2.6. □

Combining Theorems 7.2.5 and 7.2.7, we obtain the following corollary (recall that **REG** satisfies all the conditions from Theorem 7.2.5).

Corollary 7.2.8. TCFA(REG) = REG □

Context-Free-Controlled Finite Automata

Next, we consider finite automata controlled by context-free control languages.

Lemma 7.2.9. SCFA(CF) $\subseteq$ CF

Proof. Let $M = (Q, \Sigma, R, s, F)$ be a finite automaton and $C \subseteq Q^*$ be a context-free control language. Since C is context-free, there is a pushdown automaton $H = (\hat{Q}, Q, \Gamma, \hat{R}, \hat{s}, \hat{Z}, \hat{F})$ such that $L(H) = C$. Without any loss of generality, we assume that $bpa \to wq \in \hat{R}$ implies that $a \neq \varepsilon$ (see Lemma 5.2.1 in [Woo87]). We next construct a pushdown automaton M' such that $L(M') = {}_\triangleright L(M, L(H))$. Define

$$M' = (Q', \Sigma, \Gamma, R', s', Z, F')$$

where

$$Q' = \{\langle p, q \rangle \mid p \in Q, q \in \hat{Q}\}$$
$$R' = \{b\langle p, r \rangle a \to w \langle q, t \rangle \mid pa \to q \in R, bpr \to wt \in \hat{R}\}$$
$$s' = \langle s, \hat{s} \rangle$$
$$F' = \{\langle p, q \rangle \mid p \in F, q \in \hat{F}\}$$

By a similar reasoning as in Lemma 7.2.6, we can prove that $L(M') = {}_{\triangleright}L(M, L(H))$. A rigorous proof of the identity $L(M') = {}_{\triangleright}L(M, L(H))$ is left to the reader.   □

The following theorem says that even though finite automata controlled by context-free languages are more powerful than finite automata, they cannot accept any non-context-free language.

Theorem 7.2.10. SCFA(CF) = CF

Proof. The inclusion **CF** ⊆ **SCFA(CF)** is obvious. The converse inclusion follows from Lemma 7.2.9.   □

Combining Theorems 7.2.5 and 7.2.10, we obtain the following corollary (recall that **CF** satisfies all the conditions from Theorem 7.2.5).

Corollary 7.2.11. TCFA(CF) = CF   □

Program-Controlled Finite Automata

In this section, we show that there is a language family, strictly included in the family of context-sensitive languages, which significantly increases the power of finite automata. Indeed, finite automata controlled by languages generated by propagating programmed grammars with appearance checking have the same power as phrase-structure grammars. This result is of some interest because $\mathbf{P}_{ac}^{-\varepsilon} \subset \mathbf{CS}$ (see Theorem 3.5.4).

More specifically, we show how to algorithmically convert any programmed grammar with appearance checking G to a finite automaton M and a propagating programmed grammar with appearance checking G' such that ${}_{\triangleright}L(M, L(G')) = L(G)$. First, we give an insight into the algorithm. Then, we describe it formally and after that, we verify its correctness.

Let $T = \mathrm{alph}(L(G))$ and let $s \notin T$ be a new symbol. From G, we construct the propagating programmed grammar with appearance checking G' such that $w \in L(G)$ if and only if $sws^k \in L(G')$, where $k \geq 1$. Then, the set of states of M will be $T \cup \{s\}$, where s is the starting and also the only final state. For every $a, b \in T$, we introduce $aa \to b$ to M. For every $a \in T$, we introduce $s \to a$ and $aa \to s$. Finally, we add $s \to s$. An example of such a finite automaton when $T = \{a, b, c\}$ can be seen in Fig. 7.4. The key idea is that when M is in a state $a \in T$, in the next move, it has to read a. Hence, with $sws^k \in L(G')$, M moves from s to a state in T, then reads every symbol in w, ends up in s, and uses k times the rule $s \to s$.

G' works in the following way. Every intermediate sentential form is of the form xvZ, where x is a string of symbols that are not erased in the rest of the derivation, v is a string of nonterminals that are erased in the rest of the derivation, and Z is a nonterminal. When simulating a rule of G, G' non-deterministically selects symbols that are erased, appends them using Z to the end of the currently generated string, and replaces an occurrence of the left-hand side of the original rule with the not-to-be-erased symbols from the right-hand side. To differentiate the symbols in x and v, v contains barred versions of the nonterminals. If G' makes an

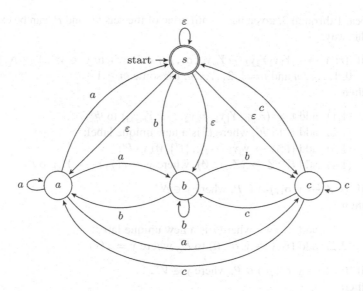

Fig. 7.4 Example of a finite automaton constructed by Algorithm 7.2.12

improper non-deterministic selection—that is, the selection does not correspond to a derivation in G—then G' is not able to generate a terminal string as explained in the notes following the algorithm.

Next, we describe the algorithm formally.

Algorithm 7.2.12.

Input: A programmed grammar with appearance checking $G = (N, T, \Psi, P, S)$.

Output: A finite automaton $M = (Q, T, R, s, F)$ and a propagating programmed grammar with appearance checking $G' = (N', Q, \Psi', P', S')$ such that $_\triangleright L(M, L(G')) = L(G)$.

Note: Without any loss of generality, we assume that $s, S', Z, \# \notin (Q \cup N \cup T)$ and $\ell_0, \bar{\ell}_0, \ell_s \notin \Psi$.

Method: Set $V = N \cup T$ and $\bar{N} = \{\bar{A} \mid A \in N\}$. Define the function τ from N^* to $\bar{N}^* \cup \{\ell_s\}$ as $\tau(\varepsilon) = \ell_s$ and $\tau(A_1 A_2 \cdots A_m) = \bar{A}_1 \bar{A}_2 \cdots \bar{A}_m$, where $A_i \in N$ for $i = 1, 2, \ldots, m$, for some $m \geq 1$. Initially, set

$$Q = T \cup \{s\}$$
$$R = \{s \to s\} \cup \{s \to a \mid a \in T\} \cup \{aa \to b \mid a \in T, b \in T \cup \{s\}\}$$
$$F = \{s\}$$
$$N' = N \cup \bar{N} \cup \{S', Z\}$$
$$\Psi' = \{\ell_0, \bar{\ell}_0, \ell_s\}$$
$$P' = \{(\ell_0 : S' \to sSZ, \{r \mid (r : S \to x, \sigma_r, \rho_r) \in P, x \in V^*\}, \emptyset)\} \cup$$
$$\quad \{(\bar{\ell}_0 : S' \to s\bar{S}Z, \{r \mid (r : S \to x, \sigma_r, \rho_r) \in P, x \in V^*\}, \emptyset)\} \cup$$
$$\quad \{(\ell_s : Z \to s, \{\ell_s\}, \emptyset)\}$$

Repeat 1 through 3, given next, until none of the sets Ψ' and P' can be extended in this way.

1. **If** $(r:A \to y_0 Y_1 y_1 Y_2 y_2 \cdots Y_m y_m, \sigma_r, \rho_r) \in P$, where $y_i \in V^*$, $Y_j \in N$, for $i = 0, 1, \ldots, m$ and $j = 1, 2, \ldots, m$, for some $m \geq 1$
   then

       (1.1) add $\ell = \langle r, y_0, Y_1 y_1, Y_2 y_2, \ldots, Y_m y_m \rangle$ to Ψ';
       (1.2) add ℓ' to Ψ', where ℓ' is a new unique label;
       (1.3) add $(\ell:A \to y_0 y_1 \cdots y_m, \{\ell'\}, \emptyset)$ to P';
       (1.4) add $(\ell':Z \to \bar{y}Z, \sigma_r, \emptyset)$, where $\bar{y} = \tau(Y_1 Y_2 \cdots Y_m)$ to P'.

2. **If** $(r:A \to y, \sigma_r, \rho_r) \in P$, where $y \in N^*$
   then

       (2.1) add $\bar{r}$ to Ψ', where $\bar{r}$ is a new unique label;
       (2.2) add $(\bar{r}:\bar{A} \to \bar{y}, \sigma_r, \emptyset)$ to P', where $\bar{y} = \tau(y)$.

3. **If** $(r:A \to y, \sigma_r, \rho_r) \in P$, where $y \in V^*$,
   then

       (3.1) add $\hat{r}$ to Ψ', where $\hat{r}$ is a new unique label;
       (3.2) add $\hat{r}'$ to Ψ', where $\hat{r}'$ is a new unique label;
       (3.3) add $(\hat{r}:A \to \#y, \emptyset, \{\hat{r}'\})$ to P';
       (3.4) add $(\hat{r}':\bar{A} \to \#y, \emptyset, \rho_r)$ to P'.

Finally, for every $t \in \Psi$, let Ψ'_t denote the set of rule labels introduced in steps (1.1), (2.1), and (3.1) from a rule labeled with t.
Replace every $(\ell':Z \to \bar{y}Z, \sigma_r, \emptyset)$ from (1.4) satisfying $\sigma_r \neq \emptyset$ with

$$(\ell':Z \to \bar{y}Z, \sigma'_r \cup \{\ell_s\}, \emptyset) \text{ where } \sigma'_r = \bigcup_{t \in \sigma_r} \Psi'_t$$

Replace every $(\bar{r}:\bar{A} \to \bar{y}, \sigma_r, \emptyset)$ from (2.2) satisfying $\sigma_r \neq \emptyset$ with

$$(\bar{r}:\bar{A} \to \bar{y}, \sigma'_r \cup \{\ell_s\}, \emptyset) \text{ where } \sigma'_r = \bigcup_{t \in \sigma_r} \Psi'_t$$

Replace every $(\hat{r}':\bar{A} \to \#y, \emptyset, \rho_r)$ from (3.4) satisfying $\rho_r \neq \emptyset$ with

$$(\hat{r}':\bar{A} \to \#y, \emptyset, \rho'_r \cup \{\ell_s\}) \text{ where } \rho'_r = \bigcup_{t \in \rho_r} \Psi'_t \qquad \square$$

Before proving that this algorithm is correct, we make some informal comments concerning the purpose of the rules of G'. The rules introduced in 1 are used to simulate an application of a rule where some of the symbols on its right-hand side are erased in the rest of the derivation while some others are not erased. The rules introduced in 2 are used to simulate an application of a rule which erases a nonterminal by making one or more derivation steps. The rules introduced in 3 are used to simulate an application of a rule in the appearance checking mode. Observe that when simulating $(r:A \to y, \sigma_r, \rho_r) \in P$ in the appearance checking mode, we have to check the absence of both A and $\bar{A}$. If some of these two nonterminals appear in the current configuration, the derivation is blocked because rules from 3 have empty success fields. Finally, notice that the final part of the algorithm ensures that after a rule $t \in P$ is applied, ℓ_s or any of the rules introduced in (1.3), (2.2), or (3.3) from rules in σ_t or ρ_t can be applied.

Reconsider 1. Notice that the algorithm works correctly although it makes no predetermination of nonterminals from which ε can be derived. Indeed, if the output grammar improperly selects a nonterminal that is not erased throughout the rest of the derivation, then this occurrence of the nonterminal never disappears so a terminal string cannot be generated under this improper selection.

We next prove that Algorithm 7.2.12 is correct.

Lemma 7.2.13. *Algorithm 7.2.12 converts any programmed grammar with appearance checking $G = (N, T, \Psi, P, S)$ to a finite automaton $M = (Q, T, R, s, F)$ and a propagating programmed grammar with appearance checking $G' = (N', Q, \Psi', P', S')$ such that $_{\triangleright}L(M, L(G')) = L(G)$.*

Proof. Clearly, the algorithm always halts. Consider the construction of G'. Observe that every string in $L(G')$ is of the form sws^k, where $k \geq 1$. From the construction of M, it is easy to see that $w \in {}_{\triangleright}L(M, L(G'))$ for $w \in T^*$ if and only if $sws^k \in L(G')$ for some $k \geq 1$. Therefore, to prove that $_{\triangleright}L(M, L(G')) = L(G)$, it is sufficient to prove that $w \in L(G)$ for $w \in T^*$ if and only if $sws^k \in L(G')$ for some $k \geq 1$.

We establish this equivalence by proving two claims. First, we prove that $w \in L(G)$ for $w \in T^+$ if and only if $sws^k \in L(G')$ for some $k \geq 1$. Then, we show that $\varepsilon \in L(G)$ if and only if $\varepsilon \in L(G')$.

In what follows, we denote a string w which is to be erased as ${}^{\varepsilon}w$; otherwise, to denote that w is not to be erased we write ${}^{\%}w$.

The next claim shows how G' simulates G.

Claim 7.2.14. If $(S, t_1) \Rightarrow^m_G ({}^{\%}x_0 {}^{\varepsilon}X_1 {}^{\%}x_1 {}^{\varepsilon}X_2 {}^{\%}x_2 \cdots {}^{\varepsilon}X_h {}^{\%}x_h, t_2) \Rightarrow^*_G (z, t_3)$, where $z \in T^+$, $t_1, t_2, t_3 \in \Psi$, $x_i \in V^*$ for $i = 0, 1, \ldots, h$, $X_j \in N$ for $j = 1, 2, \ldots, h$, for some $h \geq 0$ and $m \geq 0$, then $(S', \ell_0) \Rightarrow^*_{G'} (sx_0x_1x_2 \cdots x_hvZ, t'_2)$, where $v \in \mathrm{perm}(\tau(X_1X_2 \cdots X_h)\xi)$, $\xi \in \{s\}^*$, and t'_2 can be ℓ_s or any rule constructed from t_2 in (1.3), (2.2), or (3.3).

Proof. This claim is established by induction on $m \geq 0$.

Basis. Let $m = 0$. Then, for $(S, t_1) \Rightarrow^0_G (S, t_1) \Rightarrow^*_G (z, t_2)$, there is $(S', \ell_0) \Rightarrow_{G'} (sSZ, t'_1)$, where t'_1 can be ℓ_s or any rule constructed from t_1 in (1.3), (2.2), or (3.3). Hence, the basis holds.

Induction Hypothesis. Suppose that there exists $n \geq 0$ such that the claim holds for all derivations of length m, where $0 \leq m \leq n$.

Induction Step. Consider any derivation of the form

$$(S, t_1) \Rightarrow^{n+1}_G (w, t_3) \Rightarrow^*_G (z, t_4)$$

where $w \in V^+$ and $z \in T^+$. Since $n + 1 \geq 1$, this derivation can be expressed as

$$(S, t_1) \Rightarrow_G^n ({}^{\mathscr{G}}_{x_0}{}^{\varepsilon}X_1{}^{\mathscr{G}}_{x_1}{}^{\varepsilon}X_2{}^{\mathscr{G}}_{x_2} \cdots {}^{\varepsilon}X_h{}^{\mathscr{G}}_{x_h}, t_2) \Rightarrow_G (w, t_3) \Rightarrow_G^* (z, t_4)$$

where $x_i \in V^*$ for $i = 0, 1, \ldots, h$, $X_j \in V$ for $j = 1, 2, \ldots, h$, for some $h \geq 0$. By the induction hypothesis,

$$(S', \ell_0) \Rightarrow_{G'}^* (sx_0x_1x_2 \cdots x_h vZ, t_2')$$

where $v \in \mathrm{perm}(\tau(X_1X_2 \cdots X_h)\xi)$, $\xi \in \{s\}^*$, and t_2' can be ℓ_s or any rule constructed from t_2 in (1.3), (2.2), or (3.3).

Let $x = {}^{\mathscr{G}}_{x_0}{}^{\varepsilon}X_1{}^{\mathscr{G}}_{x_1}{}^{\varepsilon}X_2{}^{\mathscr{G}}_{x_2} \cdots {}^{\varepsilon}X_h{}^{\mathscr{G}}_{x_h}$. Next, we consider all possible forms of the derivation $(x, t_2) \Rightarrow_G (w, t_3)$, covered by the following three cases—(i) through (iii).

(i) *Application of a rule that rewrites a symbol in some x_j.* Let $x_j = x_j'Ax_j''$ and $(t_2: A \rightarrow y_0Y_1y_1Y_2y_2 \cdots Y_my_m, \sigma_{t_2}, \rho_{t_2}) \in P$ for some $j \in \{0, 1, \ldots, h\}$ and $m \geq 0$, where $y_i \in V^*$ for $i = 0, 1, \ldots, m$, $Y_i \in N$ for $i = 1, \ldots, m$, and $t_3 \in \sigma_{t_2}$ so that

$$({}^{\mathscr{G}}_{x_0}{}^{\varepsilon}X_1{}^{\mathscr{G}}_{x_1}{}^{\varepsilon}X_2{}^{\mathscr{G}}_{x_2} \cdots {}^{\varepsilon}X_j{}^{\mathscr{G}}_{x_j'}{}^{\mathscr{G}}A{}^{\mathscr{G}}_{x_j''} \cdots {}^{\varepsilon}X_h{}^{\mathscr{G}}_{x_h}, t_2) \Rightarrow_G$$
$$({}^{\mathscr{G}}_{x_0}{}^{\varepsilon}X_1{}^{\mathscr{G}}_{x_1}{}^{\varepsilon}X_2{}^{\mathscr{G}}_{x_2} \cdots {}^{\varepsilon}X_j{}^{\mathscr{G}}_{y_0}{}^{\varepsilon}Y_1{}^{\mathscr{G}}_{y_1}{}^{\varepsilon}Y_2{}^{\mathscr{G}}_{y_2} \cdots {}^{\varepsilon}Y_m{}^{\mathscr{G}}_{y_m} \cdots {}^{\varepsilon}X_h{}^{\mathscr{G}}_{x_h}, t_3)$$

By 1 and by the final step of the algorithm, P' contains

$$(\ell: A \rightarrow y_0y_1 \cdots y_m, \{\ell'\}, \emptyset)$$
$$(\ell': Z \rightarrow \bar{y}Z, \sigma_{\ell'}, \emptyset), \text{ where } \bar{y} = \tau(Y_1Y_2 \cdots Y_m)$$

By the induction hypothesis, we assume that $t_2' = \ell$. Then,

$$(sx_0x_1x_2 \cdots x_j'Ax_j'' \cdots x_hvZ, \ell) \qquad\qquad \Rightarrow_{G'}$$
$$(sx_0x_1x_2 \cdots x_j'y_0y_1y_2 \cdots y_mx_j'' \cdots x_hvZ, \ell') \Rightarrow_{G'}$$
$$(sx_0x_1x_2 \cdots x_j'y_0y_1y_2 \cdots y_mx_j'' \cdots x_hv\bar{y}Z, t_3')$$

with $t_3' \in \sigma_{\ell'}$. By the final step of the algorithm, t_3' can be ℓ_s or any rule constructed from t_3 in (1.3), (2.2), or (3.3). As $x_0x_1x_2 \cdots x_j'y_0y_1y_2 \cdots y_mx_j'' \cdots x_hv\bar{y}Z$ is of the required form, the induction step for (i) is completed.

(ii) *Application of a rule that rewrites some X_j.* Let $(t_2: X_j \rightarrow y, \sigma_{t_2}, \rho_{t_2}) \in P$ for some $j \in \{1, 2, \ldots, h\}$, where $y \in N^*$ and $t_3 \in \sigma_{t_2}$ so that

$$({}^{\mathscr{G}}_{x_0}{}^{\varepsilon}X_1{}^{\mathscr{G}}_{x_1}{}^{\varepsilon}X_2{}^{\mathscr{G}}_{x_2} \cdots {}^{\varepsilon}X_j{}^{\mathscr{G}}_{x_j} \cdots {}^{\varepsilon}X_h{}^{\mathscr{G}}_{x_h}, t_2) \Rightarrow_G$$
$$({}^{\mathscr{G}}_{x_0}{}^{\varepsilon}X_1{}^{\mathscr{G}}_{x_1}{}^{\varepsilon}X_2{}^{\mathscr{G}}_{x_2} \cdots {}^{\varepsilon}y{}^{\mathscr{G}}_{x_j} \cdots {}^{\varepsilon}X_h{}^{\mathscr{G}}_{x_h}, t_3)$$

By the induction hypothesis, $v = v_1\bar{X}_jv_2$ for some $v_1, v_2 \in \bar{N}^*$. By 2 and by the final step of the algorithm,

$$(\bar{t}_2\!:\!\bar{X}_j \rightarrow \bar{y}, \sigma_{\bar{t}_2}, \emptyset) \in P' \text{ where } \bar{y} = \tau(y)$$

By the induction hypothesis, we assume that $t_2' = \bar{t}_2$. Then,

$$(sx_0x_1x_2 \cdots x_h v_1 \bar{X}_j v_2 Z, \bar{t}_2) \Rightarrow_{G'} (sx_0x_1x_2 \cdots x_h v_1 \bar{y} v_2 Z, t_3')$$

with $t_3' \in \sigma_{\bar{t}_2}$. By the final step of the algorithm, t_3' can be ℓ_s or any rule that
was constructed from t_3 in (1.3), (2.2), or (3.3). Since $x_0x_1x_2 \cdots x_h v_1 \bar{y}_j v_2 Z$ is of
the required form, the induction step for (ii) is completed.

(iii) *Application of a rule in the appearance checking mode.* Let $(t_2\!:\!A \rightarrow$
$y, \sigma_{t_2}, \rho_{t_2}) \in P$, where $A \notin \mathrm{alph}(x)$, $y \in V^*$ and $t_3 \in \rho_{t_2}$ so that

$$(x, t_2) \Rightarrow_G (x, t_3)$$

By the induction hypothesis, we assume that t_2' was constructed from t_2 in (3.3).
By 3 and by the final step of the algorithm, P' contains

$$(\hat{r}\!:\!A \rightarrow \#y, \emptyset, \{\hat{r}'\})$$
$$(\hat{r}'\!:\!\bar{A} \rightarrow \#y, \emptyset, \rho_{\hat{r}'})$$

Since $A \notin \mathrm{alph}(x)$, $\bar{A} \notin \mathrm{alph}(x_0x_1x_2 \cdots x_h v Z)$. Then,

$$(sx_0x_1x_2 \cdots x_h v Z, \hat{r}) \Rightarrow_{G'}$$
$$(sx_0x_1x_2 \cdots x_h v Z, \hat{r}') \Rightarrow_{G'}$$
$$(sx_0x_1x_2 \cdots x_h v Z, t_3')$$

where t_3' can be ℓ_s or any rule that was constructed from t_3 in (1.3), (2.2),
or (3.3). Since $x_0x_1x_2 \cdots x_h v Z$ is of the required form, the induction step for (iii)
is completed.

Observe that cases (i) through (iii) cover all possible forms of $(x, t_2) \Rightarrow_G (w, t_3)$.
Thus, the claim holds.   □

To simplify the second claim and its proof, we define a generalization of $\Rightarrow_{G'}$. In
this generalization, we use the property that whenever a rule introduced in (1.3)
or (3.3) is applied during a successful derivation, it has to be followed by its
corresponding rule from (1.4) or (3.4), respectively. Let $V' = N' \cup T$. Define the
binary relation $\Rightarrow_{G'}$ over $V'^* \times \Psi'$ as

$$(x, r) \Rightarrow_{G'} (w, t)$$

if and only if either

$$(x, r) \Rightarrow_{G'} (w, t)$$

where $r, t \in \Psi'$ such that t is not introduced in (1.4) and (3.4), or

$$(x, r) \Rightarrow_{G'} (y, r') \Rightarrow_{G'} (w, t)$$

where r is a rule introduced in (1.3) or (3.3) and r' is its corresponding second rule introduced in (1.4) or (3.4), respectively. Define $\Rightarrow_{G'}^n$ for $n \geq 0$ and $\Rightarrow_{G'}^*$ in the usual way.

The next claim shows how G simulates G'. Define the homomorphism ι from V'^* to V^* as $\iota(X) = X$ for $X \in V$, $\iota(\bar{X}) = X$ for $X \in N$, and $\iota(s) = \iota(S') = \iota(Z) = \varepsilon$.

Claim 7.2.15. If $(S', \ell_0) \Rightarrow_{G'}^m (sxu, t) \Rightarrow_{G'}^* (z, g)$, where $x \in V^+$, $u \in (\bar{N} \cup \{s\})^* \{Z, \varepsilon\}$, $z \in Q^+$, $t, g \in \Psi'$, and $m \geq 1$, then $(S, t_1) \Rightarrow_G^* (x_0 X_1 x_1 X_2 x_2 \cdots X_h x_h, t')$, where $x = x_0 x_1 \cdots x_h$, $X_1 X_2 \cdots X_h \in \mathrm{perm}(\iota(u))$, $h \geq 0$, and t' is the rule from which t was constructed or any rule in P if t was not constructed from any rule in P.

Proof. This claim is established by induction on $m \geq 1$.

Basis. Let $m = 1$. Then, for $(S', \ell_0) \Rightarrow_{G'}^m (sSZ, t) \Rightarrow_{G'}^* (z, g)$, where $t, g \in \Psi'$ and $z \in Q^+$, there is $(S, t') \Rightarrow_G^0 (S, t')$, where t' is the rule from which t was constructed. Hence, the basis holds.

Induction Hypothesis. Suppose that there exists $n \geq 1$ such that the claim holds for all derivations of length m, where $0 \leq m \leq n$.

Induction Step. Consider any derivation of the form

$$(S', \ell_0) \Rightarrow_{G'}^{n+1} (swv, p) \Rightarrow_{G'}^* (z, g)$$

where $n \geq 1$, $w \in V^+$, $v \in (\bar{N} \cup \{s\})^* \{Z, \varepsilon\}$, $z \in Q^+$, and $p, g \in \Psi'$. Since $n + 1 \geq 1$, this derivation can be expressed as

$$(S', \ell_0) \Rightarrow_{G'}^n (sxu, t) \Rightarrow_{G'} (swv, p) \Rightarrow_{G'}^* (z, g)$$

where $x \in V^+$, $u \in (\bar{N} \cup \{s\})^* \{Z, \varepsilon\}$, and $t \in \Psi'$. By the induction hypothesis,

$$(S, t_1) \Rightarrow_G^* (x_0 X_1 x_1 X_2 x_2 \cdots X_h x_h, t')$$

where $x = x_0 x_1 \cdots x_h$, $X_1 X_2 \cdots X_h \in \mathrm{perm}(\iota(u))$, $h \geq 0$, and t' is the rule from which t was constructed or any rule in P if t was not constructed from any rule in P.

Next, we consider all possible forms of $(sxu, t) \Rightarrow_{G'} (swv, p)$, covered by the following four cases—(i) through (iv).

(i) *Application of* $(\ell_s : Z \to s, \{\ell_s\}, \emptyset)$. Let $t = \ell_s$, so

$$(swu'Z, \ell_s) \Rightarrow_{G'} (swu's, \ell_s)$$

where $u = u'Z$. Then, the induction step for (i) follows directly from the induction hypothesis (recall that $\iota(Z) = \iota(s) = \varepsilon$).

(ii) *Application of two rules introduced in 1.* Let $x = x'Ax''$ and $(\ell\colon A \to y_0y_1 \cdots y_m, \{\ell'\}, \emptyset)$, $(\ell'\colon Z \to \bar{y}Z, \sigma_{\ell'}, \emptyset) \in P'$ be two rules introduced in 1 from $(r\colon A \to y_0Y_1y_1Y_2y_2 \cdots Y_my_m, \sigma_r, \rho_r) \in P$, where $y_i \in V^*$, $Y_j \in N$, for $i = 0, 1, \ldots, m$ and $j = 1, 2, \ldots, m$, for some $m \geq 1$, and $\bar{y} = \tau(Y_1Y_2 \cdots Y_m)$. Then,

$$(sx'Ax''uZ, \ell) \Rightarrow_{G'} (sx'y_0y_1 \cdots y_mx''u\bar{y}Z, p)$$

by applying ℓ and ℓ' ($p \in \sigma_{\ell'}$). By the induction hypothesis, $t' = r$ and $x_i = x_i'Ax_i''$ for some $i \in \{0, 1, \ldots, h\}$. Then,

$$(x_0X_1x_1X_2x_2 \cdots X_ix_i'Ax_i'' \cdots X_hx_h, r) \Rightarrow_G$$
$$(x_0X_1x_1X_2x_2 \cdots X_ix_i'y_0Y_1y_1Y_2y_2 \cdots Y_my_mx_i'' \cdots X_hx_h, t'')$$

Clearly, both configurations are of the required forms, so the induction step is completed for (ii).

(iii) *Application of a rule introduced in 2.* Let $u = u'Au''$ and $(\bar{r}\colon \bar{A} \to \bar{y}, \sigma_{\bar{r}}, \emptyset) \in P'$ be a rule introduced in 2 from $(r\colon A \to y, \sigma_r, \rho_r) \in P$, where $y \in N^*$ and $\bar{y} = \tau(y)$. Then,

$$(sxu'\bar{A}u''Z, \bar{r}) \Rightarrow_{G'} (sxu'\bar{y}u''Z, p)$$

where $p \in \sigma_{\bar{r}}$. By the induction hypothesis, $t' = r$ and $X_i = A$ for some $i \in \{1, \ldots, h\}$. Then,

$$(x_0X_1x_1X_2x_2 \cdots X_ix_i \cdots X_hx_h, r) \Rightarrow_G$$
$$(x_0X_1x_1X_2x_2 \cdots yx_i \cdots X_hx_h, t'')$$

Clearly, both configurations are of the required forms, so the induction step is completed for (iii).

(iv) *Application of two rules introduced in 3.* Let $(\hat{r}\colon A \to \#y, \emptyset, \{\hat{r}'\})$, $(\hat{r}'\colon \bar{A} \to \#y, \emptyset, \rho_{\hat{r}'}) \in P'$ be two rules introduced in 3 from $(r\colon A \to y, \sigma_r, \rho_r) \in P$, where $y \in V^*$, such that $\{A, \bar{A}\} \cap alph(sxuZ) = \emptyset$. Then,

$$(sxu, \hat{r}) \Rightarrow_{G'} (sxu, p)$$

by applying $\hat{r}$ and $\hat{r}'$ in the appearance checking mode ($p \in \rho_{\hat{r}'}$). By the induction hypothesis, $t' = r$ and $A \notin alph(x_0X_1x_1X_2x_2 \cdots X_hx_h)$, so

$$(x_0X_1x_1X_2x_2 \cdots X_hx_h, r) \Rightarrow_G (x_0X_1x_1X_2x_2 \cdots X_hx_h, t'')$$

Clearly, both configurations are of the required forms, so the induction step is completed for (iv).

Observe that cases (i) through (iv) cover all possible forms of $(sxu, t) \Rightarrow_{G'} (swv, p)$. Thus, the claim holds. $\qquad\qquad\square$

Consider Claim 7.2.14 with $h = 0$. Then,

$$(S, t_1) \Rightarrow_G^* (z, r)$$

implies that

$$(S', \ell_0) \Rightarrow_{G'} (szs^k, r')$$

where $k \geq 1$, $t_1, r \in \Psi$, and $r' \in \Psi'$. Consider Claim 7.2.15 with $x \in T^+$ and $u \in \{s\}^+$. Then,

$$(S', \ell_0) \Rightarrow_{G'}^* (sxu, t)$$

implies that

$$(S, t_1) \Rightarrow_G^* (x, t')$$

Hence, we have $w \in L(G)$ for $w \in T^+$ if and only if $sws^k \in L(G')$ for some $k \geq 1$.

It remains to be shown that $\varepsilon \in L(G)$ if and only if $\varepsilon \in L(G')$. This can be proved by analogy with proving Claims 7.2.14 and 7.2.15, where G' uses $\bar{\ell}_0$ instead of ℓ_0 (see the initialization part of the algorithm). We leave this part of the proof to the reader. Hence, $\triangleright L(M, L(G')) = L(G)$, and the lemma holds. $\quad\square$

Lemma 7.2.16. $\mathbf{RE} \subseteq \mathbf{SCFA}(\mathbf{P}_{ac}^{-\varepsilon})$

Proof. Let $I \in \mathbf{RE}$ and $T = \text{alph}(I)$. By Theorem 3.5.4, there is a programmed grammar with appearance checking $G = (V, T, \Psi, P, S)$ such that $L(G) = I$. Let $M = (Q, T, R, s, F)$ and $G' = (V', Q, \Psi', P', S')$ be the finite automaton and the propagating programmed grammar with appearance checking, respectively, constructed by Algorithm 7.2.12 from G. By Lemma 7.2.13, $\triangleright L(M, L(G')) = L(G) = I$, so the lemma holds. $\quad\square$

Theorem 7.2.17. $\mathbf{SCFA}(\mathbf{P}_{ac}^{-\varepsilon}) = \mathbf{RE}$

Proof. The inclusion $\mathbf{SCFA}(\mathbf{P}_{ac}^{-\varepsilon}) \subseteq \mathbf{RE}$ follows from Turing-Church thesis. The converse inclusion $\mathbf{RE} \subseteq \mathbf{SCFA}(\mathbf{P}_{ac}^{-\varepsilon})$ follows from Lemma 7.2.16. $\quad\square$

Combining Theorems 7.2.5 and 7.2.17, we obtain the following corollary (recall that $\mathbf{P}_{ac}^{-\varepsilon}$ satisfies all the conditions from Theorem 7.2.5, see [DP89]).

Corollary 7.2.18. $\mathbf{TCFA}(\mathbf{P}_{ac}^{-\varepsilon}) = \mathbf{RE}$ $\quad\square$

Finally, we briefly investigate a reduction of the number of states in controlled finite automata. First, observe that the finite automaton $M = (Q, T, R, s, F)$ from Algorithm 7.2.12 has $\text{card}(T) + 1$ states. Therefore, we obtain the following corollary.

Corollary 7.2.19. *Let I be a recursively enumerable language, and let $T = \text{alph}(I)$. Then, there is a finite automaton $M = (Q, T, R, s, F)$ such that $\text{card}(Q) = \text{card}(T) +$*

1, *and a propagating programmed grammar with appearance checking G such that* $\triangleright L(M, L(G)) = I$. □

In a comparison to Corollary 7.2.19, a more powerful result holds in terms of transition-controlled finite automata. In this case, the number of states can be decreased to a single state as stated in the following theorem.

Theorem 7.2.20. *Let I be a recursively enumerable language. Then, there is a finite automaton* $M = (Q, T, R, s, F)$ *such that* $\text{card}(Q) = 1$, *and a propagating programmed grammar with appearance checking G such that* $\triangleright L(M, L(G)) = I$.

Proof. Reconsider Algorithm 7.2.12. We modify the construction of $M = (Q, T, R, s, F)$ in the following way. Let $G = (N, T, \Psi, P, S)$ be the input programmed grammar with appearance checking. Construct the finite automaton

$$M = (Q, \Phi, R, s, F)$$

where

$$Q = \{s\}$$
$$\Phi = \{s\} \cup T$$
$$R = \{s\colon s \to s\} \cup \{a\colon sa \to s \mid a \in T\}$$
$$F = \{s\}$$

Observe that $\text{card}(Q) = 1$ and that this modified algorithm always halts. The correctness of the modified algorithm—that is, the identity $\triangleright L(M, L(G')) = L(G)$, where G' is the propagating programmed grammar constructed by Algorithm 7.2.12—can be established by analogy with the proof of Lemma 7.2.13, so we leave the proof to the reader. The rest of the proof of this theorem parallels the proof of Lemma 7.2.16, so we omit it. □

We close this section by presenting three open problem areas that are related to the achieved results.

Open Problem 7.2.21. In general, the state-controlled and transition-controlled finite automata in Theorem 7.2.17 and Corollary 7.2.18 are non-deterministic. Do these results hold in terms of deterministic versions of these automata?

Open Problem 7.2.22. By using control languages from **CF**, we characterize **CF**. By using control languages from $\mathbf{P}_{ac}^{-\varepsilon}$, we characterize **RE**. Is there a language family $\mathcal{L}$ such that $\mathbf{CF} \subset \mathcal{L} \subset \mathbf{P}_{ac}^{-\varepsilon}$ by which we can characterize **CS**?

Open Problem 7.2.23. Theorem 7.2.5 requires $\mathcal{L}$ to contain all finite languages and to be closed under finite ε-free substitution and concatenation. Does the same result hold if there are fewer requirements placed on $\mathcal{L}$?

7.2.2   Pushdown Automata Regulated by Control Languages

Section 7.2.2 consists of four subsections. First, we define pushdown automata that regulate the application of their rules by control languages by analogy with context-free grammars regulated in this way (see Sect. 3.3). Then, we demonstrate that this regulation has no effect on the power of pushdown automata if the control languages are regular. Considering this result, we point out that pushdown automata regulated by regular languages are of little interest because their power coincides with the power of ordinary pushdown automata. Next, however, we prove that pushdown automata increase their power remarkably if they are regulated by linear languages; indeed, they characterize the family of recursively enumerable languages. Finally, we continue with the discussion of regulated pushdown automata, but we narrow our attention to their special cases, such as one-turn pushdown automata.

Definitions

Without further ado, we next define pushdown automata regulated by control languages. Recall the formalization of rule labels from Definition 2.4.10 because this formalization is often used throughout the present section.

Definition 7.2.24. Let $M = (Q, \Sigma, \Gamma, R, s, S, F)$ be a pushdown automaton, and let Ψ be an alphabet of its rule labels. Let Ξ be a *control language* over Ψ; that is, $\Xi \subseteq \Psi^*$. With Ξ, M defines the following three types of accepted languages

- $L(M, \Xi, 1)$—the *language accepted by final state*
- $L(M, \Xi, 2)$—the *language accepted by empty pushdown*
- $L(M, \Xi, 3)$—the *language accepted by final state and empty pushdown*

defined as follows. Let $\chi \in \Gamma^* Q \Sigma^*$. If $\chi \in \Gamma^* F$, $\chi \in Q$, $\chi \in F$, then χ is a 1-*final configuration*, 2-*final configuration*, 3-*final configuration*, respectively. For $i = 1, 2, 3$, we define $L(M, \Xi, i)$ as

$$L(M, \Xi, i) = \{w \mid w \in \Sigma^* \text{ and } Ssw \vdash_M^* \chi \, [\sigma] \\ \text{for an } i\text{-final configuration } \chi \text{ and } \sigma \in \Xi\}$$

The pair (M, Ξ) is called a *controlled pushdown automaton*.   □

For any family of languages $\mathscr{L}$ and $i \in \{1, 2, 3\}$, define

$$\mathbf{RPDA}(\mathscr{L}, i) = \{L \mid L = L(M, \Xi, i), \text{ where } M \text{ is a pushdown} \\ \text{automaton and } \Xi \in \mathscr{L}\}$$

We demonstrate that

$$\mathbf{CF} = \mathbf{RPDA}(\mathbf{REG}, 1) = \mathbf{RPDA}(\mathbf{REG}, 2) = \mathbf{RPDA}(\mathbf{REG}, 3)$$

and

$$RE = RPDA(LIN, 1) = RPDA(LIN, 2) = RPDA(LIN, 3)$$

Some of the following proofs involve several grammars and automata. To avoid any confusion, these proofs sometimes specify a regular grammar G as $G = (N_G, T_G, P_G, S_G)$ because this specification clearly expresses that N_G, T_G, P_G, and S_G represent the components of G. Other grammars and automata are specified analogously whenever any confusion may exist.

Regular-Controlled Pushdown Automata

This section proves that if the control languages are regular, then the regulation of pushdown automata has no effect on their power. The proof of the following lemma presents a transformation that converts any regular grammar G and any pushdown automaton K to an ordinary pushdown automaton M such that $L(M) = L(K, L(G), 1)$.

Lemma 7.2.25. *For every regular grammar G and every pushdown automaton K, there exists a pushdown automaton M such that $L(M) = L(K, L(G), 1)$.*

Proof. Let $G = (N_G, T_G, P_G, S_G)$ be any regular grammar, and let $K = (Q_K, \Sigma_K, \Gamma_K, R_K, s_K, S_K, F_K)$ be any pushdown automaton. Next, we construct a pushdown automaton M that simultaneously simulates G and K so that $L(M) = L(K, L(G), 1)$.

Let f be a new symbol. Define M as

$$M = (Q_M, \Sigma_M, \Gamma_M, R_M, s_M, S_M, F_M)$$

where

$$Q_M = \{\langle qB \rangle \mid q \in Q_K, B \in N_G \cup \{f\}\}$$
$$\Sigma_M = \Sigma_K$$
$$\Gamma_M = \Gamma_K$$
$$s_M = \langle s_K S_G \rangle$$
$$S_M = S_K$$
$$F_M = \{\langle qf \rangle \mid q \in F_K\}$$
$$R_M = \{C\langle qA \rangle b \rightarrow x\langle pB \rangle \mid a\colon Cqb \rightarrow xp \in R_K, A \rightarrow aB \in P_G\}$$
$$\quad \cup \{C\langle qA \rangle b \rightarrow x\langle pf \rangle \mid a\colon Cqb \rightarrow xp \in R_K, A \rightarrow a \in P_G\}$$

Observe that a move in M according to $C\langle qA \rangle b \rightarrow x\langle pB \rangle \in R_M$ simulates a move in K according to $a\colon Cqb \rightarrow xp \in R_K$, where a is generated in G by using $A \rightarrow aB \in P_G$. Based on this observation, it is rather easy to see that M accepts an input string w if and only if K reads w and enters a final state after using a complete string of $L(G)$; therefore, $L(M) = L(K, L(G), 1)$. A rigorous proof that $L(M) = L(K, L(G), 1)$ is left to the reader. $\square$

Theorem 7.2.26. *For* $i \in \{1, 2, 3\}$, $\mathbf{CF} = \mathbf{RPDA}(\mathbf{REG}, i)$.

Proof. To prove that $\mathbf{CF} = \mathbf{RPDA}(\mathbf{REG}, 1)$, notice that $\mathbf{RPDA}(\mathbf{REG}, 1) \subseteq \mathbf{CF}$ follows from Lemma 7.2.25. Clearly, $\mathbf{CF} \subseteq \mathbf{RPDA}(\mathbf{REG}, 1)$, so $\mathbf{RPDA}(\mathbf{REG}, 1) = \mathbf{CF}$. By analogy with the demonstration of $\mathbf{RPDA}(\mathbf{REG}, 1) = \mathbf{CF}$, we can prove that $\mathbf{CF} = \mathbf{RPDA}(\mathbf{REG}, 2)$ and $\mathbf{CF} = \mathbf{RPDA}(\mathbf{REG}, 3)$. □

Let us point out that most fundamental regulated grammars use control mechanisms that can be expressed in terms of regular control languages (see Chap. 3) to regular control languages. However, pushdown automata introduced by analogy with these grammars are of little or no interest because they are as powerful as ordinary pushdown automata (see Theorem 7.2.26 above).

Linear-Controlled Pushdown Automata

This section demonstrates that pushdown automata regulated by linear control languages are more powerful than ordinary pushdown automata. In fact, it proves that

$$\mathbf{RE} = \mathbf{RPDA}(\mathbf{LIN}, 1) = \mathbf{RPDA}(\mathbf{LIN}, 2) = \mathbf{RPDA}(\mathbf{LIN}, 3)$$

Recall the normal form for left-extended queue grammars from Definition 3.1.25, which is needed to prove the following.

Lemma 7.2.27. *Let* Q *be a left-extended queue grammar that satisfies the normal form of Definition 3.1.25. Then, there exist a linear grammar* G *and a pushdown automaton* M *such that* $L(Q) = L(M, L(G), 3)$.

Proof. Let $Q = (V_Q, T_Q, W_Q, F_Q, R_Q, g_Q)$ be a left-extended queue grammar satisfying the normal form of Definition 3.1.25. Without any loss of generality, assume that $\{@, \P, \S, \$, \lfloor, \rceil\} \cap (V_Q \cup W_Q) = \emptyset$. Define the coding ζ from V_Q^* to $\{\langle as \rangle_\psi \mid a \in V_Q\}^*$ as $\zeta(a) = \langle as \rangle_\psi$ (s is used as the start state of the pushdown automaton M defined later in this proof).

Construct the linear grammar $G = (N_G, T_G, P_G, S_G)$ in the following way. Initially, set

$$N_G = \{S_G, \langle ! \rangle, \langle !, 1 \rangle\} \cup \{\langle f \rangle \mid f \in F_Q\}$$
$$T_G = \zeta(V_Q) \cup \{\langle \S s \rangle_\psi, \langle @ \rangle_\psi\} \cup \{\langle \S f \rangle_\psi \mid f \in F_Q\}$$
$$P_G = \{S_G \to \langle \S s \rangle_\psi \langle f \rangle \mid f \in F_Q\} \cup \{\langle ! \rangle \to \langle !, 1 \rangle \langle @ \rangle_\psi\}$$

Extend N_G, T_G, and P_G by performing (1) through (3), given next.

(1) For every $(a, p, x, q) \in R_Q$, where $p, q \in W_Q$, $a \in Z$, and $x \in T^*$,

$$N_G = N_G$$
$$N_G = N_G \cup \{\langle apxqk \rangle \mid k = 0, \dots, |x|\} \cup \{\langle p \rangle, \langle q \rangle\}$$
$$T_G = T_G \cup \{\langle \mathrm{sym}(x, k) \rangle_\psi \mid k = 1, \dots, |x|\} \cup \{\langle apxq \rangle_\psi\}$$
$$P_G = P_G \cup \{\langle q \rangle \to \langle apxq|x| \rangle \langle apxq \rangle_\psi, \langle apxq0 \rangle \to \langle p \rangle\}$$
$$\cup \{\langle apxqk \rangle \to \langle apxq(k-1) \rangle \langle \mathrm{sym}(x, k) \rangle_\psi \mid k = 1, \dots, |x|\}$$

(2) For every $(a, p, x, q) \in R_Q$ with $p, q \in W_Q$, $a \in U$, and $x \in V_Q^*$,

$$N_G = N_G$$
$$N_G = N_G \cup \{\langle p, 1 \rangle, \langle q, 1 \rangle\}$$
$$P_G = P_G \cup \{\langle q, 1 \rangle \to \mathrm{reversal}(\zeta(x))\langle p, 1 \rangle \zeta(a)\}$$

(3) For every $(a, p, x, q) \in R_Q$ with $ap = q_Q$, $p, q \in W_Q$, and $x \in V_Q^*$,

$$N_G = N_G$$
$$N_G = N_G \cup \{\langle q, 1 \rangle\}$$
$$P_G = P_G \cup \{\langle q, 1 \rangle \to \mathrm{reversal}(x)\langle \$s \rangle_\psi\}$$

The construction of G is completed. Set $\Psi = T_G$. Ψ represents the alphabet of rule labels corresponding to the rules of the pushdown automaton M, defined as

$$M = \left(Q_M, \Sigma_M, \Gamma_M, R_M, s_M, S_M, \{\rceil\}\right)$$

Throughout the rest of this proof, s_M is abbreviated to s. Initially, set

$$Q_M = \{s, \langle \P! \rangle, \lfloor, \rceil\}$$
$$\Sigma_M = T_Q$$
$$\Gamma_M = \{S_M, \S\} \cup V_Q$$
$$R_M = \{\langle \S s \rangle_\psi : S_M s \to \S s\} \cup \{\langle \S f \rangle_\psi : \S \langle \P f \rangle \to \rceil \mid f \in F_M\}$$

Extend Q_M and R_M by performing (A) through (D), given next.

(A) Set $R_M = R_M \cup \{\langle bs \rangle_\psi : as \to abs \mid a \in \Gamma_M - \{S_M\}, b \in \Gamma_M\}$.
(B) Set $R_M = R_M \cup \{\langle \$s \rangle_\psi : as \to a \lfloor \mid a \in V_Q\} \cup \{\langle a \rangle_\psi : a \lfloor \to \lfloor \mid a \in V_Q\}$.
(C) Set $R_M = R_M \cup \{\langle @ \rangle_\psi : a \lfloor \to a \langle \P! \rangle \mid a \in Z\}$.
(D) For every $(a, p, x, q) \in R_Q$, where $p, q \in W_Q$, $a \in Z$, $x \in T_Q^*$, set

$$Q_M = Q_M \cup \{\langle \P p \rangle\} \cup \{\langle \P qu \rangle \mid u \in \mathrm{prefix}(x)\}$$
$$R_M = R_M \cup \{\langle b \rangle_\psi : a \langle \P qy \rangle b \to a \langle \P qyb \rangle \mid b \in T_Q, y \in T_Q^*, yb \in \mathrm{prefix}(x)\}$$
$$\cup \{\langle apxq \rangle_\psi : a \langle \P qx \rangle \to \langle \P p \rangle\}$$

The construction of M is completed. Notice that several components of G and M have this form: $\langle x \rangle_y$. Intuitively, if $y = \Psi$, then $\langle x \rangle_y \in \Psi$; or T_G, respectively. If x begins with $\P$, then $\langle x \rangle_y \in Q_M$. Otherwise, $\langle x \rangle_y \in N_G$.

First, we only sketch the reason why $L(Q)$ contains $L(M, L(G), 3)$. According to a string from $L(G)$, M accepts every string w as

$$
\begin{aligned}
\S s w_1 \cdots w_{m-1} w_m \;&\vdash_M^+\; \S b_m \cdots b_1 a_n \cdots a_1 s w_1 \cdots w_{m-1} w_m \\
&\vdash_M\; \S b_m \cdots b_1 a_n \cdots a_1 \lfloor w_1 \cdots w_{m-1} w_m \\
&\vdash_M^n\; \S b_m \cdots b_1 \lfloor w_1 \cdots w_{m-1} w_m \\
&\vdash_M\; \S b_m \cdots b_1 \langle \P q_1 \rangle w_1 \cdots w_{m-1} w_m \\
&\vdash_M^{|w_1|}\; \S b_m \cdots b_1 \langle \P q_1 w_1 \rangle w_2 \cdots w_{m-1} w_m \\
&\vdash_M\; \S b_m \cdots b_2 \langle \P q_2 \rangle w_2 \cdots w_{m-1} w_m \\
&\vdash_M^{|w_2|}\; \S b_m \cdots b_2 \langle \P q_2 w_2 \rangle w_3 \cdots w_{m-1} w_m \\
&\vdash_M\; \S b_m \cdots b_3 \langle \P q_3 \rangle w_3 \cdots w_{m-1} w_m \\
&\;\;\vdots \\
&\vdash_M\; \S b_m \langle \P q_m \rangle w_m \\
&\vdash_M^{|w_m|}\; \S b_m \langle \P q_m w_m \rangle \\
&\vdash_M\; \S \langle \P q_{m+1} \rangle \\
&\vdash_M\; \rceil
\end{aligned}
$$

where $w = w_1 \cdots w_{m-1} w_m$, $a_1 \cdots a_n b_1 \cdots b_m = x_1 \cdots x_{n+1}$, and R_Q contains (a_0, p_0, x_1, p_1), (a_1, p_1, x_2, p_2), ..., (a_n, p_n, x_{n+1}, q_1), (b_1, q_1, w_1, q_2), (b_2, q_2, w_2, q_3), ..., (b_m, q_m, w_m, q_{m+1}). According to these members of R_Q, Q makes

$$
\begin{aligned}
\# a_0 p_0 \;&\Rightarrow_Q\; a_0 \# y_0 x_1 p_1 && [(a_0, p_0, x_1, p_1)] \\
&\Rightarrow_Q\; a_0 a_1 \# y_1 x_2 p_2 && [(a_1, p_1, x_2, p_2)] \\
&\Rightarrow_Q\; a_0 a_1 a_2 \# y_2 x_3 p_3 && [(a_2, p_2, x_3, p_3)] \\
&\;\;\vdots \\
&\Rightarrow_Q\; a_0 a_1 a_2 \cdots a_{n-1} \# y_{n-1} x_n p_n && [(a_{n-1}, p_{n-1}, x_n, p_n)] \\
&\Rightarrow_Q\; a_0 a_1 a_2 \cdots a_n \# y_n x_{n+1} q_1 && [(a_n, p_n, x_{n+1}, q_1)] \\
&\Rightarrow_Q\; a_0 \cdots a_n b_1 \# b_2 \cdots b_m w_1 q_2 && [(b_1, q_1, w_1, q_2)] \\
&\Rightarrow_Q\; a_0 \cdots a_n b_1 b_2 \# b_3 \cdots b_m w_1 w_2 q_3 && [(b_2, q_2, w_2, q_3)] \\
&\;\;\vdots \\
&\Rightarrow_Q\; a_0 \cdots a_n b_1 \cdots b_{m-1} \# b_m w_1 w_2 \cdots w_{m-1} q_m && [(b_{m-1}, q_{m-1}, w_{m-1}, q_m)] \\
&\Rightarrow_Q\; a_0 \cdots a_n b_1 \cdots b_m \# w_1 w_2 \cdots w_m q_{m+1} && [(b_m, q_m, w_m, q_{m+1})]
\end{aligned}
$$

Therefore, $L(M, L(G), 3) \subseteq L(Q)$.

More formally, to demonstrate that $L(Q)$ contains $L(M, L(G), 3)$, consider any $h \in L(G)$. G generates h as

$$
\begin{aligned}
S_G \;&\Rightarrow_G\; \langle \S s \rangle_\psi \langle q_{m+1} \rangle \\
&\Rightarrow_G^{|w_m|+1}\; \langle \S s \rangle_\psi \langle q_m \rangle t_m \langle b_m q_m w_m q_{m+1} \rangle_\psi \\
&\Rightarrow_G^{|w_{m-1}|+1}\; \langle \S s \rangle_\psi \langle q_{m-1} \rangle t_{m-1} \langle b_{m-1} q_{m-1} w_{m-1} q_m \rangle_\psi t_m \langle b_m q_m w_m q_{m+1} \rangle_\psi \\
&\;\;\vdots
\end{aligned}
$$

$\Rightarrow_G^{|w_1|+1} \quad \langle \S s \rangle_\psi \langle q_1 \rangle o$

$\Rightarrow_G^{|w_1|+1} \quad \langle \S s \rangle_\psi \langle q_1, 1 \rangle \langle @ \rangle_\psi o$
$\qquad\qquad [\langle q_1 \rangle \rightarrow \langle q_1, 1 \rangle \langle @ \rangle_\psi]$

$\Rightarrow_G \quad \langle \S s \rangle_\psi \zeta(\text{reversal}(x_{n+1})) \langle p_n, 1 \rangle \langle a_n \rangle_\psi \langle @ \rangle_\psi o$
$\qquad\qquad [\langle q_1, 1 \rangle \rightarrow \text{reversal}(\zeta(x_{n+1})) \langle p_n, 1 \rangle \langle a_n \rangle_\psi \langle @ \rangle_\psi]$

$\Rightarrow_G \quad \langle \S s \rangle_\psi \zeta(\text{reversal}(x_n x_{n+1})) \langle p_{n-1}, 1 \rangle \langle a_{n-1} \rangle_\psi \langle a_n \rangle_\psi \langle @ \rangle_\psi o$
$\qquad\qquad [\langle p_n, 1 \rangle \rightarrow \text{reversal}(\zeta(x_n)) \langle p_{n-1}, 1 \rangle \langle a_{n-1} \rangle_\psi]$

$\vdots$

$\Rightarrow_G \quad \langle \S s \rangle_\psi \zeta(\text{reversal}(x_2 \cdots x_n x_{n+1})) \langle p_1, 1 \rangle \langle a_1 \rangle_\psi \langle a_2 \rangle_\psi \cdots \langle a_n \rangle_\psi \langle @ \rangle_\psi o$
$\qquad\qquad [\langle p_2, 1 \rangle \rightarrow \text{reversal}(\zeta(x_2)) \langle p_1, 1 \rangle \langle a_1 \rangle_\psi]$

$\Rightarrow_G \quad \langle \S s \rangle_\psi \zeta(\text{reversal}(x_1 \cdots x_n x_{n+1})) \langle \$ s \rangle_\psi \langle a_1 \rangle_\psi \langle a_2 \rangle_\psi \cdots \langle a_n \rangle_\psi \langle @ \rangle_\psi o$
$\qquad\qquad [\langle p_1, 1 \rangle \rightarrow \text{reversal}(\zeta(x_1)) \langle \$ s \rangle_\psi]$

where

$n, m \geq 1$
$a_i \in U$ for $i = 1, \ldots, n$
$b_k \in Z$ for $k = 1, \ldots, m$
$x_l \in V^*$ for $l = 1, \ldots, n+1$
$p_i \in W$ for $i = 1, \ldots, n$
$q_l \in W$ for $l = 1, \ldots, m+1$ with $q_1 = \,!$ and $q_{m+1} \in F$

and

$t_k = \langle \text{sym}(w_k, 1) \rangle_\psi \cdots \langle \text{sym}(w_k, |w_k| - 1) \rangle_\psi \langle \text{sym}(w_k, |w_k|) \rangle_\psi$
$\qquad$ for $k = 1, \ldots, m$;
$o = t_1 \langle b_1 q_1 w_1 q_2 \rangle_\psi \cdots t_{m-1} \langle b_{m-1} q_{m-1} w_{m-1} q_m \rangle_\psi t_m \langle b_m q_m w_m q_{m+1} \rangle_\psi$;
$h = \langle \S s \rangle_\psi \zeta(\text{reversal}(x_1 \cdots x_n x_{n+1})) \langle \$ \rangle_\psi \langle a_1 \rangle_\psi \langle a_2 \rangle_\psi \cdots \langle a_n \rangle_\psi \langle @ \rangle_\psi o$

We describe this derivation in a greater detail. Initially, G makes $S_G \Rightarrow_G$ $\langle \S s \rangle_\psi \langle q_{m+1} \rangle$ according to $S_G \rightarrow \langle \S s \rangle_\psi \langle q_{m+1} \rangle$. Then, G makes

$\qquad\qquad \langle \S s \rangle_\psi \langle q_{m+1} \rangle$
$\Rightarrow_G^{|w_m|+1} \quad \langle \S s \rangle_\psi \langle q_m \rangle t_m \langle b_m q_m w_m q_{m+1} \rangle_\psi$
$\Rightarrow_G^{|w_{m-1}|+1} \quad \langle \S s \rangle_\psi \langle q_{m-1} \rangle t_{m-1} \langle b_{m-1} q_{m-1} w_{m-1} q_m \rangle_\psi t_m \langle b_m q_m w_m q_{m+1} \rangle_\psi$
$\vdots$
$\Rightarrow_G^{|w_1|+1} \quad \langle \S s \rangle_\psi \langle q_1 \rangle o$

according to rules introduced in (1). Then, G makes

$$\langle \S s \rangle_\psi \langle q_1 \rangle o \Rightarrow_G \langle \S s \rangle_\psi \langle q_1, 1 \rangle \langle @ \rangle_\psi o$$

according to $\langle ! \rangle \to \langle !, 1 \rangle \langle @ \rangle_\psi$ (recall that $q_1 = \, !$). After this step, G makes

$$\langle \S s \rangle_\psi \langle q_1, 1 \rangle \langle @ \rangle_\psi o$$
$$\Rightarrow_G \langle \S s \rangle_\psi \zeta(\mathrm{reversal}(x_{n+1}))\langle p_n, 1 \rangle \langle a_n \rangle_\psi \langle @ \rangle_\psi o$$
$$\Rightarrow_G \langle \S s \rangle_\psi \zeta(\mathrm{reversal}(x_n x_{n+1}))\langle p_{n-1}, 1 \rangle \langle a_{n-1} \rangle_\psi \langle a_n \rangle_\psi \langle @ \rangle_\psi o$$
$$\vdots$$
$$\Rightarrow_G \langle \S s \rangle_\psi \zeta(\mathrm{reversal}(x_2 \cdots x_n x_{n+1}))\langle p_1, 1 \rangle \langle a_1 \rangle_\psi \langle a_2 \rangle_\psi \cdots \langle a_n \rangle_\psi \langle @ \rangle_\psi o$$

according to rules introduced in (2). Finally, according to $\langle p_1, 1 \rangle \to \mathrm{reversal}(\zeta(x_1))$ $\langle \$ \rangle_\psi$, which is introduced in (3), G makes

$$\langle \S s \rangle_\psi \zeta(\mathrm{reversal}(x_2 \cdots x_n x_{n+1}))\langle p_1, 1 \rangle \langle a_1 \rangle_\psi \langle a_2 \rangle_\psi \cdots \langle a_n \rangle_\psi \langle @ \rangle_\psi o$$
$$\Rightarrow_G \langle \S s \rangle_\psi \zeta(\mathrm{reversal}(x_1 \cdots x_n x_{n+1}))\langle \$ \rangle_\psi \langle a_1 \rangle_\psi \langle a_2 \rangle_\psi \cdots \langle a_n \rangle_\psi \langle @ \rangle_\psi o$$

If $a_1 \cdots a_n b_1 \cdots b_m$ differs from $x_1 \cdots x_{n+1}$, then M does not accept according to h. Assume that $a_1 \cdots a_n b_1 \cdots b_m = x_1 \cdots x_{n+1}$. At this point, according to h, M makes this sequence of moves

$$
\begin{array}{ll}
\S s w_1 \cdots w_{m-1} w_m \vdash_M^+ & \S b_m \cdots b_1 a_n \cdots a_1 s w_1 \cdots w_{m-1} w_m \\
\vdash_M & \S b_m \cdots b_1 a_n \cdots a_1 \lfloor w_1 \cdots w_{m-1} w_m \\
\vdash_M^n & \S b_m \cdots b_1 \lfloor w_1 \cdots w_{m-1} w_m \\
\vdash_M & \S b_m \cdots b_1 \langle \P q_1 \rangle w_1 \cdots w_{m-1} w_m \\
\vdash_M^{|w_1|} & \S b_m \cdots b_1 \langle \P q_1 w_1 \rangle w_2 \cdots w_{m-1} w_m \\
\vdash_M & \S b_m \cdots b_2 \langle \P q_2 \rangle w_2 \cdots w_{m-1} w_m \\
\vdash_M^{|w_2|} & \S b_m \cdots b_2 \langle \P q_2 w_2 \rangle w_3 \cdots w_{m-1} w_m \\
\vdash_M & \S b_m \cdots b_3 \langle \P q_3 \rangle w_3 \cdots w_{m-1} w_m \\
\quad \vdots & \\
\vdash_M & \S b_m \langle \P q_m \rangle w_m \\
\vdash_M^{|w_m|} & \S b_m \langle \P q_m w_m \rangle \\
\vdash_M & \S \langle \P q_{m+1} \rangle \\
\vdash_M & \rceil
\end{array}
$$

In other words, according to h, M accepts $w_1 \cdots w_{m-1} w_m$. Return to the generation of h in G. By the construction of P_G, this generation implies that R_Q contains (a_0, p_0, x_1, p_1), (a_1, p_1, x_2, p_2), ..., $(a_{j-1}, p_{j-1}, x_j, p_j)$, ..., (a_n, p_n, x_{n+1}, q_1), (b_1, q_1, w_1, q_2), (b_2, q_2, w_2, q_3), ..., (b_m, q_m, w_m, q_{m+1}).

Thus, in Q,

$$
\begin{aligned}
\#a_0p_0 &\Rightarrow_Q a_0\#y_0x_1p_1 && [(a_0,p_0,x_1,p_1)] \\
&\Rightarrow_Q a_0a_1\#y_1x_2p_2 && [(a_1,p_1,x_2,p_2)] \\
&\Rightarrow_Q a_0a_1a_2\#y_2x_3p_3 && [(a_2,p_2,x_3,p_3)] \\
&\vdots \\
&\Rightarrow_Q a_0a_1a_2\cdots a_{n-1}\#y_{n-1}x_np_n && [(a_{n-1},p_{n-1},x_n,p_n)] \\
&\Rightarrow_Q a_0a_1a_2\cdots a_n\#y_nx_{n+1}q_1 && [(a_n,p_n,x_{n+1},q_1)] \\
&\Rightarrow_Q a_0\cdots a_nb_1\#b_2\cdots b_mw_1q_2 && [(b_1,q_1,w_1,q_2)] \\
&\Rightarrow_Q a_0\cdots a_nb_1b_2\#b_3\cdots b_mw_1w_2q_3 && [(b_2,q_2,w_2,q_3)] \\
&\vdots \\
&\Rightarrow_Q a_0\cdots a_nb_1\cdots b_{m-1}\#b_mw_1w_2\cdots w_{m-1}q_m && [(b_{m-1},q_{m-1},w_{m-1},q_m)] \\
&\Rightarrow_Q a_0\cdots a_nb_1\cdots b_m\#w_1w_2\cdots w_mq_{m+1} && [(b_m,q_m,w_m,q_{m+1})]
\end{aligned}
$$

Therefore, $w_1w_2\cdots w_m \in L(Q)$. Consequently, $L(M, L(G), 3) \subseteq L(Q)$. A proof that $L(Q) \subseteq L(M, L(G), 3)$ is left to the reader. As $L(Q) \subseteq L(M, L(G), 3)$ and $L(M, L(G), 3) \subseteq L(Q)$, $L(Q) = L(M, L(G), 3)$. Therefore, Lemma 7.2.27 holds. $\square$

Theorem 7.2.28. *For $i \in \{1, 2, 3\}$, $\mathbf{RE} = \mathbf{RPDA(LIN}, i)$.*

Proof. Obviously, $\mathbf{RPDA(LIN}, 3) \subseteq \mathbf{RE}$. To prove that $\mathbf{RE} \subseteq \mathbf{RPDA(LIN}, 3)$, consider any recursively enumerable language $L \in \mathbf{RE}$. By Theorem 2.3.46, $L(Q) = L$, for a left-extended queue grammar Q. Furthermore, by Theorem 3.1.26 and Lemma 7.2.27, $L(Q) = L(M, L(G), 3)$, for a linear grammar G and a pushdown automaton M. Thus, $L = L(M, L(G), 3)$. Hence, $\mathbf{RE} \subseteq \mathbf{RPDA(LIN}, 3)$. As $\mathbf{RPDA(LIN}, 3) \subseteq \mathbf{RE}$ and $\mathbf{RE} \subseteq \mathbf{RPDA(LIN}, 3)$, $\mathbf{RE} = \mathbf{RPDA(LIN}, 3)$.

By analogy with the demonstration of $\mathbf{RE} = \mathbf{RPDA(LIN}, 3)$, we can prove that $\mathbf{RE} = \mathbf{RPDA(LIN}, i)$ for $i = 1, 2$. $\square$

One-Turn Linear-Controlled Pushdown Automata

In the present section, we continue with the discussion of regulated pushdown automata, but we narrow our attention to their special cases—*one-turn regulated pushdown automata*. To give an insight into one-turn pushdown automata, consider two consecutive moves made by an ordinary pushdown automaton M. If during the first move M does not shorten its pushdown and during the second move it does, then M makes a *turn* during the second move. A pushdown automaton is *one-turn* if it makes no more than one turn with its pushdown during any computation starting from a start configuration. Recall that one-turn pushdown automata characterize the family of linear languages (see [Har78]) while their unrestricted versions characterize the family of context-free languages (see Theorem 2.4.12). As a result, one-turn pushdown automata are less powerful than the pushdown automata.

As the most surprising result, we demonstrate that linear-regulated versions of one-turn pushdown automata characterize the family of recursively enumerable languages. Thus, as opposed to the ordinary one-turn pushdown automata, one-turn linear-regulated pushdown automata are as powerful as linear-regulated pushdown automata that can make any number of turns.

In fact, this characterization holds even for some restricted versions of one-turn regulated pushdown automata, including their atomic and reduced versions, which are sketched next.

(I) During a move, an *atomic* one-turn regulated pushdown automaton changes a state and, in addition, performs exactly one of the following three actions:

- it pushes a symbol onto the pushdown;
- it pops a symbol from the pushdown;
- it reads an input symbol.

(II) A *reduced* one-turn regulated pushdown automaton has a limited number of some components, such as the number of states, pushdown symbols, or transition rules.

We establish the above-mentioned characterization in a formal way.

Definition 7.2.29. An *atomic pushdown automaton* is a pushdown automaton (see Definition 2.4.13) $M = (Q, \Sigma, \Gamma, R, s, S, F)$, where for every rule $Apa \to wq \in R$, $|Aaw| = 1$. That is, each of the rules from R has one of the following forms.

(1) $Ap \to q$ (*popping rule*)
(2) $p \to wq$ (*pushing rule*)
(3) $pa \to q$ (*reading rule*)   □

Definition 7.2.30. Let $M = (Q, \Sigma, \Gamma, R, s, S, F)$ be a pushdown automaton. Let $x, x', x'' \in \Gamma^*$, $y, y', y'' \in \Sigma^*$, $q, q', q'' \in Q$, and $xqy \vdash_M x'q'y' \vdash_M x''q''y''$. If $|x| \le |x'|$ and $|x'| > |x''|$, then $x'q'y' \vdash_M x''q''y''$ is a *turn*. If M makes no more than one turn during any sequence of moves starting from a start configuration, then M is said to be *one-turn*.   □

One-turn pushdown automata represent an important restricted version of automata, and the formal language theory has studied their properties in detail (see Section 5.7 in [Har78] and Section 6.1 in [ABB97]).

Definition 7.2.31. Let M be a pushdown automaton. If M satisfies the conditions from Definitions 7.2.29 and 7.2.30, it is said to be *one-turn atomic pushdown automaton*. Additionally, if M is regulated (see Definition 7.2.24), it is a *one-turn atomic regulated pushdown automaton* (OA-RPDA for short).

For any family of language $\mathscr{L}$ and $i \in \{1, 2, 3\}$, define

$$\mathbf{OA\text{-}RPDA}(\mathscr{L}, i) = \{L \mid L = L(M, \Xi, i), \text{ where } M \text{ is a one-turn}$$
$$\text{atomic pushdown automaton and } \Xi \in \mathscr{L}\} \qquad □$$

We next prove that one-turn atomic pushdown automata regulated by linear languages characterize the family of recursively enumerable languages. In fact, these automata need no more than one state and two pushdown symbols to achieve this characterization.

Lemma 7.2.32. *Let Q be a left-extended queue grammar satisfying the normal form of Definition 3.1.25. Then, there is a linear grammar G and a one-turn atomic pushdown automaton $M = (\{\lfloor\}, \tau, \{0, 1\}, H, \lfloor, 0, \{\lfloor\})$ such that* $\text{card}(H) = \text{card}(\tau) + 4$ *and* $L(Q) = L(M, L(G), 3)$.

Proof. Let $Q = (V, \tau, W, F, R, g)$ be a queue grammar satisfying the normal form of Definition 3.1.25. For some $n \geq 1$, introduce a homomorphism f from R to X, where

$$X = \{1\}^*\{0\}\{1\}^*\{1\}^n \cap \{0, 1\}^{2n}$$

Extend f so it is defined from R^* to X^*. Define the substitution h from V^* to X^* as

$$h(a) = \{f(r) \mid r = (a, p, x, q) \in R \text{ for some } p, q \in W, x \in V^*\}$$

Define the coding d from $\{0, 1\}^*$ to $\{2, 3\}^*$ as $d(0) = 2$, $d(1) = 3$. Construct the linear grammar

$$G = (N, T, P, S)$$

as follows. Initially, set

$$T = \{0, 1, 2, 3\} \cup \tau$$
$$N = \{S\} \cup \{\tilde{q} \mid q \in W\} \cup \{\hat{q} \mid q \in W\}$$
$$P = \{S \to \tilde{f}2 \mid f \in F\} \cup \{\tilde{\iota} \to \hat{\iota}\}$$

Extend P by performing (1) through (3), given next.

(1) For every $r = (a, p, x, q) \in R, p, q \in w, x \in T^*$,

$$P = P \cup \{\tilde{q} \to \tilde{p}d(f(r))x\}$$

(2) For every $(a, p, x, q) \in R$,

$$P = P \cup \{\hat{q} \to y\hat{p}b \mid y \in \text{reversal}(h(x)), b \in h(a)\}$$

(3) For every $(a, p, x, q) \in R, ap = S, p, q \in W, x \in V^*$,

$$P = P \cup \{\hat{q} \to y \mid y \in \text{reversal}(h(x))\}$$

Define the atomic pushdown automaton

$$M = \left(\{\lfloor\}, \tau, \{0, 1\}, H, \lfloor, 0, \{\lfloor\}\right)$$

where H contains the following transition rules

$$0: \lfloor \to 0\lfloor$$
$$1: \lfloor \to 1\lfloor$$
$$2: 0\lfloor \to \lfloor$$
$$3: 1\lfloor \to \lfloor$$
$$a: \lfloor a \to \lfloor \text{ for every } a \in \tau$$

We next demonstrate that $L(M, L(G), 3) = L(Q)$. Observe that M accepts every string $w = w_1 \cdots w_{m-1} w_m$ as

$$0\lfloor w_1 \cdots w_{m-1} w_m \vdash_M^+ \quad 0\bar{b}_m \cdots \bar{b}_1 \bar{a}_n \cdots \bar{a}_1 \lfloor w_1 \cdots w_{m-1} w_m$$
$$\vdash_M^{|\bar{a}_n \cdots \bar{a}_1|} \quad 0\bar{b}_m \cdots \bar{b}_1 \lfloor w_1 \cdots w_{m-1} w_m$$
$$\vdash_M^{|w_1|} \quad 0\bar{b}_m \cdots \bar{b}_1 \lfloor w_2 \cdots w_{m-1} w_m$$
$$\vdash_M^{|\bar{b}_1|} \quad 0\bar{b}_m \cdots \bar{b}_2 \lfloor w_2 \cdots w_{m-1} w_m$$
$$\vdash_M^{|w_2|} \quad 0\bar{b}_m \cdots \bar{b}_2 \lfloor w_3 \cdots w_{m-1} w_m$$
$$\vdash_M^{|\bar{b}_2|} \quad 0\bar{b}_m \cdots \bar{b}_3 \lfloor w_3 \cdots w_{m-1} w_m$$
$$\vdots$$
$$\vdash_M \quad 0\bar{b}_m \lfloor w_m$$
$$\vdash_M^{|w_m|} \quad 0\bar{b}_m \lfloor$$
$$\vdash_M^{|\bar{b}_m|} \quad 0\lfloor$$
$$\vdash_M \quad \lfloor$$

according to a string of the form $\beta\alpha\alpha'\beta' \in L(G)$, where

$$\beta = \text{reversal}(f(r_m)) \, \text{reversal}(f(r_{m-1})) \cdots \text{reversal}(f(r_1))$$
$$\alpha = \text{reversal}(f(t_n)) \, \text{reversal}(f(t_{n-1})) \cdots \text{reversal}(f(t_1))$$
$$\alpha' = f(t_0)f(t_1) \cdots f(t_n)$$
$$\beta' = d(f(r_1))w_1 d(f(r_2))w_2 \cdots d(f(r_m))w_m$$

for some $m, n \geq 1$ so that for $i = 1, \ldots, m$,

$$t_i = (b_i, q_i, w_i, q_{i+1}) \in R, b_i \in V - \tau, q_i, q_{i+1} \in Q, \bar{b}_i = f(t_i)$$

and for $j = 1, \ldots, n + 1$, $r_j = (a_{j-1}, p_{j-1}, x_j, p_j)$, $a_{j-1} \in V - \tau$, $p_{j-1}, p_j \in Q - F$, $x_j \in (V - \tau)^*$, $\bar{a}_j = f(r_j)$, $q_{m+1} \in F$, $\bar{a}_0 p_0 = g$. Thus, in Q,

$$
\begin{aligned}
\#a_0 p_0 \quad &\Rightarrow_Q a_0 \# y_0 x_1 p_1 & [(a_0, p_0, x_1, p_1)] \\
&\Rightarrow_Q a_0 a_1 \# y_1 x_2 p_2 & [(a_1, p_1, x_2, p_2)] \\
&\Rightarrow_Q a_0 a_1 a_2 \# y_2 x_3 p_3 & [(a_2, p_2, x_3, p_3)]
\end{aligned}
$$

$$\vdots$$

$$
\begin{aligned}
&\Rightarrow_Q a_0 a_1 a_2 \cdots a_{n-1} \# y_{n-1} x_n p_n & [(a_{n-1}, p_{n-1}, x_n, p_n)] \\
&\Rightarrow_Q a_0 a_1 a_2 \cdots a_n \# y_n x_{n+1} q_1 & [(a_n, p_n, x_{n+1}, q_1)] \\
&\Rightarrow_Q a_0 \cdots a_n b_1 \# b_2 \cdots b_m w_1 q_2 & [(b_1, q_1, w_1, q_2)] \\
&\Rightarrow_Q a_0 \cdots a_n b_1 b_2 \# b_3 \cdots b_m w_1 w_2 q_3 & [(b_2, q_2, w_2, q_3)]
\end{aligned}
$$

$$\vdots$$

$$
\begin{aligned}
&\Rightarrow_Q a_0 \cdots a_n b_1 \cdots b_{m-1} \# b_m w_1 \cdots w_{m-1} q_m & [(b_{m-1}, q_{m-1}, w_{m-1}, q_m)] \\
&\Rightarrow_Q a_0 \cdots a_n b_1 \cdots b_m \# w_1 \cdots w_m q_{m+1} & [(b_m, q_m, w_m, q_{m+1})]
\end{aligned}
$$

Therefore, $w_1 w_2 \cdots w_m \in L(Q)$. Consequently, $L(M, L(G), 3) \subseteq L(Q)$. A proof of $L(Q) \subseteq L(M, L(G), 3)$ is left to the reader.

As $L(Q) \subseteq L(M, L(G), 3)$ and $L(M, L(G), 3) \subseteq L(Q)$, $L(Q) = L(M, L(G), 3)$. Observe that M is one-turn atomic. Furthermore, $\text{card}(H) = \text{card}(\tau) + 4$. Thus, Lemma 7.2.32 holds. $\qquad\square$

Theorem 7.2.33. *For every $L \in \textbf{RE}$, there is a linear language Ξ and a one-turn atomic pushdown automaton $M = (Q, \Sigma, \Gamma, R, s, \$, F)$ such that $\text{card}(Q) \leq 1$, $\text{card}(\Gamma) \leq 2$, $\text{card}(R) \leq \text{card}(\Sigma) + 4$, and $L(M, \Xi, 3) = L$.*

Proof. By Theorem 2.3.46, for every $L \in \textbf{RE}$, there is a left-extended queue grammar Q such that $L = L(Q)$. Thus, this theorem follows from Theorem 3.1.26 and Lemma 7.2.32. $\qquad\square$

Theorem 7.2.34. *For every $L \in \textbf{RE}$, there is a linear language Ξ and a one-turn atomic pushdown automaton $M = (Q, \Sigma, \Gamma, R, s, \$, F)$ such that $\text{card}(Q) \leq 1$, $\text{card}(\Gamma) \leq 2$, $\text{card}(R) \leq \text{card}(\Sigma) + 4$, and $L(M, \Xi, 1) = L$.*

Proof. This theorem can be proved by analogy with the proof of Theorem 7.2.33. $\qquad\square$

Theorem 7.2.35. *For every $L \in \textbf{RE}$, there is a linear language Ξ and a one-turn atomic pushdown automaton $M = (Q, \Sigma, \Gamma, R, s, \$, F)$ such that $\text{card}(Q) \leq 1$, $\text{card}(\Gamma) \leq 2$, $\text{card}(R) \leq \text{card}(\Sigma) + 4$, and $L(M, \Xi, 2) = L$.*

Proof. This theorem can be proved by analogy with the proof of Theorem 7.2.33. $\qquad\square$

From the previous three theorems, we obtain the following corollary.

Corollary 7.2.36. *For $i \in \{1, 2, 3\}$, $\textbf{RE} = \textbf{OA - RPDA}(\textbf{LIN}, i)$.* $\qquad\square$

We close this section by suggesting some open problem areas concerning regulated automata.

Open Problem 7.2.37. For $i = 1, \ldots, 3$, consider **RPDA**($\mathscr{L}, i$), where $\mathscr{L}$ is a language family satisfying **REG** $\subset \mathscr{L} \subset$ **LIN**. For instance, consider $\mathscr{L}$ as the family of *minimal linear languages* (see page 76 in [Sal73]). Compare **RE** with **RPDA**($\mathscr{L}, i$).

Open Problem 7.2.38. Investigate special cases of regulated pushdown automata, such as their deterministic versions.

Open Problem 7.2.39. By analogy with regulated pushdown automata, introduce and study some other types of regulated automata.

7.3 Self-Reproducing Pushdown Transducers

Throughout this entire book, we cover modern language-defining models. In this final section of Chap. 7, however, we make a single exception. Indeed, we explain how to modify these models in a very natural way so that they define translations (see Sect. 2.1) rather than languages. Consider the notion of a pushdown automaton (see Sect. 2.4). Based upon this notion, we next introduce and discuss the notion of a self-reproducing pushdown transducer, which defines a translation, not a language. In essence, the transducer makes its translation as follows. After a translation of an input string, x, to an output string, y, a self-reproducing pushdown transducer can make a self-reproducing step during which it moves y to its input tape and translates it again. In this self-reproducing way, it can repeat the translation n-times, for $n \geq 1$. This section demonstrates that every recursively enumerable language can be characterized by the domain of the translation obtained from a self-reproducing pushdown transducer that repeats its translation no more than three times. This characterization is of some interest because it does not hold in terms of ordinary pushdown transducers. Indeed, the domain obtained from any ordinary pushdown transducer is a context-free language (see [Har78]).

7.3.1 Definitions

Definition 7.3.1. A *self-reproducing pushdown transducer* is an 8-tuple

$$M = (Q, \Gamma, \Sigma, \Omega, R, s, S, O)$$

where Q is a finite set of states, Γ is a total alphabet such that $Q \cap \Gamma = \emptyset$, $\Sigma \subseteq \Gamma$ is an input alphabet, $\Omega \subseteq \Gamma$ is an output alphabet, R is a finite set of *translation rules* of the form $u_1 q w \rightarrow u_2 p v$ with $u_1, u_2, w, v \in \Gamma^*$ and $q, p \in Q$, $s \in Q$ is the *start state*, $S \in \Gamma$ is the *start pushdown symbol*, $O \subseteq Q$ is the set of *self-reproducing states*.

A *configuration of* M is any string of the form $\$zqy\$x$, where $x, y, z \in \Gamma^*$, $q \in Q$, and $\$$ is a special *bounding symbol* ($\$ \notin Q \cup \Gamma$). If $u_1qw \to u_2pv \in R$, $y = \$hu_1qwz\$t$, and $x = \$hu_2pz\$tv$, where $h, u_1, u_2, w, t, v, z \in \Gamma^*$, $q, p \in Q$, then M makes a *translation step* from y to x in M, symbolically written as $y_t\Rightarrow x$ $[u_1qw \to u_2pv]$ or, simply $y_t\Rightarrow x$ in M. If $y = \$hq\$t$, and $x = \$hqt\$$, where $t, h \in \Gamma^*$, $q \in O$, then M makes a *self-reproducing step* from y to x in M, symbolically written as $y_r\Rightarrow x$. Write $y \Rightarrow x$ if $y_t\Rightarrow x$ or $y_r\Rightarrow x$. In the standard manner, extend $\Rightarrow$ to $\Rightarrow^n$, where $n \geq 0$; then, based on $\Rightarrow^n$, define $\Rightarrow^+$ and $\Rightarrow^*$.

Let $w, v \in \Gamma^*$; M *translates* w to v if $\$Ssw\$ \Rightarrow^* \$q\$v$ in M. *The translation obtained from* M, $T(M)$, is defined as $T(M) = \{(w, v) : \$Ssw\$ \Rightarrow^* \$q\$v$ with $w \in \Sigma^*$, $v \in \Omega^*$, $q \in Q\}$. Set $\text{domain}(T(M)) = \{w : (w, x) \in T(M)\}$ and $\text{range}(T(M)) = \{x : (w, x) \in T(M)\}$.

Let n be a nonnegative integer; if during every translation M makes no more than n self-reproducing steps, then M is an *n-self-reproducing pushdown transducer*. Two self-reproducing transducers are equivalent if they both define the same translation. □

In the literature, there often exists a requirement that a pushdown transducer, $M = (Q, \Gamma, \Sigma, \Omega, R, s, S, O)$, replaces no more than one symbol on its pushdown and reads no more than one symbol during every move. As stated next, we can always turn any self-reproducing pushdown transducer to an equivalent self-reproducing pushdown transducer that satisfies this requirement.

Theorem 7.3.2. *Let M be a self-reproducing pushdown transducer. Then, there is an equivalent self-reproducing pushdown transducer*

$$N = (Q, \Gamma, \Sigma, \Omega, R, s, S, O)$$

in which every translation rule, $u_1qw \to u_2pv \in R$, where $u_1, u_2, w, v \in \Gamma^*$ and $q, p \in Q$, satisfies $|u_1| \leq 1$ and $|w| \leq 1$.

Proof. Basic Idea. Consider every rule $u_1qw \to u_2pv$ in M with $|u_1| \geq 2$ or $|w| \geq 2$. N simulates a move made according to this rule as follows. First, N leaves q for a new state and makes $|w|$ consecutive moves during which it reads w symbol by symbol so that after these moves, it has w recorded in a new state, $\langle qw \rangle$. From this new state, it makes $|u_1|$ consecutive moves during which it pops u_1 symbol by symbol from the pushdown so that after these moves, it has both u_1 and w recorded in another new state, $\langle u_1qw \rangle$. To complete this simulation, it performs a move according to $\langle u_1qw \rangle \to u_2pv$. Otherwise, N works as M. A detailed version of this proof is left to the reader. □

7.3.2 Results

Recall that every recursively enumerable language is generated by left-extended queue grammar (see Theorem 2.3.46).

Lemma 7.3.3. *Let Q be a left-extended queue grammar satisfying the normal form introduced in Definition 3.1.25. Then, there exists a 2-self-reproducing pushdown transducer, M, such that* $\text{domain}(T(M)) = L(Q)$ *and* $\text{range}(T(M)) = \{\varepsilon\}$.

Proof. Let $G = (V, T, W, F, s, P)$ be a left-extended queue grammar satisfying the normal form introduced in Definition 3.1.25. Without any loss of generality, assume that $\{0, 1\} \cap (V \cup W) = \emptyset$. For some positive integer, n, define an injection, ι, from P to $(\{0, 1\}^n - \{1\}^n)$ so that ι is an injective homomorphism when its domain is extended to $(VW)^*$; after this extension, ι thus represents an injective homomorphism from $(VW)^*$ to $(\{0, 1\}^n - \{1\}^n)^*$; a proof that such an injection necessarily exists is simple and left to the reader. Based on ι, define the substitution, ν, from V to $(\{0, 1\}^n - \{1\}^n)$ so that for every $a \in V$, $\nu(a) = \{\iota(p) : p \in P, \ p = (a, b, x, c)$ for some $x \in V^*; \ b, c \in W\}$. Extend the domain of ν to V^*. Furthermore, define the substitution, μ, from W to $(\{0, 1\}^n - \{1\}^n)$ so that for every $q \in W$, $\mu(q) = \{\iota(p) : p \in P, \ p = (a, b, x, c)$ for some $a \in V, \ x \in V^*; \ b, c \in W\}$. Extend the domain of μ to W^*.

Construction. Introduce the self-reproducing pushdown transducer

$$M = (Q, T \cup \{0, 1, S\}, T, \emptyset, R, z, S, O)$$

where $Q = \{o, c, f, z\} \cup \{(p, i) : p \in W \text{ and } i \in \{1, 2\}\}$, $O = \{o, f\}$, and R is constructed by performing the following steps (1) through (6).

(1) if $a_0 q_0 = s$, where $a \in V - T$ and $q \in W - F$,
    then add $Sz \to uS\langle q_0, 1\rangle w$ to R, for all $w \in \mu(q_0)$ and all $u \in \nu(a_0)$;
(2) if $(a, q, y, p) \in P$, where $a \in V - T$, $p, q \in W - F$, and $y \in (V - T)^*$,
    then add $S\langle q, 1\rangle \to uS\langle p, 1\rangle w$ to R, for all $w \in \mu(p)$ and all $u \in \nu(y)$;
(3) for every $q \in W - F$, add $S\langle q, 1\rangle \to S\langle q, 2\rangle$ to R;
(4) if $(a, q, y, p) \in P$, where $a \in V - T$, $p, q \in W - F$, and $y \in T^*$,
    then add $S\langle q, 2\rangle y \to S\langle p, 2\rangle w$ to R, for all $w \in \mu(p)$;
(5) if $(a, q, y, p) \in P$, where $a \in V - T$, $q \in W - F$, $y \in T^*$, and $p \in F$,
    then add $S\langle q, 2\rangle y \to SoS$ to R;
(6) add $o0 \to 0o$, $o1 \to 1o$, $oS \to c$, $0c \to c0$, $1c \to c1$, $Sc \to f$, $0f0 \to f$, $1f1 \to f$ to R.

For brevity, the following proofs omits some obvious details, which the reader can easily fill in. The next claim describes how M accepts each string from $L(M)$.

Claim. M accepts every $h \in L(M)$ in this way

$\$Szy_1y_2 \ldots y_{m-1}y_m\$$
$\Rightarrow \ \$g_0\langle q_0, 1\rangle y_1y_2 \ldots y_{m-1}y_m\$t_0$
$\Rightarrow \ \$g_1\langle q_1, 1\rangle y_1y_2 \ldots y_{m-1}y_m\$t_1$
$\vdots$
$\Rightarrow \ \$g_k\langle q_k, 1\rangle y_1y_2 \ldots y_{m-1}y_m\$t_k$
$\Rightarrow \ \$g_k\langle q_k, 2\rangle y_1y_2 \ldots y_{m-1}y_m\$t_k$
$\Rightarrow \ \$g_k\langle q_{k+1}, 2\rangle y_1y_2 \ldots y_{m-1}y_m\$t_{k+1}$
$\Rightarrow \ \$g_k\langle q_{k+2}, 2\rangle y_2 \ldots y_{m-1}y_m\$t_{k+2}$

$$\vdots$$

$_t\Rightarrow\ \$g_k\langle q_{k+m},2\rangle y_m\$t_{k+m}$

$_t\Rightarrow\ \$g_kSo\$t_{k+m}S$

$_r\Rightarrow\ \$g_kSot_{k+m}S\$$

$_t\Rightarrow^\iota\ \$g_kSt_{k+m}oS\$$

$_t\Rightarrow\ \$g_kSt_{k+m}c\$$

$_t\Rightarrow^\iota\ \$u_1Sc\$v_1$

$_t\Rightarrow\ \$u_1f\$v_1$

$_r\Rightarrow\ \$u_1fv_1\$$

$\Rightarrow\ \$u_2fv_2\$$

$$\vdots$$

$\Rightarrow\ \$u_lfv_l\$$

$\Rightarrow\ \$f\$$

in M, where $k,m \geq 1$; $q_0,q_1,\ldots,q_{k+m} \in W-F$; $y_1,\ldots,y_m \in T^*$; $t_i \in \mu(q_0q_1\ldots q_i)$ for $i = 0,1,\ldots,k+m$; $g_j \in \nu(d_0d_1\ldots d_j)$ with $d_1,\ldots,d_j \in (V-T)^*$ for $j = 0,1,\ldots,k$; $d_0d_1\ldots d_k = a_0a_1\ldots a_{k+m}$ where $a_1,\ldots,a_{k+m} \in V-T$, $d_0 = a_0$, and $s = a_0q_0$; $g_k = t_{k+m}$ (notice that $\nu(a_0a_1\ldots a_{k+m}) = \mu(q_0q_1\ldots q_{k+m})$); $v_i \in \mathrm{prefix}(\mu(q_0q_1\ldots q_{k+m}),|\mu(q_0q_1\ldots q_{k+m})| - i)$ for $i = 1,\ldots,\upsilon$ with $\upsilon = |\mu(q_0q_1\ldots q_{k+m})|$; $u_j \in \mathrm{suffix}(\nu(a_0a_1\ldots a_{k+m}),|\nu(a_0a_1\ldots a_{k+m})| - j)$ for $j = 1,\ldots,l$ with $l = |\nu(a_0a_1\ldots a_{k+m})|$; $h = y_1y_2\ldots y_{m-1}y_m$.

Proof. Examine steps (1) through (6) of the construction of R. Notice that during every successful computation, M uses the rules introduced in step i before it uses the rules introduced in step $i + 1$, for $i = 1,\ldots,5$. Thus, in greater detail, every successful computation $\$Szh\$ \Rightarrow^* \$f\$$ can be expressed as

$\$Szy_1y_2\ldots y_{m-1}y_m\$$

$\Rightarrow\ \$g_0\langle q_0,1\rangle y_1y_2\ldots y_{m-1}y_m\$t_0$

$\Rightarrow\ \$g_1\langle q_1,1\rangle y_1y_2\ldots y_{m-1}y_m\$t_1$

$$\vdots$$

$\Rightarrow\ \$g_k\langle q_k,1\rangle y_1y_2\ldots y_{m-1}y_m\$t_k$

$\Rightarrow\ \$g_k\langle q_k,2\rangle y_1y_2\ldots y_{m-1}y_m\$t_k$

$\Rightarrow\ \$g_k\langle q_{k+1},2\rangle y_1y_2\ldots y_{m-1}y_m\$t_{k+1}$

$\Rightarrow\ \$g_k\langle q_{k+2},2\rangle y_2y_3\ldots y_{m-1}y_m\$t_{k+2}$

$\Rightarrow\ \$g_k\langle q_{k+3},2\rangle y_3y_4\ldots y_{m-1}y_m\$t_{k+3}$

$$\vdots$$

$_t\Rightarrow\ \$g_k\langle q_{k+m},2\rangle y_m\$t_{k+m}$

$_t\Rightarrow\ \$g_kSo\$t_{k+m}S$

$\Rightarrow^*\ \$f\$$

where $k,m \geq 1$; $h = y_1y_2\ldots y_{m-1}y_m$; $q_0,q_1,\ldots,q_{k+m} \in W-F$; $y_1,\ldots,y_m \in T^*$; $t_i \in \mu(q_0q_1\ldots q_i)$ for $i = 0,1,\ldots,k+m$; $g_j \in \nu(d_0d_1\ldots d_j)$ with $d_1,\ldots,d_j \in (V-T)^*$ for $j = 0,1,\ldots,k$; $d_0d_1\ldots d_k = a_0a_1\ldots a_{k+m}$ where $a_1,\ldots,a_{k+m} \in V-T$, $d_0 = a_0$, and $s = a_0q_0$.

During $\$g_kSo\$t_{k+m}S \Rightarrow^* \$f\$$ only the rules of (6) are used. Recall these rules: $o0 \to 0o$, $o1 \to 1o$, $oS \to c$, $0c \to c0$, $1c \to c1$, $Sc \to f$, $0f0 \to f$, $1f1 \to f$. Observe that to obtain $\$f\$$ from $\$g_kSo\$t_{k+m}S$ by using these rules, M performs $\$g_kSo\$t_{k+m}S \Rightarrow^* \$f\$$ as follows

$$\$g_kSo\$t_{k+m}S$$
$$_r\Rightarrow \$g_kSot_{k+m}S\$$$
$$_l\Rightarrow^l \$g_kSt_{k+m}oS\$$$
$$_l\Rightarrow \$g_kSt_{k+m}c\$$$
$$_l\Rightarrow^l \$u_1Sc\$v_1$$
$$_l\Rightarrow \$u_1f\$v_1$$
$$_r\Rightarrow \$u_1fv_1\$$$
$$\Rightarrow \$u_2fv_2\$$$
$$\vdots$$
$$\Rightarrow \$u_lfv_l\$$$
$$\Rightarrow \$f\$$$

in M, where $g_k = t_{k+m}$; $v_i \in \mathrm{prefix}(\mu(q_0q_1 \ldots q_{k+m}), |\mu(q_0q_1 \ldots q_{k+m})| - i)$ for $i = 1, \ldots, \upsilon$ with $\upsilon = |\mu(q_0q_1 \ldots q_{k+m})|$; $u_j \in \mathrm{suffix}(\nu(a_0a_1 \ldots a_{k+m}), |\nu(a_0a_1 \ldots a_{k+m})| - j)$ for $j = 1, \ldots, l$ with $l = |\nu(a_0a_1 \ldots a_{k+m})|$. This computation implies $g_k = t_{k+m}$. As a result, the claim holds.

Let M accepts $h \in L(M)$ in the way described in the above claim. Examine the construction of R to see that at this point P contains

$$(a_0, q_0, z_0, q_1), \ldots, (a_k, q_k, z_k, q_{k+1}), (a_{k+1}, q_{k+1}, y_1, q_{k+2}), \ldots$$
$$\ldots, (a_{k+m-1}, q_{k+m-1}, y_{m-1}, q_{k+m}), (a_{k+m}, q_{k+m}, y_m, q_{k+m+1})$$

where $z_1, \ldots, z_k \in (V - T)^*$, so G makes the generation of h in the way described in Definition 3.1.25. Thus $h \in L(G)$. Consequently, $L(M) \subseteq L(G)$.

Let G generates $h \in L(G)$ in the way described in Definition 3.1.25. Then, M accepts h in the way described in the above claim, so $L(G) \subseteq L(M)$; a detailed proof of this inclusion is left to the reader.

As $L(M) \subseteq L(G)$ and $L(G) \subseteq L(M)$, $L(G) = L(M)$.

From the above Claim, it follows that M is a 2-self-reproducing pushdown transducer. Thus, Lemma 7.3.3 holds.

Theorem 7.3.4. *For every recursively enumerable language, L, there exists a 2-self-reproducing pushdown transducer, M, such that* $\mathrm{domain}(T(M)) = L$ *and* $\mathrm{range}(T(M)) = \{\varepsilon\}$.

Proof. This theorem follows from Theorems 2.3.46 and 3.1.26 and Lemma 7.3.3.

Chapter 8
Jumping Automata and Discontinuous Computation

Recall that jumping grammars (see Chap. 5) represent language-generating models for discontinuous computation. The present chapter explores their automata-based counterparts, called jumping automata. As their name suggests, they jump across their input words discontinuously, and in this way, they also formalize computation performed in a discontinuous way.

To give an insight into the notion of a jumping automaton, reconsider the basic notion of a classical finite automaton M (see Sect. 2.4). Recall that M consists of an input tape, a read head, and a finite state control. The input tape is divided into cells. Each cell contains one symbol of an input string. The symbol under the read head, a, is the current input symbol. The finite control is represented by a finite set of states together with a control relation, which is usually specified as a set of computational rules. M computes by making a sequence of moves. Each move is made according to a computational rule that describes how the current state is changed and whether the current input symbol is read. If the symbol is read, the read head is shifted precisely one cell to the right. M has one state defined as the start state and some states designated as final states. If M can read w by making a sequence of moves from the start state to a final state, M accepts w; otherwise, M rejects w.

Unfortunately, the classical versions of finite automata work so they often fail to reflect the real needs of today's informatics. Perhaps most significantly, they fail to formalize discontinuous information processing, which is central to today's computation while it was virtually unneeded and, therefore, unknown in the past. Indeed, in the previous century, most classical computer science methods were developed for continuous information processing. Accordingly, their formal models, including finite automata, work on strings, representing information, in a strictly continuous left-to-right symbol-by-symbol way. Modern information methods, however, frequently process information in a discontinuous way [GF04, BMCY06, BCC10, MRS08, BYRN11, NS05]. Within a particular running process, a typical computational step may be performed somewhere in the middle of information while the very next computational step is executed far away from it; therefore,

© Springer International Publishing AG 2017
A. Meduna, O. Soukup, *Modern Language Models and Computation*,
DOI 10.1007/978-3-319-63100-4_8

before the next step is carried out, the process has to jump over a large portion of the information to the desired position of execution. Of course, classical finite automata, which work on strings strictly continuously, inadequately and inappropriately reflect discontinuous information processing of this kind.

Formalizing discontinuous information processing adequately gives rise to the idea of adapting classical finite automata in a discontinuous way. In this way, the present chapter introduces and studies the notion of a *jumping finite automaton*, H. In essence, H works just like a classical finite automaton except it does not read the input string in a symbol-by-symbol left-to-right way. That is, after reading a symbol, H can jump over a portion of the tape in either direction and continue making moves from there. Once an occurrence of a symbol is read on the tape, it cannot be re-read again later during computation of H. Otherwise, it coincides with the standard notion of a finite automaton, and as such, it is based upon a regulated mechanism consisting in its finite state control. Therefore, we study them in detail in this book.

More precisely, concerning jumping finite automata, this chapter considers commonly studied areas of the formal language theory, such as decidability and closure properties, and establishes several results concerning jumping finite automata regarding these areas. It concentrates its attentions on results that demonstrate differences between jumping finite automata and their classical versions. As a whole, this chapter gives a systematic body of knowledge concerning jumping finite automata. At the same time, however, it points out several open questions regarding these automata, which may represent a new, attractive, significant investigation area of automata theory in the future.

This chapter is divided into two sections. First, as its title suggests, Sect. 8.1 formalizes and illustrates jumping finite automata. Then, Sect. 8.2 demonstrates their fundamental properties, including a comparison of their power with the power of well-known language-defining formal devices, closure properties, decidability. In addition, this section establishes an infinite hierarchy of language families resulting from these automata, one-directional jumps and various start configurations.

8.1 Definitions and Examples

In this section, we define a variety of jumping finite automata discussed in this chapter and illustrate them by examples.

Definition 8.1.1. A *general jumping finite automaton*, a *GJFA* for short, is general finite automaton (see Definition 2.4.1), where the binary *jumping relation*, symbolically denoted by $\curvearrowright$, over $\Sigma^* Q \Sigma^*$, is defined as follows. Let $x, z, x', z' \in \Sigma^*$ such that $xz = x'z'$ and $py \to q \in R$; then, M makes a *jump* from $xpyz$ to $x'qz'$, symbolically written as $xpyz \curvearrowright x'qz'$. In the standard manner, we extent $\curvearrowright$ to $\curvearrowright^m$, where $m \geq 0$, $\curvearrowright^+$, and $\curvearrowright^*$.

The *language accepted by M with* $\curvearrowright$, denoted by $L(M, \curvearrowright)$, is defined as $L(M, \curvearrowright) = \{uv \mid u, v \in \Sigma^*, usv \curvearrowright^* f, f \in F\}$. Let $w \in \Sigma^*$. We say that M *accepts* w if and only if $w \in L(M, \curvearrowright)$; M *rejects* w otherwise. Two GJFAs M and M' are said to be *equivalent* if and only if $L(M, \curvearrowright) = L(M', \curvearrowright)$. □

Definition 8.1.2. Let $M = (Q, \Sigma, R, s, F)$ be a GJFA. M is an *ε-free GJFA* if $py \to q \in R$ implies that $|y| \geq 1$. M is of *degree n*, where $n \geq 0$, if $py \to q \in R$ implies that $|y| \leq n$. M is a *jumping finite automaton* (a *JFA* for short) if its degree is 1. □

Definition 8.1.3. Let $M = (Q, \Sigma, R, s, F)$ be a JFA. Analogously to a GJFA, M is an *ε-free JFA* if $py \to q \in R$ implies that $|y| = 1$. M is a *deterministic JFA* (a *DJFA* for short) if (1) it is an ε-free JFA and (2) for each $p \in Q$ and each $a \in \Sigma$, there is no more than one $q \in Q$ such that $pa \to q \in R$. M is a *complete JFA* (a *CJFA* for short) if (1) it is a DJFA and (2) for each $p \in Q$ and each $a \in \Sigma$, there is precisely one $q \in Q$ such that $pa \to q \in R$. □

Definition 8.1.4. Let $M = (Q, \Sigma, R, s, F)$ be a GJFA. The *transition graph* of M, denoted by $\Delta(M)$, is a multigraph, where nodes are states from Q, and there is an edge from p to q labeled with y if and only if $py \to q \in R$. A state $q \in Q$ is *reachable* if there is a walk from s to q in $\Delta(M)$; q is *terminating* if there is a walk from q to some $f \in F$. If there is a walk from p to q, $p = q_1, q_2, \dots, q_n = q$, for some $n \geq 2$, where $q_i y_i \to q_{i+1} \in R$ for all $i = 1, \dots, n - 1$, then we write

$$p y_1 y_2 \cdots y_n \rightsquigarrow q$$ □

Next, we illustrate the previous definitions by two examples.

Example 8.1.5. Consider the DJFA

$$M = (\{s, r, t\}, \Sigma, R, s, \{s\})$$

where $\Sigma = \{a, b, c\}$ and

$$R = \{sa \to r, rb \to t, tc \to s\}$$

Starting from s, M has to read some a, some b, and some c, entering again the start (and also the final) state s. All these occurrences of a, b, and c can appear anywhere in the input string. Therefore, the accepted language is clearly

$$L(M, \curvearrowright) = \{w \in \Sigma^* \mid \#_a(w) = \#_b(w) = \#_c(w)\}$$ □

Recall that $L(M, \curvearrowright)$ in Example 8.1.5 is a well-known non-context-free context-sensitive language.

Example 8.1.6. Consider the GJFA

$$M = (\{s, t, f\}, \{a, b\}, R, s, \{f\})$$

where

$$R = \{sba \to f, fa \to f, fb \to f\}$$

Starting from s, M has to read string ba, which can appear anywhere in the input string. Then, it can read an arbitrary number of symbols a and b, including no symbols. Therefore, the accepted language is $L(M, \curvearrowright) = \{a, b\}^*\{ba\}\{a, b\}^*$.   □

8.1.1 Denotation of Language Families

Throughout the rest of this chapter, **GJFA**, **GJFA**$^{-\varepsilon}$, **JFA**, **JFA**$^{-\varepsilon}$, and **DJFA** denote the families of languages accepted by GJFAs, ε-free GJFAs, JFAs, ε-free JFAs, and DJFAs, respectively.

8.2 Properties

In this section, we discuss the generative power of GJFAs and JFAs and some other basic properties of these automata.

Theorem 8.2.1. *For every DJFA M, there is a CJFA M' such that $L(M, \curvearrowright) = L(M', \curvearrowright)$.*

Proof. Let $M = (Q, \Sigma, R, s, F)$ be a DJFA. We next construct a CJFA M' such that $L(M, \curvearrowright) = L(M', \curvearrowright)$. Without any loss of generality, we assume that $\perp \notin Q$. Initially, set

$$M' = (Q \cup \{\perp\}, \Sigma, R', s, F)$$

where $R' = R$. Next, for each $a \in \Sigma$ and each $p \in Q$ such that $pa \to q \notin R$ for all $q \in Q$, add $pa \to \perp$ to R'. For each $a \in \Sigma$, add $\perp a \to \perp$ to R'. Clearly, M' is a CJFA and $L(M, \curvearrowright) = L(M', \curvearrowright)$.   □

Lemma 8.2.2. *For every GJFA M of degree $n \geq 0$, there is an ε-free GJFA M' of degree n such that $L(M', \curvearrowright) = L(M, \curvearrowright)$.*

Proof. This lemma can be demonstrated by using the standard conversion of finite automata to ε-free finite automata (see Algorithm 3.2.2.3 in [Med00a]).   □

Theorem 8.2.3. GJFA = GJFA$^{-\varepsilon}$

Proof. **GJFA**$^{-\varepsilon} \subseteq$ **GJFA** follows from the definition of a GJFA. **GJFA** $\subseteq$ **GJFA**$^{-\varepsilon}$ follows from Lemma 8.2.2. $\qquad\square$

Theorem 8.2.4. **JFA** = **JFA**$^{-\varepsilon}$ = **DJFA**

Proof. **JFA** = **JFA**$^{-\varepsilon}$ can be proved by analogy with the proof of Theorem 8.2.3, so we only prove that **JFA**$^{-\varepsilon}$ = **DJFA**. **DJFA** $\subseteq$ **JFA**$^{-\varepsilon}$ follows from the definition of a DJFA. The converse inclusion can be proved by using the standard technique of converting ε-free finite automata to deterministic finite automata (see Algorithm 3.2.3.1 in [Med00a]). $\qquad\square$

The next theorem shows a property of languages accepted by GJFAs with unary input alphabets.

Theorem 8.2.5. *Let* $M = (Q, \Sigma, R, s, F)$ *be a GJFA such that* card$(\Sigma) = 1$. *Then,* $L(M, \curvearrowright)$ *is regular.*

Proof. Let $M = (Q, \Sigma, R, s, F)$ be a GJFA such that card$(\Sigma) = 1$. Since card$(\Sigma) = 1$, without any loss of generality, we can assume that the acceptance process for $w \in \Sigma^*$ starts from the configuration sw and M does not jump over any symbols. Therefore, we can threat M as an equivalent general finite automaton (see Definition 2.4.1). As general finite automata accept only regular languages (see Theorems 2.4.4 and 2.4.5), $L(M, \curvearrowright)$ is regular. $\qquad\square$

As a consequence of Theorem 8.2.5, we obtain the following corollary (recall that K below is not regular).

Corollary 8.2.6. *The language* $K = \{a^p \mid p$ *is a prime number*$\}$ *cannot be accepted by any GJFA.* $\qquad\square$

The following theorem gives a necessary condition for a language to be in **JFA**.

Theorem 8.2.7. *Let* K *be an arbitrary language. Then,* $K \in$ **JFA** *only if* $K = $ perm(K).

Proof. Let $M = (Q, \Sigma, R, s, F)$ be a JFA. Without any loss of generality, we assume that M is a DJFA (recall that **JFA** = **DJFA** by Theorem 8.2.4). Let $w \in L(M, \curvearrowright)$. We next prove that perm$(w) \subseteq L(M, \curvearrowright)$. If $w = \varepsilon$, then perm$(\varepsilon) = \varepsilon \in L(M, \curvearrowright)$, so we assume that $w \neq \varepsilon$. Then, $w = a_1 a_2 \cdots a_n$, where $a_i \in \Sigma$ for all $i = 1, \ldots, n$, for some $n \geq 1$. Since $w \in L(M, \curvearrowright)$, R contains

$$
\begin{aligned}
s a_{i_1} &\to s_{i_1} \\
s_{i_1} a_{i_2} &\to s_{i_2} \\
&\vdots \\
s_{i_{n-1}} a_{i_n} &\to s_{i_n}
\end{aligned}
$$

where $s_j \in Q$ for all $j \in \{i_1, i_2, \ldots, i_n\}$, $(i_1, i_2, \ldots, i_n)$ is a permutation of $(1, 2, \ldots, n)$, and $s_{i_n} \in F$. However, this implies that $a_{k_1} a_{k_2} \cdots a_{k_n} \in L(M, \curvearrowright)$, where $(k_1, k_2, \ldots, k_n)$ is a permutation of $(1, 2, \ldots, n)$, so perm$(w) \subseteq L(M, \curvearrowright)$. $\qquad\square$

From Theorem 8.2.7, we obtain the following two corollaries, which are used in subsequent proofs.

Corollary 8.2.8. *There is no JFA that accepts $\{ab\}^*$.*   □

Corollary 8.2.9. *There is no JFA that accepts $\{a, b\}^* \{ba\}\{a, b\}^*$.*   □

Consider the language of primes K from Corollary 8.2.6. Since $K = \text{perm}(K)$, the condition from Theorem 8.2.7 is not sufficient for a language to be in **JFA**. This is stated in the following corollary.

Corollary 8.2.10. *There is a language K satisfying $K = \text{perm}(K)$ that cannot be accepted by any JFA.*   □

The next theorem gives both a necessary and sufficient condition for a language to be accepted by a JFA.

Theorem 8.2.11. *Let L be an arbitrary language. $L \in$ **JFA** if and only if $L = \text{perm}(K)$, where K is a regular language.*

Proof. The proof is divided into the only-if part and the if part.
Only If. Let M be a JFA. Consider M as a finite automaton M'. Set $K = L(M')$. K is regular, and $L(M, \curvearrowright) = \text{perm}(K)$. Hence, the only-if part holds.
If. Take $\text{perm}(K)$, where K is any regular language. Let $K = L(M)$, where M is a finite automaton. Consider M as a JFA M'. Observe that $L(M', \curvearrowright) = \text{perm}(K)$, which proves the if part of the proof.   □

Finally, we show that GJFAs are stronger than JFAs.

Theorem 8.2.12. $\textbf{JFA} \subset \textbf{GJFA}$

Proof. $\textbf{JFA} \subseteq \textbf{GJFA}$ follows from the definition of a JFA. From Corollary 8.2.9, $\textbf{GJFA} - \textbf{JFA} \neq \emptyset$, because $\{a, b\}^* \{ba\}\{a, b\}^*$ is accepted by the GJFA from Example 8.1.6.   □

Open Problem 8.2.13. Is there a necessary and sufficient condition for a language to be in **GJFA**?

8.2.1 Relations with Well-Known Language Families

In this section, we establish relations between **GJFA**, **JFA**, and some well-known language families, including **FIN**, **REG**, **CF**, and **CS**.

Theorem 8.2.14. $\textbf{FIN} \subset \textbf{GJFA}$

Proof. Let $K \in \textbf{FIN}$. Since K is a finite, there exists $n \geq 0$ such that $\text{card}(K) = n$. Therefore, we can express K as $K = \{w_1, w_2, \ldots, w_n\}$. Define the GJFA

$$M = (\{s, f\}, \Sigma, R, s, \{f\})$$

where $\Sigma = \mathrm{alph}(K)$ and $R = \{sw_1 \to f, sw_2 \to f, \ldots, sw_n \to f\}$. Clearly, $L(M, \curvearrowright) = K$. Therefore, **FIN** $\subseteq$ **GJFA**. From Example 8.1.5, **GJFA** $-$ **FIN** $\neq \emptyset$, which proves the theorem. $\square$

Lemma 8.2.15. *There is no GJFA that accepts $\{a\}^*\{b\}^*$.*

Proof. By contradiction. Let $K = \{a\}^*\{b\}^*$. Assume that there is a GJFA, $M = (Q, \Sigma, R, s, F)$, such that $L(M, \curvearrowright) = K$. Let $w = a^n b$, where n is the degree of M. Since $w \in K$, during an acceptance of w, a rule, $pa^i b \to q \in R$, where $p, q \in Q$ and $0 \leq i < n$, has to be used. However, then M also accepts from the configuration $a^i b s a^{n-i}$. Indeed, as $a^i b$ is read in a single step and all the other symbols in w are just as, $a^i b a^{n-i}$ may be accepted by using the same rules as during an acceptance of w. This implies that $a^i b a^{n-i} \in K$—a contradiction with the assumption that $L(M, \curvearrowright) = K$. Therefore, there is no GJFA that accepts $\{a\}^*\{b\}^*$. $\square$

Theorem 8.2.16. **REG** *and* **GJFA** *are incomparable.*

Proof. **GJFA** $\not\subseteq$ **REG** follows from Example 8.1.5. **REG** $\not\subseteq$ **GJFA** follows from Lemma 8.2.15. $\square$

Theorem 8.2.17. **CF** *and* **GJFA** *are incomparable.*

Proof. **GJFA** $\not\subseteq$ **CF** follows from Example 8.1.5, and **CF** $\not\subseteq$ **GJFA** follows from Lemma 8.2.15. $\square$

Theorem 8.2.18. **GJFA** $\subset$ **CS**

Proof. Clearly, jumps of GJFAs can be simulated by context-sensitive grammars, so **GJFA** $\subseteq$ **CS**. From Lemma 8.2.15, it follows that **CS** $-$ **GJFA** $\neq \emptyset$. $\square$

Theorem 8.2.19. **FIN** *and* **JFA** *are incomparable.*

Proof. **JFA** $\not\subseteq$ **FIN** follows from Example 8.1.5. Consider the finite language $K = \{ab\}$. By Theorem 8.2.7, $K \notin$ **JFA**, so **FIN** $\not\subseteq$ **JFA**. $\square$

8.2.2 Closure Properties

In this section, we show the closure properties of the families **GJFA** and **JFA** under various operations.

Theorem 8.2.20. *Both* **GJFA** *and* **JFA** *are not closed under endmarking.*

Proof. Consider the language $K = \{a\}^*$. Clearly, $K \in$ **JFA**. A proof that no GJFA accepts $K\{\#\}$, where $\#$ is a symbol such that $\# \neq a$, can be made by analogy with the proof of Lemma 8.2.15. $\square$

Theorem 8.2.20 implies that both families are not closed under concatenation. Indeed, observe that the JFA

$$M = \left(\{s,f\}, \{\#\}, \{s\# \to f\}, s, \{f\}\right)$$

accepts $\{\#\}$.

Corollary 8.2.21. *Both* **GJFA** *and* **JFA** *are not closed under concatenation.*   □

Theorem 8.2.22. **JFA** *is closed under shuffle.*

Proof. Let $M_1 = (Q_1, \Sigma_1, R_1, s_1, F_1)$ and $M_2 = (Q_2, \Sigma_2, R_2, s_2, F_2)$ be two JFAs. Without any loss of generality, we assume that $Q_1 \cap Q_2 = \emptyset$. Define the JFA

$$H = \left(Q_1 \cup Q_2, \Sigma_1 \cup \Sigma_2, R_1 \cup R_2 \cup \{f \to s_2 \mid f \in F_1\}, s_1, F_2\right)$$

To see that $L(H) = \text{shuffle}(L(M_1, \curvearrowright), L(M_2, \curvearrowright))$, observe how H works. On an input string, $w \in (\Sigma_1 \cup \Sigma_2)^*$, H first runs M_1 on w, and if it ends in a final state, then it runs M_2 on the rest of the input. If M_2 ends in a final state, H accepts w. Otherwise, it rejects w. By Theorem 8.2.7, $L(M_i, \curvearrowright) = \text{perm}(L(M_i, \curvearrowright))$ for all $i \in \{1, 2\}$. Based on these observations, since H can jump anywhere after a symbol is read, we see that $L(H) = \text{shuffle}(L(M_1, \curvearrowright), L(M_2, \curvearrowright))$.   □

Notice that the construction used in the previous proof coincides with the standard construction of a concatenation of two finite automata (see [Med00a]).

Theorem 8.2.23. *Both* **GJFA** *and* **JFA** *are closed under union.*

Proof. Let $M_1 = (Q_1, \Sigma_1, R_1, s_1, F_1)$ and $M_2 = (Q_2, \Sigma_2, R_2, s_2, F_2)$ be two GJFAs. Without any loss of generality, we assume that $Q_1 \cap Q_2 = \emptyset$ and $s \notin (Q_1 \cup Q_2)$. Define the GJFA

$$H = \left(Q_1 \cup Q_2 \cup \{s\}, \Sigma_1 \cup \Sigma_2, R_1 \cup R_2 \cup \{s \to s_1, s \to s_2\}, s, F_1 \cup F_2\right)$$

Clearly, $L(H) = L(M_1, \curvearrowright) \cup L(M_2, \curvearrowright)$, and if both M_1 and M_2 are JFAs, then H is also a JFA.   □

Theorem 8.2.24. **GJFA** *is not closed under complement.*

Proof. Consider the GJFA M from Example 8.1.6. Observe that the complement of $L(M, \curvearrowright)$ (with respect to $\{a, b\}^*$) is $\{a\}^*\{b\}^*$, which cannot be accepted by any GJFA (see Lemma 8.2.15).   □

Theorem 8.2.25. **JFA** *is closed under complement.*

Proof. Let $M = (Q, \Sigma, R, s, F)$ be a JFA. Without any loss of generality, we assume that M is a CJFA (**JFA** = **DJFA** by Theorem 8.2.4 and every DJFA can be converted to an equivalent CJFA by Theorem 8.2.1). Then, the JFA

$$M' = \left(Q, \Sigma, R, s, Q - F\right)$$

accepts $\overline{L(M, \curvearrowright)}$.   □

By using De Morgan's laws, we obtain the following two corollaries of Theorems 8.2.23, 8.2.24, and 8.2.25.

Corollary 8.2.26. **GJFA** *is not closed under intersection.* ☐

Corollary 8.2.27. **JFA** *is closed under intersection.* ☐

Theorem 8.2.28. *Both* **GJFA** *and* **JFA** *are not closed under intersection with regular languages.*

Proof. Consider the language $J = \{a, b\}^*$, which can be accepted by both GJFAs and JFAs. Consider the regular language $K = \{a\}^*\{b\}^*$. Since $J \cap K = K$, this theorem follows from Lemma 8.2.15. ☐

Theorem 8.2.29. **JFA** *is closed under reversal.*

Proof. Let $K \in$ **JFA**. Since perm$(w) \subseteq K$ by Theorem 8.2.7 for all $w \in K$, also reversal$(w) \in K$ for all $w \in K$, so the theorem holds. ☐

Theorem 8.2.30. **JFA** *is not closed under Kleene star or under Kleene plus.*

Proof. Consider the language $K = \{ab, ba\}$, which is accepted by the JFA

$$M = (\{s, r, f\}, \{a, b\}, \{sa \to r, rb \to f\}, s, \{f\})$$

However, by Theorem 8.2.7, there is no JFA that accepts K^* or K^+ (notice that, for example, $abab \in K^+$, but $aabb \notin K^+$). ☐

Lemma 8.2.31. *There is no GJFA that accepts* $\{a\}^*\{b\}^* \cup \{b\}^*\{a\}^*$.

Proof. This lemma can be proved by analogy with the proof of Lemma 8.2.15. ☐

Theorem 8.2.32. *Both* **GJFA** *and* **JFA** *are not closed under substitution.*

Proof. Consider the language $K = \{ab, ba\}$, which is accepted by the JFA M from the proof of Theorem 8.2.30. Define the substitution σ from $\{a, b\}^*$ to $2^{\{a,b\}^*}$ as $\sigma(a) = \{a\}^*$ and $\sigma(b) = \{b\}^*$. Clearly, both $\sigma(a)$ and $\sigma(b)$ can be accepted by JFAs. However, $\sigma(K)$ cannot be accepted by any GJFA (see Lemma 8.2.31). ☐

Since the substitution σ in the proof of Theorem 8.2.32 is regular, we obtain the following corollary.

Corollary 8.2.33. *Both* **GJFA** *and* **JFA** *are not closed under regular substitution.* ☐

Theorem 8.2.34. **JFA** *is not closed under ε-free homomorphism.*

Proof. Define the ε-free homomorphism φ from $\{a\}$ to $\{a, b\}^+$ as $\varphi(a) = ab$, and consider the language $\{a\}^*$, which is accepted by the JFA

$$M = (\{s\}, \{a\}, \{sa \to s\}, \{s\})$$

Notice that $\varphi(L(M, \curvearrowright)) = \{ab\}^*$, which cannot be accepted by any JFA (see Corollary 8.2.8). ☐

The analogous result was recently proved for GJFAs in [Vor15].

Theorem 8.2.35 (See Theorem 2 in [Vor15]). **GJFA** *is not closed under ε-free homomorphism.*

Since ε-free homomorphism is a special case of homomorphism and since homomorphism is a special case of finite substitution, we obtain the following corollary of Theorems 8.2.34 and 8.2.35.

Corollary 8.2.36. **GJFA** *and* **JFA** *are not closed under homomorphism.*   □

Corollary 8.2.37. **GJFA** *and* **JFA** *are not closed under finite substitution.*   □

Theorem 8.2.38. **JFA** *is closed under inverse homomorphism.*

Proof. Let $M = (Q, \Gamma, R, s, F)$ be a JFA, Σ be an alphabet, and φ be a homomorphism from Σ^* to Γ^*. We next construct a JFA M' such that $L(M', \curvearrowright)$ $= \varphi^{-1}(L(M, \curvearrowright))$. Define

$$M' = (Q, \Sigma, R', s, F)$$

where

$$R' = \{pa \to q \mid a \in \Sigma, p\varphi(a) \rightsquigarrow q \text{ in } \Delta(M)\}$$

Observe that $w_1 s w_2 \curvearrowright^* q$ in M if and only if $w_1' s w_2' \curvearrowright^* q$ in M', where $w_1 w_2 = \varphi(w_1' w_2')$ and $q \in Q$, so $L(M', \curvearrowright) = \varphi^{-1}(L(M, \curvearrowright))$. A fully rigorous proof is left to the reader.   □

However, the same does not hold for GJFAs.

Theorem 8.2.39 (See Theorem 3 in [Vor15]). **GJFA** *is not closed under inverse homomorphism.*

Moreover, in [Vor15] it was shown that **GJFA** is not close under shuffle, Kleene star, and Kleene plus, while it is closed under reversal.

Theorem 8.2.40 (See Theorems 2 and 4 in [Vor15]). **GJFA** *is not closed under shuffle, Kleene star, and Kleene plus.*

Theorem 8.2.41 (See Theorem 5 in [Vor15]). **GJFA** *is closed under reversal.*

The summary of closure properties of the families **GJFA** and **JFA** is given in Fig. 8.1, where $+$ marks closure and $-$ marks non-closure. It is worth noting that **REG**, characterized by finite automata, is closed under all of these operations.

8.2.3 Decidability

In this section, we prove the decidability of some decision problems with regard to **GJFA** and **JFA**.

	GJFA	JFA
endmarking	−	−
concatenation	−	−
shuffle	−	+
union	+	+
complement	−	+
intersection	−	+
int. with regular languages	−	−
Kleene star	−	−
Kleene plus	−	−
mirror image	+	+
substitution	−	−
regular substitution	−	−
finite substitution	−	−
homomorphism	−	−
ε-free homomorphism	−	−
inverse homomorphism	−	+

Fig. 8.1 Summary of closure properties

Lemma 8.2.42. *Let $M = (Q, \Sigma, R, s, F)$ be a GJFA. Then, $L(M, \curvearrowright)$ is infinite if and only if $py \leadsto p$ in $\Delta(M)$, for some $y \in \Sigma^+$ and $p \in Q$ such that p is both reachable and terminating in $\Delta(M)$.*

Proof. If. Let $M = (Q, \Sigma, R, s, F)$ be a GJFA such that $py \leadsto p$ in $\Delta(M)$, for some $y \in \Sigma^+$ and $p \in Q$ such that p is both reachable and terminating in $\Delta(M)$. Then,

$$w_1 s w_2 \curvearrowright^* upv \curvearrowright^+ xpz \curvearrowright^* f$$

where $w_1 w_2 \in L(M, \curvearrowright)$, $u, v, x, z \in \Sigma^*$, $p \in Q$, and $f \in F$. Consequently,

$$w_1 s w_2 \curvearrowright^* upvy' \curvearrowright^+ xpz \curvearrowright^* f$$

where $y' = y^n$ for all $n \geq 0$. Therefore, $L(M, \curvearrowright)$ is infinite, so the if part holds.
Only If. Let $M = (Q, \Sigma, R, s, F)$ be a GJFA such that $L(M, \curvearrowright)$ is infinite. Without any loss of generality, we assume that M is ε-free (see Lemma 8.2.2). Then,

$$w_1 s w_2 \curvearrowright^* upv \curvearrowright^+ xpz \curvearrowright^* f$$

for some $w_1 w_2 \in L(M, \curvearrowright)$, $u, v, x, z \in \Sigma^*$, $p \in Q$, and $f \in F$. This implies that p is both terminating and reachable in $\Delta(M)$. Let $y \in \Sigma^+$ be a string read by M during $upv \curvearrowright^+ xpz$. Then, $py \leadsto p$ in $\Delta(M)$, so the only-if part holds. □

Theorem 8.2.43. *Both finiteness and infiniteness are decidable for* **GJFA**.

Proof. Let $M = (Q, \Sigma, R, s, F)$ be a GJFA. By Lemma 8.2.42, $L(M, \curvearrowright)$ is infinite if and only if $py \rightsquigarrow p$ in $\Delta(M)$, for some $y \in \Sigma^+$ and $p \in Q$ such that p is both reachable and terminating in $\Delta(M)$. This condition can be checked by any graph searching algorithm, such as breadth-first search (see page 73 in [RN02]). Therefore, the theorem holds.   □

Corollary 8.2.44. *Both finiteness and infiniteness are decidable for* **JFA**.   □

Observe that since there is no deterministic version of a GJFA, the following proof of Theorem 8.2.45 is not as straightforward as in terms of regular languages and classical deterministic finite automata.

Theorem 8.2.45. *The membership problem is decidable for* **GJFA**.

Proof. Let $M = (Q, \Sigma, R, s, F)$ be a GJFA, and let $x \in \Sigma^*$. Without any loss of generality, we assume that M is ε-free (see Theorem 8.2.3). If $x = \varepsilon$, then $x \in L(M, \curvearrowright)$ if and only if $s \in F$, so assume that $x \neq \varepsilon$. Set

$$\Gamma = \left\{ (x_1, x_2, \ldots, x_n) \mid x_i \in \Sigma^+, 1 \leq i \leq n, x_1 x_2 \cdots x_n = x, n \geq 1 \right\}$$

and

$$\Gamma_p = \left\{ (y_1, y_2, \ldots, y_n) \mid (x_1, x_2, \ldots, x_n) \in \Gamma, n \geq 1, (y_1, y_2, \ldots, y_n) \text{ is} \right.$$
$$\left. \text{a permutation of } (x_1, x_2, \ldots, x_n) \right\}$$

If there exist $(y_1, y_2, \ldots, y_n) \in \Gamma_p$ and $q_1, q_2, \ldots, q_{n+1} \in Q$, for some n, $1 \leq n \leq |x|$, such that $s = q_1$, $q_{n+1} \in F$, and $q_i y_i \rightarrow q_{i+1} \in R$ for all $i = 1, 2, \ldots, n$, then $x \in L(M, \curvearrowright)$; otherwise, $x \notin L(M, \curvearrowright)$. Since both Q and Γ_p are finite, this check can be performed in finite time.   □

Corollary 8.2.46. *The membership problem is decidable for* **JFA**.   □

Theorem 8.2.47. *The emptiness problem is decidable for* **GJFA**.

Proof. Let $M = (Q, \Sigma, R, s, F)$ be a GJFA. Then, $L(M, \curvearrowright)$ is empty if and only if no $f \in F$ is reachable in $\Delta(M)$. This check can be done by any graph searching algorithm, such as breadth-first search (see page 73 in [RN02]).   □

Corollary 8.2.48. *The emptiness problem is decidable for* **JFA**.   □

The summary of decidability properties of the families **GJFA** and **JFA** is given in Fig. 8.2, where $+$ marks decidability.

	GJFA	JFA
membership	+	+
emptiness	+	+
finiteness	+	+
infiniteness	+	+

Fig. 8.2 Summary of Decidability Properties

8.2.4 An Infinite Hierarchy of Language Families

In this section, we establish an infinite hierarchy of language families resulting from GJFAs of degree n, where $n \geq 0$. Let $\mathbf{GJFA}_n$ and $\mathbf{GJFA}_n^{-\varepsilon}$ denote the families of languages accepted by GJFAs of degree n and by ε-free GJFAs of degree n, respectively. Observe that $\mathbf{GJFA}_n = \mathbf{GJFA}_n^{-\varepsilon}$ by the definition of a GJFA and by Lemma 8.2.2, for all $n \geq 0$.

Lemma 8.2.49. *Let Σ be an alphabet such that* $\mathrm{card}(\Sigma) \geq 2$. *Then, for any $n \geq 1$, there is a GJFA of degree n, $M_n = (Q, \Sigma, R, s, F)$, such that $L(M_n)$ cannot be accepted by any GJFA of degree $n - 1$.*

Proof. Let Σ be an alphabet such that $\mathrm{card}(\Sigma) \geq 2$, and let $a, b \in \Sigma$ such that $a \neq b$. The case when $n = 1$ follows immediately from the definition of a JFA, so we assume that $n \geq 2$. Define the GJFA of degree n

$$M_n = \big(\{s, f\}, \Sigma, \{sw \to f\}, s, \{f\}\big)$$

where $w = ab(a)^{n-2}$. Clearly, $L(M_n, \curvearrowright) = \{w\}$. We next prove that $L(M_n, \curvearrowright)$ cannot be accepted by any GJFA of degree $n - 1$.

Suppose, for the sake of contradiction, that there is a GJFA of degree $n - 1$, $H = (Q, \Sigma, R, s', F)$, such that $L(H) = L(M_n, \curvearrowright)$. Without any loss of generality, we assume that H is ε-free (see Lemma 8.2.2). Since $L(H) = L(M_n, \curvearrowright) = \{w\}$ and $|w| > n - 1$, there has to be

$$us'xv \curvearrowright^m f$$

in H, where $w = uxv$, $u, v \in \Sigma^*$, $x \in \Sigma^+$, $f \in F$, and $m \geq 2$. Thus,

$$s'xuv \curvearrowright^m f$$

and

$$uvs'x \curvearrowright^m f$$

in H, which contradicts the assumption that $L(H) = \{w\}$. Therefore, $L(M_n, \curvearrowright)$ cannot be accepted by any GJFA of degree $n - 1$. $\qquad\square$

Theorem 8.2.50. $GJFA_n \subset GJFA_{n+1}$ *for all* $n \geq 0$.

Proof. $GJFA_n \subseteq GJFA_{n+1}$ follows from the definition of a GJFA of degree n, for all $n \geq 0$. From Lemma 8.2.49, $GJFA_{n+1} - GJFA_n \neq \emptyset$, which proves the theorem.   □

Taking Lemma 8.2.2 into account, we obtain the following corollary of Theorem 8.2.50.

Corollary 8.2.51. $GJFA_n^{-\varepsilon} \subset GJFA_{n+1}^{-\varepsilon}$ *for all* $n \geq 0$.   □

8.2.5 Left and Right Jumps

We define two special cases of the jumping relation.

Definition 8.2.52. Let $M = (Q, \Sigma, R, s, F)$ be a GJFA. Let $w, x, y, z \in \Sigma^*$, and $py \rightarrow q \in R$; then, (1) M makes a *left jump* from $wxpyz$ to $wqxz$, symbolically written as

$$wxpyz \,_l\!\curvearrowright wqxz$$

and (2) M makes a *right jump* from $wpyxz$ to $wxqz$, written as

$$wpyxz \,_r\!\curvearrowright wxqz$$

Let $u, v \in \Sigma^* Q \Sigma^*$; then, $u \curvearrowright v$ if and only if $u \,_l\!\curvearrowright v$ or $u \,_r\!\curvearrowright v$. Extend $\,_l\!\curvearrowright$ and $\,_r\!\curvearrowright$ to $\,_l\!\curvearrowright^m, \,_l\!\curvearrowright^*, \,_l\!\curvearrowright^+, \,_r\!\curvearrowright^m, \,_r\!\curvearrowright^*$, and $\,_r\!\curvearrowright^+$, where $m \geq 0$, by analogy with extending $\curvearrowright$. Set

$$_lL(M, \curvearrowright) = \{uv \mid u, v \in \Sigma^*, usv \,_l\!\curvearrowright^* f \text{ with } f \in F\}$$

and

$$_rL(M, \curvearrowright) = \{uv \mid u, v \in \Sigma^*, usv \,_r\!\curvearrowright^* f \text{ with } f \in F\}$$ □

Let $_lGJFA$, $_lJFA$, $_rGJFA$, and $_rJFA$ denote the families of languages accepted by GJFAs using only left jumps, JFAs using only left jumps, GJFAs using only right jumps, and JFAs using only right jumps, respectively.

Theorem 8.2.53. $_rGJFA = \,_rJFA = REG$

Proof. We first prove that $_rJFA = REG$. Consider any JFA, $M = (Q, \Sigma, R, s, F)$. Observe that if M occurs in a configuration of the form xpy, where $x \in \Sigma^*$, $p \in Q$, and $y \in \Sigma^*$, then it cannot read the symbols in x anymore because M can make only right jumps. Also, observe that this covers the situation when M starts to accept $w \in \Sigma^*$ from a different configuration than sw. Therefore, to read the whole input, M has to start in configuration sw, and it cannot jump to skips some symbols. Consequently, M behaves like an ordinary finite automaton, reading the input from the left to the right, so $L(M, \curvearrowright)$ is regular and, therefore, $_r\text{JFA} \subseteq \text{REG}$. Conversely, any finite automaton can be viewed as a JFA that starts from configuration sw and does not jump to skip some symbols. Therefore, $\text{REG} \subseteq {}_r\text{JFA}$, which proves that $_r\text{JFA} = \text{REG}$. $_r\text{GJFA} = \text{REG}$ can be proved by the same reasoning using general finite automata instead of finite automata. $\qquad\square$

Next, we show that JFAs using only left jumps accept some non-regular languages.

Theorem 8.2.54. $_l\text{JFA} - \text{REG} \neq \emptyset$

Proof. Consider the JFA

$$M = \big(\{s, p, q\}, \{a, b\}, R, s, \{s\}\big)$$

where

$$R = \big\{sa \to p, pb \to s, sb \to q, qa \to s\big\}$$

We argue that

$$_lL(M, \curvearrowright) = \big\{w \mid \#_a(w) = \#_b(w)\big\}$$

With $w \in \{a, b\}^*$ on its input, M starts over the last symbol. M reads this symbol by using $sa \to p$ or $sb \to q$, and jumps to the left in front of the rightmost occurrence of b or a, respectively. Then, it consumes it by using $pb \to s$ or $qa \to s$, respectively. If this read symbol was the rightmost one, it jumps one symbol to the left and repeats the process. Otherwise, it makes no jumps at all. Observe that in this way, every configuration is of the form urv, where $r \in \{s, p, q\}$, $u \in \{a, b\}^*$, and either $v \in \{a, \varepsilon\}\{b\}^*$ or $v \in \{b, \varepsilon\}\{a\}^*$.

Based on the previous observations, we see that

$$_lL(M, \curvearrowright) = \big\{w \mid \#_a(w) = \#_b(w)\big\}$$

Since $L(M, \curvearrowright)$ is not regular, $_l\text{JFA} - \text{REG} \neq \emptyset$, so the theorem holds. $\qquad\square$

Open Problem 8.2.55. Study the effect of left jumps to the acceptance power of JFAs and GJFAs.

8.2.6 A Variety of Start Configurations

In general, a GJFA can start its computation anywhere in the input string (see Definition 8.1.1). In this section, we consider the impact of various start configurations on the acceptance power of GJFAs and JFAs.

Definition 8.2.56. Let $M = (Q, \Sigma, R, s, F)$ be a GJFA. Set

$$^bL(M, \curvearrowright) = \{w \in \Sigma^* \mid sw \curvearrowright^* f \text{ with } f \in F\},$$
$$^aL(M, \curvearrowright) = \{uv \mid u, v \in \Sigma^*, usv \curvearrowright^* f \text{ with } f \in F\},$$
$$^eL(M, \curvearrowright) = \{w \in \Sigma^* \mid ws \curvearrowright^* f \text{ with } f \in F\}. \qquad \square$$

Intuitively, b, a, and e stand for *beginning*, *anywhere*, and *end*, respectively; in this way, we express where the acceptance process starts. Observe that we simplify $^aL(M, \curvearrowright)$ to $L(M, \curvearrowright)$ because we pay a principal attention to the languages accepted in this way in this chapter. Let $^b\mathbf{GJFA}$, $^a\mathbf{GJFA}$, $^e\mathbf{GJFA}$, $^b\mathbf{JFA}$, $^a\mathbf{JFA}$, and $^e\mathbf{JFA}$ denote the families of languages accepted by GJFAs starting at the beginning, GJFAs starting anywhere, GJFAs starting at the end, JFAs starting at the beginning, JFAs starting anywhere, and JFAs starting at the end, respectively.

We show that

(1) starting at the beginning increases the acceptance power of GJFAs and JFAs, and

(2) starting at the end does not increase the acceptance power of GJFAs and JFAs.

Theorem 8.2.57. $^a\mathbf{JFA} \subset {}^b\mathbf{JFA}$

Proof. Let $M = (Q, \Sigma, R, s, F)$ be a JFA. The JFA

$$M' = (Q, \Sigma, R \cup \{s \to s\}, s, F)$$

clearly satisfies $^aL(M, \curvearrowright) = {}^bL(M', \curvearrowright)$, so $^a\mathbf{JFA} \subseteq {}^b\mathbf{JFA}$. We prove that this inclusion is, in fact, proper. Consider the language $K = \{a\}\{b\}^*$. The JFA

$$H = (\{s, f\}, \{a, b\}, \{sa \to f, fb \to f\}, s, \{f\})$$

satisfies $^bL(H) = K$. However, observe that $^aL(H) = \{b\}^*\{a\}\{b\}^*$, which differs from K. By Theorem 8.2.7, for every JFA N, it holds that $^aL(N) \neq K$. Hence, $^a\mathbf{JFA} \subset {}^b\mathbf{JFA}$. $\qquad \square$

Theorem 8.2.58. $^a\mathbf{GJFA} \subset {}^b\mathbf{GJFA}$

Proof. This theorem can be proved by analogy with the proof of Theorem 8.2.57.
$\qquad \square$

Lemma 8.2.59. *Let M be a GJFA of degree $n \geq 0$. Then, there is a GJFA M' of degree n such that $^aL(M, \curvearrowright) = {}^eL(M', \curvearrowright)$.*

Proof. Let $M = (Q, \Sigma, R, s, F)$ be a GJFA of degree n. Then, the GJFA

$$M' = (Q, \Sigma, R \cup \{s \to s\}, s, F)$$

is of degree n and satisfies $^aL(M, \curvearrowright) = {}^eL(M', \curvearrowright)$. $\qquad \square$

Lemma 8.2.60. *Let M be a GJFA of degree $n \geq 0$. Then, there is a GJFA $\hat{M}$ of degree n such that ${}^e L(M, \curvearrowright) = {}^a L(\hat{M})$.*

Proof. Let $M = (Q, \Sigma, R, s, F)$ be a GJFA of degree n. If ${}^e L(M, \curvearrowright) = \emptyset$, then the GJFA

$$M' = (\{s\}, \Sigma, \emptyset, s, \emptyset)$$

is of degree n and satisfies ${}^a L(M', \curvearrowright) = \emptyset$. If ${}^e L(M, \curvearrowright) = \{\varepsilon\}$, then the GJFA

$$M'' = (\{s\}, \Sigma, \emptyset, s, \{s\})$$

is of degree n and satisfies ${}^a L(M'', \curvearrowright) = \{\varepsilon\}$. Therefore, assume that $w \in {}^e L(M, \curvearrowright)$, where $w \in \Sigma^+$. Then, $s \to p \in R$, for some $p \in Q$. Indeed, observe that either ${}^e L(M, \curvearrowright) = \emptyset$ or ${}^e L(M, \curvearrowright) = \{\varepsilon\}$, which follows from the observation that if M starts at the end of an input string, then it first has to jump to the left to be able to read some symbols.

Define the GJFA $\hat{M} = (Q, \Sigma, \hat{R}, s, F)$, where

$$\hat{R} = R - \{su \to q \mid u \in \Sigma^+, q \in Q, \text{ and there is no } x \in \Sigma^+ \\ \text{such that } sx \rightsquigarrow s \text{ in } \Delta(M)\}$$

The reason for excluding such $su \to q$ from $\hat{R}$ is that M first has to use a rule of the form $s \to p$, where $p \in Q$ (see the argumentation above). However, since $\hat{M}$ starts anywhere in the input string, we need to force it to use $s \to p$ as the first rule, thus changing the state from s to p, just like M does.

Clearly, $\hat{M}$ is of degree n and satisfies ${}^e L(M, \curvearrowright) = {}^a L(\hat{M})$, so the lemma holds. $\square$

Theorem 8.2.61. ${}^e\mathbf{GJFA} = {}^a\mathbf{GJFA}$ *and* ${}^e\mathbf{JFA} = {}^a\mathbf{JFA}$

Proof. This theorem follows from Lemmas 8.2.59 and 8.2.60. $\square$

We also consider combinations of left jumps, right jumps, and various start configurations. For this purpose, by analogy with the previous denotations, we define ${}^b_l\mathbf{GJFA}, {}^a_l\mathbf{GJFA}, {}^e_l\mathbf{GJFA}, {}^b_r\mathbf{GJFA}, {}^a_r\mathbf{GJFA}, {}^e_r\mathbf{GJFA}, {}^b_l\mathbf{JFA}, {}^a_l\mathbf{JFA}, {}^e_l\mathbf{JFA}, {}^b_r\mathbf{JFA}, {}^a_r\mathbf{JFA},$ and ${}^e_r\mathbf{JFA}$. For example, ${}^b_r\mathbf{GJFA}$ denotes the family of languages accepted by GJFAs that perform only right jumps and starts at the beginning.

Theorem 8.2.62. ${}^a_r\mathbf{GJFA} = {}^a_r\mathbf{JFA} = {}^b_r\mathbf{GJFA} = {}^b_r\mathbf{JFA} = {}^b_l\mathbf{GJFA} = {}^b_l\mathbf{JFA} = \mathbf{REG}$

Proof. Theorem 8.2.53, in fact, states that ${}^a_r\mathbf{GJFA} = {}^a_r\mathbf{JFA} = \mathbf{REG}$. Furthermore, ${}^b_r\mathbf{GJFA} = {}^b_r\mathbf{JFA} = \mathbf{REG}$ follows from the proof of Theorem 8.2.53 because M has to start the acceptance process of a string w from the configuration sw—that is, it starts at the beginning of w. ${}^b_l\mathbf{GJFA} = {}^b_l\mathbf{JFA} = \mathbf{REG}$ can be proved analogously. $\square$

Theorem 8.2.63. ^e_r**GJFA** $=$ ^e_r**JFA** $= \{\emptyset, \{\varepsilon\}\}$

Proof. Consider JFAs $M = (\{s\}, \{a\}, \emptyset, s, \emptyset)$ and $M' = (\{s\}, \{a\}, \emptyset, s, \{s\})$ to see that $\{\emptyset, \{\varepsilon\}\} \subseteq {}^e_r$**GJFA** and $\{\emptyset, \{\varepsilon\}\} \subseteq {}^e_r$**JFA**. The converse inclusion also holds. Indeed, any GJFA that starts the acceptance process of a string w from ws and that can make only right jumps accepts either $\emptyset$ or $\{\varepsilon\}$. $\qquad\square$

Open Problem 8.2.64. What are the properties of ^e_l**GJFA** and ^e_l**JFA**?

Notice that Open Problem 8.2.55, in fact, suggests an investigation of the properties of ^a_l**GJFA** and ^a_l**JFA**.

8.2.7 Relations Between Jumping Automata and Jumping Grammars

Next, we demonstrate that the generative power of regular and right-linear jumping grammars (see Sect. 5.1) is the same as accepting power of jumping finite automata and general jumping finite automata, respectively. Consequently, the following equivalence and the previous results in this chapter imply several additional properties of languages that are generated by regular and right-linear jumping grammars such as closure properties and decidability.

Lemma 8.2.65. GJFA $\subseteq$ JRLIN.

Proof. Construction. For every GJFA $M = (Q, \Sigma, R, s, F)$, we construct a RLG $G = (Q \cup \Sigma \cup \{S\}, \Sigma, P, S)$, where S is a new nonterminal, $S \notin Q \cup \Sigma$, such that $L(M, \curvearrowright) = L(G, {}_j\Rightarrow)$. Set $P = \{S \to f \mid f \in F\} \cup \{q \to xp \mid px \to q \in R\} \cup \{q \to x \mid sx \to q \in R\}$.
Basic Idea. The principle of the conversion is analogical to the conversion from classical lazy finite automata to equivalent RLGs with sequential derivation mode (see Section 2.6.2 in [Woo87] and Theorem 4.1 in [Sal73]).

The states of M are used as nonterminals in G. In addition, we introduce new start nonterminal S in G. The input symbols Σ are terminal symbols in G.

During the simulation of M in G there is always exactly one nonterminal symbol in the sentential form until the last jumping derivation step that produces the string of terminal symbols. If there is a sequence of jumping moves $usv \curvearrowright^* ypxy' \curvearrowright zqz'z'' \curvearrowright^* f$ in M, then G simulates it by jumping derivation $S {}_j\Rightarrow f {}_j\Rightarrow^* zz'qz'' {}_j\Rightarrow yxpy' {}_j\Rightarrow^* w$, where $yy' = zz'z''$ and $w = uv$. Firstly, S is nondeterministically rewritten to some f in G to simulate the entrance to the corresponding accepting final state of M. Then, for each rule $px \to q$ in M that processes substring x in the input string, there is x generated by the corresponding rule of the form $q \to xp$ in G. As the last jumping derivation step in G, we simulate the first jumping move of M from the start state s by rewriting the only nonterminal in the sentential form of G to a string of terminals and the simulation of M by G is completed. $\qquad\square$

Lemma 8.2.66. JRLIN $\subseteq$ GJFA.

Proof. Construction. For every RLG $G = (V, T, P, S)$, we construct a GJFA $M = (N \cup \{\sigma\}, T, R, \sigma, \{S\})$, where σ is a new start state, $\sigma \notin V$ and $N = V - T$, such that $L(G, {}_j\!\Rightarrow) = L(M, \curvearrowright)$. Set $R = \{Bx \to A \mid A \to xB \in P, A, B \in N, x \in T^*\} \cup \{\sigma x \to A \mid A \to x \in P, x \in T^*\}$.

Basic Idea. In the simulation of G in M we use nonterminals N as states, new state σ as the start state, and terminals T corresponds to input symbols of M. In addition, the start nonterminal of G corresponds to the only final state of M. Every application of a rule from P in G is simulated by a move according to the corresponding rule from R constructed above. If there is a jumping derivation $S \;{}_j\!\Rightarrow^* yy'Ay'' \;{}_j\!\Rightarrow zxBz' \;{}_j\!\Rightarrow^* w$ in G, then M simulates it by jumping moves $u\sigma v \curvearrowright^* zBxz' \curvearrowright yAy'y'' \curvearrowright^* S$, where $yy'y'' = zz'$ and $w = uv$. $\qquad\square$

Theorem 8.2.67. GJFA = JRLIN.

Proof. This theorem holds by Lemmas 8.2.65 and 8.2.66. $\qquad\square$

In the following theorem, consider jumping finite automata that processes only one input symbol in one move. We state their equivalence with jumping RGs.

Theorem 8.2.68. JFA = JREG.

Proof. Prove this statement by analogy with the proof of Theorem 8.2.67. $\qquad\square$

Figure 8.3 summarizes the achieved results on the descriptional complexity of jumping grammars and automata.

8.2.8 A Summary of Open Problems

Within the previous sections, we have already pointed out several specific open problems concerning them. We close the present chapter by pointing out some crucially important open problem areas as suggested topics of future investigations.

(I) Regarding decision problems, investigate other decision properties of **GJFA** and **JFA**, like equivalence, universality, inclusion, or regularity. Furthermore, study their computational complexity. Do there exist undecidable problems for **GJFA** or **JFA**?

(II) Section 8.2.5 has demonstrated that GJFAs and JFAs using only right jumps define the family of regular languages. How precisely do left jumps affect the acceptance power of JFAs and GJFAs?

(III) Broaden the results of Sect. 8.2.6 concerning various start configurations by investigating the properties of ${}^\varrho_f$**GJFA** and ${}^\varrho_f$**JFA**.

(IV) Determinism represents a crucially important investigation area in terms of all types of automata. In essence, the non-deterministic versions of automata can make several different moves from the same configuration while their

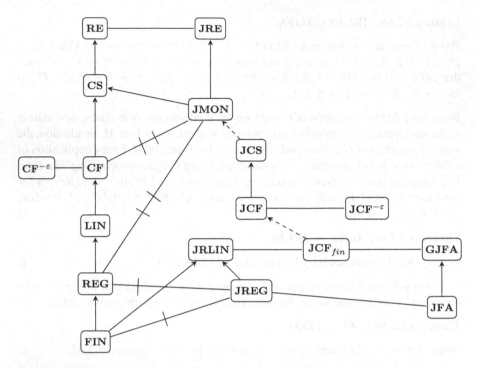

Fig. 8.3 A hierarchy of language families closely related to the language families resulting from jumping grammars and automata is shown. If there is a line or an arrow from family X to family Y in the figure, then $X = Y$ or $X \subset Y$, respectively. If there is a dashed arrow from X to Y, then $X \subseteq Y$, but $X \subset Y$ represents an open problem. A crossed line represents the incomparability between connected families. (It is noteworthy that the figure describes only some of the language-family relations that are crucially important in terms of the present section; however, by no means, it gives an exhaustive description of these relations)

deterministic counterparts cannot—that is, they make no more than one move from any configuration. More specifically, the deterministic version of classical finite automata require that for any state q and any input symbol a, there exists no more than one rule with qa on its left-hand side; in this way, they make no more than one move from any configuration. As a result, with any input string w, they make a unique sequence of moves. As should be obvious, in terms of jumping finite automata, this requirement does not guarantee their determinism in the above sense. Modify the requirement so it guarantees the determinism.

Chapter 9
Deep Pushdown Automata and New Stack Structures

Deep pushdown automata, explored in this chapter, represent language-accepting models based upon new stack structures, which can be modified deeper than on their top. As a result, these automata can make expansions deeper in their pushdown lists while ordinary pushdown automata (see Sect. 2.4) can expand only the very pushdown top.

This chapter proves that the power of deep pushdown automata is similar to the generative power of regulated context-free grammars without erasing rules (see Chap. 3). Indeed, just like these grammars, deep pushdown automata are stronger than ordinary pushdown automata but less powerful than context-sensitive grammars. More precisely, they give rise to an infinite hierarchy of language families coinciding with the hierarchy resulting from n-limited state grammars (see Sect. 3.2).

To give a more detailed insight into the concept of deep pushdown automata, consider the well-known conversion of a context-free grammar to an equivalent pushdown automaton M frequently referred to as the *general top-down parser* for the grammar (see, for instance, page 176 in [RS97a], page 148 in [Har78], page 113 in [LP81], and page 444 in [Med00a]). Recall that during every move, M either *pops* or *expands* its pushdown depending on the symbol occurring on the pushdown top. If an input symbol a occurs on the pushdown top, M compares the pushdown top symbol with the current input symbol, and if they coincide, M pops the topmost symbol from the pushdown and proceeds to the next input symbol on the input tape. If a nonterminal occurs on the pushdown top, the parser expands its pushdown so it replaces the top nonterminal with a string. M accepts an input string x if it makes a sequence of moves so it completely reads x, empties its pushdown, and enters a final state; the latter requirement of entering a final state is dropped in some books (see, for instance, Algorithm 5.3.1.1.1 in [Med00a] or Theorem 5.1 in [AU72]). In essence, a deep pushdown automaton, $_{deep}M$, represents a slight generalization of M. Indeed, $_{deep}M$ works exactly as M except that it can make expansions of depth m so $_{deep}M$ replaces the mth topmost pushdown symbol with

© Springer International Publishing AG 2017   393
A. Meduna, O. Soukup, *Modern Language Models and Computation*,
DOI 10.1007/978-3-319-63100-4_9

a string, for some $m \geq 1$. We demonstrate that the deep pushdown automata that make expansions of depth m or less, where $m \geq 1$, are equivalent to m-limited state grammars, so these automata accept a proper language subfamily of the language family accepted by deep pushdown automata that make expansions of depth $m + 1$ or less. The resulting infinite hierarchy of language families obtained in this way occurs between the families of context-free and context-sensitive languages. For every positive integer n, however, there exist some context-sensitive languages that cannot be accepted by any deep pushdown automata that make expansions of depth n or less.

The present chapter is divided into two sections—Sects. 9.1 and 9.2. The former defines and illustrates deep pushdown automata. The latter establishes their accepting power, formulates some open problem areas concerning them, and suggests introducing new deterministic and generalized versions of these automata.

9.1   Definitions and Examples

Without further ado, we define the notion of a deep pushdown automata, after which we illustrate it by an example.

Definition 9.1.1. A *deep pushdown automaton* is a septuple

$$M = (Q, \Sigma, \Gamma, R, s, S, F)$$

where

- Q is a finite set of *states*;
- Σ is an *input alphabet*;
- Γ is a *pushdown alphabet*, $\mathbb{N}$, Q, and Γ are pairwise disjoint, $\Sigma \subseteq \Gamma$, and $\Gamma - \Sigma$ contains a special *bottom symbol*, denoted by #;
- $R \subseteq (\mathbb{N} \times Q \times (\Gamma - (\Sigma \cup \{\#\}))) \times Q \times (\Gamma - \{\#\})^+)$
  $\cup$ $(\mathbb{N} \times Q \times \{\#\} \times Q \times (\Gamma - \{\#\})^* \{\#\})$ is a finite relation;
- $s \in Q$ is the *start state*;
- $S \in \Gamma$ is the *start pushdown symbol*;
- $F \subseteq Q$ is the set of *final states*.

Instead of $(m, q, A, p, v) \in R$, we write $mqA \rightarrow pv \in R$ and call $mqA \rightarrow pv$ a *rule*; accordingly, R is referred to as the *set of rules* of M. A *configuration* of M is a triple in $Q \times T^* \times (\Gamma - \{\#\})^* \{\#\}$. Let χ denote the set of all configurations of M. Let $x, y \in \chi$ be two configurations. M *pops* its pushdown from x to y, symbolically written as

$$x_p \vdash y$$

if $x = (q, au, az)$, $y = (q, u, z)$, where $a \in \Sigma$, $u \in \Sigma^*$, $z \in \Gamma^*$. M *expands* its pushdown from x to y, symbolically written as

$$x_e \vdash y$$

if $x = (q, w, uAz)$, $y = (p, w, uvz)$, $mqA \to pv \in R$, where $q, p \in Q$, $w \in \Sigma^*$, $A \in \Gamma$, $u, v, z \in \Gamma^*$, and $\#_{\Gamma - \Sigma}(u) = m - 1$. To express that M makes $x_e \vdash y$ according to $mqA \to pv$, we write

$$x_e \vdash y \, [mqA \to pv]$$

We say that $mqA \to pv$ is a *rule of depth m*; accordingly, $x_e \vdash y \, [mqA \to pv]$ is an *expansion of depth m*. M makes a *move* from x to y, symbolically written as

$$x \vdash y$$

if M makes either $x_e \vdash y$ or $x_p \vdash y$. If $n \in \mathbb{N}$ is the minimal positive integer such that each rule of M is of depth n or less, we say that M *is of depth n*, symbolically written as $_nM$. In the standard manner, we extend $_p \vdash$, $_e \vdash$, and $\vdash$ to $_p \vdash^m$, $_e \vdash^m$, and $\vdash^m$, respectively, for $m \geq 0$; then, based on $_p \vdash^m$, $_e \vdash^m$, and $\vdash^m$, we define $_p \vdash^+$, $_p \vdash^*$, $_e \vdash^+$, $_e \vdash^*$, $\vdash^+$, and $\vdash^*$.

Let M be of depth n, for some $n \in \mathbb{N}$. We define the *language accepted by* $_nM$, $L(_nM)$, as

$$L(_nM) = \{w \in \Sigma^* \mid (s, w, S\#) \vdash^* (f, \varepsilon, \#) \text{ in } _nM \text{ with } f \in F\}$$

In addition, we define the *language that* $_nM$ *accepts by empty pushdown*, $E(_nM)$, as

$$E(_nM) = \{w \in \Sigma^* \mid (s, w, S\#) \vdash^* (q, \varepsilon, \#) \text{ in } _nM \text{ with } q \in Q\} \qquad \square$$

For every $k \geq 1$, $_{deep}\mathbf{PDA}_k$ denotes the family of languages defined by deep pushdown automata of depth i, where $1 \leq i \leq k$. Analogously, $_{deep}^{empty}\mathbf{PDA}_k$ denotes the family of languages defined by deep pushdown automata of depth i by empty pushdown, where $1 \leq i \leq k$.

The following example gives a deep pushdown automaton accepting a language from

$$\left(_{deep}\mathbf{PDA}_2 \cap {}_{deep}^{empty}\mathbf{PDA}_2 \cap \mathbf{CS}\right) - \mathbf{CF}$$

Example 9.1.2. Consider the deep pushdown automaton

$$_2M = (\{s, q, p\}, \{a, b, c\}, \{A, S, \#\}, R, s, S, \{f\})$$

with R containing the following five rules

$$1sS \rightarrow qAA \qquad\qquad 1qA \rightarrow fab \qquad\qquad 1fA \rightarrow fc$$
$$1qA \rightarrow paAb \qquad\qquad 2pA \rightarrow qAc$$

On *aabbcc*, M makes

$$
\begin{aligned}
(s, aabbcc, S\#) \;&{}_e\vdash\; (q, aabbcc, AA\#) && [1sS \rightarrow qAA]\\
&{}_e\vdash\; (p, aabbcc, aAbA\#) && [1qA \rightarrow paAb]\\
&{}_p\vdash\; (p, abbcc, AbA\#)\\
&{}_e\vdash\; (q, abbcc, AbAc\#) && [2pA \rightarrow qAc]\\
&{}_e\vdash\; (q, abbcc, abbAc\#) && [1qA \rightarrow fab]\\
&{}_p\vdash\; (f, bbcc, bbAc\#)\\
&{}_p\vdash\; (f, bcc, bAc\#)\\
&{}_p\vdash\; (f, cc, Ac\#)\\
&{}_e\vdash\; (f, cc, cc\#) && [1fA \rightarrow fc]\\
&{}_p\vdash\; (f, c, c\#)\\
&{}_p\vdash\; (f, \varepsilon, \#)
\end{aligned}
$$

In brief, $(s, aabbcc, S\#) \vdash^* (f, \varepsilon, \#)$. Observe that $L(_2M) = E(_2M) = \{a^n b^n c^n \mid n \geq 1\}$, which belongs to **CS** $-$ **CF**. $\qquad\qquad\qquad\qquad\qquad\qquad\qquad\qquad\qquad\square$

9.2 Accepting Power

In the present section, we establish the main results of this chapter. That is, we demonstrate that deep pushdown automata that make expansions of depth m or less, where $m \geq 1$, are equivalent to m-limited state grammars, so these automata accept a proper subfamily of the language family accepted by deep pushdown automata that make expansions of depth $m + 1$ or less. Then, we point out that the resulting infinite hierarchy of language families obtained in this way occurs between the families of context-free and context-sensitive languages. However, we also show that there always exist some context-sensitive languages that cannot be accepted by any deep pushdown automata that make expansions of depth n or less, for every positive integer n.

To rephrase these results briefly and formally, we prove that

$$_{deep}\mathbf{PDA}_1 = {}_{deep}^{empty}\mathbf{PDA}_1 = \mathbf{CF}$$

and for every $n \geq 1$,

$$_{deep}^{empty}\mathbf{PDA}_n = {}_{deep}\mathbf{PDA}_n \subset {}_{deep}^{empty}\mathbf{PDA}_{n+1} = {}_{deep}\mathbf{PDA}_{n+1} \subset \mathbf{CS}$$

After proving all these results, we formulate several open problem areas, including some suggestions concerning new deterministic and generalized versions of deep pushdown automata.

Lemma 9.2.1. *For every state grammar G and for every $n \geq 1$, there exists a deep pushdown automaton of depth n, $_nM$, such that $L(G, n) = L(_nM)$.*

Proof. Let $G = (V, W, T, P, S)$ be a state grammar and let $n \geq 1$. Set $N = V - T$. Define the homomorphism f over $(\{\#\} \cup V)^*$ as $f(A) = A$, for every $A \in \{\#\} \cup N$, and $f(a) = \varepsilon$, for every $a \in T$. Introduce the deep pushdown automaton of depth n

$$_nM = \left(Q, T, \{\#\} \cup V, R, s, S, \{\$\} \right)$$

where

$$Q = \{S, \$\} \cup \{\langle p, u \rangle \mid p \in W, u \in N^*\{\#\}^*, |u| \leq n\}$$

and R is constructed by performing the following four steps

(1) for each $(p, S) \to (q, x) \in P, p, q \in W, x \in V^+$, add

$1sS \to \langle p, S \rangle S$ to R;

(2) if $(p, A) \to (q, x) \in P$, $\langle p, uAv \rangle \in Q$, $p, q \in W$, $A \in N$, $x \in V^+$, $u \in N^*$, $v \in N^*\{\#\}^*$, $|uAv| = n$, $p \notin_G$ states(u), add

$|uA|\langle p, uAv \rangle A \to \langle q, \text{prefix}(uf(x)v, n) \rangle x$ to R;

(3) if $A \in N, p \in W, u \in N^*, v \in \{\#\}^*, |uv| \leq n - 1, p \notin_G$ states(u), add

$|uA|\langle p, uv \rangle A \to \langle p, uAv \rangle A$ and
$|uA|\langle p, uv \rangle \# \to \langle p, uv\# \rangle \#$ to R;

(4) for each $q \in W$, add

$1\langle q, \#^n \rangle \# \to \$\#$ to R.

$_nM$ simulates n-limited derivations of G so it always records the first n nonterminals occurring in the current sentential form in its state (if there appear fewer than n nonterminals in the sentential form, it completes them to n in the state by #s from behind). $_nM$ simulates a derivation step in the pushdown and, simultaneously, records the newly generated nonterminals in the state. When G successfully completes the generation of a terminal string, $_nM$ completes reading the string, empties its pushdown, and enters the final state $. $

To establish $L(G, n) = L(_nM)$, we first prove two claims.

Claim 9.2.2. Let $(p, S)_n \Rightarrow^m (q, dy)$ in G, where $d \in T^*, y \in (NT^*)^*, p, q \in W, m \geq 0$. Then, $(\langle p, S \rangle, d, S\#) \vdash^* (\langle q, \text{prefix}(f(y\#^n), n) \rangle, \varepsilon, y\#)$ in $_nM$.

Proof. This claim is proved by induction on $m \geq 0$.

Basis. Let $m = 0$, so $(p, S)_n \Rightarrow^0 (p, S)$ in $G, d = \varepsilon$ and $y = S$. By using rules introduced in steps (1) and (4),

$$(\langle p, S \rangle, \varepsilon, S\#) \vdash^* (\langle p, \text{prefix}(f(S\#^n), n) \rangle, \varepsilon, S\#) \text{ in } {}_nM$$

so the basis holds.

Induction Hypothesis. Assume that the claim holds for all $m, 0 \leq m \leq k$, where k is a non-negative integer.

Induction Step. Let $(p, S)_n \Rightarrow^{k+1} (q, dy)$ in G, where $d \in T^*, y \in (NT^*)^*, p, q \in W$. Since $k + 1 \geq 1$, express $(p, S)_n \Rightarrow^{k+1} (q, dy)$ as

$$(p, S)_n \Rightarrow^k (h, buAo)_n \Rightarrow (q, buxo) \ [(h, A) \rightarrow (q, x)]$$

where $b \in T^*, u \in (NT^*)^*, A \in N, h, q \in W, (h, A) \rightarrow (q, x) \in P$, max-suffix$(buxo, (NT^*)^*) = y$, and max-prefix$(buxo, T^*) = d$. By the induction hypothesis,

$$(\langle p, S \rangle, w, S\#) \vdash^* (\langle h, \text{prefix}(f(uAo\#^n), n) \rangle, \varepsilon, uAo\#) \text{ in } M$$

where $w = $ max-prefix$(buAo, T^*)$. As $(h, A) \rightarrow (q, x) \in P$, step (2) of the construction introduces rule

$$|uA|\langle h, \text{prefix}(f(uAo\#^n), n) \rangle A \rightarrow \langle q, \text{prefix}(f(uxo\#^n), n) \rangle x \text{ to } R$$

By using this rule, ${}_nM$ simulates $(buAo, h)_n \Rightarrow (buxo, q)$ by making

$$(\langle h, \text{prefix}(f(uAo\#^n), n) \rangle, \varepsilon, uAo\#) \vdash (\langle q, z \rangle, \varepsilon, uxo\#)$$

where $z = \text{prefix}(f(uxo\#^n), n)$ if $x \in V^+ - T^+$ and $z = \text{prefix}(f(uxo\#^n), n-1) = \text{prefix}(f(uo\#^n), n-1)$ if $x \in T^+$. In the latter case $(z = \text{prefix}(f(uo\#^n), n-1)$, so $|z| = n-1)$, ${}_nM$ makes

$$(\langle q, \text{prefix}(f(uo\#^n), n-1) \rangle, \varepsilon, uxo\#) \vdash (\langle q, \text{prefix}(f(uo\#^n), n) \rangle, \varepsilon, uxo\#)$$

by a rule introduced in (3). If $uxo \in (NT^*)^*$, $uxo = y$ and the induction step is completed. Therefore, assume that $uxo \neq y$, so $uxo = ty$ and $d = wt$, for some $t \in T^+$. Observe that $\text{prefix}(f(uxo\#^n), n) = \text{prefix}(f(y\#^n), n)$ at this point. Then, ${}_nM$ removes t by making $|t|$ popping moves so that

$$(\langle q, \text{prefix}(f(uxo\#^n), n) \rangle, t, ty\#)_p \vdash^{|t|} (\langle q, \text{prefix}(f(y\#^n), n) \rangle, \varepsilon, y\#)$$

Thus, putting the previous sequences of moves together, we obtain

$$(p, wt, S\#^n) \vdash^* (\langle q, \text{prefix}(f(uxo\#^n), n) \rangle, t, ty\#) \ [1sS \rightarrow qAA]$$
$$_p \vdash^{|t|} (\langle q, \text{prefix}(f(y\#^n), n) \rangle, \varepsilon, y\#)$$

which completes the induction step.   □

By the previous claim for $y = \varepsilon$, if $(p, S)_n \Rightarrow^* (q, d)$ in G, where $d \in T^*, p, q \in W$, then

$$(\langle p, S \rangle, d, S\#) \vdash^* (\langle q, \mathrm{prefix}(f(\#^n), n) \rangle, \varepsilon, \#) \text{ in } {}_n M$$

As $\mathrm{prefix}(f(\#^n), n) = \#$ and R contains rules introduced in (1) and (4), we also have

$$\begin{aligned} (s, d, S\#) &\vdash & (\langle p, S \rangle, d, S\#) \\ &\vdash^* & (\langle q, \#^n, n \rangle, \varepsilon, \#) \\ &\vdash^* & (\$, \varepsilon, \#) \text{ in } {}_n M \end{aligned}$$

Thus, $d \in L(G)$ implies that $d \in L({}_n M)$, so $L(G, n) \subseteq L({}_n M)$.

Claim 9.2.3. Let $(\langle p, S\#^{n-1} \rangle, c, S\#) \vdash^m (\langle q, \mathrm{prefix}(f(y\#^n), n) \rangle, \varepsilon, by\#)$ in ${}_n M$ with $c, b \in T^*, y \in (NT^*)^*, p, q \in W$, and $m \geq 0$. Then, $(p, S)_n \Rightarrow^* (q, cby)$ in G.

Proof. This claim is proved by induction on $m \geq 0$.

Basis. Let $m = 0$. Then, $c = b = \varepsilon, y = S$, and

$$(\langle p, S\#^{n-1} \rangle, \varepsilon, S\#) \vdash^0 (\langle q, \mathrm{prefix}(f(S\#^n), n) \rangle, \varepsilon, S\#) \text{ in } {}_n M$$

As $(p, S)_n \Rightarrow^0 (p, S)$ in G, the basis holds.

Induction Hypothesis. Assume that the claim holds for all $m, 0 \leq m \leq k$, where k is a non-negative integer.

Induction Step. Let

$$(\langle p, S\#^{n-1} \rangle, c, S\#) \vdash^{k+1} (\langle q, \mathrm{prefix}(f(y\#^n), n) \rangle, \varepsilon, by\#) \text{ in } {}_n M$$

where $c, b \in T^*, y \in (NT^*)^*, p, q \in W$ in ${}_n M$. Since $k + 1 \geq 1$, we can express

$$(\langle p, S\#^{n-1} \rangle, c, S\#) \vdash^{k+1} (\langle q, \mathrm{prefix}(f(y\#^n), n) \rangle, \varepsilon, by\#)$$

as

$$\begin{aligned} (\langle p, S\#^{n-1} \rangle, c, S\#) &\vdash^k \alpha \\ &\vdash (\langle q, \mathrm{prefix}(f(y\#^n), n) \rangle, \varepsilon, by\#) \text{ in } {}_n M \end{aligned}$$

where α is a configuration of ${}_n M$ whose form depends on whether the last move is (i) a popping move or (ii) an expansion, described next.

(i) Assume that $\alpha_p \vdash (\langle q, \mathrm{prefix}(f(y\#^n), n) \rangle, \varepsilon, by\#)$ in ${}_n M$. In a greater detail, let $\alpha = (\langle q, \mathrm{prefix}(f(y\#^n), n) \rangle, a, aby\#)$ with $a \in T$ such that $c = \mathrm{prefix}(c, |c|-1)a$. Thus,

$$(\langle p, S\#^{n-1}\rangle, c, S\#) \vdash^k (\langle q, \text{prefix}(f(y\#^n), n)\rangle, a, aby\#)$$
$$_p\vdash (\langle q, \text{prefix}(f(y\#^n), n)\rangle, \varepsilon, by\#)$$

Since $(\langle p, S\#^{n-1}\rangle, c, S\#)\vdash^k(\langle q, \text{prefix}(f(y\#^n), n)\rangle, a, aby\#)$, we have

$$(\langle p, S\#^{n-1}\rangle, \text{prefix}(c, |c| - 1), S\#)\vdash^k(\langle q, \text{prefix}(f(y\#^n), n)\rangle, \varepsilon, aby\#)$$

By the induction hypothesis, $(p, S)_n\Rightarrow^*(q, \text{prefix}(c, |c| - 1)aby)$ in G. As $c = \text{prefix}(c, |c| - 1)a$, $(p, S)_n\Rightarrow^*(q, cby)$ in G.

(ii) Assume that $\alpha_e\vdash(\langle q, \text{prefix}(f(y\#^n), n)\rangle, \varepsilon, by\#)$ in $_nM$. Observe that this expansion cannot be made by rules introduced in steps (1) or (4). If this expansion is made by a rule introduced in (3), which does not change the pushdown contents at all, the induction step follows from the induction hypothesis. Finally, suppose that this expansion is made by a rule introduced in step (2). In a greater detail, suppose that

$$\alpha = (\langle o, \text{prefix}(f(uAv\#^n, n)\rangle, \varepsilon, uAv\#)$$

and $_nM$ makes

$$(\langle o, \text{prefix}(f(uAv\#^n), n)\rangle, \varepsilon, uAv\#)_e\vdash(\langle q, \text{prefix}(f(uxv\#^n), n)\rangle, \varepsilon, uxv\#)$$

by using

$$|f(uA)|\langle o, \text{prefix}(f(uAv\#^n), n)\rangle A \to \langle q, \text{prefix}(f(uxv\#^n), n)\rangle x \in R$$

introduced in step (2) of the construction, where $A \in N$, $u \in (NT^*)^*$, $v \in (N \cup T)^*$, $o \in W$, $|f(uA)| \leq n$, $by\# = uxv\#$. By the induction hypothesis,

$$(\langle p, S\#^{n-1}\rangle, c, S\#)\vdash^k(\langle o, \text{prefix}(f(uAv\#^n), n)\rangle, \varepsilon, uAv\#) \text{ in } _nM$$

implies that $(p, S)_n\Rightarrow^*(o, cuAv)$ in G. From

$$|f(uA)|\langle o, \text{prefix}(f(uAv\#^n), n)\rangle A \to \langle q, \text{prefix}(f(uxv\#^n), n)\rangle x \in R$$

it follows that $(o, A) \to (q, x) \in P$ and $A \notin_G \text{states}(f(u))$. Thus,

$$(p, S)\ _n\Rightarrow^* (o, cuAv)$$
$$_n\Rightarrow (q, cuxv) \text{ in } G$$

Therefore, $(p, S)_n\Rightarrow^*(q, cby)$ in G because $by\# = uxv\#$. $\square$

Consider the previous claim for $b = y = \varepsilon$ to see that

$$(\langle p, S\#^{n-1}\rangle, c, S\#)\vdash^*(\langle q, \text{prefix}(f(\#^n), n)\rangle, \varepsilon, \#^n) \text{ in } _nM$$

implies that $(p, S)_n \Rightarrow^* (q, c)$ in G. Let $c \in L(_nM)$. Then,

$$(s, c, S\#) \vdash^* (\$, \varepsilon, \#) \text{ in } _nM$$

Examine the construction of $_nM$ to see that $(s, c, S) \vdash^* (\$, \varepsilon, \#)$ starts by using a rule introduced in (1), so $(s, c, S) \vdash^* (\langle p, S\#^{n-1} \rangle, c, S\#)$. Furthermore, notice that this sequence of moves ends $(s, c, S) \vdash^* (\$, \varepsilon, \varepsilon)$ by using a rule introduced in step (4). Thus, we can express

$$(s, c, \#) \vdash^* (\$, \varepsilon, \#)$$

as

$$
\begin{aligned}
(s, c, \#) &\vdash^* (\langle p, S\#^{n-1} \rangle, c, S\#) \\
&\vdash^* (\langle q, \mathrm{prefix}(f(\#^n), n) \rangle, \varepsilon, \#) \\
&\vdash \ (\$, \varepsilon, \#) \text{ in } _nM
\end{aligned}
$$

Therefore, $c \in L(_nM)$ implies that $c \in L(G, n)$, so $L(_nM) \subseteq L(G, n)$.

As $L(_nM) \subseteq L(G, n)$ and $L(G, n) \subseteq L(_nM)$, $L(G, n) = L(_nM)$. Thus, Lemma 9.2.1 holds.   □

Lemma 9.2.4. *For every $n \geq 1$ and every deep pushdown automaton $_nM$, there exists a state grammar G such that $L(G, n) = L(_nM)$.*

Proof. Let $n \geq 1$ and $_nM = (Q, T, V, R, s, S, F)$ be a deep pushdown automaton. Let Z and $\$$ be two new symbols that occur in no component of $_nM$. Set $N = V - T$. Introduce sets

$$C = \{ \langle q, i, \rhd \rangle \mid q \in Q, 1 \leq i \leq n - 1 \}$$

and

$$D = \{ \langle q, i, \lhd \rangle \mid q \in Q, 0 \leq i \leq n - 1 \}$$

Moreover, introduce an alphabet W such that $\mathrm{card}(V) = \mathrm{card}(W)$, and for all i, $1 \leq i \leq n$, an alphabet U_i such that $\mathrm{card}(U_i) = \mathrm{card}(N)$. Without any loss of generality, assume that V, Q, and all these newly introduced sets and alphabets are pairwise disjoint. Set $U = \bigcup_{i=1}^n U_i$. For each i, $1 \leq i \leq n - 1$, set $C_i = \{ \langle q, i, \rhd \rangle \mid q \in Q \}$ and for each i, $0 \leq i \leq n - 1$, set $D_i = \{ \langle q, i, \lhd \rangle \mid q \in Q \}$. Introduce a bijection h from V to W. For each i, $1 \leq i \leq n$, introduce a bijection $_ig$ from N to U_i. Define the state grammar

$$G = \big(V \cup W \cup U \cup \{Z\}, Q \cup C \cup D \cup \{\$\}, T, P, Z \big)$$

where P is constructed by performing the following steps

(1) add $(s, Z) \rightarrow (\langle s, 1, \triangleright \rangle, h(S))$ to P;

(2) for each $q \in Q, A \in N, 1 \le i \le n-1, x \in V^+$, add

   (2.1) $(\langle q, i, \triangleright \rangle, A) \rightarrow (\langle q, i+1, \triangleright \rangle, {}_{i}g(A))$ and
   (2.2) $(\langle q, i, \triangleleft \rangle, {}_{i}g(A)) \rightarrow (\langle p, i-1, \triangleleft \rangle, A)$ to P;

(3) if $ipA \rightarrow qxY \in R$, for some $p, q \in Q, A \in N, x \in V^*, Y \in V, i = 1, \ldots, n$, add

   $(\langle p, i, \triangleright \rangle, A) \rightarrow (\langle q, i-1, \triangleleft \rangle, xY)$ and
   $(\langle p, i, \triangleright \rangle, h(A)) \rightarrow (\langle q, i-1, \triangleleft \rangle, xh(Y))$ to P;

(4) for each $q \in Q, A \in N$, add

   $(\langle q, 0, \triangleleft \rangle, A) \rightarrow (\langle q, 1, \triangleright \rangle, A)$ and
   $(\langle q, 0, \triangleleft \rangle, h(Y)) \rightarrow (\langle q, 1, \triangleright \rangle, h(Y))$ to P;

(5) for each $q \in F, a \in T$, add

   $(\langle q, 0, \triangleleft \rangle, h(a)) \rightarrow (\$, a)$ to P.

G simulates the application of $ipA \rightarrow qy \in R$ so it makes a left-to-right scan of the sentential form, counting the occurrences of nonterminals until it reaches the ith occurrence of a nonterminal. If this occurrence equals A, it replaces this A with y and returns to the beginning of the sentential form in order to analogously simulate a move from q. Throughout the simulation of moves of ${}_{n}M$ by G, the rightmost symbol of every sentential form is from W. G completes the simulation of an acceptance of a string x by ${}_{n}M$ so it uses a rule introduced in step (5) of the construction of P to change the rightmost symbol of x, $h(a)$, to a and, thereby, to generate x.

We next establish $L(G, n) = L({}_{n}M)$. To keep the rest of the proof as readable as possible, we omit some details in what follows. The reader can easily fill them in.

Claim 9.2.5. $L(G, n) \subseteq L({}_{n}M)$

Proof. Consider any $w \in L(G, n)$. Observe that G generates w as

$$
\begin{aligned}
(s, Z) \ {}_{n}\!\Rightarrow \ & (\langle s, 1, \triangleright \rangle, h(S)) \ [(s, Z) \rightarrow (\langle s, 1, \triangleright \rangle, h(S))] \\
{}_{n}\!\Rightarrow^* \ & (f, yh(a)) \qquad [(\langle f, 0, \triangleleft \rangle, h(a)) \rightarrow (\$, a)] \\
{}_{n}\!\Rightarrow \ & (\$, w)
\end{aligned}
$$

where $f \in F, a \in T, y \in T^*, ya = w, (s, Z) \rightarrow (\langle s, 1, \triangleright \rangle, h(S))$ in step (1) of the construction of P, $(\langle f, 0, \triangleleft \rangle, h(a)) \rightarrow (\$, a)$ in (5), every

$$
u \in \text{strings}\left(((\langle s, 1, \triangleright \rangle, h(S))_{n}\!\Rightarrow^* (f, yh(a))) \right)
$$

satisfies $u \in (V \cup U)^* W$, and every step in

$$
(\langle s, 1, \triangleright \rangle, h(S))_{n}\!\Rightarrow^* (f, yh(a))
$$

is made by a rule introduced in (2) through (4). Indeed, the rule constructed in (1) is always used in the first step and a rule constructed in (5) is always used in the very last step of any successful generation in G; during any other step, neither of them can be applied. Notice that the rule of (1) generates $h(S)$. Furthermore, examine the rules of (2) through (4) to see that by their use, G always produces a string that has exactly one occurrence of a symbol from W in any string and this occurrence appears as the rightmost symbol of the string; formally,

$$u \in \text{strings} \left((\langle s, 1, \rhd \rangle, h(S))_n \Rightarrow^* (f, yh(a)) \right)$$

implies that $u \in (V \cup U)^* W$. In a greater detail,

$$(\langle s, 1, \rhd \rangle, h(S))_n \Rightarrow^* (f, yh(a))$$

can be expressed as

$$
\begin{aligned}
(q_0, z_0) \quad &{}_n\Rightarrow^* (c_0, y_0) \ {}_n\Rightarrow (d_0, u_0) \ {}_n\Rightarrow^* (p_0, v_0) \ {}_n\Rightarrow \\
(q_1, z_1) \quad &{}_n\Rightarrow^* (c_1, y_1) \ {}_n\Rightarrow (d_1, u_1) \ {}_n\Rightarrow^* (p_1, v_1) \ {}_n\Rightarrow \\
&\quad\vdots \qquad\qquad \vdots \qquad\qquad \vdots \qquad\qquad \vdots \\
(q_m, z_m) \quad &{}_n\Rightarrow^* (c_m, y_m) \ {}_n\Rightarrow (d_m, u_m) \ {}_n\Rightarrow^* (p_m, v_m) \ {}_n\Rightarrow \\
(q_{m+1}, z_{m+1}) &
\end{aligned}
$$

for some $m \geq 1$, where $z_0 = h(S)$, $z_{m+1} = yh(a)$, $f = q_{m+1}$, and for each j, $0 \leq j \leq m$, $q_j \in C_1$, $p_j \in D_0$, $z_j \in V^* W$, and there exists $i_j \in \{1, \ldots, n\}$ so $c_j \in C_{i_j}$, $y_j \in T^* C_1 T^* C_2 \cdots T^* C_{i_j-1} V^* W$, $d_j \in D_{i_j-1}$, $u_j \in T^* C_1 T^* C_2 \cdots T^* D_{i_j-1} V^* W$, and

$$(q_j, z_j)_n \Rightarrow^* (c_j, y_j)_n \Rightarrow (d_j, u_j)_n \Rightarrow^* (p_j, v_j)_n \Rightarrow (q_{j+1}, z_{j+1})$$

satisfies (i)–(iv), given next.

For brevity, we first introduce the following notation. Let w be any string. For $i = 1, \ldots, |w|$, $\lfloor w, i, N \rfloor$ denotes the ith occurrence of a nonterminal from N in w, and if such a nonterminal does not exist, $\lfloor w, i, N \rfloor = 0$; for instance, $\lfloor ABABC, 2, \{A, C\} \rfloor$ denotes the underlined $\underline{A}$ in $AB\underline{A}BC$.

(i) $(q_j, z_j)_n \Rightarrow^* (c_i, y_i)$ consists of $i_j - 1$ steps during which G changes $\lfloor z_j, 1, N \rfloor$, $\ldots$, $\lfloor z_j, i_j - 1, N \rfloor$ to ${}_1 g(\langle \lfloor z_j, 1, N \rfloor, 2 \rangle)$, $\ldots$, ${}_{i_j} g(\langle \lfloor z_j, i_j - 1, N \rfloor, i_j - 1 \rangle)$, respectively, by using rules of (2.1) in the construction of P;

(ii) if $i_j \leq \#_N(z_j)$, then $(c_j, y_j)_n \Rightarrow (d_j, u_j)$ have to consist of a step according to $(\langle q, i, \rhd \rangle, A_j) \rightarrow (\langle q, i - 1, \lhd \rangle, x_j X_j)$, where $\lfloor z_j, i_j, N \rfloor$ is an occurrence of A_j, $x_j \in V^*$, $X_j \in V$, and if $i_j = \#_{N \cup W}(z_j)$, then $(c_j, y_j)_n \Rightarrow (d_j, u_j)$ consists of a step according to $(\langle p, i, \rhd \rangle, h(A_j)) \rightarrow (\langle q, i - 1, \lhd \rangle, x_j h(X_j))$ constructed in (3), where $\lfloor z_j, i_j, N \cup W \rfloor$ is an occurrence of $h(A_j)$, $x_j \in V^*$, $X_j \in V$;

(iii) $(d_j, u_j)_n \Rightarrow^* (p_j, v_j)$ consists of $i_j - 1$ steps during which G changes ${}_{i_j} g(\langle \lfloor z_j, i_j - 1, N \rfloor, i_j - 1 \rangle)$, $\ldots$, ${}_1 g(\langle \lfloor z_j, 1, N \rfloor, 1 \rangle)$ back to $\lfloor z_j, i_j - 1, N \rfloor$, $\ldots$, $\lfloor z_j, 1, N \rfloor$, respectively, in a right-to-left way by using rules constructed in (2.2);

(iv) $(p_j, v_j)_n \Rightarrow (q_{j+1}, z_{j+1})$ is made by a rule constructed in (4).

For every

$$(q_j, z_j) \,_n\!\Rightarrow^* (c_j, y_j)$$
$$_n\!\Rightarrow (d_j, u_j)$$
$$_n\!\Rightarrow^* (p_j, v_j)$$
$$_n\!\Rightarrow (q_{j+1}, z_{j+1}) \text{ in } G$$

where $0 \le j \le m$, $_nM$ makes

$$(q_j, o_j, \text{suffix}(z_j)t_j) \vdash^* (q_{j+1}, o_{j+1}, \text{suffix}(z_{j+1})t_{j+1})$$

with $o_0 = w$, $z_0 = S\#$, $t_{j+1} = |\max\text{-prefix}(z_{j+1}, T^*)|$, $o_{j+1} = \text{suffix}(o_j)|o_j| + t_{j+1}$, where $o_0 = w$, $z_0 = S\#$, and $t_0 = |z_0|$. In this sequence of moves, the first move is an expansion made according to $_{i_j}q_jA_j \to q_{j+1}x_jX_j$ (see steps (2) and (3) of the construction) followed by t_{j+1} popping moves (notice that $i_j \ge 2$ implies that $t_{j+1} = 0$). As $f \in F$ and $ya = w$, $w \in L(_nM)$. Therefore, $L(G, n) \subseteq L(_nM)$. □

Claim 9.2.6. $L(_nM) \subseteq L(G, n)$

Proof. This proof is simple and left to the reader. □

As $L(_nM) \subseteq L(G, n)$ and $L(G, n) \subseteq L(_nM)$, we have $L(G, n) = L(_nM)$, so this lemma holds true. □

Theorem 9.2.7. *For every $n \ge 1$ and for every language L, $L = L(G, n)$ for a state grammar G if and only if $L = L(_nM)$ for a deep pushdown automaton $_nM$.*

Proof. This theorem follows from Lemmas 9.2.1 and 9.2.4. □

By analogy with the demonstration of Theorem 9.2.7, we can establish the next theorem.

Theorem 9.2.8. *For every $n \ge 1$ and for every language L, $L = L(G, n)$ for a state grammar G if and only if $L = E(_nM)$ for a deep pushdown automaton $_nM$.* □

The main result of this chapter follows next.

Corollary 9.2.9. *For every $n \ge 1$,*

$$_{deep}^{empty}\mathbf{PDA}_n = {}_{deep}\mathbf{PDA}_n \subset {}_{deep}\mathbf{PDA}_{n+1} = {}_{deep}^{empty}\mathbf{PDA}_{n+1}$$

Proof. This corollary follows from Theorems 9.2.7 and 9.2.8 above and from Theorem 3.2.3, which says that the m-limited state grammars generate a proper subfamily of the family generated by $(m + 1)$-limited state grammars, for every $m \ge 1$. □

Finally, we state two results concerning **CF** and **CS**.

Corollary 9.2.10. $_{deep}\mathbf{PDA}_1 = {}_{deep}^{empty}\mathbf{PDA}_1 = \mathbf{CF}$

Proof. This corollary follows from Lemmas 9.2.1 and 9.2.4 for $n = 1$, and from Theorem 3.2.3, which says that one-limited state grammars characterize **CF**. □

Corollary 9.2.11. *For every* $n \geq 1$, $_{deep}\mathbf{PDA}_n = {}_{deep}^{empty}\mathbf{PDA}_n \subset \mathbf{CS}$.

Proof. This corollary follows from Lemmas 9.2.1 and 9.2.4, Theorems 9.2.7 and 9.2.8, and from Theorem 3.2.3, which says that $\mathbf{ST}_m$, for every $m \geq 1$, is properly included in **CS**. □

9.3 Open Problems

Finally, we suggest two open problem areas concerning deep pushdown automata.

9.3.1 Determinism

This chapter has discussed a general versions of deep pushdown automata, which work non-deterministically. Undoubtedly, the future investigation of these automata should pay a special attention to their deterministic versions, which fulfill a crucial role in practice. In fact, we can introduce a variety of deterministic versions, including the following two types. First, we consider the fundamental strict form of determinism.

Definition 9.3.1. Let $M = (Q, \Sigma, \Gamma, R, s, S, F)$ be a deep pushdown automaton. We say that M is *deterministic* if for every $mqA \to pv \in R$,

$$\mathrm{card}\left(\{mqA \to ow \mid mqA \to ow \in R, o \in Q, w \in \Gamma^+\} - \{mqA \to pv\}\right) = 0 \quad \square$$

As a weaker form of determinism, we obtain the following definition.

Definition 9.3.2. Let $M = (Q, \Sigma, \Gamma, R, s, S, F)$ be a deep pushdown automaton. We say that M is *deterministic with respect to the depth of its expansions* if for every $q \in Q$

$$\mathrm{card}\left(\{m \mid mqA \to pv \in R, A \in \Gamma, p \in Q, v \in \Gamma^+\}\right) \leq 1$$

because at this point from the same state, all expansions that M can make are of the same depth. □

To illustrate, consider, for instance, the deep pushdown automaton $_2M$ from Example 9.1.2. This automaton is deterministic with respect to the depth of its expansions; however, it does not satisfy the strict determinism. Notice that $_nM$ constructed in the proof of Lemma 9.2.1 is deterministic with respect to the depth of its expansions, so we obtain this corollary.

Corollary 9.3.3. *For every state grammar G and for every n ≥ 1, there exists a deep pushdown automaton $_nM$ such that $L(G, n) = L(_nM)$ and $_nM$ is deterministic with respect to the depth of its expansions.*   □

Open Problem 9.3.4. Can an analogical statement to Corollary 9.3.3 be established in terms of the strict determinism?   □

9.3.2   Generalization

Let us note that throughout this chapter, we have considered only true pushdown expansions in the sense that the pushdown symbol is replaced with a nonempty string rather than with the empty string; at this point, no pushdown expansion can result in shortening the pushdown length. Nevertheless, the discussion of moves that allow deep pushdown automata to replace a pushdown symbol with ε and, thereby, shorten its pushdown represent a natural generalization of deep pushdown automata discussed in this chapter.

Open Problem 9.3.5. What is the language family defined by deep pushdown automata generalized in this way?   □

Chapter 10
Algebra, Automata, and Computation

Traditionally, from an algebraic viewpoint, automata work over free monoids. The present chapter, however, modifies this standard approach so they work over other algebraic structures. More specifically, this chapter discusses a modification of pushdown automata that is based on two-sided pushdowns into which symbols are pushed from both ends. These pushdowns are defined over free groups, not free monoids, and they can be shortened only by the standard group reduction. We demonstrate that these automata are computational complete—that is, they characterize the family of recursively enumerable languages—even if the free groups are generated by no more than four symbols.

10.1   Two-Sided Pushdown Acceptance over Free Groups: Conceptualization

Undoubtedly, the pushdown automata fulfill a crucial role in the automata theory. Viewed as a language acceptor, pushdown automaton consists of an input tape, a read head, a pushdown and a finite state control. The input tape is divided into squares, each of which contains one symbol of an input string. The finite control is represented by a finite set of states together with a finite set of computational rules. According to these rules, pushdown automaton changes states, moves the read head on the tape to the right and replaces the top symbol on the pushdown by an arbitrary sequence of another symbols contained in the pushdown alphabet. Every next computational step is performed with respect to the current symbol under the read head on the input tape, the current state and the current symbol on the top of the pushdown. The input string is accepted by the pushdown automaton, if its last symbol is read from the input tape and (1) a state marked as final is reached, or, (2) the pushdown is empty, or, (3) a state marked as final is reached and the pushdown is empty. Note that the acceptance method is defined for every pushdown

© Springer International Publishing AG 2017   407
A. Meduna, O. Soukup, *Modern Language Models and Computation*,
DOI 10.1007/978-3-319-63100-4_10

automaton and these three methods are equivalent. In other words, every pushdown automaton with one of the three acceptance method can be transformed to another two pushdown automata which accept every string with the other two acceptance methods and define the same language as the original automaton.

The automata theory has modified pushdown automata in many different ways including various modifications concerning their pushdown stores. To give an example, recall the pushdown turn (see Definition 7.2.30). Let us introduce another modification of pushdown automata. By attaching an additional pushdown, the pushdown automaton is extended to the two-pushdown automaton. A two-pushdown automaton consists of a finite state control, an input tape with its read head, and two pushdowns. During a move, two-pushdown automaton rewrites the top symbols of both pushdowns; otherwise, it works by analogy with a pushdown automaton. It is proved that two-pushdown automata are more powerful than pushdown automata (see [Med00a]).

By putting together the previously mentioned variations of pushdown automata (i.e. one-turn pushdown automata and two-pushdown automata), the simultaneously one-turn two pushdown automata can be introduced (see [Med03c]). If the two-pushdown automaton makes a turn in both its pushdowns in one computational step, this turn is *simultaneous*. A two-pushdown automaton is simultaneously one-turn if it makes either no turn or one simultaneous turn in its pushdowns during any computation. As expected, this modification changes the power of pushdown automata.

There are another modifications of pushdown automata in the automata theory. Some of them can be found in [Cou77, Gre69, GGH67, GS68, Med03c, KM00, Sar01].

The present chapter continues with this vivid topic and introduces automata with two-sided pushdowns. As their name indicates, we can insert symbol into these pushdowns from both ends. The two-sided pushdown automaton thus consists of an input tape with read head, a finite state control, and a two-sided pushdown. Every computational step is performed according to the current symbol under the read head on the input tape, the current state and the current symbols on both tops of the two-sided pushdown.

Pushdowns are usually defined over free monoids generated by the pushdown alphabets of the pushdown automata under the operation of concatenation. However, in this chapter, we leave this concept and define these two-sided pushdowns over free groups rather than free monoids. To put it more precisely, we require that during every move, the string representing the current pushdown is over the free group generated by the pushdown alphabet under the operation of concatenation, and the standard group reduction is the only way by which the pushdown string can be shorten. We demonstrate that the pushdown automata modified in this way are as powerful as the Turing machines. In fact, they characterize the family of recursively enumerable languages. Moreover, the free groups are generated by no more than four symbols.

10.1.1 Definitions

Let us recall the notion of languages over free monoids (see Sect. 2.1) and free groups (see Sect. 6.1) and the notion of left-extended queue grammars (see Definition 2.3.44).

Definition 10.1.1. A *string-reading two-sided pushdown automaton over a free group* is an eight-tuple, $M = (Q, \Sigma, \Gamma, R, z, Z_1, Z_2, F)$, where Q is a finite set of states, Σ is an input alphabet, Γ is a pushdown alphabet, $Q \cap (\Sigma \cup \Gamma) = \emptyset$, R is a finite set of rules of the form $u_1|u_2qw \to v_1|v_2p$ with $u_1, u_2 \in \Gamma$, $v_1, v_2 \in \Gamma^\circ$, $p, q \in Q$, and $w \in \Sigma^*$, $z \in Q$ is the start state, $Z_1 \in \Gamma$ is the start symbol of the left-hand side of the pushdown, $Z_2 \in \Gamma$ is the start symbol of the right-hand side of the pushdown, and $F \subseteq Q$ is a set of final states. A configuration of M is any string of the form vqy, where $v \in \Gamma^\circ$, $y \in \Sigma^*$, and $q \in Q$. If $u_1|u_2qw \to v_1|v_2p \in R$, $y = u_1hu_2qwz$, and $x = v_1hv_2pz$, where $u_1, u_2 \in \Gamma$, $h, v_1, v_2 \in \Gamma^\circ$, $q, p \in Q$, and $w, z \in \Sigma^*$, then M makes a move from y to x in M, symbolically written as $y \vdash x[u_1|u_2qw \to v_1|v_2q]$, or, simply, $y \vdash x$. In the standard manner, extend $\vdash$ to $\vdash^n$, where $n \geq 0$; based on $\vdash^n$, define $\vdash^+$ and $\vdash^*$. We call $Z_1Z_2zw \vdash^* vqx$ a computation, where $v \in \Gamma^\circ$, $q \in Q$, $w, x \in \Sigma^*$; a computation of the form $Z_1Z_2zw \vdash^* \varepsilon f$ with $f \in F$ is a successful computation. The language of M, $L(M)$, is defined as $L(M) = \{w : Z_1Z_2zw \vdash^* \varepsilon f, \text{where } f \in F, w \in \Sigma^*\}$.

A *two-sided pushdown automaton over a free group* is a string-reading two-sided pushdown automaton over a free group, $M = (Q, \Sigma, \Gamma, R, z, Z_1, Z_2, F)$, in which every $u_1|u_2qw \to v_1|v_2p \in R$ satisfies $0 \leq |w| \leq 1$, where $u_1, u_2 \in \Gamma$, $v_1v_2 \in \Gamma^\circ$, $q, p \in Q$, and $w \in \Sigma^*$. □

10.2 Results: Computational Completeness

In this section, we study the power of the modified versions of pushdown automata defined in the previous section. We demonstrate that they are as powerful as the Turing machines. In fact, they characterize the family of recursively enumerable languages even if the free groups over which their pushdowns are defined are generated by no more than four symbols.

For the proof of the main result in this chapter presented later, we use a left-extended queue grammar, since the derivation method of these grammars is closer to the behaviour of our two-sided pushdowns. Recall that left-extended queue grammars characterize the family of recursively enumerable languages (see Theorem 2.3.46). Observe that according to the definition of left-extended queue grammars, every symbol $A \in (V - T)$ which appears in the first component of any rule used in the derivation will be moved from the place right from # to the place left from #. Every string generated right from # in the queue grammar will be inserted into the two-sided pushdown from the right-hand side. Moreover, every symbol moved to the place left from # in the queue grammar is then inserted as inverse from the left-hand side on the two-sided pushdown. At the end of every successful computation, the left half of the two-sided pushdown is equal to the inverted right half, so it can be discharged by the group reduction.

Theorem 10.2.1. *For every left-extended queue grammar,* $G = (V, T, W, F, Sq_0, P)$, *satisfying the properties of normal form described in Definition 3.1.25, there exists a string-reading two-sided pushdown automaton over a free group with the reduced pushdown alphabet,* $M = (Q, T, Z, R, z, 1, 1, F_M)$, *such that* $L(G) = L(M)$.

Proof. We construct a string-reading two-sided pushdown automaton over a free group with the reduced pushdown alphabet as follows.

Construction. Define the injections, $h : (V - T) \to \{0, 1\}^{n+2}$ and $\overline{h} : (V - T) \to \{\overline{0}, \overline{1}\}^{n+2}$, where $n = \lceil \log_2(card(V - T)) \rceil$, such that for every $A \in (V - T)$, $h(A) = \{0\}\{0, 1\}^n\{0\}$ and $\overline{h}(A) = \overline{h(A)}$. Extend the domain of h to $(V - T)^*$. After this extension, h is now an injective homomorphism from $(V - T)^*$ to $(\{0\}\{0, 1\}^n\{0\})^*$. Note that the inverses to 0 and $1 \in V - T$ are $\overline{0}$ and $\overline{1} \in V - T$, respectively.

Construct the set of states, Q, the pushdown alphabet, Z, and the set of final states, F_M, as $Q = \{f, z\} \cup \{\langle q, 1 \rangle, \langle q, 2 \rangle | q \in W\}$, $Z = \{0, \overline{0}, 1, \overline{1}\}$, and $F_M = \{f\}$, respectively.

The set of rules, R, is constructed in the following way.

(1) for the start axiom of G, Sq_0, where $S \in (V - T)$, $q_0 \in (W - F)$,
     add $1|1z \to 1|h(S)1\langle q_0, 1 \rangle$ to R
(2) for every $(A, q, x, p) \in P$, where $A \in (V - T)$, $p, q \in (W - F)$, $x \in (V - T)^*$,
     add $1|1\langle q, 1 \rangle \to 1\overline{h}(A)|h(x)1\langle p, 1 \rangle$ to R
(3) for every $q \in W$
     add $1|1\langle q, 1 \rangle \to 1|1\langle q, 2 \rangle$ to R
(4) for every $(A, q, y, p) \in P$, where $A \in (V - T)$, $p, q \in (W - F)$, $y \in T^*$,
     add $1|1\langle q, 2 \rangle y \to 1\overline{h}(A)|1\langle p, 2 \rangle$ to R
(5) for every $(A, q, y, t) \in P$, where $A \in (V - T)$, $q \in (W - F)$, $y \in T^*$, $t \in F$,
     add $1|1\langle q, 2 \rangle y \to \overline{h}(A)|\varepsilon f$ to R

The construction of M is completed. For the next parts of this proof, we introduce the following notation. If $\langle q, 1 \rangle$ is the actual state of M, we say that M is in *nonterminal-generating* mode. Similarly, if $\langle q, 2 \rangle$ is the actual state of M, we say that M is in *terminal-reading* mode, where $q \in W$.

Basic Idea. M simulates derivations in the left-extended queue grammar, G, and encodes the symbols from $V - T$ on its pushdown in a binary way. First, consider that in G, $w\#Avp$ is the actual sentential form, where $w, v \in (V - T)^*$, $A \in (V - T)$, and $p \in (W - F)$. Then, the corresponding configuration of M is $1\overline{h}(w)h(w)h(A)h(v)1\langle p, 1 \rangle \omega$, where $\omega \in T^*$. Let $(A, p, x, q) \in P$, where $x \in (V - T)^*$ Then, $w\#Avp \Rightarrow wA\#vxq$ in G. In this case, M must be in the nonterminal-generating mode and the corresponding M's rule is by construction $1|1\langle p, 1 \rangle \to 1\overline{h}(A)|h(x)1\langle q, 1 \rangle \in R$. By using this rule, M moves to a new configuration of the form $1\overline{h}(A)\overline{h}(w)h(w)h(A)h(v)h(x)1\langle q, 1 \rangle \omega$. Observe that A is encoded by $\overline{h}$ and the resulting binary string is inserted to the left-hand side of pushdown. Next, x is encoded by h and the result is inserted into the pushdown from the right-hand side.

Second, let $w\#Avup$ is the actual sentential form, where $u \in T^*$, and $(A, p, y, q) \in P$, $y \in T^*$. Then, $w\#Avup \Rightarrow wA\#vuyq$ in G. By construction, the corresponding M's rule is $1|1\langle p, 2\rangle y \rightarrow 1\overline{h}(A)|1\langle q, 2\rangle \in R$ and M makes a transition $1\overline{h}(w)h(w)h(A)h(v)1\langle p, 2\rangle y\omega' \vdash 1\overline{h}(A)\overline{h}(w)h(w)h(A)h(v)1\langle q, 2\rangle \omega'$, where $\omega' \in T^*$. Note that in this case, M must be in terminal-reading mode. In this mode, only the encoded A, $\overline{h}(A)$, is inserted into the left-hand side of the pushdown.

In other words, every $A \in (V - T)$, that is generated behind the # in G, is inserted as $h(A)$ into the right-hand side of the pushdown. Note that all these symbols in the left-extended queue grammar G satisfying the normal form from Definition 3.1.25 are moved in front of # during every successful derivation. The reason why M inserts their encoded inverses into the left-hand side of the pushdown is to correctly simulate the derivation in G. To make the pushdown empty, M uses the inverses from the free group.

Let us present an example of automaton construction to clarify the proof.

Example 10.2.2. Consider a left-extended queue grammar, $G = (V, T, W, F, s, P)$, where $V = \{S, A, B, a, b\}$, $T = \{a, b\}$, $W = \{Q, f\}$, $F = \{f\}$, $s = SQ$ and $P = \{p_1, p_2, p_3\}$, $p_1 = (S, Q, AB, Q)$, $p_2 = (A, Q, aa, Q)$ and $p_3 = (B, Q, bb, f)$. G generates sentence $aabb$ by the following derivation.

$$\#s = \#SQ \Rightarrow S\#ABQ[p_1] \Rightarrow SA\#BaaQ[p_2] \Rightarrow SAB\#aabbf[p_3]$$

We construct a string-reading two-sided pushdown automaton over a free group with the reduced pushdown alphabet, $M = (Q, T, Z, R, z, 1, 1, F_M)$, as follows:

- $Q = \{z, f, \langle Q, 1\rangle \langle Q, 2\rangle\}$,
- $T = \{a, b\}$,
- $Z = \{0, \overline{0}, 1, \overline{1}\}$,
- $R = \{$

p_1:	$1|1z \rightarrow 1|h(S)1\langle Q, 1\rangle$	for the start axiom SQ,
p_2:	$1|1\langle Q, 1\rangle \rightarrow 1\overline{h}(S)|h(AB)1\langle Q, 1\rangle$	for $(S, Q, AB, Q) \in P$,
p_3:	$1|1\langle Q, 1\rangle \rightarrow 1|1\langle Q, 2\rangle$	for $Q \in W$,
p_4:	$1|1\langle Q, 2\rangle aa \rightarrow 1\overline{h}(A)|1\langle Q, 2\rangle$	for $(A, Q, aa, Q) \in P$,
p_5:	$1|1\langle Q, 2\rangle bb \rightarrow \overline{h}(B)|\varepsilon f$	for $(B, Q, bb, f) \in P\}$,

- $F_M = \{f\}$

Encoding h of symbols from $(V - T)$:

$$h(S) = 00010 \quad \overline{h}(S) = \overline{01000}$$
$$h(A) = 00110 \quad \overline{h}(A) = \overline{01100}$$
$$h(B) = 01000 \quad \overline{h}(B) = \overline{00010}$$

Now, observe the acceptance progress of the string *aabb* by M.

two-sided pushdown	state	input	rule
11	z	*aabb*	
$1h(S)1$	$\langle Q, 1 \rangle$	*aabb*	p_1
$1\bar{h}(S)h(S)h(AB)1$	$\langle Q, 1 \rangle$	*aabb*	p_2
$1h(A)h(B)1$	$\langle Q, 1 \rangle$	*aabb*	free group reduction
$1h(A)h(B)1$	$\langle Q, 2 \rangle$	*aabb*	p_3
$1\bar{h}(A)h(A)h(B)1$	$\langle Q, 2 \rangle$	*bb*	p_4
$1h(B)1$	$\langle Q, 2 \rangle$	*bb*	free group reduction
$\bar{h}(B)h(B)$	f	ε	p_5
ε	f	ε	free group reduction

Since the two-sided pushdown is empty and the current state f is the final state, *aabb* is accepted by M.   □

Next, we prove $L(G) = L(M)$, since $L(G) \subseteq L(M)$ and $L(M) \subseteq L(G)$. First, we demonstrate Claims 10.2.3, 10.2.4 and 10.2.5 to prove $L(G) \subseteq L(M)$.

Claim 10.2.3. If

$$A_1 \ldots A_n \# B_1 \ldots B_m u \Rightarrow^i A_1 \ldots A_n B_1 \ldots B_i \# B_{i+1} \ldots B_m x_1 \ldots x_i p$$

in G, then

$$1\bar{h}(A_n) \ldots \bar{h}(A_1)h(A_1) \ldots h(A_n)h(B_1) \ldots h(B_m)1\langle u, 1 \rangle \omega \vdash^i$$

$$1\bar{h}(B_i) \ldots \bar{h}(B_1)\bar{h}(A_n) \ldots \bar{h}(A_1)h(A_1) \ldots h(A_n)h(B_1) \ldots$$

$$\ldots h(B_m)h(x_1) \ldots h(x_i)1\langle p, 1 \rangle \omega$$

in M, where $A_1, \ldots, A_n, B_1, \ldots, B_m \in (V - T)$, $x_1, \ldots, x_i \in (V - T)^*$, $u, p \in (W - F)$, $n \geq 0$, $\omega \in T^*$, $0 \leq i \leq m$.

Proof. *Basis.* Let $i = 0$. Then,

$$A_1 \ldots A_n \# B_1 \ldots B_m u \Rightarrow^0 A_1 \ldots A_n \# B_1 \ldots B_m u$$

in G. Clearly,

$$1\bar{h}(A_n) \ldots \bar{h}(A_1)h(A_1) \ldots h(A_n)h(B_1) \ldots h(B_m)1\langle u, 1 \rangle \omega \vdash^0$$

$$1\bar{h}(A_n) \ldots \bar{h}(A_1)h(A_1) \ldots h(A_n)h(B_1) \ldots h(B_m)1\langle u, 1 \rangle \omega$$

in M.

Induction Hypothesis. Assume that Claim 10.2.3 holds for every $i \leq l$, where l is a positive integer.

Induction Step. Consider any derivation of the form

$$A_1 \ldots A_n \# B_1 \ldots B_m u \Rightarrow^{l+1}$$

$$A_1 \ldots A_n B_1 \ldots B_l B_{l+1} \# B_{l+2} \ldots B_m x_1 \ldots x_l x_{l+1} q$$

Express this derivation as

$$A_1 \ldots A_n \# B_1 \ldots B_m u \Rightarrow^l$$

$$A_1 \ldots A_n B_1 \ldots B_l \# B_{l+1} \ldots B_m x_1 \ldots x_l p \Rightarrow$$

$$A_1 \ldots A_n B_1 \ldots B_l B_{l+1} \# B_{l+2} \ldots B_m x_1 \ldots x_l x_{l+1} q$$

in G, where $0 \leq l \leq m$, $q \in (W - F)$. By the induction hypothesis,

$$1\overline{h}(A_n) \ldots \overline{h}(A_1) h(A_1) \ldots h(A_n) h(B_1) \ldots h(B_m) 1 \langle u, 1 \rangle \omega \vdash^l$$

$$1\overline{h}(B_l) \ldots \overline{h}(B_1) \overline{h}(A_n) \ldots \overline{h}(A_1) h(A_1) \ldots$$

$$\ldots h(A_n) h(B_1) \ldots h(B_m) h(x_1) \ldots h(x_l) 1 \langle p, 1 \rangle \omega \vdash$$

$$1\overline{h}(B_{l+1}) \overline{h}(B_l) \ldots \overline{h}(B_1) \overline{h}(A_n) \ldots \overline{h}(A_1) h(A_1) \ldots$$

$$\ldots h(A_n) h(B_1) \ldots h(B_m) h(x_1) \ldots h(x_l) h(x_{l+1}) 1 \langle q, 1 \rangle \omega$$

in M. There is only one type of rules in P able to perform the derivation

$$A_1 \ldots A_n B_1 \ldots B_l \# B_{l+1} \ldots B_m x_1 \ldots x_l p \Rightarrow$$

$$A_1 \ldots A_n B_1 \ldots B_l B_{l+1} \# B_{l+2} \ldots B_m x_1 \ldots x_l x_{l+1} q$$

in G, namely rules of the form $(B_{l+1}, p, x_{l+1}, q) \in P$, where $B_{l+1} \in (V - T)$, $p, q \in (W - F)$ and $x_{l+1} \in (V - T)^*$. Observe that by step (2) in construction, there is a rule $1|1\langle p, 1 \rangle \rightarrow 1\overline{h}(B_{l+1})|h(x_{l+1}) 1 \langle q, 1 \rangle$ in R, so

$$1\overline{h}(B_l) \ldots \overline{h}(B_1) \overline{h}(A_n) \ldots \overline{h}(A_1) h(A_1) \ldots$$

$$\ldots h(A_n) h(B_1) \ldots h(B_m) h(x_1) \ldots h(x_l) 1 \langle p, 1 \rangle \omega \vdash$$

$$1\overline{h}(B_{l+1}) \overline{h}(B_l) \ldots \overline{h}(B_1) \overline{h}(A_n) \ldots \overline{h}(A_1) h(A_1) \ldots$$

$$\ldots h(A_n) h(B_1) \ldots h(B_m) h(x_1) \ldots h(x_l) h(x_{l+1}) 1 \langle q, 1 \rangle \omega$$

in M and Claim 10.2.3 holds.   □

Claim 10.2.4. If

$$A_1 \ldots A_n \# B_1 \ldots B_m a_1 \ldots a_k u \Rightarrow^i$$

$$A_1 \ldots A_n B_1 \ldots B_i \# B_{i+1} \ldots B_m a_1 \ldots a_k b_1 \ldots b_i p$$

in G, then

$$1\overline{h}(A_n)\ldots\overline{h}(A_1)h(A_1)\ldots h(A_n)h(B_1)\ldots h(B_m)1\langle u,2\rangle b_1\ldots b_j \vdash^i$$

$$1\overline{h}(B_i)\ldots\overline{h}(B_1)\overline{h}(A_n)\ldots\overline{h}(A_1)h(A_1)\ldots$$

$$\ldots h(A_n)h(B_1)\ldots h(B_i)h(B_{i+1})\ldots h(B_m)1\langle p,2\rangle b_{i+1}\ldots b_j$$

in M, where $A_1,\ldots,A_n,B_1,\ldots,B_m \in (V-T)$, $a_1,\ldots,a_k, b_1,\ldots,b_j \in T^*$, $u,p \in (W-F)$, $0 \le k$, $0 \le i \le j \le m$.

Proof. Basis. Let $i = 0$. Then,

$$A_1\ldots A_n\#B_1\ldots B_m a_1\ldots a_k u \Rightarrow^0 A_1\ldots A_n\#B_1\ldots B_m a_1\ldots a_k u$$

in G. Clearly,

$$1\overline{h}(A_n)\ldots\overline{h}(A_1)h(A_1)\ldots h(A_n)h(B_1)\ldots h(B_m)1\langle u,2\rangle b_1\ldots b_j \vdash^0$$

$$1\overline{h}(A_n)\ldots\overline{h}(A_1)h(A_1)\ldots h(A_n)h(B_1)\ldots h(B_m)1\langle u,2\rangle b_1\ldots b_j$$

in M.

Induction Hypothesis. Assume that Claim 10.2.4 holds for every $i \le l$, where l is a positive integer.

Induction Step. Consider any derivation of the form

$$A_1\ldots A_n\#B_1\ldots B_m a_1\ldots a_k u \Rightarrow^{l+1}$$

$$A_1\ldots A_n B_1\ldots B_l B_{l+1}\#B_{l+2}\ldots B_m a_1\ldots a_k b_1\ldots b_l b_{l+1} q$$

and express this derivation as

$$A_1\ldots A_n\#B_1\ldots B_m a_1\ldots a_k u \Rightarrow^l$$

$$A_1\ldots A_n B_1\ldots B_l\#B_{l+1}\ldots B_m a_1\ldots a_k b_1\ldots b_l p \Rightarrow$$

$$A_1\ldots A_n B_1\ldots B_l B_{l+1}\#B_{l+2}\ldots B_m a_1\ldots a_k b_1\ldots b_l b_{l+1} q$$

in G, where $0 \le k$, $0 \le l \le m$, $q \in (W-F)$. By the induction hypothesis,

$$1\overline{h}(A_n)\ldots\overline{h}(A_1)h(A_1)\ldots h(A_n)h(B_1)\ldots h(B_m)1\langle u,2\rangle b_1\ldots b_j \vdash^l$$

$$1\overline{h}(B_l)\ldots\overline{h}(B_1)\overline{h}(A_n)\ldots\overline{h}(A_1)h(A_1)\ldots$$

$$\ldots h(A_n)h(B_1)\ldots h(B_m)1\langle p,2\rangle b_{l+1}\ldots b_j \vdash$$

$$1\overline{h}(B_{l+1})\overline{h}(B_l)\ldots\overline{h}(B_1)\overline{h}(A_n)\ldots\overline{h}(A_1)h(A_1)\ldots$$

$$\ldots h(A_n)h(B_1)\ldots h(B_m)1\langle q,2\rangle b_{l+2}\ldots b_j$$

in M. In this case, there is only one possibility how G can make the derivation

$$A_1 \ldots A_n B_1 \ldots B_l \# B_{l+1} \ldots B_m a_1 \ldots a_k b_1 \ldots b_l p \Rightarrow$$

$$A_1 \ldots A_n B_1 \ldots B_l B_{l+1} \# B_{l+2} \ldots B_m a_1 \ldots a_k b_1 \ldots b_l b_{l+1} q$$

Observe that it is done by a rule of the form $(B_{l+1}, p, b_{l+1}, q) \in P$, where $B_{l+1} \in (V - T)$, $p, q \in (W - F)$, $b_{l+1} \in T^*$. By step (4) in construction, there is a rule $1|1\langle p, 2\rangle b_{l+1} \to 1\overline{h}(B_{l+1})|1\langle q, 2\rangle$ in R, so

$$1\overline{h}(B_l) \ldots \overline{h}(B_1)\overline{h}(A_n) \ldots \overline{h}(A_1)h(A_1) \ldots$$

$$\ldots h(A_n)h(B_1) \ldots h(B_m)1\langle p, 2\rangle b_{l+1} \ldots b_j \vdash$$

$$1\overline{h}(B_{l+1})\overline{h}(B_l) \ldots \overline{h}(B_1)\overline{h}(A_n) \ldots \overline{h}(A_1)h(A_1) \ldots$$

$$\ldots h(A_n)h(B_1) \ldots h(B_m)1\langle q, 2\rangle b_{l+2} \ldots b_j$$

in M and Claim 10.2.4 holds. $\qquad\square$

Claim 10.2.5. If

$$A_1 \ldots A_{n-1} \# A_n y q \Rightarrow A_1 \ldots A_{n-1} A_n \# y z t$$

in G, where $A_1, \ldots, A_n \in (V - T)$, $y, z \in T^*$, $q \in (W - F)$, $t \in F$, then

$$1\overline{h}(A_{n-1}) \ldots \overline{h}(A_1)h(A_1) \ldots h(A_n)1\langle q, 2\rangle z \vdash$$

$$\overline{h}(A_n) \ldots \overline{h}(A_1)h(A_1) \ldots h(A_n)f = \varepsilon f$$

in M, where $f \in F_M$.

Proof. Grammar G performs the described derivation by a rule of the form $(A_n, q, z, t) \in P$, where $A_n \in (V - T)$, $z \in T^*$, $q \in (W - F)$, $t \in F$. By step (5) of the construction, there is a rule $1|1\langle q, 2\rangle z \to \overline{h}(A_n)|\varepsilon f$ in R, so the corresponding computational step described in Claim 10.2.5 indeed occurs in M, so Claim 10.2.5 holds. $\qquad\square$

Claims 10.2.3, 10.2.4, and 10.2.5 prove that $L(G) \subseteq L(M)$. Next, we establish Claims 10.2.6, 10.2.7, and 10.2.8 to prove $L(M) \subseteq L(G)$.

Claim 10.2.6. Automaton M accepts every $w \in L(M)$ in this way

$$1 1 z w_1 w_2 \ldots w_r \vdash$$

$$1 h(S) 1 \langle q_0, 1\rangle w_1 w_2 \ldots w_r \vdash$$

$$1\overline{h}(S)h(S)h(X_1^1)h(X_2^1) \ldots h(X_{n_1}^1)1\langle q_1, 1\rangle w_1 w_2 \ldots w_r \vdash$$

$$1\overline{h}(X_1^1)\overline{h}(S)h(S)h(X_1^1)h(X_2^1) \ldots h(X_{n_1}^1)h(X_1^2)h(X_2^2) \ldots$$

$$\ldots h(X_{n_2}^2)1\langle q_2,1\rangle w_1 w_2 \ldots w_r \vdash$$

$$1\overline{h}(X_2^1)\overline{h}(X_1^1)\overline{h}(S)h(S)h(X_1^1)h(X_2^1)\ldots h(X_{n_1}^1)h(X_1^2)h(X_2^2)\ldots$$

$$\ldots h(X_{n_2}^2)h(X_1^3)h(X_2^3)\ldots h(X_{n_3}^3)1\langle q_3,1\rangle w_1 w_2 \ldots w_r \vdash$$

$$\vdots$$

$$1\overline{h}(X_j^k)\ldots\overline{h}(X_2^1)\overline{h}(X_1^1)\overline{h}(S)h(S)h(X_1^1)h(X_2^1)\ldots$$

$$\ldots h(X_{n_1}^1)h(X_1^2)h(X_2^2)\ldots h(X_{n_2}^2)h(X_1^3)h(X_2^3)\ldots$$

$$\ldots h(X_{n_3}^3)\ldots h(X_1^m)h(X_2^m)\ldots h(X_{n_m}^m)1\langle q_m,1\rangle w_1 w_2 \ldots w_r \vdash$$

$$1\overline{h}(X_j^k)\ldots\overline{h}(X_2^1)\overline{h}(X_1^1)\overline{h}(S)h(S)h(X_1^1)h(X_2^1)\ldots$$

$$\ldots h(X_{n_1}^1)h(X_1^2)h(X_2^2)\ldots h(X_{n_2}^2)h(X_1^3)h(X_2^3)\ldots$$

$$\ldots h(X_{n_3}^3)\ldots h(X_1^m)h(X_2^m)\ldots h(X_{n_m}^m)1\langle q_m,2\rangle w_1 w_2 \ldots w_r \vdash$$

$$1\overline{h}(X_{j+1}^k)\overline{h}(X_j^k)\ldots\overline{h}(X_2^1)\overline{h}(X_1^1)\overline{h}(S)h(S)h(X_1^1)h(X_2^1)\ldots$$

$$\ldots h(X_{n_1}^1)h(X_1^2)h(X_2^2)\ldots h(X_{n_2}^2)h(X_1^3)h(X_2^3)\ldots$$

$$\ldots h(X_{n_3}^3)\ldots h(X_1^m)h(X_2^m)\ldots h(X_{n_m}^m)1\langle q_{m+1},2\rangle w_2 \ldots w_r \vdash$$

$$1\overline{h}(X_{j+2}^k)\overline{h}(X_{j+1}^k)\overline{h}(X_j^k)\ldots\overline{h}(X_2^1)\overline{h}(X_1^1)\overline{h}(S)h(S)h(X_1^1)h(X_2^1)\ldots$$

$$\ldots h(X_{n_1}^1)h(X_1^2)h(X_2^2)\ldots h(X_{n_2}^2)h(X_1^3)h(X_2^3)\ldots$$

$$\ldots h(X_{n_3}^3)\ldots h(X_1^m)h(X_2^m)\ldots h(X_{n_m}^m)1\langle q_{m+2},2\rangle w_3 \ldots w_r \vdash$$

$$\vdots$$

$$1\overline{h}(X_{n_m-1}^m)\ldots\overline{h}(X_{j+2}^k)\overline{h}(X_{j+1}^k)\overline{h}(X_j^k)\ldots$$

$$\ldots\overline{h}(X_2^1)\overline{h}(X_1^1)\overline{h}(S)h(S)h(X_1^1)h(X_2^1)\ldots$$

$$\ldots h(X_{n_1}^1)h(X_1^2)h(X_2^2)\ldots h(X_{n_2}^2)h(X_1^3)h(X_2^3)\ldots$$

$$\ldots h(X_{n_3}^3)\ldots h(X_1^m)h(X_2^m)\ldots h(X_{n_m}^m)1\langle q_{m+r-1},2\rangle w_r \vdash$$

$$\overline{h}(X_{n_m}^m)\overline{h}(X_{n_m-1}^m)\ldots\overline{h}(X_{j+2}^k)\overline{h}(X_{j+1}^k)\overline{h}(X_j^k)\ldots$$

$$\ldots \bar{h}(X_2^1)\bar{h}(X_1^1)\bar{h}(S)h(S)h(X_1^1)h(X_2^1)\ldots$$

$$\ldots h(X_{n_1}^1)h(X_1^2)h(X_2^2)\ldots h(X_{n_2}^2)h(X_1^3)h(X_2^3)\ldots$$

$$\ldots h(X_{n_3}^3)\ldots h(X_1^m)h(X_2^m)\ldots h(X_{n_m}^m)f = \varepsilon f$$

where $w = w_1 w_2 \ldots w_r$, $r \geq 1$, $w_1, \ldots, w_r \in T^*$, $q_0, q_1, \ldots, q_{m+r-1} \in (W - F)$, $X_1^1, \ldots, X_{n_1}^1, X_1^2, \ldots, X_{n_2}^2, \ldots, X_1^m, \ldots, X_{n_m}^m \in (V - T)$, $n_1, n_2, \ldots, n_m \geq 0$, $0 \leq k \leq m$.

Proof. We examine steps (1) through (5) of the construction of R. Note that in every successful computation, M uses rules created in step b before it uses rules created in step $b + 1$, for $b = 1, \ldots, 4$.

In the first computational step, the rule $1|1z \to 1|h(S)1\langle q_0, 1\rangle$ introduced in (1) is applied, where Sq_0 is the axiom of G. This is the only way by which M can make the transition

$$11zw_1w_2 \ldots w_r \vdash 1h(S)1\langle q_0, 1\rangle w_1w_2 \ldots w_r$$

Observe that this rule is used exactly once during one successful computation. By this step, automaton is switched to the nonterminal-generating mode.

In the next part of computation, namely

$$1h(S)1\langle q_0, 1\rangle w_1 w_2 \ldots w_r \vdash^*$$

$$1\bar{h}(X_j^k)\ldots \bar{h}(X_2^1)\bar{h}(X_1^1)\bar{h}(S)h(S)h(X_1^1)h(X_2^1)\ldots$$

$$\ldots h(X_{n_1}^1)h(X_1^2)h(X_2^2)\ldots h(X_{n_2}^2)h(X_1^3)h(X_2^3)\ldots$$

$$\ldots h(X_{n_3}^3)\ldots h(X_1^m)h(X_2^m)\ldots h(X_{n_m}^m)1\langle q_m, 1\rangle w_1 w_2 \ldots w_r$$

M uses rules of the form $1|1\langle q, 1\rangle \to 1\bar{h}(A)|h(x)1\langle p, 1\rangle$ constructed in (2), where $A \in (V-T)$, $x \in (V-T)^*$, $p, q \in (W-F)$. This part of computation is characterized by M's states of the form $\langle q, 1\rangle$, $q \in (W - F)$. For the more detailed proof of this part, see Claim 10.2.7.

By the next computational step,

$$1\bar{h}(X_j^k)\ldots \bar{h}(X_2^1)\bar{h}(X_1^1)\bar{h}(S)h(S)h(X_1^1)h(X_2^1)\ldots$$

$$\ldots h(X_{n_1}^1)h(X_1^2)h(X_2^2)\ldots h(X_{n_2}^2)h(X_1^3)h(X_2^3)\ldots$$

$$\ldots h(X_{n_3}^3)\ldots h(X_1^m)h(X_2^m)\ldots h(X_{n_m}^m)1\langle q_m, 1\rangle w_1 w_2 \ldots w_r \vdash$$

$$1\bar{h}(X_j^k)\ldots \bar{h}(X_2^1)\bar{h}(X_1^1)\bar{h}(S)h(S)h(X_1^1)h(X_2^1)\ldots$$

$$\ldots h(X_{n_1}^1)h(X_1^2)h(X_2^2)\ldots h(X_{n_2}^2)h(X_1^3)h(X_2^3)\ldots$$

$$\ldots h(X_{n_3}^3) \ldots h(X_1^m) h(X_2^m) \ldots h(X_{n_m}^m) 1 \langle q_m, 2 \rangle w_1 w_2 \ldots w_r$$

M switches to the terminal-reading mode by a rule of the form $1|1\langle q, 1\rangle \to 1|1\langle q, 2\rangle$ constructed in (2). Observe that this rule is used exactly once during one successful computation. Since this rule changes an actual state of automaton of the form $\langle q, 1\rangle$ to the state of the form $\langle q, 2\rangle$, $q \in (W - F)$, there is no further possibility of using any rules constructed in parts (1) through (3).

In the next part of computation, namely

$$1\overline{h}(X_j^k) \ldots \overline{h}(X_2^1)\overline{h}(X_1^1)\overline{h}(S)h(S)h(X_1^1)h(X_2^1) \ldots$$

$$\ldots h(X_{n_1}^1)h(X_1^2)h(X_2^2) \ldots h(X_{n_2}^2)h(X_1^3)h(X_2^3) \ldots$$

$$\ldots h(X_{n_3}^3) \ldots h(X_1^m)h(X_2^m) \ldots h(X_{n_m}^m)1 \langle q_m, 2 \rangle w_1 w_2 \ldots w_r \vdash^*$$

$$1\overline{h}(X_{n_m-1}^m) \ldots \overline{h}(X_{j+2}^k)\overline{h}(X_{j+1}^k)\overline{h}(X_j^k) \ldots$$

$$\ldots \overline{h}(X_2^1)\overline{h}(X_1^1)\overline{h}(S)h(S)h(X_1^1)h(X_2^1) \ldots$$

$$\ldots h(X_{n_1}^1)h(X_1^2)h(X_2^2) \ldots h(X_{n_2}^2)h(X_1^3)h(X_2^3) \ldots$$

$$\ldots h(X_{n_3}^3) \ldots h(X_1^m)h(X_2^m) \ldots h(X_{n_m}^m)1 \langle q_{m+r-1}, 2 \rangle w_r$$

M uses rules constructed in 4 and reads input strings of terminals. The detailed proof of this part of computation is described in Claim 10.2.8.

The last computational step switches M to the final state. It is done by a rule of the form $1|1\langle q, 2\rangle y \to \overline{h}(A)|\varepsilon f$ constructed in (5), where $q \in (W - T)$, $y \in T^*$, $A \in (V - T)$ and $f \in F_M$. After that, if the two-sided pushdown is empty by a group reduction and the input string is read, then M accepts the input string. Otherwise, the input string is not accepted, since there is no rule with the left-hand side of the form $1|1fy$, where $f \in F_M$, $y \in T^*$, so Claim 10.2.6 holds.   □

Claim 10.2.7. If

$$1\overline{h}(A_n) \ldots \overline{h}(A_1)h(A_1) \ldots h(A_n)h(B_1) \ldots h(B_m)1\langle u, 1\rangle \omega \vdash^i$$

$$1\overline{h}(B_i) \ldots \overline{h}(B_1)\overline{h}(A_n) \ldots \overline{h}(A_1)h(A_1) \ldots$$

$$\ldots h(A_n)h(B_1) \ldots h(B_m)h(x_1) \ldots h(x_i)1\langle p, 1\rangle \omega$$

in M, then

$$A_1 \ldots A_n \# B_1 \ldots B_m u \Rightarrow^i A_1 \ldots A_n B_1 \ldots B_i \# B_{i+1} \ldots B_m x_1 \ldots x_i p$$

in G, where $A_1, \ldots, A_n, B_1, \ldots, B_m \in (V - T)$, $x_1, \ldots, x_i \in (V - T)^*$, $u, p \in (W - F)$, $0 \le i \le m$.

Proof. Basis. Let $i = 0$. Then,

$$1\overline{h}(A_n)\ldots\overline{h}(A_1)h(A_1)\ldots h(A_n)h(B_1)\ldots h(B_m)1\langle u, 1\rangle\omega \vdash^0$$

$$1\overline{h}(A_n)\ldots\overline{h}(A_1)h(A_1)\ldots h(A_n)h(B_1)\ldots h(B_m)1\langle u, 1\rangle\omega$$

in M. Clearly,

$$A_1\ldots A_n\#B_1\ldots B_m u \Rightarrow^0 A_1\ldots A_n\#B_1\ldots B_m u$$

in G.

Induction Hypothesis. Assume that Claim 10.2.7 holds for every $i \leq l$, where l is a positive integer.

Induction Step. Consider any computation of the form

$$1\overline{h}(A_n)\ldots\overline{h}(A_1)h(A_1)\ldots h(A_n)h(B_1)\ldots h(B_m)1\langle u, 1\rangle\omega \vdash^{l+1}$$

$$1\overline{h}(B_{l+1})\overline{h}(B_l)\ldots\overline{h}(B_1)\overline{h}(A_n)\ldots\overline{h}(A_1)h(A_1)\ldots$$

$$\ldots h(A_n)h(B_1)\ldots h(B_m)h(x_1)\ldots h(x_l)h(x_{l+1})1\langle q, 1\rangle\omega$$

and express this derivation as

$$1\overline{h}(A_n)\ldots\overline{h}(A_1)h(A_1)\ldots h(A_n)h(B_1)\ldots h(B_m)1\langle u, 1\rangle\omega \vdash^l$$

$$1\overline{h}(B_l)\ldots\overline{h}(B_1)\overline{h}(A_n)\ldots\overline{h}(A_1)h(A_1)\ldots$$

$$\ldots h(A_n)h(B_1)\ldots h(B_m)h(x_1)\ldots h(x_l)1\langle p, 1\rangle\omega \vdash$$

$$1\overline{h}(B_{l+1})\overline{h}(B_l)\ldots\overline{h}(B_1)\overline{h}(A_n)\ldots\overline{h}(A_1)h(A_1)\ldots$$

$$\ldots h(A_n)h(B_1)\ldots h(B_m)h(x_1)\ldots h(x_l)h(x_{l+1})1\langle q, 1\rangle\omega$$

in M, where $q \in (W - F)$, $0 \leq l \leq m$. By the induction hypothesis,

$$A_1\ldots A_n\#B_1\ldots B_m u \Rightarrow^l$$

$$A_1\ldots A_nB_1\ldots B_l\#B_{l+1}\ldots B_m x_1\ldots x_l p \Rightarrow$$

$$A_1\ldots A_nB_1\ldots B_lB_{l+1}\#B_{l+2}\ldots B_m x_1\ldots x_l x_{l+1} q$$

in G. There is only one type of rules in R able to perform the computation

$$1\overline{h}(B_l)\ldots\overline{h}(B_1)\overline{h}(A_n)\ldots\overline{h}(A_1)h(A_1)\ldots$$

$$\ldots h(A_n)h(B_1)\ldots h(B_m)h(x_1)\ldots h(x_l)1\langle p, 1\rangle\omega \vdash$$

$$1\overline{h}(B_{l+1})\overline{h}(B_l)\ldots\overline{h}(B_1)\overline{h}(A_n)\ldots\overline{h}(A_1)h(A_1)\ldots$$

$$\ldots h(A_n)h(B_1)\ldots h(B_m)h(x_1)\ldots h(x_l)h(x_{l+1})1\langle q, 1\rangle\omega$$

in M, namely rules of the form

$$1|1\langle p, 1\rangle \to 1\overline{h}(B_{l+1})|h(x_{l+1})1\langle q, 1\rangle \in R$$

Observe that by construction, there is a rule (B_{l+1}, p, x_{l+1}, q) in P, so

$$A_1 \ldots A_n \# B_1 \ldots B_m u \Rightarrow^l$$

$$A_1 \ldots A_n B_1 \ldots B_l \# B_{l+1} \ldots B_m x_1 \ldots x_l p \Rightarrow$$

$$A_1 \ldots A_n B_1 \ldots B_l B_{l+1} \# B_{l+2} \ldots B_m x_1 \ldots x_l x_{l+1} q$$

in G and Claim 10.2.7 holds.   □

Claim 10.2.8. If

$$1\overline{h}(A_n) \ldots \overline{h}(A_1)h(A_1) \ldots h(A_n)h(B_1) \ldots h(B_m)1\langle u, 2\rangle b_1 \ldots b_j \vdash^i$$

$$1\overline{h}(B_i) \ldots \overline{h}(B_1)\overline{h}(A_n) \ldots \overline{h}(A_1)h(A_1) \ldots$$

$$\ldots h(A_n)h(B_1) \ldots h(B_i)h(B_{i+1}) \ldots h(B_m)1\langle p, 2\rangle b_{i+1} \ldots b_j$$

in M, then

$$A_1 \ldots A_n \# B_1 \ldots B_m a_1 \ldots a_k u \Rightarrow^i$$

$$A_1 \ldots A_n B_1 \ldots B_i \# B_{i+1} \ldots B_m a_1 \ldots a_k b_1 \ldots b_i p$$

in G, where $A_1, \ldots, A_n, B_1, \ldots, B_m \in V - T$, $a_1, \ldots, a_k, b_1, \ldots, b_j \in T^*$ and $p, u \in W - F$, $0 \le i \le m$.

Proof. Basis. Let $i = 0$. Then,

$$1\overline{h}(A_n) \ldots \overline{h}(A_1)h(A_1) \ldots h(A_n)h(B_1) \ldots h(B_m)1\langle u, 2\rangle b_1 \ldots b_j \vdash^0$$

$$1\overline{h}(A_n) \ldots \overline{h}(A_1)h(A_1) \ldots h(A_n)h(B_1) \ldots h(B_m)1\langle u, 2\rangle b_1 \ldots b_j$$

in M. Clearly,

$$A_1 \ldots A_n \# B_1 \ldots B_m a_1 \ldots a_k u \Rightarrow^0 A_1 \ldots A_n \# B_1 \ldots B_m a_1 \ldots a_k u$$

in G.

Induction Hypothesis. Assume that Claim 10.2.8 holds for every $i \le l$, where l is a positive integer.

Induction Step. Consider any computation of the form

$$1\overline{h}(A_n) \ldots \overline{h}(A_1)h(A_1) \ldots h(A_n)h(B_1) \ldots h(B_m)1\langle u, 2\rangle b_1 \ldots b_j \vdash^{l+1}$$

$$1\overline{h}(B_{l+1})\overline{h}(B_l) \ldots \overline{h}(B_1)\overline{h}(A_n) \ldots \overline{h}(A_1)h(A_1) \ldots$$

$$\ldots h(A_n)h(B_1) \ldots h(B_m)1\langle q, 2\rangle b_{l+2} \ldots b_j$$

and express this derivation as

$$1\overline{h}(A_n)\ldots\overline{h}(A_1)h(A_1)\ldots h(A_n)h(B_1)\ldots h(B_m)1\langle u,2\rangle b_1\ldots b_j \vdash^l$$

$$1\overline{h}(B_l)\ldots\overline{h}(B_1)\overline{h}(A_n)\ldots\overline{h}(A_1)h(A_1)\ldots$$

$$\ldots h(A_n)h(B_1)\ldots h(B_m)1\langle p,2\rangle b_{l+1}\ldots b_j \vdash$$

$$1\overline{h}(B_{l+1})\overline{h}(B_l)\ldots\overline{h}(B_1)\overline{h}(A_n)\ldots\overline{h}(A_1)h(A_1)\ldots$$

$$\ldots h(A_n)h(B_1)\ldots h(B_m)1\langle q,2\rangle b_{l+2}\ldots b_j$$

in M, where $0 \le l \le j \le m$, $q \in (W - F)$. By the induction hypothesis,

$$A_1\ldots A_n\#B_1\ldots B_m u_1\ldots a_k u \Rightarrow^l$$

$$A_1\ldots A_n B_1\ldots B_l\#B_{l+1}\ldots B_m a_1\ldots a_k b_1\ldots b_l p \Rightarrow$$

$$A_1\ldots A_n B_1\ldots B_l B_{l+1}\#B_{l+2}\ldots B_m a_1\ldots a_k b_1\ldots b_l b_{l+1} q$$

in G, where $0 \le l \le m$. In this case, there is the only way by which M can make the computational step

$$1\overline{h}(B_l)\ldots\overline{h}(B_1)\overline{h}(A_n)\ldots\overline{h}(A_1)h(A_1)\ldots$$

$$\ldots h(A_n)h(B_1)\ldots h(B_m)1\langle p,2\rangle b_{l+1}\ldots b_j \vdash$$

$$1\overline{h}(B_{l+1})\overline{h}(B_l)\ldots\overline{h}(B_1)\overline{h}(A_n)\ldots\overline{h}(A_1)h(A_1)\ldots$$

$$\ldots h(A_n)h(B_1)\ldots h(B_m)1\langle q,2\rangle b_{l+2}\ldots b_j$$

Observe that it is done by a rule of the form

$$1|1\langle p,2\rangle b_{l+1} \to 1\overline{h}(B_{l+1})|1\langle q,2\rangle \in R$$

By step (4) of the construction, there is a rule $(B_{l+1}, p, b_{l+1}, q) \in P$ where $B_{l+1} \in (V - T)$, $p, q \in (W - F)$, $b_{l+1} \in T^*$, so

$$A_1\ldots A_n B_1\ldots B_l\#B_{l+1}\ldots B_m a_1\ldots a_k b_1\ldots b_l p \Rightarrow$$

$$A_1\ldots A_n B_1\ldots B_l B_{l+1}\#B_{l+2}\ldots B_m a_1\ldots a_k b_1\ldots b_l b_{l+1} q$$

in G and Claim 10.2.8 holds.   □

By Claims 10.2.6, 10.2.7 and 10.2.8, we proved that $L(M) \subseteq L(G)$. As a result, $L(G) = L(M)$, so Theorem 10.2.1 is proved.   □

Theorem 10.2.9. *For every string-reading two-sided pushdown automaton over a free group with the reduced pushdown alphabet, Q', there exists a two-sided pushdown automaton over a free group with the reduced pushdown alphabet, Q, such that $L(Q') = L(Q)$.*

Proof. The formal proof of this theorem is simple and left to the reader.   □

10.3   Conclusion

In this chapter, we proved that the power of two-sided pushdown automata with pushdowns defined over free groups is equal to the power of Turing machines, so these automata generate the whole family of recursively enumerable languages. Moreover, the pushdown alphabet contains no more than four symbols. Note that the same result can be also reached with two-sided pushdowns defined over free monoids. This modification affects only the set of rules with their construction, and it is left to the reader.

Another modifications of pushdown automata have been studied in theory of automata and formal languages. We can mention simultaneously one-turn two-pushdown automata introduced in [Med03c], regulated pushdown automata described in [KM00], or finite-turn pushdown automata (see [GS68]). Very simple and natural modification of pushdown automata is also presented in [Cou77] and [ABB97], where there is the ability for pushdown reversal added. The main goal of all these modifications is to increase the generative power of ordinary pushdown automata. In our chapter, we significantly increased the power and moreover, the number of transition rules was reduced by defining of the two-sided pushdowns over free groups.

Part IV
Languages Defined in Combined Ways

This part, consisting of Chaps. 11 and 12, covers important language-defining devices that are based on combinations of other language models as well as the ways they work. As a result, these devices actually reflect and formalize a cooperating way of computation. More specifically, Chap. 11 untraditionally combines grammars and automata in terms of the way they operate. Indeed, it studies how to generate languages by automata although languages are traditionally generated by grammars. Chapter 12 studies the generation of languages by several grammars that work in a simultaneously cooperative way, thus formalizing computational cooperation in a quite straightforward way.

Chapter 11
Language-Generating Automata and State-Controlled Computation

Traditionally, computation controlled by finitely many states is formalized by finite state automata, which accept their languages (see Sect. 2.4). Untraditionally, however, the present chapter explains how to adapt these automata in a very natural way so they act as language generators just like grammars. Consequently, the formalization of state-controlled computation can be based on the language-generating automata resulting from this adaptation.

We first give a conceptual insight into adapting automata so they generate languages. Consider, for instance, the notion of a context-free grammar G (see Sect. 2.3). Recall that G contains an alphabet of terminal symbols and an alphabet of nonterminal symbols, one of which represents the start symbol. Starting from this symbol, G rewrites nonterminal symbols in the sentential forms by its rules until it generates a string of terminals. The set of all terminal strings generated in this way is the language that G defines. To illustrate automata, the notion of a finite-state automaton M (see Definition 2.4.1). M has a finite set of states, one of which is defined as the start state. In addition, some states are specified as final states. M works by making moves. During a move, it changes its current state and reads an input symbol. If with an input string, M makes a sequence of moves according to its rules so it starts from the start state, reads the input string, and reaches a final state, then M accepts the input string. The set of all strings accepted in this way represents the language that M defines.

Although it is obviously quite natural to design language-defining models based on a combination of grammars and automata and, thereby, make their scale much broader, only a tiny minority of these devices is designed in this combined way (see [BF94, BF95, Kas70, MHHO05]). To support this combined design, the present chapter introduces new rewriting systems, called #-*rewriting systems*, having features of both grammars and automata. Indeed, like grammars, they are generative devices. However, like automata, they use finitely many states without any nonterminals. As its main result, this chapter characterizes the well-known infinite hierarchy of language families resulting from programmed grammars of

© Springer International Publishing AG 2017
A. Meduna, O. Soukup, *Modern Language Models and Computation*,
DOI 10.1007/978-3-319-63100-4_11

finite index by the #-rewriting systems (see Theorems 3.1.2i and 3.1.7 in [DP89]). From a broader perspective, this result thus demonstrates that rewriting systems based on a combination of grammars and automata are naturally related to some classical topics and results concerning formal languages, on which they can shed light in an alternative way.

11.1 Definitions and Examples

Definition 11.1.1. A *#-rewriting system* is a quadruple $M = (Q, \Sigma, s, R)$, where Q is a finite set of states, Σ is an alphabet containing # called a *bounder*, $Q \cap \Sigma = \emptyset$, $s \in Q$ is a start state and $R \subseteq Q \times \mathbb{N} \times \{\#\} \times Q \times \Sigma^*$ is a finite relation whose members are called *rules*. A rule $(p, n, \#, q, x) \in R$, where $n \in \mathbb{N}$, $q, p \in Q$ and $x \in \Sigma^*$, is usually written as $r: p_n\# \to q\,x$ hereafter, where r is its unique label.

A *configuration* of M is a pair from $Q \times \Sigma^*$. Let χ denote the set of all configurations of M. Let $pu\#v, quxv \in \chi$ be two configurations, $p, q \in Q$, $u, v \in \Sigma^*$, $n \in \mathbb{N}$ and $\#_\#(u) = n - 1$. Then, M makes a *derivation step* from $pu\#v$ to $quxv$ by using $r: p_n\# \to q\,x$, symbolically written $pu\#v \Rightarrow quxv$ [r] in M or simply $pu\#v \Rightarrow quxv$.

In the standard manner, we extend $\Rightarrow$ to $\Rightarrow^m$, for $m \geq 0$; then, based on $\Rightarrow^m$, we define $\Rightarrow^+$ and $\Rightarrow^*$ in the standard way. The *language generated* by M, $L(M)$, is defined as

$$L(M) = \{w \mid s\# \Rightarrow^* qw, \ q \in Q, w \in (\Sigma - \{\#\})^*\}$$

Let k be a positive integer. A #-rewriting system M is of *index k* if for every configuration $x \in \chi$, $s\# \Rightarrow^* qy = x$ implies $\#_\#(y) \leq k$. $_k\#\mathbf{RS}$ denote the families of languages derived by #-rewriting systems of index k. □

Notice that M of index k cannot derive a string containing more than k #s; in this sense, this notion differs from the corresponding notion in terms of programmed grammars (see Sect. 3.5), which can derive strings containing more than k nonterminals. Remark that for a positive integer k, $_k\mathbf{P}$ denote the families of programmed languages of index k (see definition 3.5.2).

Example 11.1.2. $M = (\{s, p, q, f\}, \{a, b, c, \#\}, s, R)$, where R contains

1: $s_1\# \to p\,\#\#$
2: $p_1\# \to q\,a\#b$
3: $q_2\# \to p\,\#c$
4: $p_1\# \to f\,ab$
5: $f_1\# \to f\,c$ □

Obviously, M is of index 2, and $L(M) = \{a^n b^n c^n \mid n \geq 1\}$. For instance, M generates $aaabbbccc$ as

$$s\# \Rightarrow p\#\# \qquad\qquad [1]$$

$$\Rightarrow qa\#b\# \qquad\qquad [2]$$

$$\Rightarrow pa\#b\#c \qquad\qquad [3]$$

$$\Rightarrow qaa\#bb\#c \qquad\qquad [2]$$

$$\Rightarrow paa\#bb\#cc \qquad\qquad [3]$$

$$\Rightarrow faaabbb\#cc \qquad\qquad [4]$$

$$\Rightarrow faaabbbccc \qquad\qquad [5]$$

11.2 Results

This section establishes an infinite hierarchy of language families resulting from the #-rewriting systems defined in the previous section.

In what follows, for a programmed grammar $G = (V, T, P, S)$ and a rule $(p: S \rightarrow \alpha, \varphi) \in P$, $g(p) = \varphi$ denotes the set of all rules applicable after the rule p.

Lemma 11.2.1. *For every $k \geq 1$, ${}_k\mathbf{P} \subseteq {}_k\#\mathbf{RS}$.*

Proof. Construction. Let $k \geq 1$ be a positive integer. Let $G = (V, T, P, S)$ be a programmed grammar of index k, where $N = V - T$. We construct a #-rewriting system of index k, $H = (Q, T \cup \{\#\}, s, R)$, where $\# \notin T$, $s = \langle \sigma \rangle$, σ is a new symbol, and R and Q are constructed by performing the following steps:

(1) For each $(p: S \rightarrow \alpha, g(p)) \in P$, $\alpha \in V^*$, add $\langle \sigma \rangle_1 \# \rightarrow \langle [p] \rangle \#$ to R, where $\langle [p] \rangle$ is new state in Q.

(2) If $A_1 A_2 \ldots A_j \ldots A_h \in N^*$, $h \in \{1, 2, \ldots, k\}$,

$$(p: A_j \rightarrow x_0 B_1 x_1 B_2 x_2 \ldots x_{n-1} B_n x_n, g(p)) \in P$$

$j \in \{1, 2, \ldots, h\}$ for $n \geq 0$, $x_0, x_t \in T^*$, $B_t \in N$, $1 \leq t \leq n$ and $n + h - 1 \leq k$, then

(2.1) if $g(p) = \emptyset$, then $\langle A_1 A_2 \ldots A_{j-1} [p] A_{j+1} \ldots A_h \rangle$, $\langle A_1 A_2 \ldots B_1 \ldots B_n \ldots A_h \rangle$ are new states in Q and the rule

$$\langle A_1 A_2 \ldots A_{j-1} [p] A_{j+1} \ldots A_h \rangle_j \# \rightarrow$$

$$\langle A_1 A_2 \ldots B_1 \ldots B_n \ldots A_h \rangle x_0 \# x_1 \ldots x_{n-1} \# x_n$$

is added to R;

(2.2) for every $q \in g(p)$, $(q: D_d \to \alpha, g(q)) \in P$, $\alpha \in V^*$ add new states $\langle A_1 A_2 \ldots A_{j-1} [p] A_{j+1} \ldots A_h \rangle$ and $\langle D_1 D_2 \ldots [q] \ldots D_{n+h-1} \rangle$ to Q and add the following rule to R:

$$\langle A_1 A_2 \ldots A_{j-1}[p]A_{j+1} \ldots A_h \rangle_j \# \to$$

$$\langle D_1 D_2 \ldots [q] \ldots D_{n+h-1} \rangle x_0 \# x_1 \ldots x_n$$

where $A_1 \ldots A_{j-1} B_1 \ldots B_n A_{j+1} \ldots A_h = D_1 \ldots D_{h+n-1}$, $B_1 \ldots B_n = D_j \ldots D_{j+n-1}$ for some $d \in \{1, 2, \ldots, n + h - 1\}$.

Basic Idea. H simulates G's derivations. The information necessary for this simulation is recorded inside of states. Each state in Q carries string of nonterminals from N^*, where one symbol of this string is replaced with the label of a rule in P.

Let $x_0 A_1 x_1 \ldots x_{h-1} A_h x_h$ be a sentential form derived by G, where $x_i \in T^*$ for $0 \le i \le h$ and $A_l \in V - T$ for $1 \le l \le h$, and let $(p: A_j \to \alpha, g(p))$ be a rule in P applicable in the next step to A_j, $1 \le j \le h$. Then, H's new configuration is of the form $\langle A_1 A_2 \ldots A_{j-1}[p]A_{j+1} \ldots A_h \rangle x_0 \# x_1 \ldots x_{h-1} \# x_h$, which encodes the nonterminals in G's sentential form and the next applicable rule label.

We establish the proof of Lemma 11.2.1 by proving the following two claims. Remark, that we give a simplified proof; the fully rigorous proof is left to the reader.

Claim 11.2.2. If $S \Rightarrow^m x_0 A_1 x_1 A_2 x_2 \ldots x_{h-1} A_h x_h$ in G, then $\langle \sigma \rangle \# \Rightarrow^r \langle A_1 A_2 \ldots A_h \rangle$ $x_0 \# x_1 \ldots x_h [q_1 q_2 \ldots q_r]$ in H, for $m \ge 0$. If $g(q_r) \ne \emptyset$, then there exists a rule $(q_{r+1}: A_j \to y_0 B_1 y_1 \ldots y_{h-1} B_n y_n, g(q_{r+1}))$, $n + h - 1 \le k$, $q_{r+1} \in g(q_r)$ and $A_j = [q_{r+1}]$, $q_1, \ldots, q_r, q_{r+1} \in lab(R)$.

We prove Claim 11.2.2 by induction on $m \ge 0$.

Proof. We omit the basis of the induction. Supposing that Claim 11.2.2 holds for all derivations of length m or less for some $m \ge 0$, we consider $S \Rightarrow^{m+1} x$, where $x \in V^*$. Express $S \Rightarrow^{m+1} x$ as $S \Rightarrow^m y [p_1 p_2 \ldots p_m]$, where $y = x_0 A_1 x_1 \ldots x_{h-1} A_h x_h$ and $p_1, \ldots, p_m, p_{m+1} \in lab(P)$, $y \Rightarrow x [p_{m+1}]$. If $m = 0$, then $p_{m+1} \in \{p \mid lhs(p) = S, p \in lab(P)\}$; otherwise, $p_{m+1} \in g(p_m)$. For $(p_{m+1}: A_j \to y_0 B_1 y_1 \ldots y_{n-1} B_n y_n, g(p_{m+1}))$, x is of the form $x = x_0 A_1 x_1 \ldots A_{j-1} x_{j-1} y_0 B_1 y_1 \ldots y_{n-1} B_n y_n x_j A_{j+1} \ldots x_{h-1} A_h x_h$, where $x_0, \ldots, x_h \in T^*$ and $y_0, \ldots, y_n \in T^*$. By the induction hypothesis, there exists

$$\langle \sigma \rangle \# \Rightarrow^r$$

$$\langle A_1 A_2 \ldots A_{j-1}[p_{m+1}]A_{j+1} \ldots A_h \rangle x_0 \# x_1 \ldots x_{h-1} \# x_h [q_1 q_2 \ldots q_r] \Rightarrow$$

$$\langle A_1 A_2 \ldots A_{j-1} B_1 \ldots B_n A_{j+1} \ldots A_h \rangle x_0 \# \ldots \# x_{j-1} y_0 \# \ldots$$

$$\ldots \# y_n x_j \# \ldots \# x_h [q_{r+1}]$$

$r \geq 1$, $q_i \in lab(R)$, $1 \leq i \leq r+1$. If $g(p_{m+1}) \neq \emptyset$, then there exists a rule $p_{m+2} \in g(p_{m+1})$ and a sequence $D_1 \ldots D_{n+h-1}$ so that

$$A_1 A_2 \ldots A_{j-1} B_1 \ldots B_n A_{j+1} \ldots A_h = D_1 D_2 \ldots D_{n+h-1}$$

where for at most one $d \in \{1, 2, \ldots, n+h-1\}$ is $D_d = [q_{r+2}]$, $q_{r+2} \in g(q_{r+1})$. □

Claim 11.2.3. If $S \Rightarrow^z x$ in G, then $\langle \sigma \rangle \# \Rightarrow^* \langle \rangle x$ in H, where $z \geq 0$, $x \in T^*$.

Proof. Consider the claim for $h = 0$; then, $x_0 A_1 x_1 A_2 \ldots A_h x_h = x_0$ and $A_1 A_2 \ldots A_h = \varepsilon$. At this point, if $S \Rightarrow^z x_0$, then $\langle \sigma \rangle \# \Rightarrow^* \langle \rangle x_0$, so $x_0 = x$. □

Thus, Lemma 11.2.1 holds true. □

Lemma 11.2.4. *For every* $k \geq 1$, $_k \#\mathbf{RS} \subseteq {}_k\mathbf{P}$.

Proof. Construction. Let $k \geq 1$ be a positive integer. Let $H = (Q, T \cup \{\#\}, s, R)$ be a #-rewriting system of index k, where $\Sigma = T \cup \{\#\}$. We construct the programmed grammar of index k, $G = (V, T, P, S)$, where the sets of nonterminals $N = V - T$ and the rules of P are constructed as follows:

(1) $S = \langle s, 1, 1 \rangle$;
(2) $N = \{\langle p, i, h \rangle \mid p \in Q, 1 \leq i \leq h, i \leq h \leq k\} \cup \{\langle q', i, h \rangle \mid q \in Q, 1 \leq i \leq h,$
$i \leq h \leq k\} \cup \{\langle q'', i, h \rangle \mid q \in Q, 1 \leq i \leq h, i \leq h \leq k\} \cup \{\langle q'', 1, 0 \rangle \mid q \in Q\}$;
(3) For every rule $r: p_i \# \to qy \in R$, $y = y_0 \# y_1 \ldots y_{m-1} \# y_m$, $y_0, y_1, y_2 \ldots y_m \in T^*$, add the following set to P:

  (i) $\{\langle p, j, h \rangle \to \langle q', j, h+m-1 \rangle,$
    $\{r' \mid$ if $j+1 = i$ then $r': \langle p, i, h \rangle \to \langle q'', i, h+m-1 \rangle$
          else $r': \langle p, j+1, h \rangle \to \langle q', j+1, h+m-1 \rangle \}$
  $\mid 1 \leq j < i, i \leq h \leq h_{max}\}$
    $\cup$
  (ii) $\{\langle p, i, h \rangle \to \langle q'', i, h+m-1 \rangle,$
    $\{r' \mid$ if $i = h$, then $r': \langle q'', i, h+m-1 \rangle \to$
          $y_0 \langle q', i, h+m-1 \rangle y_1 \langle q', i+1, h+m-1 \rangle y_2 \ldots$
          $\ldots y_{m-1} \langle q', i+m-1, h+m-1 \rangle y_m$
          else $r': \langle p, i+1, h \rangle \to$
          $\langle q', i+1+m-1, h+m-1 \rangle \}$
  $\mid i \leq h \leq h_{max}\}$
    $\cup$
  (iii) $\{\langle p, j, h \rangle \to \langle q', j+m-1, h+m-1 \rangle,$
    $\{r' \mid$ if $j = h$, then $r': \langle q'', i, h+m-1 \rangle \to$
          $y_0 \langle q', i, h+m-1 \rangle y_1 \langle q', i+1, h+m-1 \rangle y_2 \ldots$
          $\ldots y_{m-1} \langle q', i+m-1, h+m-1 \rangle y_m$
          else $r': \langle p, j+1, h \rangle \to \langle q', j+1+m-1, h+m-1 \rangle \}$
  $\mid i < j \leq h, i \leq h \leq h_{max}\}$
    $\cup$

(iv) $\{\langle q'', i, h + m - 1\rangle \rightarrow$
$$y_0\langle q', i, h + m - 1\rangle y_1 \langle q', i + 1, h + m - 1\rangle y_2 \ldots$$
$$\ldots y_{m-1}\langle q', i + m - 1, h + m - 1\rangle y_m,$$
$$\{r' \mid r': \langle q', 1, h + m - 1\rangle \rightarrow \langle q, 1, h + m - 1\rangle\}$$
$$\mid i \leq h \leq h_{max}\}$$
$$\cup$$

(v) $\{\langle q', j, h + m - 1\rangle \rightarrow \langle q, j, h + m - 1\rangle,$
$$\{r' \mid \text{ if } j < h + m - 1, \text{ then } r': \langle q', j + 1, h + m - 1\rangle \rightarrow$$
$$\langle q, j + 1, h + m - 1\rangle$$
$$\text{else } r': \langle \tilde{p}, 1, h + m - 1\rangle \rightarrow$$
$$\langle \tilde{q}', 1, h + m - 1 + \tilde{m} - 1\rangle,$$

where
$$\tilde{p}_{\tilde{i}}\# \rightarrow \tilde{q}\tilde{y}_0\#\tilde{y}_1 \ldots \tilde{y}_{\tilde{m}-1}\#\tilde{y}_{\tilde{m}} \in R, \tilde{y}_0, \tilde{y}_1, \ldots, \tilde{y}_{\tilde{m}} \in T^*,$$
$$\text{if } \tilde{i} = 1, \text{ then } \tilde{q}' := \tilde{q}''\}$$
$$\mid 1 \leq j \leq h + m - 1, i \leq h \leq h_{max}\},$$

where $h_{max} = k$ if $m = 0$; otherwise $h_{max} = k - m + 1$.

Basic Idea. By several derivation steps, G simulates a single step in H. Inside of every nonterminal of the form $\langle p, i, h\rangle$ occurring in a sentential form of G, we record

(a) p—the current state of H;
(b) i—the position of the occurrence of # in H's current configuration;
(c) h—the total number of all #s in the current configuration.

From these three pieces of information and the set $g(p)$ associated with p, we find out whether p is applicable in the next step and if so, we simulate the step by rules introduced in the third step of the above construction as follows:

1. inside of all nonterminals in the sentential form, change h to $h + m - 1$, where m is the number of nonterminals occurring on the right-hand side of p, so $h + m - 1$ is the number of nonterminals after the application of p (see (i) through (iii));
2. in the nonterminals that follow the rewritten nonterminal, change their position so it corresponds to the position after the application of p (see (iii));
3. apply p and select a rule label q from $g(p)$ to be applied in the next step (see (iv));
4. complete the simulated derivation step in H by rules introduced in (v).

We leave a rigorous version of this proof to the reader.   □

Theorem 11.2.5. *For every $k \geq 1$, $_k\#\mathbf{RS} \subset {}_{k+1}\#\mathbf{RS}$.*

Proof. $_k\mathbf{P} = {}_k\#\mathbf{RS}$ follows from Lemmas 11.2.1 and 11.2.4. Thus, from Theorems 3.5.5, 11.2.5 holds.   □

Before closing this chapter, we suggest some open problem areas related to its subject.

Open Problem 11.2.6. The present chapter has concentrated its attention on the #-rewriting systems of index k, where $k \geq 1$. Consider these systems without this index-based restriction. What is their generative power?

Finally, we suggest three variants of the #-rewriting systems to be discussed in the future. Let $M = (Q, \Sigma, s, R)$ be a #-rewriting system.

(1) M is *deterministic* if for every $p \in Q$ and every positive integer i, $p_i\#$ is the left-hand side of no more than one rule in M.

(2) Let $\alpha, \beta \in \chi$. If $\alpha \Rightarrow \beta$ in M, then M *directly reduces* β to α, denoted by $\beta \vdash \alpha$. The *language reduced by M*, $_rL(M)$, is defined as

$$_rL(M) = \{w \mid qw \vdash^* s\#, q \in Q, w \in (\Sigma - \{\#\})^*\}$$

where $\vdash^*$ is the transitive and reflexive closure of $\vdash$.

(3) M works in a *parallel way* if it simultaneously rewrites all #s in the current configuration during a single derivation step.

Chapter 12
Multigenerative Grammar Systems and Parallel Computation

Today's environment of cooperating multiprocessor computers allows us to base modern information technologies on a large combination of simultaneously running processes, which make use of this powerful environment as much as possible. Consequently, parallel computation plays a crucially important role in computer science at present as already pointed out in the beginning of Chap. 4.

Recall that parallel computation is conceptually accomplished by breaking a large computational task into many independent subtasks, which are simultaneously executed. Once they are completed, their results are combined together. Of course, to obtain solid knowledge about this way of computation, we need formal models that adequately formalize computational parallelism, including the final combination of the achieved results. The present section describes multigenerative grammar systems, which can serve for this purpose very well. Indeed, they consist of several simultaneously working components represented by context-free grammars, so they reflect and formalize concurrent computation in a natural and proper way.

The chapter consists of two sections. The first section introduces the basic versions of multigenerative grammar systems. During one generation step, each of their grammatical components rewrites a nonterminal in its sentential form. After this simultaneous generation is completed, all the generated strings are composed into a single string by some common string operation, such as union and concatenation. More precisely, for a positive integer n, an n-generative grammar system works with n context-free grammatical components, each of which makes a derivation, and these n derivations are simultaneously controlled by a finite set of n-tuples consisting of rules. In this way, the grammar system generates n terminal strings, which are combined together by operation union, concatenation or the selection of the first generated string. We show that these systems characterize the family of matrix languages. In addition, we demonstrate that multigenerative grammar systems with any number of grammatical components can be transformed to equivalent two-component versions of these systems. Section 12.2 discusses leftmost versions of multigenerative grammar systems in which each generation step is performed in the

© Springer International Publishing AG 2017
A. Meduna, O. Soukup, *Modern Language Models and Computation*,
DOI 10.1007/978-3-319-63100-4_12

leftmost manner. That is, all the grammatical components of these versions rewrite the leftmost nonterminal occurrence in their sentential forms; otherwise, they work as the basic versions. We prove that leftmost multigenerative grammar systems are more powerful than their basic versions. Indeed, they generate the family of recursively enumerable languages, which properly contains the family of matrix languages (see Theorems 3.3.6 and 3.4.5). We also consider regulation by n-tuples of nonterminals, rather than rules, and prove that leftmost multigenerative grammar systems regulated by rules or nonterminals have the same generative power. In addition, like for the basic versions, we demonstrate that leftmost multigenerative grammar systems with any number of grammatical components can be transformed to equivalent two-component versions of these systems.

12.1 Multigenerative Grammar Systems

In this section, we define multigenerative grammar systems and demonstrate that they are as powerful as matrix grammars. We also show that any multigenerative grammar system can be transformed to an equivalent two-component multigenerative grammar system.

Definition 12.1.1. An *n-generative rule-synchronized grammar system* (an *n-MGR* for short) is an $n + 1$ tuple

$$\Gamma = (G_1, G_2, \ldots, G_n, Q)$$

where

- $G_i = (V_i, T_i, P_i, S_i)$ is a context-free grammar, for each $i = 1, \ldots, n$;
- Q is a finite set of n-tuples of the form $(p_1, p_2, \ldots, p_n)$, where $p_i \in P_i$, for all $i = 1, \ldots, n$.

A *sentential n-form* is an n-tuple of the form $\chi = (x_1, x_2, \ldots, x_n)$, where $x_i \in V_i^*$, for all $i = 1, \ldots, n$. Let $\chi = (u_1 A_1 v_1, u_2 A_2 v_2, \ldots, u_n A_n v_n)$ and $\bar{\chi} = (u_1 x_1 v_1, u_2 x_2 v_2, \ldots, u_n x_n v_n)$ be two sentential n-forms, where $A_i \in N_i$ and $u_i, v_i, x_i \in V_i^*$, for all $i = 1, \ldots, n$. Let $(p_i: A_i \rightarrow x_i) \in P_i$, for all $i = 1, \ldots, n$ and $(p_1, p_2, \ldots, p_n) \in Q$. Then, χ directly derives $\bar{\chi}$ in Γ, denoted by

$$\chi \Rightarrow_\Gamma \bar{\chi}$$

In the standard way, we generalize $\Rightarrow_\Gamma$ to $\Rightarrow_\Gamma^k$, for all $k \geq 0$, $\Rightarrow_\Gamma^*$, and $\Rightarrow_\Gamma^+$.

The *n-language of* Γ, denoted by $n\text{-}L(\Gamma)$, is defined as

$$n\text{-}L(\Gamma) = \{(w_1, w_2, \ldots, w_n) \mid (S_1, S_2, \ldots, S_n) \Rightarrow_\Gamma^* (w_1, w_2, \ldots, w_n),$$
$$w_i \in T_i^*, \text{ for all } i = 1, \ldots, n\}$$

The *language generated by* Γ *in the union mode*, $L_{union}(\Gamma)$, is defined as

$$L_{union}(\Gamma) = \bigcup_{i=1}^{n} \{w_i \mid (w_1, w_2, \ldots, w_n) \in n\text{-}L(\Gamma)\}$$

The *language generated by* Γ *in the concatenation mode*, $L_{conc}(\Gamma)$, is defined as

$$L_{conc}(\Gamma) = \{w_1 w_2 \ldots w_n \mid (w_1, w_2, \ldots, w_n) \in n\text{-}L(\Gamma)\}$$

The *language generated by* Γ *in the first mode*, $L_{first}(\Gamma)$, is defined as

$$L_{first}(\Gamma) = \{w_1 \mid (w_1, w_2, \ldots, w_n) \in n\text{-}L(\Gamma)\} \qquad \square$$

We illustrate the above definition by an example.

Example 12.1.2. Consider the 2-MGR $\Gamma = (G_1, G_2, Q)$, where

- $G_1 = (\{S_1, A_1, a, b, c\}, \{a, b, c\}, \{1\colon S_1 \to aS_1, 2\colon S_1 \to aA_1, 3\colon A_1 \to bA_1c, 4\colon A_1 \to bc\}, S_1)$,
- $G_2 = (\{S_2, A_2, d\}, \{d\}, \{1\colon S_2 \to S_2A_2, 2\colon S_2 \to A_2, 3\colon A_2 \to d\}, S_2)$,
- $Q = \{(1, 1), (2, 2), (3, 3), (4, 3)\}$.

Observe that

- $2\text{-}L(\Gamma) = \{(a^n b^n c^n, d^n) \mid n \geq 1\}$,
- $L_{union}(\Gamma) = \{a^n b^n c^n \mid n \geq 1\} \cup \{d^n \mid n \geq 1\}$,
- $L_{conc}(\Gamma) = \{a^n b^n c^n d^n \mid n \geq 1\}$, and
- $L_{first}(\Gamma) = \{a^n b^n c^n \mid n \geq 1\}$. $\qquad \square$

Next, we prove that multigenerative grammar systems under all of the defined modes are equivalent to matrix grammars. Throughout the rest of this section, as indicated in Sect. 2.3.1, for brevity, we use rules and rule labels interchangeably.

Algorithm 12.1.3. *Conversion of an n-MGR in the union mode to an equivalent matrix grammar.*

Input: An n-MGR, $\Gamma = (G_1, G_2, \ldots, G_n, Q)$.
Output: A matrix grammar, $H = (G, M)$, satisfying $L_{union}(\Gamma) = L(H)$.
Method: Let $G_i = (V_i, T_i, P_i, S_i)$, for all $i = 1, \ldots, n$, and without any loss of generality, we assume that N_1 through N_n are pairwise disjoint. Let us choose arbitrary S satisfying $S \notin \bigcup_{j=1}^{n} N_j$. Then, construct

$$G = (V, T, P, S)$$

where

- $N = \{S\} \cup (\bigcup_{i=1}^{n} N_i) \cup (\bigcup_{i=1}^{n} \{\bar{A} \mid A \in N_i\})$;
- $T = \bigcup_{i=1}^{n} T_i$;

- $P = \{(s_1: S \rightarrow S_1 h(S_2) \ldots h(S_n)),$
  $(s_2: S \rightarrow h(S_1) S_2 \ldots h(S_n)),$

  $\vdots$

  $(s_n: S \rightarrow h(S_1) h(S_2) \ldots S_n)\}$
  $\cup (\bigcup_{i=1}^{n} P_i)$
  $\cup (\bigcup_{i=1}^{n} \{h(A) \rightarrow h(x) \mid A \rightarrow x \in P_i\}),$
  where h is a homomorphism from $((\bigcup_{i=1}^{n} V_i))^*$ to $(\bigcup_{i=1}^{n} \{\bar{A} \mid A \in N_i\})^*$,
  defined as $h(a) = \varepsilon$, for all $a \in \bigcup_{i=1}^{n} T_i$, and $h(A) = \bar{A}$, for all $A \in \bigcup_{i=1}^{n} N_i$.
- $M = \{s_1, s_2, \ldots, s_n\}$
  $\cup \{p_1 \bar{p}_2 \ldots \bar{p}_n \mid (p_1, p_2, \ldots, p_n) \in Q\}$
  $\cup \{\bar{p}_1 p_2 \ldots \bar{p}_n \mid (p_1, p_2, \ldots, p_n) \in Q\}$

  $\vdots$

  $\cup \{\bar{p}_1 \bar{p}_2 \ldots p_n \mid (p_1, p_2, \ldots, p_n) \in Q\}.$ □

Theorem 12.1.4. *Let $\Gamma = (G_1, G_2, \ldots, G_n, Q)$ be an n-MGR. With Γ as its input, Algorithm 12.1.3 halts and correctly constructs a matrix grammar, $H = (G, M)$, such that $L_{union}(\Gamma) = L(H)$.*

Proof. Let $(p: A \rightarrow x)$ be a rule. Then, for simplicity and brevity, $\bar{p}$ denotes the rule $h(A) \rightarrow h(x)$. To prove this theorem, we first establish Claims 12.1.5 and 12.1.6.

Claim 12.1.5. Let $(S_1, S_2, \ldots, S_n) \Rightarrow_{\Gamma}^{m} (y_1, y_2, \ldots, y_n)$, where $m \geq 0, y_i \in V_i^*$, for all $i = 1, \ldots, n$. Then, $S \Rightarrow_{H}^{m+1} h(y_1) h(y_2) \ldots h(y_{j-1}) y_j h(y_{j+1}) \ldots h(y_n)$, for any $j = 1, \ldots, n$.

Proof. This claim is proved by induction on $m \geq 0$.

Basis. Let $m = 0$. Then, $(S_1, S_2, \ldots, S_n) \Rightarrow_{\Gamma}^{0} (S_1, S_2, \ldots, S_n)$. Notice that

$$S \Rightarrow_H h(S_1) h(S_2) \ldots h(S_{j-1}) S_j h(S_{j+1}) \ldots h(S_n)$$

for any $j = 1, \ldots, n$, because

$$(s_j: S \rightarrow h(S_1) h(S_2) \ldots h(S_{j-1}) S_j h(S_{j+1}) \ldots h(S_n)) \in M$$

Induction Hypothesis. Assume that the claim holds for all m-step derivations, where $m = 0, \ldots, k$, for some $k \geq 0$.

Induction Step. Consider any derivation of the form

$$(S_1, S_2, \ldots, S_n) \Rightarrow_{\Gamma}^{k+1} (y_1, y_2, \ldots, y_n)$$

Then, there exists a sentential n-form $(u_1 A_1 v_1, u_2 A_2 v_2, \ldots, u_n A_n v_n)$, where $u_i, v_i \in V_i^*, A_i \in N_i$ such that

$$(S_1, S_2, \ldots, S_n) \Rightarrow_{\Gamma}^{k} (u_1 A_1 v_1, u_2 A_2 v_2, \ldots, u_n A_n v_n)$$
$$\Rightarrow_{\Gamma} (u_1 x_1 v_1, u_2 x_2 v_2, \ldots, u_n x_n v_n)$$

where $u_i x_i v_i = y_i$, for all $i = 1, \ldots, n$. First, observe that

$$(S_1, S_2, \ldots, S_n) \Rightarrow_\Gamma^k (u_1 A_1 v_1, u_2 A_2 v_2, \ldots, u_n A_n v_n)$$

implies that

$$S \Rightarrow_H^{k+1} h(u_1 A_1 v_1) h(u_2 A_2 v_2) \ldots h(u_{j-1} A_{j-1} v_{j-1})$$
$$u_j A_j v_j h(u_{j+1} A_{j+1} v_{j+1}) \ldots h(u_n A_n v_n)$$

for any $j = 1, \ldots, n$ by the induction hypothesis. Furthermore, let

$$(u_1 A_1 v_1, u_2 A_2 v_2, \ldots, u_n A_n v_n) \Rightarrow_\Gamma$$
$$(u_1 x_1 v_1, u_2 x_2 v_2, \ldots, u_n x_n v_n)$$

Then, $((p_1 : A_1 \rightarrow x_1), (p_2 : A_2 \rightarrow x_2), \ldots, (p_n : A_n \rightarrow x_n)) \in Q$. Algorithm 12.1.3 implies that $\bar{p}_1 \bar{p}_2 \ldots p_{j-1}^- p_j p_{j+1}^- \ldots \bar{p}_n \in M$, for any $j = 1, \ldots, n$. Hence,

$$h(u_1 A_1 v_1) h(u_2 A_2 v_2) \ldots h(u_{j-1} A_{j-1} v_{j-1}) u_j A_j v_j h(u_{j+1} A_{j+1} v_{j+1}) \ldots h(u_n A_n v_n) \Rightarrow_H$$
$$h(u_1 x_1 v_1) h(u_2 x_2 v_2) \ldots h(u_{j-1} x_{j-1} v_{j-1}) u_j x_j v_j h(u_{j+1} x_{j+1} v_{j+1}) \ldots h(u_n x_n v_n)$$

by matrix $\bar{p}_1 \bar{p}_2 \ldots p_{j-1}^- p_j p_{j+1}^- \ldots \bar{p}_n$, for any $j = 1, \ldots, n$. As a result, we obtain

$$S \Rightarrow_H^{k+2} h(u_1 x_1 v_1) h(u_2 x_2 v_2) \ldots h(u_{j-1} x_{j-1} v_{j-1})$$
$$u_j x_j v_j h(u_{j+1} x_{j+1} v_{j+1}) \ldots h(u_n x_n v_n)$$

for any $j = 1, \ldots, n$. □

Claim 12.1.6. Let $S \Rightarrow_H^m y$, where $m \geq 1, y \in V^*$. Then, there exist $j \in \{1, \ldots, n\}$ and $y_i \in V_i^*$, for $i = 1, \ldots, n$, such that $(S_1, \ldots, S_n) \Rightarrow_\Gamma^{m-1} (y_1, \ldots, y_n)$ and $y = h(y_1) \ldots h(y_{j-1}) y_j h(y_{j+1}) \ldots h(y_n)$.

Proof. This claim is proved by induction on $m \geq 1$.

Basis. Let $m = 1$. Then, there exists exactly one of the following one-step derivations in H:

$$S \Rightarrow_H S_1 h(S_2) \ldots h(S_n) \text{ by matrix } s_1, \text{ or}$$
$$S \Rightarrow_H h(S_1) S_2 \ldots h(S_n) \text{ by matrix } s_2, \text{ or}$$
$$\ldots, \text{ or}$$
$$S \Rightarrow_H h(S_1) h(S_2) \ldots S_n \text{ by matrix } s_n$$

Notice that trivially $(S_1, S_2, \ldots, S_n) \Rightarrow_\Gamma^0 (S_1, S_2, \ldots, S_n)$.

Induction Hypothesis. Assume that the claim holds for all m-step derivations, where $m = 1, \ldots, k$, for some $k \geq 1$.

Induction Step. Consider any derivation of the form

$$S \Rightarrow_H^{k+1} y$$

Then, there exists a sentential form w such that

$$S \Rightarrow_H^k w \Rightarrow_H y$$

where $w, y \in (N \cup T)^*$. As $w \Rightarrow_H y$, this derivation step can use only a matrix of the form $p_1 p_2 \ldots p_{j-1} p_j p_{j+1} \ldots p_n \in Q$, where p_j is a rule from P_j and $\bar{p}_i \in h(P_i)$, for $i = 1, \ldots, j-1, j+1, \ldots, n$. Hence, $w \Rightarrow_H y$ can be written as

$$h(w_i) \ldots h(w_{j-1}) w_j h(w_{j+1}) \ldots h(w_n) \Rightarrow_H z_1 \ldots z_n$$

where $w_j \Rightarrow_H z_j$ by the rule p_j and $h(w_i) \Rightarrow_H z_i$ by $\bar{p}_i$, for $i = 1, \ldots, j-1, j+1, \ldots, n$. Each rule $\bar{p}_i$ rewrites a barred nonterminal $\bar{A}_i \in h(N_i)$. Of course, then each rule p_i can be used to rewrite the respective occurrence of a non-barred nonterminal A_i in w_i in such a way that $w_i \Rightarrow_H y_i$ and $h(y_i) = z_i$, for all $i = 1, \ldots, j-1, j+1, \ldots, n$. By setting $y_j = z_j$, we obtain

$$(w_1, \ldots, w_n) \Rightarrow_\Gamma (y_1, \ldots, y_n)$$

and $y = h(y_1) \ldots h(y_{j-1}) y_j h(y_{j+1}) \ldots h(y_n)$. As a result, we obtain

$$(S_1, S_2, \ldots, S_{j-1}, S_j, S_{j+1}, \ldots, S_n) \Rightarrow_\Gamma^k$$
$$(u_1 x_1 v_1, u_2 x_2 v_2, \ldots, u_{j-1} x_{j-1} v_{j-1}, u_j x_j v_j, u_{j+1} x_{j+1} v_{j+1}, \ldots, u_n x_n v_n)$$

so $y = u_1 x_1 v_1 u_2 x_2 v_2 \ldots u_{j-1} x_{j-1} v_{j-1} u_j x_j v_j u_{j+1} x_{j+1} v_{j+1} \ldots u_n x_n v_n$.   □

Consider Claim 12.1.5 for $y_i \in T_i^*$, for all $i = 1, \ldots, n$. Notice that $h(a) = \varepsilon$, for all $a \in T_i$. We obtain an implication of the form

$$\text{if } (S_1, S_2, \ldots, S_n) \Rightarrow_\Gamma^* (y_1, y_2, \ldots, y_n)$$
$$\text{then } S \Rightarrow_H^* y_j, \text{ for any } j = 1, \ldots, n$$

Hence, $L_{union}(\Gamma) \subseteq L(H)$. Consider Claim 12.1.5 for $y \in T^*$. Notice that $h(a) = \varepsilon$, for all $a \in T_i$. We obtain an implication of the form

$$\text{if } S \Rightarrow_H^* y$$
$$\text{then } (S_1, S_2, \ldots, S_n) \Rightarrow_\Gamma^* (y_1, y_2, \ldots, y_n)$$

and there exist an index $j = 1, \ldots, n$ such that $y = y_j$. Hence, $L(H) \subseteq L_{union}(\Gamma)$.   □

Algorithm 12.1.7. *Conversion of an n-MGR in the concatenation mode to an equivalent matrix grammar.*

Input: An n-MGR, $\Gamma = (G_1, G_2, \ldots, G_n, Q)$.
Output: A matrix grammar, $H = (G, M)$, satisfying $L_{conc}(\Gamma) = L(H)$.
Method: Let $G_i = (V_i, T_i, P_i, S_i)$, for all $i = 1, \ldots, n$, and without any loss of generality, we assume that N_1 through N_n are pairwise disjoint. Let us choose arbitrary S satisfying $S \notin \bigcup_{j=1}^{n} N_j$. Construct

$$G = (V, T, P, S)$$

where

- $N = \{S\} \cup (\bigcup_{i=1}^{n} N_i)$;
- $T = \bigcup_{i=1}^{n} T_i$;
- $P = \{(s\colon S \to S_1 S_2 \ldots S_n)\} \cup (\bigcup_{i=1}^{n} P_i)$.

Finally, set $M = \{s\} \cup \{p_1 p_2 \ldots p_n \mid (p_1, p_2, \ldots, p_n) \in Q\}$. $\square$

Theorem 12.1.8. *Let $\Gamma = (G_1, G_2, \ldots, G_n, Q)$ be an n-MGR. On input Γ, Algorithm 12.1.7 halts and correctly constructs a matrix grammar, $H = (G, M)$, such that $L_{conc}(\Gamma) = L(H)$.*

Proof. To prove this theorem, we first establish Claims 12.1.9 and 12.1.10.

Claim 12.1.9. Let $(S_1, S_2, \ldots, S_n) \Rightarrow_{\Gamma}^{m} (y_1, y_2, \ldots, y_n)$, where $m \geq 0, y_i \in V_i^*$, for all $i = 1, \ldots, n$. Then, $S \Rightarrow_{H}^{m+1} y_1 y_2 \ldots y_n$.

Proof. This claim is proved by induction on $m \geq 0$.

Basis. Let $m = 0$. Then, $(S_1, S_2, \ldots, S_n) \Rightarrow_{\Gamma}^{0} (S_1, S_2, \ldots, S_n)$. Notice that $S \Rightarrow_H S_1 S_2 \ldots S_n$, because $(s\colon S \to S_1 S_2 \ldots S_n) \in M$.

Induction Hypothesis. Assume that the claim holds for all m-step derivations, where $m = 0, \ldots, k$, for some $k \geq 0$.

Induction Step. Consider any derivation of the form

$$(S_1, S_2, \ldots, S_n) \Rightarrow_{\Gamma}^{k+1} (y_1, y_2, \ldots, y_n)$$

Then, there exists a sentential n-form $(u_1 A_1 v_1, u_2 A_2 v_2, \ldots, u_n A_n v_n)$, where $u_i, v_i \in V_i^*, A_i \in N_i$ such that

$$(S_1, S_2, \ldots, S_n) \Rightarrow_{\Gamma}^{k} (u_1 A_1 v_1, u_2 A_2 v_2, \ldots, u_n A_n v_n)$$
$$\Rightarrow_{\Gamma} (u_1 x_1 v_1, u_2 x_2 v_2, \ldots, u_n x_n v_n)$$

where $u_i x_i v_i = y_i$, for all $i = 1, \ldots, n$. First, observe that

$$(S_1, S_2, \ldots, S_n) \Rightarrow_{\Gamma}^{k} (u_1 A_1 v_1, u_2 A_2 v_2, \ldots, u_n A_n v_n)$$

implies that

$$S \Rightarrow_H^{k+1} u_1 A_1 v_1 u_2 A_2 v_2 \ldots u_n A_n v_n$$

by the induction hypothesis. Furthermore, let

$$(u_1 A_1 v_1, u_2 A_2 v_2, \ldots, u_n A_n v_n) \Rightarrow_\Gamma$$
$$(u_1 x_1 v_1, u_2 x_2 v_2, \ldots, u_n x_n v_n)$$

Then, it holds that $((p_1 : A_1 \rightarrow x_1), (p_2 : A_2 \rightarrow x_2), \ldots, (p_n : A_n \rightarrow x_n)) \in Q$. Algorithm 12.1.7 implies that $p_1 p_2 \ldots p_n \in M$. Hence,

$$u_1 A_1 v_1 u_2 A_2 v_2 \ldots u_n A_n v_n \Rightarrow_H$$
$$u_1 x_1 v_1 u_2 x_2 v_2 \ldots u_n x_n v_n$$

by matrix $p_1 p_2 \ldots p_n$. As a result, we obtain

$$S \Rightarrow_H^{k+2} u_1 x_1 v_1 u_2 x_2 v_2 \ldots u_n x_n v_n \qquad \square$$

Claim 12.1.10. Let $S \Rightarrow_H^m y$, where $m \geq 1, y \in V^*$. Then, $(S_1, S_2, \ldots, S_n) \Rightarrow_\Gamma^{m-1}$ $(y_1, y_2, \ldots, y_n)$ such that $y = y_1 y_2 \ldots y_n$, where $y_i \in V_i^*$, for all $i = 1, \ldots, n$.

Proof. This claim is proved by induction on $m \geq 1$.

Basis. Let $m = 1$. Then, there exists exactly one one-step derivation in H: $S \Rightarrow_H$ $S_1 S_2 \ldots, S_n$ by matrix s. Notice that $(S_1, S_2, \ldots, S_n) \Rightarrow_\Gamma^0 (S_1, S_2, \ldots, S_n)$ trivially.

Induction Hypothesis. Assume that the claim holds for all m-step derivations, where $m = 1, \ldots, k$, for some $k \geq 1$.

Induction Step. Consider any derivation of the form

$$S \Rightarrow_H^{k+1} y$$

Then, there exists a sentential form w such that

$$S \Rightarrow_H^k w \Rightarrow_H y$$

where $w, y \in V^*$. First, observe that $S \Rightarrow_H^k w$ implies that

$$(S_1, S_2, \ldots, S_n) \Rightarrow_\Gamma^{k-1} (w_1, w_2, \ldots, w_n)$$

so that $w = w_1 w_2 \ldots w_n$, where $w_i \in V_i^*$, for all $i = 1, \ldots, n$, by the induction hypothesis. Furthermore, let $w \Rightarrow_H y$ by matrix $p_1 p_2 \ldots p_n \in M$, where $w = w_1 w_2 \ldots w_n$. Let p_i be a rule of the form $A_i \rightarrow x_i$. The rule p_i can be applied only inside substring w_i, for all $i = 1, \ldots, n$. Assume that $w_i = u_i A_i v_i$, where $u_i, v_i \in V^*, A_i \in N_i$, for all $i = 1, \ldots, n$. There exist a derivation step

$$u_1A_1v_1u_2A_2v_2\ldots u_nA_nv_n \Rightarrow_H$$
$$u_1x_1v_1u_2x_2v_2\ldots u_nx_nv_n$$

by matrix $p_1p_2\ldots p_n \in M$. Algorithm 12.1.7 implies that

$$\big((p_1\!:\!A_1 \to x_1), (p_2\!:\!A_2 \to x_2), \ldots, (p_n\!:\!A_n \to x_n)\big) \in Q$$

because $p_1p_2\ldots p_n \in M$. Hence,

$$(u_1A_1v_1, u_2A_2v_2, \ldots, u_nA_nv_n \Rightarrow_\Gamma$$
$$(u_1x_1v_1, u_2x_2v_2, \ldots, u_nx_nv_n)$$

As a result, we obtain

$$(S_1, S_2, \ldots, S_n) \Rightarrow_\Gamma^k (u_1x_1v_1, u_2x_2v_2, \ldots, u_nx_nv_n)$$

so that $y = u_1x_1v_1u_2x_2v_2\ldots u_nx_nv_n$. □

Consider Claim 12.1.9 for $y_i \in T_i^*$, for all $i = 1, \ldots, n$. We obtain an implication of this form

$$\text{if } (S_1, S_2, \ldots, S_n) \Rightarrow_\Gamma^* (y_1, y_2, \ldots, y_n)$$
$$\text{then } S \Rightarrow_H^* y_1y_2\ldots y_n$$

Hence, $L_{conc}(\Gamma) \subseteq L(H)$. Consider Claim 12.1.10 for $y \in T^*$. We obtain an implication of the form

$$\text{if } S \Rightarrow_H^* y \text{ then } (S_1, S_2, \ldots, S_n) \Rightarrow_\Gamma^* (y_1, y_2, \ldots, y_n), \text{ such that } y = y_1y_2\ldots y_n$$

Hence, $L(H) \subseteq L_{conc}(\Gamma)$. □

Algorithm 12.1.11. *Conversion of an n-MGR in the first mode to an equivalent matrix grammar.*

Input: An n-MGR, $\Gamma = (G_1, G_2, \ldots, G_n, Q)$.
Output: A matrix grammar, $H = (G, M)$, satisfying $L_{first}(\Gamma) = L(H)$.
Method: Let $G_i = (V_i, T_i, P_i, S_i)$, for all $i = 1, \ldots, n$, and without any loss of generality, we assume that N_1 through N_n are pairwise disjoint. Let us choose arbitrary S satisfying $S \notin \bigcup_{j=1}^n N_j$. Construct

$$G = (V, T, P, S)$$

where

- $N = \{S\} \cup N_1 \cup (\bigcup_{i=2}^n \{\bar{A}\!:\!A \in N_i\})$;
- $T = T_1$;

- $P = \{(s: S \to S_1 h(S_2) \ldots h(S_n))\} \cup P_1$
  $\cup \; (\bigcup_{i=2}^{n} \{h(A) \to h(x) \mid A \to x \in P_i\})$,
  where h is a homomorphism from $((\bigcup_{i=2}^{n} V_i))^*$ to $(\bigcup_{i=2}^{n} \{\bar{A} \mid A \in N_i\})^*$
  defined as $h(a) = \varepsilon$, for all $a \in \bigcup_{i=2}^{n} T_i$ and $h(A) = \bar{A}$, for all $A \in \bigcup_{i=2}^{n} N_i$.

Finally, set $M = \{s\} \cup \{p_1 \bar{p}_2 \ldots \bar{p}_n \mid (p_1, p_2, \ldots, p_n) \in Q\}$.   □

Theorem 12.1.12. *Let* $\Gamma = (G_1, G_2, \ldots, G_n, Q)$ *be an n-MGR. With* Γ *as its input, Algorithm 12.1.11 halts and correctly constructs a matrix grammar,* $H = (G, M)$, *such that* $L_{first}(\Gamma) = L(H)$.

Proof. Let $(p: A \to x)$ be a rule. Then, for simplicity and brevity, $\bar{p}$ denotes the rule $h(A) \to h(x)$. To prove this theorem, we first establish Claims 12.1.13 and 12.1.14.

Claim 12.1.13. Let $(S_1, S_2, \ldots, S_n) \Rightarrow_\Gamma^m (y_1, y_2, \ldots, y_n)$, where $m \geq 0$, $y_i \in V_i^*$, for all $i = 1, \ldots, n$. Then, $S \Rightarrow_H^{m+1} y_1 h(y_2) \ldots h(y_n)$.

Proof. This claim is proved by induction on $m \geq 0$.

Basis. Let $m = 0$. Then, $(S_1, S_2, \ldots, S_n) \Rightarrow_\Gamma^0 (S_1, S_2, \ldots, S_n)$. Notice that $S \Rightarrow_H S_1 h(S_2) \ldots h(S_n)$, because $(s: S \to S_1 h(S_2) \ldots h(S_n)) \in M$.

Induction Hypothesis. Assume that the claim holds for all m-step derivations, where $m = 0, \ldots, k$, for some $k \geq 0$.

Induction Step. Consider any derivation of the form

$$(S_1, S_2, \ldots, S_n) \Rightarrow_\Gamma^{k+1} (y_1, y_2, \ldots, y_n)$$

Then, there exists a sentential n-form $(u_1 A_1 v_1, u_2 A_2 v_2, \ldots, u_n A_n v_n)$, where $u_i, v_i \in V_i^*, A_i \in N_i$ such that

$$(S_1, S_2, \ldots, S_n) \Rightarrow_\Gamma^k (u_1 A_1 v_1, u_2 A_2 v_2, \ldots, u_n A_n v_n)$$
$$\Rightarrow_\Gamma (u_1 x_1 v_1, u_2 x_2 v_2, \ldots, u_n x_n v_n)$$

where $u_i x_i v_i = y_i$, for all $i = 1, \ldots, n$. First, observe that

$$(S_1, S_2, \ldots, S_n) \Rightarrow_\Gamma^k (u_1 A_1 v_1, u_2 A_2 v_2, \ldots, u_n A_n v_n)$$

implies that

$$S \Rightarrow_H^{k+1} u_1 A_1 v_1 h(u_2 A_2 v_2) \ldots h(u_n A_n v_n)$$

by the induction hypothesis. Furthermore, let

$$(u_1 A_1 v_1, u_2 A_2 v_2, \ldots, u_n A_n v_n) \Rightarrow_\Gamma$$
$$(u_1 x_1 v_1, u_2 x_2 v_2, \ldots, u_n x_n v_n)$$

Then, it holds $((p_1: A_1 \to x_1), (p_2: A_2 \to x_2), \ldots, (p_n: A_n \to x_n)) \in Q$. Algorithm 12.1.11 implies that $p_1 \bar{p}_2 \ldots \ldots \bar{p}_n \in M$. Hence,

$$u_1 A_1 v_1 h(u_2 A_2 v_2) \ldots h(u_n A_n v_n) \Rightarrow_H$$
$$u_1 x_1 v_1 h(u_2 x_2 v_2) \ldots h(u_n x_n v_n)$$

by matrix $p_1 \bar{p}_2 \ldots \bar{p}_n$. As a result, we obtain

$$S \Rightarrow_H^{k+2} u_1 x_1 v_1 h(u_2 x_2 v_2) \ldots h(u_n x_n v_n) \qquad \square$$

Claim 12.1.14. Let $S \Rightarrow_H^m y$, where $m \geq 1, y \in V^*$. Then, $(S_1, S_2, \ldots, S_n) \Rightarrow_\Gamma^{m-1}$ $(y_1, y_2, \ldots, y_n)$, where $y_i \in V_i^*$, for all $i = 1, \ldots, n$ so that $y = y_1 h(y_2) \ldots h(y_n)$.

Proof. This claim is proved by induction on $m \geq 1$.

Basis. Let $m = 1$. Then, there exists exactly one one-step derivation in H: $S \Rightarrow_H$ $S_1 h(S_2) \ldots h(S_n)$ by matrix s. Notice that $(S_1, S_2, \ldots, S_n) \Rightarrow_\Gamma^0 (S_1, S_2, \ldots, S_n)$ trivially.

Induction Hypothesis. Assume that the claim holds for all m-step derivations, where $m = 1, \ldots, k$, for some $k \geq 1$.

Induction Step. Consider any derivation of the form

$$S \Rightarrow_H^{k+1} y$$

Then, there is w such that

$$S \Rightarrow_H^k w \Rightarrow_H y$$

where $w, y \in V^*$. First, observe that $S \Rightarrow_H^k w$ implies that

$$(S_1, S_2, \ldots, S_n) \Rightarrow_\Gamma^{k-1} (w_1, w_2, \ldots, w_n)$$

so that $w = w_1 h(w_2) \ldots h(w_n)$, where $w_i \in V_i^*$, for all $i = 1, \ldots, n$, by the induction hypothesis. Furthermore, let $w \Rightarrow_H y$, where $w = w_1 h(w_2) \ldots h(w_n)$. Let p_1 be a rule of the form $A_1 \rightarrow x_1$. Let $\bar{p}_i$ be a rule of the form $h(A_i) \rightarrow h(x)$, for all $i = 2, \ldots, n$. The rule p_1 can be applied only inside substring w_1, the rule $\bar{p}_i$ can be applied only inside substring w_i, for all $i = 2, \ldots, n$. Assume that $w_i = u_i A_i v_i$, where $u_i, v_i \in V_i^*, A_i \in N_i$, for all $i = 1, \ldots, n$. There exists a derivation step

$$u_1 A_1 v_1 h(u_2 A_2 v_2) \ldots h(u_n A_n v_n) \Rightarrow_H$$
$$u_1 x_1 v_1 h(u_2 x_2 v_2) \ldots h(u_n x_n v_n)$$

by matrix $p_1 \bar{p}_2 \ldots \bar{p}_n \in M$. Algorithm 12.1.11 implies that

$$((p_1 : A_1 \rightarrow x_1), (p_2 : A_2 \rightarrow x_2), \ldots, (p_n : A_n \rightarrow x_n)) \in Q$$

because $p_1 \bar{p}_2 \ldots \bar{p}_n \in M$. Hence,

$$(u_1 A_1 v_1, u_2 A_2 v_2, \ldots, u_n A_n v_n) \Rightarrow_{\Gamma}$$
$$(u_1 x_1 v_1, u_2 x_2 v_2, \ldots, u_n x_n v_n)$$

As a result, we obtain

$$(S_1, S_2, \ldots, S_n) \Rightarrow_{\Gamma}^{k} (u_1 x_1 v_1, u_2 x_2 v_2, \ldots, u_n x_n v_n)$$

so that $y = u_1 x_1 v_1 h(u_2 x_2 v_2) \ldots h(u_n x_n v_n)$. $\qquad\square$

Consider Claim 12.1.13 for $y_i \in T_i^*$, for all $i = 1, \ldots, n$. Notice that $h(a) = \varepsilon$, for all $a \in T_i$. We obtain an implication of the form

$$\text{if } (S_1, S_2, \ldots, S_n) \Rightarrow_{\Gamma}^{*} (y_1, y_2, \ldots, y_n)$$
$$\text{then } S \Rightarrow_{H}^{*} y_1$$

Hence, $L_{first}(\Gamma) \subseteq L(H)$. Consider Claim 12.1.14 for $y \in T^*$. Notice that $h(a) = \varepsilon$, for all $a \in T_i$. We obtain an implication of the form

$$\text{if } S \Rightarrow_{H}^{*} y$$
$$\text{then } (S_1, S_2, \ldots, S_n) \Rightarrow_{\Gamma}^{*} (y_1, y_2, \ldots, y_n), \text{ such that } y = y_1$$

Hence, $L(H) \subseteq L_{first}(\Gamma)$. Therefore, $L(H) = L_{first}(\Gamma)$. $\qquad\square$

Algorithm 12.1.15. *Conversion of a matrix grammar to an equivalent 2-MGR.*

Input: A matrix grammar, $H = (G, M)$, and a string, $\bar{w} \in \bar{T}^*$, where $\bar{T}$ is any alphabet.

Output: A 2-MGR, $\Gamma = (G_1, G_2, Q)$, satisfying $\{w_1 \mid (w_1, \bar{w}) \in 2\text{-}L(\Gamma)\} = L(H)$.

Method: Let $G = (V, T, P, S)$. Then, set $G_1 = G$ and construct

$$G_2 = (V_2, T_2, P_2, S_2)$$

where

- $N_2 = \{\langle p_1 p_2 \ldots p_k, j \rangle \mid p_1, \ldots, p_k \in P, p_1 p_2 \ldots p_k \in M,$
  $1 \le j \le k-1\} \cup \{S_2\}$;
- $T_2 = \bar{T}$;
- $P_2 = \{S_2 \to \langle p_1 p_2 \ldots p_k, 1 \rangle \mid p_1, \ldots, p_k \in P, p_1 p_2 \ldots p_k \in M, k \ge 2\}$
  $\cup \{\langle p_1 p_2 \ldots p_k, j \rangle \to \langle p_1 p_2 \ldots p_k, j+1 \rangle \mid p_1 p_2 \ldots p_k \in M, k \ge 2,$
  $1 \le j \le k-2\}$
  $\cup \{\langle p_1 p_2 \ldots p_k, k-1 \rangle \to S_2 \mid p_1, \ldots, p_k \in P, p_1 p_2 \ldots p_k \in M, k \ge 2\}$
  $\cup \{S_2 \to S_2 \mid p_1 \in M, |p_1| = 1\}$

$\cup \{\langle p_1p_2 \ldots p_k, k-1\rangle \rightarrow \bar{w} \mid p_1, \ldots, p_k \in P, p_1p_2 \ldots p_k \in M, k \geq 2\}$
$\cup \{S_2 \rightarrow \bar{w} \mid p_1 \in M, |p_1| = 1\}$;
- $Q = \{(p_1, S_2 \rightarrow \langle p_1p_2 \ldots p_k, 1\rangle)) \mid p_1, \ldots, p_k \in P, p_1p_2 \ldots p_k \in M,$
  $k \geq 2\}$
  $\cup \{(p_{j+1}, \langle p_1p_2 \ldots p_k, j\rangle \rightarrow \langle p_1p_2 \ldots p_k, j+1\rangle)) \mid p_1p_2 \ldots p_k \in M,$
  $k \geq 2, 1 \leq j \leq k-2\}$
  $\cup \{(p_k, \langle p_1p_2 \ldots p_k, k-1\rangle \rightarrow S_2) \mid p_1, \ldots, p_k \in P, p_1p_2 \ldots p_k \in M,$
  $k \geq 2\}$
  $\cup \{(p_1, S_2 \rightarrow S_2) \mid p_1 \in M, |p_1| = 1\}$
  $\cup \{(p_k, \langle p_1p_2 \ldots p_k, k-1\rangle \rightarrow \bar{w}) \mid p_1, \ldots, p_k \in P, p_1p_2 \ldots p_k \in M,$
  $k \geq 2\}$
  $\cup \{(p_1, S_2 \rightarrow \bar{w}) \mid p_1 \in M, |p_1| = 1\}$. $\quad\quad\Box$

Theorem 12.1.16. *Let H be a matrix grammar and $\bar{w}$ be a string. With H and $\bar{w}$ as its input, Algorithm 12.1.15 halts and correctly constructs a 2-MGR, $\Gamma = (G_1, G_2, Q)$, such that $\{w_1 \mid (w_1, \bar{w}) \in 2\text{-}L(\Gamma)\} = L(H)$.*

Proof. To prove this theorem, we first establish Claims 12.1.17 through 12.1.20.

Claim 12.1.17. Let $x \Rightarrow_H y$, where $x, y \in V^*$. Then, $(x, S_2) \Rightarrow_\Gamma^* (y, S_2)$ and $(x, S_2) \Rightarrow_\Gamma^* (y, \bar{w})$.

Proof. In this proof, we distinguish two cases—I. and II. In I., we consider any derivation step of the form $x \Rightarrow_H y$ by a matrix consisting of a single rule. In II., we consider $x \Rightarrow_H y$ by a matrix consisting of several rules.

I. Consider any derivation step of the form $x \Rightarrow_H y$ by a matrix which contains only one rule $(p_1: A_1 \rightarrow x_1)$. It implies that $uA_1v \Rightarrow_G ux_1v$ $[p_1]$, where $uA_1v = x, ux_1v = y$. Algorithm 12.1.15 implies that $(A_1 \rightarrow x_1, S_2 \rightarrow S_2) \in Q$ and $(A_1 \rightarrow x_1, S_2 \rightarrow \bar{w}) \in Q$. Hence, $(uA_1v, S_2) \Rightarrow_\Gamma (ux_1v, S_2)$ and $(uA_1v, S_2) \Rightarrow_\Gamma (ux_1v, \bar{w})$.

II. Let $x \Rightarrow_H y$ by a matrix of the form $p_1p_2 \ldots p_k$, where $p_i, \ldots, p_k \in P, k \geq 2$. It implies that

$$
\begin{aligned}
x &\Rightarrow_H y_1 & [p_1] \\
&\Rightarrow_H y_2 & [p_2] \\
&\;\;\vdots \\
&\Rightarrow_H y_{k-1} & [p_{k-1}] \\
&\Rightarrow_H y_k & [p_k]
\end{aligned}
$$

where $y_k = y$. Algorithm 12.1.15 implies that $(p_1, S_2 \rightarrow \langle p_1p_2 \ldots p_k, 1\rangle) \in Q$, $(p_{j+1}, \langle p_1p_2 \ldots p_k, j\rangle \rightarrow \langle p_1p_2 \ldots p_k, j+1\rangle) \in Q$, where $j = 1, \ldots, k-2$, $(p_k, \langle p_1p_2 \ldots p_k, k-1\rangle \rightarrow S_2) \in Q$, $(p_k, \langle p_1p_2 \ldots p_k, k-1\rangle \rightarrow \bar{w}) \in Q$. Hence,

$$(x, S_2) \Rightarrow_\Gamma (y_1, \langle p_1 p_2 \ldots p_k, 1 \rangle)$$
$$\Rightarrow_\Gamma (y_2, \langle p_1 p_2 \ldots p_k, 2 \rangle)$$
$$\vdots$$
$$\Rightarrow_\Gamma (y_{k-1}, \langle p_1 p_2 \ldots p_k, k-1 \rangle)$$
$$\Rightarrow_\Gamma (y_k, S_2)$$

and

$$(x, S_2) \Rightarrow_\Gamma (y_1, \langle p_1 p_2 \ldots p_k, 1 \rangle)$$
$$\Rightarrow_\Gamma (y_2, \langle p_1 p_2 \ldots p_k, 2 \rangle)$$
$$\vdots$$
$$\Rightarrow_\Gamma (y_{k-1}, \langle p_1 p_2 \ldots p_k, k-1 \rangle)$$
$$\Rightarrow_\Gamma (y_k, \bar{w})$$

where $y_k = y$. □

Claim 12.1.18. Let $x \Rightarrow_H^m y$, where $m \geq 1, y \in V^*$. Then, $(x, S_2) \Rightarrow_\Gamma^* (y, \bar{w})$.

Proof. This claim is proved by induction on $m \geq 1$.

Basis. Let $m = 1$ and let $x \Rightarrow_H y$. Claim 12.1.17 implies that $(x, S_2) \Rightarrow_\Gamma^* (y, \bar{w})$.

Induction Hypothesis. Assume that the claim holds for all m-step derivations, where $m = 1, \ldots, k$, for some $k \geq 1$.

Induction Step. Consider any derivation of the form

$$S \Rightarrow_H^{k+1} y$$

Then, there exists w such that

$$S \Rightarrow_H w \Rightarrow_H^k y$$

where $w, y \in V^*$. First, observe that $w \Rightarrow_H^k y$ implies that

$$(w, S_2) \Rightarrow_\Gamma^* (y, \bar{w})$$

by the induction hypothesis. Furthermore, let $x \Rightarrow_H w$. Claim 12.1.17 implies that

$$(x, S_2) \Rightarrow_\Gamma^* (w, S_2)$$

As a result, we obtain

$$(x, S_2) \Rightarrow_\Gamma^* (y, \bar{w})$$ □

Claim 12.1.19. Let

$$(y_0, S_2) \Rightarrow_\Gamma (y_1, z_1) \quad \text{or} \quad (y_0, S_2) \Rightarrow_\Gamma (y_1, z_1)$$
$$\Rightarrow_\Gamma (y_2, z_2) \qquad\qquad \Rightarrow_\Gamma (y_2, z_2)$$
$$\vdots \qquad\qquad\qquad \vdots$$
$$\Rightarrow_\Gamma (y_{k-1}, z_{k-1}) \qquad \Rightarrow_\Gamma (y_{k-1}, z_{k-1})$$
$$\Rightarrow_\Gamma (y_k, S_2) \qquad\qquad \Rightarrow_\Gamma (y_k, \bar{w})$$

where $z_i \neq S_2$, for all $i = 1, \ldots, k - 1$. Then, there exists a direct derivation step $y_0 \Rightarrow_H y_k$.

Proof. In this proof, we distinguish two cases—I. and II. In I., we consider any derivation step of the form $x \Rightarrow_H y$ by a matrix consisting of a single rule. In II., we consider $x \Rightarrow_H y$ by a matrix consisting of several rules.

I. Consider any derivation step of the form

$$(uA_1v, S_2) \Rightarrow_\Gamma (ux_1v, S_2)$$

or

$$(uA_1v, S_2) \Rightarrow_\Gamma (ux_1v, \bar{w})$$

where $uA_1v = y_0, ux_1v = y_1$. Then, $(A_1 \to x_1, S_2 \to S_2) \in Q$ or $(A_1 \to x_1, S_2 \to \bar{w}) \in Q$. Algorithm 12.1.15 implies that there exists a matrix of the form $(p_1 : A_1 \to x_1) \in M$. Hence,

$$uA_1v \Rightarrow_H ux_1v$$

II. Let

$$(y_0, S_2) \Rightarrow_\Gamma (y_1, z_1) \quad \text{or} \quad (y_0, S_2) \Rightarrow_\Gamma (y_1, z_1)$$
$$\Rightarrow_\Gamma (y_2, z_2) \qquad\qquad \Rightarrow_\Gamma (y_2, z_2)$$
$$\vdots \qquad\qquad\qquad \vdots$$
$$\Rightarrow_\Gamma (y_{k-1}, z_{k-1}) \qquad \Rightarrow_\Gamma (y_{k-1}, z_{k-1})$$
$$\Rightarrow_\Gamma (y_k, S_2) \qquad\qquad \Rightarrow_\Gamma (y_k, \bar{w})$$

where $z_i \neq S_2$, for all $i = 1, \ldots, k - 1$ and $k \geq 2$. Algorithm 12.1.15 implies that there exists a matrix $p_1 p_2 \ldots p_k \in M$ and that $z_i = \langle p_1 p_2 \ldots p_k, i \rangle$, for all $i = 1, \ldots k - 1$. Hence,

$$y_0 \Rightarrow_H y_k \qquad\qquad\qquad \square$$

Claim 12.1.20. Let

$$(y_0, S_2) \Rightarrow_\Gamma (y_1, z_1)$$
$$\Rightarrow_\Gamma (y_2, z_2)$$
$$\vdots$$
$$\Rightarrow_\Gamma (y_{r-1}, z_{r-1})$$
$$\Rightarrow_\Gamma (y_r, \bar{w})$$

Set $m = \mathrm{card}(\{i \mid 1 \le i \le r - 1, z_i = S_2\})$. Then, $y_0 \Rightarrow_H^{m+1} y_r$.

Proof. This claim is proved by induction on $m \ge 0$.

Basis. Let $m = 0$. Then, $z_i \ne S_2$, for all $i = 1, \dots, k - 1$. Claim 12.1.19 implies that there exists a derivation step $y_0 \Rightarrow_H y_r$.

Induction Hypothesis. Assume that the claim holds for all m-step derivations, where $m = 0, \dots, k$, for some $k \ge 0$.

Induction Step. Consider any derivation of the form

$$(y_0, S_2) \Rightarrow_\Gamma (y_1, z_1)$$
$$\Rightarrow_\Gamma (y_2, z_2)$$
$$\vdots$$
$$\Rightarrow_\Gamma (y_{r-1}, z_{r-1})$$
$$\Rightarrow_\Gamma (y_r, \bar{w})$$

where $\mathrm{card}(\{i \mid 1 \le i \le r-1, z_i = S_2\}) = k+1$. Then, there exists $p \in \{1, \dots, r-1\}$ such that $z_p = S_2$, $\mathrm{card}(\{i \mid 1 \le i \le p - 1, z_i = S_2\}) = 0$, $\mathrm{card}(\{i \mid p + 1 \le i \le r - 1, z_i = S_2\}) = k$, and

$$(y_0, z_0) \Rightarrow_\Gamma (y_1, z_1)$$
$$\vdots$$
$$\Rightarrow_\Gamma (y_p, z_p)$$
$$\vdots$$
$$\Rightarrow_\Gamma (y_{r-1}, z_{r-1})$$
$$\Rightarrow_\Gamma (y_r, \bar{w})$$

First, observe that from

$$(y_p, z_p) \Rightarrow_\Gamma (y_{p+1}, z_{p+1})$$
$$\vdots$$
$$\Rightarrow_\Gamma (y_{r-1}, z_{r-1})$$
$$\Rightarrow_\Gamma (y_r, \bar{w})$$

where $z_p = S_2$ and $\mathrm{card}(\{i \mid p+1 \le i \le r-1, z_i = S_2\}) = k$, it follows

$$y_p \Rightarrow_H^{k+1} y_r$$

by the induction hypothesis. Furthermore, let

$$(y_0, z_0) \Rightarrow_\Gamma (y_1, z_1)$$
$$\vdots$$
$$\Rightarrow_\Gamma (y_p, z_p)$$

$\mathrm{card}(\{i \mid 1 \le i \le p-1, z_i = S_2\}) = 0$ implies that $z_i \ne S_2$, for all $i = 1, \dots, p-1$. Claim 12.1.19 implies that there exists a derivation step $y_0 \Rightarrow_H y_p$. As a result, we obtain

$$y_0 \Rightarrow_H^{k+2} y_r \qquad \qquad \Box$$

We next prove the following two identities, (1) and (2).

(1) $\{w_1 \mid (w_1, \bar{w}) \in 2\text{-}L(\Gamma)\} = L(H)$. Consider Claim 12.1.18 for $x = S$ and $y \in T^*$. We obtain an implication of the form

$$\text{if } S \Rightarrow_H^* y$$
$$\text{then } (S, S_2) \Rightarrow_\Gamma^* (y, \bar{w})$$

Hence, $L(H) \subseteq \{w_1 \mid (w_1, \bar{w}) \in 2\text{-}L(\Gamma)\}$. Consider Claim 12.1.20 for $y_0 = S$ and $y_r \in T^*$. We see that

$$\text{if } (S, S_2) \Rightarrow_\Gamma^* (y_r, \bar{w})$$
$$\text{then } S \Rightarrow_H^* y_r$$

Hence, $\{w_1 \mid (w_1, \bar{w}) \in 2\text{-}L(\Gamma)\} \subseteq L(H)$.

(2) $\{(w_1, w_2) \mid (w_1, w_2) \in 2\text{-}L(\Gamma), w_2 \ne \bar{w}\} = \emptyset$. Notice that Algorithm 12.1.15 implies that $G_2 = (N_2, T_2, P_2, S_2)$ contains only rules of the form $A \to B$ and $A \to \bar{w}$, where $A, B \in N_2$. Hence, G_2 generates $\emptyset$ or $\{\bar{w}\}$. Γ contains G_2 as a second component; hence, $\{(w_1, w_2) \mid (w_1, w_2) \in 2\text{-}L(\Gamma), w_2 \ne \bar{w}\} = \emptyset$. $\qquad \Box$

Theorem 12.1.21. *For every matrix grammar H, there is a 2-MGR Γ such that $L(H) = L_{union}(\Gamma)$.*

Proof. To prove this theorem, we make use of Algorithm 12.1.15 with matrix grammar H and $\bar{w}$ as input, where $\bar{w}$ is any string in $L(H)$, provided that $L(H)$ is nonempty. Otherwise, if $L(H)$ is empty, let $\bar{w}$ be any string. We prove that $L(H) = L_{union}(\Gamma)$.

1. If $L(H) = \emptyset$, take any string $\bar{w}$ and use Algorithm 12.1.15 to construct Γ. Observe that $L_{union}(\Gamma) = \emptyset = L(H)$.
2. If $L(H) \neq \emptyset$, take any $\bar{w} \in L(H)$ and use Algorithm 12.1.15 to construct Γ. As obvious, $L_{union}(\Gamma) = L(H) \cup \bar{w} = L(H)$.   □

Theorem 12.1.22. *For every matrix grammar H, there is a 2-MGR Γ such that $L(H) = L_{conc}(\Gamma)$.*

Proof. To prove this theorem, we make use of Algorithm 12.1.15 with matrix grammar H and $\bar{w} = \varepsilon$ as input. We prove that $L(H) = L_{conc}(\Gamma)$. Theorem 12.1.16 says that

$$\{w_1 \mid (w_1, \bar{w}) \in 2\text{-}L(\Gamma)\} = L(H)$$

and

$$\{(w_1, w_2) \mid (w_1, w_2) \in 2\text{-}L(\Gamma), w_2 \neq \bar{w}\} = \emptyset$$

Then,

$$
\begin{aligned}
L_{conc}(\Gamma) &= \{w_1 w_2 \mid (w_1, w_2) \in 2\text{-}L(\Gamma)\} \\
&= \{w_1 w_2 \mid (w_1, w_2) \in 2\text{-}L(\Gamma), w_2 = \bar{w}\} \\
&\quad \cup \{w_1 w_2 \mid (w_1, w_2) \in 2\text{-}L(\Gamma), w_2 \neq \bar{w}\} \\
&= \{w_1 \bar{w} \mid (w_1, \bar{w}) \in 2\text{-}L(\Gamma)\} \cup \emptyset \\
&= \{w_1 \bar{w} \mid (w_1, \bar{w}) \in 2\text{-}L(\Gamma)\} \\
&= L(H)
\end{aligned}
$$

because $\bar{w} = \varepsilon$.   □

Theorem 12.1.23. *For every matrix grammar H, there is a 2-MGR Γ such that $L(H) = L_{first}(\Gamma)$.*

Proof. To prove this theorem, we make use of Algorithm 12.1.15 with matrix grammar H and any $\bar{w}$ as input. We prove that $L(H) = L_{first}(\Gamma)$. Theorem 12.1.16 says that

$$\{w_1 \mid (w_1, \bar{w}) \in 2\text{-}L(\Gamma)\} = L(H)$$

and

$$\{(w_1, w_2) \mid (w_1, w_2) \in 2\text{-}L(\Gamma), w_2 \neq \bar{w}\} = \emptyset$$

Then,

$$
\begin{aligned}
L_{first}(\Gamma) &= \{w_1 \mid (w_1, w_2) \in 2\text{-}L(\Gamma)\} \\
&= \{w_1 \mid (w_1, w_2) \in 2\text{-}L(\Gamma), w_2 = \bar{w}\} \\
&\quad \cup \{w_1 \mid (w_1, w_2) \in 2\text{-}L(\Gamma), w_2 \neq \bar{w}\} \\
&= \{w_1 \mid (w_1, \bar{w}) \in 2\text{-}L(\Gamma)\} \cup \emptyset \\
&= \{w_1 \mid (w_1, \bar{w}) \in 2\text{-}L(\Gamma)\} \\
&= L(H)
\end{aligned}
$$

Hence, the theorem holds. □

Let $\mathbf{MGR}_{n,X}$ denote the language families defined by n-MGRs in the X mode, where $X \in \{union, conc, first\}$. From the previous results, we obtain the following corollary.

Corollary 12.1.24. $\mathbf{M} = \mathbf{MGR}_{n,X}$, where $n \geq 2$, $X \in \{union, conc, first\}$. □

To summarize all the results, multigenerative grammar systems with any number of grammatical components are equivalent with two-component versions of these systems. Perhaps even more importantly, these systems are equivalent with matrix grammars, which generate a proper subfamily of the family of recursively enumerable languages (see Theorem 3.4.5).

We close this section by suggesting two open problem areas.

Open Problem 12.1.25. Consider other operations, like intersection, and study languages generated in this way by multigenerative grammar systems.

Open Problem 12.1.26. Study multigenerative grammar systems that are based on other grammars than context-free grammars. Specifically, determine the generative power of multigenerative grammar systems with regular or right-linear grammars as components.

12.2 Leftmost Multigenerative Grammar Systems

In this section, we study leftmost versions of multigenerative grammar systems, whose basic versions were defined and investigated in the previous section of this chapter. We prove that they characterize the family of recursively enumerable languages, which properly contains the family of matrix languages (see Theorems 3.3.6 and 3.4.5). We also consider regulation by n-tuples of nonterminals rather than rules, and we prove that leftmost multigenerative grammar systems regulated by rules or nonterminals have the same generative power. Just like for multigenerative grammar systems in the previous section, we explain how to reduce the number of grammatical components in leftmost multigenerative grammar systems to two.

Definition 12.2.1. A *leftmost n-generative rule-synchronized grammar system* (an *n-LMGR* for short) is an *n*-MGR (see Definition 12.1.1), where for two sentential *n*-forms $\chi = (u_1 A_1 v_1, u_2 A_2 v_2, \ldots, u_n A_n v_n)$ and $\bar{\chi} = (u_1 x_1 v_1, u_2 x_2 v_2, \ldots, u_n x_n v_n)$, where $A_i \in N_i$, $u_i, v_i, x_i \in V_i^*$, and an *n*-tuple $(p_1, p_2, \ldots, p_n) \in Q$, where $(p_i : A_i \to x_i) \in P_i$,

$$\chi \Rightarrow_\Gamma \bar{\chi}$$

if and only if $u_i \in T_i^*$, for all $i = 1, \ldots, n$.   □

Next, we introduce regulation by *n*-tuples of nonterminals rather than rules.

Definition 12.2.2. A *leftmost n-generative nonterminal-synchronized grammar system* (an *n-LMGN* for short) is an $n + 1$ tuple

$$\Gamma = (G_1, G_2, \ldots, G_n, Q)$$

where

- $G_i = (V_i, T_i, P_i, S_i)$ is a context-free grammar, for each $i = 1, \ldots, n$;
- Q is a finite set of *n*-tuples of the form $(A_1, A_2, \ldots, A_n)$, where $A_i \in N_i$, for all $i = 1, \ldots, n$.

A *sentential n-form* is defined as a sentential *n*-form of an *n*-LMGR. Let $\chi = (u_1 A_1 v_1, u_2 A_2 v_2, \ldots, u_n A_n v_n)$ and $\bar{\chi} = (u_1 x_1 v_1, u_2 x_2 v_2, \ldots, u_n x_n v_n)$ be two sentential *n*-forms, where $A_i \in N_i$, $u_i \in T^*$, and $v_i, x_i \in V_i^*$, for all $i = 1, \ldots, n$. Let $(p_i : A_i \to x_i) \in P_i$, for all $i = 1, \ldots, n$ and $(A_1, A_2, \ldots, A_n) \in Q$. Then, χ directly derives $\bar{\chi}$ in Γ, denoted by

$$\chi \Rightarrow_\Gamma \bar{\chi}$$

In the standard way, we generalize $\Rightarrow_\Gamma$ to $\Rightarrow_\Gamma^k$, for all $k \geq 0$, $\Rightarrow_\Gamma^*$, and $\Rightarrow_\Gamma^+$.

An *n-language* for *n*-LMGN is defined as the *n*-language for *n*-LMGR, and a language generated by *n*-LMGN in the X mode, for each $X \in \{union, conc, first\}$, is defined as the language generated by *n*-LMGR in the X mode.   □

Example 12.2.3. Consider the 2-LMGN $\Gamma = (G_1, G_2, Q)$, where

- $G_1 = (\{S_1, A_1, a, b, c\}, \{a, b, c\}, \{S_1 \to aS_1, S_1 \to aA_1, A_1 \to bA_1 c, A_1 \to bc\}, S_1)$,
- $G_2 = (\{S_2, A_2, d\}, \{d\}, \{S_2 \to S_2 A_2, S_2 \to A_2, A_2 \to d\}, S_2)$,
- $Q = \{(S_1, S_2), (A_1, A_2)\}$.

Observe that

- $2\text{-}L(\Gamma) = \{(a^n b^n c^n, d^n) \mid n \geq 1\}$,
- $L_{union}(\Gamma) = \{a^n b^n c^n \mid n \geq 1\} \cup \{d^n \mid n \geq 1\}$,
- $L_{conc}(\Gamma) = \{a^n b^n c^n d^n \mid n \geq 1\}$, and
- $L_{first}(\Gamma) = \{a^n b^n c^n \mid n \geq 1\}$.   □

Lemma 12.2.4. *Let Γ be an n-LMGN and let $\bar{\Gamma}$ be an n-LMGR such that $n\text{-}L(\Gamma)$ $= n\text{-}L(\bar{\Gamma})$. Then, $L_X(\Gamma) = L_X(\bar{\Gamma})$, for each $X \in \{union, conc, first\}$.*

Proof.

I. First, we prove that $L_{union}(\Gamma) = L_{union}(\bar{\Gamma})$ as follows:

$$L_{union}(\Gamma) = \{w \mid (w_1, \ldots, w_n) \in n\text{-}L(\Gamma), w \in \{w_i \mid 1 \le i \le n\}\}$$
$$= \{w \mid (w_1, \ldots, w_n) \in n\text{-}L(\bar{\Gamma}), w \in \{w_i \mid 1 \le i \le n\}\}$$
$$= L_{union}(\bar{\Gamma})$$

II. Second, we prove that $L_{conc}(\Gamma) = L_{conc}(\bar{\Gamma})$ as follows:

$$L_{conc}(\Gamma) = \{w_1 \cdots w_n \mid (w_1, \ldots, w_n) \in n\text{-}L(\Gamma)\}$$
$$= \{w_1 \cdots w_n \mid (w_1, \ldots, w_n) \in n\text{-}L(\bar{\Gamma})\}$$
$$= L_{conc}(\bar{\Gamma})$$

III. Finally, we prove that $L_{first}(\Gamma) = L_{first}(\bar{\Gamma})$ as follows:

$$L_{first}(\Gamma) = \{w_1 \mid (w_1, \ldots, w_n) \in n\text{-}L(\Gamma)\}$$
$$= \{w_1 \mid (w_1, \ldots, w_n) \in n\text{-}L(\bar{\Gamma})\}$$
$$= L_{first}(\bar{\Gamma}) \qquad \qquad \square$$

Algorithm 12.2.5. *Conversion of an n-LMGN to an equivalent n-LMGR.*

Input: An n-LMGN, $\Gamma = (G_1, G_2, \ldots, G_n, Q)$.
Output: An n-LMGR, $\bar{\Gamma} = (G_1, G_2, \ldots, G_n, \bar{Q})$, such that $n\text{-}L(\Gamma) = n\text{-}L(\bar{\Gamma})$.
Method: Let $G_i = (V_i, T_i, P_i, S_i)$, for all $i = 1, \ldots, n$, and set

$$\bar{Q} = \{(A_1 \to x_1, A_2 \to x_2, \ldots, A_n \to x_n) \mid A_i \to x_i \in P_i,$$
$$\text{for all } i = 1, \ldots, n, \text{ and } (A_1, A_2, \ldots, A_n) \in Q\} \qquad \square$$

Theorem 12.2.6. *Let $\Gamma = (G_1, G_2, \ldots, G_n, Q)$ be an n-LMGN. With Γ as its input, Algorithm 12.2.5 halts and correctly constructs an n-LMGR, $\bar{\Gamma} = (G_1, G_2, \ldots, G_n, \bar{Q})$, such that $n\text{-}L(\Gamma) = n\text{-}L(\bar{\Gamma})$, and $L_X(\Gamma) = L_X(\bar{\Gamma})$, for each $X \in \{union, conc, first\}$.*

Proof. To prove this theorem, we first establish Claims 12.2.7 and 12.2.8.

Claim 12.2.7. Let $(S_1, S_2, \ldots, S_n) \Rightarrow_{\Gamma}^m (y_1, y_2, \ldots, y_n)$, where $m \ge 0$, $y_i \in V_i^*$, for all $i = 1, \ldots, n$. Then, $(S_1, S_2, \ldots, S_n) \Rightarrow_{\bar{\Gamma}}^m (y_1, y_2, \ldots, y_n)$.

Proof. This claim is proved by induction on $m \ge 0$.

Basis. The basis is clear.

Induction Hypothesis. Assume that Claim 12.2.7 holds for all m-step derivations, where $m = 0, \ldots, k$, for some $k \ge 0$.

Induction Step. Consider any derivation of the form

$$(S_1, S_2, \ldots, S_n) \Rightarrow_{\Gamma}^{k+1} (y_1, y_2, \ldots, y_n)$$

Then, there exists a sentential n-form $(u_1 A_1 v_1, u_2 A_2 v_2, \ldots, u_n A_n v_n)$, where $u_i \in T_i^*$, $A_i \in N_i$, and $v_i \in V_i^*$, such that

$$(S_1, S_2, \ldots, S_n) \Rightarrow_{\Gamma}^{k} (u_1 A_1 v_1, u_2 A_2 v_2, \ldots, u_n A_n v_n)$$
$$\Rightarrow_{\Gamma} (u_1 x_1 v_1, u_2 x_2 v_2, \ldots, u_n x_n v_n)$$

where $u_i x_i v_i = y_i$, for all $i = 1, \ldots, n$. Then, by the induction hypothesis, we have

$$(S_1, S_2, \ldots, S_n) \Rightarrow_{\bar{\Gamma}}^{k} (u_1 A_1 v_1, u_2 A_2 v_2, \ldots, u_n A_n v_n)$$

Since

$$(u_1 A_1 v_1, u_2 A_2 v_2, \ldots, u_n A_n v_n) \Rightarrow_{\Gamma}$$
$$(u_1 x_1 v_1, u_2 x_2 v_2, \ldots, u_n x_n v_n)$$

$(A_1, A_2, \ldots, A_n) \in Q$ and $A_i \to x_i \in P_i$, for all $i = 1, \ldots, n$. Algorithm 12.2.5 implies that $(A_1 \to x_1, A_2 \to x_2, \ldots, A_n \to x_n) \in \bar{Q}$, so

$$(u_1 A_1 v_1, u_2 A_2 v_2, \ldots, u_n A_n v_n) \Rightarrow_{\bar{\Gamma}}$$
$$(u_1 x_1 v_1, u_2 x_2 v_2, \ldots, u_n x_n v_n)$$

which proves the induction step. Therefore, Claim 12.2.7 holds.   □

Claim 12.2.8. Let $(S_1, S_2, \ldots, S_n) \Rightarrow_{\bar{\Gamma}}^{m} (y_1, y_2, \ldots, y_n)$, where $m \geq 0$, $y_i \in V_i^*$, for all $i = 1, \ldots, n$. Then, $(S_1, S_2, \ldots, S_n) \Rightarrow_{\Gamma}^{m} (y_1, y_2, \ldots, y_n)$.

Proof. This claim is proved by induction on $m \geq 0$.

Basis. The basis is clear.

Induction Hypothesis. Assume that Claim 12.2.8 holds for all m-step derivations, where $m = 0, \ldots, k$, for some $k \geq 0$.

Induction Step. Consider any derivation of the form

$$(S_1, S_2, \ldots, S_n) \Rightarrow_{\bar{\Gamma}}^{k+1} (y_1, y_2, \ldots, y_n)$$

Then, there exists a sentential n-form $(u_1 A_1 v_1, u_2 A_2 v_2, \ldots, u_n A_n v_n)$, where $u_i \in T_i^*$, $A_i \in N_i$, $v_i \in V_i^*$, such that

$$(S_1, S_2, \ldots, S_n) \Rightarrow_{\bar{\Gamma}}^{k} (u_1 A_1 v_1, u_2 A_2 v_2, \ldots, u_n A_n v_n)$$
$$\Rightarrow_{\bar{\Gamma}} (u_1 x_1 v_1, u_2 x_2 v_2, \ldots, u_n x_n v_n)$$

where $u_i x_i v_i = y_i$, for all $i = 1, \ldots, n$. Then, by the induction hypothesis, we have

$$(S_1, S_2, \ldots, S_n) \Rightarrow_{\Gamma}^{k} (u_1 A_1 v_1, u_2 A_2 v_2, \ldots, u_n A_n v_n)$$

Since

$$(u_1 A_1 v_1, u_2 A_2 v_2, \ldots, u_n A_n v_n) \Rightarrow_{\bar{\Gamma}}$$
$$(u_1 x_1 v_1, u_2 x_2 v_2, \ldots, u_n x_n v_n)$$

$(A_1 \to x_1, A_2 \to x_2, \ldots, A_n \to x_n) \in \bar{Q}$, for all $i = 1, \ldots, n$. Algorithm 12.2.5 implies that $(A_1, A_2, \ldots, A_n) \in Q$ and $A_i \to x_i \in P_i$, so

$$(u_1 A_1 v_1, u_2 A_2 v_2, \ldots, u_n A_n v_n) \Rightarrow_{\Gamma}$$
$$(u_1 x_1 v_1, u_2 x_2 v_2, \ldots, u_n x_n v_n)$$

which proves the induction step. Therefore, Claim 12.2.8 holds.   □

Consider Claim 12.2.7 for $y_i \in T_i^*$, for all $i = 1, \ldots, n$. At this point, if

$$(S_1, S_2, \ldots, S_n) \Rightarrow_{\Gamma}^{*} (y_1, y_2, \ldots, y_n)$$

then

$$(S_1, S_2, \ldots, S_n) \Rightarrow_{\bar{\Gamma}}^{*} (y_1, y_2, \ldots, y_n)$$

Hence, $n\text{-}L(\Gamma) \subseteq n\text{-}L(\bar{\Gamma})$. Consider Claim 12.2.8 for $y_i \in T_i^*$, for all $i = 1, \ldots, n$. At this point, if

$$(S_1, S_2, \ldots, S_n) \Rightarrow_{\bar{\Gamma}}^{*} (y_1, y_2, \ldots, y_n)$$

then

$$(S_1, S_2, \ldots, S_n) \Rightarrow_{\Gamma}^{*} (y_1, y_2, \ldots, y_n)$$

Hence, $n\text{-}L(\bar{\Gamma}) \subseteq n\text{-}L(\Gamma)$. As $n\text{-}L(\Gamma) \subseteq n\text{-}L(\bar{\Gamma})$ and $n\text{-}L(\bar{\Gamma}) \subseteq n\text{-}L(\Gamma)$, $n\text{-}L(\Gamma)$ $= n\text{-}L(\bar{\Gamma})$. By Lemma 12.2.4, this identity implies that $L_X(\Gamma) = L_X(\bar{\Gamma})$, for each $X \in \{union, conc, first\}$. Therefore, Theorem 12.2.6 holds.   □

Algorithm 12.2.9. *Conversion of an n-LMGR to an equivalent n-LMGN.*

Input: An n-LMGR, $\Gamma = (G_1, G_2, \ldots, G_n, Q)$.
Output: An n-LMGN, $\bar{\Gamma} = (\bar{G}_1, \bar{G}_2, \ldots, \bar{G}_n, \bar{Q})$, such that $n\text{-}L(\Gamma) = n\text{-}L(\bar{\Gamma})$.
Method: Let $G_i = (V_i, T_i, P_i, S_i)$, for all $i = 1, \ldots, n$, and set

- $\bar{G}_i = (\bar{V}_i, T_i, \bar{P}_i, S_i)$, for all $i = 1, \ldots, n$, where
  $\bar{N}_i = \{\langle A, x \rangle \mid A \to x \in P_i\} \cup \{S_i\}$;

$\bar{P}_i = \{\langle A, x \rangle \to y \mid A \to x \in P_i, y \in \tau_i(x)\} \cup \{S_i \to y \mid y \in \tau_i(S_i)\}$, where τ_i is a finite substitution from V_i^* to $\bar{V}_i^*$ defined as $\tau_i(a) = \{a\}$, for all $a \in T$, and $\tau_i(A) = \{\langle A, x \rangle \mid A \to x \in P_i\}$, for all $A \in N_i$;

- $\bar{Q} = \{(\langle A_1, x_1 \rangle, \langle A_2, x_2 \rangle, \ldots, \langle A_n, x_n \rangle) \mid (A_1 \to x_1, A_2 \to x_2, \ldots, A_n \to x_n) \in Q\}$
  $\cup \{(S_1, S_2, \ldots, S_n)\}$. $\qquad\qquad\qquad\qquad\qquad\qquad\qquad$ □

Theorem 12.2.10. *Let* $\Gamma = (G_1, G_2, \ldots, G_n, Q)$ *be an n-LMGR. With* Γ *as its input, Algorithm 12.2.9 halts and correctly constructs an n-LMGN,* $\bar{\Gamma} = (\bar{G}_1, \bar{G}_2, \ldots, \bar{G}_n, \bar{Q})$, *such that* $n\text{-}L(\Gamma) = n\text{-}L(\bar{\Gamma})$, *and* $L_X(\Gamma) = L_X(\bar{\Gamma})$, *for each* $X \in \{union, conc, first\}$.

Proof. To prove this theorem, we first establish Claims 12.2.11 and 12.2.12.

Claim 12.2.11. Let $(S_1, S_2, \ldots, S_n) \Rightarrow_\Gamma^m (z_1, z_2, \ldots, z_n)$, where $m \geq 0$, $z_i \in V_i^*$, for all $i = 1, \ldots, n$. Then, $(S_1, S_2, \ldots, S_n) \Rightarrow_{\bar{\Gamma}}^{m+1} (\bar{z}_1, \bar{z}_2, \ldots, \bar{z}_n)$, for any $\bar{z}_i \in \tau_i(z_i)$.

Proof. This claim is proved by induction on $m \geq 0$.
Basis. Let $m = 0$. Then,

$$(S_1, S_2, \ldots, S_n) \Rightarrow_\Gamma^0 (S_1, S_2, \ldots, S_n)$$

Observe that

$$(S_1, S_2, \ldots, S_n) \Rightarrow_{\bar{\Gamma}}^1 (\bar{z}_1, \bar{z}_2, \ldots, \bar{z}_n)$$

for any $\bar{z}_i \in \tau_i(z_i)$, because Algorithm 12.2.9 implies that $(S_1, S_2, \ldots, S_n) \in \bar{Q}$ and $S_i \to \bar{z}_i \in \bar{P}_i$, for any $\bar{z}_i \in \tau_i(z_i)$, for all $i = 1, \ldots, n$. Thus, the basis holds.

Induction Hypothesis. Assume that the claim holds for all m-step derivations, where $m = 0, \ldots, k$, for some $k \geq 0$.

Induction Step. Consider any derivation of the form

$$(S_1, S_2, \ldots, S_n) \Rightarrow_\Gamma^{k+1} (y_1, y_2, \ldots, y_n)$$

Then, there exists a sentential n-form $(u_1 A_1 v_1, u_2 A_2 v_2, \ldots, u_n A_n v_n)$, where $u_i \in T_i^*$, $A_i \in N_i$, $v_i \in V_i^*$, such that

$$(S_1, S_2, \ldots, S_n) \Rightarrow_\Gamma^k (u_1 A_1 v_1, u_2 A_2 v_2, \ldots, u_n A_n v_n)$$
$$\Rightarrow_\Gamma (u_1 x_1 v_1, u_2 x_2 v_2, \ldots, u_n x_n v_n)$$

where $u_i x_i v_i = y_i$, for all $i = 1, \ldots, n$. Then, by the induction hypothesis, we have

$$(S_1, S_2, \ldots, S_n) \Rightarrow_{\bar{\Gamma}}^{k+1} (\bar{w}_1, \bar{w}_2, \ldots, \bar{w}_n)$$

for any $\bar{w}_i \in \tau_i(u_i A_i v_i)$, for all $i = 1, \ldots, n$. Since

$$(u_1 A_1 v_1, u_2 A_2 v_2, \ldots, u_n A_n v_n) \Rightarrow_\Gamma (u_1 x_1 v_1, u_2 x_2 v_2, \ldots, u_n x_n v_n)$$

$(A_1 \rightarrow x_1, A_2 \rightarrow x_2, \ldots, A_n \rightarrow x_n) \in Q$. Algorithm 12.2.9 implies that $(\langle A_1, x_1 \rangle,$ $\langle A_2, x_2 \rangle, \ldots, \langle A_n, x_n \rangle) \in \bar{Q}$ and $\langle A_i, x_i \rangle \rightarrow \bar{y}_i \in \bar{P}_i$, for any $\bar{y}_i \in \tau_i(x_i)$, for all $i = 1, \ldots, n$. Let $\bar{w}_i$ be any sentential form of the form $\bar{u}_i \langle A_i, x_i \rangle \bar{v}_i$, for all $i = 1, \ldots, n$, where $\bar{u}_i \in \tau_i(u_i)$ and $\bar{v}_i \in \tau_i(v_i)$. Then,

$$(\bar{u}_1 \langle A_1, x_1 \rangle \bar{v}_1, \bar{u}_2 \langle A_2, x_2 \rangle \bar{v}_2, \ldots, \bar{u}_n \langle A_n, x_n \rangle \bar{v}_n) \Rightarrow_{\bar{\Gamma}}$$
$$(\bar{u}_1 \bar{y}_1 \bar{v}_1, \bar{u}_2 \bar{y}_2 \bar{v}_2, \ldots, \bar{u}_n \bar{y}_n \bar{v}_n)$$

where $\bar{u}_i \bar{y}_i \bar{v}_i$ is any sentential form, $\bar{u}_i \bar{y}_i \bar{v}_i \in \tau_i(u_i y_i v_i)$, for all $i = 1, \ldots, n$, which proves the induction step. Therefore, Claim 12.2.11 holds.   □

Claim 12.2.12. Let $(S_1, S_2, \ldots, S_n) \Rightarrow_{\bar{\Gamma}}^m (\bar{z}_1, \bar{z}_2, \ldots, \bar{z}_n)$, where $m \geq 1$, $\bar{z}_i \in \bar{V}_i^*$, for all $i = 1, \ldots, n$. Then, $(S_1, S_2, \ldots, S_n) \Rightarrow_{\Gamma}^{m-1} (z_1, z_2, \ldots, z_n)$, where $\bar{z}_i \in \tau_i(z_i)$, for all $i = 1, \ldots, n$.

Proof. This claim is proved by induction on $m \geq 1$.

Basis. Let $m = 1$. Then,

$$(S_1, S_2, \ldots, S_n) \Rightarrow_{\bar{\Gamma}} (\bar{z}_1, \bar{z}_2, \ldots, \bar{z}_n)$$

implies that $S_i \rightarrow \bar{z}_i \in \bar{P}_i$, for all $i = 1, \ldots, n$. Algorithm 12.2.9 implies that $\bar{z}_i \in \tau_i(S_i)$, for all $i = 1, \ldots, n$, so

$$(S_1, S_2, \ldots, S_n) \Rightarrow_{\bar{\Gamma}}^0 (S_1, S_2, \ldots, S_n)$$

Since $\bar{z}_i \in \tau_i(S_i)$, for all $i = 1, \ldots, n$, the basis holds.

Induction Hypothesis. Assume that the claim holds for all m-step derivations, where $m = 1, \ldots, k$, for some $k \geq 1$.

Induction Step. Consider any derivation of the form

$$(S_1, S_2, \ldots, S_n) \Rightarrow_{\bar{\Gamma}}^{k+1} (\bar{y}_1, \bar{y}_2, \ldots, \bar{y}_n)$$

Then, there exists a sentential n-form

$$(\bar{u}_1 \langle A_1, x_1 \rangle \bar{v}_1, \bar{u}_2 \langle A_2, x_2 \rangle \bar{v}_2, \ldots, \bar{u}_n \langle A_n, x_n \rangle \bar{v}_n)$$

where $\bar{u}_i \in T_i^*$, $\langle A_i, x_i \rangle \in \bar{N}_i$, $\bar{v}_i \in \bar{V}_i^*$, such that

$$(S_1, S_2, \ldots, S_n) \Rightarrow_{\bar{\Gamma}}^k (\bar{u}_1 \langle A_1, x_1 \rangle \bar{v}_1, \bar{u}_2 \langle A_2, x_2 \rangle \bar{v}_2, \ldots, \bar{u}_n \langle A_n, x_n \rangle \bar{v}_n)$$
$$\Rightarrow_{\bar{\Gamma}} (\bar{u}_1 \bar{x}_1 \bar{v}_1, \bar{u}_2 \bar{x}_2 \bar{v}_2, \ldots, \bar{u}_n \bar{x}_n \bar{v}_n)$$

where $\bar{u}_i \bar{x}_i \bar{v}_i = \bar{y}_i$, for all $i = 1, \ldots, n$. Then, by the induction hypothesis, we have

$$(S_1, S_2, \ldots, S_n) \Rightarrow_{\Gamma}^{k-1} \Rightarrow_{\Gamma} (w_1, w_2, \ldots, w_n)$$

where $\bar{u}_i \langle A_i, x_i \rangle \bar{v}_i \in \tau_i(w_i)$, for all $i = 1, \ldots, n$. Since

$$(\bar{u}_1 \langle A_1, x_1 \rangle \bar{v}_1, \bar{u}_2 \langle A_2, x_2 \rangle \bar{v}_2, \ldots, \bar{u}_n \langle A_n, x_n \rangle \bar{v}_n) \Rightarrow_{\bar{\Gamma}}$$

$$(\bar{u}_1 \bar{x}_1 \bar{v}_1, \bar{u}_2 \bar{x}_2 \bar{v}_2, \ldots, \bar{u}_n \bar{x}_n \bar{v}_n)$$

there are $(\langle A_1, x_1 \rangle, \langle A_2, x_2 \rangle, \ldots, \langle A_n, x_n \rangle) \in \bar{Q}$ and $\langle A_i, x_i \rangle \to \bar{x}_i \in \bar{P}_i$, for all $i = 1, \ldots, n$. Algorithm 12.2.9 implies that $(A_1 \to x_1, A_2 \to x_2, \ldots, A_n \to x_n) \in Q$ and $A_i \to x_i \in P_i$, where $\bar{x}_i \in \tau_i(x_i)$, for all $i = 1, \ldots, n$. We can express w_i as $w_i = u_i A_i v_i$, where $\bar{u}_i \in \tau_i(u_i)$, $\bar{v}_i \in \tau_i(v_i)$, and observe that $\langle A_i, x_i \rangle \in \tau_i(A_i)$ holds by the definition of τ_i, for all $i = 1, \ldots, n$. Then,

$$(u_1 A_1 v_1, u_2 A_2 v_2, \ldots, u_n A_n v_n) \Rightarrow_{\Gamma} (u_1 x_1 v_1, u_2 x_2 v_2, \ldots, u_n x_n v_n)$$

where $\bar{u}_i \in \tau_i(u_i)$, $\bar{v}_i \in \tau_i(v_i)$, and $\bar{x}_i \in \tau_i(x_i)$, for all $i = 1, \ldots, n$, which means that $\bar{u}_i \bar{x}_i \bar{v}_i \in \tau_i(u_i x_i v_i)$, for all $i = 1, \ldots, n$. Therefore,

$$(S_1, S_2, \ldots, S_n) \Rightarrow_{\Gamma}^{k+1} (u_1 A_1 v_1, u_2 A_2 v_2, \ldots, u_n A_n v_n)$$
$$\Rightarrow_{\Gamma} (u_1 x_1 v_1, u_2 x_2 v_2, \ldots, u_n x_n v_n)$$

where $\bar{u}_i \bar{x}_i \bar{v}_i \in \tau_i(u_i x_i v_i)$, for all $i = 1, \ldots, n$. Let $\bar{z}_i = \bar{u}_i \bar{x}_i \bar{v}_i$ and $z_i = u_i x_i v_i$, for all $i = 1, \ldots, n$. Then,

$$(S_1, S_2, \ldots, S_n) \Rightarrow_{\Gamma}^{k+2} (z_1, z_2, \ldots, z_n)$$

for all $\bar{z}_i \in \tau_i(z_i)$, which proves the induction step. Therefore, Claim 12.2.12 holds.   □

Consider Claim 12.2.11 when $z_i \in T_i^*$, for all $i = 1, \ldots, n$. At this point, if

$$(S_1, S_2, \ldots, S_n) \Rightarrow_{\Gamma}^* (z_1, z_2, \ldots, z_n)$$

then

$$(S_1, S_2, \ldots, S_n) \Rightarrow_{\bar{\Gamma}}^* (\bar{z}_1, \bar{z}_2, \ldots, \bar{z}_n)$$

where $\bar{z}_i \in \tau_i(z_i)$, for all $i = 1, \ldots, n$. Since $\tau_i(a_i) = a_i$, for all $a_i \in T_i$, $\bar{z}_i = z_i$. Hence, $n\text{-}L(\Gamma) \subseteq n\text{-}L(\bar{\Gamma})$. Consider Claim 12.2.12 when $\bar{z}_i \in T_i^*$, for all $i = 1, \ldots, n$. At this point, if

$$(S_1, S_2, \ldots, S_n) \Rightarrow_{\bar{\Gamma}}^m (\bar{z}_1, \bar{z}_2, \ldots, \bar{z}_n)$$

then

$$(S_1, S_2, \ldots, S_n) \Rightarrow_{\Gamma}^{m-1} (z_1, z_2, \ldots, z_n)$$

where $\bar{z}_i \in \tau_i(z_i)$, for all $i = 1,\ldots,n$. Since $\tau_i(a_i) = a_i$, for all $a_i \in T_i$, $z_i = \bar{z}_i$. Hence, $n\text{-}L(\bar{\Gamma}) \subseteq n\text{-}L(\Gamma)$. As $n\text{-}L(\Gamma) \subseteq n\text{-}L(\bar{\Gamma})$ and $n\text{-}L(\bar{\Gamma}) \subseteq n\text{-}L(\Gamma)$, $n\text{-}L(\Gamma)$ $= n\text{-}L(\bar{\Gamma})$. By Lemma 12.2.4, this identity implies that $L_X(\Gamma) = L_X(\bar{\Gamma})$, for each $X \in \{union, conc, first\}$. Therefore, Theorem 12.2.10 holds. □

From the achieved results, we immediately obtain the following corollary.

Corollary 12.2.13. *The family of languages generated by n-LMGN in the X mode coincides with the family of languages generated by n-LMGR in the X mode, where* $X \in \{union, conc, first\}$. □

Theorem 12.2.14. *For every recursively enumerable language L over some alphabet T, there exits a 2-LMGR,* $\Gamma = ((\bar{V}_1, T, \bar{P}_1, S_1), (\bar{V}_2, T, \bar{P}_2, S_2), Q)$*, such that*

(i) $\{w \in T^* \mid (S_1, S_2) \Rightarrow_\Gamma^* (w, w)\} = L$,
(ii) $\{w_1 w_2 \in T^* \mid (S_1, S_2) \Rightarrow_\Gamma^* (w_1, w_2), w_1 \neq w_2\} = \emptyset$.

Proof. Recall that for every recursive enumerable language L over some alphabet T, there exist two context-free grammars, $G_1 = (V_1, \bar{T}, P_1, S_1)$, $G_2 = (V_2, \bar{T}, P_2, S_2)$, and a homomorphism h from $\bar{T}^*$ to T^* such that $L = \{h(x) \mid x \in L(G_1) \cap L(G_2)\}$ (see Theorem 2.3.18). Furthermore, by Theorem 3.1.22, for every context-free grammar, there exists an equivalent context-free grammar in the Greibach normal form (see Definition 3.1.21). Hence, without any lost of generality, we assume that G_1 and G_2 are in the Greibach normal form. Consider the 2-LMGR

$$\Gamma = (G_1, G_2, Q)$$

where

- $G_i = (\bar{V}_i, T, \bar{P}_i, S_i)$, where

    $\bar{N}_i = N_i \cup \{\bar{a} \mid a \in \bar{T}\}$;
    $\bar{P}_i = \{A \to \bar{a}x \mid A \to ax \in P_i, a \in \bar{T}, x \in N_i^*\} \cup \{\bar{a} \to h(a) \mid a \in \bar{T}\}$,

    for $i = 1, 2$;
- $Q = \{(A_1 \to \bar{a}x_1, A_2 \to \bar{a}x_2) \mid A_1 \to \bar{a}x_1 \in P_1, A_2 \to \bar{a}x_2 \in P_2, a \in \bar{T}\}$
    $\cup \{(\bar{a} \to h(a), \bar{a} \to h(a)) \mid a \in \bar{T}\}$.

Consider properties (i) and (ii) in Theorem 12.2.14. Next, Claims 12.2.15 and 12.2.16 establish (i) and (ii), respectively.

Claim 12.2.15. $\{w \in T^* \mid (S_1, S_2) \Rightarrow_\Gamma^* (w, w)\} = L$

Proof.

I. We prove that $L \subseteq \{w \in T^* \mid (S_1, S_2) \Rightarrow_\Gamma^* (w, w)\}$. Let w be any string. Then, there exists a string, $a_1 a_2 \cdots a_n \in \bar{T}^*$, such that

- $a_1 a_2 \cdots a_n \in L(G_1)$,
- $a_1 a_2 \cdots a_n \in L(G_2)$, and
- $h(a_1 a_2 \cdots a_n) = w$.

This means that there exist the following derivations in G_1 and G_2

$$
\begin{aligned}
S_1 &\Rightarrow_{G_1} a_1 x_1 & [p_1] \\
&\Rightarrow_{G_1} a_1 a_2 x_2 & [p_2] \\
&\ \ \vdots \\
&\Rightarrow_{G_1} a_1 a_2 \cdots a_n & [p_n] \\
S_2 &\Rightarrow_{G_2} a_1 y_1 & [r_1] \\
&\Rightarrow_{G_2} a_1 a_2 y_2 & [r_2] \\
&\ \ \vdots \\
&\Rightarrow_{G_2} a_1 a_2 \cdots a_n & [r_n]
\end{aligned}
$$

where $a_i \in \bar{T}$, $x_i \in N_1^*$, $y_i \in N_2^*$, $p_i \in P_1$, $r_i \in P_2$, for all $i = 1, \ldots, n$. Observe that that $\mathrm{sym}(\mathrm{rhs}(p_i), 1) = \mathrm{sym}(\mathrm{rhs}(r_i), 1) = a_i$, for all $i = 1, \ldots, n$. The construction of Q implies the following two statements.

- Let $p_i \colon A_i \to a_i u_i \in \bar{P}_1$, $r_i \colon B_i \to a_i v_i \in \bar{P}_2$. Then, $(A_i \to \bar{a}_i u_i, B_i \to \bar{a}_i v_i) \in Q$, for all $i = 1, \ldots, n$.
- Q contains $(\bar{a}_i \to h(a_i), \bar{a}_i \to h(a_i))$, for all $i = 1, \ldots, n$.

Therefore, there exists

$$
\begin{aligned}
(S_1, S_2) &\Rightarrow_{\Gamma} (\bar{a}_1 x_1, \bar{a}_1 y_1) \\
&\Rightarrow_{\Gamma} (h(a_1) x_1, h(a_1) y_1) \\
&\Rightarrow_{\Gamma} (h(a_1) \bar{a}_2 x_2, h(a_1) \bar{a}_2 y_2) \\
&\Rightarrow_{\Gamma} (h(a_1) h(a_2) x_2, h(a_1) h(a_2) y_2) \\
&\ \ \vdots \\
&\Rightarrow_{\Gamma} (h(a_1) h(a_2) \cdots h(a_n), h(a_1) h(a_2) \cdots h(a_n)) \\
&= (h(a_1 a_2 \cdots a_n), h(a_1 a_2 \cdots a_n)) \\
&= (w, w)
\end{aligned}
$$

In brief, $(S_1, S_2) \Rightarrow_{\Gamma}^* (w, w)$. Hence, $L \subseteq \{ w \in T^* \mid (S_1, S_2) \Rightarrow_{\Gamma}^* (w, w) \}$.

II. We prove that $\{ w \in T^* \mid (S_1, S_2) \Rightarrow_{\Gamma}^* (w, w) \} \subseteq L$. Let $(S_1, S_2) \Rightarrow_{\Gamma}^* (w, w)$. Then, there exists

$$
\begin{aligned}
(S_1, S_2) &\Rightarrow_{\Gamma} (\bar{a}_1 x_1, \bar{a}_1 y_1) \\
&\Rightarrow_{\Gamma} (h(a_1) x_1, h(a_1) y_1) \\
&\Rightarrow_{\Gamma} (h(a_1) \bar{a}_2 x_2, h(a_1) \bar{a}_2 y_2) \\
&\Rightarrow_{\Gamma} (h(a_1) h(a_2) x_2, h(a_1) h(a_2) y_2) \\
&\ \ \vdots \\
&\Rightarrow_{\Gamma} (h(a_1) h(a_2) \cdots h(a_n), h(a_1) h(a_2) \cdots h(a_n)) \\
&= (h(a_1 a_2 \cdots a_n), h(a_1 a_2 \cdots a_n)) \\
&= (w, w)
\end{aligned}
$$

By analogy with part I., we can prove that there exist derivations in G_1 and G_2 of the forms

$$S_1 \Rightarrow_{G_1} a_1 x_1 \qquad [p_1]$$
$$\Rightarrow_{G_1} a_1 a_2 x_2 \qquad [p_2]$$
$$\vdots$$
$$\Rightarrow_{G_1} a_1 a_2 \cdots a_n \ [p_n]$$

$$S_2 \Rightarrow_{G_2} a_1 y_1 \qquad [r_1]$$
$$\Rightarrow_{G_2} a_1 a_2 y_2 \qquad [r_2]$$
$$\vdots$$
$$\Rightarrow_{G_2} a_1 a_2 \cdots a_n \ [r_n]$$

This implies that $a_1 a_2 \cdots a_n \in L(G_1)$, $a_1 a_2 \cdots a_n \in L(G_2)$, and $h(a_1 a_2 \cdots a_n) = w$, so $w \in L$. Hence, $\{w \in T^* \mid (S_1, S_2) \Rightarrow_r^* (w, w)\} \subseteq L$. Therefore, Claim 12.2.15 holds. $\square$

Claim 12.2.16. $\{w_1 w_2 \in T^* \mid (S_1, S_2) \Rightarrow_r^* (w_1, w_2), w_1 \neq w_2\} = \emptyset$

Proof. By contradiction. Let $\{w_1 w_2 \in T^* \mid (S_1, S_2) \Rightarrow_r^* (w_1, w_2), w_1 \neq w_2\} \neq \emptyset$. Then, there have to exist two different strings, $w_1 = h(a_1)h(a_2) \cdots h(a_n)$ and $w_2 = h(b_1)h(b_2) \cdots h(b_n)$, such that $(S_1, S_2) \Rightarrow_r^* (w_1, w_2)$.

I. Assume that $a_i = b_i$, for all $i = 1, \ldots, n$. Then, $w_1 = h(a_1)h(a_2) \cdots h(a_n) = h(b_1)h(b_2) \cdots h(b_n) = w_2$, which contradicts $w_1 \neq w_2$.
II. Assume that there exists some $k \leq n$ such that $a_k \neq b_k$. Then, there exists a derivation of the form

$$(S_1, S_2) \Rightarrow_r (\bar{a}_1 x_1, \bar{a}_1 y_1)$$
$$\Rightarrow_r (h(a_1) x_1, h(a_1) y_1)$$
$$\Rightarrow_r (h(a_1)\bar{a}_2 x_2, h(a_1)\bar{a}_2 y_2)$$
$$\Rightarrow_r (h(a_1)h(a_2) x_2, h(a_1)h(a_2) y_2)$$
$$\vdots$$
$$\Rightarrow_r (h(a_1)h(a_2) \cdots h(a_{k-1}) x_{k-1}, h(a_1)h(a_2) \cdots h(a_{k-1}) y_{k-1})$$

Then, there has to exist a derivation

$$(x_{k-1}, y_{k-1}) \Rightarrow_r (\bar{a}_k x_k, \bar{b}_k y_k)$$

where $\bar{a}_k \neq \bar{b}_k$. By the definition of Q, there has to be $(p, r) \in Q$ such that

$$\text{sym}\big(\text{rhs}(p), 1\big) = \text{sym}\big(\text{rhs}(r), 1\big)$$

Therefore, the next derivation has to be of the form

$$(x_{k-1}, y_{k-1}) \Rightarrow_\Gamma (\bar{a}_k x_k, \bar{b}_k y_k)$$

where $\bar{a}_k = \bar{b}_k$, which is a contradiction. Therefore, Claim 12.2.16 holds. $\square$

Claims 12.2.15 and 12.2.16 imply that Theorem 12.2.14 holds. $\square$

Theorem 12.2.17. *For any recursively enumerable language L over an alphabet T, there exists a 2-LMGR, $\Gamma = (G_1, G_2, Q)$, such that $L_{union}(\Gamma) = L$.*

Proof. By Theorem 12.2.14, for every recursively enumerable language L over an alphabet T, there exits a 2-LMGR

$$\bar{\Gamma} = \big((V_1, T, P_1, S_1), (V_2, T, P_2, S_2), Q\big)$$

such that

$$\{w \in T^* \mid (S_1, S_2) \Rightarrow_\Gamma^* (w, w)\} = L$$

and

$$\{w_1 w_2 \in T^* \mid (S_1, S_2) \Rightarrow_\Gamma^* (w_1, w_2), w_1 \neq w_2\} = \emptyset$$

Let $\Gamma = \bar{\Gamma}$. Then,

$$
\begin{aligned}
L_{union}(\Gamma) &= \{w \mid (S_1, S_2) \Rightarrow_\Gamma^* (w_1, w_2), w_i \in T^*, \text{ for } i = 1, 2, \\
&\qquad w \in \{w_1, w_2\}\} \\
&= \{w \mid (S_1, S_2) \Rightarrow_\Gamma^* (w, w)\} \cup \{w \mid (S_1, S_2) \Rightarrow_\Gamma^* (w_1, w_2), \\
&\qquad w_i \in T^*, \text{ for } i = 1, 2, w \in \{w_1, w_2\}, w_1 \neq w_2\} \\
&= \{w \mid (S_1, S_2) \Rightarrow_\Gamma^* (w, w)\} \cup \emptyset \\
&= \{w \mid (S_1, S_2) \Rightarrow_\Gamma^* (w, w)\} \\
&= L
\end{aligned}
$$

Therefore, Theorem 12.2.17 holds. $\square$

Theorem 12.2.18. *For any recursively enumerable language L over an alphabet T, there exists a 2-LMGR, $\Gamma = (G_1, G_2, Q)$, such that $L_{first}(\Gamma) = L$.*

Proof. By Theorem 12.2.14, for every recursively enumerable language L over an alphabet T, there exits a 2-LMGR

$$\bar{\Gamma} = \big((V_1, T, P_1, S_1), (V_2, T, P_2, S_2), Q\big)$$

such that

$$\{w \in T^* \mid (S_1, S_2) \Rightarrow_\Gamma^* (w, w)\} = L$$

and

$$\{w_1 w_2 \in T^* \mid (S_1, S_2) \Rightarrow_{\Gamma}^* (w_1, w_2), w_1 \neq w_2\} = \emptyset$$

Let $\Gamma = \bar{\Gamma}$. Then,

$$
\begin{aligned}
L_{first}(\Gamma) &= \{w_1 \mid (S_1, S_2) \Rightarrow_{\Gamma}^* (w_1, w_2), w_i \in T^*, \text{ for } i = 1, 2\} \\
&= \{w \mid (S_1, S_2) \Rightarrow_{\Gamma}^* (w, w)\} \cup \{w_1 \mid (S_1, S_2) \Rightarrow_{\Gamma}^* (w_1, w_2), \\
&\quad\quad w_i \in T^*, \text{ for } i = 1, 2, w_1 \neq w_2\} \\
&= \{w \mid (S_1, S_2) \Rightarrow_{\Gamma}^* (w, w)\} \cup \emptyset \\
&= \{w \mid (S_1, S_2) \Rightarrow_{\Gamma}^* (w, w)\} \\
&= L
\end{aligned}
$$

Therefore, Theorem 12.2.18 holds.   □

Theorem 12.2.19. *For any recursively enumerable language L over an alphabet T, there exists a 2-LMGR, $\Gamma = (G_1, G_2, Q)$, such that $L_{conc}(\Gamma) = L$.*

Proof. By Theorem 12.2.14, we have that for every recursively enumerable language L over an alphabet T, there exits a 2-LMGR

$$\bar{\Gamma} = ((V_1, T, P_1, S_1), (V_2, T, P_2, S_2), Q)$$

such that

$$\{w \in T^* \mid (S_1, S_2) \Rightarrow_{\Gamma}^* (w, w)\} = L$$

and

$$\{w_1 w_2 \in T^* \mid (S_1, S_2) \Rightarrow_{\Gamma}^* (w_1, w_2), w_1 \neq w_2\} = \emptyset$$

Let $G_1 = (V_1, T, P_1, S_1)$ and $G_2 = (V_2, \emptyset, \bar{P}_2, S_2)$, where $\bar{P}_2 = \{A \rightarrow g(x) \mid A \rightarrow x \in P_2\}$, where g is a homomorphism from V_2^* to N_2^* defined as $g(X) = X$, for all $X \in N_2$, and $g(a) = \varepsilon$, for all $a \in T$. We prove that $L_{conc}(\Gamma) = L$.

I. We prove that $L \subseteq L_{conc}(\Gamma)$. Let $w \in L$. Then, there exists a derivation of the form

$$(S_1, S_2) \Rightarrow_{\Gamma}^* (w, w)$$

Thus, there exist a derivation of the form

$$(S_1, S_2) \Rightarrow_{\Gamma}^* (w, g(w))$$

Since $g(a) = \varepsilon$, for all $a \in T$, $g(w) = \varepsilon$, for all $w \in T^*$. Thus,

$$(S_1, S_2) \Rightarrow_{\Gamma}^* (w, \varepsilon)$$

Hence, $w\varepsilon = w$ and $w \in L_{conc}(\Gamma)$.

II. We prove that $L_{conc}(\Gamma) \subseteq L$. Let $w \in L$. Then, there exists a derivation of the form

$$(S_1, S_2) \Rightarrow^*_\Gamma (w, \varepsilon)$$

because $L(G_2) = \{\varepsilon\}$. Since $g(x) = \varepsilon$ in Γ, for all $x \in T^*$, there is a derivation of the form

$$(S_1, S_2) \Rightarrow^*_\Gamma (w, x)$$

where x is any string. Theorem 12.2.14 implies that $x = w$. Thus,

$$(S_1, S_2) \Rightarrow^*_\Gamma (w, w)$$

Hence, $w \in L$.

By I. and II., Theorem 12.2.19 holds.   □

We close this section by suggesting the next open problem area.

Open Problem 12.2.20. By analogy with leftmost n-generative nonterminal-synchronized grammar systems, discussed in this section, introduce n-generative nonterminal-synchronized grammar systems and study their generative power.

Part V
Modern Language Models Applied to Computation

This part, consisting of Chaps. 13 through 15, discusses computational applications of modern language models studied earlier in the book. Chapter 13 covers these applications and their perspectives in computer science from a rather general standpoint. Then, more specifically, Chaps. 14 and 15 describe many applications in computational linguistics and computational biology, respectively. Both chapters contain several case studies of real-world applications described in detail.

Chapter 13
Applications and Their Perspectives in General

This chapter makes several general remarks about computational applications of modern language models covered earlier in this book. It also discusses their application perspectives in computer science in the near future.

As we know by now, however, all these models represent an enormously large variety of grammars and automata. Therefore, we narrow our attention only to some of them. Specifically, we choose regulated grammars (see Chap. 3), scattered context grammars (see Sect. 4.1), grammar systems (see Sect. 12), and regulated pushdown automata (see Chap. 7) for this purpose. Regarding the computer science application areas, we focus our principle attention on two areas—computational linguistics and computational biology.

13.1 General Comments on Applications in Computational Linguistics

In terms of English syntax, grammatical regulation can specify a number of relations between individual syntax-related elements of sentences in natural languages. For instance, relative clauses are introduced by *who* or *which* depending on the subject of the main clause. If the subject in the main clause is a person, the relative clause is introduced by *who*; otherwise, it starts by *which*. We encourage the reader to design a regulated grammar that describes this dependency (consult [HP05]).

In other natural languages, there exist syntax relations that can be elegantly handled by regulated grammars, too. To illustrate, in Spanish, all adjectives inflect according to gender of the noun they characterize. Both the noun and the adjective may appear at different parts of a sentence, which makes their syntactical dependency difficult to capture by classical grammars; obviously, regulated grammars, discussed in Chap. 3, can describe this dependency in a more elegant and simple way. As a result, parsing is expected as their principle application field.

© Springer International Publishing AG 2017
A. Meduna, O. Soukup, *Modern Language Models and Computation*,
DOI 10.1007/978-3-319-63100-4_13

Ordinary parsers represent crucially important components of translators, and they are traditionally underlined by ordinary context-free grammars. As their name indicates, regulated parsers are based upon regulated context-free grammars. Considering their advantages, including properties (I) through (IV) listed next, it comes as no surprise that they become increasingly popular in modern design of language translators.

(I)   Regulated parsers work in a faster way than classical parsers do. Indeed, ordinary parsers control their parsing process so they consult their parsing tables during every single step. As opposed to this exhaustively busy approach, in regulated parsers, regulated grammatical mechanisms take control over the parsing process to a large extent; only during very few pre-determined steps, they consult their parsing tables to decide how to continue the parsing process under the guidance of regulating mechanism. Such a reduction of communication with the parsing tables obviously results into a significant acceleration of the parsing process as a whole.

(II) Regulated context-free grammars are much stronger than ordinary context-free grammars. Accordingly, parsers based upon regulated grammars are more powerful than their ordinary versions. As an important practical consequence, they can parse syntactical structures that cannot be parsed by ordinary parsers.

(III) Regulated parsers make use of their regulation mechanisms to perform their parsing process in a deterministic way.

(IV) Compared to ordinary parsers, regulated parsers are often written more succinctly and, therefore, readably as follows from reduction-related results concerning the number of their components, such as nonterminals and rules, achieved earlier in this book (see Sects. 3.1.4, 3.1.6, 4.1.4, 4.2.2, 4.2.3, and 4.2.4).

From a general point of view, some fundamental parts of translators, such as syntax-directed translators, run within the translation process under the parser-based regulation. Furthermore, through their symbol tables, parsers also regulate exchanging various pieces of information between their components, further divided into several subcomponents. Indeed, some parts of modern translators may be further divided into various subparts, which are run in a regulated way, and within these subparts, a similar regulation can be applied again, and so on. As a matter of fact, syntax-directed translation is frequently divided into two parts, which work concurrently. One part is guided by a precedence parser that works with expressions and conditions while the other part is guided by a predictive parser that processes the general program flow. In addition, both parts are sometimes further divided into several subprocesses or threads. Of course, this two-parser design of syntax-directed translation requires an appropriate regulation of translation as a whole. Indeed, prior to this syntax-directed translation, a pre-parsing decomposition of the tokenized source program separates the syntax constructs for both parsers. On the other hand, after the syntax-directed translation based upon the two parsers is successfully completed, all the produced fragments of the intermediate code are carefully composed together so the resulting intermediate code is functionally equivalent to the source program. Of course, handling translation like this requires a proper regulation of all these translation subphases.

To give one more example in terms of modern translator design, various optimization methods are frequently applied to the generation of the resulting target code to speed the code up as much as possible. This way of code generation may result from an explicit requirement in the source program. More often, however, modern translators themselves recognize that a generation like this is appropriate within the given computer framework, so they generate the effective target code to speed up its subsequent execution. Whatever they do, however, they always have to guarantee that the generated target code is functionally equivalent to the source program. Clearly, this design of translators necessitates an extremely careful control over all the optimization routines involved, and this complicated control has to be based upon a well developed theory of computational regulation. Within formal language theory, which has always provided translation techniques with their formal models, this control can be accomplished by regulated grammars, which naturally and elegantly formalize computational regulation.

Apart from description, specification, and transformation of language syntax, regulated grammars can be applied to other linguistically oriented fields, such as *morphology* (see [Bau03, AF04]).

13.2   General Comments on Applications in Computational Biology

Because the grammar-based information processing fulfills a crucially important role in biology as a whole, it is literally impossible to cover all these applications in this scientific filed. Therefore, we restrict our attention only to microbiology, which also makes use of the systematically developed knowledge concerning these grammars significantly. Even more specifically, we narrow our attention to *molecular genetics* (see [SR10, WBB+07, Rus09]). A solidly developed control of information processing is central to this scientific field although it approaches this processing in a specific way. Indeed, genetically oriented studies usually investigate how to prescribe the modification of several symbols within strings that represent a molecular organism. To illustrate a modification like this, consider a typical molecular organism consisting of several groups of molecules; for instance, take any organism consisting of several parts that slightly differ in behavior of DNA molecules made by specific sets of enzymes. During their development, these groups of molecules communicate with each other, and this communication usually influences the future behavior of the whole organism. A simulation of such an organism might be formally based upon regulated grammars, which can control these changes at various places. Consequently, genetic dependencies of this kind represent another challenging application area of regulated grammars in the future.

To sketch the applicability of regulated grammars in this scientific area in a greater detail, consider forbidding grammars, studied earlier in Sect. 3.1.4. These grammars can formally and elegantly simulate processing information in molecular genetics, including information concerning macromolecules, such as DNA, RNA, and polypeptides. For instance, consider an organism consisting of DNA molecules made by enzymes. It is a common phenomenon that a molecule m made by a specific enzyme can be modified unless molecules made by some other enzymes occur either to the left or to the right of m in the organism. Consider a string w that formalizes this organism so every molecule is represented by a symbol. As obvious, to simulate a change of the symbol a that represents m requires forbidding occurrences of some symbols that either precede or follow a in w. As obvious, forbidding grammars can provide a string-changing formalism that can capture this forbidding requirement in a very succinct and elegant way. To put it more generally, forbidding grammars can simulate the behavior of molecular organisms in a rigorous and uniform way. Application-oriented topics like this obviously represent a future investigation area concerning forbidding grammars.

In the near future, highly regulated information processing is expected to intensify rapidly and significantly. Indeed, to take advantage of highly effective parallel and mutually connected computers as much as possible, a modern software product simultaneously run several processes, each of which gather, analyze and modify various elements occurring within information of an enormous size, largely spread and constantly growing across the virtually endless and limitless computer environment. During a single computational step, a particular running process selects a finite set of mutually related information elements, from which it produces new information as a whole and, thereby, completes the step. In many respects, the newly created information affects the way the process performs the next computational step, and from a more broadly perspective, it may also significantly change the way by which the other processes work as well. Clearly, a product conceptualized in this modern way requires a very sophisticated regulation of its computation performed within a single process as well as across all the processes involved.

As already explained in "Preface" section, computer science urgently needs to express regulated computation by appropriate mathematical models in order to express its fundamentals rigorously. Traditionally, formal language theory provides computer science with various automata and grammars as formal models of this kind. However, classical automata and grammars, such as ordinary finite automata or context-free grammars, represent unregulated formal models because they were introduced several decades ago when hardly any highly regulated computation based upon parallelism and distribution occurred in computer science. As an inescapable consequence, these automata and grammars fail to adequately formalize highly regulated computation. Consequently, so far, most theoretically oriented computer science areas whose investigation involve this computation simplify their investigation so they reduce their study to quite specific areas in which they work with various ad-hoc simplified models without any attempt to formally describe highly regulated computation generally and systematically. In this sense, theoretical computer science based upon unregulated formal models is endangered by approaching computation in an improper way, which does not reflect the expected regulated computation in the future at all. Simply put, rather than shed some light on fundamental ideas of this processing, this approach produces little or no relevant results concerning future computation.

Taking into account this unsatisfactory and dangerous situation occurring in the very heart of computational theory, the present book has paid an explicit attention to modifying automata and grammars so they work in a regulated way. As a result of this modification, the resulting regulated versions of grammars and automata can properly and adequately underlie a systematized theory concerning general ideas behind future regulated information processing. Out of all these regulated grammars and automata, we next select three types and demonstrate the way they can appropriately act as formal models of regulated computation. Namely, we choose

(1) scattered context grammars (see Sect. 4.1);
(2) grammar systems (see Chap. 12);
(3) regulated pushdown automata (see Sect. 7.2.2).

(1) In general, the heart of every grammar consists of a finite set of rules, according to which the grammar derives sentences. The collection of all sentences derived by these rules forms the language generated by the grammar. Most classical grammars perform their derivation steps in a strictly sequential way. To illustrate, context-free grammars work in this way because they rewrite a single symbol of the sentential form during every single derivation step (see [Med00a, Sal73, Har78, Woo87, RS97a]).

As opposed to strictly sequential grammars, the notion of a scattered context grammar is based upon finitely many sequences of context-free rules that are simultaneously applied during a single derivation step. Beginning from its start symbol, the derivation process, consisting of a sequence of derivation steps, successfully ends when the derived strings contain only terminal symbols. A terminal word derived in this successful way is included into the language of this grammar, which contains all strings derived in this way. As obvious, this way of rewriting makes scattered context grammars relevant to regulated information processing as illustrated next in terms of computational linguistics.

Consider several texts such that (a) they all are written in different natural languages, but (b) they correspond to the same syntactical structure, such as the structure of basic clauses. With respect to (b), these texts are obviously closely related, yet we do not tend to compose them into a single piece of information because of (a). Suppose that a multilingual processor simultaneously modifies all these texts in their own languages so all the modified texts again correspond to the same syntactical structure, such as a modification of basic clauses to the corresponding interrogative clauses; for instance, *I said that* would be changed to *Did I say that?* in English. At this point, a processor like this needs to regulate its computation across all these modified texts in mutually different languages. As obvious, taking advantage of their simultaneous way of rewriting, scattered context grammars can handle changes of this kind while ordinary unregulated context-free grammars cannot.

(2) Classical grammar systems combine several grammars (see [CVDKP94]). All the involved grammars cooperate according to some protocol during their derivations. Admittedly, compared to isolated grammars, these grammar systems show several significant advantages, including an increase of the generative power and, simultaneously, a decrease of their descriptional complexity. In essence, the classical grammar systems can be classified into cooperating distributed (CD) and parallel communicating (PC) grammar systems (see [CVDKP94]). CD grammar systems work in a sequential way. Indeed, all the grammars that form components of these systems have a common sentential form, and every derivation step is performed by one of these grammars. A cooperation protocol dictates the way by which the grammars cooperate. For instance, one grammar performs precisely k derivation steps, then another grammar works in this way, and so on, for a positive integer k. In addition, some stop conditions are given to determine when the grammar systems become inactive and produce their sentences. For example, a stop condition of this kind says that no grammar of the system can make another derivation step. Many other cooperating protocols and stop conditions are considered in the literature (see [CVDKP94] and Chapter 4 in [RS97b] for an overview). As opposed to a CD grammar system, a PC grammar system works in parallel. The PC grammatical components have their own sentential forms, and every derivation step is performed by each of the components with its sentential form. A cooperation protocol is based on a communication between the components through query symbols. More precisely, by generating these query symbols, a component specifies where to insert the sentential form produced by another component. Nevertheless, even PC grammar systems cannot control their computation across all their grammatical components simultaneously and globally.

Multigenerative grammars, discussed in Chap. 12, are based upon classical grammar systems, sketched above, because they also involve several grammatical components. However, these multigenerative versions can control their computation across all these components by finitely many sequences of nonterminals or rules while their ordinary counterparts cannot. As illustrated next, since the ordinary grammar systems cannot control information processing across all the grammatical components, they may be inapplicable under some circumstances while multigenerative grammar systems are applicable.

Consider regulated information processing concerning digital images. Suppose that the processor composes and transforms several fragments of these images into a single image according to its translation rules. For instance, from several digital images that specify various parts of a face, the processor produces a complete digital image of the face. Alternatively, from a huge collection of files containing various image data, the translator selects a set of images satisfying some prescribed criteria and composes them into a single image-data file. Of course, the processor makes a multi-composition like this according to some compositional rules. As obvious, a proper composition-producing process like this necessities a careful regulation of all the simultaneously applied rules, which can be elegantly accomplished by regulated grammar systems that control their computation by sequences of rules. On the other hand, a regulation like this is hardly realizable based upon unregulated grammar systems, which lack any rule-controlling mechanism.

(3) Classical pushdown automata work by making moves during which they change states (see [Med00a, Sal73, Har78, Woo87, RS97a]). As a result, this state mechanism is the only way by which they can control their computation. In practice, however, their applications may require a more sophisticated regulation, which cannot be accomplished by state control. Frequently, however, the regulated versions of pushdown automata (see Sect. 7.2.2) can handle computational tasks like this by their control languages, so under these circumstance, they can act as appropriate computational models while their unregulated counterparts cannot as illustrated next in terms of parsing.

Consider a collection of files, each of which contains a portion of a source program that should be parsed as a whole by a syntax analyzer, underlined by a pushdown automaton. By using a simple control language, we can prescribe the order in which the syntax analyzer should properly compose all these fragmented pieces of code stored in several different files, after which the entire code composed in this way is parsed. As obvious, we cannot prescribe any global composition like this over the collection of files by using any classical pushdown automata, which does not regulate its computation by any control language.

To summarize this chapter, regulated grammars and automata represent appropriate formal models of highly regulated computation, which is likely to fulfill a central role in computer science as a whole in the near future. As such, from a theoretical perspective, they will allow us to express the theoretical fundamentals of this computation rigorously and systematically. From a more pragmatic perspective, based upon them, computer science can create a well-designed methodology concerning regulated information processing. Simply put, as their main perspective in near future, regulated grammars and automata allow us to create (a) a systematized body of knowledge representing an in-depth theory of highly regulated computation and (b) a sophisticated methodology concerning regulated information processing, based upon this computation.

Chapter 14
Applications in Computational Linguistics

This chapter gives several specific case studies concerning linguistics. Specifically, it demonstrates applications of scattered context grammars in this scientific field. It concentrates its attention to many complicated English syntactical structures and demonstrates how scattered context grammars allow us to explore them clearly, elegantly, and precisely.

Clearly, scattered context grammars are useful to every linguistic field that formalizes its results by strings in which there exist some scattered context dependencies spread over the strings. Since numerous linguistic areas, ranging from discourse analysis, through psycholinguistics up to neurolinguistics, formalize and study their results by using strings involving dependencies of this kind, describing applications of scattered context grammars in all these areas would be unbearably sketchy and, therefore, didactically inappropriate. Instead of an encyclopedic approach like this, we narrow our attention to the investigation of *English syntax* (see [Bak95, HP05]), which describes the rules concerning how words relate to each other in order to form well-formed grammatical English sentences. We have selected syntax of this language because the reader is surely familiar with English very well. Nevertheless, analogical ideas can be applied to members of other language families, including Indo-European, Sino-Tibetan, Niger-Congo, Afro-Asiatic, Altaic, and Japonic families of languages. We explore several common linguistic phenomena involving scattered context in English syntax and explain how to express these phenomena by scattered context grammars.

However, even within the linguistics concerning English syntax, we cannot be exhaustive in any way. Rather, we consider only selected topics concerning English syntax and demonstrate how scattered context grammars allow us to explore them clearly, elegantly, and precisely. Compared to the previous parts of this book, which are written in a strictly mathematical way, we discuss and describe scattered context grammars less formally here because we are interested in demonstrating real applications rather than theoretical properties. Specifically, we primarily use scattered context grammars to transform and, simultaneously, verify that the English sentences under discussion are grammatical.

© Springer International Publishing AG 2017
A. Meduna, O. Soukup, *Modern Language Models and Computation*,
DOI 10.1007/978-3-319-63100-4_14

The present section consists of Sects. 14.1, 14.2, and 14.3. Section 14.1 connects the theoretically oriented discussion of scattered context grammars given earlier in this book and the pragmatically oriented discussion of these grammars applied to English syntax in the present section. Then, Sect. 14.2 modifies scattered context grammars to their transformational versions, which are easy to apply to syntax-related modifications of sentences. Most importantly, Sect. 14.3 describes English syntax and its transformations by methods based upon the transformational versions of scattered context grammars.

14.1　Syntax and Related Linguistic Terminology

In the linguistic study concerning English syntax, we discuss and describe the principles and rules according to which we correctly construct and transform grammatical English sentences. To give an insight into the discussion of English syntax, we open this section by some simple examples that illustrate how we connect the theoretically oriented discussion of scattered context grammars with the application-oriented discussion of English syntax. Then, we introduce the basic terminology used in syntax-oriented linguistics.

14.1.1　Introduction

Observe that many common English sentences contain expressions and words that mutually depend on each other although they are not adjacent to each other in the sentences. For example, consider this sentence

He usually goes to work early.

The subject (*he*) and the predicator (*goes*) are related; sentences

∗He usually go to work early.

and

∗I usually goes to work early.

are ungrammatical because the form of the predicator depends on the form of the subject, according to which the combinations *∗he...go* and *∗I...goes* are illegal (throughout this section, ∗ denotes ungrammatical sentences or their parts). Clearly, any change of the subject implies the corresponding change of the predicator as well. Linguistic dependencies of this kind can be easily and elegantly captured by scattered context grammars. Let us construct a scattered context grammar that contains this rule

$$(He, goes) \rightarrow (We, go)$$

This rule checks whether the subject is the pronoun *he* and whether the verb *go* is in third person singular. If the sentence satisfies this property, it can be transformed to the grammatically correct sentence

We usually go to work early.

Observe that the related words may occur far away from each other in the sentence in question. In the above example, the word *usually* occurs between the subject and the predicator. While it is fairly easy to use context-sensitive grammars to model context dependencies where only one word occurs between the related words, note that the number of the words appearing between the subject and the predicator can be virtually unlimited. We can say

He almost regularly goes to work early.

but also

He usually, but not always, goes to work early.

and many more grammatical sentences like this. To model these context dependencies by ordinary context-sensitive grammars, many auxiliary rules have to be introduced to send the information concerning the form of a word to another word, which may occur at the opposite end of the sentence. As opposed to this awkward and tedious description, the single scattered context rule above is needed to perform the same job regardless of the number of the words appearing between the subject and the predicator.

We next give another example that illustrates the advantage of scattered context grammars over classical context-sensitive grammars under some circumstances. Consider these two sentences

John recommended it.

and

Did John recommend it?

There exists a relation between the basic clause and its interrogative counterpart. Indeed, we obtain the second, interrogative clause by adding *did* in front of *John* and by changing *recommended* to *recommend* while keeping the rest of the sentence unchanged. In terms of scattered context grammars, this transformation can be described by the scattered context rule

$$(\text{John, recommended}) \rightarrow (\text{Did John, recommend})$$

Clearly, when applied to the first sentence, this rule performs exactly the same transformation as we have just described. Although this transformation is possible by using an ordinary context rule, the inverse transformation is much more difficult to achieve. The inverse transformation can be performed by a scattered context rule

$$(\text{Did, recommend}) \rightarrow (\varepsilon, \text{recommended})$$

Obviously, by erasing *did* and changing *recommend* to *recommended*, we obtain the first sentence from the second one. Again, instead of *John* the subject may consist of a noun phrase containing several words, which makes it difficult to capture this context dependency by ordinary context-sensitive grammars.

Considering the examples above, the advantage of scattered context grammars is more than obvious: scattered context grammars allow us to change only some words during the transformation while keeping the others unchanged. On the other hand, context-sensitive grammars are inconvenient to perform transformations of this kind. A typical context-sensitive grammar that performs this job usually needs many more context-sensitive rules by which it repeatedly traverses the transformed sentence in question just to change very few context dependent words broadly spread across the sentence.

14.1.2 Terminology

Taking into account the intuitive insight given above, we see that there are structural rules and regularities underlying syntactically well-formed English sentences and their transformations. Although we have already used some common linguistic notions, such as subject or predicator, we now introduce this elementary linguistic terminology more systematically so we can express these English sentences in terms of their syntactic structure in a more exact and general way. However, we restrict this introduction only to the very basic linguistic notions, most of which are taken from [HP02, HP05].

Throughout the rest of this section, we narrow our discussion primarily to verbs and personal pronouns, whose proper use depends on the context in which they occur. For instance, *is, are, was*, and *been* are different forms of the same verb *be*, and their proper use depends on the context in which they appear. We say that words in these categories *inflect* and call this property *inflection*. Verbs and personal pronouns often represent the key elements of a clause—the *subject* and the *predicate*. In simple clauses like

She loves him.

we can understand the notion of the subject and the predicate so that some information is "predicated of" the subject (*she*) by the predicate (*loves him*). In more complicated clauses, the best way to determine the subject and the predicate is the examination of their syntactic properties (see [HP02] for more details). The predicate is formed by a *verb phrase*—the most important word of this phrase is the verb, also known as the *predicator*. In some verb phrases, there occur several verbs. For example, in the sentence

He has been working for hours.

the verb phrase contains three verbs—*has, been*, and *working*. The predicator is, however, always the first verb of a verb phrase (*has* in the above example). In this study, we focus on the most elementary clauses—*canonical clauses*. In these clauses, the subject always precedes the predicate, and these clauses are positive, declarative, and without subordinate or coordinate clauses.

Next, we describe the basic categorization of verbs and personal pronouns, and further characterize their inflectional forms in a greater detail.

14.1.3   Verbs

We distinguish several kinds of verbs based upon their grammatical properties. The set of all verbs is divided into two subsets—the set of *auxiliary verbs*, and the set of *lexical verbs*. Further, the set of auxiliary verbs consists of *modal verbs* and *non-modal verbs*. The set of modal verbs includes the following verbs—*can, may, must, will, shall, ought, need, dare*; the verbs *be, have*, and *do* are non-modal. All the remaining verbs are lexical. In reality, the above defined classes overlap in certain situations; for example, there are sentences, where *do* appears as an auxiliary verb, and in different situations, *do* behaves as a lexical verb. For simplicity, we do not take into account these special cases in what follows.

Inflectional forms of verbs are called *paradigms*. In English, every verb, except for the verb *be*, may appear in each of the six paradigms described in Table 14.1 (see [HP02]). Verbs in *primary form* may occur as the only verb in a clause and form the head of its verb phrase (predicator); on the other hand, verbs in *secondary form* have to be accompanied by a verb in primary form.

The verb *be* has nine paradigms in its neutral form. All primary forms have, in addition, their negative contracted counterparts. Compared to other verbs, there is one more verb paradigm called *irrealis*. The irrealis form *were* (and *weren't*) is used in sentences of an unrealistic nature, such as

I wish I were rich.

All these paradigms are presented in Table 14.2.

Table 14.1 Paradigms of English verbs

Form	Paradigm	Person	Example
Primary	Present	3rd sg	*She* walks *home.*
		Other	*They* walk *home.*
	Preterite		*She* walked *home.*
Secondary	Plain form		*They should* walk *home.*
	Gerund-participle		*She is* walking *home.*
	Past participle		*She has* walked *home.*

Table 14.2 Paradigms of the verb *be*

Form	Paradigm	Person	Neutral	Negative
Primary	Present	1st sg	*am*	*aren't*
		3rd sg	*is*	*isn't*
		other	*are*	*aren't*
	Preterite	1st sg, 3rd sg	*was*	*wasn't*
		other	*were*	*weren't*
	Irrealis	1st sg, 3rd sg	*were*	*weren't*
Secondary	Plain form		*be*	–
	Gerund-participle		*being*	–
	Past participle		*been*	–

Table 14.3 Personal pronouns

Non-reflexive				
Nominative	Accusative	Genitive		
Plain		Dependent	Independent	Reflexive
I	*me*	*my*	*mine*	*myself*
you	*you*	*your*	*yours*	*yourself*
he	*him*	*his*	*his*	*himself*
she	*her*	*her*	*hers*	*herself*
it	*it*	*its*	*its*	*itself*
we	*us*	*our*	*ours*	*ourselves*
you	*you*	*your*	*yours*	*yourselves*
they	*them*	*their*	*theirs*	*themselves*

14.1.4 Personal Pronouns

Personal pronouns exhibit a great amount of inflectional variation as well. Table 14.3
summarizes all their inflectional forms. The most important for us is the class of
pronouns in *nominative* because these pronouns often appear as the subject of a
clause.

14.2 Transformational Scattered Context Grammars

As we have already mentioned, we primarily apply scattered context grammars to transform grammatical English sentences to other grammatical English sentences. To do so, we next slightly modify scattered context grammars so they start their derivations from a language rather than a single start symbol. Even more importantly, these grammars define transformations of languages, not just their generation.

Definition 14.2.1. A *transformational scattered context grammar* is a quadruple

$$G = (V, T, P, I)$$

where

- V is the *total vocabulary*;
- $T \subset V$ is the set of terminals (or the *output vocabulary*);
- P is a finite set of scattered context rules;
- $I \subset V$ is the *input vocabulary*.

The derivation step is defined as in scattered context grammars (see Definition 4.1.1). The *transformation t that G defines from* $K \subseteq I^*$ is denoted by $\tau(G, K)$ and defined as

$$\tau(G, K) = \{(x, y) \mid x \Rightarrow_G^* y, x \in K, y \in T^*\}$$

If $(x, y) \in \tau(G, K)$, we say that x *is transformed to* y by G; x and y are called the *input* and the *output sentence*, respectively. □

As already pointed out, while scattered context grammars generate strings, transformational scattered context grammars translate them. In a sense, however, the language generated by any scattered context grammar $G = (V, T, P, S)$ can be expressed by using a transformational scattered context grammar $H = (V, T, P, \{S\})$ as well. Observe that

$$L(G) = \{y \mid (S, y) \in t(H, \{S\})\}$$

Before we make use of transformational scattered context grammars in terms of English syntax in the next section, we give two examples to demonstrate a close relation of these grammars to the theoretically oriented studies given previously in this book. To link the theoretical discussions given earlier in this book to the present section, the first example presents a transformational scattered context grammar that works with purely abstract languages. In the second example, we discuss a transformational scattered context grammar that is somewhat more linguistically oriented.

Example 14.2.2. Define the transformational scattered context grammar

$$G = (V, T, P, I)$$

where $V = \{A, B, C, a, b, c\}$, $T = \{a, b, c\}$, $I = \{A, B, C\}$, and

$$P = \{(A, B, C) \rightarrow (a, bb, c)\}$$

For example, for the input sentence *AABBCC*,

$$AABBCC \Rightarrow_G aABbbcC \Rightarrow_G aabbbbcc$$

Therefore, the input sentence $AABBCC \in I^*$ is transformed to the output sentence $aabbbbcc \in T^*$, and

$$(AABBCC, aabbbbcc) \in \tau(G, I^*)$$

If we restrict the input sentences to the language $L = \{A^n B^n C^n \mid n \geq 1\}$, we get

$$\tau(G, L) = \{(A^n B^n C^n, a^n b^{2n} c^n) \mid n \geq 1\}$$

so every $A^n B^n C^n$, where $n \geq 1$, is transformed to $a^n b^{2n} c^n$.   □

In the following example, we modify strings consisting of English letters by a transformational scattered context grammar, and in this way, we relate these grammars to lexically oriented linguistics—that is, the area of linguistics that concentrates its study on vocabulary analysis and dictionary design.

Example 14.2.3. We demonstrate how to lexicographically order alphabetic strings and, simultaneously, convert them from their uppercase versions to lowercase versions. More specifically, we describe a transformational scattered context grammar G that takes any alphabetic strings that consist of English uppercase letters enclosed in angle brackets, lexicographically orders the letters, and converts them to the corresponding lowercases. For instance, G transforms $\langle XXUY \rangle$ to *uxxy*.

More precisely, let J and T be alphabets of English uppercases and English lowercases, respectively. Let $\prec$ denote the *lexical order* over J; that is, $A \prec B \prec \cdots \prec Z$. Furthermore, let h be the function that maps the uppercases to the corresponding lowercases; that is, $h(A) = a$, $h(B) = b$, ..., $h(Z) = z$. Let i denote the inverse of h, so $i(a) = A$, $i(b) = B$, ..., $i(z) = Z$. Let $N = \{\hat{a} \mid a \in T\}$. We define the transformational scattered context grammar

$$G = (V, T, P, I)$$

where T is defined as above, $I = J \cup \{\langle, \rangle\}$, $V = I \cup N \cup T$, and P is constructed as follows:

(1) for each $A, B \in I$, where $A \prec B$, add $(B, A) \rightarrow (A, B)$ to P;
(2) for each $a \in T$, add $(\langle) \rightarrow (\hat{a})$ to P;

(3) for each $a \in T$ and $A \in J$, where $i(a) = A$, add $(\hat{a}, A) \rightarrow (a, \hat{a})$ to P;
(4) for each $a, b \in T$, where $i(a) \prec i(b)$, add $(\hat{a}) \rightarrow (\hat{b})$ to P;
(5) for each $a \in T$, add $(\hat{a}, \rangle) \rightarrow (\varepsilon, \varepsilon)$ to P.

Set $K = \{\langle \} J^* \{ \rangle\}$. For instance, G transforms $\langle ORDER \rangle \in K$ to $deorr \in T^*$ as

$$\langle ORDER \rangle \Rightarrow_G \langle OEDRR \rangle \Rightarrow_G \langle DEORR \rangle$$
$$\Rightarrow_G \hat{d}DEORR \rangle \Rightarrow_G d\hat{d}EORR \rangle \Rightarrow_G d\hat{e}EORR \rangle \Rightarrow_G de\hat{e}ORR \rangle$$
$$\Rightarrow_G de\hat{o}ORR \rangle \Rightarrow_G deo\hat{o}RR \rangle \Rightarrow_G deo\hat{r}RR \rangle \Rightarrow_G deor\hat{r}R \rangle$$
$$\Rightarrow_G deorr\hat{r} \rangle \Rightarrow_G deorr$$

so $(\langle ORDER \rangle, deorr) \in \tau(G, K)$. Clearly, G can make the same transformation in many more ways; on the other hand, notice that the set of all transformations of $\langle ORDER \rangle$ to $deorr$ is finite.

More formally, we claim that G transforms every string $\langle A_1 \dots A_n \rangle \in K$ to $b_1 \dots b_n \in T^*$, for some $n \geq 0$, so that $i(b_1) \dots i(b_n)$ represents a permutation of $A_1 \dots A_n$, and for all $1 \leq j \leq n - 1$, $i(b_j) \prec i(b_{j+1})$ (the case when $n = 0$ means that $A_1 \dots A_n = b_1 \dots b_n = \varepsilon$). To see why this claim holds, notice that $T \cap I = \emptyset$, so every successful transformation of a string from K to a string from T^* is performed so that all symbols are rewritten during the computation. By rules introduced in (1), G lexicographically orders the input uppercases. By a rule of the form $(\langle) \rightarrow (\hat{a})$ introduced in (2), G changes the leftmost symbol $\langle$ to $\hat{a}$. By rules introduced in (3) and (4), G verifies that the alphabetic string is properly ordered and, simultaneously, converts its uppercase symbols into the corresponding lowercases in a strictly left-to-right one-by-one way. Observe that a rule introduced in (2) is applied precisely once during every successful transformation because the left-to-right conversion necessities its application, and on the other hand, no rule can produce $\langle$. By a rule from (5), G completes the transformation; notice that if this completion is performed prematurely with some uppercases left, the transformation is necessary unsuccessful because the uppercases cannot be turned to the corresponding lowercases. Based upon these observations, it should be obvious that G performs the desired transformation. $\square$

Having illustrated the lexically oriented application, we devote the next section solely to the applications of transformational scattered context grammars in English syntax.

14.3   Scattered Context in English Syntax

In this section, we apply transformational scattered context grammars to English syntax. Before opening this topic, let us make an assumption regarding the set of all English words. We assume that this set, denoted by T, is finite and fixed. From a practical point of view, this is obviously a reasonable assumption because we all commonly use a finite and fixed vocabulary of words in everyday English (purely hypothetically, however, this may not be the case as illustrated by the study that closes this section). Next, we subdivide this set into subsets with respect to the classification of verbs and pronouns described in Sect. 14.1:

- T is the set of all words including all their inflectional forms;
- $T_{vrbs} \subset T$ is the set of all verbs including all their inflectional forms;
- $T_{vaux} \subset T_{vrbs}$ is the set of all auxiliary verbs including all their inflectional forms;
- $T_{vplain} \subset T_{vrbs}$ is the set of all verbs in plain form;
- $T_{ppron} \subset T$ is the set of all personal pronouns in nominative.

To describe all possible paradigms of a verb $v \in T_{vplain}$, we use the following notation

- $\pi_{vthird}(v)$ is the verb v in third person singular present;
- $\pi_{vpres}(v)$ is the verb v in present (other than third person singular);
- $\pi_{vpret}(v)$ is the verb v in preterite.

There are several conventions we use throughout this section in order to simplify the presented case studies, given next.

- We do not take into account capitalization and punctuation. Therefore, according to this convention,

He is your best friend.

and

he is your best friend

are equivalent.
- To make the following studies as simple and readable as possible, we expect every input sentence to be a canonical clause. In some examples, however, we make slight exceptions to this rule; for instance, sometimes we permit the input sentence to be negative. The first example and the last example also demonstrate a simple type of coordinated canonical clauses.
- The input vocabulary is the set $I = \{\langle x \rangle \mid x \in T\}$, where T is the set of all English words as stated above. As a result, every transformational scattered context grammar in this section takes an input sentence over I and transforms it to an output sentence over T. For instance, in the case of the declarative-to-interrogative transformation,

$$\langle he \rangle \langle is \rangle \langle your \rangle \langle best \rangle \langle friend \rangle$$

is transformed to

is he your best friend

As we have already mentioned, we omit punctuation and capitalization, so the above sentence corresponds to

$$Is\ he\ your\ best\ friend?$$

Next, we give several studies that describe how to transform various kinds of grammatical sentences to other grammatical sentences by using transformational scattered context grammars.

14.3.1 Clauses with neither and nor

The first example shows how to use transformational scattered context grammars to negate clauses that contain the pair of the words *neither* and *nor*, such as

$$Neither\ Thomas\ nor\ his\ wife\ went\ to\ the\ party.$$

Clearly, the words *neither* and *nor* are related, but there is no explicit limit of the number of the words appearing between them. The following transformational scattered context grammar G converts the above sentence to

$$Both\ Thomas\ and\ his\ wife\ went\ to\ the\ party.$$

In fact, the constructed grammar G is general enough to negate every grammatical clause that contains the pair of the words *neither* and *nor*.

Set $G = (V, T, P, I)$, where $V = T \cup I$, and P is defined as follows:

$$P = \{((\langle neither \rangle), \langle nor \rangle)) \to (both, and)\}$$
$$\cup \{((\langle x \rangle)) \to (x) \mid x \in T - \{neither, nor\}\}$$

For example, for the above sentence, the transformation can proceed in this way

$$\langle neither \rangle \langle thomas \rangle \langle nor \rangle \langle his \rangle \langle wife \rangle \langle went \rangle \langle to \rangle \langle the \rangle \langle party \rangle$$
$$\Rightarrow_G both\ \langle thomas \rangle\ and\ \langle his \rangle \langle wife \rangle \langle went \rangle \langle to \rangle \langle the \rangle \langle party \rangle$$
$$\Rightarrow_G both\ thomas\ and\ \langle his \rangle \langle wife \rangle \langle went \rangle \langle to \rangle \langle the \rangle \langle party \rangle$$
$$\Rightarrow_G both\ thomas\ and\ his\ \langle wife \rangle \langle went \rangle \langle to \rangle \langle the \rangle \langle party \rangle$$
$$\Rightarrow_G^5 both\ thomas\ and\ his\ wife\ went\ to\ the\ party$$

The rule

$$((\langle neither \rangle), \langle nor \rangle)) \to (both, and)$$

replaces *neither* and *nor* with *both* and *and*, respectively. Every other word $\langle w \rangle \in I$ is changed to $w \in T$. Therefore, if we denote all possible input sentences, described in the introduction of this example, by K, $\tau(G, K)$ represents the set of all negated sentences from K, and

$$\big(\langle \text{neither} \rangle \langle \text{thomas} \rangle \langle \text{nor} \rangle \langle \text{his} \rangle \langle \text{wife} \rangle \langle \text{went} \rangle \langle \text{to} \rangle \langle \text{the} \rangle \langle \text{party} \rangle,$$
$$\text{both thomas and his wife went to the party} \big) \in \tau(G, K)$$

14.3.2 Existential Clauses

In English, clauses that indicate an existence are called *existential*. These clauses are usually formed by the dummy subject *there*; for example,

There was a nurse present.

However, this dummy subject is not mandatory in all situations. For instance, the above example can be rephrased as

A nurse was present.

We construct a transformational scattered context grammar G that converts any canonical existential clause without the dummy subject *there* to an equivalent existential clause with *there*.

Set $G = (V, T, P, I)$, where $V = T \cup I \cup \{X\}$ (X is a new symbol such that $X \notin T \cup I$), and P is defined as follows:

$$
\begin{aligned}
P = \big\{ &(\langle x \rangle, \langle \text{is} \rangle) \rightarrow (\text{there is } xX, \varepsilon), \\
&(\langle x \rangle, \langle \text{are} \rangle) \rightarrow (\text{there are } xX, \varepsilon), \\
&(\langle x \rangle, \langle \text{was} \rangle) \rightarrow (\text{there was } xX, \varepsilon), \\
&(\langle x \rangle, \langle \text{were} \rangle) \rightarrow (\text{there were } xX, \varepsilon) \mid x \in T \big\} \\
\cup\ &\big\{ (X, \langle x \rangle) \rightarrow (X, x) \mid x \in T \big\} \\
\cup\ &\big\{ (X) \rightarrow (\varepsilon) \big\}
\end{aligned}
$$

For the above sample sentence, we get the following derivation

$$
\begin{aligned}
&\langle \text{a} \rangle \langle \text{nurse} \rangle \langle \text{was} \rangle \langle \text{present} \rangle \\
\Rightarrow_G\ &\text{there was a } X \langle \text{nurse} \rangle \langle \text{present} \rangle \\
\Rightarrow_G\ &\text{there was a } X \text{ nurse } \langle \text{present} \rangle \\
\Rightarrow_G\ &\text{there was a } X \text{ nurse present} \\
\Rightarrow_G\ &\text{there was a nurse present}
\end{aligned}
$$

A rule from the first set has to be applied first because initially there is no symbol X in the sentential form and all other rules require X to be present in the sentential form. In our case, the rule

$$(\langle a \rangle, \langle was \rangle) \rightarrow (\text{there was a } X, \varepsilon)$$

is applied; the use of other rules from this set depends on what tense is used in the input sentence and whether the subject is in singular or plural. The rule non-deterministically selects the first word of the sentence, puts *there was* in front of it, and the symbol X behind it; in addition, it erases *was* in the middle of the sentence. Next, all words $\langle w \rangle \in I$ are replaced with $w \in T$ by rules from the second set. These rules also verify that the previous non-deterministic selection was made at the beginning of the sentence; if not, there remains a word $\langle w \rangle \in I$ in front of X that cannot be rewritten. Finally, the derivation ends by erasing X from the sentential form.

This form of the derivation implies that if we denote the input existential clauses described in the introduction of this example by K, $\tau(G, K)$ represents the set of these clauses with the dummy subject *there*. As a result,

$$(\langle a \rangle \langle nurse \rangle \langle was \rangle \langle present \rangle, \text{there was a nurse present}) \in \tau(G, K)$$

14.3.3 Interrogative Clauses

In English, there are two ways of transforming declarative clauses into interrogative clauses depending on the predicator. If the predicator is an auxiliary verb, the interrogative clause is formed simply by swapping the subject and the predicator. For example, we get the interrogative clause

Is he mowing the lawn?

by swapping *he*, which is the subject, and *is*, which is the predicator, in

He is mowing the lawn.

On the other hand, if the predicator is a lexical verb, the interrogative clause is formed by adding the dummy *do* to the beginning of the declarative clause. The dummy *do* has to be of the same paradigm as the predicator in the declarative clause and the predicator itself is converted to its plain form. For instance,

She usually gets up early.

is a declarative clause with the predicator *gets*, which is in third person singular, and the subject *she*. By inserting *do* in third person singular to the beginning of the sentence and converting *gets* to its plain form, we obtain

Does she usually get up early?

To simplify the following transformational scattered context grammar G, which performs this conversion, we assume that the subject is a personal pronoun in nominative.

Set $G = (V, T, P, I)$, where $V = T \cup I \cup \{X\}$ (X is a new symbol such that $X \notin T \cup I$), and P is defined as follows:

$$P = \{(\langle p \rangle, \langle v \rangle) \to (vp, X) \mid v \in T_{\text{vaux}}, p \in T_{\text{ppron}}\}$$
$$\cup \{(\langle p \rangle, \langle \pi_{\text{vpret}}(v) \rangle) \to (\text{did } p, vX),$$
$$(\langle p \rangle, \langle \pi_{\text{vthird}}(v) \rangle) \to (\text{does } p, vX),$$
$$(\langle p \rangle, \langle \pi_{\text{vpres}}(v) \rangle) \to (\text{do } p, vX) \mid v \in T_{\text{vplain}} - T_{\text{vaux}}, p \in T_{\text{ppron}}\}$$
$$\cup \{(\langle x \rangle, X) \to (x, X),$$
$$(X, \langle y \rangle) \to (X, y) \mid x \in T - T_{\text{vrbs}}, y \in T\}$$
$$\cup \{(X) \to (\varepsilon)\}$$

For sentences whose predicator is an auxiliary verb, the transformation made by G proceeds as follows:

$$\langle \text{he} \rangle \langle \text{is} \rangle \langle \text{mowing} \rangle \langle \text{the} \rangle \langle \text{lawn} \rangle$$
$$\Rightarrow_G \text{is he } X \langle \text{mowing} \rangle \langle \text{the} \rangle \langle \text{lawn} \rangle$$
$$\Rightarrow_G \text{is he } X \text{ mowing } \langle \text{the} \rangle \langle \text{lawn} \rangle$$
$$\Rightarrow_G \text{is he } X \text{ mowing the } \langle \text{lawn} \rangle$$
$$\Rightarrow_G \text{is he } X \text{ mowing the lawn}$$
$$\Rightarrow_G \text{is he mowing the lawn}$$

The derivation starts by the application of a rule from the first set, which swaps the subject and the predicator, and puts X behind them. Next, rules from the third set rewrite every word $\langle w \rangle \in I$ to $w \in T$. Finally, X is removed from the sentential form.

The transformation of the sentences in which the predicator is a lexical verb is more complicated:

$$\langle \text{she} \rangle \langle \text{usually} \rangle \langle \text{gets} \rangle \langle \text{up} \rangle \langle \text{early} \rangle$$
$$\Rightarrow_G \text{does she } \langle \text{usually} \rangle \text{ get } X \langle \text{up} \rangle \langle \text{early} \rangle$$
$$\Rightarrow_G \text{does she usually get } X \langle \text{up} \rangle \langle \text{early} \rangle$$
$$\Rightarrow_G \text{does she usually get } X \text{ up } \langle \text{early} \rangle$$
$$\Rightarrow_G \text{does she usually get } X \text{ up early}$$
$$\Rightarrow_G \text{does she usually get up early}$$

As the predicator is in third person singular, a rule from

$$\left\{ (\langle p \rangle, \langle \pi_{\text{vthird}}(v) \rangle) \to (\text{does } p, vX) \mid v \in T_{\text{vplain}} - T_{\text{vaux}}, p \in T_{\text{ppron}} \right\}$$

is applied at the beginning of the derivation. It inserts *does* to the beginning of the sentence, converts the predicator *gets* to its plain form *get*, and puts X behind it. Next, rules from

$$\left\{ (\langle x \rangle, X) \to (x, X) \mid x \in T - T_{\text{vrbs}} \right\}$$

rewrite every word $\langle w \rangle \in I$ appearing in front of the predicator to $w \in T$. Notice that they do not rewrite verbs—in this way, the grammar verifies that the first verb in the sentence was previously selected as the predicator. For instance, in the sentence

He has been working for hours.

has must be selected as the predicator; otherwise, the derivation is unsuccessful. Finally, the grammar rewrites all words behind X, and erases X in the last step as in the previous case.

Based on this intuitive explanation, we can see that the set of all input sentences K described in the introduction of this example is transformed by G to $\tau(G, K)$, which is the set of all interrogative sentences constructed from K. Therefore,

$$(\langle \text{he} \rangle \langle \text{is} \rangle \langle \text{mowing} \rangle \langle \text{the} \rangle \langle \text{lawn} \rangle, \text{is he mowing the lawn}) \in \tau(G, K),$$
$$(\langle \text{she} \rangle \langle \text{usually} \rangle \langle \text{gets} \rangle \langle \text{up} \rangle \langle \text{early} \rangle, \text{does she usually get up early}) \in \tau(G, K)$$

14.3.4 Question Tags

Question tags are special constructs that are primarily used in spoken language. They are used at the end of declarative clauses, and we customarily use them to ask for agreement or confirmation. For instance, in

Your sister is married, isn't she?

isn't she is a question tag, and we expect an answer stating that she is married. The polarity of question tags is always opposite to the polarity of the main clause—if the main clause is positive, the question tag is negative, and vice versa. If the predicator is an auxiliary verb, the question tag is formed by the same auxiliary verb. For lexical verbs, the question tag is made by using *do* as

He plays the violin, doesn't he?

There are some special cases that have to be taken into account. First, the verb *be* has to be treated separately because it has more paradigms than other verbs and the question tag for first person singular is irregular:

I am always right, aren't I?

Second, for the verb *have*, the question tag depends on whether it is used as an auxiliary verb, or a lexical verb. In the first case, *have* is used in the question tag as

He has been working hard, hasn't he?

in the latter case, the auxiliary *do* is used as

They have a dog, don't they?

To explain the basic concepts as simply as possible, we omit the special cases of the verb *have* in the following transformational scattered context grammar G, which supplements a canonical clause with a question tag. For the same reason, we only sketch its construction and do not mention all the created rules explicitly. In addition, we suppose that the subject is represented by a personal pronoun.

Set $G = (V, T, P, I)$, where $V = T \cup I \cup \{X, Y\}$ (X, Y are new symbols such that $X, Y \notin T \cup I$), and P is defined as follows:

$$
\begin{aligned}
P = \{ & (\langle p \rangle, \langle \text{will} \rangle, \langle x \rangle) \to (p, \text{will } X, Yx \text{ won't } p), \\
& (\langle p \rangle, \langle \text{won't} \rangle, \langle x \rangle) \to (p, \text{won't } X, Yx \text{ will } p), \\
& \cdots \mid p \in T_{\text{ppron}}, x \in T \} \\
\cup \{ & (\langle \text{I} \rangle, \langle \text{am} \rangle, \langle x \rangle) \to (\text{I, am } X, Yx \text{ aren't I}), \\
& (\langle \text{you} \rangle, \langle \text{are} \rangle, \langle x \rangle) \to (\text{you, are } X, Yx \text{ aren't you}), \\
& \cdots \mid x \in T \} \\
\cup \{ & (\langle p \rangle, \langle v \rangle, \langle x \rangle) \to (p, vX, Yx \text{ doesn't } p), \\
& (\langle q \rangle, \langle v \rangle, \langle x \rangle) \to (q, vX, Yx \text{ don't } q) \mid \\
& p \in \{\text{he, she, it}\}, q \in T_{\text{ppron}} - \{\text{he, she, it}\}, v \in T_{\text{vrbs}} - T_{\text{vaux}}, x \in T \} \\
& \quad\quad \vdots \\
\cup \{ & (\langle x \rangle, X) \to (x, X), \\
& (X, \langle y \rangle, Y) \to (X, y, Y) \mid x \in T - T_{\text{vrbs}}, y \in T \} \\
\cup \{ & (X, Y) \to (\varepsilon, \varepsilon) \}
\end{aligned}
$$

First, we describe the generation of question tags for clauses whose predicator is an auxiliary verb:

$$\langle I\rangle\langle am\rangle\langle always\rangle\langle right\rangle$$
$$\Rightarrow_G I\ am\ X\langle always\rangle Y\ right\ aren't\ I$$
$$\Rightarrow_G I\ am\ X\ always\ Y\ right\ aren't\ I$$
$$\Rightarrow_G I\ am\ always\ right\ aren't\ I$$

Here, the rule

$$(\langle I\rangle, \langle am\rangle, \langle right\rangle) \to (I, am\ X, Y\ right\ aren't\ I)$$

initiates the derivation. When it finds *I am* at the beginning of the sentence, it generates the question tag *aren't I* at its end. In addition, it adds X behind *I am* and Y in front of *right aren't I*. Next, it rewrites all words from $\langle w\rangle \in I$ to $w \in T$. It makes sure that the predicator was chosen properly by rules from

$$\left\{(\langle x\rangle, X) \to (x, X) \mid x \in T - T_{\text{vrbs}}\right\}$$

similarly to the previous example. In addition, rules from

$$\left\{(X, \langle y\rangle, Y) \to (X, y, Y) \mid y \in T\right\}$$

check whether the question tag was placed at the very end of the sentence. If not, there remains some symbol from the input vocabulary behind Y that cannot be rewritten. Finally, the last rule removes X and Y from the sentential form.

When the predicator is a lexical verb in present, the question tag is formed by *does* or *do* depending on person in which the predicator occurs:

$$\langle he\rangle\langle plays\rangle\langle the\rangle\langle violin\rangle$$
$$\Rightarrow_G he\ plays\ X\langle the\rangle Y\ violin\ doesn't\ he$$
$$\Rightarrow_G he\ plays\ X\ the\ violin\ Y\ doesn't\ he$$
$$\Rightarrow_G he\ plays\ the\ violin\ doesn't\ he$$

The rest of the derivation is analogous to the first case.

Based on these derivations, we can see that the set of all input sentences K described in the introduction of this example is transformed by G to $\tau(G, K)$, which is the set of all sentences constructed from K that are supplemented with question tags. Therefore,

$$(\langle I\rangle\langle am\rangle\langle always\rangle\langle right\rangle, I\ am\ always\ right\ aren't\ I) \in \tau(G, K),$$
$$(\langle he\rangle\langle plays\rangle\langle the\rangle\langle violin\rangle, he\ plays\ the\ violin\ doesn't\ he) \in \tau(G, K)$$

14.3.5   *Generation of Grammatical Sentences*

The purpose of the next discussion, which closes this section, is six-fold—(1) through (6), stated below.

(1) We want to demonstrate that ordinary scattered context grammars, discussed earlier in this book, can be seen as a special case of transformational scattered context grammars, whose applications are discussed in the present section.

(2) As pointed out in the notes following the general definition of a transformational scattered context grammar (see Definition 14.2.1), there exists a close relation between ordinary scattered context grammars and transformational scattered context grammars. That is, for every scattered context grammar $G = (V, T, P, S)$, there is a transformational scattered context grammar $H = (V, T, P, \{S\})$ satisfying

$$L(G) = \left\{ y \mid (S, y) \in t\big(H, \{S\}\big) \right\}$$

and in this way, $L(G)$ is defined by H. Next, we illustrate this relation by a specific example.

(3) From a syntactical point of view, we want to show that scattered context grammars can generate an infinite non-context-free grammatical subset of English language in a very succinct way.

(4) In terms of morphology—that is, the area of linguistics that studies the structure of words and their generation—we demonstrate how to use transformational scattered context grammars to create complicated English words within English sentences so that the resulting words and sentences are grammatically correct.

(5) As stated in the beginning of the present section, so far we have assumed that the set of common English words is finite. Next, we want to demonstrate that from a strictly theoretical point of view, the set of all possible well-formed English words, including extremely rare words in everyday English, is infinite. Indeed, L, given next, includes infinitely many words of the form

$$(great-)^i grandparents$$
$$(great-)^i grandfathers$$
$$(great-)^i grandmothers$$

for all $i \geq 0$, and purely theoretically speaking, they all represent well-formed English words. Of course, most of them, such as

$$great - great - great - great - great - great$$
$$-great - great - great - grandfathers$$

cannot be considered as common English words because most people never use them during their lifetime.

(6) We illustrate that the language generation based upon scattered context grammars may have significant advantages over the generation based upon classical grammars, such as context-sensitive grammars.

Without further ado, consider the language L consisting of these grammatical English sentences:

> *Your grandparents are all your grandfathers and all your grandmothers.*
> *Your great-grandparents are all your great-grandfathers and all your great-grandmothers.*
> *Your great-great-grandparents are all your great-great-grandfathers and all your great-great-grandmothers.*
> $\vdots$

In brief,

$$L = \{\text{your } \{\text{great-}\}^i \text{grandparents are all your } \{\text{great-}\}^i \text{grandfathers} \\ \text{and all your } \{\text{great-}\}^i \text{grandmothers} \mid i \geq 0\}$$

Introduce the scattered context grammar $G = (V, T, P, S)$, where

$$T = \{\text{all, and, are, grandfathers, grandmothers, grandparents, great-, your}\}$$

$V = T \cup \{S, \#\}$, and P consists of these three rules

$$(S) \rightarrow (\text{your \#grandparents are all your \#grandfathers} \\ \text{and all your \#grandmothers})$$
$$(\#, \#, \#) \rightarrow (\text{\#great-, \#great-, \#great-})$$
$$(\#, \#, \#) \rightarrow (\varepsilon, \varepsilon, \varepsilon)$$

Obviously, this scattered context grammar generates L; formally, $L = L(G)$. Consider the transformational scattered context grammar $H = (V, T, P, \{S\})$. Notice that

$$L(G) = \left\{y \mid (S, y) \in t(H, \{S\})\right\}$$

Clearly, L is not context-free, so its generation is beyond the power of context-free grammars. It would be possible to construct a context-sensitive grammar that generates L. However, a context-sensitive grammar like this would have to keep traversing across its sentential forms to guarantee the same number of occurrences of *great-* in the generated sentences. Compared to this awkward way of generating L, the scattered context grammar G generates L in a more elegant, economical, and effective way.

In this section, we have illustrated how to transform and generate grammatical sentences in English by using transformational scattered context grammars, which represent a very natural linguistic apparatus straightforwardly based on scattered context grammars. However, from a more general perspective, we can apply these grammars basically in any area of science that formalizes its results by strings containing some scattered context dependencies.

Chapter 15
Applications in Computational Biology

This chapter presents some case studies concerning biology. It consists of Sects. 15.1, 15.2, and 15.3. Section 15.1 introduces simple case study using jumping scattered context derivation in DNA processing. Section 15.2 presents two case studies of biological organisms whose development is affected by some abnormal conditions, such as a virus infection. From a more practical point of view, Sect. 15.3 discusses parametric 0L grammars (see [PL90b]), which represent a powerful and elegant implementation tool in the area of biological simulation and modeling today. More specifically, we extend parametric 0L grammars by context conditions and demonstrate their use in models of growing plants.

15.1 DNA Processing with Jumping Scattered Context Derivations

In this section, we add some remarks concerning application-related perspectives of jumping scattered context grammars (see Sect. 5.2) in terms of molecular biology—namely, DNA processing.

As already sketched, jumping grammars serve as grammatical models that allow us to explore information processing performed in a discontinuous way adequately and rigorously. Taking into account the way these grammars are conceptualized, we see that they are particularly useful and applicable under the circumstances that primarily concern the number of occurrences of various symbols or substrings rather than their mutual context.

Case Study 15.1.1. Recall that a DNA is a molecule encoding genetic information by a repetition of four basic units called nucleotides—namely, guanine, adenine, thymine, and cytosine, denoted by letters G, A, T, and C, respectively. In terms

© Springer International Publishing AG 2017
A. Meduna, O. Soukup, *Modern Language Models and Computation*,
DOI 10.1007/978-3-319-63100-4_15

of formal language theory, a DNA is described as a string over $\{G, A, T, C\}$; for instance,

$$GGGGAGTGGGATTGGGAGAGGGGTTTGCCCCGCTCCC$$

Suppose that a DNA-computing-related investigation needs to study all the strings that contain the same number of As and Cs so all As precede Cs; for instance, $AGGAATCGCGTC$ is a proper string, but $CGCACCGGTA$ is not. Consider the jumping scattered context grammar

$$G = (\{1, 2, 3, 4, G, A, T, C\}, \{G, A, T, C\}, P, 1)$$

with P containing rules

$$
\begin{array}{ll}
(1) \rightarrow (23) & (3) \rightarrow (G3) \\
(3) \rightarrow (T3) & (3) \rightarrow (4) \\
(2, 4) \rightarrow (A2, 4C) & (2)|(4) \rightarrow (\varepsilon)
\end{array}
$$

Assume that the grammar works under $_2\Rightarrow$. It first generates an arbitrary string of Gs and Ts, in which there are no restrictions, by classical regular rules, since $_2\Rightarrow$ does not change the behaviour of context-free rules. However, then it comes the essential phase generating As and Cs. Indeed, the only context-sensitive rule under $_2\Rightarrow$ generates the equal number of As and Cs randomly scattered through the resulting sentence, but always with As preceding Cs. For instance, previously mentioned string $AGGAATCGCGTC$ can be generated by the following derivation.

$$
\begin{array}{ll}
1 \Rightarrow 23 & \Rightarrow 2G3 \\
\Rightarrow 2GG3 & \Rightarrow 2GGT3 \\
\Rightarrow 2GGTG3 & \Rightarrow 2GGTGG3 \\
\Rightarrow 2GGTGGT3 & \Rightarrow 2GGTGGT4 \\
\Rightarrow A2GGTGGT4C & \Rightarrow AGGA2TG4CGTC \\
\Rightarrow AGGAA2T4CGCGTC & \Rightarrow^2 AGGAATCGCGTC \qquad\qquad \square
\end{array}
$$

As obvious, under $_2\Rightarrow$, the grammar generates the language consisting of all the strings satisfying the above-stated requirements. Therefore, as we can see, jumping grammars may fulfill a useful role in studies related to DNA computing.

15.2 Biological Development and Its Grammatical Simulation

Case Study 15.2.1. Consider a cellular organism in which every cell divides itself into two cells during every single step of healthy development. However, when a virus infects some cells, all of the organism stagnates until it is cured again. During

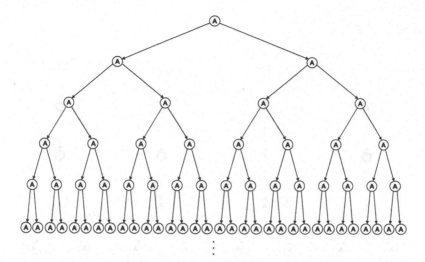

Fig. 15.1 Healthy development

the stagnation period, all of the cells just reproduce themselves without producing any new cells. To formalize this development by a suitable simple semi-conditional L grammar (see Sect. 4.2.3), we denote a healthy cell and a virus-infected cell by A and B, respectively, and introduce the simple semi-conditional 0L grammar

$$G = (\{A, B\}, P, A)$$

where P contains the following rules

$$(A \to AA, 0, B) \qquad (B \to B, 0, 0)$$
$$(A \to A, B, 0) \qquad (B \to A, 0, 0)$$
$$(A \to B, 0, 0)$$

Figure 15.1 describes G simulating a healthy development while Fig. 15.2 gives a development with a stagnation period caused by the virus. $\square$

In the next case study, we discuss an 0L grammar that simulates the developmental stages of a red alga (see [Sal73]). Using context conditions, we can modify this grammar so that it describes some unhealthy development of this alga that leads to its partial death or degeneration.

Case Study 15.2.2. Consider an 0L grammar

$$G = (V, P, 1)$$

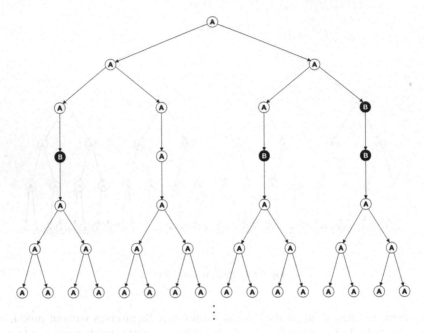

Fig. 15.2 Development with a stagnating period

where

$$V = \{1, 2, 3, 4, 5, 6, 7, 8, [,]\}$$

and the set of rules P contains

$$
\begin{array}{lllll}
1 \to 23 & 2 \to 2 & 3 \to 24 & 4 \to 54 & [\, \to\, [\\
5 \to 6 & 6 \to 7 & 7 \to 8[1] & 8 \to 8 &]\, \to\,]
\end{array}
$$

From a biological viewpoint, expressions in fences represent branches whose position is indicated by 8s. These branches are shown as attached at alternate sides of the branch on which they are born. Figure 15.3 gives a biological interpretation of the developmental stages formally specified by the next derivation, which contains 13 strings corresponding to stages (a) through (m) in this figure.

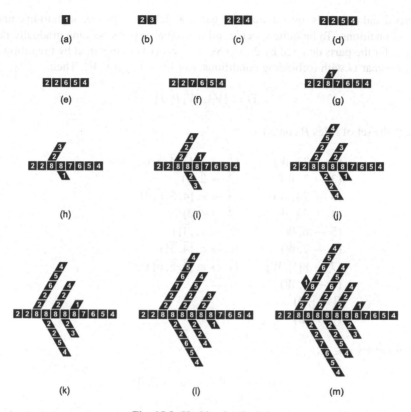

Fig. 15.3 Healthy development

$1 \Rightarrow_G 23$

$ \Rightarrow_G 224$

$ \Rightarrow_G 2254$

$ \Rightarrow_G 22654$

$ \Rightarrow_G 227654$

$ \Rightarrow_G 228[1]7654$

$ \Rightarrow_G 228[23]8[1]7654$

$ \Rightarrow_G 228[224]8[23]8[1]7654$

$ \Rightarrow_G 228[2254]8[224]8[23]8[1]7654$

$ \Rightarrow_G 228[22654]8[2254]8[224]8[23]8[1]7654$

$ \Rightarrow_G 228[227654]8[22654]8[2254]8[224]8[23]8[1]7654$

$ \Rightarrow_G 228[228[1]7654]8[227654]8[22654]8[2254]8[224]8[23]8[1]7654$

Death Let us assume that the red alga occurs in some unhealthy conditions under which only some of its parts survive while the rest dies. This dying process starts from the newly born, marginal parts of branches, which are too young and weak to survive, and proceeds toward the older parts, which are strong enough to live under these conditions. To be quite specific, all the red alga parts become gradually dead except for the parts denoted by 2s and 8s. This process is specified by the following OL grammar G with forbidding conditions. Let $W = \{\bar{a} \mid a \in V\}$. Then,

$$G = (V \cup W, P, 1)$$

where the set of rules P contains

$$
\begin{array}{ll}
(1 \to 23, W) & (\bar{1} \to \bar{2}, \{\bar{3}, \bar{4}, \bar{5}, \bar{6}, \bar{7}\}) \\
(2 \to 2, W) & (\bar{2} \to \bar{2}, \emptyset) \\
(3 \to 24, W) & (\bar{3} \to \varepsilon, \{\bar{4}, \bar{5}, \bar{6}, \bar{7}\}) \\
(4 \to 54, W) & (\bar{4} \to \varepsilon, \emptyset) \\
(5 \to 6, W) & (\bar{5} \to \varepsilon, \{\bar{4}\}) \\
(6 \to 7, W) & (\bar{6} \to \varepsilon, \{\bar{4}, \bar{5}\}) \\
(7 \to 8[1], W) & (\bar{7} \to \varepsilon, \{\bar{4}, \bar{5}, \bar{6}\}) \\
(8 \to 8, W) & \\
([\to [, \emptyset) & \\
(] \to], \emptyset) &
\end{array}
$$

and for every $a \in V$,

$$(a \to \bar{a}, \emptyset) \qquad (\bar{a} \to \bar{a}, \emptyset)$$

Figure 15.4 pictures the dying process corresponding to the next derivation, whose last eight strings correspond to stages (a) through (h) in the figure.

$$
\begin{aligned}
1 &\Rightarrow_G^* 228[228[1]7654]8[227654]8[22654]8[2254]8[224]8[23]8[1]7654 \\
&\Rightarrow_G \overline{228[228[1]7654]8[227654]8[22654]8[2254]8[224]8[23]8[1]7654} \\
&\Rightarrow_G \overline{228[228[1]765]8[22765]8[2265]8[225]8[22]8[23]8[1]765} \\
&\Rightarrow_G \overline{228[228[1]76]8[2276]8[226]8[22]8[22]8[23]8[1]76} \\
&\Rightarrow_G \overline{228[228[1]7]8[227]8[22]8[22]8[22]8[23]8[1]7} \\
&\Rightarrow_G \overline{228[228[1]]8[22]8[22]8[22]8[22]8[23]8[1]} \\
&\Rightarrow_G \overline{228[228[1]]8[22]8[22]8[22]8[22]8[2]8[1]} \\
&\Rightarrow_G \overline{228[228[2]]8[22]8[22]8[22]8[22]8[2]8[2]}
\end{aligned}
$$

Degeneration Consider circumstances under which the red alga has degenerated. During this degeneration, only the main stem was able to give a birth to new branches while all the other branches lengthened themselves without any branching out. This degeneration is specified by the forbidding OL grammar $G = (V \cup \{D, E\}, P, 1)$, with P containing

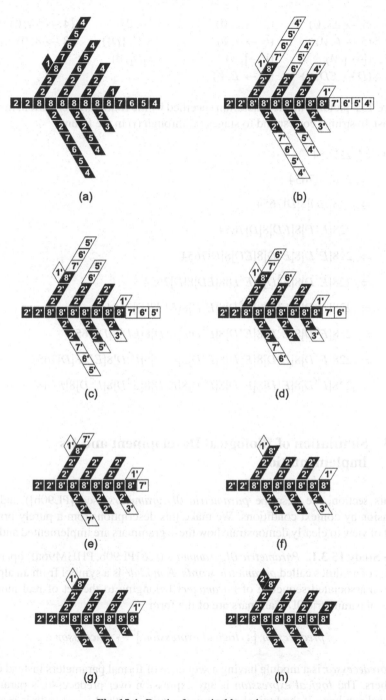

Fig. 15.4 Death of marginal branch parts

$$(1 \rightarrow 23, \emptyset) \qquad (2 \rightarrow 2, \emptyset) \qquad (3 \rightarrow 24, \emptyset) \qquad (4 \rightarrow 54, \emptyset)$$
$$(5 \rightarrow 6, \emptyset) \qquad (6 \rightarrow 7, \emptyset) \qquad (7 \rightarrow 8[1], \{D\}) \qquad (8 \rightarrow 8, \emptyset)$$
$$([\rightarrow [, \emptyset) \qquad (] \rightarrow], \emptyset) \qquad (7 \rightarrow 8[D], \emptyset)$$
$$(D \rightarrow ED, \emptyset) \qquad (E \rightarrow E, \emptyset)$$

Figure 15.5 pictures the degeneration specified by the following derivation, in which the last 10 strings correspond to stages (a) through (j) in the figure:

$$1 \Rightarrow_G^* 227654$$

$$\Rightarrow_G 228[D]7654$$

$$\Rightarrow_G 228[ED]8[D]7654$$

$$\Rightarrow_G 228[E^2D]8[ED]8[D]7654$$

$$\Rightarrow_G 228[E^3D]8[E^2D]8[ED]8[D]7654$$

$$\Rightarrow_G 228[E^4D]8[E^3D]8[E^2D]8[ED]8[D]7654$$

$$\Rightarrow_G 228[E^5D]8[E^4D]8[E^3D]8[E^2D]8[ED]8[D]7654$$

$$\Rightarrow_G 228[E^6D]8[E^5D]8[E^4D]8[E^3D]8[E^2D]8[ED]8[D]7654$$

$$\Rightarrow_G 228[E^7D]8[E^6D]8[E^5D]8[E^4D]8[E^3D]8[E^2D]8[ED]8[D]7654$$

$$\Rightarrow_G 228[E^8D]8[E^7D]8[E^6D]8[E^5D]8[E^4D]8[E^3D]8[E^2D]8[ED]8[D]7654 \qquad \Box$$

15.3   Simulation of Biological Development and Its Implementation

In this section, we describe *parametric 0L grammars* (see [PL90b]) and their extension by context conditions. We make this description from a purely practical point of view to clearly demonstrate how these grammars are implemented and used.

Case Study 15.3.1. *Parametric 0L grammars* (see [PL90b, PHHM96a]) operate on strings of modules called *parametric words*. A *module* is a symbol from an alphabet with an associated sequence of *parameters* belonging to the set of real numbers. Rules of parametric 0L grammars are of the form

$$predecessor\ [:\ logical\ expression\ ] \rightarrow successor$$

The *predecessor* is a module having a sequence of formal parameters instead of real numbers. The *logical expression* is any expression over predecessor's parameters and real numbers. If the logical expression is missing, the logical truth is assumed.

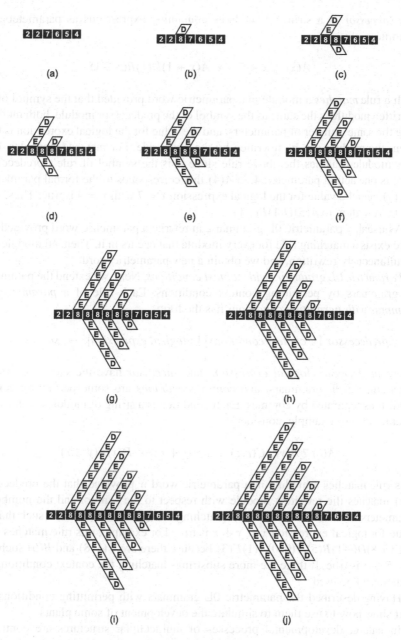

Fig. 15.5 Degeneration

The *successor* is a string of modules containing expressions as parameters; for example,

$$A(x) \; : \; x < 7 \; \rightarrow \; A(x+1)D(1)B(3-x)$$

Such a rule *matches* a module in a parametric word provided that the symbol of the rewritten module is the same as the symbol of the predecessor module, both modules have the same number of parameters, and the value for the logical expression is true. Then, the module can be rewritten by the given rule. For instance, consider $A(4)$. This module matches the above rule since A is the symbol of rule's predecessor, there is one actual parameter, 4, in $A(4)$ that corresponds to the formal parameter x in $A(x)$, and the value for the logical expression $x < 7$ with $x = 4$ is true. Thus, $A(4)$ can be rewritten to $A(5)D(1)B(-1)$.

As usual, a parametric 0L grammar can rewrite a parametric word provided that there exists a matching rule for every module that occurs in it. Then, all modules are simultaneously rewritten, and we obtain a new parametric word.

Parametric 0L grammars with context conditions. Next, we extend the parametric 0L grammars by permitting context conditions. Each rule of a *parametric 0L grammar with permitting conditions* has the form

$$predecessor \; [\; ? \; context \; conditions] \; [\; : \; logical \; expression] \; \rightarrow \; successor$$

where *predecessor*, *logical expression*, and *successor* have the same meaning as in parametric 0L grammars, and *context conditions* are some permitting context conditions separated by commas. Each condition is a string of modules with formal parameters. For example, consider

$$A(x) \; ? \; B(y), \; C(r,z) \; : \; x < y + r \; \rightarrow \; D(x)E(y+r)$$

This rule matches a module in a parametric word w provided that the predecessor $A(x)$ matches the rewritten module with respect to the symbol and the number of parameters and there exist modules matching to $B(y)$ and $C(r,z)$ in w such that the value for logical expression $x < y + r$ is true. For example, this rule matches $A(1)$ in $C(3,8)D(-1)B(5)H(0,0)A(1)F(3)$ because there are $C(3,8)$ and $B(5)$ such that $1 < 5 + 3$ is true. If there are more substrings matching the context condition, any of them can be used.

Having described the parametric 0L grammars with permitting conditions, we next show how to use them to simulate the development of some plants.

In nature, developmental processes of multicellular structures are controlled by the quantity of substances exchanged between modules. In the case of plants, growth depends on the amount of water and minerals absorbed by the roots and carried upward to the branches. The model of branching structures making use of the resource flow was proposed by Borchert and Honda in [BH84]. The model is controlled by a *flux* of resources that starts at the base of the plant and propagates the substances toward the apexes. An apex accepts the substances, and when the quantity of accumulated resources exceeds a predefined threshold value, the apex bifurcates and initiates a new lateral branch. The distribution of the flux depends on the number of apexes that the given branch supports and on the type of the branch—plants usually carry greater amount of resources to straight branches than to lateral branches (see [BH84] and [PHHM96a]).

The following two examples (I) and (II) illustrate the idea of plants simulated by parametric 0L grammars with permitting conditions.

(I) Consider the model

$$\text{start} : I(1, 1, e_{root}) A(1)$$
$$p_1 : \quad A(id) \ ? \ I(id_p, c, e) \ : \ id == id_p \ \wedge \ e \geq e_{th}$$
$$\rightarrow \ [+(\alpha) I(2 * id + 1, \gamma, 0)$$
$$A(2 * id + 1)]/(\pi) I(2 * id, 1 - \gamma, 0) A(2 * id)$$
$$p_2 : \quad I(id, c, e) \ ? \ I(id_p, c_p, e_p) \ : \ id_p == \lfloor id/2 \rfloor$$
$$\rightarrow \ I(id, c, c * e_p)$$

This L grammar describes a simple plant with a constant resource flow from its roots and with a fixed distribution of the stream between lateral and straight branches. It operates on the following types of modules.

- $I(id, c, e)$ represents an internode with a unique identification number id, a distribution coefficient c, and a flux value e.
- $A(id)$ is an apex growing from the internode with identification number equal to id.
- $+(\varphi)$ and $/(\varphi)$ rotate the segment orientation by angle φ (for more information, consult [PHHM96a]).
- [and] enclose the sequence of modules describing a lateral branch.

We assume that if no rule matches a given module $X(x_1, \ldots, x_n)$, the module is rewritten by an implicit rule of the form

$$X(x_1, \ldots, x_n) \rightarrow X(x_1, \ldots, x_n)$$

That is, it remains unchanged.

At the beginning, the plant consists of one internode $I(1, 1, e_{root})$ with apex $A(1)$, where e_{root} is a constant flux value provided by the root. The first rule, p_1, simulates the bifurcation of an apex. If an internode preceding the apex $A(id)$ reaches a sufficient flux $e \geq e_{th}$, the apex creates two new internodes I terminated by apexes A. The lateral internode is of the form $I(2*id+1, \gamma, 0)$ and the straight internode is of the form $I(2 * id, 1 - \gamma, 0)$. Clearly, the identification numbers of these internodes are unique. Moreover, every child internode can easily calculate the identification number of its parent internode; the parent internode has $id_p = \lfloor id/2 \rfloor$. The coefficient γ is a fraction of the parent flux to be directed to the lateral internode. The second rule, p_2, controls the resource flow of a given internode. Observe that the permitting condition $I(id_p, c_p, e_p)$

with $id_p = \lfloor id/2 \rfloor$ matches only the parent internode. Thus, p_2 changes the flux value e of $I(id, c, e)$ to $c * e_p$, where e_p is the flux of the parent internode, and c is either γ for lateral internodes or $1 - \gamma$ for straight internodes. Therefore, p_2 simulates the transfer of a given amount of parent's flux into the internode. Figure 15.6 pictures 12 developmental stages of this plant with e_{root}, e_{th}, and γ set to 12, 0.9, and 0.4, respectively. The numbers indicate the flow values of internodes.

It is easy to see that this model is unrealistically simple. Since the model ignores the number of apexes, its flow distribution does not depend on the size of branches, and the basal flow is set to a constant value. However, it sufficiently illustrates the technique of communication between adjacent internodes. Thus, it can serve as a template for more sophisticated models of plants, such as the following model.

(II) We discuss a plant development with a resource flow controlled by the number of apexes. This example is based on Example 17 in [PHHM96a].

$$
\begin{aligned}
&start : N(1)\, I(1, straight, 0, 1)\, A(1) \\
&p_1 : \quad N(k) \to N(k + 1) \\
&p_2 : \quad I(id, t, e, c)\; ?\, N(k),\, A(id) \\
&\qquad\quad : id == 1 \\
&\qquad\quad \to I(id, t, \sigma_0 2^{(k-1)\eta^k}, 1) \\
&p_3 : \quad I(id, t, e, c)\; ?\, N(k),\, I(id_s, t_s, e_s, c_s),\, I(id_l, t_l, e_l, c_l) \\
&\qquad\quad : id == 1 \;\wedge\; id_s == 2 * id \;\wedge\; id_l == 2 * id + 1 \\
&\qquad\quad \to I(id, t, \sigma_0 2^{(k-1)\eta^k}, c_s + c_l) \\
&p_4 : \quad I(id, t, e, c)\; ?\, I(id_p, t_p, e_p, c_p),\, I(id_s, t_s, e_s, c_s),\, I(id_l, t_l, e_l, c_l) \\
&\qquad\quad : id_p == \lfloor id/2 \rfloor \;\wedge\; id_s == 2 * id \;\wedge\; id_l == 2 * id + 1 \\
&\qquad\quad \to I(id, t, \delta(t, e_p, c_p, c), c_s + c_l) \\
&p_5 : \quad Id(id, t, e, c)\; ?\, I(id_p, t_p, e_p, c_p),\, A(id_a) \\
&\qquad\quad : id_p == \lfloor id/2 \rfloor \;\wedge\; id_a == id \\
&\qquad\quad \to I(id, t, \delta(t, e_p, c_p, c), 1) \\
&p_6 : \quad A(id)\; ?\, I(id_p, t_p, e_p, c_p) \\
&\qquad\quad : id == id_p \;\wedge\; e_p \geq e_{th} \\
&\qquad\quad \to [+(\alpha)\, I(2 * id + 1, lateral, e_p * (1 - \lambda), 1)\, A(2 * id + 1)] \\
&\qquad\qquad /(\pi)\, I(2 * id, straight, e_p * \lambda, 1)\, A(2 * id)
\end{aligned}
$$

This L grammar uses the following types of modules.

- $I(id, t, e, c)$ is an internode with a unique identification number id, where t is a type of this internode, $t \in \{straight, lateral\}$, e is a flux value, and c is a number of apexes the internode supports.
- $A(id)$ is an apex terminating the internode id.
- $N(k)$ is an auxiliary module, where k is the number of a developmental cycle to be done by the next derivation.
- $+(\varphi)$, $/(\varphi)$, [and] have the same meaning as in the previous example.

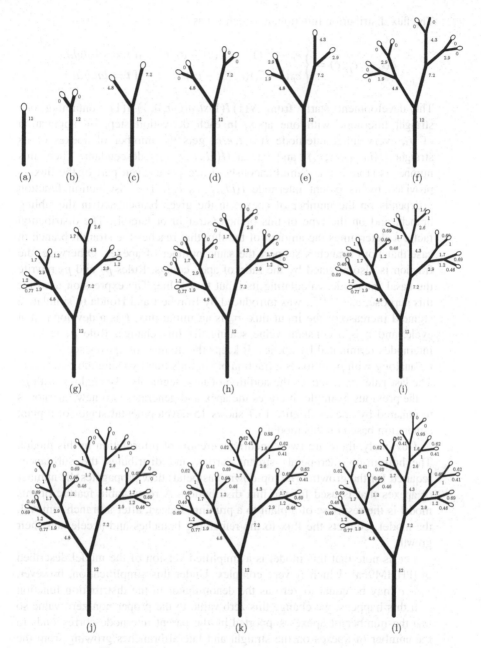

Fig. 15.6 Developmental stages of the plant generated by (I)

The flux distribution function, δ, is defined as

$$\delta(t, e_p, c_p, c) = \begin{cases} e_p - e_p(1-\lambda)((c_p - c)/c) & \text{if } t = straight, \\ e_p(1-\lambda)(c/(c_p - c)) & \text{if } t = lateral \end{cases}$$

The development starts from $N(1) I(1, straight, 0, 1) A(1)$ containing one straight internode with one apex. In each derivation step, by application of p_4, every inner internode $I(id, t, e, c)$ gets the number of apexes of its straight $(I(id_s, t_s, e_s, c_s))$ and lateral $(I(id_l, t_l, e_l, c_l))$ descendant. Then, this number is stored in c. Simultaneously, it accepts a given part of the flux e_p provided by its parent internode $I(id_p, t_p, e_p, c_p)$. The distribution function δ depends on the number of apexes in the given branch and in the sibling branch, and on the type of this branch (straight or lateral). The distribution factor λ determines the amount of the flux that reaches the straight branch in case that both branches support the same number of apexes. Otherwise, the fraction is also affected by the ratio of apex counts. Rules p_2 and p_3 rewrite the basal internode, calculating its input flux value. The expression used for this purpose, $\sigma_0 2^{(k-1)\eta^k}$, was introduced by Borchert and Honda to simulate a sigmoid increase of the input flux; σ_0 is an initial flux, k is a developmental cycle, and η is a constant value scaling the flux change. Rule p_5 rewrites internodes terminated by apexes. It keeps the number of apexes set to 1, and by analogy with p_4, it loads a fraction of parent's flux by using the δ function. The last rule, p_6, controls the addition of new segments. By analogy with p_1 in the previous example, it erases the apex and generates two new internodes terminated by apexes. Figure 15.7 shows 15 developmental stages of a plant simulation based on this model.

Obviously, there are two concurrent streams of information in this model. The bottom-up (acropetal) stream carries and distributes the substances required for the growth. The top-down (basipetal) flow propagates the number of apexes that is used for the flux distribution. A remarkable feature of this model is the response of a plant to a pruning. Indeed, after a branch removal, the model redirects the flux to the remaining branches and accelerates their growth.

Let us note that this model is a simplified version of the model described in [PHHM96a], which is very complex. Under this simplification, however, $c_p - c$ may be equal to zero as the denominator in the distribution function δ. If this happens, we change this zero value to the proper non-zero value so that the number of apexes supported by the parent internode corresponds to the number of apexes on the straight and lateral branches growing from the parent internode. Consult [PHHM96a] for a more appropriate, but also more complicated solution of this problem.

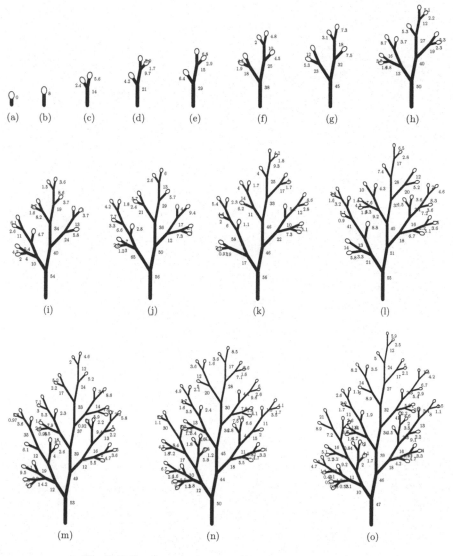

Fig. 15.7 Developmental stages of the plant generated by (II)

From the presented examples, we see that with permitting conditions, parametric 0L grammars can describe sophisticated models of plants in a very natural way. Particularly, compared to the context-sensitive L grammars, they allow one to refer to modules that are not adjacent to the rewritten module, and this property makes them more adequate, succinct, and elegant.   □

Part VI
Conclusion

This concluding part closes the entire book by adding several remarks concerning its coverage. It consists of a single chapter—Chap. 16. This chapter first briefly summarizes all the material covered in the text. Furthermore, it sketches many brand new investigation trends and long-time open problems. Finally, it makes several bibliographical remarks.

Chapter 16
Concluding Remarks

This three-section chapter closes the book by adding several remarks concerning its coverage. Section 16.1 briefly summarizes all the material covered in the text. Furthermore, Sect. 16.2 sketches many brand new investigation trends as well as points out long-time open problems. Finally, it makes several bibliographical and historical remarks.

16.1 Summary

This book deals with formal language theory, which represents a branch of mathematics that formalizes languages and devices that define them (see [Med14]). In other words, this theory represents a mathematically systematized body of knowledge concerning languages in general. It defines languages as sets of finite sequences consisting of symbols. As a result, this general definition encompasses almost all languages, including natural languages as well as artificial languages, such as programming languages.

The strictly mathematical approach to languages necessitates an introduction of mathematical systems that define them. Traditionally, these systems are based upon finitely many rules by which they sequentially rewrite strings, and that is why they are called language models. They are classified into two basic categories— grammars and automata. Grammars define strings of their language so their rewriting process generates them from a special start symbol. Automata define strings of their language by rewriting process that starts from these strings and ends in a special set of strings, usually called final configurations. However, apart from these traditional versions of grammars and automata, language theory have also developed several systems that rewrite words in a non-traditional way. These non-traditional versions of rewriting systems used as modern language models represent the principal subject of this book.

© Springer International Publishing AG 2017
A. Meduna, O. Soukup, *Modern Language Models and Computation*,
DOI 10.1007/978-3-319-63100-4_16

Many of the modern language models have their great advantages over their traditional counterparts. From a practical viewpoint, an important advantage of these modern language models consists in controlling their language-defining process and, therefore, operating in a more deterministic way than classical language models, which perform their derivations in a quite traditional way. Indeed, in an ever-changing environment in which real language processors work, the modern language models adequately reflect and simulate real communication technologies applied in such real-world areas as various engineering techniques for language analysis. Most importantly, the modern versions of language models are stronger than their traditional counterparts. Considering these significant advantages and properties, modern language models fulfill a highly beneficial role in many kinds of language-related work conducted by a broad variety of scientists, ranging from mathematicians through computer scientists up to linguists and biologists.

This book restricted its principal attention to these crucially important investigation areas concerning modern language models–their properties, transformations and applications. Next, we recall their significance, and in general, we sum up their coverage in the book.

First, concerning properties, the power of the language models under consideration represents perhaps the most important property concerning them, so we always determined the language family defined by these models. A special attention was also paid to algorithms that arrange modern language models so they satisfy some prescribed properties while the generated languages remain unchanged because many language processors strictly require their satisfaction. From a theoretical viewpoint, these properties frequently simplified proofs demonstrating results about the models.

Second, transformations of language models were central to this book, too. Specifically, transformations that reduce the models represent one of its important investigation areas because reduced versions of these models define languages in a succinct and easy-to-follow way. As obvious, this reduction simplifies the development of language processing technologies, which then work economically and effectively. Of course, the same languages can be defined by different language models. We obviously tend to define them by the most appropriate models under given circumstances. Therefore, whenever discussing different types of equally powerful language models, we also presented transformations that converted them to each other. More specifically, given a language model of one type, we explained how to convert it to a language model of another equally powerful type so both the original model and the model produced by this conversion define the same language.

Finally, the book demonstrated applications of modern language models. It narrowed its attention to grammars rather than automata. First, it described these applications and their perspectives from a general viewpoint. Then, it gave many case studies to show quite specific real-world applications concerning computational linguistics and biology.

Part I, consisting of Chaps. 1 through 2, gives an introduction to this book in order to express all its discussion clearly and make the book completely self-contained. It places all the coverage of the book into scientific context and reviews important mathematical concepts with a focus on formal language theory. Chapter 1 gives the mathematical background of this book. It reviews all the necessary mathematical concepts to grasp the topics covered in the book. These concepts primarily include fundamental areas of discrete mathematics. First, this chapter reviews basic concepts from set theory. Then, it gives the essentials concerning relations and their crucially important special cases, namely, functions. Finally, this chapter reviews fundamental concepts from graph theory. Chapter 2 covers selected areas of formal language theory needed to follow the rest of this book. It introduces the basic terminology concerning strings, languages, operations, and closure properties. Furthermore, it overviews a large variety of grammars, automata and language families resulting from them. Apart from the classical rudiments of formal language theory, Chap. 2 covers several lesser-known areas of this theory, such as fundamentals concerning parallel grammars, because these areas are also needed to grasp some topics of this book.

Part II consists of Chaps. 3 through 6. It deals with the most important modern versions of grammars. Chapter 3 gives the fundamentals of grammars that regulate their generation process by additional mechanisms, based upon simple mathematical concepts, such as finite sets of symbols. Chapter 4 studies grammars that generate their languages in parallel and, thereby, accelerate this generation significantly. First, it studies partially parallel generation of languages, after which it investigates the totally parallel generation of languages. Chapter 5 explores grammars that work on their words in a discontinues way. Chapter 6 studies the generation of languages based on algebraic restrictions. In particular, it examines grammatical generation defined over free groups.

Part III consists of Chaps. 7 through 10. It covers the essential modern versions of automata. In many respects, it parallels what Part II covers in terms of grammars. Chapter 7 gives the fundamentals of regulated automata. Similarly to grammars discussed in Chap. 5, Chap. 8 studies automata that jump across the words they work on discontinuously. Chapter 9 discusses automata with deep pushdown lists, which can be modified deeper than on their top. Chapter 10 studies automata that work over free groups.

Part IV, which consists of Chaps. 11 and 12, covers important language-defining devices that combine other language models. Chapter 11 untraditionally combines grammars and automata in terms of the way they operate. Specifically, it studies how to generate languages by automata although traditionally, languages are always generated by grammars. Chapter 12 studies the generation of languages by several grammars that work in a simultaneously cooperative way.

Part V consists of Chaps. 13 through 15. It discusses applications of language models studied earlier in the book. First, Chap. 13 covers these applications and their perspectives from a rather general viewpoint. Then, more specifically, Chaps. 14 and 15 describe applications in computational linguistics and molecular biology, respectively. Both chapters contain several case studies of real-world applications described in detail.

Part VI consists of a single chapter—Chap. 16, which closes the entire book by adding several remarks concerning its coverage. It briefly summarizes all the material covered in the text. Furthermore, it sketches many brand new investigation trends and long-time open problems. Finally, it makes several bibliographical and historical remarks.

16.2   Modern Trends

In this section, we point out three new directions in the investigation of modern language models. In its conclusion, we make several suggestions regarding their future investigation. We also point out many open problems.

16.2.1   An Algebraic Approach to Modern Versions of Grammars and Automata

As demonstrated in Chaps. 6 and 10, from an algebraic viewpoint, various kinds of modern language models can be viewed as restrictions placed upon relations by which the language models define their languages. Indeed, several modern versions of grammars are based on restrictions placed upon derivations while modern versions of automata restrict the way they make moves. From this point of view, the investigation of modern grammars is closely related to many algebraically oriented studies in formal language theory. Investigate how to replace some of the previous regulating mechanisms by suitable relation-domain restrictions and vice versa. Furthermore, study how some well-known special cases of these relations affect the resulting language-defining power. Specifically, perform this study under the assumptions that these relations represent functions, injections, or surjections. The algebraic theory of formal languages and their automata is discussed in a great number of articles and books, some of which are summarized in Chapters 6 through 11 of [RS97a]. Furthermore, [Gin75, vzGG03, GN03, Ito03, Tru98] represent a systematic introduction to this area of formal language theory.

16.2.2   Combining Grammars and Automata

In formal language theory, the overwhelming majority of language-defining devices is based on language models that represent either grammars or automata. Although it is obviously quite natural to design language-defining devices based on a combination of both grammars and automata and, thereby, make the scale of language-defining models much richer and broader, only a tiny minority of these models is designed in this combined way; some of them are covered in Sect. 3.2 and Chap. 11.

In terms of modern language-defining models, state grammars and so-called #-rewriting systems represent language models that have features of both grammars and automata. Introduced several decades ago, state grammars (see Sect. 3.2) represent a classical grammatical model of regulation, which has been covered in this monograph in detail. On the other hand, #-rewriting systems have been introduced relatively recently (see [Kři08, KMS07, KM07, Kři07, KMS06a]). These systems generate languages just like any grammars. On the other hand, like automata, they use simple state-based regulation during their language-generation process. These systems characterize an infinite hierarchy of language families resulting from programmed grammars of finite index (see [KMS06a]). As a result, to put it from a broader perspective, systems of this combined kind are naturally related to some classical results about formal languages, on which they can shed light in an modern way. Therefore, it is highly expectable that formal language theory will introduce and investigate many more language models based upon a combination of modern versions of grammars and automata.

16.2.3 Modern Translation-Defining Models

Modern versions of grammars and automata discussed in this book generate languages. As obvious, they can be easily and naturally modified to modern translation-defining models by analogy with the modification of context-free grammars and pushdown automata to context-free translation grammars and pushdown transducers, respectively (see [AU72]). Most probably, formal language theory will open their investigation of modern translation-defining models by studying their properties from a theoretical point of view by analogy with other well-known studies of formal translation, including [AU72, ALSU06, Bro89, Cho02, Gri71, LRS76, Pag81, SSS87]. Simultaneously, however, we can expect a struggle to apply them to the translation of programming as well as natural languages. As a matter of fact, to some extent, [ČHM12, HM12, Hor12, HM11] have already sketched applications concerning the specification and translation of natural languages in this way.

16.2.4 Open Problem Areas

Throughout this book, we have already formulated many open problems. We close the present section by selecting and repeating the most important questions, which deserve our special attention. To see their significance completely, however, we suggest that the reader returns to the referenced parts of the book in order to view these questions in the full context of their formulation and discussion in detail.

I. Over the last four decades, formal language theory has struggled to determine the precise impact of erasing rules to the power of modern versions of grammars. Indeed, it is still an open question whether regular-controlled, matrix, programmed, and forbidding grammars are equivalent to their propagating versions (see Chap. 3). Sections 3.3 presents a partial solutions to this problem in terms of regular-controlled grammars. However, in general, this important question has not been answered yet. For some very recent results regarding this topic, see [Zet09, Zet10, Zet11b, Zet11a].

II. By Theorem 4.1.6, we can convert any propagating scattered context grammar into an equivalent context-sensitive grammar. However, it is a long-standing open problem whether these two types of grammars are, in fact, equivalent.

III. Consider the results in Sect. 4.1.4 concerning the reduction of scattered context grammars. While one-nonterminal versions of scattered context grammars do not generate the entire family of recursively enumerable languages (see Theorem 4.1.13), their two-nonterminal versions do (see Theorem 4.1.20). Therefore, regarding the number of nonterminals, this open problem area has been completely solved. By Theorem 4.1.21, the two-context-sensitive rule versions of scattered context grammars characterize the family of recursively enumerable languages. On the other hand, the generative power of their one-context-sensitive rule versions has not been determined yet.

IV. All the uniform rewriting discussed in Chap. 3 is obtained for grammars with erasing rules. In the proof techniques by which we have achieved this rewriting, these rules fulfill a crucial role. Indeed, these techniques cannot be straightforwardly adapted for grammars without erasing rules. Can we achieve some uniform rewriting for grammars without erasing rules in a different way?

V. Return to LRC-ET0L grammars and their variants, discussed in Sect. 4.2.4. Recall that ET0L and EPT0L grammars have the same generative power (see Theorem 2.3.41). Do LF-E0L and LF-EP0L grammars have the same power? Are LP-E0L and LP-EP0L grammars equally powerful? What is the relation between the language families generated by ET0L grammars and by LP-E0L grammars? What is the generative power of LF-E0L grammars?

VI. Chapter 12 gives the basics of multigenerative grammar systems. Recall that they are based upon a simultaneous generation of several strings, which are composed together by some basic operation, such as concatenation, after their generation is completed. Consider other operations, like intersection, and study languages generated in this way by multigenerative grammar systems. Furthermore, study multigenerative grammar systems based on special cases of context-free grammars. Specifically, what is the generative power of multigenerative grammar systems based upon regular or linear grammars?

VII. In Sect. 7.2.1, we have proved that state-controlled and transition-controlled finite automata regulated by languages generated by propagating programmed grammars with appearance checking characterize the family of recursively enumerable languages (see Theorem 7.2.17 and Corollary 7.2.18). Let us point out, however, that these automata are, in a general case, non-deterministic. Does this characterization hold in terms of their deterministic versions, too? Furthermore, try to achieve an analogical characterization of the family of context-sensitive languages.

VIII. Consider jumping finite automata and their general versions, discussed in Chap. 8. Theorem 8.2.11 gives a necessary and sufficient condition for a language to belong to the family defined by jumping finite automata. Does there exist a similar necessary and sufficient condition for general jumping finite automata as well? Furthermore, how precisely do left jumps affect the power of these automata? Are there any undecidable problems concerning the family of languages accepted by these automata?

IX. Reconsider deep pushdown automata, discussed in Chap. 9. In its conclusion, this section discusses two special types of these automata—deterministic deep pushdown automata and deep pushdown automata whose expansions can erase symbols inside of their pushdowns. Determine the language families defined by these two variants.

16.3 Bibliographical Remarks

This section gives an overview of the crucially important studies published on the subject of this book from a historical perspective. As this book represents primarily a theoretically oriented treatment, we concentrate our attention primarily on theoretical studies.

Although the present treatment of modern versions of grammars and automata is self-contained, some background in formal language theory is definitely helpful to grasp the material of this book easily. As an introduction to formal language theory, we recommend [Med00a, FB94, Gin75, Har78, HU69, Kel95, LP81, Mar02, MAK88, Sal73, Sip06]. The three-volume *Handbook of Formal Languages* (see [RS97a, RS97b, RS97c]) gives an overview of the recent important trends in formal language theory.

For a summary of the fundamental knowledge about modern rewriting published by 1989, consult [DP89]. Furthermore, [MVMP04] and Chapter 3 of [RS97b] give a brief overview of recent results concerning regulated grammars. Reference [MZ10] summarizes recent results concerning various transformations of regulated grammars. More specifically, it concentrates its attention on algorithms that transform these grammars and some related modern language-defining models so the resulting transformed models are equivalent and, in addition, satisfy some prescribed properties.

16.3.1 Context-Based Grammatical Models

The classical normal forms from Sect. 3.1.1 were established in [Pen74, Kur64, Gef91]. The two new normal forms appearing in this book were recently introduced in [MZ14]. Consult page 180 in [RS97a] for a summary of normal forms of phrase-structure grammars.

The uniform generation of sentences by phrase-structure grammars, discussed in Sect. 3.1.1, has been investigated in [Med98b].

Conditional grammars were introduced in [Fri68]. Several variants of these grammars were discussed in [CV92, DPS93, EKR85, EPR94, Med91a, Kel84, Kel89, KR81, Krá73, Nav70, P79, P85, Roz77, RS78, Urb83, Vas03]. The crucial concepts of these grammars and results concerning them are summarized in [MCV93].

Random context grammars were introduced in [vdW70]. Strictly speaking, in [vdW70], their definition coincides with the definition of permitting grammars in this book. Forbidding grammars, also known as *N-grammars* (see [Pen75]), together with other variants of random context grammars were originally studied by Lomkovskaya in [Lom72a, Lom72b, Lom72c]. After these studies, many more papers discussed these grammars, including [vdWE00, vdWE02, Mas09a, AES06, EW03, Mv05, Zet10, EW13]. In [DM12, MM09, Mas10b], simplified versions of random context grammars, called *restricted context-free grammars*, were studied. Moreover, [GMM10, CVMV09, Mas09a, KM11] studied grammar systems with their components represented by random context grammars.

Generalized forbidding grammars were introduced in [Med90b] and further investigated in [Mv03a, MM07a, Mv05].

Semi-conditional and simple semi-conditional grammars were introduced and investigated in [P85] and [MG94], respectively. Their descriptional complexity was studied in [Mv02, Oku09, MM07b, Mv05, Vas03, Vas05, Mas06].

Originally, scattered context grammars were defined in [GH69]. Their original version disallowed erasing rules, however. Four years later, [Vir73] generalized them to scattered context grammars with erasing rules (see also [Med95a]). The following studies represent the most important studies that have discussed these grammars: [GH69, Vir73, Cre73, Mas07b, Med95a, MR71, P82, MT08a, May72, Fer96, GW89, Vas05, ER79, Med91b, Med93, FM03b, FM03a, Krá69, Mas09a, MM08, Med97, Med98a, Med00c, Med00b, Med01, Med02, Med03a, Med03b, Mv05, MT05, MT07a, KMv05, MT07c, MT07b, MT08b, MMv08, MT09, Tec07, Tec08, CVV10, Mas10a, Mv11]. Uniform generation of languages by scattered context grammars was investigated in [Med01]. For an in-depth overview of scattered context grammars and their applications, consult [MT10] and the references given therein.

Sequential rewriting over word monoids has been studied in [Med90a, Med96]. Moreover, [BBM07a, BBM05, BBM07b] investigate sequential rewriting over free groups.

16.3.2   Rule-Based Grammatical Regulation

Grammars regulated by regular control languages over the set of rules were introduced in [GS68]. Their workspace conditions were established in [MZ11] (see also [Zem10]). Generation of sentences with their parses by these grammars was investigated in [MZ13b].

Matrix grammars were first defined and studied in [Abr65]. For some very recent results regarding the elimination of erasing rules from these grammars, see [Zet09, Zet10, Zet11b, Zet11a]. This book originally introduces even matrix grammars; however, they generalize the concept of simple matrix grammars which were introduced in [Iba70] and simultaneously in [KM70] as tuple grammars. Their leftmost versions were studied in [Mau73].

Programmed grammars were introduced in [Ros69]. Their non-determinism has been investigated in [MVZ11, BBD+06, BH06, Vrá11, Vrá12]. Some other recent papers include [FS97, Fer03, FFOR07].

State grammars were defined by Kasai in [Kas70]. A generalized version of these grammars with erasing rules was originally studied in [HM88].

16.3.3 Modern Parallel Grammars

In general, modern versions of ET0L grammars have been studied in [Mv03b, RS78, Sol76, Š03, DP89, Mv05, Das07, BCVHV05, Das07, Sos03, DP89]. Context-conditional ET0L grammars were studied in Section 4.2.1 in [Mv05]. Forbidding ET0L grammars were introduced and investigated in [Mv03b]. Simple semi-conditional ET0L grammars were introduced in [Š03] and further studied in [KM04]. Left versions of ET0L grammars were introduced and studied in [MZ13a]. Their nonterminal complexity was investigated in [Zem11]. Parallel rewriting over word monoids was studied in [Med92, KMv05].

Let us finally add that there also exist modern versions of (uniformly) limited ET0L grammars (see [W93, W94, W95, W96, FW98]) and ET0L grammars regulated by other mechanisms, such as mechanisms based upon control languages (see [Asv77, GR74, DF84] and Chapter 8 in [DP89]).

16.3.4 Modern Versions of Grammar Systems

Multigenerative grammar systems based upon leftmost derivations (see Sect. 12.2) were introduced in [LM06]. Their general versions were studied in [LM10b]. Controlled pure grammar systems were introduced and investigated in [MZ12b]. Moreover, [MVZ14] gives a preliminary solution to four open problems raised in [MZ12b].

16.3.5 Modern Versions of Automata

Self-regulated finite and pushdown automata were introduced in [MM07c]. Finite automata regulated by control languages were introduced in [MZ14]. For a study of finite automata controlled by Petri nets, see [JKZ07]. Regulated pushdown automata were introduced in [KM00]. Their special versions, referred to as one-turn linear-regulated pushdown automata, were studied in [KM01] (see also [KM05, Ryc09]). *Blackhole pushdown automata*, which are closely related to regulated pushdown automata, were introduced and investigated in [ECV10, CVMV11]. Deep pushdown automata were proposed and studied in [Med06]. For more results related to these automata, consult [KMS06c, KMS06b, LM10a, Sol12, QS09]. Finite automata over free groups were studied in [DM00, MS01]. Pushdown automata with pushdowns defined over free groups were introduced and studied in [BB06]. #-rewriting systems were recently introduced and studied in [Kři08, KMS07, KM07, Kři07, KMS06a].

16.3.6 Discontinuous Rewriting

Jumping grammars, discussed in Chap. 5, were introduced in [KM15]. Jumping finite automata from Chap. 8 were introduced in [MZ12a]. Several open problems stated there were solved in [Mad16, Vor15]. Other related models involving discontinuity include *nested word automata* [AM09], *bag automata* [DEM03], and *input-revolving finite automata* [BBHK09].

Bibliography

[ABB97] J. Autebert, J. Berstel, L. Boasson, (eds.), *Context-Free Languages and Pushdown Automata*, in *Handbook of Formal Languages*, chapter 3 (Springer, Berlin, 1997), pp. 111–174

[Abr65] S. Abraham, Some questions of language theory, in *Proceedings of the 1965 Conference on Computational Linguistics* (Association for Computational Linguistics, 1965), pp. 1–11

[AES06] B. Atcheson, S. Ewert, D. Shell, A note on the generative capacity of random context. South Afr. Comput. J. **36**, 95–98 (2006)

[AF04] M. Aronoff, K. Fudeman, *What Is Morphology (Fundamentals of Linguistics)* (Wiley-Blackwell, New Jersey, 2004)

[ALSU06] A.V. Aho, M.S. Lam, R. Sethi, J.D. Ullman, *Compilers: Principles, Techniques, and Tools*, 2nd edn. (Addison-Wesley, Boston, 2006)

[AM09] R. Alur, P. Madhusudan, Adding nesting structure to words. J. ACM **56**(3), 16:1–16:43 (2009)

[Asv77] P.R.J. Asveld, Controlled iteration grammars and full hyper-AFL's. Inf. Control **34**(3), 248–269 (1977)

[AU72] A.V. Aho, J.D. Ullman, *The Theory of Parsing, Translation and Compiling, Volume I: Parsing* (Prentice-Hall, New Jersey, 1972)

[Bak95] C.L. Baker, *English Syntax*, 2nd edn. (MIT Press, Cambridge, 1995)

[Bau03] L. Bauer, *Introducing Linguistic Morphology*, 2nd edn. (Georgetown University Press, Washington, DC, 2003)

[BB06] R. Bidlo, P. Blatný, Two-sided pushdown automata over free groups, in *Proceedings of the 12th Conference Student EEICT 2006 Volume 4* (Faculty of Electrical Engineering and Communication, BUT, 2006), pp. 352–355

[BBD⁺06] M. Barbaiani, C. Bibire, J. Dassow, A. Delaney, S. Fazekas, M. Ionescu, G. Liu, A. Lodhi, B. Nagy, The power of programmed grammars with graphs from various classes. J. Appl. Math. Comput. **22**(1–2), 21–38 (2006)

[BBHK09] S. Bensch, H. Bordihn, M. Holzer, M. Kutrib, On input-revolving deterministic and nondeterministic finite automata. Inf. Comput. **207**(11), 1140–1155 (2009)

[BBM05] R. Bidlo, P. Blatný, A. Meduna, Formal models over free groups, in *1st Doctoral Workshop on Mathematical and Engineering Methods in Computer Science* (Faculty of Information Technology BUT, Brno, CZ, 2005), pp. 193–199

[BBM07a] R. Bidlo, P. Blatný, A. Meduna, Automata with two-sided pushdowns defined over free groups generated by reduced alphabets. Kybernetika **2007**(1), 21–35 (2007)

© Springer International Publishing AG 2017
A. Meduna, O. Soukup, *Modern Language Models and Computation*,
DOI 10.1007/978-3-319-63100-4

[BBM07b]   R. Bidlo, P. Blatný, A. Meduna, Context-free and E0L derivations over free groups.
            Schedae Informaticae **2007**(16), 14–24 (2007)
[BCC10]   S. Buettcher, C.L.A. Clarke, G.V. Cormack, *Information Retrieval: Implementing
            and Evaluating Search Engines* (MIT Press, Cambridge, 2010)
[BCVHV05]   M. Beek, E. Csuhaj-Varjú, M. Holzer, G. Vaszil, On competence in CD grammar
            systems, in *Developments in Language Theory*, vol. 3340 of *Lecture Notes in
            Computer Science* (Springer, Berlin/Heidelberg, 2005), pp. 3–14
[BCVHV07]   M. Beek, E. Csuhaj-Varjú, M. Holzer, G. Vaszil, On competence in CD grammar
            systems with parallel rewriting. Int. J. Found. Comput. Sci. **18**(6), 1425–1439
            (2007)
[BF94]   H. Bordihn, H. Fernau, Accepting grammars with regulation. Int. J. Comput.
            Math. **53**(1), 1–18 (1994)
[BF95]   H. Bordihn, H. Fernau, Accepting grammars and systems: an overview, in *Proc.
            of Development in Language Theory Conf.*, Magdeburg (1995), pp. 199–208
[BH84]   R. Borchert, H. Honda, Control of development in the bifurcating branch system
            of tabebuia rosa : a computer simulation. Bot. Gaz. **145**, 184–195 (1984)
[BH00]   H. Bordihn, M. Holzer, Grammar systems with negated conditions in their
            cooperation protocols. J. Universal Comput. Sci. **6**(12), 1165–1184 (2000)
[BH06]   H. Bordihn, M. Holzer, Programmed grammars and their relation to the LBA
            problem. Acta Informatica **43**(4), 223–242 (2006)
[BH08]   H. Bordihn, M. Holzer, Random context in regulated rewriting versus cooperating
            distributed grammar systems, in *LATA'08: Proceedings of the 2nd International
            Conference on Language and Automata Theory and Applications* (Springer, 2008),
            pp. 125–136
[BMCY06]   B. Bouchon-Meunier, G. Coletti, R.R. Yager (eds.), *Modern Information Process-
            ing: From Theory to Applications* (Elsevier Science, New York, 2006)
[Bro89]   J.G. Brookshear, *Theory of Computation* (Benjamin Cummings Publishing, San
            Francisco, 1989)
[BYRN11]   R. Baeza-Yates, B. Ribeiro-Neto, *Modern Information Retrieval: The Concepts
            and Technology behind Search*, 2nd edn. (Addison-Wesley Professional, Boston,
            2011)
[ČHM12]   M. Čermák, P. Horáček, A. Meduna, Rule-restricted automaton-grammar trans-
            ducers: Power and linguistic applications. Math. Appl. **1**(1), 13–35 (2012)
[Cho56]   N. Chomsky, Three models for the description of language. IRE Trans. Inf. Theory
            2(3), 113–124 (1956)
[Cho59]   N. Chomsky, On certain formal properties of grammars. Inf. Control **2**, 137–167
            (1959)
[Cho02]   N. Chomsky, *Syntactic Structures* (Mouton, New York, 2002)
[Chu36a]   A. Church, A note on the entscheidungsproblem. J. Symb. Log. **1**(1), 40–41 (1936)
[Chu36b]   A. Church, An unsolvable problem of elementary number. Am. J. Math. **58**(2),
            345–363 (1936)
[CLR90]   T.H. Cormen, Ch.E. Leiserson, R.L. Rivest, *Introduction to Algorithms* (MIT
            Press, Cambridge, 1990)
[Cou77]   B. Courcelle, On jump deterministic pushdown automata. Math. Syst. Theory **11**,
            87–109 (1977)
[Cre73]   A.B. Cremers, Normal forms for context-sensitive grammars. Acta Informatica **3**,
            59–73 (1973)
[CV92]   E. Csuhaj-Varjú, On grammars with local and global context conditions. Int. J.
            Comput. Math. **47**, 17–27 (1992)
[CVDKP94]   E. Csuhaj-Varju, J. Dassow, J. Kelemen, G. Păun, *Grammar Systems: A Grammat-
            ical Approach to Distribution and Cooperation* (Gordon and Breach, Yverdon,
            1994)

[CVDV08]   E. Csuhaj-Varjú, J. Dassow, G. Vaszil, Some new modes of competence-based
            derivations in CD grammar systems, in *Developments in Language Theory*,
            vol. 5257 of *Lecture Notes in Computer Science* (Springer, Berlin/Heidelberg,
            2008), pp. 228–239
[CVMV09]   E. Csuhaj-Varjú, T. Masopust, G. Vaszil, Cooperating distributed grammar
            systems with permitting grammars as components. Rom. J. Inf. Sci. Technol.
            12(2), 175–189 (2009)
[CVMV11]   E. Csuhaj-Varjú, T. Masopust, G. Vaszil, Blackhole pushdown automata. Funda-
            menta Informaticae **112**(2–3), 137–156 (2011)
[CVPS95]   E. Csuhaj-Varjú, G. Păun, A. Salomaa, Conditional tabled eco-grammar systems.
            J. Universal Comput. Sci. **1**(5), 252–268 (1995)
[CVV10]   E. Csuhaj-Varjú, G. Vaszil, Scattered context grammars generate any recursively
            enumerable language with two nonterminals. Inf. Process. Lett. **110**, 902–907
            (2010)
[Das07]   J. Dassow, On cooperating distributed grammar systems with competence based
            start and stop conditions. Fundamenta Informaticae **76**, 293–304 (2007)
[DEM03]   M. Daley, M. Eramian, I. McQuillan, Bag automata and stochastic retrieval
            of biomolecules in solution, in *Implementation and Application of Automata,
            Eighth International Conference CIAA 2003, Santa Barbara, CA, July 16–18,
            2003*, vol. 2759 in *Lecture Notes in Computer Science*, ed. by O.H. Ibarra, Z. Dang
            (Springer, New York, 2003), pp. 239–250
[DF84]   J. Dassow, U. Fest, On regulated L systems. Rostock. Math. Kolloqu. **25**, 99–118
            (1984)
[DM00]   J. Dassow, V. Mitrana, Finite automata over free groups. Int. J. Algebra Comput.
            10(6), 725 (2000)
[DM12]   J. Dassow, T. Masopust, On restricted context-free grammars. J. Comput. Syst.
            Sci. **78**(1), 293–304 (2012)
[DP89]   J. Dassow, G. Păun, *Regulated Rewriting in Formal Language Theory* (Springer,
            Berlin, 1989)
[DPS93]   J. Dassow, G. Păun, A. Salomaa, Grammars based on patterns. Int. J. Found.
            Comput. Sci. **4**(1), 1–14 (1993)
[ECV10]   G. Vaszil, E. Csuhaj-Varjú, T. Masopust, Blackhole state-controlled regulated
            pushdown automata, in *Second Workshop on Non-Classical Models for Automata
            and Applications* (2010), pp. 45–56
[EKR85]   A. Ehrenfeucht, H.C.M. Kleijn, G. Rozenberg, Adding global forbidding context
            to context-free grammars. Theor. Comput. Sci. **37**, 337–360 (1985)
[EPR94]   A. Ehrenfeucht, P. Pasten, G. Rozenberg, Context-free text grammars. Acta
            Informatica **31**, 161–206 (1994)
[ER79]   A. Ehrenfeucht, G. Rozenberg, An observation on scattered grammars. Inf.
            Process. Lett. **9**(2), 84–85 (1979)
[EW03]   S. Ewert, A. Walt, The power and limitations of random context, in *Grammars
            and Automata for String Processing: From Mathematics and Computer Science to
            Biology* (Taylor and Francis, 2003), pp. 33–43
[EW13]   S. Ewert, A. Walt, Necessary conditions for subclasses of random context
            languages. Theor. Comput. Sci. **475**, 66–72 (2013)
[FB94]   R.W. Floyd, R. Beigel, *The Language of Machines: An Introduction to Com-
            putability and Formal Languages* (Computer Science Press, New York, 1994)
[Fer96]   H. Fernau, Scattered context grammars with regulation. Ann. Buchar. Univ. Math.
            Inf. Ser. **45**(1), 41–49 (1996)
[Fer03]   H. Fernau, Nonterminal complexity of programmed grammars. Theor. Comput.
            Sci. **296**(2), 225–251 (2003)

[FFOR07]   H. Fernau, R. Freund, M. Oswald, K. Reinhardt, Refining the nonterminal complexity of graph-controlled, programmed, and matrix grammars. J. Automata Lang. Comb. **12**(1–2), 117–138 (2007)

[FHF01]   H. Fernau, M. Holzer, R. Freund, Hybrid modes in cooperating distributed grammar systems: internal versus external hybridization. Theor. Comput. Sci. **259**(1–2), 405–426 (2001)

[FM03a]   H. Fernau, A. Meduna, On the degree of scattered context-sensitivity. Theor. Comput. Sci. **290**(3), 2121–2124 (2003)

[FM03b]   H. Fernau, A. Meduna, A simultaneous reduction of several measures of descriptional complexity in scattered context grammars. Inf. Process. Lett. **86**(5), 235–240 (2003)

[FR68]   P.C. Fischer, A.L. Rosenberg, Multitape one-way nonwriting automata. J. Comput. Syst. Sci. **2**, 38–101 (1968)

[Fri68]   I. Fris, Grammars with partial ordering of the rules. Inf. Control **12**, 415–425 (1968)

[FS97]   H. Fernau, F. Stephan, How powerful is unconditional transfer? — When UT meets AC, in *Developments in Language Theory* (1997), pp. 249–260

[FW98]   H. Fernau, D. Wätjen, Remarks on regulated limited ET0L systems and regulated context-free grammars. Theor. Comput. Sci. **194**(1–2), 35–55 (1998)

[Gef91]   V. Geffert, Normal forms for phrase-structure grammars. Theor. Inf. Appl. **25**(5), 473–496 (1991)

[GF04]   D.A. Grossman, O. Frieder, *Information Retrieval: Algorithms and Heuristics*, 2nd edn. (Springer, Berlin, 2004)

[GGH67]   S. Ginsburg, S.A. Greibach, M. Harrison, One-way stack automata. J. ACM **14**(2), 389–418 (1967)

[GH69]   S.A. Greibach, J.E. Hopcroft, Scattered context grammars. J. Comput. Syst. Sci. **3**(3), 233–247 (1969)

[Gin66]   S. Ginsburg, *The Mathematical Theory of Context-free Languages* (McGraw Hill, New York, 1966)

[Gin75]   S. Ginsburg, *Algebraic and Automata-Theoretic Properties of Formal Languages* (Elsevier Science, New York, 1975)

[GMM10]   F. Goldefus, T. Masopust, A. Meduna, Left-forbidding cooperating distributed grammar systems. Theor. Comput. Sci. **411**(40–42), 3661–3667 (2010)

[GN03]   W.J. Gilbert, W.K. Nicholson, *Modern Algebra with Applications (Pure and Applied Mathematics: A Wiley Series of Texts, Monographs and Tracts)*, 2nd edn. (Wiley-Blackwell, New Jersey, 2003)

[GR74]   S. Ginsburg, G. Rozenberg, T0L schemes and control sets. Inf. Control **27**, 109–125 (1974)

[Gre65]   S.A. Greibach, A new normal-form theorem for context-free phrase structure grammars. J. ACM **12**(1), 42–52 (1965)

[Gre69]   S.A. Greibach, Checking automata and one-way stack languages. J. Comput. Syst. Sci. **3**, 196–217 (1969)

[Gri71]   D. Gries, *Compiler Construction for Digital Computers* (Wiley, New York, 1971)

[GS68]   S. Ginsburg, E. Spanier, Finite-turn pushdown automata. SIAM J. Control **4**, 429–453 (1968)

[GW89]   J. Gonczarowski, M.K. Warmuth, Scattered versus context-sensitive rewriting. Acta Informatica **27**, 81–95 (1989)

[Har78]   M.A. Harrison, *Introduction to Formal Language Theory* (Addison-Wesley, Boston, 1978)

[HM88]   G. Horváth, A. Meduna, On state grammars. Acta Cybernetica **1988**(8), 237–245 (1988)

[HM11]   P. Horáček, A. Meduna, Regulated rewriting in natural language translation, in *7th Doctoral Workshop on Mathematical and Engineering Methods in Computer Science* (Brno University of Technology, Brno, CZ, 2011), pp. 35–42

[HM12]   P. Horáček, A. Meduna, Synchronous versions of regulated grammars: Generative power and linguistic applications. Theor. Appl. Inf. **24**(3), 175–190 (2012)

[Hor12] P. Horáček, On generative power of synchronous grammars with linked rules, in *Proceedings of the 18th Conference STUDENT EEICT 2012 Volume 3* (Brno University of Technology, Brno, CZ, 2012), pp. 376–380

[HP02]   R. Huddleston, G. Pullum, *The Cambridge Grammar of the English Language* (Cambridge University Press, New York, 2002)

[HP05]   R. Huddleston, G. Pullum, *A Student's Introduction to English Grammar* (Cambridge University Press, New York, 2005)

[HU69]   J.E. Hopcroft, J.D. Ullman, *Formal Languages and Their Relation to Automata* (Addison-Wesley, Boston, 1969)

[Iba70] O.H. Ibarra, Simple matrix languages. Inf. Control **17**, 359–394 (1970)

[Ito03] M. Ito, *Algebraic Theory of Automata and Languages*, 2nd edn. (World Scientific Publishing Company, Singapore, 2003)

[JKZ07] M. Jantzen, M. Kudlek, G. Zetzsche, Finite automata controlled by Petri nets, in *Proceedings of the 14th Workshop; Algorithmen und Werkzeuge für Petrinetze*, ed. by S. Philippi, A. Pini. Technical Report Nr. 25/2007 (University Koblenz-Landau, 2007), pp. 57–62

[Kas70] T. Kasai, A hierarchy between context-free and context-sensitive languages. J. Comput. Syst. Sci. **4**, 492–508 (1970)

[Kel84] J. Kelemen, Conditional grammars: Motivations, definition, and some properties, in *Proceedings on Automata, Languages and Mathematical Systems* (K. Marx University of Economics, Budapest, 1984), pp. 110–123

[Kel89] J. Kelemen, Measuring cognitive resources use (a grammatical approach). Comput. Artif. Intell. **8**(1), 29–42 (1989)

[Kel95] D. Kelley, *Automata and Formal Languages* (Prentice-Hall, New Jersey, 1995)

[KM70]   W. Kuich, H.A. Maurer, Tuple languages, in *International Computing Symposium, Bonn* (1970), pp. 881–891

[KM00]   D. Kolář, A. Meduna, Regulated pushdown automata. Acta Cybernetica **2000**(4), 653–664 (2000)

[KM01]   D. Kolář, A. Meduna, One-turn regulated pushdown automata and their reduction. Fundamenta Informaticae **2001**(21), 1001–1007 (2001)

[KM04]   T. Kopeček, A. Meduna, Simple-semi-conditional versions of matrix grammars with a reduced regulating mechanism. Comput. Inf. **2004**(23), 287–302 (2004)

[KM05]   D. Kolář, A. Meduna, Regulated automata: from theory towards applications, in *Proceeding of 8th International Conference on Information Systems Implementation and Modelling (ISIM'05)* (2005), pp. 33–48

[KM07]   Z. Křivka, A. Meduna, Generalized #-rewriting systems of finite index, in *Information Systems and Formal Models (Proceedings of 2nd International Workshop on Formal Models (WFM'07))* (Silesian University, Opava, CZ, 2007), pp. 197–204

[KM11]   Z. Křivka, T. Masopust, Cooperating distributed grammar systems with random context grammars as components. Acta Cybernetica **20**(2), 269–283 (2011)

[KM15]   Z. Křivka, A. Meduna, Jumping grammars. Int. J. Found. Comput. Sci. **26**(6), 709–731 (2015)

[KMS06a] Z. Křivka, A. Meduna, R. Schönecker, Generation of languages by rewriting systems that resemble automata. Int. J. Found. Comput. Sci. **17**(5), 1223–1229 (2006)

[KMS06b] Z. Křivka, A. Meduna, R. Schönecker, General top-down parsers based on deep pushdown expansions, in *Proceedings of 1st International Workshop on Formal Models (WFM'06)*, Ostrava, CZ (2006), pp. 11–18

[KMS06c] Z. Křivka, A. Meduna, R. Schönecker, Reducing deep pushdown automata and infinite hierarchy, in *2nd Doctoral Workshop on Mathematical and Engineering Methods in Computer Science* (Brno University of Technology, Brno, CZ, 2006), pp. 214–221

[KMS07]   Z. Křivka, A. Meduna, J. Smrček, n-right-linear #-rewriting systems, in *Third Doctoral Workshop on Mathematical and Engineering Methods in Computer Science (MEMICS 2007)*, Znojmo, CZ (2007), pp. 105–112

[KMv05]   T. Kopeček, A. Meduna, M. Švec, Simulation of scattered context grammars and phrase-structured grammars by symbiotic E0L grammars, in *Proceeding of 8th International Conference on Information Systems, Implementation and Modelling (ISIM'05)*, Brno, CZ (2005), pp. 59–66

[KMVP00]   M. Kudlek, C. Martín-Vide, Gh. Păun, Toward FMT (formal macroset theory), in *Pre-proceedings of the Workshop on Multiset Processing*, Curtea de Arges (2000), pp. 149–158

[KR81]   H.C.M. Kleijn, G. Rozenberg, Context-free-like restrictions on selective rewriting. Theor. Comput. Sci. **16**, 237–239 (1981)

[KR83]   H.C.M. Kleijn, G. Rozenberg, On the generative power of regular pattern grammars. Acta Informatica **20**, 391–411 (1983)

[Krá69]   J. Král, On multiple grammars. Kybernetika **1**, 60–85 (1969)

[Krá73]   J. Král, A note on grammars with regular restrictions. Kybernetika **9**(3), 159–161 (1973)

[Kři07]   Z. Křivka, Rewriting systems with restricted configurations, PhD thesis, Faculty of Information Technology, Brno University of Technology, 2007

[Kři08]   Z. Křivka, Rewriting systems with restricted configurations, PhD thesis, Faculty of Information Technology, BUT, 2008

[Kur64]   S.Y. Kuroda, Classes of languages and linear-bounded automata. Inf. Control **7**(2), 207–223 (1964)

[LM06]   R. Lukáš, A. Meduna, Multigenerative grammar systems. Schedae Informaticae **2006**(15), 175–188 (2006)

[LM10a]   P. Leupold, A. Meduna, Finitely expandable deep PDAs, in *Automata, Formal Languages and Algebraic Systems: Proceedings of AFLAS 2008* (Hong Kong University of Scinece and Technology, 2010), pp. 113–123

[LM10b]   R. Lukáš, A. Meduna, Multigenerative grammar systems and matrix grammars. Kybernetika **46**(1), 68–82 (2010)

[Lom72a]   M.V. Lomkovskaya, Conditional grammars and intermediate classes of languages (in Russian). Sov. Math. Doklady **207**, 781–784 (1972)

[Lom72b]   M.V. Lomkovskaya, On c-conditional and other commutative grammars (in Russian). Nauchno-Tekhnicheskaya Informatsiya **2**(2), 28–31 (1972)

[Lom72c]   M.V. Lomkovskaya, On some properties of c-conditional grammars (in Russian). Nauchno-Tekhnicheskaya Informatsiya **2**(1), 16–21 (1972)

[LP81]   H.R. Lewis, C.H. Papadimitriou, *Elements of the Theory of Computation* (Prentice-Hall, New Jersey, 1981)

[LRS76]   P.M. Lewis, D.J. Rosenkrantz, R.E. Stearns, *Compiler Design Theory* (Addison-Wesley, Boston, 1976)

[Mad16]   G. Madejski, Jumping and pumping lemmas and their applications, in *Proceedings of NCMA 2016 - 8th Workshop on Non-Classical Models of Automata and Applications* (2016), pp. 25–34

[MAK88]   R.N. Moll, M.A. Arbib, A.J. Kfoury, *An Introduction to Formal Language Theory* (Springer, New York, 1988)

[Mar60]   A.A. Markov, The theory of algorithms. Am. Math. Soc. Trans. **2**(15), 1–14 (1960)

[Mar02]   J.C. Martin, *Introduction to Languages and the Theory of Computation*, 3rd edn. (McGraw-Hill, New York, 2002)

[Mas06]   T. Masopust, An improvement of the descriptional complexity of grammars regulated by context conditions, in *2nd Doctoral Workshop on Mathematical and Engineering Methods in Computer Science* (Faculty of Information Technology BUT, Brno, CZ, 2006), pp. 105–112

[Mas07a]   T. Masopust, Formal models: regulation and reduction, PhD thesis, Faculty of Information Technology, Brno University of Technology, 2007

[Mas07b]   T. Masopust, Scattered context grammars can generate the powers of 2, in
             Proceedings of the 13th Conference and Competition EEICT 2007, vol. 4 (Faculty
             of Electrical Engineering and Communication, Brno University of Technology,
             2007), pp. 401–404

[Mas09a]   T. Masopust, On the descriptional complexity of scattered context grammars.
             Theor. Comput. Sci. **410**(1), 108–112 (2009)

[Mas09b]   T. Masopust, On the terminating derivation mode in cooperating distributed
             grammar systems with forbidding components. Int. J. Found. Comput. Sci. **20**(2),
             331–340 (2009)

[Mas10a]   T. Masopust, Bounded number of parallel productions in scattered context
             grammars with three nonterminals. Fundamenta Informaticae **99**(4), 473–480
             (2010)

[Mas10b]   T. Masopust, Simple restriction in context-free rewriting. J. Comput. Syst. Sci.
             76(8), 837–846 (2010)

[Mau73]   H.A. Maurer, Simple matrix languages with a leftmost restriction. Inf. Control **23**,
             128–139 (1973)

[May72]   O. Mayer, Some restrictive devices for context-free grammars. Inf. Control **20**,
             69–92 (1972)

[MCV93]   A. Meduna, E. Csuhaj-Varjú, Grammars with context conditions. EATCS Bull.
             32, 112–124 (1993)

[Med90a]   A. Meduna, Context free derivations on word monoids. Acta Informatica **27**,
             781–786 (1990)

[Med90b]   A. Meduna, Generalized forbidding grammars. Int. J. Comput. Math. **36**(1–2),
             31–38 (1990)

[Med91a]   A. Meduna, Global context conditional grammars. J. Automata Lang. Comb.
             1991(27), 159–165 (1991)

[Med91b]   A. Meduna, Scattered rewriting in the formal language theory, in *Missourian
             Annual Conference on Computing*, Columbia, US, (1991), pp. 26–36

[Med92]   A. Meduna, Symbiotic E0L systems. Acta Cybernetica **10**, 165–172 (1992)

[Med93]   A. Meduna, Canonical scattered rewriting. Int. J. Comput. Math. **51**, 122–129
             (1993)

[Med95a]   A. Meduna, Symbiotic e0l systems. artificial life: Gramatical models. Bucharest,
             pp. 122–129, 1995

[Med95b]   A. Meduna, Syntactic complexity of scattered context grammars. Acta Informatica
             32, 285–298 (1995)

[Med96]   A. Meduna, Syntactic complexity of context-free grammars over word monoids.
             Acta Informatica **1996**(33), 457–462 (1996)

[Med97]   A. Meduna, Four-nonterminal scattered context grammars characterize the family
             of recursively enumerable languages. Int. J. Comput. Math. **63**, 67–83 (1997)

[Med98a]   A. Meduna, Economical transformations of phrase-structure grammars to scat-
             tered context grammars. Acta Cybernetica **13**, 225–242 (1998)

[Med98b]   A. Meduna, Uniform rewriting based on permutations. Int. J. Comput. Math.
             69(1–2), 57–74 (1998)

[Med00a]   A. Meduna, *Automata and Languages: Theory and Applications* (Springer,
             London, 2000)

[Med00b]   A. Meduna, Generative power of three-nonterminal scattered context grammars.
             Theor. Comput. Sci. **2000**(246), 279–284 (2000)

[Med00c]   A. Meduna, Terminating left-hand sides of scattered context grammars. Theor.
             Comput. Sci. **2000**(237), 424–427 (2000)

[Med01]   A. Meduna, Uniform generation of languages by scattered context grammars.
             Fundamenta Informaticae **44**, 231–235 (2001)

[Med02]   A. Meduna, Descriptional complexity of scattered rewriting and multirewriting:
             An overview. J. Automata Lang. Comb. **2002**(7), 571–577 (2002)

[Med03a]   A. Meduna,   Coincidental extension of scattered context languages.   Acta
            Informatica **39**(5), 307–314 (2003)
[Med03b]   A. Meduna, Erratum: Coincidental extension of scattered context languages. Acta
            Informatica **39**(9), 699 (2003)
[Med03c]   A. Meduna, Simultaneously one-turn two-pushdown automata. Int. J. Comput.
            Math. **2003**(80), 679–687 (2003)
[Med04]   A. Meduna, Two-way metalinear PC grammar systems and their descriptional
            complexity. Acta Cybernetica **2004**(16), 385–397 (2004)
[Med06]   A. Meduna, Deep pushdown automata. Acta Informatica **2006**(98), 114–124
            (2006)
[Med14]   A. Meduna, *Formal Languages and Computation: Models and Their Applications*
            (Taylor & Francis, New York, 2014)
[MG94]   A. Meduna, A. Gopalaratnam, On semi-conditional grammars with productions
            having either forbidding or permitting conditions. Acta Cybernetica **11**, 307–323
            (1994)
[MHHO05]   E. Moriya, D. Hofbauer, M. Huber, F. Otto, On state-alternating context-free
            grammars. Theor. Comput. Sci. **337**, 183–216 (2005)
[MK02]   A. Meduna, D. Kolář, Homogenous grammars with a reduced number of non-
            context-free productions. Inf. Process. Lett. **81**, 253–257 (2002)
[MM07a]   T. Masopust, A. Meduna, Descriptional complexity of grammars regulated by
            context conditions, in *LATA '07 Pre-proceedings*. *Reports of the Research Group
            on Mathematical Linguistics 35/07, Universitat Rovira i Virgili* (2007), pp. 403–
            411
[MM07b]   T. Masopust, A. Meduna, Descriptional complexity of semi-conditional gram-
            mars. Inf. Process. Lett. **104**(1), 29–31 (2007)
[MM07c]   T. Masopust, A. Meduna, Self-regulating finite automata. Acta Cybernetica **18**(1),
            135–153 (2007)
[MM08]   T. Masopust, A. Meduna, On descriptional complexity of partially parallel
            grammars. Fundamenta Informaticae **87**(3), 407–415 (2008)
[MM09]   T. Masopust, A. Meduna, On context-free rewriting with a simple restriction
            and its computational completeness. RAIRO Theor. Inf. Appl. **43**(2), 365–378
            (2009)
[MMv08]   T. Masopust, A. Meduna, J. Šimáček, Two power-decreasing derivation restric-
            tions in generalized scattered context grammars. Acta Cybernetica **18**(4), 783–793
            (2008)
[MR71]   D. Milgram, A. Rosenfeld, A note on scattered context grammars. Inf. Process.
            Lett. **1**, 47–50 (1971)
[MRS08]   C.D. Manning, P. Raghavan, H. Schütze, *Introduction to Information Retrieval*
            (Cambridge University Press, New York, 2008)
[MS01]   V. Mitrana, R. Stiebe, Extended finite automata over groups. Discrete Appl. Math.
            108(3), 287–300 (2001)
[MT05]   A. Meduna, J. Techet, Generation of sentences with their parses: the case of
            propagating scattered context grammars. Acta Cybernetica **17**, 11–20 (2005)
[MT07a]   A. Meduna, J. Techet, Canonical scattered context generators of sentences with
            their parses. Theor. Comput. Sci. **2007**(389), 73–81 (2007)
[MT07b]   A. Meduna, J. Techet, Maximal and minimal scattered context rewriting, in *FCT
            2007 Proceedings*, Budapest (2007), pp. 412–423
[MT07c]   A. Meduna, J. Techet, Reduction of scattered context generators of sentences
            preceded by their leftmost parses, in *DCFS 2007 Proceedings*, High Tatras, SK
            (2007), pp. 178–185
[MT08a]   T. Masopust, J. Techet, Leftmost derivations of propagating scattered context
            grammars: A new proof. Discrete Math. Theor. Comput. Sci. **10**(2), 39–46 (2008)
[MT08b]   A. Meduna, J. Techet, Scattered context grammars that erase nonterminals in a
            generalized *k*-limited way. Acta Informatica **45**(7), 593–608 (2008)

[MT09]   A. Meduna, J. Techet, An infinite hierarchy of language families generated by scattered context grammars with n-limited derivations. Theor. Comput. Sci. **410**(21), 1961–1969 (2009)

[MT10]   A. Meduna, J. Techet, *Scattered Context Grammars and Their Applications* (WIT Press, Southampton, 2010)

[Mv02]   A. Meduna, M. Švec, Reduction of simple semi-conditional grammars with respect to the number of conditional productions. Acta Cybernetica **15**, 353–360 (2002)

[Mv03a]   A. Meduna, M. Švec, Descriptional complexity of generalized forbidding grammars. Int. J. Comput. Math. **80**(1), 11–17 (2003)

[Mv03b]   A. Meduna, M. Švec, Forbidding ET0L grammars. Theor. Comput. Sci. **2003**(306), 449–469 (2003)

[Mv05]   A. Meduna, M. Švec, *Grammars with Context Conditions and Their Applications* (Wiley, New Jersey, 2005)

[Mv11]   A. Meduna, S. Židek, Scattered context grammars generating sentences followed by derivation trees. Theor. Appl. Inf. **2011**(2), 97–106 (2011)

[MVMP04] C. Martín-Vide, V. Mitrana, G. Păun (eds.), *Formal Languages and Applications*, chapter 13 (Springer, Berlin, 2004), pp. 249–274

[MVZ11]   A. Meduna, L. Vrábel, P. Zemek, On nondeterminism in programmed grammars, in *13th International Conference on Automata and Formal Languages* (Computer and Automation Research Institute, Hungarian Academy of Sciences, Debrecen, HU, 2011), pp. 316–328

[MVZ14]   A. Meduna, L. Vrábel, P. Zemek, Solutions to four open problems concerning controlled pure grammar systems. Int. J. Comput. Math. **91**(6), 1156–1169 (2014)

[MZ10]   A. Meduna, P. Zemek, *Regulated Grammars and Their Transformations* (Faculty of Information Technology, Brno University of Technology, Brno, CZ, 2010)

[MZ11]   A. Meduna, P. Zemek, Workspace theorems for regular-controlled grammars. Theor. Comput. Sci. **412**(35), 4604–4612 (2011)

[MZ12a]   A. Meduna, P. Zemek, Jumping finite automata. Int. J. Found. Comput. Sci. **23**(7), 1555–1578 (2012)

[MZ12b]   A. Meduna, P. Zemek, One-sided random context grammars with leftmost derivations, in *LNCS Festschrift Series: Languages Alive*, vol. 7300 (Springer, 2012), pp. 160–173

[MZ13a]   A. Meduna, P. Zemek, Left random context ET0L grammars. Fundamenta Informaticae **123**(3), 289–304 (2013)

[MZ13b]   A. Meduna, P. Zemek, On the generation of sentences with their parses by propagating regular-controlled grammars. Theor. Comput. Sci. **477**(1), 67–75 (2013)

[MZ14]   A. Meduna, P. Zemek, *Regulated Grammars and Automata* (Springer US, New York, 2014)

[Nav70]   E. Navrátil, Context-free grammars with regular conditions. Kybernetika **6**(2), 118–125 (1970)

[NS05]   N. Nisan, S. Schocken, *The Elements of Computing Systems: Building a Modern Computer from First Principles* (MIT Press, Cambridge, 2005)

[Oku09]   F. Okubo, A note on the descriptional complexity of semi-conditional grammars. Inf. Process. Lett. **110**(1), 36–40 (2009)

[Pag81]   F.G. Pagen, *Formal Specifications of Programming Language: A Panoramic Primer* (Prentice-Hall, New Jersey, 1981)

[Pen74]   M. Penttonen, One-sided and two-sided context in formal grammars. Inf. Control **25**(4), 371–392 (1974)

[Pen75]   M. Penttonen, ET0L-grammars and N-grammars. Inf. Process. Lett. **4**(1), 11–13 (1975)

[PHHM96a]   P. Prusinkiewicz, M. Hammel, J. Hanan, R. Mech, *Visual Models of Plant Development*, in *Handbook of Formal Languages*, ed. by G. Rozenberg, A. Salomaa (Springer, New York, 1996)

[PHHM96b]   P. Prusinkiewicz, M. Hammel, J. Hanan, R. Měch, L-systems: From the theory to visual models of plants, in *Proceedings of the 2nd CSIRO Symposium on Computational Challenges in Life Sciences* (CSIRO Publishing, Collingwood, Victoria, Australia, 1996)

[PL90a]   P. Prusinkiewicz, A. Lindenmayer, *The Algorithmic Beauty of Plants* (Springer, New York, 1990)

[PL90b]   P. Prusinkiewicz, A. Lindenmayer, *The Algorithmic Beauty of Plants* (Springer, New York, 1990)

[Pos43]   E. Post, Formal reductions of the general combinatorial decision problem. Am. J. Math. **65**(2), 197–215 (1943)

[Pǎ79]   G. Pǎun, On the generative capacity of conditional grammars. Inf. Control **43**, 178–186 (1979)

[Pǎ82]   G. Pǎun, On simple matrix languages versus scattered context languages. Inf. Théor. Appl. **16**(3), 245–253 (1982)

[Pǎ85]   G. Pǎun, A variant of random context grammars: semi-conditional grammars. Theor. Comput. Sci. **41**(1), 1–17 (1985)

[QS09]   A.A. Quesada, I.A. Stewart, On the power of deep pushdown stacks. Acta Informatica **46**(7), 509–531 (2009)

[RN02]   S. Russell, P. Norvig, *Artificial Intelligence: A Modern Approach*, 2nd edn. (Prentice-Hall, New Jersey, 2002)

[Rog87]   H. Rogers, *Theory of Recursive Functions and Effective Computability* (MIT Press, Cambridge, 1987)

[Ros69]   D.J. Rosenkrantz, Programmed grammars and classes of formal languages. J. ACM **16**(1), 107–131 (1969)

[Roz77]   G. Rozenberg, Selective substitution grammars (towards a framework for rewriting systems). Part 1: Definitions and examples. Elektronische Informationsverarbeitung und Kybernetik **13**(9), 455–463 (1977)

[RS78]   G. Rozenberg, S.H. Solms, Priorities on context conditions in rewriting systems. Inf. Sci. **14**(1), 15–50 (1978)

[RS80]   G. Rozenberg, A. Salomaa, *Mathematical Theory of L Systems* (Academic Press, Orlando, 1980)

[RS86]   G. Rozenberg, A. Salomaa, *The Book of L* (Springer, New York, 1986)

[RS97a]   G. Rozenberg, A. Salomaa (eds.), *Handbook of Formal Languages, Vol. 1: Word, Language, Grammar* (Springer, New York, 1997)

[RS97b]   G. Rozenberg, A. Salomaa (eds.), *Handbook of Formal Languages, Vol. 2: Linear Modeling: Background and Application* (Springer, New York, 1997)

[RS97c]   G. Rozenberg, A. Salomaa (eds.), *Handbook of Formal Languages, Vol. 3: Beyond Words* (Springer, New York, 1997)

[Rus09]   P.J. Russel, *iGenetics: A Molecular Approach*, 3rd edn. (Benjamin Cummings Publishing, San Francisco, 2009)

[RW73]   R.D. Rosebrugh, D. Wood, A characterization theorem for *n*-parallel right linear languages. J. Comput. Syst. Sci. **7**, 579–582 (1973)

[RW75]   R.D. Rosebrugh, D. Wood, Restricted parallelism and right linear grammars. Utilitas Mathematica **7**, 151–186 (1975)

[Ryc09]   L. Rychnovský, Regulated pushdown automata revisited, in *Proceedings of the 15th Conference STUDENT EEICT 2009* (Brno University of Technology, Brno, CZ, 2009), pp. 440–444

[Sal73]   A. Salomaa, *Formal Languages* (Academic Press, London, 1973)

[Sar01]   P. Sarkar, Pushdown automaton with the ability to flip its stack, in *TR01-081, Electronic Colloquium on Computational Complexity (ECCC)*, 2001

[Sip06]   M. Sipser, *Introduction to the Theory of Computation*, 2nd edn. (PWS Publishing Company, Boston, 2006)

[Sir69]   R. Siromoney, Studies in the mathematical theory of grammars and its applications, PhD thesis, University of Madras, Madras, India, 1969

[Sir71]   R. Siromoney, Finite-turn checking automata. J. Comput. Syst. Sci. **5**, 549–559 (1971)

[Sol76]   S.H. Solms, Some notes on ET0L languages. Int. J. Comput. Math. **5**, 285–296 (1976)

[Sol12]   P. Solár, Parallel deep pushdown automata, in *Proceedings of the 18th Conference STUDENT EEICT 2012*, vol. 3 (Brno University of Technology, Brno, CZ, 2012), pp. 410–414

[Sos03]   P. Sosík, The power of catalysts and priorities in membrane systems. Grammars **6**(1), 13–24 (2003)

[SR10]   T. Strachan, A. Read, *Human Molecular Genetics*, 4th edn. (Garland Science, New York, 2010)

[SSS87]   S. Sippu, E. Soisalon-Soininen, *Parsing Theory* (Springer, New York, 1987)

[Tec07]   J. Techet, A note on scattered context grammars with non-context-free components, in *3rd Doctoral Workshop on Mathematical and Engineering Methods in Computer Science* (Brno University of Technology, Brno, CZ, 2007), pp. 225–232

[Tec08]   J. Techet, *Scattered context in formal languages*, PhD thesis, Faculty of Information Technology, Brno University of Technology, 2008

[Tru98]   J. Truss, *Discrete Mathematics for Computer Scientists (International Computer Science Series)*, 2nd edn. (Addison-Wesley, Boston, 1998)

[Tur36]   A.M. Turing, On computable numbers, with an application to the entscheidungsproblem. Proc. Lond. Math. Soc. **42**(2), 230–265 (1936)

[Urb83]   F.J. Urbanek, A note on conditional grammars. Revue Roumaine de Mathématiques Pures at Appliquées **28**, 341–342 (1983)

[Vas03]   G. Vaszil, On the number of conditional rules in simple semi-conditional grammars, in *Descriptional Complexity of Formal Systems* (MTA SZTAKI, Budapest, HU, 2003), pp. 210–220

[Vas05]   G. Vaszil, On the descriptional complexity of some rewriting mechanisms regulated by context conditions. Theor. Comput. Sci. **330**(2), 361–373 (2005)

[vdW70]   A.P.J. van der Walt, Random context grammars, in *Proceedings of Symposium on Formal Languages* (1970), pp. 163–165

[vdWE00]   A.P.J. van der Walt, S. Ewert, A shrinking lemma for random forbidding context languages. Theor. Comput. Sci. **237**(1–2), 149–158 (2000)

[vdWE02]   A.P.J. van der Walt, S. Ewert, A pumping lemma for random permitting context languages. Theor. Comput. Sci. **270**(1–2), 959–967 (2002)

[Vir73]   V. Virkkunen, On scattered context grammars. Acta Universitatis Ouluensis **20**(6), 75–82 (1973)

[Vor15]   V. Vorel, On basic properties of jumping finite automata. CoRR, abs/1511.08396, 2015

[Vrá11]   L. Vrábel, A new normal form for programmed grammars, in *Proceedings of the 17th Conference STUDENT EEICT 2011* (Brno University of Technology, Brno, CZ, 2011)

[Vrá12]   L. Vrábel, A new normal form for programmed grammars with appearance checking, in *Proceedings of the 18th Conference STUDENT EEICT 2012* (Brno University of Technology, Brno, CZ, 2012), pp. 420–425

[Š03]   M. Švec, Simple semi-conditional ET0L grammars, in *Proceedings of the International Conference and Competition Student EEICT 2003* (Brno University of Technology, Brno, CZ, 2003), pp. 283–287

[vzGG03]   J. von zur Gathen, J. Gerhard, *Modern Computer Algebra*, 2nd edn. (Cambridge University Press, New York, 2003)

[Wä93]   D. Wätjen, Regulation of k-limited ET0L systems. Int. J. Comput. Math. **47**,
            29–41 (1993)

[Wä94]   D. Wätjen, Regulation of uniformly k-limited T0L systems. J. Inf. Process.
            Cybern. **30**(3), 169–187 (1994)

[Wä95]   D. Wätjen, On regularly controlled k-limited T0L systems. Int. J. Comput. Math.
            55(1–2), 57–66 (1995)

[Wä96]   D. Wätjen, Regulations of uniformly k-limited ET0L systems and their relations to
            controlled context-free grammars. J. Automata Lang. Comb. **1**(1), 55–74 (1996)

[WBB$^+$07] J.D. Watson, T.A. Baker, S.P. Bell, A. Gann, M. Levine, R. Losick, *Molecular
            Biology of the Gene*, 6th edn. (Benjamin Cummings Publishing, San Francisco,
            2007)

[Woo73]   D. Wood, Properties of n-parallel finite state languages. Technical report,
            McMaster University, 1973

[Woo75]   D. Wood, m-parallel n-right linear simple matrix languages. Utilitas Mathematica
            8, 3–28 (1975)

[Woo87]   D. Wood, *Theory of Computation: A Primer* (Addison-Wesley, Boston, 1987)

[Zem10]   P. Zemek, k-limited erasing performed by regular-controlled context-free gram-
            mars, in *Proceedings of the 16th Conference STUDENT EEICT 2011*, vol. 3 (Brno
            University of Technology, Brno, CZ, 2010), pp. 42–44

[Zem11]   P. Zemek, On the nonterminal complexity of left random context E0L grammars,
            in *Proceedings of the 17th Conference STUDENT EEICT 2011*, vol. 3 (Brno
            University of Technology, Brno, CZ, 2011), pp. 510–514

[Zet09]   G. Zetzsche, Erasing in Petri net languages and matrix grammars, in *DLT '09:
            Proceedings of the 13th International Conference on Developments in Language
            Theory* (Springer, 2009), pp. 490–501

[Zet10]   G. Zetzsche, On erasing productions in random context grammars, in *ICALP'10:
            Proceedings of the 37th International Colloquium on Automata, Languages and
            Programming* (Springer, 2010), pp. 175–186

[Zet11a]   G. Zetzsche, A sufficient condition for erasing productions to be avoidable,
            in *DLT'11: Developments in Language Theory*, vol. 6795 of *Lecture Notes in
            Computer Science* (Springer, Berlin/Heidelberg, 2011), pp. 452–463

[Zet11b]   G. Zetzsche, Toward understanding the generative capacity of erasing rules in
            matrix grammars. Int. J. Comput. Math. **22**(2), 411–426 (2011)

Index to Models and Language Families They Define

Used Abbreviations

prop.   Propagating

a.c.   Appearance checking

0L   Zero-sided Lindenmayer

Family	Page	Formal model
FIN	17	–
RE	26	Phrase-structure grammar
MON	27	Monotone phrase-structure grammar
CS	27	Context-sensitive grammar
CF	27	Context-free grammar
LIN	28	Linear grammar
$CF^{-\varepsilon}$	28	Prop. context-free grammar
REG	28	Regular grammar
RLIN	28	Right-linear grammar
CF_{fin}	29	Context-free grammar of finite index
0L	47	0L grammar
E0L	47	Extended 0L grammar
EP0L	47	Extended prop. 0L grammar
ET0L	47	Extended tabled 0L grammar
EPT0L	47	Extended prop. tabled 0L grammar
$EPDA_f$	52	Extended pushdown automaton accepting by final state
$EPDA_e$	52	Extended pushdown automaton accepting by empty pushdown

(continued)

© Springer International Publishing AG 2017

A. Meduna, O. Soukup, *Modern Language Models and Computation*,

DOI 10.1007/978-3-319-63100-4

Family	Page	Formal model
EPDA$_{ef}$	52	Extended pushdown automaton accepting by final state and empty pushdown
DPDA	53	Deterministic pushdown automaton
PDA$_f$	53	Pushdown automata accepting by final state
PDA$_e$	53	Pushdown automata accepting by empty pushdown
PDA$_{ef}$	53	Pushdown automata accepting by final state and empty pushdown
PS$[.j]$	68	Phrase-structure grammar with j-l-uniform rewriting
PS$[j.]$	68	Phrase-structure grammar with j-r-uniform rewriting
CG	77	Conditional grammar
CG$^{-\varepsilon}$	77	Prop. conditional grammar
RC	84	Random context grammar
RC$^{-\varepsilon}$	84	Prop. random context grammar
For	84	Forbidding grammar
For$^{-\varepsilon}$	84	Prop. forbidding grammar
Per	84	Permitting grammar
Per$^{-\varepsilon}$	84	Prop. permitting grammar
GF	89	Generalized forbidding grammar
GF$^{-\varepsilon}$	89	Prop. generalized forbidding grammar
SSC	107	Simple semi-conditional grammar
SSC$^{-\varepsilon}$	107	Prop. simple semi-conditional grammar
ST	140	State grammar
ST$_n$	140	n-limited state grammar
rC	145	Regular-controlled grammar
rC$^{-\varepsilon}$	145	Prop. regular-controlled grammar
rC$_{ac}$	145	Regular-controlled grammar with a.c.
rC$_{ac}^{-\varepsilon}$	145	Prop. regular-controlled grammar with a.c.
M	147	Matrix grammar
M$^{-\varepsilon}$	147	Prop. matrix grammar
M$_{ac}$	147	Matrix grammar with a.c.
M$_{ac}^{-\varepsilon}$	147	Prop. matrix grammar with a.c.
$_n$**EM**$_m$	150	Even matrix grammar of mode m and degree n
EM$_m$	150	Even matrix grammar of mode m
P	163	Programmed grammar
P$^{-\varepsilon}$	163	Prop. programmed grammar
P$_{ac}$	163	Programmed grammar with a.c.
P$_{ac}^{-\varepsilon}$	163	Prop. programmed grammar with a.c.
$_k$**P**	163	Programmed grammar of index k
$_k$**P**$^{-\varepsilon}$	163	Prop. programmed grammar of index k
$_k$**P**$_{ac}$	163	Programmed grammar of index k with a.c.

(continued)

Family	Page	Formal model
$_k\mathbf{P}_{ac}^{-\varepsilon}$	163	Prop. programmed grammar of index k with a.c.
SC	170	Scattered context grammar
$\mathbf{SC}^{-\varepsilon}$	170	Prop. scattered context grammar
C - E0L	196	Context-conditional extended 0L grammar
C - EP0L	196	Context-conditional extended prop. 0L grammar
C - ET0L	196	Context-conditional extended tabled 0L grammar
C - EPT0L	196	Context-conditional extended prop. tabled 0L grammar
F - E0L	203	Forbidding extended 0L grammar
F - EP0L	203	Forbidding extended prop. 0L grammar
F - ET0L	203	Forbidding extended tabled 0L grammar
F - EPT0L	203	Forbidding extended prop. tabled 0L grammar
SSC - E0L	224	Simple semi-conditional extended 0L grammar
SSC - EP0L	224	Simple semi-conditional extended prop. 0L grammar
SSC - ET0L	224	Simple semi-conditional extended tabled 0L grammar
SSC - EPT0L	224	Simple semi-conditional extended prop. tabled 0L grammar
LRC - E0L	240	Left random context extended 0L grammar
LRC - EP0L	240	Left random context extended prop. 0L grammar
LRC - ET0L	240	Left random context extended tabled 0L grammar
LRC - EPT0L	240	Left random context extended prop. tabled 0L grammar
LF - E0L	240	Left forbidding extended 0L grammar
LF - EP0L	240	Left forbidding extended prop. 0L grammar
LF - ET0L	240	Left forbidding extended tabled 0L grammar
LF - EPT0L	240	Left forbidding extended prop. tabled 0L grammar
LP - E0L	240	Left permitting extended 0L grammar
LP - EP0L	240	Left permitting extended prop. 0L grammar
LP - ET0L	240	Left permitting extended tabled 0L grammar
LP - EPT0L	240	Left permitting extended prop. tabled 0L grammar
RC - ET0L	253	Random context extended tabled 0L grammar
RC - EPT0L	253	Random context extended prop. tabled 0L grammar
JMON	260	Jumping monotone phrase-structure grammar
JCS	260	Jumping context-sensitive grammar
JREG	260	Jumping regular grammar
JRLIN	262	Jumping right-linear grammar
JLIN	262	Jumping linear grammar
$\mathbf{JCF}_{fin}$	262	Jumping context-free grammar of finite index
$\mathbf{JCF}^{-\varepsilon}$	263	Prop. jumping context-free grammar
JRE	265	Jumping phrase-structure grammar
JCF	271	Jumping context-free grammar
$\mathbf{JSC}_X$	276	X-mode jumping scattered context grammar
FSFA	321	First-move self-regulated finite automaton
ASFA	322	All-move self-regulated finite automaton

(continued)

Subject Index

© Springer International Publishing AG 2017
A. Meduna, O. Soukup, *Modern Language Models and Computation*,
DOI 10.1007/978-3-319-63100-4

Printed in the United States
By Bookmasters